Handbook of Experimental Pharmacology

Volume 83

Calcium in Drug Actions

Contributors

D. M. Bers, P. J. R. Bevis, M. P. Blaustein, R. D. Bukoski, B. Ceccarelli, R. A. Chapman, S. Cockcroft, M. Crompton, A.W. Cuthbert, S. Ebashi, C. H. Evans, H. Fleisch, M. Fosset, M. L. Garcia, T. Godfraind, B. D. Gomperts, T. R. Hinds, P. Honerjäger, M. Hugues, G. J. Kaczorowski, U. Kikkawa, V. F. King, M. Lazdunski, B. A. Levine, I. MacIntyre, K.T. MacLeod, D. A. McCarron, J. Meldolesi, C. Milet, C. Mourre, H. Nagamoto, T. Narahashi, Y. Nishizuka, Y. Ogawa, T. Pozzan, J. F. Renaud, G. Romey, H. Schmid-Antomarchi, M. Schramm, I. Schulz, T. J. B. Simons, R. S. Slaughter, R. Towart, D. J. Triggle, J. Tunstall, F. F. Vincenzi, H. J. Vogel, R. J. P. Williams, M. Zaidi

Editor

P. F. Baker

Springer-Verlag
Berlin Heidelberg New York
London Paris Tokyo

PETER F. BAKER, Professor Sc. D., F.R.S. (†)

Department of Physiology
King's College, University of London, Strand,
London WC2R 2LS, Great Britain

With 123 Figures

ISBN-13: 978-3-642-71808-3 e-ISBN-13: 978-3-642-71806-9
DOI: 10.1007/978-3-642-71806-9

Library of Congress Cataloging-in-Publication Data. Calcium in drug actions/contributors. D. M. Bers ... [et al.]: editor, P. F. Baker. p. cm. - (Handbook of experimental pharmacology: v. 83) Includes bibliographies and index. 1. Calcium - Therapeutic use - Testing. 2. Calcium channels - Effect of drugs on. 3. Calcium - Receptors - Effect of drugs on. 4. Drugs - Physiological effect. I. Bers, D. M. II. Baker, Peter F. (Peter Frederick), 1939– . III. Series. [DNLM: 1. Calcium - physiology. W1HA51L v. 83/ QV 276 C1433] QP905.H3 vol. 83 [RM666.C6] 615'.1 s - cc19 [615'.2393] DNLM/DLC for Library of Congress

Softcover reprint of the hardcover 1st edition 1988

2122/3130-543210

List of Contributors

D. M. BERS, Division of Biomedical Science, University of California, Riverside, CA 92521, USA

P. J. R. BEVIS, Endocrine Unit, Department of Chemical Pathology, Royal Postgraduate Medical School, Hammersmith Hospital, Ducane Road, London W12 OHS, Great Britain

M. P. BLAUSTEIN, University of Maryland School of Medicine, Department of Physiology, 655 West Baltimore Street, Baltimore, MD 21201, USA

R. D. BUKOSKI, Division of Nephrology, Oregon Health Sciences University, 3181 SW Sam Jackson Park Rd., Portland, OR 97201, USA

B. CECCARELLI, Center for the Study of Peripheral Neuropathies and Neuromuscular Diseases, Department of Pharmacology, University of Milan, Via Vanvitelli, 32, I-20129 Milan

R. A. CHAPMAN, Department of Physiology, School of Veterinary Science, University of Bristol, Park Row, Bristol BS1 5LS, Great Britain

S. COCKCROFT, Department of Experimental Pathology, University College London, University Street, London WC1E 6JJ, Great Britain

M. CROMPTON, Department of Biochemistry, University College London, Gower Street, London WC1E 6BT, Great Britain

A. W. CUTHBERT, Department of Pharmacology, University of Cambridge, Hills Road, Cambridge CB2 2QD, Great Britain

S. EBASHI, National Institute for Physiological Sciences, Myodaiji, Okazaki 444, Japan

C. H. EVANS, Ferguson Laboratory for Orthopedic Research, University of Pittsburgh, School of Medicine, 986 Scaife Hall, Pittsburgh, PA 15261, USA

H. FLEISCH, Department of Pathophysiology, University of Berne, Murtenstrasse 35, CH-3010 Berne

M. FOSSET, Centre de Biochimie, Université de Nice, Centre National de la Recherche Scientifique, Parc Valrose, 28, Ave. Valrose, F-06034 Nice Cedex

M. L. GARCIA, Department of Membrane Biochemistry and Biophysics, Merck Institute for Therapeutic Research, P.O. Box 2000, Rahway, NJ 07065, USA

T. Godfraind, Laboratoire de Pharmacodynamie Générale et de Pharmacologie, Université Catholique de Louvain, UCL 7350, Ave. Emmanuel Mounier 73, B-1200 Brussels

B. D. Gomperts, Department of Pathology, University College London, Gower Street, London WC1E 6BT, Great Britain

T. R. Hinds, Department of Pharmacology, SJ-30, School of Medicine, University of Washington, Seattle, WA 98195, USA

P. Honerjäger, Institut für Pharmakologie und Toxikologie der Technischen Universität München, Biedersteiner Strasse 29, D-8000 München 40

M. Hugues, Centre de Biochimie, Université de Nice, Centre National de la Recherche Scientifique, Parc Valrose, 28, Ave. Valrose, F-06034 Nice Cedex

G. J. Kaczorowski, Department of Membrane Biochemistry and Biophysics, Room 80N-32C, Merck Institute for Therapeutic Research, P.O. Box 2000, Rahway, NJ 07065, USA

U. Kikkawa, Department of Biochemistry, Kobe University School of Medicine, Chuo-ku, Kobe 650, Japan

V. F. King, Department of Membrane Biochemistry and Biophysics, Merck Institute for Therapeutic Research, P.O. Box 2000, Rahway, NJ 07065, USA

M. Lazdunski, Centre de Biochimie, Université de Nice, Centre National de la Recherche Scientifique, Parc Valrose, 28, Ave. Valrose, F-06034 Nice Cedex

B. A. Levine, University of Oxford, Inorganic Chemistry Laboratory, South Parks Road, Oxford OX1 3QR, Great Britain

I. MacIntyre, Endocrine Unit, Department of Chemical Pathology, Royal Postgraduate Medical School, Hammersmith Hospital, Ducane Road, London W12 OHS, Great Britain

K. T. MacLeod, Division of Biomedical Science, University of California, Riverside, CA 92521, USA

D. A. McCarron, Division of Nephrology and Hypertension, Oregon Health Sciences University, 3181 SW Sam Jackson Park Road, Portland, OR 97201, USA

J. Meldolesi, University of Milan, Department of Pharmacology, Scientific Institute S. Raffaele, Via Olgettina, 60, I-20132 Milan

C. Milet, Laboratoire de Physiologie Generale et Comparee, Centre National de la Recherche Scientifique, F-75231 Paris Cedex 5

C. Mourre, Centre de Biochimie, Université de Nice, Centre National de la Recherche Scientifique, Parc Valrose, 28, Ave. Valrose, F-06034 Nice Cedex

H. Nagamoto, Tokushima Research Institute, Otsuka Pharmaceutical Co. Ltd., Tokushima, 771-01, Japan

T. Narahashi, The Medical and Dental Schools, Department of Pharmacology, Northwestern University, 303 East Chicago Ave., Chicago, IL 60611, USA

Y. Nishizuka, Department of Biochemistry, Kobe University School of Medicine, Chuo-ku, Kobe 650, Japan

Y. Ogawa, Department of Pharmacology, Juntendo University School of Medicine, Hongo, Bunkyo-ku, Tokyo 113, Japan

T. Pozzan, Department of General Pathology and CNR Center Membrane Physiology, University of Padova, Via Loredan, 16, I-Padova

J. F. Renaud, Centre de Biochimie, Université de Nice, Centre National de la Recherche Scientifique, Parc Valrose, 28, Ave. Valrose, F-06034 Nice Cedex

G. Romey, Centre de Biochimie, Université de Nice, Centre National de la Recherche Scientifique, Parc Valrose, 28, Ave. Valrose, F-06034 Nice Cedex

H. Schmid-Antomarchi, Centre de Biochimie, Université de Nice, Centre National de la Recherche Scientifique, Parc Valrose, 28, Ave. Valrose, F-06034 Nice Cedex

M. Schramm, Bayer AG, Institut für Pharmakologie, Aprather Weg 18a, Postfach 101 709, D-5600 Wuppertal

I. Schulz, Max-Planck Institut für Biophysik, Kennedyallee 70, D-6000 Frankfurt/Main 70

T. J. B. Simons, Department of Physiology, King's College, University of London, Strand, London WC2R 2LS, Great Britain

R. S. Slaughter, Department of Membrane Biochemistry and Biophysics, Merck Institute for Therapeutic Research, P.O. Box 2000, Rahway, NJ 07065, USA

R. Towart, Miles Laboratories Ltd., Stoke Court, Stoke Poges, Bucks. SL2 4LY, Great Britain

D. J. Triggle, Department of Biochemical Pharmacology, School of Pharmacy, Faculty of Health Sciences, State University of New York at Buffalo, 313 Hochstetter Hall, Buffalo, NY 14260, USA

J. Tunstall, Department of Physiology, University of Leicester, Leicester, Great Britain

F. F. Vincenzi, Department of Pharmacology, SJ-30, School of Medicine, University of Washington, Seattle, WA 98195, USA

H. J. Vogel, University of Calgary, Division of Biochemistry, Department of Biological Sciences, 2500 University Drive, N.W., Calgary, Alberta T2N 1N4, Canada

R. J. P. WILLIAMS, University of Oxford, Inorganic Chemistry Laboratory, South Parks Road, Oxford OX1 3QR, Great Britain

M. ZAIDI, Endocrine Unit, Department of Chemical Pathology, Royal Postgraduate Medical School, Hammersmith Hospital, Ducane Road, London W12 OHS, Great Britain

Foreword

The Editorial Board and the Publishers of the Handbook of Experimental Pharmacology wish to express their profound grief at the untimely death of Professor Peter Baker. Aware of his international recognition as an expert on the ubiquitous role of calcium in physiological processes and their pharmacological control, the Board was gratified when Professor Baker accepted its invitation to edit a new Handbook volume on "Calcium in Drug Actions". He went about this task with his usual energy and effectiveness so that, in the few months before his unexpected death, Professor Baker had mustered his distinguished contributors, got them to provide their manuscripts, and seen almost the entire material into the press. This achievement is all the more remarkable when one bears in mind the extraordinary number of his other commitments during the same time; they are mentioned in Sir Alan Hodgkin's preface to this volume.

With so many other professional and personal responsibilities upon him, the Board of the Handbook wishes to record its grateful appreciation for the admirable way in which Professor Baker took on and carried out the additional work of bringing this fine book into existence; and the Board wishes it to be dedicated to the memory of Professor Peter Frederick Baker.

The Editorial Board: G. V. R. Born, P. Cuatrecasas, H. Herken, A. Schwartz

PETER FREDERICK BAKER
11 March 1939 to 10 March 1987

PETER BAKER's sudden death from a heart attack has deprived the international scientific community of one of its most gifted and versatile biologists. He had just finished editing the contributions to this volume of the *Handbook of Experimental Pharmacology* and it is sad that he will not be here to see the printed book or to take part in the discussion that publication of a major scientific treatise is likely to generate.

BAKER's research activities developed at an early age and probably arose from his interest in natural history. While still at school in Lincoln he published an ecological note on the common earwig and in the interval between school and university he carried out research at Rothamsted which resulted in an article on aphid behaviour on virus-affected sugar beet. After a brilliant undergraduate career at Cambridge, studying Natural Sciences and specialising in Biochemistry in his third year, he decided to take a doctorate in Physiology working on the relationship between phosphorus metabolism and ion transport in nerve. To begin with it was arranged that he would divide his time between Cambridge and Plymouth where there were many interesting experiments to be done with giant nerve fibres that can be obtained from the common squid.

At Plymouth BAKER started to work with the late Dr. TREVOR SHAW and almost immediately made a major advance by showing that after the protoplasm had been squeezed out of a giant nerve fibre, conduction of impulses could be restored by perfusing the remaining membrane and sheath with an appropriate solution. With SHAW and the writer, BAKER worked out a method of changing internal solutions while recording with an internal electrode from a perfused axon. It turned out that it did not much matter what internal solution was employed so long as it contained potassium and not much sodium. Provided this condition was satisfied a perfused nerve fibre was able to conduct nearly a million impulses without the intervention of any biochemical processes. ATP is needed for the Na-K pump, but not for the conduction of impulses. There were also some unexpected findings of which one of the most interesting was that reducing the internal ionic strength led to a dramatic shift in the operating characteristics of the membrane. This effect which has a simple physical explanation helps to explain some puzzling findings which were sometimes thought to be inconsistent with the ionic theory of nerve conduction.

It did not take PETER BAKER long to finish off the main project he had selected for a Ph.D. thesis, namely the relation of phosphorus metabolism to the active transport of sodium and potassium. In both crab and squid nerve it seems that 3 Na^+ are ejected per ATP split. The lower figure obtained previously by CALDWELL, KEYNES, SHAW, and myself was explained by the fact that ¾ of the ATP breakdown occurs in the protoplasm and only ¼ at the membrane.

BAKER's work on the Na-K pump in different tissues led to experiments on the way in which living cells maintain an extremely low level of calcium and the importance of this phenomenon for various kinds of cellular activity. A new finding was the demonstration of a Na-Ca exchange system which helps to maintain internal calcium at a low level at the expense of the sodium gradient. The first evidence for this mechanism which has since been seen in many cells was provided in the autumn and early winter of 1966–67 by BAKER and colleagues working on squid nerve at Plymouth and independently by REUTER and SEITZ at Bern, on heart muscle.

PETER BAKER's interest in calcium was by no means confined to Na-Ca exchange. Some of the many subjects he and his colleagues tackled were the use of aequorin to detect calcium movements in giant nerve fibres, the uncoupled, "Schatzmann" calcium pump, the state of calcium in axoplasm and finally exocytosis.

BAKER's early research was so successful that he soon became a college tutor and university lecturer at Cambridge. He was in fact a brilliant teacher, both in the classroom and as a lecturer. It was, therefore, not surprising that in 1975 he was offered and accepted the Professorship of Physiology at King's College London, where he built up an active centre of research as well as a fine teaching laboratory. At King's College BAKER's main research interest was in the way in which neural transmitters are discharged by exocytosis. Characteristically, BAKER, and his close colleague, DEREK KNIGHT, chose to attack this fundamental problem in its simplest situation, namely the adrenal medulla, which may be regarded as an extension of the nervous system. Again characteristically, BAKER developed drastic but highly effective ways of getting large molecules into cells.

PETER BAKER was never one to shirk the world's work. He acted as secretary and chairman of the Editorial Board of The Journal of Physiology; was an energetic member of the Medical Research Council's Board on Neurosciences and became a member of the Agricultural and Food Research Council, a post which he greatly enjoyed as he basically felt himself more a biologist than a physiologist. He also served on several Royal Society Committees and one or two University working parties.

As if this was not enough he managed to be Chairman of the Parish Council and one of the Governors of the village school at his home in the country in Bourn near Cambridge.

In 1966 PETER BAKER married PHYLLIS LIGHT, a geneticist with whom he shared many interests – gardening, natural history and biology among them. They had four children, LUCY, ALEXANDER, SARAH and CHARLOTTE who meant a great deal to him.

A. L. HODGKIN

Contents

CHAPTER 3

Troponin C and Calmodulin as Calcium Receptors: Mode of Action and Sensitivity to Drugs

CHAPTER 4

Ligand-Binding Sites on Calmodulin

CHAPTER 5

Calcium Channels as Drug Receptors

CHAPTER 11

Pharmacology of Calcium Uptake and Release from the Sarcoplasmic Reticulum: Sensitivity to Methylxanthines and Ryanodine

CHAPTER 12

Effect of Lithium in Stimulus-Response Coupling

CHAPTER 16

Some New Questions Concerning the Role of Ca^{2+} in Exocytosis

CHAPTER 17

Exo-Endocytosis: Mechanisms of Drug and Toxin Action

CHAPTER 18

Pharmacology of Calcium Metabolism in Smooth Muscle

CHAPTER 19

Drugs Affecting Cardiac Calcium Metabolism

CHAPTER 21

Bisphosphonates: A New Class of Drugs in Diseases of Bone and Calcium Metabolism

CHAPTER 22

Calcium and Hypertension

Drugs and Toxicological Agents that Either Mimic Calcium or Elements of Intracellular Calcium Metabolism

CHAPTER 23

Calcium Chelators and Calcium Ionophores

CHAPTER 24

Lead-Calcium Interactions and Lead Toxicity

CHAPTER 25

Alkaline Earths, Transition Metals, and Lanthanides

CHAPTER 1

The Multiple Physiological Roles of Calcium: Possible Sites for Pharmacological Intervention

A. W. CUTHBERT

It is not at all surprising that calcium is an important constituent of living cells since it is one of the more readily available elements, comprising some 3.6% of the lithosphere. Furthermore evidence for the most ancient forms of life comes from the algal limestones of Central Africa, which are apparently 2.7×10^9 years old (OPARIN 1969). In one of the earliest symposia on the cellular functions of calcium, the late Peter Baker made some remarks which are just as pertinent today as they were in 1970. At that time the first estimates of the intracellular ionised calcium concentration were being made both by direct measurement, using aequorin, and indirectly, by using calcium buffer systems. It was discovered that muscle contraction required the intracellular ionised calcium concentration to rise to the micromolar range and that resting levels were perhaps only one tenth of this value. Baker pointed out that most cells would accumulate calcium if this was dependent only upon the Donnan ratio. Further with intracellular potentials of around -60 mV and an external calcium concentration of 1–10 mM, intracellular concentrations of 0.1–1.0 M could be expected. Clearly this was not the case; living cells had developed a variety of mechanisms to keep the intracellular ionised calcium at very low values indeed. Given this scenario transient increases in intracellular ionised calcium might then act as a trigger for a variety of cellular functions and these functions would be terminated by the activities which maintained the normal resting low level of intracellular calcium ions. It must not be forgotten that the total calcium concentration in living cells may equal or exceed the external concentration but the majority is stored in a sequestered or unionised form. It is the concentration of intracellular ionised calcium which is relevant in cell function. A pivotal role for calcium in muscle contraction was apparent from the classical studies of Sidney Ringer in 1883, who showed that the in vitro frog heart would continue beating for prolonged periods if bathed in a calcium-containing salt solution. Later it was found that smooth muscle was more uniquely dependent on external calcium, indeed action potentials currents in smooth muscle are carried largely by calcium ions. On the other hand, skeletal muscle can tolerate deprivation of external calcium more readily because release and reuptake from intracellular calcium stores play a major role in contraction. Binding studies with calcium antagonists have confirmed these differences in relation to the numbers and location of calcium channels in various muscle types. Thus in the 1960s the notion of the role of excitation-contraction coupling was consolidated, that is, ways in which signals arriving at the membrane surface were coupled to the contraction process through the agency of raised intracellular calcium – the first of the second messengers. Following rapidly upon the heels of these discoveries was the concept of excitation-secretion coupling, in which secretory processes were also shown to be dependent upon raised intracellular calcium triggered by excitatory signals delivered to the cell surface.

What was undreamt of in these early days was the diversity of ways in which the concentration of ionised calcium within cells could be regulated. This volume is a testament to that diversity. As the intracellular calcium concentration within a cell at any instant is the resultant of a multitude of different processes, the possibilities of chemical intervention with these processes are legion. Only a few of these strategies have yet been tried but already the possibilities of distinct therapeutic advantage are a reality, as in the cardiovascular arena.

What I shall do in this overview is to give a foretaste of what can be found in the following chapters, indicating where pharmacological agents are already available but, more importantly, pointing to possible targets for future pharmacological endeavour. However the format of this overview will not follow the organisation of this volume, rather it will follow three main themes. The ways in which Ca_i can be raised will be discussed first, followed by a consideration of agencies and processes which lead to a reduction in Ca_i. Finally, various intracellular processes which are switched on by a rise in Ca_i will be catalogued together with an account of chemical agents which interfere with those processes.

A wide variety of hormones, neurotransmitters, autacoids and growth factors bring about their cellular effects by reacting first with membrane receptors to bring about an increase in Ca_i. Obviously this increase in Ca_i can arise either from increased entry or from release from internal stores. The principal cellular mechanisms involved are the activation of calcium channels in the plasmalemma or activation of the "PI system". The latter involves primarily the activation of a membrane phospholipase C and the subsequent hydrolysis of phosphatidylinositol 4,5-bisphosphate to inositol 1,4,5-trisphosphate, which binds to receptors in the endoplasmic reticulum to release stored calcium. The other product of the hydrolysis reaction is diacylglycerol, a potent activator of protein kinase C.

At least four main types of calcium channel are known: leak channels, stretch sensitive channels, voltage sensitive channels and receptor operated channels. There are at least three subdivisions of voltage operated channels which are recognised, and voltage sensitive and receptor operated channels may be different stimulation modes of the same channel types.

There is a considerable history of the use of metal ions to block calcium entry into cells, particularly so with lanthanum, europium, cadmium etc. However not until Fleckenstein working with verapamil and D600 (FLECKENSTEIN et al. 1969) and Godfraind with the antihistamine cinnarizine (GODFRAIND and KABA 1969) was it realised that organic calcium antagonists were a reality. These are now more properly called calcium channel antagonists, and the three main groups (phenylalkylamines, 1,4-dihydropyridines and the benzothiazepines) are adequately reviewed in this volume. They present a pharmacologically fascinating group which appear to have receptors coupled to but separate from the calcium ionophore. Some show use dependence and some have already found a place in the therapy of cardiovascular disease. Calcium channel agonists, of which the best known is BAY K8644, have also joined the pharmacologists armoury. Both types of agent appear to affect the probability of channel opening. An increasing number of known drugs are now found to affect the operation of calcium channels; for example, pyrethroids and opiates block type I and II voltage gated calcium channels, whereas phenytoin blocks only the type 1 channel. The long known effect of opiates on reducing transmitter release now has a probable explanation in terms of reduced calcium entry. We can expect many new agents which block calcium channels to be discovered following the successful application of these agents to therapy. For example, a natural sesterterpenoid, manoalide, has recently been found that inhibits calcium channels in a variety of cells (WHEELER et al. 1987).

In contrast there are as yet no agents which affect the release of calcium by IP_3, and the receptor in the endoplasmic reticulum is an obvious target. Similarly inhibitors of protein kinase C would not only provide useful investigative tools, but may give new insights for developing antitumour drugs.

While the discovery of IP_3 as a messenger is relatively new, agents which release calcium from the endoplasmic reticulum have been known for several decades. The methylxanthines cause a positive inotropic effect, prolong the active state in skeletal muscle, and have even more complex actions in cardiac muscle. Caffeine is the most widely studied and has multiple effects, including the inhibition of phosphodiesterase, but contracture in skeletal muscle is a result of calcium release from the sarcoplasmic reticulum. This latter effect is blocked by local anaesthetics, such as procaine. Ryanodine is similar to caffeine in some respects but with more complex actions affecting multiple sites in excitation-contraction coupling mechanisms.

While there is clearly much scope for developing other agents that affect calcium release from internal stores, it must be remembered that specificity resides in the membrane receptors which trigger the response.

There are two principal mechanisms for maintaining low Ca_i; these are the ubiquitous Na–Ca exchanger and Ca-ATPase. It is to Peter Baker (1972) that we owe a deal of our knowledge of these two. The exchanger can move calcium ions in either direction across the plasmalemma, but under resting conditions movement is usually outwards. This may reverse in other conditions, for example during the plateau of the cardiac action potential. The exchanger is sodium dependent, voltage sensitive and electrogenic, exchanging one calcium ion for three sodium ions. The exchanger can work in the Ca–Ca and Na–Na exchange modes and Li can substitute for Na. Pyrazine carboxamides, developed as potassium sparing diuretics, and acting by blocking epithelial sodium channels, were found to inhibit Na–Ca exchange at high concentration. New analogues, such as naphthylmethylamiloride, are more active (EC_{50} 10 μM) but really potent drugs for the exchanger are still to be developed. Amiloride-like drugs are competitive with respect to sodium, as indeed they are for some sodium channels. Amiloride-calcium competition is demonstrated only under some conditions with the exchanger. Of course, the exchanger can be affected by a multitude of subtle effects which change either intracellular Ca or Na and which can be brought about by a variety of drugs, but these effects on the exchanger are indirect.

Ca-ATPase directs calcium out of the cell across the plasmalemma countering the leak influx, as first described by SCHATZMANN and VINCENZI (1969) for the red blood cell. This ion pump is modulated by CaM which disinhibits the enzyme.

The enzyme is also present in mitochondria and the endoplasmic reticulum, but in these sites it is not CaM sensitive. CaM antagonists will have an effect on calcium extrusion from cells, but as many other systems are CaM sensitive the effect will not be specific. There are, however, prospects for developing specific inhibitors of the calcium pump. For example, it is known that phenylglyoxal binds to the low affinity site for ATP to prevent hydrolysis. Also it is known that this membrane pump is sensitive to the phospholipid environment and its activity can be modified by the cholesterol-phospholipid ratio. Ca-ATPase activity is inhibited by kappa opiate agonists, divicine and N-acetyl-p-benzoquinone imine, by unknown mechanisms. The latter is a metabolite of paracetamol and the hepatotoxicity caused by this agent is likely due to increased Ca_i leading to cell death, possibly as a result of the metabolite acting on the pump.

Mitochondria are often considered an important intracellular store for calcium, but in general calcium uptake occurs only under pathological conditions and when Ca_i rises above 5 μM. Uptake into mitochondria is blocked by ruthenium red and gentamycin acting by unknown mechanisms. Amiloride analogues inhibit the Na–Ca exchange mechanism present in mitochondria.

Finally the concentration of ionised calcium within cells is partly controlled by cytosolic buffers in the form of calcium-binding proteins. Among these may be listed troponin C, calmodulin, parvalabumin and vitamin D calcium binding protein.

Raised intracellular concentrations of ionised calcium are known to be the crucial trigger for a variety of cellular functions, yet precise knowledge of the mechanisms involved are far from complete. In the future all these mechanisms will be better understood and provide targets for pharmacological intervention by drugs.

The functional responsibilities of troponin C and calmodulin in muscle contraction are well understood. Each has four calcium binding sites, and the two demonstrate a significant degree of homology. The troponin C-calcium complex triggers muscle contraction in skeletal muscle, while calmodulin has a comparable role in smooth muscle. Calmodulin, however, has many other functions via protein kinases, phosphatases and membrane ATPases; therefore Ca–M antagonists, such as trifluoperazine and related phenothiazines, are not very specific as they inhibit many calcium dependent processes. Calcium channels contain a calcium binding protein similar to calmodulin, yet relatively specific calcium channel blockers have been developed indicating that inhibition of single calmodulin dependent processes may be possible once the interactions of Ca–M with target proteins are better understood. A number of endogenous peptides (ACTH, VIP and gastric inhibitory peptide) bind to calmodulin in a calcium dependent manner but the physiological relevance of these observations is unclear.

In a number of physiological situations Ca_i may exert direct action on effector processes. For example, in neurotransmitter release binding of Ca to the vesicle surface will reduce surface charge and allow the approach of the vesicle to the cell membrane prior to exocytosis. However this is likely a simplistic view of exocytosis. Synexin, a calcium binding protein, and synapsin I, which can be phosphorylated by CaM or cAMP dependent protein kinase, are claimed to have a role in vesicle fusion and to influence exocytotic processes. Once these mechanisms are clarified, then strategies to interfere with them selectively and specifically can be considered.

A number of ion channels appear to be activated or inhibited by Ca_i, for example calcium activated potassium channels and the calcium inhibited GABA activated chloride conductance. Specific blockers are already available for some of these calcium activated channels, such as the apamin sensitive potassium channel. In the sarcoplasmic reticulum calcium release can be brought about by calcium ions themselves. Calcium release from isolated microsomes can be blocked by some agents, such as TMB 8, but regrettably this effect is not very specific.

Receptor activation of the PI mechanism not only generates IP_3 but also diacylglycerol. Protein kinase C, a calcium activated phospholipid dependent enzyme, is markedly activated by DAG. This it does by increasing the affinity for Ca, so much so that DAG alone can, in some circumstances, stimulate calcium dependent effects at basal intracellular calcium levels. Quite a lot is known of the structural requirements for DAG. Unsaturated fatty acids give more active DAGs than saturated ones and the 1,2-Sn configuration appears to be an absolute requirement. The 2–3 Sn and 1–3 Sn diacyl congeners neither activate nor inhibit protein kinase C. Synthetic DAGs are already available for researchers, such as 1-oleoyl-2-acetyl glycerol, 1,2-dioctanoyl glycerol and 1,2-didecanoyl glycerol. Phorbol esters, such as 12-O-tetradecanoylphorbol-13-acetate, are natural analogues of DAG, while tumour promotors unrelated to phorbol esters in structure (for example mezerein, teleocidin and aplysia toxin) also simulate DAG. Inhibitors of protein kinase C or of DAG are awaited with interest.

Perhaps the most interesting functions of raised Ca_i are yet to be discovered. Provocative suggestions in relation to memory mechanisms can be found in the pages of this book. For example, in PC12 cells, derived from a phaeochomocytoma, raised Ca_i leads to activation of the proto-oncogene c-*fos*. Elsewhere it argued that elevation of Ca_i activates a protease (calpain) that degrades a spectrin-like protein (fodrin) which leads to an increase in glutamate receptors in the plasma membrane. Whether or not these findings are the basis for memory mechanisms may be known by the time the next major volume on calcium and cellular function is published. The possibility of pharmacological intervention in distressing situations of mental deterioration is a worthy goal. The recent announcement of successful cloning of the receptor for dihydropyridine calcium channel antagonists (Tanabe et al. 1987) opens up totally new prospects for drug design based upon the tertiary structure of the active site.

References

Baker PF (1970) Sodium-Calcium exchange across the nerve cell membrane. In Cuthbert AW (ed) Calcium and Cellular Function. MacMillan, London, pp 96–107

Baker PF (1972) Transport and metabolism of calcium ions in nerve. Prog Biophys Mol Biol 24:177–223

Fleckenstein A, Tritthart H, Fleckenstein B, Herbst A, Grun G (1969) A new group of competitive Ca-antagonists (Proveratil, D 600, prenylamine) with highly potent inhibiting effects on excitation contraction coupling in mammalian myocardium. Pflügers Arch 307:R25

Godfraind T, Kaba A (1969) Blockade and reversal of contraction induced by calcium and adrenaline in depolarised arterial smooth muscle. Br J Pharmacol 36:549–560
Oparin AI (1969) Chemistry and the origin of life. Roy. Inst Chem Rev 2:1–12
Ringer S (1883) A further contribution regarding the influence of the different constituents of blood on the contraction of the heart. J Physiol 4:29–42
Schatzmann HJ, Vincenzi FF (1969) Calcium movements acros the membrane of human red cells. J Physiol 201:369–395
Tanabe T, Takeshima H, Mikami A, Flockerzi V, Takahashi H, Kangawa K, Kojima M, Matsuo H, Hirose T, Numa S (1987) Primary structure of the receptor for calcium channel blockers from skeletal muscle. Nature 328:313–318
Wheeler LA, Sachs G, Vries G De, Goodrum D, Woldemussie E, Muallem S (1987) Manoalide, a natural sesterterpenoid that inhibits calcium channels. J Biol Chem 262:6531–6538

Calcium Receptors and Calcium Metabolism

CHAPTER 2

Chemical Factors Determining the Affinity of a Receptor for Calcium

B. A. LEVINE and R. J. P. WILLIAMS

A. Introduction

In this chapter we shall give a somewhat different account of calcium chemistry in relation to biology than we have done previously. Our previous papers on the topic have described a detailed examination of the energetics of calcium–ligand interactions, including both thermodynamic and kinetic considerations, and of the structures of calcium complexes (LEVINE and WILLIAMS 1982a, b). The discussion was made in the light of a detailed knowledge of model, small molecule calcium complexes and a comparison with parallel magnesium chemistry. While some of this earlier work will be reported again in the present chapter, we shall describe here mainly the features of biological calcium chemistry directly related to the physiological situations of pH, Na^+, K^+, Mg^{2+} and ligand concentrations. Inevitably we shall be looking at large protein ligands for much of the account, but we shall refer to data on small ligands from time to time. This reflects new and growing knowledge of the precise nature of calcium binding to large molecules of direct biological interest.

The chapter will start with a discussion of competition between calcium, protons and other metal ions for these ligands. This is not only a thermodynamic, but also a kinetic problem since we deal with transient, time-dependent effects as well as steady state situations in physiology. Later we turn to structural knowledge.

B. Concentrations in Physiological Conditions

Table 1 gives the concentrations of the major cations and anions inside and outside cells in different conditions. The first point to observe is that most of the Na^+ and K^+ ions are free both inside and outside cells. By "free" we do not imply that there is no loose association with polyelectrolytes such as DNA. The condition of calcium and magnesium is quite different since a larger proportion of these ions are directly bound to ligands (coordinated) and therefore we need to look at the concentrations of different ligands and at concentrations of both ligands and metal ions under different conditions.

Outside cells free Mg^{2+} and Ca^{2+} ions are at approximately 10^{-3} M and there is not a large amount of bound calcium or magnesium associated with freely diffusing, large or small molecules in this part of space. We turn to cell outer surfaces later. Inside cells the concentration of free plus bound magnesium can approach 10 mM and the amount of free plus bound calcium is also high, probably

Table 1. The properties of some cations[a]

Cation	Na^+	K^+	Mg^{2+}	Ca^{2+}	Mn^{2+}	Ln^{3+}
Size (Å)	1.0	1.33	0.6	0.9	0.75	1.05–0.85
Coordination number	6(+)	8(±)	6	7(±)	6(+)	7(+)
Bond lengths	Irregular	Irregular	Regular	Irregular	Regular	Irregular
Bond angles	Irregular	Irregular	Regular	Irregular	Regular[b]	Irregular
H_2O exchange (s)	10^{-10}	10^{-10}	10^{-6}	10^{-9}	10^{-7}	10^{-9}
Concentration (M)						
cellular	10^{-2}	10^{-1}	10^{-3}	10^{-7}	10^{-8}	
blood	10^{-1}	10^{-2}	10^{-3}	10^{-3}	10^{-8}	

[a] Mn^{2+} and Ln^{3+} are useful probes for Ca^{2+} activation. All the cations bind to oxygen ligands almost exclusively with the exception of Mn^{2+}. Na^+ and K^+ bind weakly to all donors except for those in special cavities.
[b] Variability in angles reported.

always exceeding 1 mM. However while the free magnesium is fairly evenly distributed in different organelles, vesicular systems and in the cytoplasm at 10^{-3} M, the calcium concentration distribution is very different. Free calcium is below 10^{-7} M in the cell cytoplasm, which contrasts with 10^{-3} M free magnesium, but free calcium can be quite high, approximately 10^{-3} M, in mitochondria and some other vesicles, e.g. chromaffin granules and platelet vesicles. Calcium is much more frequently found in insoluble matrices than is magnesium. The exact complexes which can be formed in different cellular compartments are then very different. We need to look at the concentrations of ligands which are contained in each compartment. This type of analysis is far from complete at present. There are ligands in cell cytoplasm and organelles such as ATP^{4-}, ADP^{3-} and P_2^{4-} which are likely to be 10^{-3} M and are almost totally in the form of magnesium complexes. However nearly all the small anions of biology – both carboxylates and phosphates – do not have binding constants to calcium or magnesium great enough to give stable complexes either inside or outside cells, given their concentration levels, e.g. HPO_4^{2-}, malonate and sugar phosphates. Turning to larger ligands or ligands such as proteins giving relatively rigid cavities of a particular size, there are specific combinations to which we turn later. However, in general there are many proteins in the cytoplasm of cells at effective concentrations greater than 10^{-7} M so that potential ligand concentrations exceed free levels of calcium, but not that of magnesium. Of course, the total bound calcium greatly exceeds the free, especially in reticula and organelles (Table 2). We need to look at each type of cell separately. Again the bathing liquids of cells are specific to particular organisms and we shall have to consider a variety of problems there.

When we turn to vesicular systems we must always be aware of the local very high concentrations of anion centres and of the fact that calcium is usually pumped into these vesicles. Some typical systems are described in Table 2. Notice that there are peculiar associations such as those of calcium with pyrophosphate (in platelets), with calsequestrin (in the sarcoplasmic reticulum), with phosphate (in mitochondrial granules) and so on. The stability constants and solubility

Table 2. Vesicle systems containing high calcium

System	Calcium-binding centre
Sarcoplasmic reticulum	Calsequestrin
Mitochondria	As phosphate
$CaCO_3$ precipitation pH = 7	Shell minerals
Ca_2OHPO_4 precipitation pH = 7	Bone minerals
Chromaffin granules	Associated with ATP and chromagranin A
Vesicles of sperm	Unknown
Platelets	Pyrophosphate

products of these associations indicate that free calcium must exceed 10^{-4} *M* and free ligand must be similar in concentration. We do not always know the anions associated with the vesicular calcium of course. We must also note that occasionally the concentrations must be very high as in platelet vesicles. Another curiosity here is that in different species the pyrophosphate is associated with magnesium (rat) rather than calcium (man).

All these reactions are also under the control of protons which vary in concentration between 10^{-8} and 10^{-5} *M* and are sometimes as high as 10^{-1} *M*. It is usual for the cytoplasmic pH to be 7 and also the external liquid is usually at this pH, but vesicles often have a pH as low as 5. We turn immediately to problems caused by the pH.

C. Calcium–Proton Competition

The proton concentration in a compartment greatly affects metal binding. The binding *association* constant of calcium for a group L will be written K_{Ca} and the *dissociation* constant of the proton complex, K_a, as is conventional. If the total ligand is T_L we have

$$\begin{aligned} T_L &= [CaL]+[L]+[HL] \\ &= [Ca]K_{Ca}[L]+[L]\left(1+\frac{[H^+]}{K_a}\right) \end{aligned}$$

(A similar equation can be written for total calcium.)

$$\frac{[CaL]}{T_L} = \text{degree of formation of CaL}$$

assuming that $T_{Ca} < T_L$ (and a similar equation can be written for the fraction of L bound). Calling this F_{Ca} (*F* for fraction)

$$F_{Ca} = \frac{K_{Ca}[Ca][L]}{K_{Ca}[Ca]\,[L]+[L]\left(1+\frac{[H^+]}{K_a}\right)}$$

$$F_{Ca} = \frac{K_{Ca}[Ca]}{K_{Ca}[Ca]+\left(1+\frac{[H^+]}{K_a}\right)}$$

For free calcium ion concentrations of 10^{-6} *M* and $K_{Ca} \gg 10^6$, F_{Ca} approaches unity so long as $[H^+]/K_a < 1.0$, i.e. for $K_a = 10^{-6}$, at pH > 6. These considerations are quite general and apply to all large and small ligand reactions and must be obeyed in all cell compartments at rest. In this chapter we wish to look in particular at the calcium trigger proteins, calmodulin and troponin. The binding constant $K_{Ca} = 10^7$ for calcium and a pK_a of 6 is quite close to the values in these proteins. In stimulated cells calcium is 10^{-6} *M*. It is then obvious that at pH = 7 there is no problem of proton competition for calcium binding in the troponin or calmodulin reactions. At pH = 5 the problem would be very real however since proton binding is at least equal to calcium binding. The problem is made more acute since probably more than one proton can bind. For two protons binding and keeping all the same illustrative numbers, calcium can only bind to 1.0% of the protein ligands at pH = 5.0. There are obvious biological consequences (which others will discuss in this volume) as the pH of muscle cytoplasm drops with prolonged exercise. This is confirmed by the observations on crystal forms of calmodulin and troponin (SUDHAKAR et al. 1985; SUNDRALINGAM et al. 1985). These crystals were grown at pH = 5.6 and 5.0, respectively. Troponin C was found to be bound to only two of four possible calcium ions. There are then two very different pairs of binding sites or "hands". One, the weak-binding hands, has low affinity for calcium, log $K \sim 6$ and must have $pK_a \sim 6$, and suffers from proton competition while the high affinity sites have a log $K \sim 7$ and an unknown, but probably similar proton affinity of $pK_a \sim 6$.

Outside cells the pH is usually fixed at around 7. (We can not generalise since there are conditions when the pH ~ 1.0). The proton does not protect any phosphates or carboxylates from calcium binding at this high pH, especially since the free calcium concentration is high outside cells – generally 10^{-3} *M* – and there is then the general condition that there will be binding to ligands for $K_{Ca} \geqq 10^3$. Now we have already seen that K_{Ca} does not exceed 10^3 for simple phosphates or carboxylates. Biological systems have therefore devised more complex ligands such as heavily phosphorylated proteins, e.g. β-casein, and multicarboxylated proteins such as prothrombin which contains multiple units of Gla (γ-carboxyglutamate) and particular lowly charged sites of the correct size to accept calcium such as phospholipase A_2. These complexes will be broken down in the acid part of the digestive tract of course.

The acidity of vesicles may also help to keep calcium away from ligands to which it would bind at pH = 7. On the other hand these vesicles sometimes hold proteins in a low calcium solution, but vesicle contents are ejected to the external solution by stimulation. Many such proteins have $K_{Ca} \sim 10^3$ and are then activated by external calcium, Table 3. External calcium is associated in many ways with stimulated systems outside cells, e.g. blood clotting, digestion, fertilisation, viral invasion, complement systems, and organisation (of sponges for example).

Finally there is the cell surface. Here the problem of binding is complicated by the surface potential, the field in which the binding groups lie as well as the pH. Most cell surfaces have an *effective* binding constant for some calcium of $K_{Ca} > 10^3$ owing to the negative surface potential and the presence of anionic substituents of lipids, saccharides and proteins. pH is not usually a problem since it is around 7.0 so generally calcium is bound to cell surfaces.

Table 3. Calcium sites in some proteins

	$\log K_{Ca}$	Sequence of ligands	sequential?
Extracellular			
Phospholipase A_2	3.5	Tyr-28.Glu-30.Gly-32.Asp-49.$2H_2O$	No
Staphylococcal nuclease	~3	Asp-19.Asp-21.Asp-40.Thr-41.Glu-43.$1H_2O$	No
Trypsin	3.5	Glu-70. Asn-72. Glu-80.$2H_2O$	Yes
Thermolysin	4.7	Asp-138.Glu-177.Asp-185.Glu-187.Glu-190.$1H_2O$	No
		Asp-57.Asp-59.Glu-61.$3H_2O$	Yes
Intracellular			
Parvalbumin	8	Asp-51.Asp-53.Ser-55.Phe-57.Glu-59.Glu-62	Yes
		Asp-90.Asp-92.Asp-94.Lys-96.Glu-101.$1H_2O$	Yes
Calmodulin	~6.5	Asp-20.Asp-22.Asn-24.Thr-26.Thr-28.Glu-31	Yes
		Asp-56.Asp-58.Asn-60.Thr-60.Thr-62.Asp-64.Glu-67	Yes
		Asp-93.Asp-95.Asn-97.Tyr-99.Ser-101.Glu-104	Yes
		Asn-129.Asp-131.Asp-133.Glu-135.Asn-137.Glu-140	Yes
Intestinal calcium-binding protein	~6	Ala-15.Glu-17.Asp-19.Gln-22.Ser-24.Glu-27	Yes
		Asp-54.Asn-56.Asp-58.Glu-60.Ser-62.Glu-65	Yes
Troponin C	5.5	Asp-27.Asp-29.Gly-31.Asp-33.Ser-35.Glu-38	Yes
	7.3	Asp-63.Asp-65.Ser-67.Thr-69.Asp-71.Glu-74	Yes
		Asp-103.Asn-105.Asp-107.Tyr-109.Asp-111.Glu-114	Yes
		Asp-139.Asn-141.Asp-143.Arg-145.Asp-147.Glu-150	Yes

D. The Calcium–Magnesium Problem

The same equations can be written for the competition between two metal ions as for proton–metal competition. The fraction of calcium bound is now

$$F_{Ca} = \frac{K_{Ca}[Ca][L]}{[Ca]K_{Ca}[L] + [Mg]K_{Mg}[L] + [L]\left(1 + \frac{[H^+]}{K_a}\right)}$$

and leaving aside the proton competition at pH = 7

$$F_{Ca} = \frac{K_{Ca}[Ca]}{[Ca]K_{Ca} + [Mg]K_{Mg} + 1}$$

Putting $[Ca^{2+}] = 10^{-6}$ M, $K_{Ca} = 10^7$ and $[Mg^{2+}] = 10^{-3}$ M means that K_{Mg} must be less than 10^4 if heavy competition for calcium sites is to be avoided in stimulated cells but in resting cells where $[Ca] < 10^{-7}$ competition is always likely. We ask the following questions. Which ligands provide a difference of 10^4 units between Ca and Mg binding to protect stimulation and which ligands have $K_{Mg} < 10^3$ to make certain that resting cells do not bind magnesium? What guarantee have we that the strong-binding hands of troponin, for example, are not bound by magnesium in vivo at rest? Once again these questions are important since we have structures for calcium-bound proteins only and we do not know from crystal structure data how the structure varies on changing from calcium to magnesium. The study by NMR of both calcium and magnesium proteins in solution can be very helpful here (LEVINE and DALGARNO 1983).

We have already pointed out that the smaller molecules inside cells are not affected by the calcium ion, but this may not be true in organelles which pump in calcium. We do not know the Mg^{2+}–Ca^{2+} antagonisms in mitochondria. In the internal volumes of reticula there is a similar problem since calcium concentration is high, for example in the sarcoplasmic reticulum where there is a calcium-binding protein, calsequestrin.

One other group of proteins is worth mentioning – the "internal" proteases which appear to depend on calcium ion concentrations of around 10^{-5} *M* for activation. These proteases appear to be suicide reagents causing rapid cell death if by accident $[Ca^{2+}]$ rises above 10^{-5} *M*. We know little about Mg^{2+} binding to them.

E. Other Metal Ions

The other metal ions which are in high enough concentration to cause concern are Na^+, K^+ and possibly Mn^{2+}. Manganese could be as high as 10^{-8} *M* in cell cytoplasm, but its binding to calmodulin-type hands is weaker than that of calcium. We conclude that manganese will not interfere with calcium under physiological conditions. Sodium and potassium are present at 10^{-2} *M* and 10^{-1} *M*, respectively in cells and we suspect that there will generally be some binding of these elements to calmodulin-like hands in the resting cells. However, these bindings are not likely to give $K_{Na}[Na]$ or $K_K[K]$ a value as great as 10 and therefore they will not swamp the $K_{Ca}[Ca]$ which exceeds 10 in the stimulated cell.

A general problem usually referred to in discussions of "acid rain" is the effect of elevated aluminium ion levels. Since most of this effect is not physiological, but is really due to pollution, we just give a reference to the considerations of aluminium–calcium competition (TAM and WILLIAMS 1986), while noting that the ion content of drinking waters may need careful watching, e.g. with reference to Alzheimer's disease.

F. Cooperative Interactions

Finally there can be cooperative interactions. A possible case is illustrated by the calcium-binding protein S-100 which may also bind to zinc, see Sect. O.

G. Binding and Conformational Energy

When a ligand which has internal mobility in its bond structure binds to a calcium ion, we can look upon the reaction in two steps in order to appreciate the overall energetics though this is not likely to be the order in which they occur in practice. In the first step the ensemble of states of the free ligand, assuming a mobile ligand, is reduced to a more limited group of states which represents the form of the molecule which binds to calcium. There is therefore an energy of conformational

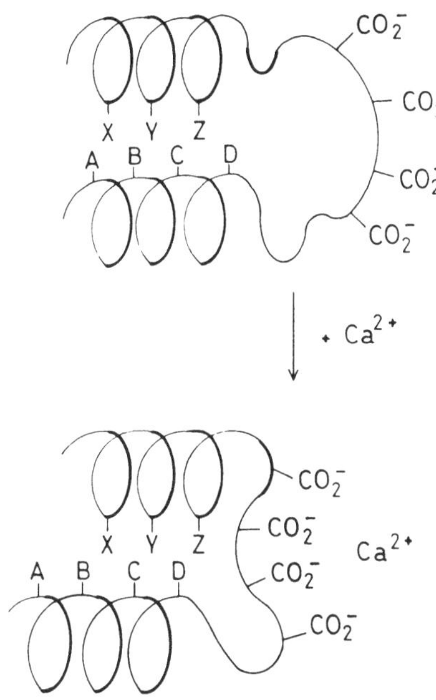

Fig. 1. Schematic representation of the Ca^{2+}–ligand framework able to wrap around the cation. The induced structural reorganization is depicted

change and it is a positive free energy, i.e. against metal ion binding. The second step is calcium binding to the constrained state. The overall binding is a sum of the first (disadvantageous) change and the second (advantageous) change

$$-\Delta G = -(\Delta G_2 + \Delta G_1) = RT \log K_{Ca}$$

The first change, ΔG_1, is made up of an entropic component including a loss of configurational entropy as well as an enthalpic component which includes any strain in thc ligand. Thc valuc of log K_{Ca} is then not just a problem of binding of Ca^{2+} ions to a fixed ligand as, for example, in the reaction of Ca^{2+} with oxalate or phosphate ions, but involves the opposing rearrangement energy of the ligand, as is the case for a ligand like EGTA (Fig. 1). It is the second term which controls the subtle changes of binding constant from one calcium protein to another while the binding groups give a more gross control. The fold of the protein is involved in the first term. An interesting example is the protein, lactalbumin.

Now, just because the ligand in both the bound and free states has many different possible conformations both can further interact with molecules and ions other than calcium. The cations such as Mg^{2+}, H^+, La^{3+} and Na^+ will tend to bind competitively at the calcium-binding site, but other metal ions and organic molecules and anions will bind to the proteins at remote sites. As a consequence there will be cooperative and anticooperative effects between calcium and other organic molecules (and perhaps metal ions) linked via the ligand through noncooperative binding at remote sites on the ligand (protein). While we often use two-state models to describe these allosteric effects all recent evidence indicates that in the real situation the calcium protein ligands are ensembles of many states. The ensembles are biased by all kinds of bindings. We shall use a two-state model here because of its simplicity. Notice that we are saying that it is not the calcium bond energies, ΔG_2, which are necessarily altered, although they could be, but it is the change in ΔG_1 by binding to the free or bound ligand which can bias the conformational energy for or against calcium binding. The constant $-\Delta G = -(\Delta G_2 + \Delta G_1)$ is then open to many controls. We return to the structural part of the description later. It must also be clear that we have widened the scope of the discussion of the observed binding constant for calcium and the effects of acid and metal ions upon it to the much wider problem of the cooperation or anticooperation between calcium and other binding agents such as other proteins or small organic molecules such as drugs. The first interaction is a part of triggering. Notice that the fold energy change on binding calcium can be manipulated by the position in the sequence of the calcium-binding amino acid side chains (see Table 3).

H. Kinetic Constraints

We can often express the effect in a biological system by the general equation which applies in any steady state

$$\text{Effect} = [\text{M}]\, K_{\text{Ca}}\, \text{f (structure)}$$

where [M] is the free ion concentration, K_{Ca} is the binding constant (discussed in Sect. G) and f (structure) expresses the fact that no matter how much of a metal ion is bound its functional effect depends upon the structure it induces on binding. However, this equation applies only to an equilibrium situation of metal bound to ligand. In the case of calcium activation of cellular or extracellular processes it is almost certainly the case that equilibrium is not reached and that we deal with a fluctuation in calcium (or protein) levels. In this case the rate at which calcium enters and leaves the site of action is of great concern. There are three processes of interest: (a) the rates of binding to the ligand, i.e. the on–off contants; and the rates of diffusion (b) to the site and (c) from the site.

$$\text{Effect} = (k)_{\text{Ca}}(\text{time})[\text{M}]\ \text{f (structure)}$$

where (k) refers to overall rate constant considerations; $(k)_{\text{Ca}}(\text{time})$ replaces K_{Ca}. We shall now analyse various rate steps.

J. On–Off Binding Constants

The rate at which a metal binds initially to a ligand is usually controlled by the rate of replacement of water of hydration. Table 1 gives some rate data when we see that Ca^{2+}, Na^+, K^+ as well as H^+ react very rapidly, but Mg^{2+} is at least 10^3 times slower. If conformational rearrangement of the ligand (protein) follows then this may be the overall rate-limiting process. Such conformation changes can be followed by NMR and studies with the hands of the calcium-binding proteins have given the data in Table 4. The off-rate of calcium and magnesium is around 10^3–10^4 for all the hands of binding constant around 10^6 (slower for tighter binding), but notice that this similarity hides a binding constant difference between calcium and magnesium of 10^3. These biological trigger systems of magnesium/calcium/proteins are then tuned to an overall rate constant (relaxation) of around 10^{-3} s which is probably close to an optimal limit for reversible triggering. We see that even if a ligand is partially bound in a resting situation by Na^+ or K^+, these ions do not act as a kinetic constraint even if they act as a small thermodynamic limitation. Mg^{2+} introduces a very different situation.

A consequence of the different rates of dissociation of calcium and magnesium is that the following situation can arise. Consider a solution containing two chelating agents L_1 and L_2 in the presence of 10^{-3} *M* Mg^{2+} ions. The ligand L_1 has a binding constant of $10^{2.5}$ for Mg^{2+} and is but poorly bound while the ligand L_2 has a binding constant of $10^{3.5}$ and is virtually saturated with Mg^{2+}. None of this binding is affected by the presence of 10^{-8} *M* calcium if the Ca^{2+} binding constants are respectively $10^{6.5}$ and $10^{7.5}$ for L_1 and L_2. Let calcium ions now be added to the solution rapidly so that $[Ca^{2+}]$ becomes 10^{-6} *M* in 1 ms. L_1 binds calcium within the time of the addition and is then virtually CaL_1. The binding of calcium to L_2 has to wait for the dissociation of Mg^{2+} which may well take 10 ms when L_2 becomes saturated with calcium. Thus, the order of binding of calcium is the reverse of the binding constants.

In terms of triggering L_1 is a good trigger while L_2 is a poor trigger and can only assist relaxation and removal of calcium. The difference is that between calmodulin and parvalbumin, see Sects. K.I, N. A similar discussion can be formulated for the extracellular proteins (see Table 3).

K. Diffusion of Ions

I. In Water

Diffusion of ions in water is relatively fast. Although buffering accelerates proton diffusion, the diffusion of other ions such as calcium is more complicated since in general there is a huge reservoir of *bound* protons, but a very small reservoir of bound calcium. The situation is different in some muscle cells and especially muscle cells with a fast response. Here parvalbumin is high in concentration and we must ask if such a protein can assist calcium diffusion to the membrane after a pulse. The core of the argument (Birdsall et al. 1979) rests with the relative effects of free ion diffusion

$$\text{Transfer of calcium} = k_T[\text{Ca}]$$

and

$$\text{Transfer of calcium by parvalbumin} = k_{Pn}[\text{CaPn}]$$

where Pn indicates parvalbumin. We must note that k_T is a simple diffusion coefficient, but k_{Pn} is diffusion of CaPn plus its dissociation to free Ca^{2+} and Pn. It does appear that the parvalbumin can distribute calcium faster than calcium itself could diffuse. The situation could be further affected by interaction of CaPn with membrane calcium pumps. The same argument applies to the vitamin D-dependent calcium-binding protein.

II. Through Membranes

The problem of diffusion in a membrane looked on as a rather rigid matrix is best described through an analysis of solid state diffusion. Ions can diffuse in solid lattices provided that the lattice: (a) forms sites which the ion can occupy and a proportion of which are unoccupied at any one time; and (b) has the occupied and unoccupied sites so structured that there is a "window" through which the ions can move from site to site. The energy of an ion sitting in the window must not greatly differ from the energy of the ion in the site since this energy difference is the activation energy for motion. Motion is then described within the channel (Fig. 2) by the equation

$$\text{Rate of motion} = ki \exp\left(-\frac{\Delta G_c}{RT}\right)$$

where ΔG_c is the channel excitation energy, i is the degeneracy of the motion, i.e. the number of equivalent hopping possibilities from occupied to unoccupied sites, T is the temperature and k is an intrinsic rate describing linear motion in the ab-

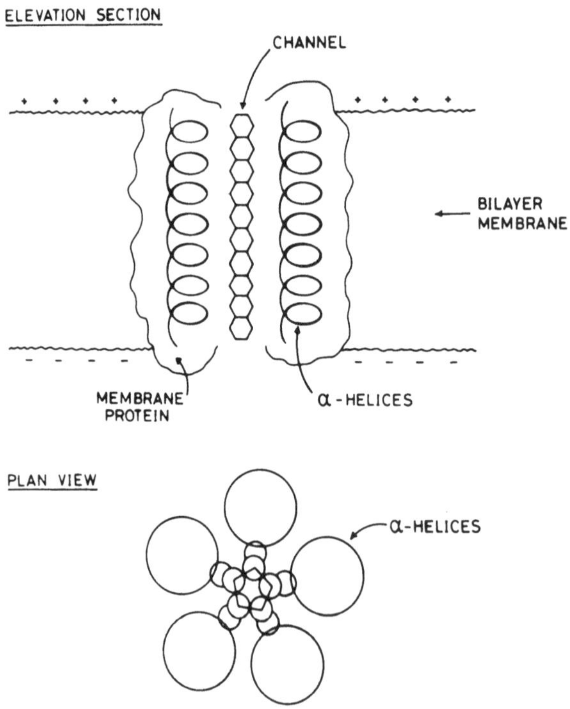

Fig. 2. Membrane protein channel protein groups containing an assembly of α-helices extend roughly perpendicular to the membrane, enclosing a water channel (e.g. EDMONDS 1979)

sence of barriers. Now, such a channel may need to be selective and yet to be able to allow fast motion from one aqueous phase to another. Thus, $\Delta G_c(Ca^{2+})$ must be close to zero, and the release of Ca^{2+} at both ends of the channel should involve a very low binding energy too.

These conditions can be met by proper consideration of the radius ratio effect, i.e. the ratio of the size of the ion, here calcium, to the effective size of the hole in the membrane channel (WILLIAMS 1953). There will be some circumstances where these effective sizes closely match so that calcium ions do not feel a change in free energy on entering the channel. A smaller ion will not enter because its hydration energy opposes entry – water molecules collapse around an ion, but a channel can not. A larger ion will not enter because it is too large to get in. Thus, selectivity can be achieved even when binding is virtually nonexistent.

III. Pumps

While a channel does not have to involve binding at all, a pump must involve binding so that the energy state of the calcium ion can be changed. The ion is trapped selectively, its binding is reduced by conformational change and the ion is ejected. If the conformational change alters access for the ion to the protein so that at first it reaches the site only from inside the cell, but after conformational change the ion only sees the site from outside the cell, then the conformation change, strong to weak binding, pumps the calcium ion. It can be seen in outline that if the pump is to be driven by the conformation change some energy input is required. This is usually done by ATP or an ion travelling, being pumped, in an opposite sense (Fig. 3). The ATP (or other ion) does not need to bind near to the calcium since transmission of energy from remote sites is possible as in calmodulin (Sect. N).

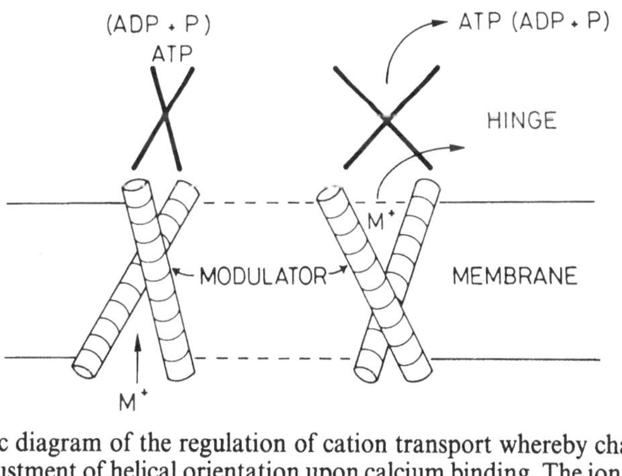

Fig. 3. Schematic diagram of the regulation of cation transport whereby channel opening results from adjustment of helical orientation upon calcium binding. The ion-induced helix twisting can thereby be related to chemical reactions

L. Selective Binding to Unstructured Molecules of High Anionic Charge Density

In this section we wish to make again a simple point about electrostatics within *assemblies of molecules* which is lost when more general *bulk* physical models of electrostatic interactions are used. The major source of selectivity in electrostatics does not appear in the attractive term z_1z_2/rD where z_1 is the charge on the cation, z_2 is the charge on the anion, r is the distance between them, i.e. the sum of their radii, and D is the dielectric constant of the supposed continuous medium. This term states simply that selectivity depends on the value of r, smaller ions binding better, provided that no changes are made in z_1 and z_2. Ions in solution do not show such simple selectivity (TAM and WILLIAMS 1985). Now, we can make an *apparent* extension of this picture if we try to account for the observed variation of binding constants between anions and cations by describing the energy of interaction in terms of the difference between hydrated free ions and the bound complex as follows

$$-\Delta G = \frac{-C'z_1z_2}{rD} + \frac{C''z_1}{r_1D} + \frac{C'''z_2}{r_2D} \tag{1}$$

where C refers to appropriately chosen constants to describe the ion pair energy and the hydration of the individual ions. As PHILLIPS and WILLIAMS (1966) showed, this expression does give one the observed selectivity orders – namely cations and anions of similar sizes bind best. For example Mg^{2+} binds better to OH^- than does Ca^{2+}, but Ca^{2+} binds better to SO_4^{2-}. This shows that different selectivity orders arise through the use of chosen values of r_1, r_2 and the constants C. However, these choices are based not so much upon fundamental electrostatic theory, but upon empirical knowledge of crystal energies and hydration energies which automatically take into account the repulsive terms at close approach between molecules or ions. Repulsion is then taken care of without explicitly mentioning it in Eq. 1. If we look at the structures of the hydrates or of the complexes we often find no simple connection between so-called crystal radii and distances so that r in Eq. 1 is not equal to $r_1 + r_2$. There is a packing problem. The packing includes the repulsions. In fact, space filling, which is totally absent from bulk electrostatic theory, gives the major part of selectivity. Consider an example.

$EDTA^{4-}$ binds Ca^{2+} with roughly the same strength as it binds to Mg^{2+}, $\log K = 10$. $EGTA^{4-}$ binds to calcium 10^4 times more strongly than to Mg^{2+}. $EDTA^{4-}$ and $EGTA^{4-}$ carry the same charge. We can only get these results from Eq. 1 by manipulating the constants. Inspection of structures reveals that these organic molecules bind the two ions Mg^{2+} and Ca^{2+} in different ways. Note that these anions do not have rigid structures.

We turn back now to some of the observed bindings in biological systems. We do not want to analyse again the binding to cavities in proteins which match the sizes of the individual ions. This is simply described even by Eq. 1 since $r \leqq r_1 + r_2$ for an ion which is too small to fit and for $r \leqq r_1 + r_2$ the ion can not enter. It is independent of the nature of the ligand donors. This is the same problem as channel selectivity. There is selectivity by size in an obvious way, e.g. in phospholipase A_2. We wish to examine the more open chain systems of charges such as several carboxylate (Glu), malonate (Gla) or phosphate head groups when they are side chains of proteins. We observe that calcium can be selectively bound over magnesium, e.g. in EGTA, but that this selectivity is absent in the binding to all individual head groups per se, e.g. glutamate, malonate or phosphate. For these simple monodentate ligands the binding of ions is similar almost independent of size, $Mg \geqq Ca$, and it is too weak to be of significance in biology. The proteins carry several such groups all of which are on flexible chains $-CH_2-CH_2-X$ for Glu and Gla and $-CH_2-O-PO_3^{2-}$ for phosphate. Selectivity of cation binding must arise from the use of packing repulsions, acting against overall attraction, such that the packing repulsions are much greater for magnesium than for calcium. Magnesium tends to remain as a partially or fully hydrated ion. This means that there is a stable conformation or ensemble of conformations into which the anionic protein can fold in order to bind calcium, but that this or any other conformation or ensemble of conformations, even approaching equal stability, does not exist for magnesium. The fold here includes the disposition of the side chains. The *absolute* stability of binding can be regulated so that at the relative concentration of magnesium and calcium in a particular biological compartment magnesium is not an effective competitor for the calcium-binding site. This, we surmise, is the situation for the glutamate centres of the calcium-ATPase (see Fig. 4), for the Gla centres of prothrombin and for the phosphate centres of casein. On the other hand the arrangement of charge in ATP itself is such that the binding is only to Mg^{2+}, given the concentration differential of 10^3 between Mg^{2+} and Ca^{2+} in the cytoplasm of cells. It is then the **charge density pattern** that decides the selectivity and not the magnitude of the anionic charge. The description is readily extended to charged surfaces of polysaccharides and membranes or to combinations of any of the components, proteins, polysaccharides and surfaces. We note that it is calcium not magnesium that is implicated in the cross-linking of all these components. Cross-linking involves a spread-out set of anion charges. A new example is provided by the modified amino acid, hydroxyaspartate.

M. The Function of Neutral Donors

Neutral donors such as carbonyl and hydroxyl groups are not obviously suitable as sites of binding for cations in water. Yet in various molecules they are observed to bind. Again reference to simple molecules shows that it is not to the electrostatic terms of dipole–charge interactions that we must look since we find that the dipoles of monodentate ketones and alcohols do not bind any cations in water. It is structured series of dipoles which bind (see Table 3). This is explicable on the basis of entropy gain on loss of hydration, and of cavity fitting giving rise to selectivity. We can go on to ask if open flexible chains of carbonyls or hydroxyls could bind to, say, magnesium or calcium in water. The answer is no, within the limitations imposed by biological concentrations. The implication is that binding constants of 10 could arise, but not of 10^3 without an anionic centre, e.g. between M^{2+} and polysaccharides, including DNA and RNA.

There is another question however. Could the binding become important if the loosely assembled dipoles are trapped in a hydrophobic (membrane) region?

We can imagine a relatively mobile set of carbonyl or alcohol (serine) donors retained in a membrane. What is the possibility of interacting with them, given that the M^{2+} ion is initially in the surrounding aqueous phase? Effectively we have changed Eq. 1 so that D is not constant in all three terms. In the transfer of the calcium ion from water to a membrane phase the bulk dielectric constant drops tenfold. We are asking if dipoles can screen the charge effectively in the low dielectric medium. There is of course the gain in the stability of the dipoles immersed in the organic medium by the neutralisation of their field. Such extractions of ions into organic solvents by dipolar reagents are well known in fact and they are usually much easier the lower the charge on the cation and the larger it is.

The importance of this analysis is that there can be effective transfer of ions such as Ca^{2+}, Na^{+} and K^{+} into membranes by dipolar fields, but this transfer is less likely for H^{+} and Mg^{2+}. Selectivity is altered by the density of charges on more and more rigid frameworks. We do not believe that extraction itself into the membrane phase is commonly brought about by such neutral but polar molecules. However, the interaction is a very plausible one in the case of the design of channels, as stressed by URRY et al. (1983).

The advantage of channels of neutral dipoles in a membrane is that they do not generate local deep potential energy minima so that ions are not trapped

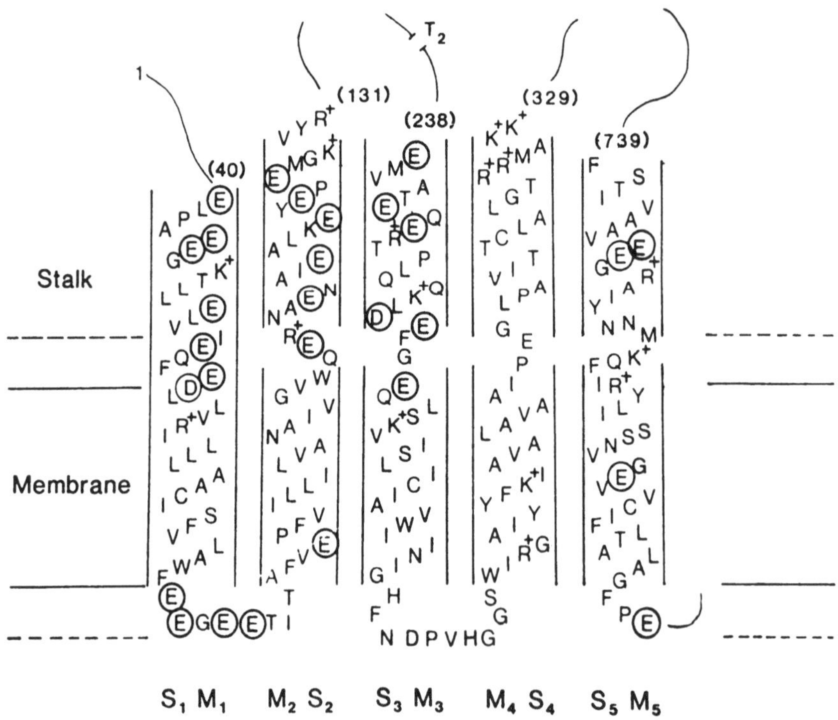

Fig. 4. Representation of the structure of the sarcoplasmic reticulum Ca^{2+}-ATPase in the membrane showing the helical segments that extend into the membrane deduced from the primary structure. (GREEN 1985)

within the channels, but flow freely. However, if channels are required in aqueous media it is doubtful if a mobile dipolar construction will give sufficient interaction or selectivity. Anionic surfaces in water do not give deep potential energy traps and in fact there are several immediately obvious cases where anionic surfaces may act as flow surfaces. The flow of ions along polyelectrolytes such as DNA must occur easily; the flow along the proposed surface of the calcium-ATPase helices must be easy (Fig. 4); the flow along the phosphoproteins of teeth and bone may also be the *raison d'être* for these proteins (COOKSON et al. 1980). Immediately we see that dynamics not so much of the protein backbone, but of the protein side chains may well assist ion flow. This is to be contrasted with the impingement of a cation at a site to generate flow of mechanical stress. In all of these we refer mainly to calcium and other inorganic cations, but the discussion is general in electrostatic terms and therefore applies to organic cations, e.g. polyamines, histones, and DNA regulators moving the surface of DNA. What is required is a very extended trap and not a very local one on a molecular scale (TAM and WILLIAMS 1985). In the final sections we give two specific examples in order to illustrate our general theme that the function of the calcium ion has to be understood in terms of the relationships.

Composition of ligand (protein) : structure : dynamics.

N. Calmodulin: An Example of a Calcium Trigger

Calmodulin (approximate molecular weight 17000) has been found in a structurally conserved form in a wide range of eukaryotic cells (KLEE and VANAMAN 1982). The design and operation of the calcium-modulated conformational trigger has been recognised by NMR studies (LEVINE and DALGARNO 1983; DALGARNO et al. 1984 b) and the recently determined solid state structure (SUDHAKAR et al. 1985). The calmodulin molecule is essentially bimodular and is made up of two closely homologous domains each containing a pair of calcium-binding sites of binding constant, $K_b \sim 5 \times 10^5$. The individual sites are flanked by short helical segments and are paired in each domain in a structural motif that brings the calcium-binding regions (loops) back to back to form a short two-strand, β-sheet (Fig. 5). The relative orientation and mutual disposition of the helical segments that flank the binding loops is determined by the backbone torsion angles within the β-sheet. A flow of conformational change throughout each (helix–binding loop–helix)$_2$ domain is therefore possible whereby adjustment of the helical elements is induced by changes at the binding loops joined by an array of hydrogen bonds to form the small β-sheet.

The movements in the β-sheet initiated by calcium binding can be readily analysed by NMR since the variation in energy of a resonance of a given backbone $C_\alpha H$ signal can be correlated with an alteration in the twist of the β-sheet. By contrast with an extended series of strands that impart rigidity to the β-sheet configuration, the short two-strand segment between pairs of calcium-binding sites can readily adopt a variety of somewhat different conformational states. Examination of the binding of protons, magnesium and calcium to the protein has shown that the β-sheet strands vary in contact angles for the H^+-, Mg^{2+}- and Ca^{2+}-

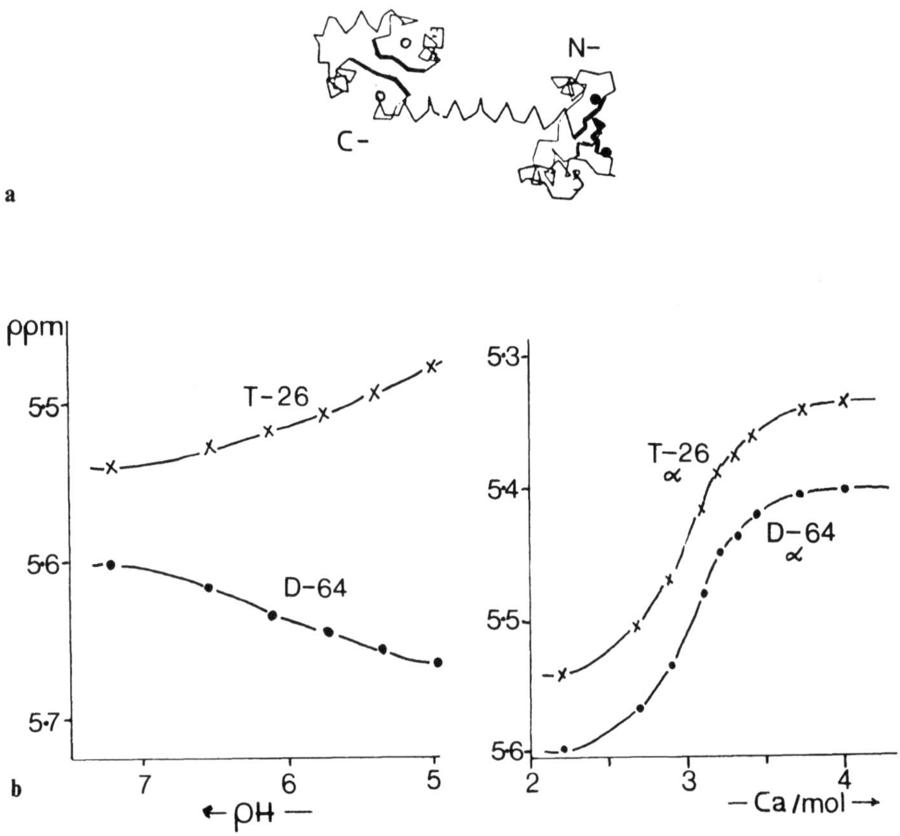

Fig. 5. a The bilobed structure of calmodulin, pH 5.6. The short β-sheet regions between each pair of calcium-binding sites are indicated by *bold lines*. Note the different twist of the two strands in the COOH and NH_2 terminal domains. **b** Variation of chemical shift of two NH_2-terminal β-sheet backbone proton signals with H^+ (left) and Ca^{2+} (right) binding, the latter at pH 7.1. The differences reflect alterations in the twist of the β-sheet upon binding of the two cations. The binding of calcium to each pair of sites is cooperative, calmodulin existing free in solution in the apo, Ca_2-bound and/or Ca_4-bound states

bound conditions (Fig. 5). The local reorganisation of the β-sheet configuration upon binding by these cations at the loop sequences gives a different adjustment of the side chain disposition at the terminal helical interface in each domain of the molecule. The exposure of the helical faces thus differs not only from calcium-free to calcium-bound forms, but also varies upon substitution of Mg^{2+} or H^+ for Ca^{2+}.

The helical surfaces exposed upon calcium coordination contribute to the sites of binding of noncompetitive drugs as well as to the surfaces of contact with partner enzymes whose activity is modulated by calmodulin (KLEE and VANAMAN 1982; DALGARNO et al. 1984a, b). The binding of drugs that antagonise calcium activation occurs differentially to the calcium-bound and calcium-free states and causes a reduction in the calcium off-rate. The kinetics of calcium dissociation are paralleled by the small adjustments to the β-strand linking the calcium-binding loops remote from the site of drug binding. Such drugs are therefore allosteric effectors of calcium binding and underline the cooperativity between local and global fold energy (see Sects. G, H, J) that determines the affinity of a site for calcium. Transmission of the drug-binding effect operates in the reverse fashion to the mechanical adjustment of the conformation induced by calcium – a relative reorientation of the crossed terminal helical segments in each domain so that interaction through the helices runs to the calcium-binding loops.

Antagonism of the calcium trigger by drug binding operates by restricting the accessibility of the protein to particular conformational states. Although the tertiary fold of the calmodulin molecule is stabilised by calcium binding, the solution structure is observed to encompass a range of conformations with side chain mobility (rotational and vibrational) that is relatively unrestricted in each domain. N*H* exchange rates indicate that even in the presence of calcium the overall fold possesses relatively easy internal relaxation as would be expected from the functional role of calmodulin as a transducer. Similarly, considerable rotational mobility is found to be available to the cluster of three aromatic residues which contribute to the interhelical contacts in the tertiary fold of each domain. Greater mobility is found in the calcium-free protein. These characteristic contacts provide the fulcrum of the structural change that occurs upon calcium binding. The flanking structural elements swivel about the binding loops whose β-sheet configuration is perturbed by cation coordination. Although the outline fold is similar upon substitution of calcium by H^+ or Mg^{2+}, the local conformation of the hydrophobic cluster is altered to a small degree (as in the β-sheet) with a resulting distortion in the disposition of the helices. The kinetics of conformational adjustment (Table 4) are closely matched to the role of calcium as an on–off switch through the very flexibility of the protein conformation that enables easy relaxation of any structural strain between the binding loops and the rest of the molecule.

As in the case of calcium coordination, structural flexibility of the loops at the opposite ends of the helices enables dynamic yet specific interaction with target enzymes. Such binding to the surface of the calmodulin molecule interacts cooperatively with the calcium-binding regions of the protein. Directed transmission of information is achieved by rotation and lateral motion of helical segments enabled by the intrinsic conformational mobility of the protein structure. In this way it can be seen how the transduction of the calcium signal can be modulated by any molecule (e.g. a drug) that binds to the surface of the protein. Information input and information output are therefore mutually adjusting.

Table 4. Kinetic parameters of calcium binding to calmodulin and parvalbumin

Calmodulin[a]	
COOH terminal sites	3– 10 s^{-1}
NH_2 terminal sites	250–400 s^{-1}
Parvalbumin	$\leqq$ 20 s^{-1}

[a] The calcium off-rate is decreased approximately tenfold when calmodulin is complexed with drugs

Apart from the general accord between NMR and X-ray structure data the recently determined solid state structure shows the two domains of the calmodulin molecule to be linked by a length of exposed helix, the molecule resembling a dumbbell in overall conformation. No confirmation of the existence of the linker helix is available from the solution data which indicates that agreement with the crystal data is otherwise very good when comparison is made under solution conditions typical of the crystallographic study (pH 5.6). At this pH protons compete for the calcium sites and small changes in the twist of the β-sheet and the helical disposition are observed as protons bind (Fig. 5). Further, little evidence for structural cooperativity between the two domains of calmodulin has been obtained from solution studies of the isolated molecule. These show positive cooperativity for calcium binding only within each of the two two-handed individual domains. The possibility that the two domains do interact is raised by the variety of calcium-dependent contact sites of the molecule relevant to its role as a control protein in eukaryotic cells. The elongated shape may enable the maximisation of free energy coupling and the tunable adjustment of the protein configuration, the free energy of calcium binding being dependent upon the state of the organised assembly triggered by the calcium–calmodulin complex.

Clearly, these studies illustrate cases of cation–cation competition. We have also shown how indirect competition between drug and cation arises. In the next section we consider a possible case of cation–cation long-range cooperativity. Cation cooperativity with protein–protein interaction is known for calmodulins (e.g. KELLER et al. 1984).

O. S-100b: An Example of Ion–Ion Cooperativity

S-100b is a putative calcium-binding protein. In solution it is a dimer of two identical subunits. We shall treat it as a single subunit protein. We have studied its NMR spectrum, but there is little other direct evidence apart from CD data about its structure (MANI et al. 1983). The circular dichroism and NMR spectrum together with the sequence indicate that the protein has a typical four-helix fold, i.e. helix–hand–helix–loop–helix–hand–helix. An outline of the structure is given in Fig. 6. The points of interest here are that the hands which bind calcium are normal and we draw them forming a small β-sheet as indicated by the NMR spectrum. The diagnostic down-field shifted α-protons are present. Again we show the typical interlocking three aromatic rings identically placed in sequence to those in other calcium binding proteins and showing a similar NMR pattern. These rings form a fulcrum about which the helices twist and move on adding calcium, assuming the same description as we have given for the conformation change of other calcium-triggered proteins. The major differences between the S-100b protein and other calcium proteins is in the large number of histidines in the sequence. There are five. This gives the protein a special proton–calcium interaction which is unlike that in all the other proteins from which histidine is absent. The proton binding to the calcium hands has a pK_a in the region below 6.0, as is usual, and this binding is competitive with calcium. The pK_a values of the histidines divide into two groups. The first two are around 6.0, but the remaining three are

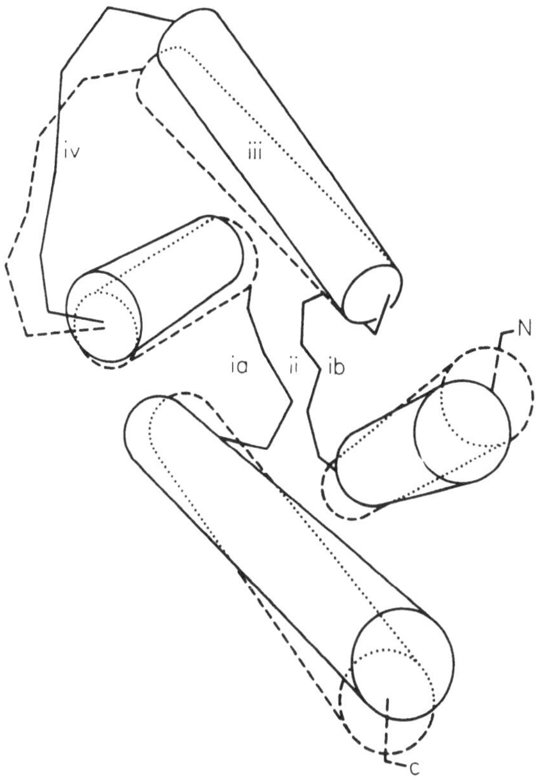

Fig. 6. Outline configuration of the fold of one pair of calcium-binding regions of the S-100b protein, showing the relative (vectorial) motion of the helical segments induced by calcium binding; as for calmodulin, calcium binding at the loops (*ia, ib*) alters the β-sheet (*ii*). This information is transferred via helices (*iii*) to the segments (*N*) that link the helices

above 7.5. The protein alters conformation during the pH titration of the histidines as shown by NMR. Thus, there is competition between the conformational demands of proton binding and calcium binding so that one adjusts the other. It is the helical structure of the protein that generates this feature. We do not know if it is of physiological significance.

The binding of metal ions is also intriguing since it is known that the protein binds zinc at the histidine sites. The binding is relatively strong. Thus, there is direct proton–zinc competition and indirect zinc–calcium and proton–calcium competition. We need a modified series of equations to handle even this situation, but the problem is further aggravated in that both magnesium and alkali metal ions bind to the calcium sites in direct competition. The protein then becomes a general sensor for the levels of H^+, Mg^{2+}, K^+, Na^+, Zn^{2+} and Ca^{2+}. The significance is not known.

A final example of mutual interaction between two regulators for one protein is provided by anion binding – phosphorylation.

P. A Look at Phosphorylation Regulation

It is not amiss here to contrast the way in which calcium is used as a control with the way in which the anion, phosphate, is used since it may well be that the use of these two doubly charged ions, one positive the other negative, is extremely similar in the sense that both alter the electrostatic interactions within and between polymers. In this sense phosphate acts as "negative calcium", i.e. compare the formulae P^{2-} with Ca^{2+}, one reducing the positive charges of certain regions of the protein surface and the other reducing the negative charge on such surfaces. This superficial comparison hides the very interesting differences in the way the negative P^{2-} is used relative to the positive Ca^{2+}. The differences are largely due to kinetic not thermodynamic factors. Inorganic phosphate as such could have been used as a control, but it seems that it is not so used.

Essentially P^{2-} is bound *covalently* to a protein (usually a serine or a threonine) close to a group of arginines or lysines. Covalent binding is open to control by enzymes – the serine–OP^{2-} or threonine–OP^{2-} are actually unstable covalent bonds and must be made from energised compounds such as acyl phosphates or ATP, e.g. by the kinases. Moreover, the release of P^{2-} from a covalent bond in a protein is also controlled by enzymes, e.g. the phosphatases. We see then that P^{2-} is delivered by a covalent carrier ATP or acylP, in which the phosphate is in an energised condition which can diffuse readily, but the specificity of its attachment to proteins is due to protein–protein recognition not phosphate recognition. Its release is also due to protein–protein recognition not to the undoing of phosphate interaction with positive charges. Thus, all the processes involved with P^{2-} are *kinetically specifically controlled* by catalysis (enzymic reactions) not by electrostatics. It is only in the effect of the P^{2-} on the protein to which it is bound that P^{2-} interactions could be described in electrostatic terms, that is the conformation change of the protein P^{2-} can be likened to that of Ca^{2+} protein.

Now, just because both the on- and off-reactions of P^{2-} are enzymatic and kinetically specific the attachment of P^{2-} must be on an *open surface*. The contrast with calcium could not be greater. Apart from the different regions of a protein which are used we see that calcium triggers in a local though not necessarily deep cavity in a protein and can not act in cross-linking while triggering. However, these differences in structure must not hide the even greater differences in the kinetics. Calcium binding and specific site action is controlled by thermodynamics (governed by electrostatics) limited by the rates of diffusion of the ion in the matrix. The off-reaction is limited then by the thermodynamic binding constants. These points are collected in Tables 3 and 4. These data show that while calcium effects are fast and fast relaxing, controlled by the inflow and pump-out reactions in large part, i.e. physical processes, phosphate on–off reactions are slow regulators controlled by chemistry.

Q. Conclusion

Perhaps the main lesson of this study is that the interaction of small molecules, drugs and ions with proteins is interesting not just because of the binding constants and structures. There is also the effect of binding on a whole range of time-dependent phenomena. The protein matrix is the "solvent" for many biological reactions and there are equivalent properties to solvent viscosity, solvent cooperativity, and so on which are familiar in fluid dynamics and which must now be put into protein activities. The dynamics also concern the control over flow along given field directions through channels and pumps and along surfaces. Thus, mechanical energies become important in many activities. All of which adds new dimensions to such topics as drug design.

References

Birdsall WJ, Levine BA, Williams RJP, Demaille JG, Haiech J, Pechere JF (1979) Calcium and magnesium binding by parvalbumin. Biochimie 61:741–750

Cookson DJ, Levine BA, Williams RJP, Jontell M, Linde A, de Bernard B (1980) Cation binding by the rat incisor dentine phosphoprotein. Eur J Biochem 110:273–278

Dalgarno DC, Klevit RE, Levine BA, Scott GMM, Williams RJP, Gergely J, Grabarek Z, Leavis PC, Grand RJA, Drabikowski W (1984a) Nature of the trifluoperazine binding sites on calmodulin and troponin-C. Biochim Biophys Acta 791:164–172

Dalgarno DC, Klevit RE, Levine BA, Williams RJP (1984b) The calcium receptor and trigger. Trends Pharmacol Sci 5:266–271

Edmonds DT (1979) A reversible electrostatic channel for ion transport. Chem Phys Lett 65:429–436

Green M (1985) Primary structure of the calcium ATPase. Ciba Found Symp 122:93–107

Keller CH, Bradley BD, LaPorte DC, Storm DR (1984) Determination of the free energy coupling for binding of calcium ions and troponin-I to calmodulin. Biochem 21:156–162

Klee CB, Vanaman TC (1982) Calmodulin. Adv Protein Chem 35:213–321

Levine BA, Dalgarno DC (1983) The dynamics and function of calcium binding proteins. Biochim Biophys Acta 726:187–204

Levine BA, Williams RJP (1982a) Calcium binding to proteins and other large biological anion centres. In: Cheung WJ (ed) Calcium and cell function, II. Academic, New York, pp 1–38

Levine BA, Williams RJP (1982b) The chemistry of the calcium ion and its biological relevance. In: Anghileri LJ (ed) The role of calcium in biological systems. CRC, Boca Raton, pp 4–26

Mani RS, Shelling JG, Sykes BD, Kay CM (1983) Spectral studies on the calcium binding properties of bovine brain S-100b protein. Biochem 22:1734–1740

Phillips CSG, Williams RJP (1966) Inorganic chemistry. Oxford University Press, Oxford

Sudhakar YB, Sack JS, Greenhough TJ, Bugg CE, Means AR, Cook WJ (1985) Three-dimensional structure of calmodulin. Nature 31:37–40

Sundralingam M, Bergstrom R, Strasburg G, Rao ST, Roychowdhury P, Greaser M, Wang CB (1985) Molecular structure of troponin-C from chicken skeletal muscle. Science 227:945–948

Tam SC, Williams RJP (1985) Electrostatics and biological systems. Structure Bonding 63:105–151

Tam SC, Williams RJP (1986) One problem of acid rain: aluminium. J Inorg Biochem 26:35–44

Urry DW, Trapane TL, Prasad KU (1983) Is the Granicidin A transmembrane channel single stranded? Science 221:1064–1067

Williams RJP (1953) A systematic approach to the choice of organic reagents for metal ions. Analyst 78:586–593

CHAPTER 3

Troponin C and Calmodulin as Calcium Receptors: Mode of Action and Sensitivity to Drugs

S. EBASHI and Y. OGAWA

A. Introduction

The first suggestion of the essential role of Ca^{2+} was made in cardiac muscle research more than a century ago, in 1883, and the second, more direct indication came from a study on skeletal muscle in 1940. Studies of Ca^{2+} as the intracellular regulator were, until recently, almost exclusively confined to muscle, so the history of Ca research in muscle is per se that of the early stage of Ca research in general. A brief history is given in Sect. B. The discovery of troponin (TN) in 1965, the first Ca receptor protein, provided a firm basis for the Ca concept in muscle. The subsequent discovery of calmodulin (CaM) around 1970 emancipated Ca^{2+} from muscle. Now Ca^{2+} is the common interest of all biological scientists.

A number of review articles have been published concerning Ca regulation of muscle and various tissues by TN (EBASHI 1974; OHTSUKI and NAGANO 1982; POTTER and JOHNSON 1982; LEAVIS and GERGELY 1984; OHTSUKI et al. 1986), and CaM (CHEUNG 1980; KAKIUCHI 1981; KELLER et al. 1982a; KLEE and VANAMAN 1982; ROUFOGALIS 1982; COX et al. 1984). To avoid overlap with these articles, it may be reasonable to focus this chapter on the Ca-binding properties of TNC and CaM and their physiologic and pharmacologic significance.

B. Brief Historical Survey of Ca Receptor Proteins

The development of the Ca concept in muscle research was far from straightforward. The essential point of RINGER's revolutionary discovery (1883) was masked by the elegant explanation, given by himself and his followers, that Ca^{2+} was necessary for maintaining the physiologic excitability not only of cardiac muscle, but also of skeletal (RINGER 1886) and smooth muscle (STILES 1901). Since then, more than half a century passed quietly, until HEILBRUNN (1940) intuitively sensed the importance of Ca^{2+} in muscle contraction. Prior to his proposal, several scientists had already observed a vigorous contraction-inducing action of Ca^{2+} on injured or somewhat denatured fibers, but none of them claimed that Ca^{2+} was a physiologic factor. Therefore, though the Ca injection experiments carried out later by KAMADA and KINOSITA (1943) and HEILBRUNN and WIERCINSKI (1947) were important, HEILBRUNN's courageous conclusion made in 1940 should be seen as the most crucial contribution to Ca research.

In spite of the effort of HEILBRUNN and his followers, the Ca concept was overshadowed by the brilliant success of the actomyosin–ATP system initiated and developed by SZENT-GYÖRGYI (1942) and his colleagues. In the 1950s, studies on the actomyosin system, which appeared to have nothing to do with Ca^{2+}, were almost synonymous with muscle research for those who wished to inquire into the biochemical mechanism of muscle contraction.

The modern Ca concept of muscle contraction was initially developed through studies which were subsidiary to the mainstream of muscle research at that time. The discovery of natural relaxing factor (MARSH 1951) and its identification with Kielley and Meyerhof's granular ATPase (KUMAGAI et al. 1955; KIELLEY and MEYERHOF 1948a, b) were the crucial steps in this direction, but the recognition of the relaxing effect of EDTA (BOZLER 1954; WATANABE 1955) and other chelating agents, including EGTA (EBASHI 1959) was no less important in finally establishing the role of Ca^{2+} in the contraction–relaxation cycle at the molecular level (WEBER 1959; EBASHI 1960, 1961; EBASHI and LIPMANN 1962; WEBER and HERZ 1963). Discovery of native tropomyosin (EBASHI 1963) and then TN (EBASHI and KODAMA 1965) answered the question why Ca^{2+} did not work on the pure actomyosin system (WEBER and WINICUR 1961); for those who remained unconvinced, this question had appeared to be a strong challenge to the Ca^{2+} hypothesis.

Nevertheless, Ca^{2+} was still confined to muscle. Even the work showing the effect of Ca^{2+} on excitation–metabolism coupling (i.e., Ca^{2+} in the same concentration range as that for muscle contraction activated phosphorylase *b* kinase; the key enzyme in glycolysis (OZAWA et al. 1967) did not evoke much interest in Ca^{2+} by biochemists outside the muscle field (the history up to this time has been described in more detail in the review article by EBASHI and ENDO 1968).

Influenced deeply by the work on TN, Kakiuchi, working at that time in Sutherland's laboratory, drew up a plan to investigate the role of Ca^{2+} in the brain. Encouraged further by the work on Ca activation of phosphorylase *b* kinase, he discovered the accelerating effect of Ca^{2+} on brain phosphodiesterase (KAKIUCHI et al. 1970) and found that a factor, which he named modulator protein (KAKIUCHI et al. 1969; KAKIUCHI and YAMAZAKI 1970a, b), was responsible for this effect. Meanwhile, investigations by CHEUNG (1970) of the activator of phosphodiesterase enabled him to identify a factor which was shown to be identical with that of Kakiuchi. Cheung's factor was later related to Ca^{2+} by TEO and WANG (1973); without this work the determination of the chemical nature of the factor, named calmodulin, would have taken much longer.

In the meantime, HARTSHORNE and MUELLER (1968) succeeded in separating TN into two fractions, troponin A, now called troponin C (TNC), and troponin B, a mixture of troponin I and troponin T. Influenced by the development of TN research, PECHÈRE et al. (1971) demonstrated the Ca-binding property of parvalbumin, a protein which had been discovered by DEUTICKE (1934) and crystallized by HENROTTE (1955), but its function has remained undecided (GILLIS 1980; ROBERTSON et al. 1981; GILLIS et al. 1982; OGAWA and TANOKURA 1986a, b). Thus, three Ca-binding proteins of relatively low molecular weight were known at the beginning of the 1970s.

However, it must be emphasized that, though TN was certainly the first protein recognized as the Ca receptor, actin was the first protein studied as the Ca-binding protein (FEUER et al. 1948); in its monomer form (G-actin) it has an affinity for Ca^{2+} of the order of 10^5 M^{-1} or larger, comparable to that of CaM

(KASAI and OOSAWA 1968) and in its aggregated form (F-actin) retains Ca so firmly that it can not be removed by ordinary procedures. Later, this Ca binding was shown to be a mere in vitro artifact and Mg is the main physiologic constituent of F-actin (WEBER 1966; KASAI 1969). This has suggested that we should discriminate between Ca-binding proteins and Ca receptor proteins; the latter should be used only for those which have physiologic significance. Since Ca research is now very popular, a cautious attitude may be necessary to distinguish true Ca receptor proteins from in vitro Ca-binding proteins.

C. General Views of TNC and CaM

Although physicochemically related, TNC and CaM are quite distinct from each other as Ca receptor proteins. First of all, TNC is firmly included in the TN molecule, just like the light chains in the myosin molecule. In all circumstances, the three subunits behave as a unit, functionally as well as physicochemically, none of them being detached from the parent particle (as myosin is not called the "myosin complex," the term "TN complex" is not justified, whereas the term "TN–tropomyosin complex" may be a reasonable one). On the other hand, CaM is characterized by its being easily dissociated from and associated with the partner proteins. The "flip-flop" mechanism, which can be seen in many CaM-related reactions (KAKIUCHI 1981; KAKIUCHI and SOBUE 1981), may be considered as an inherent property of CaM. In this connection, it is interesting that parvalbumin exists as a lone protein, independent of other proteins.

This may be reflected in the sensitivity of proteins to drugs. So far, no marked inhibitor of the function of TNC has been found, whereas CaM has various kinds of drugs which inhibit its function and the "CaM antagonists" are becoming popular, a new genre of pharmacologic agents, some of which are useful not only in cell-free experiments, but also for intact cells (see Sect. G).

Second, TN, like myosin, is one of the most differentiated proteins, as represented by its immunochemical reactivity (TOYOTA and SHIMADA 1981) as well as its functional variety (cf. the review articles mentioned in Sect. A). TNC and other TN subunits are specialized for the function of the muscle that contains them. For example, cardiac TNC provides cardiac TN and hybrid TNs with higher Sr^{2+} sensitivity than does skeletal TNC (EBASHI et al. 1968; YAMAMOTO 1983; MORIMOTO and OHTSUKI 1987). In contrast with this, CaM is well known as one of the most conservative proteins and can adapt itself to cooperation with various proteins.

D. Ca Binding of TNC and CaM

The general physicochemical properties of TNC and CaM are listed in Table 1 and the properties related to Ca binding in Table 2.

Both proteins have four Ca-binding sites. One of the important questions is whether or not there is cooperativity among the four Ca-binding sites. Virtually no cooperativity is found in the case of TN and TNC. In the case of CaM, it is

Table 1. Comparative properties of troponin C and calmodulin[a]

	Troponin C	Calmodulin
Amino acid residues	159	148/151
EF-hand domain	4	4
Bound Ca (mol/mol)	4	4
Molecular weight	18000	16700/17000
$s_{20,w}$	1.9	1.85/1.9
Stokes radius (Å)	24.0	20.9/23.2
$D_{20,w}$ ($cm^2/s \times 10^7$)	8.9	10.9/9.2
f/f_0	1.37	1.20/1.34
$\bar{v}$ (ml/g)	0.72	0.71, 0.72/0.72
Isoelectric pH	4.2	4.3/3.9
Percentage α-helix (circular dichroism)		
−Ca	19	28–39/45
+Ca	40	42–57/54
Enhancement of tyrosine fluorescence (+Ca)	1.5-fold	2.5-fold
3-Dimensional structure	Unusual dumbbell shape about 70 Å long with the two lobes connected by a long nine-turn α-helix	Dumbbell (about 65 Å long) with the two lobes (25 × 20 × 20 Å) connected by an eight-turn α-helix[b]

[a] The results shown here are mostly from rabbit skeletal muscle for TNC and from bovine brain and rat testis (indicated as 148/151 etc.) from CaM.
[b] Small-angle X-ray scattering experiments reveal that the Ca^{2+}-free form of CaM has a smaller maximum vector length, by approximately 4 Å, than the Ca^{2+}-containing form, which corresponds to an increase in Stokes radius from 20.6 to 21.5 Å upon Ca binding (SEATON et al. 1985). In contrast to this, spectroscopic (circular dichroism) and sedimentation velocity studies indicate an increase in ellipticity and a reduced Stokes radius from 21.4 to 20.9 Å upon Ca binding to the CaM molecules (CROUGH and KLEE 1980).
References for TNC: DEDMAN et al. (1977); SUNDARALINGAM et al. (1985); HERZBERG and JAMES (1985). For CaM: DEDMAN et al. (1977) LIN et al. (1974); LIU and CHEUNG (1976); WOLFF et al. (1977); WATTERSON et al. (1980); CROUCH and KLEE (1980); BABU et al. (1985); Seaton et al. (1985).

not easy to give an exact answer to this question, but no serious trouble will occur even if we treat the four Ca-binding sites as being independent of one another (for details see Sect. F).

The second important question may be whether, and how, Mg^{2+} interferes with Ca binding. Since the free Mg^{2+} concentration is kept constant around 1 m*M* in cytoplasm, it can not take part directly in the regulation mechanism. However, if the affinity of the Ca-binding site for Mg^{2+} is high enough to be affected by 1 m*M* Mg^{2+} and the rate of Mg release from the sites is low, the availability of the site for Ca^{2+} is more or less restricted by Mg^{2+} under physiologic conditions. There is general agreement that the high affinity sites of TN and TNC have some affinity for Mg^{2+}, but it has been controversial whether or not the low affinity sites are influenced by Mg^{2+}. It was claimed that the low affinity site had no affinity for Mg^{2+}; Mg^{2+} certainly represses Ca binding, but this is interpreted as the result of an allosteric effect (POTTER and GERGELY 1975; JOHNSON et al.

Table 2. Comparison of calcium binding properties of calcium-binding proteins and EGTA[a]

Ca (Mg) binding	Parvalbumins		Troponin C		Cal-modulin	EGTA
	PA-1	PA-2	High	Low		
Binding sites (mol/mol)	2	2	2	2	4	1
K_{Ca} (M^{-1})	1×10^7	6×10^6	4.5×10^6	6.4×10^4	2.0×10^5	1.0×10^6
k_{on} (Ca) $(M^{-1}s^{-1})$[b]	$\geqq 1.5 \times 10^7$	7×10^6	$\geqq 4 \times 10^7$		1×10^7	1.4×10^6
K_{Mg} (M^{-1})	900	830	1000	520	130[c]	ND
Ionic strenght dependence[d]	0.4–0.8	0.35–0.65	1.0	2.9	3.2	1.0
$\Delta \log K_{Ca}/\Delta$ pH (pH=6.5–7.2)	0		0	0	0	2.0
Temperature dependence	Marked		PA-like	CaM-like	Negli-gible	Weak

[a] Experiments were carried out in the reaction mixture containing 100 mM KCl, 20 mM MOPS-KOH pH 6.80, and 0.13 mM tetramethylmurexide (ionic strength 0.106) at 20 °C. When Mg^{2+} was to be added, the ionic strength was kept constant by reducing KCl. Parvalbumins from bullfrog skeletal muscle (OGAWA and TANOKURA 1986a, b), troponin C from rabbit skeletal muscle (OGAWA 1985), calmodulin from bovine brain (OGAWA and TANOKURA 1984, 1985), and EGTA (HARAFUJI and OGAWA 1980).
[b] Association rate constant per binding site. POTTER and his associates assume a value of $1 \times 10^8\ M^{-1}\ s^{-1}$ for k_{on} for Ca^{2+} of diffusion-limited reactions (e.g. JOHNSON et al. 1981; POTTER and JOHNSON 1982). However, reported values for biologic molecules which have actually been determined range from 10^5–$10^9\ M^{-1}\ s^{-1}$ because of the marked effect of net charge and steric configuration of macromolecules as well as the diffusion coefficient and dehydration of water molecules (HIROMI 1978).
[c] As Mg^{2+} concentration was increased, not only the apparent affinity for Ca^{2+}, but also the total number of Ca-binding sites were decreased. K_{Mg} was calculated from apparent affinities for Ca^{2+} in the presence of various concentrations of Mg^{2+} under the assumption that Mg^{2+} and Ca^{2+} would compete with the binding sites.
[d] The results are values of the coefficient B in the equation $\log K_{Ca} = A - B[2\sqrt{I}/(1+\sqrt{I}) - 0.4\,I]$ where I is the ionic strenght.
ND = not detectable.

1980; POTTER et al. 1980, 1981). Reviewing all the reports, however, it is more likely that Mg^{2+} also competes with Ca^{2+} in binding to the low affinity sites (Tables 2 and 3). Ca binding of CaM is only slightly affected by Mg^{2+} (Table 2).

The third problem, not so physiologically important as the first two items, but often crucial for in vitro experiments, is the influence of ionic strength on Ca binding. CaM is far more susceptible to this factor than the other two proteins. An interesting conclusion may be deduced from Table 2 that the properties of the high affinity sites of TNC resemble those of parvalbumin, whereas those of low affinity sites resemble those of CaM.

Finally, the difficulties in estimating Ca-binding properties of these proteins will be listed:

1. The rates of Ca binding to TNC and CaM are extremely fast compared with EGTA, more than 100 times that of EGTA on molar basis (Table 2). TNC

Table 3. Calcium binding to troponin Cs and troponins[a]

Protein	High affinity site			Low affinity site			References
	N (mol/mol)	K_{Ca} (M^{-1})	K_{Mg} (M^{-1})	N (mol/mol)	K_{Ca} (M^{-1})	K_{Mg} (M^{-1})	
Skeletal							
troponin C	2	2×10^7	5×10^3	2	2×10^5	ND	POTTER and GERGELY (1975)
	2	4.5×10^6	10^3	2	6.4×10^4	520	OGAWA (1985)
troponin	2	5×10^8	5×10^4	2	5×10^6	ND	POTTER and GERGELY (1975)
	2	$\geqq 10^7$	384[b] 2×10^4[c]	2	$\geqq 10^5$	90[b]	KOHAMA (1980)
	2	2.5×10^8	2×10^4	2	3.5×10^6	440	OGAWA (1985)
Cardiac							
troponin C	3	7×10^6	3.5×10^3	ND	ND	ND	BURTNICK and KAY (1977)
	2	1.4×10^7	$(0.7–3.5) \times 10^3$	1	$(2.5–4.5) \times 10^5$	ND	JOHNSON et al. (1980); POTTER and JOHNSON (1982)
	2	1.4×10^7	0.7×10^3	1	2.5×10^5	ND	HOLROYDE et al. (1980)
troponin	1	4×10^8	384[b] 2×10^4[c]	1	10^6	45[b]	KOHAMA (1980)
	2	3.7×10^8	$(3–20) \times 10^3$	1	$(2.5–3) \times 10^6$	ND	JOHNSON et al. (1980); POTTER and JOHNSON (1982)
	2	4.2×10^8	3×10^3	1	1.7×10^6	ND	HOLROYDE et al. (1980)
	2	4×10^7	1.6×10^3	ND	ND	ND	STULL and BUSS (1978)

[a] Most experiments, but not those by OGAWA (1985) and (in part) STULL and BUSS (1978), were carried out in the presence of an EGTA-Ca^{2+} buffer system. The results shown in this table are uncorrected for the difference in the apparent binding constant of EGTA for Ca^{2+} which may be as large as fourfold. It should be noted that the results were obtained under different experimental conditions.
[b] Calculated from Ca binding in the presence of various concentrations of Mg^{2+}.
[c] Determined in the absence of Ca^{2+}.
ND = not detectable.

shows a somewhat faster rate than does CaM. This makes it very difficult to use EGTA to measure the rates of Ca binding to and Ca release from TN, TNC, and CaM.

2. To determine the Ca-binding properties of these proteins, measurements of the conformational change associated with Ca binding (isomerization), e.g., fluorescence (JOHNSON and POTTER 1978; JOHNSON et al. 1978, 1979), are often used (e.g., KLEE and VANAMAN 1982; POTTER and JOHNSON 1982). There may be no serious problem for equilibrium states (GRABAREK et al. 1983), but, in the case of transient phases, the situation will become very complicated (HIROMI 1978; see also JOHNSON et al. 1979; MALENICK et al. 1981). Unless precise analysis of the rates of forward and back reactions of all steps, including isomerization, is made, the values obtained will be meaningless.

As a whole, to study the kinetic properties of Ca-binding sites, it is most desirable to use the method of measuring the decrease of Ca^{2+} concentration in the medium containing Ca-binding proteins first developed by OHNISHI and EBASHI (1963), using a fast-reacting indicator. So far as the present authors' experiences are concerned, tetramethylmurexide (OHNISHI 1978) is the most reliable indicator, but its sensitivity to Ca^{2+} is low (this is no disadvantage in this kind of analysis) so that it requires a highly sensitive detector (for details of the method see OGAWA et al. 1980; OGAWA 1985).

Some Ca analogs such as lanthanides or transition metals have been used to assess Ca-binding properties of proteins. However, the results thus obtained are often at variance with those derived from experiments using Ca^{2+} itself (e.g., WANG et al. 1984; MILLS and JOHNSON 1985). In principle, we should avoid the use of such substitutes unless there is a specified purpose.

E. Ca Binding to TNC in Relation to Tension Development

Table 3 shows the general features of Ca binding to TN and TNC reported by different authors. The number of Ca-binding sites of cardiac TNC has been a matter for some argument. From the primary amino acid sequence in comparison with skeletal TNC, cardiac TNC has three EF-hand structure domains (VAN EERD and TAKAHASHI 1976). Therefore, it seemed plausible to assume three Ca-binding sites for cardiac TNC. Indeed, a few reports asserted that cardiac TNC had three binding sites, two of high affinity and one of low affinity. However, EF-hand structure (KRETSINGER and NOCKOLDS 1973) by itself can not give a high affinity for Ca^{2+} without the cooperation of other steric arrangements. A recent result (Y. OGAWA 1980–1981, unpublished work) has favored the opinion that it has only two sites, one high and one low (KOHAMA 1979).

There has been some controversy whether the affinity for Ca^{2+} is influenced by pH. Although some reports showed a slight increase in the affinity by elevating pH of skeletal TNC or TN (KOHAMA 1980; ROBERTSON et al. 1978), it is more probable that pH variation does not cause a substantial change in Ca binding around neutral pH (STULL and BUSS 1978; OGAWA 1985). This is in accord with the observations that H^+ release upon binding of Ca to TN and TNC is very small, around 0.01 mol/mol (POTTER et al. 1977; YAMADA and KOMETANI 1982).

Table 4. Comparison of calcium-binding properties between rabbit and bullfrog skeletal muscle troponin Cs[a]

	High affinity site		Low affinity site		Reference
	K_{Ca} (M^{-1})	K_{Mg} (M^{-1})	K_{Ca} (M^{-1})	K_{Mg} (M^{-1})	
Rabbit at 20 °C	4.5×10^6	1000	6.4×10^4	520	OGAWA (1985)
Frog at 20 °C	6×10^6	1100	8.6×10^4	700	Y. OGAWA and M. TANOKURA (1985–1986) (unpublished work)
Frog at 5.7 °C	3×10^7	2000	7.8×10^4	350	Y. OGAWA and M. TANOKURA (1985–1986) (unpublished work)

[a] The experimental conditions were similar to those in Table 2.

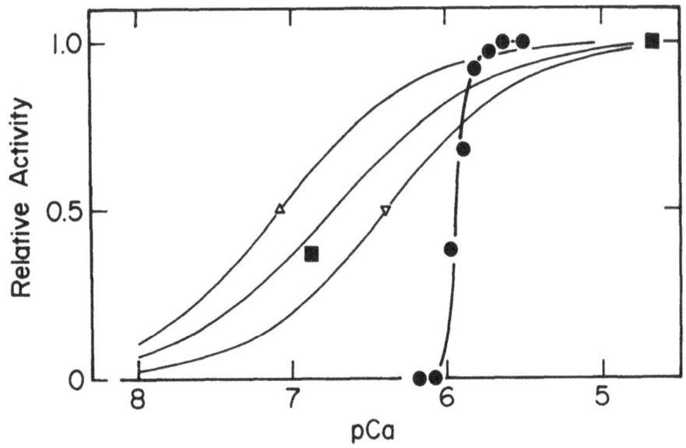

Fig. 1. Relationships between Ca binding to troponin and Ca^{2+}-induced tension. Ca binding at various pCa values is calculated on the basis of the results obtained with rabbit skeletal muscle troponin (OGAWA 1985), i.e., with binding constants of 1.19×10^7 and 2.43×10^6 M^{-1} for high and low affinity sites, respectively, considering the presence of 1 mM Mg^{2+} in the solution (ionic strength 0.11). *Thin curve with upward triangle* high affinity sites; *thin curve with downward triangle* low affinity sites; *thin curve with no symbol* TN as a whole. Unit values of these sites are 2, 2, and 4 mol/mol, respectively. The results by electron probe microanalysis of Ca binding to the thin filament of frog skeletal muscle (*full squares*) (KITAZAWA et al. 1982) are consistent with the calculated binding curve for TN. Ca^{2+}-induced tension (*thick curve with full circles*) developed by a skinned fiber from frog skeletal muscle under similar conditions is derived from the results of KUREBAYASHI and OGAWA (1985)

It is said that the affinity of TNC is strongly dependent on temperature (WNUK et al. 1984). As shown in Tables 2 and 4, this change is limited to the high affinity site, and virtually no temperature dependence is observed in the low affinity site, like CaM. This is consistent with the enthalpy changes reported by YAMADA and KOMETANI (1982), who showed that the high affinity sites show a large heat liberation upon Ca binding, but virtually no change is exhibited by the low affinity sites. Table 4 also indicates close similarities between frog and rabbit TNC.

As seen in Table 3, a definite increase in the affinity of TNC for Ca^{2+} can be observed when it is included in TN. Troponin I, which has both Ca-dependent

Table 5. Effects of various agents on Ca binding to troponin C and Ca^{2+}-induced tension development. Bracketed numbers refer to the references in the list under the table

Agents	Ca binding to troponin C	Ca^{2+}-induced tension
Ca^{2+}	No cooperativity [1–4]	Hill coefficient ~7 [5–7], see also [8–12]
H^+	See Table 2	Decrease[a] [13, 14]
Mg^{2+}	See Tables 2 and 4	Decrease[a] [8–10, 15]
Ionic strength	See Table 2	Decrease[a], decrease also in maximum tension [16–18]
Temperature	See Table 4, see also reference [19]	Decrease[a] [20, 21]
Sarcomere length		Increase[a] [22–26]
DMSO 1%	Unchanged [6]	Decrease[a], less steep pCa-tension relationship [6]
Quercetin 120 μ*M*	Unchanged [6]	Increase[a], less steep pCa-tension relationship [6]
Trifluoperazine 10–100 μ*M*	Increase in K_{Ca} [4]	Increase[a], less steep pCa-tension relationship [27], cf. [28]
Caffeine, dibucaine, halothane, quinidine, quinine, Triton X-100		Increase[a] [17, 27, 29–32]

[a] Decrease or increase in the Ca^{2+} sensitivity of the contractile apparatus with increase in the agent listed or upon its addition.

References: [1] Potter and Gergely (1975); [2] Kohama (1980); [3] Grabarek et al. (1983); [4] Ogawa (1985); [5] Brandt et al. (1980); [6] Kurebayashi and Ogawa (1985); [7] Gulati and Babu (1985); [8] Ebashi and Endo (1968); [9] Kerrick and Donaldson (1972, 1975); [10] Ashley and Moisescu (1977); [11] Moss et al. (1985); [12] Fabiato (1981); [13] Fabiato and Fabiato (1978a); [14] Donaldson and Hermansen (1978); [15] Fabiato and Fabiato (1975); [16] Gordon et al. (1973); [17] Thames et al. (1974); [18] Gulati and Podolsky (1978); [19] Wnuk et al. (1984); [20] Stephenson and Williams (1981); [21] Godt and Lindley (1982); [22] Endo (1972, 1973); [23] Fabiato and Fabiato (1978b); [24] Stephenson and Williams (1982); [25] Hibberd and Jewell (1982); [26] Moss et al. (1983); [27] N. Kurebayashi and Y. Ogawa (1985–1986, unpublished work); [28] Kerrick et al. (1981); [29] Endo and Kitazawa (1978); [30] Takagi (1976); [31] Yagi and Endo (1980); [32] Wendt and Stephenson (1983).

and virtually Ca-independent affinities for TNC, is responsible for this increase (Potter and Gergely 1975; Ogawa 1985). No agreement has been reached as to the influences of tropomyosin, F-actin, and myosin on the affinity of TN for Ca^{2+} (Bremel and Weber 1972; Murray et al. 1975; Fuchs 1977; Fuchs and Fox 1982; Grabarek et al. 1983; Zot et al. 1983). It is very probable that the affinity of the actomyosin system for Ca^{2+} is essentially the same as that of TN (Fig. 1).

In Fig. 1, the relationship of Ca^{2+} concentrations with Ca binding to TN is illustrated in comparison with the relationship to tension development. It is interesting that crayfish fiber, which has only one Ca-binding site per TN molecule, also shows a sharp pCa–tension relationship, almost the same as that of rabbit muscle fibers (Wnuk et al. 1984), indicating that the cooperativity is not determined by TN but by the nature of the actin–myosin–ATP interaction.

Physical and chemical influences on Ca binding to TNC and tension development are summarized in Table 5. There are many cases where the pCa–tension

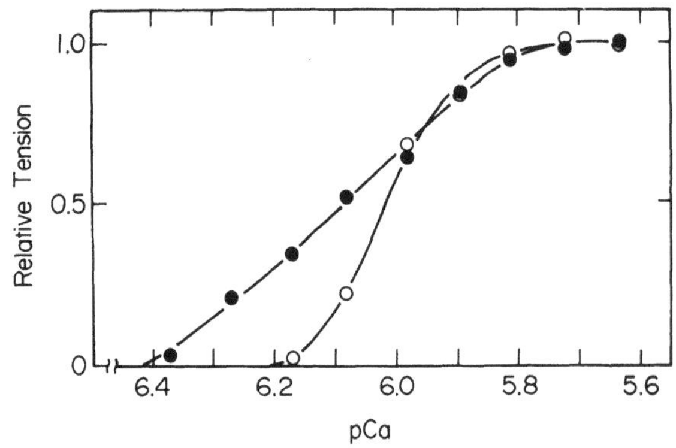

Fig. 2. Effect of quercetin on tension–pCa relationship for frog skinned skeletal muscle fiber. A mechanically skinned fiber was treated with Triton X-100 to remove the function of the sarcoplasmic reticulum. Ca binding to troponin C or EGTA is unaffected by quercetin. *Full circles* with 120 *M* quercetin; *open circles* without quercetin. The effect of quercetin is almost completely reversible. Unity for relative tension of the ordinate is defined as the tension developed at pCa 5.4 without the drug. For details, see KUREBAYASHI and OGAWA (1985)

relationship is altered without change in the affinity of TNC for Ca^{2+} (see legend to Fig. 2). Thus, Ca binding to TNC is not straightforwardly expressed as the tension.

The common properties of the agents that increase the Ca sensitivity of tension are:

1. Their effects are more marked at low Ca^{2+} concentrations and do not increase the maximum tension, giving rise to a less sharp pCa–tension relationship (Fig. 2; this indicates the possibility of exploring new types of contraction-modulating drugs for cardiac muscle where physiologic contraction is operating at low Ca^{2+} concentrations).
2. The rate of tension development is also enhanced.

F. Ca Binding to CaM and Enzyme Activities

CaM has four Ca-binding sites, the capacities of which are dependent on the ionic strength and influenced to some extent by Mg^{2+}, but not by pH and temperature. While the apparent binding constant of CaM for Ca^{2+} (ionic strength 0.15, 1 m*M* Mg^{2+}, neutral pH) is only 1.2×10^5 M^{-1}, CaM exhibits its multiple functions in the same range of Ca^{2+} concentration as that for muscle contraction, i.e., 1×10^{-7} to $2–3 \times 10^{-6}$ *M* Ca^{2+}, and the CaM-bound enzymes are activated by Ca^{2+} in a cooperative manner. The question then arises whether or not the mode of Ca binding to CaM is cooperative, which might be the basis of the cooperativity of the CaM-dependent enzyme actions.

A number of papers dealing with this matter have been published. For instance, CROUCH and KLEE (1980) explained their results by positive cooperativity and HAIECH et al. (1981) proposed the idea of ordered sequence binding to explain

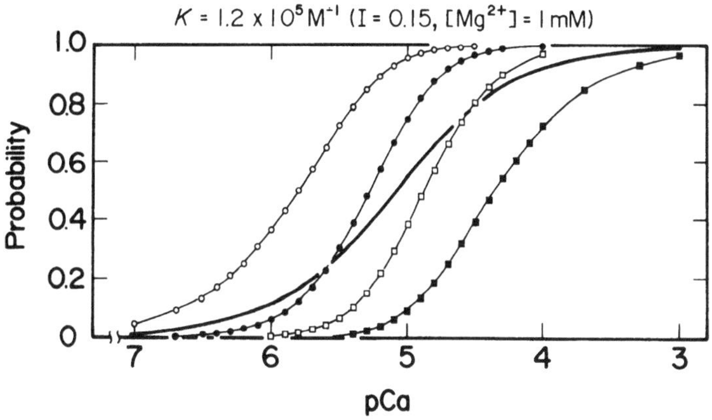

Fig. 3. Ca binding to calmodulin; probability of occupancy of calmodulin molecule with Ca. Calmodulin is assumed to have four independent Ca-binding sites per molecule, of which the intrinsic affinities for Ca^{2+} are estimated to be 1.2×10^5 M^{-1} in the solution containing 1 m*M* Mg^{2+} (ionic strength 0.15). The *bold continuous curve* represents Ca binding to calmodulin. From the binding curve, the probabilities of calmodulin molecules are calculated assuming the occupancy of binding sites with Ca: *open circles* at least one; *full circles* two or more; *open squares* three or four; *full squares* four sites occupied. (OGAWA and TANOKURA 1984)

their results, but BURGER et al. (1984) indicated that the same results could also be interpreted by assuming four independent sites. Using proteolytic fragments, MINOWA and YAGI (1984) showed some examples of cooperativity, a weakly to moderately positive one within each of the two fragments. However, it is generally agreed that the Scatchard plot of Ca binding to CaM is linear in the range 0.5–4 mol/mol under varied conditions, indicating no cooperativity in this range. Therefore, the arguments are concerned with the range 0.0–0.5 mol/mol, where experimental errors allow no clear conclusion.

Thus, even if such a cooperativity did exist, it could not form the basis for the cooperativity of the whole CaM–enzyme complex. As indicated by the calculation shown in Fig. 3, the amount of Ca-CaM complex in relation to the Ca^{2+} concentration shows a considerable cooperativity under some conditions, if we assume that the occupancy of two or three of the Ca-binding sites with Ca is enough for the function of CaM. However, these considerations can not give a satisfactory explanation for the fact that CaM activates enzymes at low Ca^{2+} concentrations, comparable to those at which TN exerts its physiologic function.

One of the characteristics of CaM is its abundant presence in many tissues. The content of CaM is (see Table 3 in KLEE and VANAMAN 1982; KAKIUCHI et al. 1982): very high (20–60 μ*M*) in brain, testis, gizzard, uterus, etc.; moderate (10–20 μ*M*) in aorta, lung, prostate, adrenal gland, kidney medulla, erythrocytes, etc.; or low (less than 10 μ*M*) in liver, kidney cortex, spleen, heart, and skeletal muscle. On the other hand, the concentration of most abundant CaM-dependent enzymes such as phosphodiesterase or myosin light chain kinase is about 1 μ*M*, much lower than the CaM concentrations in most tissues. Since the equilibrium constant defined by the following equation

$$K_{\mathrm{Enz}} = \frac{[\mathrm{Ca}_n \cdot \mathrm{CaM} \cdot \mathrm{Enz}^*]}{[\mathrm{Ca}_n \cdot \mathrm{CaM}][\mathrm{Enz}]}$$

has been reported to be 10^7–10^{10} M^{-1} (YAZAWA and YAGI 1978; LAPORTE et al. 1979; BLUMENTHAL and STULL 1980; CHEUNG et al. 1981; COX et al. 1981; CROUCH et al. 1981; HUANG et al. 1981; KLEE and VANAMAN 1982; MALENCIK et al. 1982; see also references cited in Table 7), the ratio of activated enzyme to free enzyme

$$\frac{[\mathrm{Ca}_n \cdot \mathrm{CaM} \cdot \mathrm{Enz}^*]}{[\mathrm{Enz}]} = K_{\mathrm{Enz}} [\mathrm{Ca}_n \cdot \mathrm{CaM}]$$

can readily be larger than 1, even at $\mathrm{Ca}^{2+} = 10^{-6}$ M, owing to high CaM concentrations, close to or higher than the reciprocal of the affinity of CaM for Ca^{2+}, viz., 10 μM.

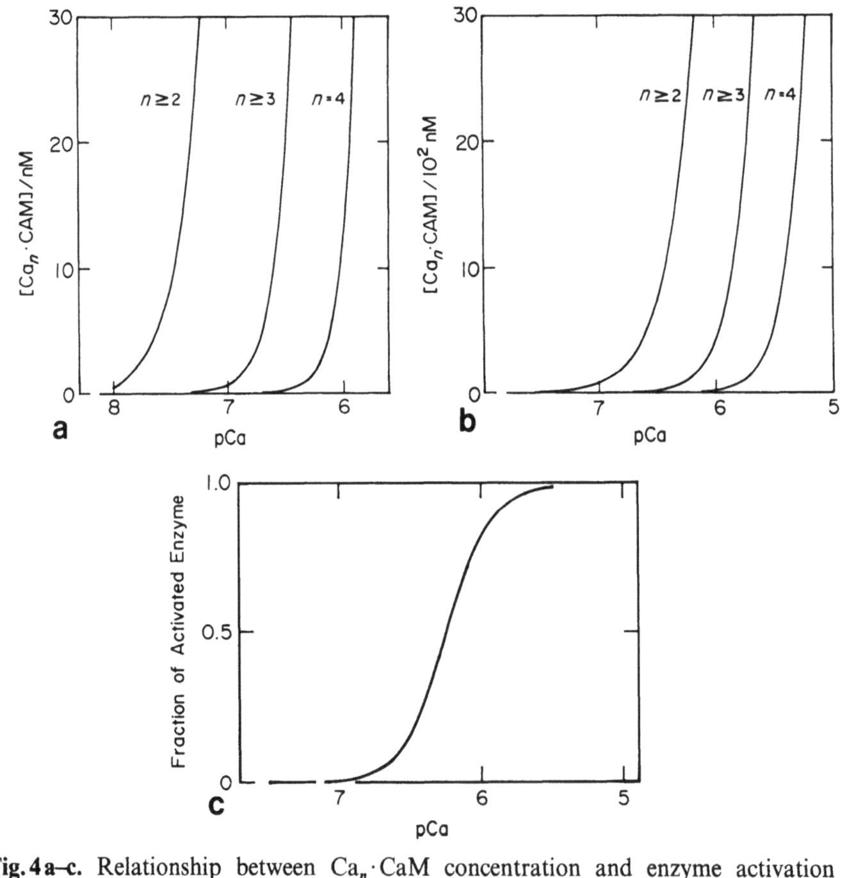

Fig. 4a–c. Relationship between $\mathrm{Ca}_n \cdot \mathrm{CaM}$ concentration and enzyme activation by $\mathrm{Ca}_n \cdot \mathrm{CaM}$. $[\mathrm{Ca}_n \cdot \mathrm{CaM}]$, calculated as in Fig. 3 using $K_{\mathrm{Ca}} = 1.2 \times 10^5$ M^{-1} and total concentration of calmodulin $[\mathrm{CaM}]_\mathrm{T} = 100$ μM, where K_{Enz} is taken as 10^9 M^{-1} (**a**) and 10^7 M^{-1} (**b**). The fraction of activated enzyme is calculated (**c**), assuming Ca occupancy of an active CaM molecule and K_{Enz} to be $n \geqq 3$ and 10^7 M^{-1}, respectively. As $[\mathrm{Enz}]_\mathrm{T}$ is much smaller than $[\mathrm{CaM}]_\mathrm{T}$ in most in vivo cases, the equilibrium between Ca^{2+} and CaM is scarcely perturbed by the presence of the enzyme in the two-step model of activation

In Fig. 4, the concentrations of activated CaM, [$Ca_n \cdot CaM$], are calculated ($K_{Enz} = 10^9\ M^{-1}$ in Fig. 4a; $10^7\ M^{-1}$ in Fig. 4b). One example of the amount of activated enzyme is shown in Fig. 4c, assuming K_{Enz}, n, total [CaM], and K_{Ca} to be $10^7\ M^{-1}$, $\geqq 3$, 100 μM, and $1.2 \times 10^5\ M^{-1}$, respectively. The Hill coefficient of this curve is 2.8, so altogether the profile is in accord with the actual activation curves of phosphodiesterase (HUANG et al. 1981) and myosin light chain kinase (BLUMENTHAL and STULL 1980). Thus, the cooperativity of CaM-activated enzyme action can be demonstrated by a simple scheme without assuming cooperativity of Ca-binding sites. This means that the CaM concentration in a given cell is a decisive factor in its CaM-related activity (BLUMENTHAL and STULL 1980; HUANG et al. 1981; COX et al. 1984; RÜEGG et al. 1984).

If K_{Enz} is $10^9\ M^{-1}$ or more, the dissociation rate of $Ca_n \cdot CaM$ from $Ca_n \cdot CaM \cdot Enz^*$ is expected to be very slow. This tendency may be enhanced if the affinity of complexed CaM for Ca^{2+} is apparently higher than that of free CaM. This is very probable, so it is expected that the off-rate of a CaM-dependent reaction is very much slower than the on-rate [1] (see Sect. G).

G. CaM Ligands: Their Pharmacologic and Physiologic Significance

In the early part of this section, the development of inhibitory drugs on CaM function, so-called CaM antagonists, will be briefly described, and in the later part, the nature of CaM ligands in relation to Ca-binding properties of CaM will be discussed.

I. Discovery of CaM-Inhibiting Drugs (CaM Antagonists)

Some psychotropic drugs, especially phenothiazine derivatives, such as chlorpromazine and trifluoperazine, were shown to have a strong inhibitory action on CaM-activated enzymes (LEVIN and WEISS 1976). Later experiments have shown that many drugs other than psychotropics have the same effect and, therefore, we are concerned only with their chemical affinity for CaM.

A high degree of parallelism was found between the hydrophobicity of phenothiazine derivatives and their inhibitory actions on CaM (NORMAN et al. 1979; PROZIALECK and WEISS 1982); the hydrophobic part of CaM would be exposed by its binding Ca and the hydrophobic drugs would bind to the exposed hydrophobic area (LAPORTE et al. 1980; TANAKA and HIDAKA 1980), resulting in the abolition of the function of CaM. The use of phenothiazine columns for chromatographic separation of CaM seemed to have some justification (JAMIESON and

---

[1] The inactivation of CaM-activated enzyme can be induced not only by this reverse reaction, but also by Ca^{2+} release from the complex

$$Ca_n \cdot CaM \cdot Enz \text{ (active form)} \rightarrow Ca^{2+} + Ca_{n-1} \cdot CaM \cdot Enz \text{ (inactive form)}$$

However, this process has not been thoroughly investigated yet; even the value of n for activation is unknown. So this problem, though physiologically important, will not be discussed here.

VANAMAN 1979; CHARBONNEAU and CORMIER 1979). This tendency appears to be recognized to some extent in the series of naphthalenesulfonamide complexes (TANAKA et al. 1982). Recent studies have revealed, however, that a considerable interaction of CaM and phenothiazines exists, even in the absence of Ca^{2+} (cf. Table 6; see also SHIMIZU and HATANO 1983, 1984; SHIMIZU et al. 1984).

However, the discovery of the strong inhibitory action of endorphin on CaM (SELLINGER-BARNETTE and WEISS 1982) has indicated that hydrophobicity is not the sole basis for the antagonism of CaM function, e.g., even among non-peptide antagonists, their activities are not related to hydrophobicity if comparison is made between different chemical series (PROZIALECK and WEISS 1982). As is seen in Table 6, quite different kinds of chemicals have inhibitory actions on CaM. The CaM antagonist is rather characterized by its diversity in chemical structure. Perhaps the only common property among these chemicals is their basic nature. Since CaM is an acidic protein, ionic interactions may also be involved (e.g., OLIVER et al. 1986).

An interesting finding in this connection is that a proteolytic fragment of CaM (amino acid sequence 78–148) sometimes shows CaM-like action, sometimes antagonistic action, and sometimes no action (NEWTON et al. 1984). As a whole, the interactions of CaM with its inhibitors are complicated and their modes of action vary from case to case.

Table 6. Calmodulin ligands which inhibit calmodulin-dependent enzyme activities[a]. Bracketed numbers refer to references in the list below the table

Drugs and reagents	Inhibitory potency (IC_{50} (μM))				Binding to CaM				
	PDE	Ca^{2+}, Mg^{2+}-ATPase	MLCK	References	+Ca		−Ca		References
					$K_{0.5}$ (μM)	N	$K_{0.5}$ (μM)	N	
Chlorpromazine	42			[1]	5.0	3	150	17	[2]
	47	38	50	[3]	2.4				[3]
	6			[4]					
					22		100		[5]
					20	8	30	1	[6]
	19–42			[7]	17	5	30–40	0.6	[7]
		22–135		[8]					
Trifluoperazine	10	9–50		[1, 8]					
	17			[12]	1.5	2	5000	24	[9]
					1.5	2	500	27	[10]
					5.0				[11]
W7	67	27	50	[13]	11	3	200	9	[13]
Felodipine					1–10	1–2			[14]
					2.8				[11]
					22	2			[15]
Diltiazem					80				[11]
Tetracaine	440		1010	[16]	920				[16]

Table 6 (continued)

Drugs and reagents	Inhibitory potency (IC_{50} (μM))				Binding to CaM				
	PDE	Ca^{2+}, Mg^{2+}-ATPase	MLCK	References	+Ca		−Ca		References
					$K_{0.5}$ (μM)	N	$K_{0.5}$ (μM)	N	
Troponin I	0.03			[17]					
					0.02	1	70	1	[18, 19]
					0.03–0.04				[20]
					0.003	1	5.7	1	[27]
β-Endorphin 1–31	3.4			[5]	4.6	1.3			[5]
					2				[20]
	10			[21]		2			[21]
β-Endorphin fragments 14–31	10			[21]		2			[21]
ACTH					1.5–2.5				[20]
Substance P					2				[20]
Glucagon					3.4				[20]
Dynorphin 1–17	10			[5]					
					4.5				[20]
					0.6	1			[22]
VIP					0.08				[20]
GIP					0.08–0.14				[20]
Secretin					0.14–0.2				[20]
Melittin	0.0059			[23]	0.003	1			[23]
							10	2[c]	[24]
Mastoparan					0.0003				[25]
Mastoparan X					0.0009				[25]
Polistes mastoparan					0.0035	1			[25]
LK 1[b]	0.25			[25]	0.15				[26]
LK 2[b]	0.005			[25]	0.003				[26]

[a] A23187 and nicardipine bind to calmodulin in the presence of Ca^{2+}, but they do not inhibit CaM-activated enzyme activity (INAGAKI et al. 1985).
[b] LK n is 9-fluorenylmethoxycarbonyl-(Leu-Lys-Lys-Leu-Leu-Lys-Lys)$_n$ ($n = 1$, 2).
[c] Up to 5, if denatured.
References: [1] LEVIN and WEISS (1976); [2] LEVIN and WEISS (1979); [3] HIDAKA et al. (1980); [4] NORMAN et al. (1979); [5] SELLINGER-BARNETTE and WEISS (1982); [6] GIEDROC et al. (1985); [7] MARSHAK et al. (1985); [8] ROUFOGALIS (1982); [9] LEVIN and WEISS (1977); [10] LEVIN and WEISS (1978); [11] JOHNSON and FUGMAN (1983); [12] PROZIALECK and WEISS (1982); [13] HIDAKA et al. (1980); [14] BOSTRÖM et al. (1981); [15] MILLS et al. (1985); [16] TANAKA and HIDAKA (1981); [17] LAPORTE et al. (1980); [18] OLWIN et al. (1982); [19] KELLER et al. (1982b); [20] MALENCIK and ANDERSON (1982, 1983a); [21] GIEDROC et al. (1985); [22] MALENCIK and ANDERSON (1984); [23] COMTE et al. (1983); [24]MAULET and COX (1983); [25] MALENCIK and ANDERSON (1983b); [26] COX et al. (1985); [27] OLWIN and STORM (1985).

Table 6 summarizes the inhibitory actions of various chemicals on CaM-activated enzymes and their affinities for CaM. There is fairly good correlation between both activities (considerable discrepancies can be seen in some phenothiazine derivatives, but they may be ascribed to the measurement of the activities under different conditions, e.g., at different ionic strengths; see Table 2).

A more important aspect from the physiologic point of view is how a drug is able to exhibit its intracellular action, even when it is applied to intact cells from outside. Phenothiazine derivatives are disqualified in this respect because of their strong pharmacologic actions on other sites at the surface membrane. Peptides, e.g., melittin, a specific antagonist, hardly permeate the cell membrane. W7, one of the naphthalenesulfonamide complexes, has no other prominent pharmacologic action and is relatively permeable through surface membranes (HIDAKA et al. 1981), but requires a high concentration. Thus, many pharmacologic tasks remain to be done.

II. CaM Ligands

The discovery of the antagonistic action of some peptides against CaM has suggested that the three groups of CaM-related substances, i.e., CaM antagonists represented by phenothiazine derivatives, the peptides that inhibit CaM-related functions, and the enzymes to be activated by CaM, should have something in common in their binding to CaM. Hence, it is more convenient to include these substances in the category of CaM ligands.

This concept was justified by the following interesting fact. Binding of these three groups to CaM results in an enormous increase in their affinity for Ca^{2+} (KELLER et al. 1982a, b; OGAWA and TANOKURA 1985) as summarized in Table 7. Two kinds of mechanisms can be postulated for this increase. One is that the complex of CaM and ligands has higher affinity for Ca^{2+}. The other is that the affinity of Ca · CaM for ligands is very strong so that the CaM exists mostly as the complex of Ca · CaM and the ligand; as a result, the apparent affinity of CaM for Ca^{2+} increases. The explanation in Sects. F and G.I followed this line. Although no linear relationship can be found between the increase in the affinity for Ca^{2+} and the inhibitory action on CaM-activated enzyme activities, both actions seem inseparably related.

In this connection, two interesting findings have been presented: (a) addition of troponin I to CaM at a molar ratio of 1:2 results in the coexistence of high and low affinity sites; and (b) the measurement of the rates of Ca binding and Ca release in the presence of trifluoperazine by the stopped-flow method has revealed that the apparent increase in the affinity for Ca^{2+} is mainly due to the decrease in the rate of Ca release, without a substantial increase of the binding rate (OGAWA and TANOKURA 1985) this is in accord with the assumption made for the CaM–enzyme complex in Sect. F.

Table 7. Enhanced apparent affinity of calmodulin for Ca^{2+} in the presence of calmodulin ligands. Bracketed numbers refer to references in the list below the table

Ligand	Amount	Increase in apparent affinity of CaM for Ca^{2+} [a]	Positive cooperativity	Remarks and References
Trifluoperazine	0.1 mM	58.8	+	Reduced affinity of the fourth Ca^{2+}-binding site [1]
	0.3 mM	76.3	+	
	0.1 mM	56.7	$n_H \geqq 2$ [b]	[2]
Chlorpromazine	0.1 mM	14.7	+	[2]
W7	0.1 mM	9.9	+	[2]
Felodipine	0.03 mM	4.2	−	[2]
Tetracaine	3 mM	12.7	+	[2]
Halothane	0.2%	2.2	−	[2]
Diltiazem	0.1 mM	1.3	−	[8]
Melittin	1:1	34.8	$n_H = 1.38$	[3]
Troponin I	0.5:1	6.0 [c]	−	[2]
	1:1	5	−	[2]
	1:1	6.7	$n_H = 1.45$	$\overline{K_{Enz}} = 20$ nM (+Ca), 70 μM (−Ca) [d] [4]
	3:1	8.2	$n_H = 1.36$	
	1:1	7.4	−	$\overline{K_{Enz}} = 3$ nM (+Ca), 5.7 μM (−Ca) [5]
Phosphorylase kinase		73	+	$\overline{K_{Enz}} = 2.3$ nM (+Ca) [6], markedly enhanced affinity of the third Ca^{2+}-binding site
δ-Subunit of phosphorylase kinase		6.9		[7]
Myosin light chain kinase, skeletal muscle	1:1	14.3	+	$\overline{K_{Enz}} = 15$ nM (+Ca), 3 nM (+Ca, substrates), 30 mM (−Ca) [5]
Phosphodiesterase	1:1	25.5	+	$\overline{K_{Enz}} = 3$ nM (+Ca), 80 mM (−Ca) [5]

[a] Ratio of averaged apparent affinity of CaM for Ca^{2+} in the presence of ligand to that in its absence.
[b] Hill coefficient.
[c] One-half of the Ca-binding sites of CaM unchanged.
[d] $\overline{K_{Enz}} = 1/K_{Enz}$.
References: [1] KELLER et al. (1982); [2] OGAWA and TANOKURA (1985); [3] MAULET and COX (1983); [4] KELLER et al. (1982a, b); [5] OLWIN et al. (1984); OLWIN and STORM (1985); [6] BURGER et al. (1983); [7] BURGER et al. (1982); [8] Y. OGAWA (1983, unpublished work).

H. Concluding Remarks

Although related to each other chemically and perhaps evolutionarily, TNC and CaM are distinct from each other from the physiologic and pharmacologic points of view. TNC is a highly differentiated protein, specialized for muscle contraction, and its function is always performed as a part of the whole TN molecule, never independent of it. On the other hand, CaM can cooperate with an incredibly large number of proteins, but the modes of interaction of CaM and partner proteins are subtly different from one another. Furthermore, the cellular content of CaM, which is distinctly different from tissue to tissue, plays a crucial role in activating the CaM-dependent systems. Consequently, this simple, common molecule can confer characteristic features on the cell that contains it.

Other Ca receptor proteins may occupy a position between these two distinct proteins. Therefore, research into these two proteins provides us with representative patterns of Ca regulation, which is the most fundamental mechanism of regulating cell function and is the most widely investigated topic in biological science today.

References

Ashley CC, Moisescu DG (1977) Effect of changing the composition of the bathing solutions upon the isometric tension-pCa relationship in bundles of crustacean myofibrils. J Physiol 270:627–652

Babu YS, Sack JS, Greenhough TJ, Bugg CE, Means AR, Cook WJ (1985) Three-dimensional structure of calmodulin. Nature 315:37–40

Blumenthal DK, Stull JT (1980) Activation of skeletal muscle myosin light chain kinase by calcium (2+) and calmodulin. Biochemistry 19:5608–5614

Boström S-L, Ljung B, Mårdh S, Forsen S, Thulin E (1981) Interaction of the antihypertensive drug felodipine with calmodulin. Nature 292:777–778

Bozler E (1954) Relaxation in extracted muscle fibers. J Gen Physiol 38:149–159

Brandt PW, Cox RN, Kawai M (1980) Can the binding of Ca^{2+} to two regulatory sites on troponin C determine the steep pCa/tension relationship of skeletal muscle? Proc Natl Acad Sci USA 77:4717–4720

Bremel RD, Weber A (1972) Cooperation within actin filament in vertebrate skeletal muscle. Nature New Biol 238:97–101

Burger D, Cox JA, Fischer EH, Stein EA (1982) The activation of rabbit skeletal muscle phosphorylase kinase requires the binding of 3 Ca^{2+} per δ subunit. Biochem Biophys Res Commun 105:632–638

Burger D, Stein EA, Cox JA (1983) Free energy coupling in the interactions between Ca^{2+}, calmodulin, and phosphorylase kinase. J Biol Chem 258:14733–14739

Burger D, Cox JA, Comte M, Stein EA (1984) Sequential conformational changes in calmodulin upon binding of calcium. Biochemistry 23:1966–1971

Burtnick LD, Kay CM (1977) The calcium-binding properties of bovine cardiac troponin C. FEBS Lett 75:105–110

Charbonneau H, Cormier MJ (1979) Purification of plant calmodulin by fluphenazine-sepharose affinity chromatography. Biochem Biophys Res Commun 90:1039–1047

Cheung WY (1970) Cyclic 3′,5′-nucleotide phosphodiesterase: demonstration of an activator. Biochem Biophys Res Commun 38:533–538

Cheung WY (ed) (1980) Calcium and cell function, vol I. Calmodulin. Academic, New York

Cheung WY, Lynch TJ, Wallace RW, Tallant EA (1981) cAMP renders Ca^{2+}-dependent phosphodiesterase refractory to inhibition by a calmodulin-binding protein (calcineurin). J Biol Chem 256:4439–4443

Comte M, Maulet Y, Cox JA (1983) Ca^{2+}-dependent high-affinity complex formation between calmodulin and melittin. Biochem J 209:269–272

Cox JA, Malnoë A, Stein EA (1981) Regulation of brain cyclic nucleotide phosphodiesterase by calmodulin: a quantitative analysis. J Biol Chem 256:3218–3222

Cox JA, Comte M, Malnoë A (1984) Mode of action of the regulatory protein calmodulin. In: Sigel H (ed) Metal ions in biological systems, vol 17. Calcium and its role in biology. Dekker, New York, pp 215–273

Cox JA, Comte M, Fitton JE, DeGrado WF (1985) The interaction of calmodulin with amphiphilic peptides. J Biol Chem 260:2527–2534
Crouch TH, Klee CB (1980) Positive cooperative binding of calcium to bovine brain calmodulin. Biochemistry 19:3692–3698
Crouch TH, Holroyde MJ, Collins JH, Solaro J, Potter JD (1981) Interaction of calmodulin with skeletal muscle myosin light chain kinase. Biochemistry 20:6318–6325
Dedman JR, Potter JD, Jackson RL, Johnson JD, Means AR (1977) Physicochemical properties of rat testis Ca^{2+}-dependent regulator protein of cyclic nucleotide phosphodiesterase: relationship of Ca^{2+}-binding, conformational changes, and phosphodiesterase activity. J Biol Chem 252:8415–8422
Deuticke HJ (1934) Über die Sedimentationskonstante von Muskelproteinen. Hoppe Seylers Z Physiol Chem 224:216–228
Donaldson SKB, Hermansen L (1978) Differential, direct effects of H^+ on Ca^{2+}-activated force of skinned fibers from the soleus, cardiac and adductor magnus muscles of rabbits. Pflugers Arch 376:55–65
Ebashi S (1959) The mechanism of relaxation in glycerinated muscle fiber (in Japanese). In: Natori R (ed) IVth symposium on physicochemistry of biomacromolecules, 1958. Nanko-do, Tokyo, pp 25–34
Ebashi S (1960) Calcium binding and relaxation in the actomyosin system. J Biochem 48:150–151
Ebashi S (1961) Calcium binding activity of vesicular relaxing factor. J Biochem 50:236–244
Ebashi S (1963) Third component participating in the superprecipitation of "natural actomyosin". Nature 200:1010–1011
Ebashi S (1974) Regulatory mechanism of muscle contraction with special reference to the Ca-troponin-tropomyosin system. Essays Biochem 10:1–36
Ebashi S, Endo M (1968) Calcium ion and muscle contraction. Prog Biophys Mol Biol 18:123–183
Ebashi S, Kodama A (1965) A new protein factor promoting aggregation of tropomyosin. J Biochem 58:107–108
Ebashi S, Lipmann F (1962) Adenosine triphosphate-linked concentration of calcium ions in a particulate fraction of rabbit muscle. J Cell Biol 14:389–400
Ebashi S, Kodama A, Ebashi F (1968) Troponin. I. Preparation and physiological function. J Biochem 64:465–477
Endo M (1972) Stretch-induced increase in activation of skinned muscle fibres by calcium. Nature New Biol 237:211–213
Endo M (1973) Length dependence of activation of skinned muscle fibres by calcium. Cold Spring Harbor Symp Quant Biol 37:505–510
Endo M, Kitazawa T (1978) Excitation-contraction coupling in chemically skinned fibers of cardiac muscle. In: Hayase S, Murao S (eds) Proceedings of the VIII world congress on cardiology. Excerpta Medica, Amsterdam, pp 800–803
Fabiato A (1981) Myoplasmic free calcium concentration reached during the twitch of an intact isolated cardiac cell and during calcium-induced release of calcium from the sarcoplasmic reticulum of a skinned cardiac cell from the adult rat or rabbit ventricle. J Gen Physiol 78:457–497
Fabiato A, Fabiato F (1975) Effects of magnesium on contractile activation of skinned cardiac cells. J Physiol 249:497–517
Fabiato A, Fabiato F (1978a) Effects of pH on the myofilaments and the sarcoplasmic reticulum of skinned cells from cardiac and skeletal muscles. J Physiol 276:233–255
Fabiato A, Fabiato F (1978b) Myofilament-generated tension oscillations during partial calcium activation and activation dependence of the sarcomere length-tension relation of skinned cardiac cells. J Gen Physiol 72:667–699
Feuer G, Molnar F, Pettko E, Straub FB (1948) Studies on the composition and polymerisation of actin. Hungar Acta Physiol 1:150–163
Fuchs F (1977) Cooperative interactions between calcium binding sites on glycerinated muscle fibres: the influence of cross-bridge attachments. Biochim Biophys Acta 462:314–322

Fuchs F, Fox C (1982) Parallel measurements of bound calcium and force in glycerinated rabbit psoas muscle fibers. Biochim Biophys Acta 679:110–115
Giedroc DP, Keravis TM, Staros JV, Ling N, Wells JN, Puet D (1985) Functional properties of covalent β-endorphin peptide/calmodulin complexes. Chlorpromazine binding and phosphodiesterase activation. Biochemistry 24:1203–1211
Gillis JM (1980) The biological significance of muscle parvalbumins. In: Siegel FL, Carafoli E, Kretzinger RH, MacLennan DH, Wasserman RH (eds) Calcium binding proteins: structure and function. Elsevier, Amsterdam, pp 309–311
Gillis JM, Thomason D, Lefèvre J, Kretsinger RH (1982) Parvalbumins and muscle relaxation: a computer simulation study. J Muscle Res Cell Motil 3:377–398
Godt RE, Lindley BD (1982) Influence of temperature upon contractile activation and isometric force production in mechanically skinned muscle fibers of the frog. J Gen Physiol 80:279–297
Gordon AM, Godt RE, Donaldson SKB, Harris CE (1973) Tension in skinned frog muscle fibers in solutions of varying ionic strength and neutral salt composition. J Gen Physiol 62:550–574
Grabarek Z, Grabarek J, Leavis PC, Gergely J (1983) Cooperative binding to the Ca^{2+}-specific sites of troponin C in regulated actin and actomyosin. J Biol Chem 258:14098–14102
Gulati J, Babu A (1985) Contraction kinetics of intact and skinned frog muscle fibers and degree of activation. Effects of intracellular Ca^{2+} on unloaded shortening. J Gen Physiol 86:479–500
Gulati J, Podolsky RJ (1978) Contraction transients of skinned muscle fibers: effects of calcium and ionic strength. J Gen Physiol 72:701–716
Haiech J, Klee CB, Demaille JG (1981) Effects of cations on affinity of calmodulin for calcium: ordered binding of calcium ions allows the specific activation of calmodulin-stimulated enzymes. Biochemistry 20:3890–3897
Harafuji H, Ogawa Y (1980) Re-examination of the apparent binding constant of ethylene glycol bis(β-aminoethylether)-*N*,*N*,*N'*,*N'*-tetraacetic acid with calcium around neutral pH. J Biochem 87:1305–1312
Hartshorne DJ, Mueller H (1968) Fractionation of troponin into two distinct proteins. Biochem Biophys Res Commun 31:647–653
Heilbrunn LV (1940) The action of calcium on muscle protoplasm. Physiol Zool 13:88–94
Heilbrunn LV, Wiercinski FL (1947) The action of various cations on muscle protoplasm. J Cell Comp Physiol 29:15–32
Henrotte JG (1955) A crystalline component of carp myogen precipitating at high ionic strength. Nature 176:1221
Herzberg O, James MNG (1985) Structure of the calcium regulatory muscle protein troponin-C at 2.8 Å resolution. Nature 313:653–659
Hibberd MG, Jewell BR (1982) Calcium- and length-dependent force production in rat ventricle muscle. J Physiol 329:527–540
Hidaka H, Yamaki T, Naka M, Tanaka T, Hayashi H, Kobayashi R (1980) Calcium-regulated modulator protein interacting agents inhibit smooth muscle calcium-stimulated protein kinase and ATPase. Mol Pharmacol 17:66–72
Hidaka H, Sasaki Y, Tanaka T, Endo T, Ohno S, Fujii Y, Nagata T (1981) *N*-(6-aminohexyl)-5-chloro-1-naphthalenesulfonamide, a calmodulin antagonist, inhibits cell proliferation. Proc Natl Acad Sci USA 78:4354–4357
Hiromi K (1978) Kinetics of fast enzyme reactions: theory and practice. Kodansha, Tokyo
Holroyde MJ, Robertson SP, Johnson JD, Solaro RJ, Potter JD (1980) The calcium and magnesium binding sites on cardiac troponin and their role in the regulation of myofibrillar adenosine triphosphatase. J Biol Chem 255:11688–11693
Huang CY, Chau V, Chock PB, Wang JH, Sharma RK (1981) Mechanism of activation of cyclic nucleotide phosphodiesterase: requirement of the binding of four Ca^{2+} to calmodulin for activation. Proc Natl Acad Sci USA 78:871–874

Inagaki M, Tanaka T, Sasaki Y, Hidaka H (1985) Calcium-dependent interactions of an ionophore A23187 with calmodulin. Biochem Biophys Res Commun 130:200–206
Jamieson GA Jr, Vanaman TC (1979) Calcium dependent affinity chromatography of calmodulin on an immobilized phenothiazine. Biochem Biophys Res Commun 90:1048–1056
Johnson JD, Fugman DA (1983) Calcium and calmodulin antagonists binding to calmodulin and relaxation of coronary segments. J Pharmacol Exp Ther 226:330–334
Johnson JD, Potter JD (1978) Detection of two classes of Ca^{2+} binding sites in troponin C with circular dichroism and tyrosine fluorescence. J Biol Chem 253:3775–3777
Johnson JD, Collins JH, Potter JD (1978) Dansylaziridine-labeled troponin C: a fluorescent probe of Ca^{2+} binding to the Ca^{2+}-specific regulatory sites. J Biol Chem 253:6451–6458
Johnson JD, Charlton SC, Potter JD (1979) A fluorescence stopped flow analysis of Ca^{2+} exchange with troponin C. J Biol Chem 254:3497–3502
Johnson JD, Collins JH, Robertson SP, Potter JD (1980) A fluorescent probe study of Ca^{2+} binding to the Ca^{2+}-specific sites of cardiac troponin and troponin C. J Biol Chem 255:9635–9640
Johnson JD, Robinson DE, Robertson SP, Schwartz A, Potter JD (1981) Ca^{2+} exchange with troponin and the regulation of muscle contraction. In: Grinnell AD, Brazier MAB (eds) The regulation of muscle contraction: excitation-contraction coupling. Academic, New York, pp 241–259
Kakiuchi S (1981) Calmodulin and cytoskeletal system. Seikagaku 53:1267–1289 (in Japanese)
Kakiuchi S, Sobue K (1981) Ca^{2+}- and calmodulin-dependent flip-flop mechanism in microtubule assembly-disassembly. FEBS Lett 132:141–143
Kakiuchi S, Yamazaki R (1970a) Stimulation of the activity of cyclic 3′,5′-nucleotide phosphodiesterase by calcium ion. Proc Jpn Acad 46:387–392
Kakiuchi S, Yamazaki R (1970b) Calcium dependent phosphodiesterase activity and its activating factor (PAF) from brain. III. Studies on cyclic 3′,5′-nucleotide phosphodiesterase. Biochem Biophys Res Commun 41:1104–1110
Kakiuchi S, Yamazaki R, Nakajima H (1969) Studies on brain phosphodiesterase (2). Bull Jpn Neurochem Soc 8:17–20 (in Japanese)
Kakiuchi S, Yamazaki R, Nakajima H (1970) Properties of a heat-stable phosphodiesterase activating factor isolated from brain extract. Proc Jpn Acad 46:587–592
Kakiuchi S, Yasuda S, Yamazaki R, Teshima Y, Kanda K, Kakiuchi R, Sobue K (1982) Quantitative determinations of calmodulin in the supernatant and particulate fractions of mammalian tissues. J Biochem 92:1041–1048
Kamada T, Kinosita H (1943) Disturbances initiated from naked surface of muscle protoplasm. Jpn J Zool 10:469–493
Kasai M (1969) The divalent cation bound to actin and thin filament. Biochim Biophys Acta 172:171–173
Kasai M, Oosawa F (1968) The exchangeability of actin-bound calcium with various divalent cations. Biochim Biophys Acta 154:520–528
Keller CH, Olwin BB, Heideman W, Storm DR (1982a) The energetics and chemistry for interactions between calmodulin and calmodulin-binding proteins. In: Cheung WY (ed) Calcium and cell function, vol III. Academic, New York, pp 103–127
Keller CH, Olwin BB, LaPorte DC, Storm DR (1982b) Determination of the free-energy coupling for binding of calcium ions and troponin I to calmodulin. Biochemistry 21:156–162
Kerrick WGL, Donaldson SKB (1972) The effects of Mg^{2+} on submaximum Ca^{2+}-activated tension in skinned fibers of frog skeletal muscle. Biochim Biophys Acta 275:117–122
Kerrick WGL, Donaldson SKB (1975) The comparative effects of $[Ca^{2+}]$ and $[Mg^{2+}]$ on tension generation in the fibers of skinned frog skeletal muscle and mechanically disrupted rat ventricular cardiac muscle. Pflugers Arch 358:195–201

Kerrick WGL, Hoar PE, Cassidy PS, Malencik DA (1981) Ca^{2+} regulation of contraction in skinned muscle fibers. In: Grinnell AD, Brazier MAB (eds) The regulation of muscle contraction: excitation-contraction coupling. Academic, New York, pp 227–239
Kielley WW, Meyerhof O (1948 a) A new magnesium-activated adenosinetriphosphatase from muscle. J Biol Chem 174:387–388
Kielley WW, Meyerhof O (1948 b) Studies on adenosinetriphosphatase of muscle. II. a new magnesium activated adenosinetriphosphatase. J Biol Chem 183:391–401
Kitazawa T, Shuman H, Somlyo AP (1982) Calcium and magnesium binding to thin and thick filaments in skinned muscle fibres: electron probe analysis. J Muscle Res Cell Motil 3:437–454
Klee CB, Vanaman TC (1982) Calmodulin. Adv Protein Chem 35:213–321
Kohama K (1979) Divalent cation binding properties of slow skeletal muscle troponin in comparison with those of cardiac and fast skeletal muscle troponins. J Biochem 86:811–820
Kohama K (1980) Role of the high affinity Ca binding sites of cardiac and fast skeletal troponins. J Biochem 88:591–599
Kretsinger RH, Nockolds CE (1973) Carp muscle calcium-binding protein. II. Structure determination and general description. J Biol Chem 248:3313–3326
Kumagai H, Ebashi S, Takeda F (1955) Essential relaxing factor in muscle other than myokinase and creatine phosphokinase. Nature 176:166
Kurebayashi N, Ogawa Y (1985) Effect of quercetin on tension development by skinned fibres from frog skeletal muscle. J Muscle Res Cell Motil 6:189–195
LaPorte DC, Toscano WA Jr, Storm DR (1979) Cross-linking of iodine-125-labeled, calcium-dependent regulatory protein to the Ca^{2+}-sensitive phosphodiesterase purified from bovine heart. Biochemistry 18:2820–2825
LaPorte DC, Wierman BM, Storm DR (1980) Calcium induced exposure of a hydrophobic surface on calmodulin. Biochemistry 19:3814–3819
Leavis PC, Gergely J (1984) Thin filament proteins and thin filament-linked regulation of vertebrate muscle contraction. CRC Crit Rev Biochem 16:235–305
Levin RM, Weiss B (1976) Mechanism by which psychotropic drugs inhibit adenosine cyclic 3′,5′-monophosphate phosphodiesterase of brain. Mol Pharmacol 12:581–589
Levin RM, Weiss B (1977) Binding of trifluoperazine to the calcium-dependent activator of cyclic nucleotide phosphodiesterase. Mol Pharmacol 13:690–697
Levin RM, Weiss B (1978) Specificity of the binding of trifluoperazine to the calcium-dependent activator of phosphodiesterase and to a series of other calcium-binding proteins. Biochim Biophys Acta 540:197–204
Levin RM, Weiss B (1979) Selective binding of antipsychotics and other psychoactive agents to the calcium-dependent activator of cyclic nucleotide phosphodiesterase. J Pharmacol Exp Ther 208:454–459
Lin YM, Liu YP, Cheung WY (1974) Cyclic 3′:5′-nucleotide phosphodiesterase. Purification, characterization and active form of the protein activator from bovine brain. J Biol Chem 249:4943–4954
Liu YP, Cheung WY (1976) Cyclic 3′,5′-nucleotide phosphodiesterase. Ca^{2+} confers more helical conformation to the protein activator. J Biol Chem 251:4193–4198
Malencik DA, Anderson SR (1982) Binding of simple peptides, hormones, and neurotransmitters by calmodulin. Biochemistry 21:3480–3486
Malencik DA, Anderson SR (1983 a) Binding of hormones and neuropeptides by calmodulin. Biochemistry 22:1995–2001
Malencik DA, Anderson SR (1983 b) High affinity binding of the mastoparans by calmodulin. Biochem Biophys Res Commun 114:50–56
Malencik DA, Anderson SR (1984) Peptide binding by calmodulin and its proteolytic fragments and by troponin C. Biochemistry 23:2420–2428
Malencik DA, Anderson SR, Shalitin Y, Shimerlik MI (1981) Rapid kinetic studies on calcium interactions with native and fluorescently labeled calmodulin. Biochem Biophys Res Commun 101:390–395

Malencik DA, Anderson SR, Bohnert JL, Shalitin Y (1982) Functional interactions between smooth muscle myosin light chain kinase and calmodulin. Biochemistry 21:4031–4039
Marsh BB (1951) A factor modifying muscle fibre syneresis. Nature 167:1065–1066
Marshak DR, Lukas TJ, Watterson DM (1985) Drug-protein interactions: binding of chlorpromazine to calmodulin, calmodulin fragments, and related calcium binding proteins. Biochemistry 24:144–150
Maulet Y, Cox JA (1983) Structural changes in melittin and calmodulin upon complex formation and their modulation by calcium. Biochemistry 22:5680–5686
Mills JS, Johnson JD (1985) Metal ions as allosteric regulators of calmodulin. J Biol Chem 260:15100–15105
Mills JS, Bailey BL, Johnson JD (1985) Cooperativity among calmodulin's drug binding sites. Biochemistry 24:4897–4902
Minowa O, Yagi K (1984) Calcium binding to tryptic fragments of calmodulin. J Biochem 96:1175–1182
Morimoto S, Ohtsuki I (1987) Ca^{2+}- and Sr^{2+}-sensitivity of the ATPase activity of rabbit skeletal myofibrils. Effect of the complete substitution of troponin C with cardiac troponin C, calmodulin and parvalbumins. J Biochem 101:291–301
Moss RL, Swinford AE, Greaser ML (1983) Alterations in the Ca^{2+} sensitivity of tension development by single skeletal muscle fibres at stretched lengths. Biophys J 43:115–119
Moss RL, Giulian GG, Greaser ML (1985) The effects of partial extraction of TnC upon the tension-pCa relationship in rabbit skinned skeletal muscle fibers. J Gen Physiol 86:585–600
Murray JM, Weber A, Bremel A (1975) Could cooperativity in the actin filament play a role in muscle contraction? In: Carafoli E, Clementi F, Drabikowski W, Margreth A (eds) Calcium transport in contraction and secretion. Elsevier, Amsterdam, pp 489–496
Newton DL, Oldewurted MD, Krinks MH, Shiloach J, Klee CB (1984) Agonist and antagonist properties of calmodulin fragments. J Biol Chem 259:4419–4426
Norman JA, Drummond AH, Moser P (1979) Inhibition of calcium-dependent regulator-stimulated phosphodiesterase activity by neuroleptic drugs is unrelated to their clinical efficacy. Mol Pharmacol 16:1089–1094
Ogawa Y (1985) Calcium binding to troponin C and troponin: effect of Mg^{2+}, ionic strength and pH. J Biochem 97:1011–1023
Ogawa Y, Tanokura M (1984) Calcium binding to calmodulin: effects of ionic strength, Mg^{2+}, pH and temperature. J Biochem 95:19–28
Ogawa Y, Tanokura M (1985) Calcium binding to calmodulin and its modification by calmodulin-ligands in view of the regulatory role in vivo of the calmodulin system. In: Ebashi S (ed) Cellular regulation and malignant growth. Japan Sci Soc Press, Tokyo/Springer, Berlin Heidelberg New York Tokyo, pp 250–258
Ogawa Y, Tanokura M (1986a) Steady-state properties of calcium binding to parvalbumins from bullfrog skeletal muscle: effects of Mg^{2+}, pH, ionic strength, and temperature. J Biochem 99:73–80
Ogawa Y, Tanokura M (1986b) Kinetic studies of calcium binding to parvalbumins from bullfrog skeletal muscle. J Biochem 99:81–89
Ogawa Y, Harafuji H, Kurebayashi N (1980) Comparison of the characteristics of four metallochromic dyes as potential calcium indicators for biological experiments. J Biochem 87:1293–1303
Ohnishi ST (1978) Characterization of the murexide method: dual wavelength spectrophotometry of cations under physiological conditions. Anal Biochem 85:165–179
Ohnishi T, Ebashi S (1963) Spectrophotometrical measurement of instantaneous calcium binding of the relaxing factor of muscle. J Biochem 54:506–511
Ohtsuki I, Nagano K (1982) Molecular arrangement of troponin-tropomyosin in the thin filament. Adv Biophys 15:93–130
Ohtsuki I, Maruyama K, Ebashi S (1986) Regulatory and cytoskeletal protein of vertebrate skeletal muscle. Adv Protein Chem 38:1–67

Oliver JL, Rainteau D, Bereziat G, Wolf C (1986) Interaction between calmodulin and five different spin-labelled chlorophenothiazines. Biochem J 233:853–857
Olwin BB, Storm DR (1985) Calcium binding to complexes of calmodulin and calmodulin binding proteins. Biochemistry 24:8081–8086
Olwin BB, Keller CH, Storm DR (1982) Interaction of a fluorescent *N*-dansylaziridine derivative of troponin I with calmodulin in the absence and presence of calcium. Biochemistry 21:5669–5675
Olwin BB, Edelman AM, Krebs EG, Storm DR (1984) Quantitation of energy coupling between Ca^{2+}, calmodulin, skeletal muscle myosin light chain kinase, and kinase substrates. J Biol Chem 259:10949–10955
Ozawa E, Hosoi K, Ebashi S (1967) Reversible stimulation of muscle phosphorylase b kinase by low concentration of calcium ions. J Biochem 61:531–533
Pechère J-F, Capony J-P, Ryden L (1971) The primary structure of the major parvalbumin from hake muscle. Isolation and general properties of the protein. Eur J Biochem 23:421–428
Potter JD, Gergely J (1975) The calcium and magnesium binding sites on troponin and their role in the regulation of myofibrillar adenosine triphosphatase. J Biol Chem 250:4628–4633
Potter JD, Johnson JD (1982) Troponin. In: Cheung WY (ed) Calcium and cell function, vol II. Academic, New York, pp 145–173
Potter JD, Hsu F-J, Pownall HJ (1977) Thermodynamics of Ca^{2+} binding to troponin-C. J Biol Chem 252:2452–2454
Potter JD, Robertson SP, Collins JH, Johnson JD (1980) The role of the Ca^{2+} and Mg^{2+} binding sites on troponin and other myofibrillar proteins in the regulation of muscle contraction. In: Siegel FL, Carafoli E, Kretsinger RH, MacLennan DH, Wasserman RH (eds) Calcium-binding proteins: structure and function. Elsevier, Amsterdam, pp 279–288
Potter JD, Robertson SP, Johnson JD (1981) Magnesium and the regulation of muscle contraction. Fed Proc 40:2653–2656
Proziakeck WC, Weiss B (1982) Inhibition of calmodulin by phenothiazines and related drugs: structure-activity relationships. J Pharmacol Exp Ther 222:509–516
Ringer S (1883) A further contribution regarding the influence of the different constituents of the blood on the contraction of the heart. J Physiol 4:29–42
Ringer S (1886) Further experiments regarding the influence of small quantities of lime potassium and other salts on muscular tissue. J Physiol 7:291–308
Robertson SP, Johnson JD, Potter JD (1978) The effect of pH on calcium binding to the Ca^{2+}-Mg^{2+} and Ca^{2+}-specific sites of rabbit skeletal TnC. Biophys J 21:16a
Robertson SP, Johnson JD, Potter JD (1981) The time-course of Ca^{2+} exchange with calmodulin, troponin, parvalbumin, and myosin in response to transient increases in Ca^{2+}. Biophys J 34:559–569
Roufogalis BD (1982) Specificity of trifluoperazine and related phenothiazines for calcium-binding proteins. In: Cheung WY (ed) Calcium and cell function, vol III. Academic, New York, pp 129–159
Rüegg JC, Pfitzer G, Zimmer M, Hofmann F (1984) The calmodulin fraction responsible for contraction in an intestinal smooth muscle. FEBS Lett 170:383–386
Seaton BA, Head JF, Engelman DM, Richards FM (1985) Calcium-induced increase in the radius of gyration and maximum dimension of calmodulin measured by small-angle X-ray scattering. Biochemistry 24:6740–6743
Sellinger-Barnette M, Weiss B (1982) Interaction of β-endorphin and other opioid peptides with calmodulin. Mol Pharmacol 21:86–91
Shimizu T, Hatano M (1983) Interaction of trifluoperazine with porcine calmodulin ^{19}F NMR and induced CD spectral studies. FEB Lett 160:182–186
Shimizu T, Hatano M (1984) Effects of metal cations on trifluoperazine-calmodulin interactions: induced circular dichroism studies. Biochemistry 23:6403–6409
Shimizu T, Hatano M, Muto Y, Nozawa Y (1984) Interaction of trifluoperazine with *Tetrahymena* calmodulin: a ^{19}F NMR study. FEBS Lett 166:373–377

Stephenson DG, Williams DA (1981) Calcium-activated force responses in fast- and slow-twitch skinned muscle fibres of the rat at different temperatures. J Physiol 317:281–302
Stephenson DG, Williams DA (1982) Effects of sarcomere length on the force-pCa relations in fast- and slow-twitch skinned muscle fibres from the rat. J Physiol 333:637–653
Stiles PG (1901) On the rhythmic activity of the oesophagus and the influence upon it of various media. Am J Physiol 5:338–357
Stull JT, Buss JE (1978) Calcium binding properties of beef cardiac troponin. J Biol Chem 253:5932–5938
Sundaralingam M, Bergstrom R, Strasburg G, Rao ST, Roychowdhury P, Greaser M, Wang BC (1985) Molecular structure of troponin C from chicken skeletal muscle at 3-angstrom resolution. Science 227:945–948
Szent-Györgyi A (1942) The reversibility of the contraction of myosin threads. Studies Inst Med Chem Univ Szeged 2:25–26
Takagi A (1976) Abnormality of sarcoplasmic reticulum in malignant hyperthermia. Adv Neurol 20:109–113 (in Japanese)
Tanaka T, Hidaka H (1980) Hydrophobic regions function in calmodulin-enzyme(s) interactions. J Biol Chem 255:11078–11080
Tanaka T, Hidaka H (1981) Interaction of local anesthetics with calmodulin. Biochem Biophys Res Commun 101:447–453
Tanaka T, Ohmura T, Hidaka H (1982) Hydrophobic interaction of the Ca^{2+}-calmodulin complex with calmodulin antagonists: naphthalenesulfonamide derivatives. Mol Pharmacol 22:403–407
Teo TS, Wang JH (1973) Mechanism of activation of a cyclic adenosine 3′:5′-monophosphate phosphodiesterase from bovine heart by calcium ions: identification of the protein activator as a Ca^{2+} binding protein. J Biol Chem 248:5950–5955
Thames MD, Teichholz LE, Podolsky RJ (1974) Ionic strength and the contraction kinetics of skinned muscle fibers. J Gen Physiol 63:509–530
Toyota N, Shimada Y (1981) Differentiation of troponin in cardiac and skeletal muscles in chicken embryos as studied by immunofluorescence microscopy. J Cell Biol 91:497–504
Van Eerd JP, Takahashi K (1976) Determination of the complete amino acid sequence of bovine cardiac troponin C. Biochemistry 15:1171–1180
Wang C-LA, Leavis PC, Gergely J (1984) Kinetic studies that Ca^{2+} and Tb^{3+} have different binding preferences toward the four Ca^{2+}-binding sites of calmodulin. Biochemistry 23:6410–6415
Watanabe S (1955) Relaxing effects of EDTA on glycerol-treated muscle fibers. Arch Biochem Biophys 54:559–562
Watterson DM, Sharief F, Vanaman TC (1980) The complete amino acid sequence of calmodulin of bovine brain. J Biol Chem 255:962–975
Weber A (1959) On the role of calcium in the activity of adenosine 5′-triphosphate hydrolysis by actomyosin. J Biol Chem 234:2764–2769
Weber A (1966) Energized calcium transport and relaxing factors. In: Sanadi DR (ed) Current topics in bioenergetics, vol 1. Academic, New York, pp 203–254
Weber A, Herz R (1963) The binding of calcium to actomyosin systems in relation to their biological activity. J Biol Chem 238:599–605
Weber A, Winicur S (1961) The role of calcium in the superprecipitation of actomyosin. J Biol Chem 236:3198–3202
Wendt IR, Stephenson DG (1983) Effect of caffeine on Ca-activated force production in skinned cardiac and skeletal muscle fibres of the rat. Pflugers Arch 398:213–219
Wnuk W, Schoechlin M, Stein EA (1984) Regulation of actomyosin ATPase by a single calcium-binding site on troponin C from crayfish. J Biol Chem 259:9017–9023
Wolff DJ, Poirier PG, Brostrom CO, Brostrom MA (1977) Divalent cation binding properties of bovine brain Ca^{2+}-dependent regulator protein. J Biol Chem 252:4108–4117
Yagi S, Endo M (1980) Effects of dibucaine on skinned skeletal muscle fibers. An example of multiple actions of a drug on a single subcellular structure. Biomed Res 1:269–272

Yamada K, Kometani K (1982) The changes in heat capacity and entropy of troponin C induced by calcium binding. J Biochem 92:1505–1517
Yamamoto K (1983) Sensitivity of actomyosin ATPase to calcium and strontium ions. Effect of hybrid troponins. J Biochem 93:1061–1069
Yazawa M, Yagi K (1978) Purification of modulator-deficient myosin light-chain kinase by modulator protein-sepharose affinity chromatography. J Biochem 84:1259–1265
Zot HG, Iida S, Potter JD (1983) Thin filament interactions and Ca^{2+} binding to Tn. Chemica Scripta 21:133–136

CHAPTER 4

Ligand-Binding Sites on Calmodulin

H. J. VOGEL

A. Introduction

Calcium plays a pivotal role as an intracellular messenger in all eukaryotic cells. In resting cells the Ca^{2+} level is kept as low as 10^{-7} *M*. A series of membrane-bound exchange proteins and ATP-driven pump proteins are responsible for excreting the Ca^{2+} into the extracellular environment or into intracellular organelles that function as calcium stores. Taken together, these proteins produce a 10000-fold gradient of Ca^{2+} over the plasma membrane, thus creating a situation that lends itself to a regulatory function (RASMUSSEN 1983). A hormonal or nerval impulse can generate an influx of Ca^{2+} into the cytoplasm through the calcium channels. This influx gives rise to a transiently increased Ca^{2+} level inside the cell. Although the Ca^{2+} concentration in an activated cell may differ widely between various cell types, it is generally around 10^{-6} *M*, which represents a tenfold increase from the resting level of 10^{-7} *M* (RASMUSSEN 1983). This situation is schematically depicted in Fig. 1. As a second step, the transient increase in intracellular Ca^{2+} needs to be translated into metabolic or contractile responses. To this end, nature has deployed a unique class of calcium-binding proteins. Upon binding Ca^{2+} they undergo a large conformational change which allows them to interact with and activate specific target proteins. For example, in skeletal and cardiac muscles the protein troponin C is the calcium-binding component which triggers contraction. In smooth muscle tissues the homologous protein calmodulin (CaM) is responsible for the Ca^{2+} triggering of the initiation of contraction (see Chap. 3). Relaxation is in both cases simply brought about by the removal of the Ca^{2+} stimulus. Whereas troponin C only seems to play a role in muscle contraction, CaM exerts an enormous variety of effects. It can interact with many different enzymes and proteins, including protein kinases, phosphatases, and

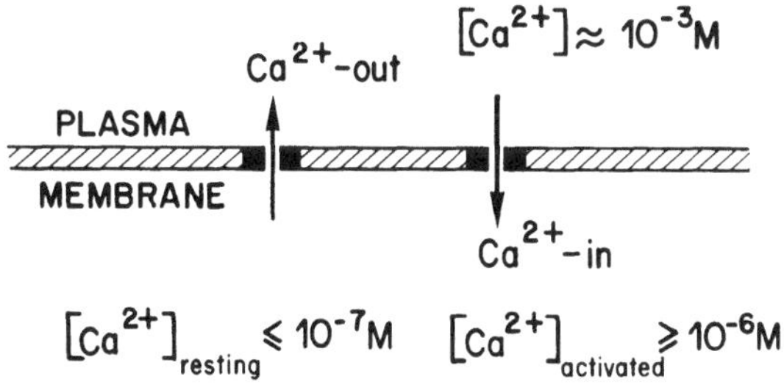

Fig. 1. Schematic diagram indicating the differences in intra- and extracellular calcium levels

Table 1. Examples of Enzymes and proteins whose activity can be affected by binding of calmodulin. (Adapted from JOHNSON and MILLS 1986)

Protein/enzyme	$K_{0.5}$ CaM (M)
Cyclic nucleotide metabolism	
Phosphodiesterase	10^{-9}
Adenylate cyclase	10^{-9}
Kinases and phosphatases	
Smooth muscle myosin light chain kinase	10^{-9}
Skeletal muscle myosin light chain kinase	6×10^{-8}
Calmodulin-dependent kinase	10^{-8}
Phosphorylase kinase	4×10^{-8}
Glycogen synthase kinase	ND
Calcineurin	10^{-8}
Membrane cation pumps	
Calcium-ATPase	10^{-7}–10^{-9}
Sodium, potassium-ATPase	10^{-7}
Other proteins	
Troponin I	2×10^{-8}
Caldesmon (SOBUE et al. 1981)	ND
Phosphofructokinase (MAYR and HEILMEYER 1983b)	ND
Glycogen phosphorylase (VILLAR-PALASI et al. 1983)	ND
NAD-kinase	3×10^{-11}

membrane ATPases (see Table 1). The variety of protein and enzyme systems that are thought to be regulated through the calcium–CaM complex is impressive (for review see KLEE and VANAMAN 1982) and one often wonders how selectivity and specificity can be accomplished inside a cell. It should be stressed at this point that the vast majority of proteins only interact with CaM after it has bound Ca^{2+} ions. The apo form of CaM is incapable of forming complexes with most target proteins, thus allowing CaM to function as a true receptor protein for the second messenger Ca^{2+}.

Despite their ubiquitous nature and their important functions, it has taken considerable time and effort to obtain a detailed molecular picture of both CaM and troponin C. When the amino acid sequence of CaM was completed, it became immediately apparent that it had a high degree of homology with troponin C (VANAMAN 1981). The crystal structures for both proteins were reported in 1985 (HERZBERG and JAMES 1985a, b; SUNDRALINGAM et al. 1985a, b; BABU et al. 1985; KRETSINGER et al. 1986) and this demonstrated that the sequence homology extended into a large structural homology. Figure 2 shows a schematic representation of troponin C, which clearly demonstrates the α-helical nature of the protein. The most intriguing aspect of the structure is the unusual dumbbell structure. The protein is clearly composed of two almost independently folded domains that are connected by an α-helical loop. Both halves contain binding sites for two calcium

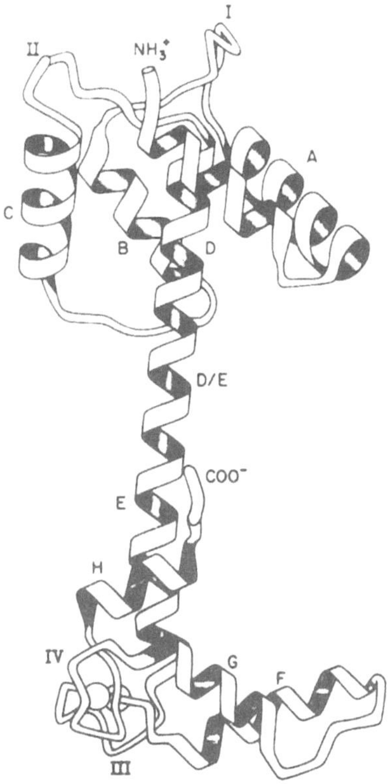

Fig. 2. Schematic picture of the structure of turkey skeletal muscle troponin C, which clearly indicates the α-helical nature of the protein. Two calcium ions are bound to the *III* and *IV* calcium-binding domain, whereas no calcium is bound to the first two sites in the top of the molecule. Note that the protein is shaped like a dumbbell with two domains connected by an α-helical loop. (HERZBERG and JAMES 1985a)

ions each. The structure of CaM is virtually identical to that of troponin C. CaM can interact with metal ions (in particular Ca^{2+}), target proteins, and a variety of drugs and other hydrophobic substances. The purpose of this chapter is to discuss our present understanding of the location of the binding sites for these different ligands on CaM. In addition the structural requirements for ligands that CaM recognizes will be discussed.

B. Metal Ion-Binding Sites

I. Calcium

CaM has an impressive ability to recognize Ca^{2+} specifically over other metal ions. The concentration of Mg^{2+}, for example, is between 1 and 5 m*M* in most cells, the level of K^+ is around 100 m*M* whereas the intracellular Ca^{2+} concentration during activation rises from 10^{-7} to 10^{-6} *M*. Few proteins are capable of such a specificity and a unique molecular architecture of the calcium-binding sites is used to accomplish this. The specific features of these sites were first recognized when the crystal structure for the intracellular protein parvalbumin became available (KRETSINGER 1980). It showed that two calcium ions were bound to this protein in a loop of amino acids that wraps itself around the metal ion and follows an octahedral pattern. The ligands coordinating to the Ca^{2+} are all oxygen atoms, mainly from Asp and Glu side chains, but also from Ser and Thr hydroxyl groups, carbonyl groups from the protein backbone, or water molecules. These calcium-binding loops are flanked by two α-helices, thus each Ca^{2+}-binding site is contained in a helix–loop–helix segment. Screening of the amino acid sequences of a variety of intracellular calcium-binding proteins revealed that similar patterns of helix-forming and acidic amino acid could be recognized. This led to the suggestion that these proteins would have very similar structures (KRETSINGER 1980), which has now been confirmed (SZEBENYI et al. 1981; HERZBERG and JAMES 1985a; BABU et al. 1985). Attempts to determine the solution structure of CaM are also in agreement with this notion (AULABAUGH et al. 1984; SEATON et al. 1985).

Since CaM contains four calcium-binding sites, it was of interest to know if the Ca^{2+} ions bound in a specific sequence to the protein or if this was a random process. There are in principle two different approaches to this question. The indirect method is to follow the conformational changes in the protein with spectroscopic techniques and deduce from this the mode of metal ion binding. A more direct approach is through observing the Ca^{2+} directly. A problem with the latter approach is that there are no known isotopes for Ca^{2+} which have optical or ESR properties. Although one can substitute other cations which do have such properties, there is never a guarantee that these will bind in a similar manner to CaM (see Sect. B) and thus the direct study remains preferable. There are essentially two ways in which this can be done. First of all, one can perform equilibrium dialysis measurements using the radioactive isotope ^{45}Ca. This strategy has been used by various groups (for a summary of the results see BURGER et al. 1984). These studies did not seem to indicate a specific sequence of binding and it was suggested that the four sites are identical and independent. A second direct approach is through ^{43}Ca NMR measurements (VOGEL et al. 1983b; VOGEL and FORSÉN 1987). Such experiments on CaM provided evidence for a biphasic binding pattern with two strong sites and two weaker ones (TELEMAN et al. 1986). Binding to each pair of sites appears to be positive cooperative. The same pattern of Ca^{2+} binding was found in studies where the off-rates of Ca^{2+} from CaM were measured with the help of a fluorescent chelator (BAYLEY et al. 1984; MARTIN et al. 1985). In addition, in all indirect spectroscopic studies (UV, circular dichroism, fluorescence, NMR) where the protein conformational change was followed when Ca^{2+} was added, a biphasic pattern was observed (DRABIKOWSKI et al. 1982; SEAMON 1980; ANDERSSON et al. 1983; THULIN et al. 1984; AULABAUGH et al. 1984; IKURA et al. 1984; MINOWA and YAGI 1984; for review see FORSÉN et al. 1986). The conflict between the outcome of the equilibrium dialysis data and all other data was not resolved until WANG (1985) reported that the former data could equally well be fitted with a biphasic model with 2×2 cooperative sites as well as by a model of random binding with no cooperativity, i.e., the accuracy of the equilibrium dialysis method does not allow one to distinguish between these two modes of binding. In view of the large body of evidence in support of a biphasic pattern with positive cooperativity, it is now almost generally accepted that CaM functions in this manner.

Of particular help in settling the question of metal ion affinities were studies in which proteolytic fragments of CaM were used. CaM can be clipped by limited

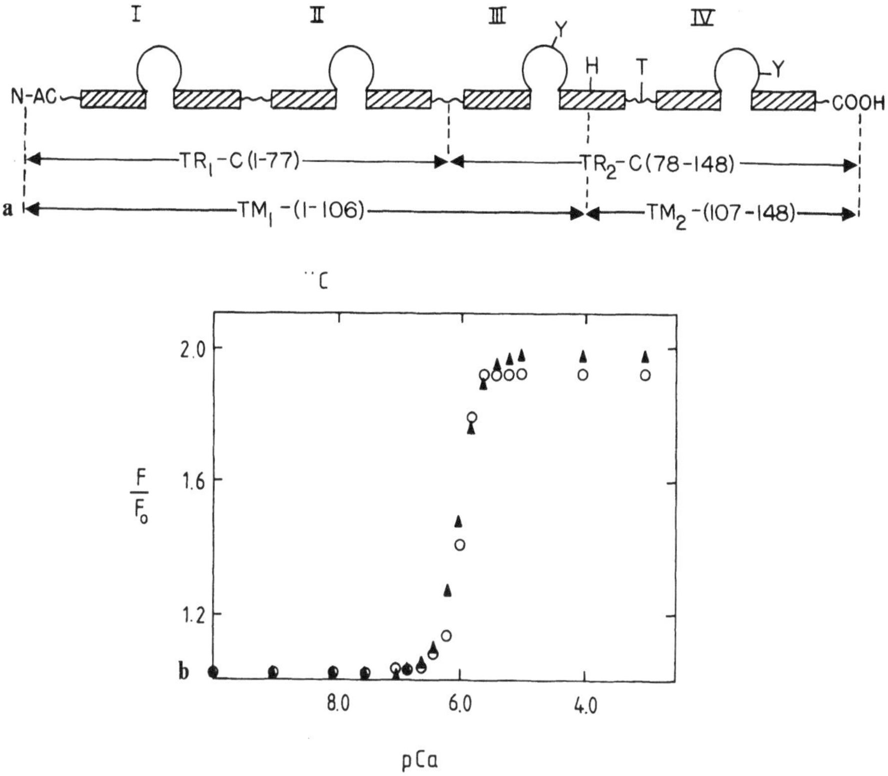

Fig. 3. **a** Schematic diagram indicating the selective proteolytic cleavage of calmodulin by trypsin (*TR*) or thrombin (*TM*). **b** Change in intrinsic tyrosine fluorescence for calmodulin (*triangles*) and its proteolytic fragment TR_2C (78–148) (*circles*) upon titration with Ca^{2+}

tryptic proteolysis into two halves, as shown in Fig. 3a (Drabikowski et al. 1977; Walsh et al. 1977; Thulin et al. 1984). The two fragments that are released in this manner still retain their native structure and their metal ion-binding properties (Drabikowski et al. 1982; Thulin et al. 1984; Tsalkova and Privalov 1985). For example, Fig. 3a shows that the change in intrinsic fluorescence for the two tyrosine residues of CaM which are located in the third and fourth calcium-binding loop is not altered by the proteolytic cleavage. Using these proteolytic fragments, it has been possible to establish that the two strong calcium-binding sites are sites III and IV and the two weaker ones are in the NH_2 terminal half (Andersson et al. 1983; Aulabaugh et al. 1984; Thulin et al. 1984; Ikura et al. 1984; Dalgarno et al. 1984). The positive cooperativity between the two sites is probably transmitted by a set of hydrogen bonds between the two calcium-binding loops (Ikura et al. 1985). In porcine intestinal calcium-binding protein, a proline residue has been inserted in one of the loops which would disrupt this pattern of hydrogen bonds (Szebenyi et al. 1981). As one would predict, this has also altered the cooperative properties in metal ion binding of this protein (Vogel et al. 1985).

II. Calcium Probe Cations

Because of the difficulties in directly observing protein-bound calcium ions, a variety of other cations with spectroscopic properties have been used as probes for the calcium-binding sites. Their use seemed to be justified in view of the fact that many di- and trivalent cations can substitute for Ca^{2+} and confer biologic activity in assays of CaM-stimulated enzymes (CHAO et al. 1984). Trivalent lanthanide ions have enjoyed particular popularity in this regard as the luminescent properties of Eu^{3+} or the fluorescent properties of Tb^{3+} can provide many clues about the structure of proteinaceous calcium-binding sites (MARTIN and RICHARDSON 1979; HORROCKS and SUDNICK 1981). Thus, these two ions appeared ideally suited to probe the calcium-binding sites of CaM and to study the order of binding of metal ions to CaM. This strategy has been used by a number of researchers (KILHOFFER et al. 1980a, b; WANG et al. 1982; WALLACE et al. 1982). They showed that the most probable sequence of Tb^{3+} binding to CaM was that sites I and II were filled before sites III and IV. It should be noted that this is the opposite sequence from what has been found for Ca^{2+}, where the two NH_2 terminal sites are filled last. Although these papers originally suggested that Ca^{2+} would bind in the same fashion, it has now been proposed, based on detailed kinetic experiments and ^{1}H NMR experiments, that Tb^{3+} and Ca^{2+} have a different sequence of binding from CaM (WANG et al. 1984; ANDERSSON 1986; BUCCIGROS et al. 1986). This is a very surprising observation as no such discrepancies have been observed for the homologous protein troponin C (LEAVIS et al. 1978; WANG et al. 1981). It raises questions about the use of lanthanides as a general probe for calcium-binding sites on proteins.

Another cation which has been used to a large extent for the study of CaM is Cd^{2+}. The Cd^{2+} ion has an ionic radius which is almost indistinguishable from that of Ca^{2+} (see Table 2). Its use has been attractive as the ^{113}Cd isotope has very useful NMR properties. As such it has been used successfully in studies of a wide variety of calcium-binding proteins such as parvalbumin, troponin C, lactalbumin, intestinal calcium-binding protein, trypsin, etc. (VOGEL and FORSÉN 1987). Judging from spectroscopic (MARTIN and BAYLEY 1986) and ^{1}H NMR studies (VOGEL and FORSÉN 1987), titrations with Cd^{2+} or Ca^{2+} give rise to al-

Table 2. Ionic radiis of a series of metal ions

Metal ion	Ionic radius (nm)
Ca^{2+}	0.099
La^{3+} [a]	0.1016
Lu^{3+} [a]	0.085
Cd^{2+}	0.097
Mn^{2+}	0.080
Mg^{2+}	0.066
Na^{+}	0.097
K^{+}	0.133

[a] The ionic radius of the lanthanide series of trivalent metal ions varies gradually between lanthanum and lutetium.

most identical conformational changes in CaM and the majority of these proteins, although some important differences have been found for bovine α-lactalbumin and porcine intestinal calcium-binding protein. The binding of Cd^{2+} to CaM as followed by ^{113}Cd NMR clearly showed a biphasic pattern. Moreover, it could be demonstrated that binding of drugs and peptides to CaM gives rise to an enhanced affinity for Cd^{2+}. In fact, surprising as it may seem, ^{113}Cd has turned out to be one of the most sensitive methods for monitoring the binding of drugs and peptides to CaM (ANDERSSON et al. 1985; LINSE et al. 1986; VOGEL et al. 1987). Some of these results will be discussed in Sects. C, D.

Yet another divalent probe cation that has been used with CaM is Mn^{2+}, an ion with ESR properties. Mn^{2+} binds in the same sequence to CaM as Ca^{2+}. This was deduced in part by estimating its distance to an ESR spin label attached at the single Cys-25 of wheat germ CaM (YOSHIDA et al. 1983).

III. Other Metal Ions

Na^{+} has been used in ^{23}Na NMR studies as a monovalent probe for Ca^{2+} and it appears to bind in a different sequence from Ca^{2+} (DELVILLE et al. 1980). K^{+} has been used in flow dialysis experiments in competition with Ca^{2+} (HAIECH et al. 1981). Its mode of binding is complex and it does not appear to follow the same pattern as for Ca^{2+}. Both monovalent cations fail to induce the large-scale conformational changes in CaM that are brought about by Ca^{2+}, Cd^{2+}, or lanthanides. Nevertheless, they may play an important physiologic role by combining with other sites on the protein as K^{+} does exert considerable effects on the off-rates of Ca^{2+} determined in stopped flow experiments (MARTIN et al. 1985). In addition, a series of other metal ions, including Zn^{2+}, Hg^{2+}, Pb^{2+}, and Sr^{2+} have been shown to bind to CaM. Most of these can also replace Ca^{2+} to a certain degree in assays of CaM-stimulated enzymes. They also produce the characteristic tyrosine fluorescence enhancement. Their effectiveness in replacing Ca^{2+} in both measurements appears to be directly related to their ionic radius, i.e., the closer this radius resembles that of Ca^{2+} the higher its effect (CHAO et al. 1984). These measurements are all consistent with the idea that these cations bind to the calcium-binding sites of CaM. In agreement with this notion, it has been shown that Pb^{2+} and lanthanides can displace Ca^{2+} from the protein (FULLMER et al. 1985). Recently, however, MILLS and JOHNSON (1985) produced evidence that one or more additional metal ion-binding sites must exist on CaM that have a high affinity for ions such as Zn^{2+}, Pb^{2+}, or lanthanides, but a low affinity for Ca^{2+}. Occupation of this site by these cations produces an allosterically potentiated conformer of CaM. In contrast, if these sites are occupied by Cu^{2+} or Hg^{2+} an inactive conformer is produced. The exact localization of this site is unknown at present. As its occupation by specific metal ions may produce effects on the way in which the Ca_4^{2+}–CaM complex interacts with its targets, it seems that it may play a role in fine tuning the response which is mediated through the four calcium-binding sites of CaM.

C. Binding of Drugs

I. Trifluoperazine

In 1974 WEISS et al. reported that CaM activation of phosphodiesterase was blocked by trifluoperazine (TFP), a clinically effective antipsychotic agent. Subsequently, they demonstrated that TFP and related phenothiazines bound selectively to CaM in a Ca^{2+}-dependent manner (LEVIN and WEISS 1979). From equilibrium dialysis studies, it was concluded that CaM contained two high affinity sites for TFP ($K_d \sim 1\ \mu M$) and a series of weaker ones ($K_d \sim 5\ mM$) (LEVIN and WEISS 1979). Although these numbers have been questioned in later equilibrium dialysis studies (MARSHAK et al. 1985), various spectroscopic studies are consistent with the presence of not more than two high affinity sites for TFP (FORSÉN et al. 1980; KREBS and CARAFOLI 1982; VOGEL et al. 1984). These observations have raised a lot of interest as they suggested that certain drugs could act by interfering with CaM-stimulated events. Later studies showed, however, that drugs without any clinical effect bound equally well to CaM as did TFP, thus suggesting that the physiologic site of action of phenothiazines is not through its binding to CaM (NORMAN et al. 1979; ROUFOGOLIS 1981). Nevertheless, the phenothiazines have been popular as experimental tools for in vitro studies to inhibit a large series of CaM-modulated events. Of particular value has been the development of phenothiazine affinity columns for use in the purification of CaM (CHARBONNEAU and CORMIER 1979; JAMIESON and VANAMAN 1979). By running these in the presence and absence of Ca^{2+}, it was easy to purify CaM on a large scale.

Because TFP has been so effective in inhibiting so many different CaM-regulated enzymes, the view emerged that the drug bound to the same surface on CaM, as where the target systems would bind. Thus, it became of interest to study the localization of the two strong binding sites on CaM for TFP. Two approaches have been used. The first one relied on the use of proteolytic fragments of CaM in combination with a phenothiazine affinity column (VOGEL et al. 1983a). These studies showed convincingly that one site for TFP was located in each domain of CaM, as both halves of the molecule retained the capacity to bind to the phenothiazine column in a Ca^{2+}-dependent manner. The second approach was through NMR spectroscopy. Both in ^{1}H and ^{113}Cd NMR spectra it was observed that addition of up to 2 equiv-TFP to CaM caused large changes in the environment of all four metal ion-binding sites, and some methionine and phenylalanine residues (FORSÉN et al. 1980; KREBS and CARAFOLI 1982). NMR studies with proteolytic fragments of CaM demonstrated clearly that the binding of TFP to its two strong sites is sequential, the first equivalent of TFP binds to the COOH terminal half, the second equivalent binds to the NH_2 terminal half (THULIN et al. 1984). Based on fluorescence energy transfer measurements it has been proposed that TFP and a hydrophobic probe molecule will bind to the NH_2 terminal helix at domain III (STEINER 1984; STEINER and MARSHALL 1985). Also GARIEPY and HODGES (1983) have suggested, based on studies with the fragment CB9 (84–135) from troponin C, that TFP binds to the NH_2 terminal helix of the third calcium-binding domain of this protein. However, DALGARNO et al. (1984) used ^{1}H NMR to identify two TFP sites in troponin C and CaM that were formed by: (a) the NH_2 terminal helix of site I and the COOH terminal helix of site II; and (b) the NH_2 terminal helix of site III and the COOH terminal helix of site IV. These findings are surprising in view of the fact that both CB9 (84–135) and TR_2C (89–159) of troponin C do *not* bind to a phenothiazine affinity column, in contrast to TR_1C (9–84) (Vogel et al. 1983 a). Moreover, the CaM fragment 77–124 appears to retain a complete TFP-binding site (Head et al. 1982) which seems to preclude the involvement of the COOH terminal helix of site IV. From these conflicting results, we have to conclude that it is not clear at present where exactly the TFP molecules are bound to CaM and troponin C.

The forces that stabilize the interactions between CaM and TFP are probably mainly hydrophobic in nature. This became especially clear after LaPorte et al. (1980) and Tanaka and Hidaka (1980) demonstrated that hydrophobic surfaces were exposed on CaM upon binding Ca^{2+}. This was confirmed in later studies using hydrophobic photoaffinity probes for CaM (Krebs et al. 1984) as well as in fluorescence studies where the binding of the hydrophobic probe TNS was followed (Johnson et al. 1986). In both studies, it was concluded that the two domains of CaM have the capacity to expose a hydrophobic surface upon binding Ca^{2+}. This was in good agreement with other studies on TNS binding to CaM where it was shown that the intact protein contained two strong sites for this probe molecule (Follenius and Gerard 1984). Studies using proteolytic fragments have also shown that the surface in the NH_2 terminal part of CaM is more hydrophobic than the one in the COOH terminal half (Drabikowski and Brzeska 1983; Johnson et al. 1986). Probably inspired by the discovery of these Ca^{2+}-exposed hydrophobic surfaces, Gopalakrishna and Andersson (1982) showed that commercially available phenyl-Sepharose could be substituted for the phenothiazine column in the purification scheme of CaM. As was to be expected, both proteolytic halves of CaM can be retained on this material, thus providing further evidence for the capacity of each half of CaM to expose a hydrophobic patch upon binding of Ca^{2+} (Vogel et al. 1983 a; Brzeska et al. 1983; Newton et al. 1984).

The binding of TFP to Ca_4^{2+}–CaM enhances the affinity of the protein for metal ions (Forsén et al. 1980; Vogel et al. 1984; Tanokura and Yamada 1985; Suko et al. 1986; Inagaki et al. 1983). In particular, the two sites in the NH_2 terminal domain are affected and this gives rise to a situation where the four sites have very similar affinities (Martin et al. 1985).

II. Calmodulin Antagonists

Following TFP, a whole series of other hydrophobic substances have been shown to interact with CaM (see Table 3). Most of these successfully inhibit the CaM stimulation of enzymes which has given them their name: CaM antagonists (Johnson and Wittenauer 1983; Van Belle 1981). Others have been used for the preparation of affinity columns for the purification of CaM (Hidaka et al. 1981; Endo et al. 1981; Hart et al. 1983; Tanaka et al. 1984). It is not clear where exactly these drugs bind to CaM; nevertheless, it has been shown that oxidation of methionine residues of CaM resulted in a decrease in the binding of naphthalenesulfonamides and TFP (Tanaka et al. 1983). A survey of the structures of a

Table 3. Drugs that can bind to calmodulin and block enzyme activation. (Compiled from data taken from JOHNSON and WITTENAUER 1983; FOLLENIUS and GERARD 1984)

Compound	Number of strong binding sites	Dissociation constant (μ*M*)
Calmodulin antagonists		
Trifluoperazine (TFP)	2	5
Naphthalenesulfonamides (W7)	ND	4
Calmidazolium (R24571)	ND	2×10^{-3}
Toluidinylnaphthalene sulfonamide (TNS)	2	100
Calcium antagonists		
Felodipine	2	2.8
Verapamil and D600	ND	30
Diltiazam	ND	80
Prenylamine	ND	0.6

ND, not determined.

variety of these drugs shows that compounds such as TFP, W7, and R24571 all possess one positive charge at physiologic pH. Thus, in addition to hydrophobic forces that stabilize the CaM–drug interaction, it is likely that ionic charges are involved as well. In this respect, the Glu-81, -82, and -83 residues have been pinpointed as a potential site for neutralizing the positive charge on these drugs (JOHNSON et al. 1986).

One other feature of the CaM antagonists should be mentioned here as well. Generally, they are not very specific and thus effects observed upon using these compounds do not necessarily reflect an action of CaM. For example, W7, TFP, and R24571 can bind directly to myosin light chain kinase (MLCK) (ZIMMER and HOFMANN 1984) and the plasma membrane Ca^{2+}-ATPase (ADUNYAH et al. 1982); whereas W7 even inhibits the CaM-independent protein kinase C activity (SCHATZMANN et al. 1983). Thus, caution is needed in interpreting results obtained with so-called CaM antagonists.

III. Calcium Antagonists

Calcium antagonists are pharmacologic agents that are thought to act by binding to calcium channels and blocking the influx of calcium into a cell (see Chaps. 5 and 6). Nevertheless, it has been shown that a whole series of such compounds can bind to CaM in a Ca^{2+}-dependent manner (BOSTRÖM et al. 1981, 1985; ANDERSSON et al. 1985; JOHNSON and WITTENAUER 1983). This property is of considerable interest as the binding to the calcium channel proteins is also a Ca^{2+}-dependent event. Figure 4 shows the structures of some calcium antagonists. Most of them appear to be structurally unrelated and hence one can distinguish dihydropyridines (felodipine), verapamil, diltiazem, and prenylamine as separate agents. Of these different classes felodipine and other dihydropyridines have been studied in some detail. Felodipine is a very potent and efficient antihypertensive agent. It has excellent vascular selectivity, i.e., it has no effects on the contractility of the heart in the concentration ranges where it affects the smooth muscle activ-

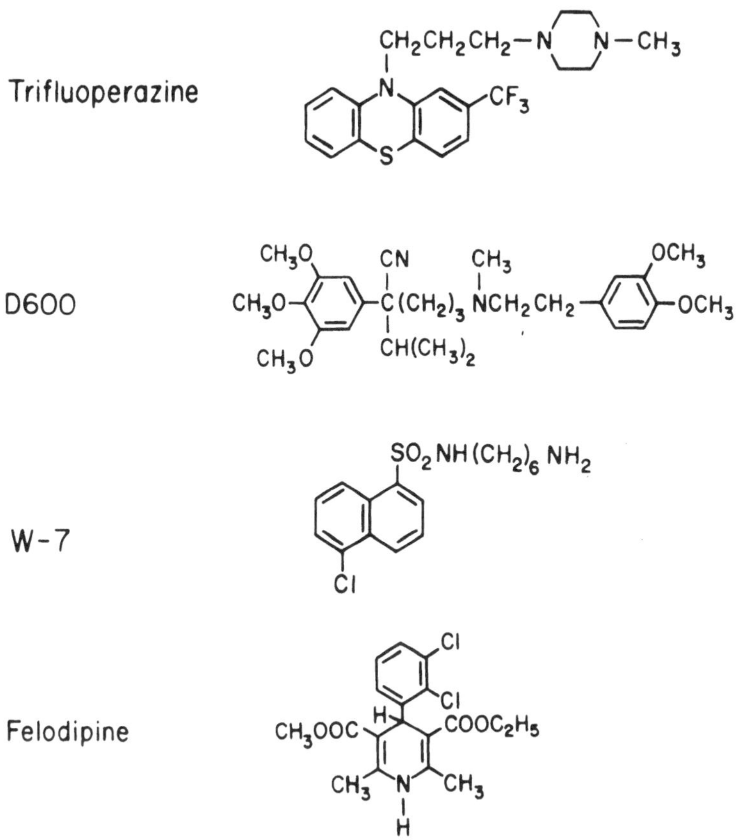

Fig. 4. Structures of some of the drugs that are discussed in the text and in Table 2

ity (Fig. 5). This selectivity is rare among calcium antagonists (LJUNG et al. 1987). Also, the structure–activity relationship for 1,4-dihydropyridines has been probed by X-ray crystallographic studies (see Chap. 6) and statistical analysis of physicochemical data (BERNTSSON et al. 1987; BERNTSSON and WOLD 1987). Based on the latter studies, it has been suggested that physicochemical substitution parameters give a better prediction of the biologic activity than X-ray data.

A property of felodipine which has been useful experimentally is its intrinsic fluorescence which increases upon binding to Ca_4^{2+}–CaM (JOHNSON 1983; MILLS et al. 1985). This has allowed us to study the localization of a felodipine-binding site on CaM. It was found that the latter does *not* bind to the COOH terminal half, but that it binds to the NH_2 terminal half and that it requires part of the central loop to be intact (JOHNSON et al. 1986; BOSTRÖM et al. 1987). Despite this different behavior when compared with TFP, felodipine is still capable of inhibiting CaM activation of the phosphodiesterase and MLCK (BOSTROM et al. 1985). A recent detailed study (ZIMMER and HOFMANN 1987) of the inhibition of CaM's enzyme activation by calmodulin- and calcium antagonists led to the conclusion that there are at least three distinct binding sites on calmodulin for the different antagonists. Most interesting, however, was the observation that felodipine bind-

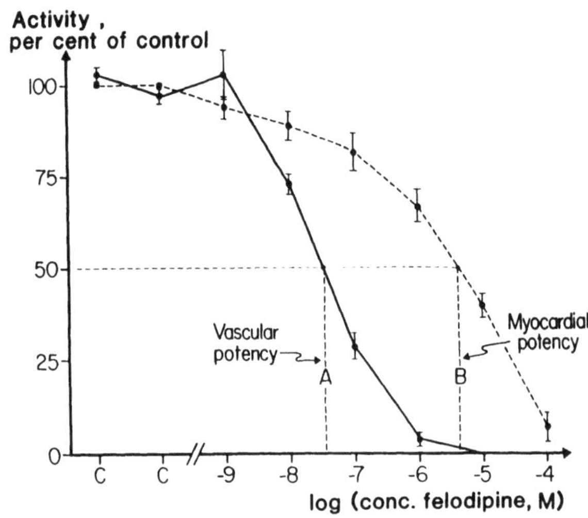

Fig. 5. The activity of felodipine in inhibiting isolated portal vein (*full line*) and a paced papillary muscle of the left ventricle from the rat (*broken line*). Note the large difference in vascular (*A*) and myocardial (*B*) potency. This varied across a series of calcium antagonists as follows: felodipine ≫ nifedipine > diltiazem > verapamil. (LJUNG et al. 1987)

ing can be potentiated by the binding of other drugs to the COOH terminal half of CaM (MILLS et al. 1985; JOHNSON et al. 1986). This provided the first evidence that the two domains of CaM are not independent, but that they communicate with each other in the presence of drugs. Similar cooperativity has been observed by others (NEWTON et al. 1983). In this respect, it is of interest that dihydropyridine binding to calcium channels can also be potentiated by the binding of other calcium antagonists (see for example MURPHY et al. 1983).

D. Binding of Peptides

I. Structural Requirements

Many target proteins of CaM contain a separate active site and CaM-binding domains (see Sect. E.II). Ideally, one would of course like to study a complex between CaM and a proteolytic fragment containing the CaM-binding domain of the target enzyme. However, until quite recently it has proven a rather difficult task to purify such a CaM-binding domain. Therefore, much interest was generated when several reports appeared in the literature showing that CaM could interact in a Ca^{2+}-dependent manner with a variety of small peptide hormones and toxins which have a relatively well understood structure. Regardless of the physiologic relevance of such interactions, it was felt that such studies could provide useful insight into the otherwise elusive problem of the CaM-binding domain. By now an enormous variety of peptides have been screened for their inhibitory potential in a CaM-stimulated phosphodiesterase assay. A listing of some of these peptides is provided in Table 4.

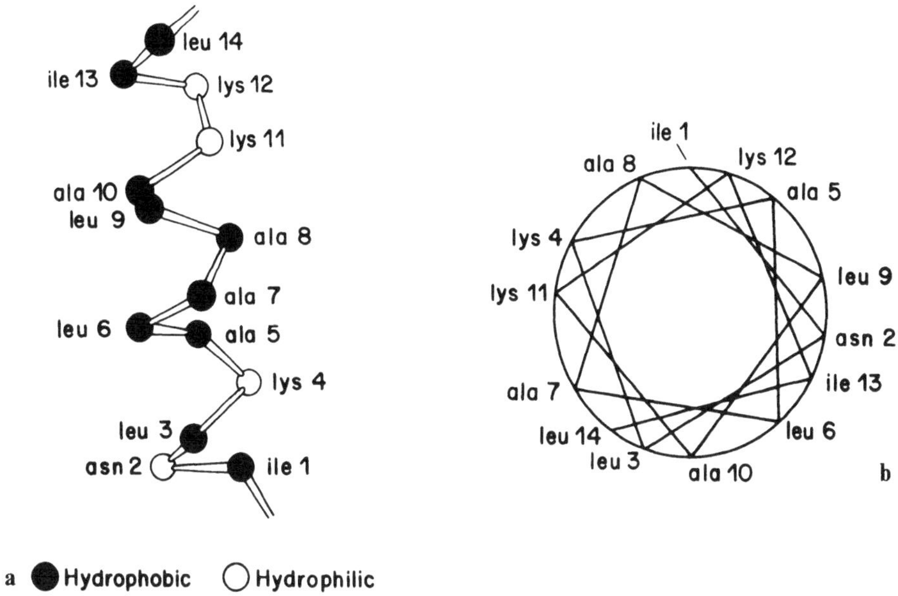

Fig. 6. a An α-helical structure for mastoparan. Note that the three lysine residues are all on one face of the helix, whereas the other side of the helix contains only hydrophobic amino acids. Such a helix is called amphiphilic. **b** A convenient way of looking at the amphiphilicity of a helix is the so-called helical wheel projection which was developed by SCHIFFER and EDMUNDSON (1967). By looking down the center of a helix one can observe if all the residues on one side are hydrophobic and if all the hydrophilic chains are on the other side. This is shown for the mastoparan helix. (Adapted from MCDOWELL et al. 1984)

One of the first peptides to be shown to interact with CaM was β-endorphin (SELLINGER-BARNETTE and WEISS 1981). Through cross-linking studies it was demonstrated that a 2:1 complex was formed (GIEDROC et al. 1983). By analysis of a series of deletion peptides it was shown that the central region of β-endorphin (residues 14–25) comprised the CaM-binding domain. This part of the molecule is thought to form an amphiphilic helix (Fig. 6; GIEDROC et al. 1985b). Another feature which appears to be important in such CaM-binding peptides is the presence of some basic positively charged groups. This was noted when some arginine- and lysine-rich proteins such as the myelin basic protein and histones formed complexes with CaM in a Ca^{2+}-dependent manner (GRAND and PERRY 1980). These results demonstrate the importance of ionic interactions in addition to hydrophobic forces in stabilizing a CaM–protein complex. Interestingly, the introduction of some negative charges on histone 2B through phosphorylation reduces the affinity for CaM (IWASA et al. 1981).

In a series of papers (MALENCIK and ANDERSON 1982, 1983, 1984; MALENCIK et al. 1982a, b), the interactions between CaM and a number of peptides have been probed by fluorescence measurements. It was shown that a large series of peptides, such as ACTH, endorphin, substance P, and glucagon were able to interact with CaM (MALENCIK and ANDERSON 1982). Amino acid sequence com-

Table 4. Peptide hormones and other peptides that bind to calmodulin. (Compiled from data taken from MALENCIK and ANDERSON 1983, 1984; COY et al. 1985)

Peptide	Number of strong binding sites	Dissociation constant (μM)
β-Endorphin	2	2000
Adrenocorticotropic hormone (ACTH)	ND	1500
Glucagon	ND	3400
Histone 2B	2[a]	~5000[a]
Vasoactive intestinal peptide (VIP)	1	50
Gastric inhibitory peptide (GIP)	1	80
Melittin	1	3
Mastoparans	1	0.3
Leu-Lys amphiphilic model peptide	1	5

[a] Unpublished observations.

Mastoparan	[1]I	N	L	K	A	L	A	A	L	A	K	K	I	L		
Mastoparan X	[1]I	N	W	K	G	I	A	A	M	A	K	K	L	L		
P. Mastoparan	[1]V	D	W	K	K	I	G	Q	H	I	L	S	V	L		
LK2			L	K	K	L	L	K	L	L	K	K	L	L	K	L
Melittin	[3]G	A	V	L	K	V	L	T	T	G	L	P	A	L		
VIP	[17]M	A	V	K	K	Y	L	N	S	I	L	N				
GIP	[17]I	R	Q	Q	D	F	V	N	W	L	L	A	Q	Q		
δ-Hemolysin	[7]S	T	I	G	D	L	V	K	W	I	I	D	T	V	N	K

Fig. 7. The amino acid sequences of peptides which interact tightly with calmodulin. The invariant hydrophobic residues are *boxed*. (COX et al. 1985)

parisons of the peptides that showed affinity for CaM demonstrated that the regions that CaM recognizes bear a structural resemblance to the recognition sequence for the cAMP-dependent protein kinase. Nevertheless, although a synthetic protein kinase substrate was able to bind to CaM, it did so rather poorly ($K_d \sim 2$–7 mM) (MALENCIK et al. 1982a), thus suggesting that this peptide lacks certain features which determine high affinity binding to CaM. Also, a peptide derived from a heat-stable protein kinase inhibitor will bind to CaM (MALENCIK et al. 1986). Most interesting, however, was the observation by these authors that peptides such as VIP and GIP, and mastoparan could from 1:1 complexes with CaM with submicromolar affinity (Table 4; MALENCIK and ANDERSON 1983, 1984). This stoichiometry and these dissociation constants closely resembled the ones observed for CaM-regulated enzyme systems. Also melittin, an amphiphilic peptide from bee venom, has been found to have a very high affinity for CaM (COMTE et al. 1983; MAULET and COX 1983). Concomitant with complex formation, the helical content of these peptides increases dramatically. Based on this information, Cox et al. (1985) tested a series of synthetic peptides containing only Leu and Lys amino acids to test the hypothesis that the minimum structural requirement for high affinity peptide binding is a basic amphiphilic helix approximately 15 Å in length. Indeed, they observed that these peptides bound to CaM with nanomolar affinities. α-Helix formation takes place both in CaM and in the peptide upon complex formation. The amino acid sequences of all peptides can be aligned such that the hydrophobic resonances occupy invariant positions (Fig. 7). A second feature is the clustering of a set of basic residues. Amphiphilic peptides with a negative charge do not bind to CaM, those that are neutral have a reduced affinity for CaM (Cox et al. 1985). Thus, there seems to be a parallel with the binding of drugs, as discussed in the previous section. Also there, we noticed that the driving forces for binding of some drugs were hydrophobic and furthermore that a positive charge seemed to be involved.

II. Localization of Binding Sites

From the previous studies, it is not clear where exactly on CaM's surface the peptide will bind. In order to address this question the binding of proteolytic fragments of CaM with the various model peptides has been studied. Mastoparan for example, binds both the fragment TM_1 (1–106) and TR_2C (78–148). This is consistent with binding of this peptide to part of the central helix of CaM (MALENCIK and ANDERSON 1984). From fluorescence measurements it was concluded that the site on CaM was either hydrophobic or rigid hydrophilic. The latter would be consistent with binding to the central helix. Moreover, in ^{113}Cd NMR studies it was demonstrated that the calcium-binding sites in both halves of CaM were affected by the addition of 1 equiv. mastoparan (LINSE et al. 1986) or melittin (VOGEL et al. 1987). In an attempt to further localize the site of attachment of melittin we have used affinity chromatography of a series of CaM-proteolytic fragments on a melittin affinity column (VOGEL et al. 1987). The results demonstrated that melittin can interact with both halves of the protein although it seems to have a preference for the NH_2 terminal half of CaM. Melittin also displaces the drug felodipine, which is known to require the central loop of CaM for binding (JOHNSON et al. 1986; BOSTRÖM et al. 1987).

These conclusions are in agreement with fluorescence studies on melittin binding to CaM, where it was shown that the peptide interacts with both halves and that Trp-19 of melittin is close to the COOH terminal half (STEINER et al. 1986). A different approach for studying the localization of the two β-endorphin sites on CaM was pursued by GIEDROC et al. (1985a). They utilized ^3H-labeled acetic anhydride as a reagent which is specific for the ε-amino group of the seven lysyl residues on CaM. Addition of 2 equiv. β-endorphin greatly reduced the reactivity of Lys-75 and Lys-148. Similar results were obtained by additions of TFP, suggesting that β-endorphin resembles the phenothiazines rather than mastoparan or melittin. Lys-75 is located in the long helix that connects the two domains of CaM. Thus, part of the central helix appears to be involved in one of the binding sites for β-endorphin. It is of interest in this regard that Lys-77 which is close to Lys-75 and is also part of the central helix does *not* have an altered reactivity toward acetic anhydride in the presence of β-endorphin.

E. Interactions with Target Proteins

In order to get a complete description of the interactions between CaM and a target protein it will be necessary to determine the crystal structures of a CaM–target protein complex. Unfortunately, it will be a while yet before such studies will be available for CaM. In the meantime a large body of information has been gathered through biochemical and chemical studies. These will be discussed in this section.

Once it became clear that CaM had more widespread effects than simply regulating the activity of the cyclic nucleotide phosphodiesterase, the search was on for proteins that could interact with this unique regulatory protein. A wide range of experimental strategies has been devised to identify putative CaM-binding proteins. These approaches have included the use of CaM affinity columns, gel overlay techniques, bifunctional cross-linking reagents, photoaffinity labeling strategies, etc., (for review see Klee and Vanaman 1982). From these studies it became clear that the interaction of CaM with its target proteins was an exclusive one, i.e., the binding of CaM to one target system prevented the interaction with another (Manalan and Klee 1983a). Also, cross-linking studies showed conclusively that 1:1 complexes were formed between CaM and its targets (Andreasen et al. 1981; Kincaid 1984; Klevitt and Vanaman 1984). Thus Ca_4^{2+}–CaM appears to have one surface available to which all proteins bind to some extent. Cross-linking studies can be a first step on the way to identifying the location of a CaM-binding site on the target protein. Unfortunately, no such studies have been reported to date. Perhaps the experimental strategy could be improved if cross-linking reagents were used that are designed in such a way that they introduce a radioactive label in the target protein while the two proteins can be dissociated before the analysis of the attachment site is initiated. Such reagents have been used to advantage in the study of the interactions between troponin C and troponin I (Chong and Hodges 1981) and gelatin and fibronectin, for example (Schwartz et al. 1982).

I. Chemical Modifications and Affinity Labeling

It should be possible to identify the groups on CaM that are involved in the interactions with target proteins through chemical modification studies. First attempts in this direction were made by Walsh and Stevens (1977). They reported that the modification of the single histidine, the two tyrosines, and four of the six arginines did not cause a loss of activity in a phosphodiesterase assay.[1] Carbamylation (but *not* guanidation) of one or more lysines resulted in inactivity. This observation is of interest as the former modification does abolish the positive charge on the lysines, but in the latter the positive charge is retained. In addition, it has been reported that oxidation of Met-71, Met-72, and Met-76 with *N*-chlorosuccinimide or chloramine-T completely abolished activity (Walsh and Stevens 1978), leading to the suggestion that the nuthionine residues form part of the interaction surface. This interpretation has been questioned by Gopalakrishna

[1] In contrast, Arg-modified CaM has a reduced stimulatory activity toward the plasma membrane Ca^{2+}-ATPase and carbamylated CaM displayed full activation (J. Krebs 1986, personal communication).

and ANDERSON (1984). They argued that if these methionines are directly involved in the Ca^{2+}-dependent interaction between CaM and its target enzymes, Met-71, Met-72, and Met-76 should only be modified after addition of Ca^{2+} to CaM. This was not observed experimentally; modification of CaM in the presence and absence of Ca^{2+} apparently gives rise to the same result. We have also studied this question by ^1H NMR. Modification of the three methionine residues with *N*-chlorosuccinimide and chloramine-T resulted in drastic changes in the structure of the homologous protein rabbit parvalbumin (HIRAOKI et al. 1987). Preliminarly studies with CaM have shown that this protein's structure is also rather sensitive to methionine modification. It has also been reported that the affinity for Ca^{2+} of two of the sites is drastically reduced upon modification of the methionines (WALSH and STEVENS 1978; GOPALAKRISHNA and ANDERSON 1984). Thus, the available data can hardly be taken as strong evidence for the methionines being directly involved in the binding of target systems. Rather, they suggest that the methionine oxidation results in a partial disruption of CaM's structure which in turn makes it impossible for target proteins to recognize CaM.

Affinity labeling studies have been attempted by various groups. NEWTON et al. (1983) produced an adduct of CaM with an isothiocyanate phenothiazine derivative. Interestingly, the major product is a 1:1 complex although normally CaM strongly binds two phenothiazine molecules. The site of modification is not clear at present, but is presumably on the Lys-75 residue (BABU et al. 1985). The 1:1 adduct, known as $CAPP_1$-CaM has some interesting properties. It lacks the ability to stimulate the phosphodiesterase, but it still binds to the enzyme as it is capable of inhibiting the action of intact CaM. Similarly, it acts as an antagonist with MLCK, but it can partially activate calcineurin while it completely activates phosphorylase kinase and glycogen synthase kinase (NEWTON and KLEE 1984; NEWTON et al. 1985). These data indicate that the occupancy of one of the two phenothiazine-binding sites on CaM does not abolish the affinity of CaM for binding to its target proteins, but that it does affect its ability to activate some enzymes. In later studies the authors also studied $CAPP_2$–CaM. This adduct is inactive as an agonist or antagonist with almost all enzymes tested. These observations seem to imply that both domains of CaM must be involved in the interaction with the majority of all CaM-regulated enzymes.[2] Moreover, they also suggest that the mode of interaction has to be different with all enzymes. JARRET (1984, 1986) synthesized a succinimide along of TFP and demonstrated that it could label two sites in CaM but that it was particularily reactive with Lys-148 (FAUST et al. 1987). Adducts containing 1–2 equiv. of this analog were incapable of activating either NAD kinase or phosphodiesterase. Similarly, covalent modification of CaM with up to 2 equiv. phenoxybenzamine (LUKAS et al. 1985a) or a tricyclic fluorene ring ESR spin label (JACKSON and PUETT 1984) prevented CaM from activating the phosphodiesterase and MLCK. Phenoxybenzamine was mainly found attached to residues 38–75, 107–126, and 127–148. The other reagent almost exclusively modified Lys-75 and Lys-148. The combination of all these results clearly implicates the regions around the lysine residues 75 and 148 in the interaction between CaM and these hydrophobic compounds. Neverthe-

[2] A possible exception is phosphorylase kinase. However, this enzyme seems to have a low requirement for specificity as troponin C can substitute as well.

less, it would be worthwhile if it could also be shown in these instances that the structural integrity and the calcium-binding properties of CaM have not been adversely affected by the modifications. Only when this condition is fulfilled can it be assumed that these regions are directly implicated. In this respect it is encouraging that the fluorenylmethyloxy group can be partially removed from CaM with an almost complete recovery of CaM's capacity to activate enzymes. Thus, apparently, Lys-75 and Lys-148 are not directly involved, but it is mainly the bulky tricyclic ring which blocks the interactions between CaM and its target proteins (JACKSON and PUETT 1984). In other studies it could also be shown that the covalent adducts between CaM and β-endorphin (with 1:1 and 1:2 stoichiometry) were incapable of activating target proteins (GIEDROC et al. 1985b).

The relevance of the unique posttranslationally modified trimethyl-Lys-115 residue has long puzzled many investigators. It has been assumed for some time that it could play a role in the interactions between CaM and its targets. Nevertheless, studies using the unmethylated form of CaM showed that there were no differences in the capacity of either form to activate a variety of enzymes (PUTKEY et al. 1985). Recently, it has been shown that unmethylated CaM is an excellent substrate for the ubiquitin-associated protein degradation system, whereas the trimethylated form is not. Thus, the role of this unusual amino acid appears to be more in protecting CaM against proteolytic attack than in being involved in enzyme activation (GREGORY et al. 1985).

II. Activation and Binding with Calmodulin Fragments

Another way in which it should be possible to study what parts of CaM are involved in the interaction with target proteins is to determine the potency of proteolytic fragments to act as activators or as inhibitors of CaM in assays of CaM-regulated enzymes. Phosphorylase kinase seems to be poorly discriminatory as it can be stimulated by both TR_1C (1–77) and TR_2C (78–148) (KUZNICKI et al. 1981; NEWTON et al. 1984, 1985). In contrast, phosphodiesterase, calcineurin, MLCK, and glycogen synthase are *not* activated by any of the peptides. In fact, judged by the fragments' ability to interfere with CaM activation, some of the fragments may not bind strongly to some of these enzymes, with the exception of TR_2C (78–148) which is capable of inhibiting CaM in phosphodiesterase and MLCK assays. The activation of erythrocyte Ca^{2+}-ATPase has also been tested (GUERINI et al. 1984). This membrane cation pump can be stimulated by the fragments TR_2C (78–148) and TM_1 (1–106), but not by TR_1C (1–77), TR_2E (1–90), or TM_2 (107–148). These observations clearly implicate the region around calcium-binding loop III as a requirement for the activation of the Ca^{2+}-ATPase. A second means of studying binding between proteolytic fragments of CaM and target proteins is by affinity chromatography. NI and KLEE (1985) produced three columns containing immobilized TM_2 (107–148), TR_1C (1–77), and TR_2C (78–148). The former did not retain any of the CaM-binding proteins in brain. However, binding did occur to the other two columns. Based on their behavior on the latter two columns, three classes of CaM-binding proteins could be identified. The first group interacts with both domains of CaM and includes the phosphodiesterase. The second group interacts only with the COOH terminal portion of CaM. This comprises the majority of the brain CaM-binding proteins and it includes, for example, calcineurin. The third group differs from the second in that the interaction with the COOH terminal part can be reversed at high ionic strength, even in the presence of Ca^{2+}. What these experiments clearly demonstrate is that the COOH terminal domain of CaM contains a Ca^{2+}-dependent interaction site for all the enzymes tested.

III. Calmodulin-Binding Domains

It has been shown for a variety of CaM-regulated enzymes such as phosphodiesterase (KINCAID et al. 1985), calcineurin (MANALAN and KLEE 1983 b; TALLANT and CHEUNG 1984), plasma membrane Ca^{2+}-ATPase (ZURINI et al. 1984), and skeletal and smooth muscle MLCK (WALSH et al. 1982) that they can be activated by limited proteolysis. This had led to the idea that most of these enzymes contain a separate CaM-binding domain which is distinct from the active site domain. Proteolytic removal of the CaM domain releases the inhibition, giving rise to an enzyme which is continuously active and no longer requires Ca^{2+} and CaM for activation. These data suggest that an inhibitory peptide is arranged in such a way that it could block the entrance to the active site. CaM removes this blockage, either by binding directly to the peptide, or by causing conformational changes in the target enzyme that are propagated to the inhibitory peptide and cause it to be released from its blocking position. In the latter case it should in theory be possible to isolate a proteolytic product of the protein which would still bind Ca_4^{2+}–CaM, but this interaction would no longer be required for activity. In the former case both Ca_4^{2+}–CaM activation and binding would disappear simultaneously. It should be possible to differentiate between these two models by checking the ability of the activated proteolyzed target enzyme to interact with a CaM affinity column, for example. In the cases where this has been studied to date, usually the inhibition of enzymatic activity and the capacity to bind CaM are released simultaneously. Thus, it is most likely that the CaM-binding domain and the inhibitory peptide are closely related and perhaps even in the same stretch of amino acids.

In a few recent studies the purification of a peptide that retains both the CaM-binding activity and the inhibitory activity has been successfully completed. For example, an 18000 dalton fragment of caldesmon still retains both activities (M. WALSH 1986, personal communication). For skeletal muscle MLCK, a much smaller peptide has been characterized which seems to be the CaM-binding domain. This peptide was isolated from a CNBr digest of MLCK and it was purified on its ability to inhibit the CaM activation of MLCK. The peptide is 27 residues in length and it represents the COOH terminus of MLCK (BLUMENTAL et al. 1985; EDELMAN et al. 1985). It has a high percentage of basic and hydrophobic residues and it has a high probability of helix formation. Upon complexation with CaM, the formation of α-helices is induced, moreover, NMR data show conclusively that structural changes occur in both domains of CaM (KLEVITT et al. 1985). These studies are in agreement with earlier measurements by MAYR and HEILMEYER (1983 a) who showed that CaM induces α-helix formation in much larger proteolytic fragments of MLCK (37000 dalton). A further careful analysis of the complexes of CaM with these CaM-binding domains by NMR and perhaps X-ray crystallography could provide a much better picture of the residues involved at the protein–peptide interface.

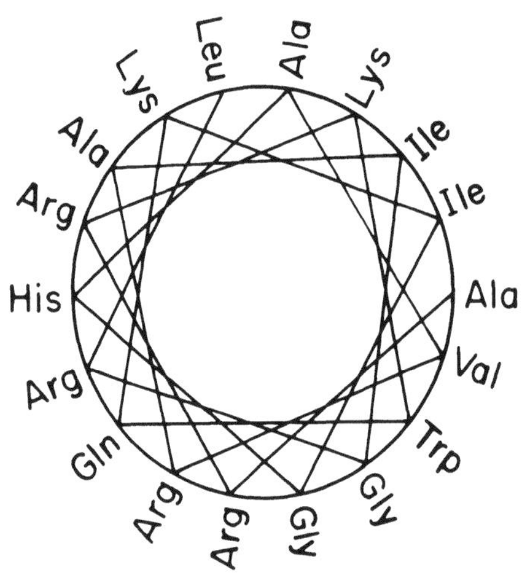

Fig. 8. Helical wheel projection of the peptide of smooth muscle myosin light chain kinase that contains both the CaM-binding and inhibitory properties. The amphiphilic nature can clearly be seen. (Adapted from LUKAS et al. 1985b)

Also, the CaM-binding peptide of smooth muscle MLCK has been purified and subsequently synthesized. The strategy employed in this case involved the purification of the phosphorylated peptide (LUKAS et al. 1985b) as it was known that CaM strongly inhibited phosphorylation (IKEBE et al. 1985). Indeed, one peptide could be isolated which contained a site that could be phosphorylated by protein kinase and which also bound strongly to CaM ($K_d \sim 1$ n*M*). Complex formation follows a 1:1 stoichiometry and α-helix formation is induced as the peptide interacts with CaM. These results have been confirmed and extended by KEMP et al. (1987) who showed that the site was between residues 480 and 501. On the basis of the characteristics of this peptide, the authors identified a similar peptide in another CaM-regulated enzyme, phosphorylase kinase (LUKAS et al. 1985b). The helical structure of the peptide is shown in Fig. 8. It is of considerable interest that the two CaM-binding peptides from skeletal and smooth muscle MLCK have very similar properties to those of the peptides that showed high affinity binding to CaM that were discussed in the previous section. All of them are capable of forming amphiphilic helices and have a cluster of basic residues. Moreover, helix formation is induced by addition of CaM and Ca^{2+}. However, they are more hydrophilic than mastoparan or melittin. For example, these peptides contain 35% hydrophilic and 20% hydrophobic residues, whereas mastoparan has 25% and 45%, respectively. Thus, the interactions between CaM and its CaM-binding domains may involve stronger electrostatic elements than previously appreciated. In addition, it is still not clear if the formation of an α-helix is a prerequisite as the troponin I inhibitory peptide does not adopt an α-helical conformation upon binding to CaM or troponin C (CACHIA et al. 1986).

F. Epilogue

Although we know presently what the structural requirements are for drugs and peptides to show high affinity binding to CaM and we have some idea about where they bind, we still know very little about the exact nature of the surface on CaM to which they bind. Another question that remains is: do these compounds bind to CaM and leave the protein structure relatively unperturbed or do drugs intercalate in the protein and cause large structural perturbations? What we know presently is that CaM's affinity for Ca^{2+} increases in such complexes, we also know that usually α-helix formation in CaM is induced upon ligand binding. These data seem to suggest that relatively large structural perturbations may occur upon ligand interaction with CaM and that inspection of the potential surfaces available for interaction in the crystal structure of Ca_4^{2+}–CaM may not necessarily provide the right answer (O'NEILL and DEGRADO 1985). We would also like to learn how CaM can act as such an effective switch for Ca^{2+}. Upon binding of Ca^{2+} to CaM, amino acid side chains become available for interaction with ligands that were previously inaccessible. We would like to know what these groups are, and how binding of Ca^{2+} to the Ca^{2+}-specific sites can bring about their exposure. One approach which seems powerful in attempting to answer some of these questions is site-specific modification of CaM. To this end the protein will need to be cloned and expressed in a bacterial strain. This has been accomplished now by several laboratories (PUTKEY et al. 1985; ROBERTS et al. 1985). This has set the stage for studies in which specific amino acids can be replaced or deleted at will and several interesting results have already been forth coming (PUTKEY et al. 1986; CRAIG et al. 1987). This experimental approach should eventually allow us to pinpoint the residues that are of critical importance to CaM's functioning.

In some other areas we feel that we are relatively close to understanding this unique protein. First of all, we have learned that most target proteins contain a CaM-binding domain, which often appears to be a peptide that can form an amphiphilic α-helical structure and contains a cluster of basic amino acids. We have also learned that both halves of CaM and the central helix play a role in the binding to target proteins and thus in enzyme activation. This agrees with recent NMR studies where it can be shown that both domains of CaM undergo structural changes upon binding 1 equiv. of a peptide resembling a target protein. Also, the mode of metal ion binding to CaM is now well established. Obviously, positive cooperative metal ion binding occurs within each domain of CaM. First, the two sites in the COOH terminal half are saturated before the two sites in the NH_2 terminal half are filled. Furthermore, cooperativity exists between the two interaction domains in the two halves of CaM. If the site in the COOH terminal half is occupied by a drug, a site which is on the helical linker and the NH_2 terminal half becomes exposed. All these observations have been combined in the model depicted in Fig. 9. It is shown that the COOH terminal half of CaM after binding two Ca^{2+} ions exposes an interaction surface, which can then initiate the binding to the CaM-binding domain of the target protein. Once this binding has been established, this information is propagated through the central helix to the NH_2 terminal half. The latter undergoes a conformational change which increases its affinity for Ca^{2+}. Thus, in the presence of a target system, all four metal ion-binding sites have become strongly positive cooperative. This has been confirmed experimentally by a variety of techniques (OLWIN and STORM 1985; HUANG et al. 1981; FORSÉN et al. 1986). In the presence of a target CaM will become fully saturated when going from 10^{-7}–10^{-6} M Ca^{2+}, whereas in the absence of a target a much shallower curve is obtained and complete saturation is

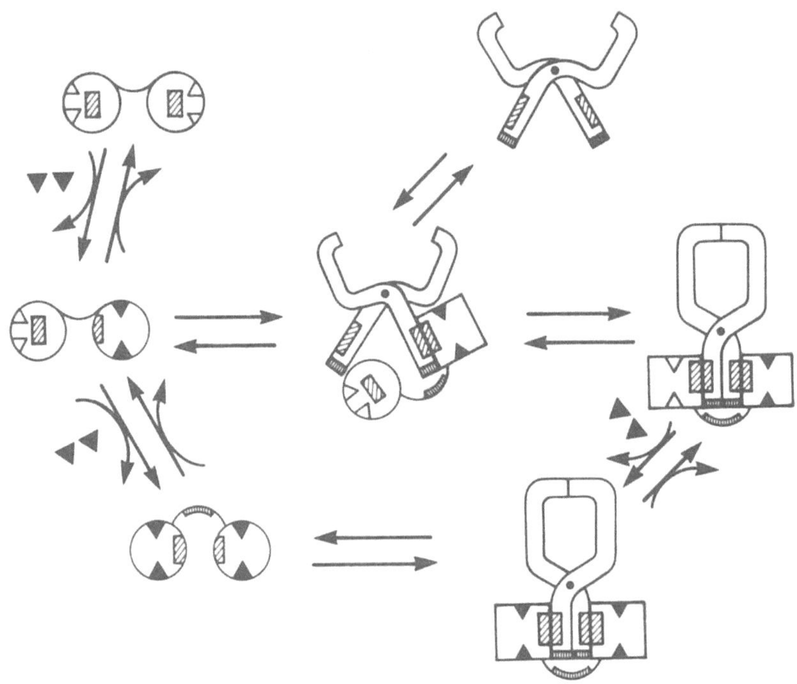

Fig. 9. Schematic representation of the mode of interaction of calmodulin with a typical target protein which contains a separate active site and CaM-binding domain. For further explanation see text. (Adapted from HIRAOKI and VOGEL 1987)

only reached over the range 10^{-7}–10^{-4} *M*. Obviously, the latter numbers are not in agreement with a physiologic function as during stimulus the intracellular Ca^{2+} level changes from 10^{-7}–10^{-6} only. It is this target protein-mediated positive cooperativity that allows CaM to function effectively over such a narrow Ca^{2+} concentration range.

This model also provides a rationale for the fact that CaM contains four calcium-binding sites. In principle, nature could have used a protein with only one calcium-binding site. However, such a protein would not be able to display positive cooperativity. Thus, intracellular Ca^{2+} would have to increase to 10^{-4} *M* during stimulation in order to accomplish complete activation. By having four metal ion-binding sites a cooperative system is created. The principle of such positive cooperativity is not unique to CaM. Nature has utilized the same principle in other situations where regulation over a very narrow concentration range is required. A well-known example is the protein hemoglobin whose main function it is to deliver oxygen to the cell. This protein binds the oxygen at the lungs and releases it at the tissues. Nevertheless, the difference in O_2 tension between these two environments is not very large and a protein with only one O_2-binding site would have problems carrying out this function. Also, hemoglobin is a protein containing four sites which possesses strong cooperative properties that are modulated in part through ligand binding. It is this property that allows hemoglobin to function properly over a very narrow range of O_2 concentrations, just as it allows CaM to function over a very narrow range of Ca^{2+} concentrations.

The studies with the drugs, phenyl-Sepharose and the model peptides have drawn attention to the importance of hydrophobic interactions in complex formation with CaM. Nevertheless, the two CaM-binding peptides from skeletal and smooth muscle MLCK that have been isolated and studied to date suggest that these domains may be more hydrophilic than previously considered. Does this mean that the role of the hydrophobic interactions has been overestimated? In contrast, studies aimed at elucidating the forces between CaM and target proteins have clearly established that hydrophobic interactions are involved (Gopalakrishna and Andersson 1983; Blumenthal and Stull 1982). In fact, it has been proposed that the hydrophobic interactions play a major role as an initial step by establishing the complex between CaM and a target protein. A second step would be the formation of hydrogen bonds, ionic and van der Waals interactions which would only take place after the binding has been established (Blumenthal and Stull 1982). Thus, two kinds of conformational changes can take place in CaM, one upon binding of Ca^{2+} and a second after the ligand has bound. The first would mainly expose the hydrophobic surfaces, the second would provide for the other interactions. This two-step model is consistent with available spectroscopic data (Krebs and Carafoli 1982; Klevitt et al. 1985; Linse et al. 1986; Vogel et al. 1987).

Finally, a few more words could be said about the future of studying interactions between drugs and calcium-binding proteins such as CaM. On one hand it could be argued that, since we now have peptides available that resemble CaM-binding domains of target proteins, there is no longer a need to use drugs as models of target proteins. Of course, the drugs retain their usefulness as experimental tools where one wants to inhibit CaM-activated processes, but it should be recalled that even so-called calmodulin antagonists are not always specific and they have been shown to inhibit other activities as well. There are, however, two reasons why such studies remain useful for the development of new pharmacologic agents to treat hypertension, angina, and related diseases. Calcium antagonists act by binding to the calcium channels. These channel proteins may contain a calcium-binding protein which closely resembles CaM (Johnson 1984) or a homologous calcium-binding domain similar to the one found in calcium-dependent proteases (Ohno et al. 1984). Attempts to purify this protein have not been successful to date and thus CaM remains for the time being the best model system to study interactions between the channel and calcium antagonist. The second reason relates to a potentially new strategy for treating hypertension. Rather than inhibiting smooth muscle contraction, hypertension could potentially also be regulated by developing agents that would specifically increase the heart function by combining with cardiac troponin C. Bepridil is a first example of a drug that exerts such an action (Solaro et al. 1986). As such it could become a prototype for a whole new class of antihypertension drugs.

Acknowledgments. The work on calcium-binding proteins in the author's laboratory is presently made possible through support from the Alberta Heritage Foundation for Medical Research (AHFMR) and the Medical Research Council of Canada (MRC). Valuable comments on the manuscript were made by Dr. M. Walsh, Dr. J. Krebs and Dr. B. Ljung. The secretarial assistance of Sylvia Thome is greatly appreciated.

References

Adunyah ES, Niggli V, Carafoli E (1982) The anticalmodulin drugs trifluoperazine and R24571 remove the activation of purified erythrocyte Ca^{2+}-ATPase by acidic phospholipids and by controlled proteolysis. FEBS Lett 143:65–68

Andersson A (1986) Calmodulin: binding of cations and drugs studied by NMR and stopped-flow techniques. Ph D Thesis, University of Lund, Lund, Sweden

Andersson A, Forsén S, Thulin E, Vogel HJ (1983) Cadmium-113 nuclear magnetic resonance studies of proteolytic fragments of calmodulin: assignment of strong and weak cation binding sites. Biochemistry 22:2309–2313

Andersson A, Drakenberg T, Forsén S (1985) The interaction of various drugs with calmodulin as monitored by cadmium-113 NMR. In: Hidaka T, Hartshorne D (eds) Calmodulin antagonists and cellular physiology. Academic, New York, pp 27–45

Andreasen TJ, Keller CH, LaPorte DC, Edelman AM, Storm DR (1981) Preparation of azidocalmodulin: a photoaffinity label for calmodulin-binding proteins. Biochemistry 78:2782–2785

Aulabaugh A, Niemczyra WP, Blundell TL, Gibbons WA (1984) A study of the interactions between residues in the C-terminal half of calmodulin by one and two-dimensional NMR methods and computer modelling. Eur J Biochem 143:409–418

Babu YS, Sack JS, Greenhough TJ, Bugg CE, Means AR, Cook WJ (1985) Three-dimensional structure of calmodulin. Nature 315:37–40

Bayley P, Ahlstrom P, Martin SR, Forsén S (1984) The kinetics of calcium binding to calmodulin: QUIN-2 and ANS stopped flow fluorescence. Biochem Biophys Res Commun 120:185–191

Berntsson P, Wold S (1986) Comparison between X-ray crystallographic data and physicochemical parameters with respect to their information about the calcium channel antagonist activity of 4-phenyl-1,4-dihydropyridines. Quant Struct Act Relat (to be published)

Berntsson P, Johansson E, Westerlund C (1987) Felodipine analogues, structure activity relationships. J Cardiovasc Pharm 10:560–565

Blumenthal DK, Stull JT (1982) Effects of pH, ionic strength, and temperature on activation by calmodulin and catalytic activity of myosin light chain kinase. Biochemistry 21:2386–2391

Blumenthal DK, Takio K, Edelman AM, Charbonneau H, Titani K, Walsh KA, Krebs EG (1985) Identification of the calmodulin-binding domain of skeletal muscle myosin light chain kinase. Proc Natl Acad Sci USA 82:3187–3191

Boström SL, Ljung B, Mardh S, Forsén S, Thulin E (1981) Interaction of the antihypertensive drug felodipine with calmodulin. Nature 292:777–778

Boströom SL, Ljung B, Nordlander M, Johanssen B (1985) Actions of felodipine in vascular smooth muscle. In: Hidaka T, Hartshorn D (eds) Calmodulin antagonists and cellular physiology. Academic, New York, pp 273–286

Boström SL, Westerlund C, Rochester S, Vogel HJ (1987) Binding of the dihydropyridine felodipine to calmodulin and related calcium-binding proteins (to be published)

Brzeska H, Szynklowicz J, Drabikowski W (1983) Localization of hydrophobic sites in calmodulin and skeletal muscle troponin C studies using tryptic fragments. A simple method of their preparation. Biochem Biophys Res Commun 115:87–93

Buccigros JM, O'Donnel CL, Nelson DJ (1986) A flow-dialysis method for obtaining relative measures of association constants in calmodulin metal-ion systems. Biochem J 235:677–684

Burger D, Cox JA, Comte M, Stein EA (1984) Sequential conformational changes in calmodulin upon binding of Ca^{2+}. Biochemistry 23:1966–1971

Cachia PJ, Van Eyk J, Ingraham RH, McCubbin WD, Kay CM, Hodges RS (1986) Calmodulin and troponin C: a comparative study of the interaction of mastoparan and troponin I inhibitory peptide. Biochemistry 25:3553–3562

Chao S-H, Suzuki Y, Zysk JR, Cheung WY (1984) Activation of calmodulin by various metal cations as a function of ionic radius. Mol Pharmacol 26:75–82

Charbonneau H, Cormier MJ (1979) Purification of calmodulin on immobilized phenothiazines. Biochem Biophys Res Commun 90:1039–1047
Chong PCS, Hodges RS (1981) A new heterobifunctional cross-linking reagent for the study of biological interactions between proteins. J Biol Chem 256:5064–5076
Comte M, Maulet Y, Cox JA (1983) Ca^{2+}-dependent high-affinity complex formation between calmodulin and melittin. Biochem J 209:269–272
Cox JA, Comte M, Fitton JE, DeGrado WF (1985) The interaction of calmodulin with amphiphilic peptides. J Biol Chem 260:2527–2534
Craig TA, Watterson DM, Prendergast FG, Haiech J, Roberts DM (1987) Site-specific mutagenesis of the α-helices of calmodulin. J Biol Chem 262:3278–3284
Dalgarno DC, Klevit RE, Levine BA, Scott GMM, Williams RJP, Gergely J, Grabarek Z, Leavis PC, Grand RJA, Drabikowski W (1984) The nature of the trifluoperazine binding sites on calmodulin and troponin-C. Biochim Biophys Acta 791:164–172
Delville A, Grandjean J, Laszlo P, Gerday C, Brzeska H, Drabikowski W (1980) Sodium-23 nuclear magnetic resonance as an indicator of sodium binding to calmodulin and tryptic fragments, in relation to calcium content. Eur J Biochem 109:515–522
Drabikowski W, Brzeska H (1983) Proteolytic fragments in the study of calcium binding proteins. In: de Bernard B et al. (eds) Calcium-binding proteins. Elsevier, Amsterdam, pp 227–235
Drabikowski W, Kuznicki J, Grabarek Z (1977) Similarity in Ca^{2+}-induced changes between troponin-C and protein activator of 3′:5′-cyclic nucleotide phosphodiesterase and their tryptic fragments. Biochim Biophys Acta 485:124–133
Drabikowski W, Brzeska H, Venyaminov SY (1982) Tryptic fragments of calmodulin. Ca^{2+}- and Mg^{2+}-induced conformational changes. J Biol Chem 257:11584–11590
Edelman AM, Takio K, Blumenthal DK, Hansen RS, Walsh KA, Titani K, Krebs EG (1985) Characterization of the calmodulin-binding and catalytic domains in skeletal muscle myosin light chain kinase. J Biol Chem 260:11275–11285
Endo T, Tanaka T, Isobe T, Kasai H, Okuyama T, Hidaka H (1981) Calcium-dependent affinity chromatography of S-100 and calmodulin on calmodulin antagonist-coupled sepharose. J Biol Chem 256:12485–12489
Faust FM, Slisz M, Jarrett HW (1987) Calmodulin in labelled at lysine-148 by a chemically reactive phenothiazine. J Biol Chem 262:1938–1941
Follenius A, Gerard D (1984) Fluorescence investigations of calmodulin hydrophobic sites. Biochem Biophys Res Commun 119:1154–1160
Forsén S, Thulin E, Drakenberg T, Krebs J, Seamon K (1980) A ^{113}Cd NMR study of calmodulin and its interaction with calcium, magnesium and trifluoperazine. FEBS Lett 117:189–194
Forsén S, Vogel HJ, Drakenberg T (1986) Biophysical studies on calmodulin. In: Cheung WY (ed) Calcium and cell function, vol VI. Academic, New York, pp 113–157
Fullmer CS, Edelstein S, Wasserman RH (1985) Lead-binding properties of intestinal calcium-binding proteins. J Biol Chem 260:6816–6819
Gariepy J, Hodges RS (1983) Localization of a trifluoperazine binding site on troponin C. Biochemistry 22:1586–1594
Giedroc DP, Puett D, Ling N, Staros JV (1983) Demonstration by covalent cross-linking of a specific interaction between β-endorphin and calmodulin. J Biol Chem 258:16–19
Giedroc DP, Sinha SK, Brew K, Puett D (1985a) Differential trace labeling of calmodulin: investigation of binding sites and conformational states by individual lysine reactivities. J Biol Chem 260:13406–13413
Giedroc DP, Keravis TM, Staros JV, Ling N, Wells JN, Puett D (1985b) Functional properties of covalent β-endorphin peptide/calmodulin complexes. Chlorpromazine binding and phosphodiesterase activation. Biochemistry 24:1203–1211
Gopalakrishna R, Anderson WB (1982) Calcium-induced hydrophobic site on calmodulin, application for purification by phenyl-sepharose chromatography. Biochem Biophys Res Commun 104:830
Gopalakrishna R, Anderson WB (1983) Calmodulin interacts with cyclic nucleotide phosphodiesterase and calcineurin by binding to a metal ion-independent hydrophobic region on these proteins. J Biol Chem 258:2405–2409

Gopalakrishna R, Anderson WB (1984) The effects of chemical modification of calmodulin on Ca^{2+}-induced exposure of a hydrophobic region. Biochim Biophys Acta 844:265–269
Grand RJA, Perry SV (1980) The binding of calmodulin to myelin basic protein and histone H2B. Biochem J 189:227–240
Gregori L, Marriott D, West CM, Chau V (1985) Specific recognition of calmodulin from dictyostelium discoideum by the ATP, ubiquitin-dependent degradative pathway. J Biol Chem 260:5232–5235
Guerini D, Krebs J, Carafoli E (1984) Stimulation of the purified erythrocyte Ca^{2+}-ATPase by tryptic fragments of calmodulin. J Biol Chem 259:15172–15177
Haiech J, Klee CB, Demaille JG (1981) Effects of cations on affinity of calmodulin for calcium: ordered binding of calcium ions allows the specific activation of calmodulin-stimulated enzymes. Biochemistry 20:3890–3897
Hart RC, Hice RE, Charbonneau H, Putnam-Evans C, Cormier MJ (1983) Preparation and properties of calcium-dependent resins with increased selectivity of calmodulin. Anal Biochem 135:208–220
Head JF, Masure HR, Kaminer B (1982) Identification and purification of a phenothiazine binding fragment from bovine brain calmodulin. FEBS Lett 137:71–74
Herzberg O, James MNG (1985a) Structure of the calcium regulatory muscle protein troponin-C at 2.8 Å resolution. Nature 313:653–659
Herzberg O, James MNG (1985b) Common structural framework of the two Ca^{2+}/Mg^{2+} binding loops of troponin C and other Ca^{2+} binding proteins. Biochemistry 24:5298–5302
Hidaka H, Asano M, Tanaka T (1981) Activity-structure relationship of calmodulin antagonists. Mol Pharmacol 20:571–578
Hiraoki T, Vogel HJ (1987) Structure and function of calcium-binding proteins. J Cardiovasc Pharmacol 10:514–531
Hiraoki T, Vogel HJ (1987) Modification of methionine residues in proteins followed by NMR (to be published)
Horrocks W De W, Sudnick DR (1981) Lanthanide ion luminescence probes of the structure of biological macromolecules. Acc Chem Res 14:384–392
Huang CH, Cau V, Chock PB, Wang JH, Sharma RK (1981) Mechanism of activation of cyclic nucleotide phosphodiesterase: requirement of the binding of four Ca^{2+} to calmodulin for activation. Proc Natl Acad Sci USA 78:871–874
Ikebe M, Inagaki M, Kanamura K, Hidaka H (1985) Phosphorylation of smooth muscle myosin light chain kinase by protein kinase C. J Biol Chem 256:4547–4550
Ikura M, Hiraoki T, Hikichi K, Minowa O, Yagi K (1984) NMR studies on calmodulin: Ca^{2+}-dependent spectral change of proteolytic fragments. Biochemistry 23:3124–3128
Ikura M, Minowa O, Hikichi K (1985) Hydrogen bonding in the carboxyl-terminal half-fragment 78–148 of calmodulin as studied by two-dimensional nuclear magnetic resonance. Biochemistry 24:4264–4269
Inagaki M, Tanaka T, Hidaka H (1983) Calmodulin antagonists enhance calcium binding to calmodulin. Pharmacology 27:125–129
Iwasa Y, Iwasa T, Higashi K, Matsui K, Miyamoto E (1981) Modulation by phosphorylation of interaction between calmodulin and histones. FEBS Lett 133:95–98
Jackson AE, Puett D (1984) Specific acylation of calmodulin – synthesis and adduct formation with a fluorenyl-based spin label. J Biol Chem 259:14985–14992
Jamieson GA Jr, Vanaman TC (1979) Calcium-dependent affinity chromatography of calmodulin on an immobilized phenothiazine. Biochem Biophys Res Commun 90:1048–1056
Jarrett HW (1984) The synthesis and reaction of a specific affinity label for the hydrophobic drug-binding domains of calmodulin. J Biol Chem 259:10136–10144
Jarrett HW (1986) Response of three enzymes to oleic acid, trypsin and calmodulin chemically modified with a reactive phenothiazine. J Biol Chem 261:4967–4972
Johnson JD (1983) Allosteric interactions among drug binding sites on calmodulin. Biochem Biophys Res Commun 112:787–793

Johnson JD (1984) A calmodulin-like Ca^{2+} receptor in the Ca^{2+}-channel. Biophys J 45:134–136
Johnson JD, Mills JS (1986) Calmodulin. Med Res Rev 6:341–363
Johnson JD, Wittenauer LA (1983) A fluorescent calmodulin that reports the binding of hydrophobic inhibitory ligands. Biochem J 211:473–479
Johnson JD, Wittenauer LA, Thulin E, Forsén S, Vogel HJ (1986) Localization of a felodipine (dihydropyridine) binding site on calmodulin. Biochemistry 25:2226–2231
Kemp BE, Pearson RB, Guerriero V, Bagchi IC, Means AR (1987) The calmodulin binding domain of chicken smooth muscle myosin light chain kinase contains a pseudosubstrate sequence. J Biol Chem 262:2542–2548
Kilhoffer MC, Gerard D, Demaille JG (1980a) Terbium binding to octopus calmodulin provides the complete sequence of ion binding. FEBS Lett 120:99–103
Kilhoffer MC, Demaille JG, Gerard D (1980b) Terbium as a luminescent probe of calmodulin calcium-binding sites. FEBS Lett 116:269–272
Kincaid RL (1984) Preparation of an enzymatically active cross-linked complex between brain cyclic nucleotide phosphodiesterase and 3-(2-pyridyldithio)propionyl-substituted calmodulin. Biochemistry 23:1143–1147
Kincaid RL, Stith-Coleman IE, Vaughan M (1985) Proteolytic activation of calmodulin-dependent cyclic nucleotide phosphodiesterase. J Biol Chem 260:9009–9015
Klee CB, Vanaman TC (1982) Calmodulin. Adv Prot Chem 35:213–321
Klevitt RE, Vanaman TC (1984) Azidotyrosylcalmodulin derivatives – specific probes for protein-binding domains. J Biol Chem 259:15414–15424
Klevitt RE, Blumenthal DK, Wemmer DE, Krebs EG (1985) Interaction of calmodulin and a calmodulin-binding peptide from myosin light chain kinase: major spectral changes in both occur as the result of complex formation. Biochemistry 24:8152–8157
Krebs J, Carafoli E (1982) Influence of Ca^{2+} and trifluoperazine on the structure of calmodulin. A ^{1}H-nuclear magnetic resonance study. Eur J Biochem 124:619–627
Krebs J, Buerkler J, Guerini D, Brunner J, Carafoli E (1984) 3-(Trifluoromethyl)-3-(m-[^{125}I]iodophenyl)diazirine, a hydrophobic, photoreactive probe, labels calmodulin and calmodulin fragments in a Ca^{2+}-dependent way. Biochemistry 23:400–403
Kretsinger RH (1980) Structure and evolution of calcium-modulated proteins. CRC Crit Rev Biochem 8:119–174
Kretsinger RH, Rudnick SE, Weissman LJ (1986) J Inorg Biochem 28:289–302
Kuznicki J, Grabarek Z, Brzeska H, Drabikowski W, Cohen P (1981) Stimulation of enzyme activities by fragments of calmodulin. FEBS Lett 130:141–145
LaPorte DC, Wierman BM, Storm DR (1980) Calcium-induced exposure of a hydrophobic surface on calmodulin. Biochemistry 19:3814–3819
Leavis PC, Rosenfeld SS, Gergely J, Grabarek Z, Drabikowski W (1978) Proteolytic fragments of troponin C. Localization of high and low affinity Ca^{2+} binding sites and in teractions with troponin I and troponin T. J Biol Chem 253:5452–5459
Levin RM, Weiss B (1979) Binding of trifluoperazine to the calcium dependent activator of cyclic nucleotide phosphodiesterase. Mol Pharmacol 13:690–697
Linse S, Drakenberg T, Forsén S (1986) Mastoparan binding induces a structural change affecting both the N-terminal and C-terminal domains of calmodulin. A ^{113}Cd NMR study. FEBS Lett 199:28–32
Ljung B, Kjellstadt A, Oreback D (1987) Vascular versus myocardial selectivity of calcium antagonists as studied by concentration-time-effect relations. J Cardiovasc Pharmacol 10:534–539
Lukas TJ, Marshak DR, Watterson DM (1985a) Drug-protein interactions: isolation and characterization of covalent adducts of phenoxybenzamine and calmodulin. Biochemistry 24:151–157
Lukas TJ, Burgess WH, Prendergast FG, Lau W, Watterson DM (1985) Calmodulin binding domains: characterization of a phosphorylation and calmodulin binding site from myosin light chain kinase. Biochemistry 25:1458–1464
Malencik DA, Anderson SR (1982) Binding of simple peptides, hormones, and neurotransmitters by calmodulin. Biochemistry 21:3480–3486

Malencik DA, Anderson SR (1983) Binding of hormones and neuropeptides by calmodulin. Biochemistry 22:1995–2001
Malencik DA, Anderson SR (1984) Peptide binding by calmodulin and its proteolytic fragments and by troponin C. Biochemistry 23:2420–2428
Malencik DA, Huang T-S, Anderson SR (1982a) Binding of protein kinase substrates by fluorescently labeled calmodulin. Biochem Biophys Res Commun 108:266–272
Malencik DA, Anderson SR, Bohnert JL, Shalitin Y (1982b) Functional interactions between smooth muscle myosin light chain kinase and calmodulin. Biochemistry 21:4031–4039
Malencik DA, Scott JD, Fischer EH, Krebs EG, Anderson SR (1986) Association of calmodulin with peptide analogues of the inhibitory region of the heat stable protein inhibitor of adenosine cyclic 3′,5′-phosphate dependent protein kinase. Biochemistry 25:3502–3508
Manalan AS, Klee CB (1983a) Interaction of calmodulin with its target proteins. Chem Scrip 21:137–142
Manalan AS, Klee CB (1983b) Activation of calcineurin by limited proteolysis. Proc Natl Acad Sci USA 80:4291–4295
Marshak DR, Lukas TJ, Watterson DM (1985) Drug-protein interactions: binding of chlorpromazine to calmodulin, calmodulin fragments, and related calcium binding proteins. Biochemistry 24:144–150
Martin RB, Richardson FS (1979) Lanthanides as probes for calcium in biological systems. Quart Rev Biophys 12:181–209
Martin SR, Bayley PM (1986) The effects of Ca^{2+} and Cd^{2+} on the secondary and tertiary structure of bovine testis calmodulin. Biochem J 238:485–490
Martin SR, Andersson-Teleman A, Bayley PM, Drakenberg T, Forsén S (1985) Kinetics of calcium dissociation from calmodulin and its tryptic fragments. A stopped-flow fluorescence study using Quin 2 reveals a two-domain structure. Eur J Biochem 151:543–550
Maulet Y, Cox JA (1983) Structural changes in melittin and calmodulin upon complex formation and their modulation by calcium. Biochemistry 22:5680–5686
Mayr GW, Heilmeyer LMG Jr (1983a) Shape and substructure of skeletal muscle myosin light chain kinase. Biochemistry 22:4316–4326
Mayr GW, Heilmeyer LMG Jr (1983b) Phosphofructokinase is a calmodulin binding protein. FEBS Lett 159:51–57
McDowell L, Sanyal G, Prendergast FG (1984) Probable role of amphiphilicity in the binding of mastoparan to calmodulin. Biochemistry 24:2979–2984
Mills JS, Johnson JD (1985) Metal ions as allosteric regulators of calmodulin. J Biol Chem 260:15100–15105
Mills JS, Bailey BL, Johnson JD (1985) Cooperativity among calmodulin's drug binding sites. Biochemistry 24:4897–4902
Minowa O, Yagi K (1984) Calcium binding to tryptic fragments of calmodulin. J Biochem 96:1175–1182
Murphy KMM, Gould RJ, Largent BL, Snyder SH (1983) A unitary mechanism of calcium antagonist drug action. Proc Natl Acad Sci USA 80:860–864
Newton DL, Klee CB (1984) CAPP-calmodulin: a potent competitive inhibitor of calmodulin actions. FEBS Lett 165:269–272
Newton DL, Burke TR Jr, Rice KC, Klee CB (1983) Calcium ion dependent covalent modification of calmodulin with norchlorpromazine isothiocyanate. Biochemistry 22:5472–5476
Newton DL, Oldewurtel MD, Krinks MH, Shiloach J, Klee CB (1984) Agonist and antagonist properties of calmodulin fragments. J Biol Chem 259:4419–4426
Newton D, Klee C, Woodgett J, Cohen P (1985) Selective effects of $CAPP_1$-calmodulin on its target proteins. Biochim Biophys Acta 845:533–539
Ni W-C, Klee CB (1985) Selective affinity chromatography with calmodulin fragments coupled to sepharose. J Biol Chem 260:6974–6981
Norman JA, Drummond AH, Moser P (1979) Inhibition of calcium-dependent regulator stimulated phosphodiesterase activity by neuroleptic drugs is unrelated to their chemical efficacy. Mol Pharmacol 16:1089–1094

Olwin DD, Storm DR (1985) Calcium binding to complexes of calmodulin and calmodulin-binding proteins. Biochemistry 24:8081–8086
O'Neil KT, DeGrado WF (1985) A predicted structure of calmodulin suggests an electrostatic basis for its function. Proc Natl Acad Sci USA 82:4954–4958
Putkey JA, Slaughter GR, Means AR (1985) Bacterial expression and characterization of proteins derived from the chicken calmodulin cDNA and a calmodulin processed gene. J Biol Chem 260:4704–4712
Putkey JA, Draetta GF, Slaughter GR, Klee CB, Cohen P, Stull JT, Means AR (1986) Genetically engineered calmodulins differentially activate target enzymes. J Biol Chem 261:9896–9903
Rasmussen H (1983) Cellular calcium metabolism. Ann Intern Med 98:809–816
Roberts DM, Crea R, Malecha M, Alvarado-Urbina G, Chiarello RH, Watterson DM (1985) Chemical synthesis and expression of a calmodulin gene designed for site-specific mutagenesis. Biochemistry 24:5090–5098
Roufogalis BD (1981) Phenothiazine antagonism of calmodulin: a structurally-nonspecific interaction. Biochem Biophys Res Commun 98:607–613
Schatzman RC, Raynor RL, Kuo JF (1983) N-(6-aminohexyl)-5-chloro-1-naphthalenesulfonamide(W-7), a calmodulin antagonist, also inhibits phospholipid-sensitive calcium-dependent protein kinase. Biochim Biophys Acta 755:144–147
Schiffer M, Edmundson AB (1967) Use of helical wheels to represent the structures of proteins and to identify segments with helical potential. Biophys J 7:121–128
Schwartz MA, Das OP, Hynes RO (1982) A new radioactive cross-linking reagent for studying the interactions of proteins. J Biol Chem 257:2343–2349
Seamon KB (1980) Calcium and magnesium dependent conformational states of calmodulin as determined by NMR. Biochemistry 19:207–215
Seaton BA, Head JF, Engelman DM, Richards FM (1985) Calcium-induced increase in the radius of gyration and maximum dimension of calmodulin measured by small-angle X-ray scattering. Biochemistry 24:6740–6743
Sellinger-Barnette M, Weiss B (1981) Interaction of β-endorphin and other opioid peptides with calmodulin. Mol Pharmacol 21:86–91
Sobue K, Muramoto Y, Fujita M, Kakiuchi S (1981) Purification of a calmodulin-binding protein from chicken gizzard that interacts with F-actin. Proc Natl Acad Sci USA 78:5652–5655
Solaro RJ, Bousquet P, Johnson JD (1986) Stimulation of cardiac myofilament force, ATPase activity and troponin C Ca^{2+} binding by bepridil. J Pharmacol Exp Ther 238:502–507
Steiner RF (1984) Location of a binding site for 1-anilinaphthalene-8-sulfonate on calmodulin. Arch Biochem Biophys 228:105–112
Steiner RF, Marshall L (1985) Sites of interaction of calmodulin with trifluoperazine and glucagon. Arch Biochem Biophys 240:297–311
Steiner RF, Marshall L, Needleman D (1986) The interaction of melittin with its tryptic fragments. Arch Biochem Biophys 246:286–300
Suko J, Wyskovsky W, Pidlich J, Hamptner R, Plank B, Helmann G (1986) Calcium release from calmodulin antagonists phenoxybenzamine and melittin measured by stopped-flow fluorescence with Quin-2 and intrinsic fluorescence. Eur J Biochem 159:425–434
Sundralingam M, Bergstrom R, Strasburg G, Rao ST, Roychowdhury P, Greaser M, Wang BC (1985a) Molecular structure of troponin C from chicken skeletal muscle at 3 angstrom resolution. Science 227:945–948
Sundralingam M, Drendel W, Greaser M (1985b) Stabilization of the long central helix of troponin C by intrahelical salt bridges between charged amino acid side chains. Proc Natl Acad Sci USA 82:7944–7947
Szebenyi DHE, Ohlendorf SK, Moffat K (1981) Structure of the vitamin D dependent calcium binding protein from bovine intestine. Nature 294:327–332
Tallant EA, Cheung WY (1984) Activation of bovine brain calmodulin-dependent protein phosphatase by limited trypsinization. Biochemistry 23:973–979
Tanaka T, Hidaka H (1980) Hydrophobic regions function in calmodulin-enzyme(s) interactions. J Biol Chem 255:11078–11080

Tanaka T, Ohmura T, Hidaka H (1983) Calmodulin antagonists' binding sites on calmodulin. Pharmacology 26:249–257
Tanaka T, Umekawa H, Ohmura T, Hidaka H (1984) Calcium-dependent hydrophobic chromatography of calmodulin, S-100 protein and troponin-C. Biochim Biophys Acta 787:158–164
Tanokura M, Yamada K (1985) Effects of trifluoperazine on calcium binding by calmodulin – a microcalorimetric study. J Biol Chem 260:8680–8682
Teleman A, Drakenberg T, Forsén S (1986) Kinetics of calcium binding to calmodulin and its tryptic fragments studied by calcium-43 NMR. Biochim Biophys Acta 873:204–213
Thulin E, Andersson A, Drakenberg T, Forsén S, Vogel HJ (1984) Metal ion and drug binding to proteolytic fragments of calmodulin: proteolytic, cadmium-113, and proton nuclear magnetic resonance studies. Biochemistry 23:1862–1870
Tsalkova TN, Privalov P (1985) Thermodynamic study of domain organization in troponin C and calmodulin. J Mol Biol 181:533–544
Vanaman TC (1981) Structure, function and evolution of calmodulin. In: Cheung WY (ed) Calcium and cell function, vol II. Academic, New York, pp 41–57
Van Belle H (1981) R 24571: a potent inhibitor of calmodulin-activated enzymes. Cell Calcium 2:483–494
Villar-Palasi C, Oshiro DL, Kretsinger RH (1983) Interaction of calmodulin and glycogen phosphorylase. Biochim Biophys Acta 757:40–46
Vogel HJ, Forsén S (1987) NMR studies of calcium binding proteins. In: Berliner LJ (ed) Biological magnetic resonance, vol VII. Plenum, New York
Vogel HJ, Lindahl L, Thulin E (1983a) Calcium-dependent hydrophobic interaction chromatography of calmodulin, troponin C and their proteolytic fragments. FEBS Lett 157:241–246
Vogel HJ, Drakenberg T, Forsén S (1983b) Calcium binding proteins. In: Laszlo P (ed) NMR of newly accessible nuclei. Academic, New York, pp 157–192
Vogel HJ, Andersson T, Braunlin WH, Drakenberg T, Forsén S (1984) Trifluoperazine binding to calmodulin: a shift reagent ^{43}Ca NMR study. Biochem Biophys Res Commun 122:1350–1356
Vogel HJ, Drakenberg T, Forsén S, O'Neill JDG, Hofmann J (1985) Structural differences in the two calcium-binding sites of the porcine intestinal calcium binding protein: a multinuclear NMR study. Biochemistry 24:3870–3876
Vogel HJ, Rochester S, Johnson JD (1987) Localization of a melittin binding site on calmodulin (to be published)
Wallace RA, Tallant EA, Dockter ME, Cheung WY (1982) Calcium binding domains of calmodulin. J Biol Chem 257:1845–1854
Walsh M, Stevens FC (1977) Chemical modification studies on the Ca^{2+}-dependent protein modulator of cyclic nucleotide phosphodiesterase. Biochemistry 16:2742–2749
Walsh M, Stevens FC (1978) Chemical modification studies on the Ca^{2+}-dependent protein modulator: the role of methionine residues in the activation of cyclic nucleotide phosphodiesterase. Biochemistry 17:3924–3930
Walsh M, Stevens FC, Kuznicki J, Drabikowski W (1977) Charactérization of tryptic fragments obtained from bovine brain protein modulator of cyclic nucleotide phosphodiesterase. J Biol Chem 252:7440–7443
Walsh MP, Dabrowska R, Hinkins S, Hartshore DJ (1982) Calcium-independent myosin light chain kinase of smooth muscle. Preparation by limited chymotryptic digestion of the calcium ion dependent enzyme, purification, and characterization. Biochemistry 21:1919–1925
Wang C-L (1985) A note on Ca^{2+} binding to calmodulin. Biochem Biophys Res Commun 130:426–430
Wang C-L, Leavis PC, Horrocks W DeW Jr, Gergely J (1981) Binding of lanthanide ions to troponin C. Biochemistry 20:2439–2444
Wang CLA, Aquaron RR, Leavis PC, Gergely J (1982) Metal-binding properties of calmodulin. Eur J Biochem 124:7–12

Wang CLA, Leavis PC, Gergely J (1984) Kinetic studies show that Ca^{2+} and Tb^{3+} have different binding preferences towards the four Ca^{2+}-binding sites of calmodulin. Biochemistry 23:4
Weiss B, Fertel R, Figlin R, Uzunov P (1974) Selective alteration of the multiple forms of phosphodiesterase of rat cerebrum. Mol Pharmacol 10:615–625
Yoshida M, Minowa O, Yagi K (1983) Divalent cation binding to wheat germ calmodulin. J Biochem (Tokyo) 94:1925–1933
Zimmer M, Hofmann F (1984) Calmodulin antagonists inhibit activity of myosin light-chain kinase independent of calmodulin. Eur J Biochem 142:393–397
Zimmer M, Hofmann F (1987) Differentiation of the drug-binding sites of calmodulin. Eur J Biochem 164:411–420
Zurini M, Krebs J, Penniston JT, Carafoli E (1984) Activation of the plasma-membrane Ca^{2+}-ATPase by limited proteolysis. J Biol Chem 259:618–627

CHAPTER 5

Calcium Channels as Drug Receptors

M. SCHRAMM and R. TOWART

A. Introduction

Of all the many aspects of the biological handling of calcium in this volume, the story of the calcium channel, and of the substances which modulate its function, is among the most fascinating. Although our knowledge of the necessity of calcium for cardiac contraction goes back to the year 1883 in which Sydney RINGER published his famous paper, the concept of voltage-sensitive calcium entry did not start until 1953. Based on the pioneering work of HODGKIN and HUXLEY (1952) on the role of the sodium channel in the propagation of excitation, the existence of calcium channels was hinted at by FATT and KATZ (1953) who investigated the electrical properties of crustacean muscle.

By the 1960s it was realized that the cardiac action potential depended not only on the entry of sodium ions (the "fast" current), but also upon the entry of calcium ions (the "slow" current) (see REUTER 1973). The entry of calcium ions through these slow channels has been shown to play a vital role in the excitation–contraction coupling of the heart, and Fleckenstein's group became interested in the effects of calcium removal or of inorganic cations such as nickel or cobalt on these processes. When testing the effects of new antianginal drugs such as verapamil, prenylamine or nifedipine (VATER et al. 1972) on cardiac excitation and contraction, they found that those drugs mimicked in many ways the effect of calcium-free solution: the upstroke velocity and overshoot of the action potential were unaffected, but both the action potential duration and the force of contraction were markedly reduced. Fleckenstein therefore suggested that the drugs' mechanism of action was to inhibit the entry of calcium ions through the activated calcium channels, and termed the drugs "calcium antagonists" (FLECKENSTEIN et al. 1969; FLECKENSTEIN 1977, 1983). His group also showed that smooth muscle (especially coronary arteries) was relaxed by these drugs and that these actions could be reversed by increasing the extracellular calcium concentration, suggesting that an inhibition of calcium ion entry was operating here also (FLECKENSTEIN 1983).

Similar conclusions were reached by Godfraind's group, using the peripheral vasodilator drug cinnarizine (GODFRAIND and KABA 1969). The beneficial effects of drugs such as verapamil or nifedipine on peripheral vascular resistance or coronary spasm led to their widespread use in Europe as antianginal and antihypertensive drugs. FLECKENSTEIN (1977) reviewed the mechanism of action of the calcium antagonists, and emphasized their selectivity, both as specific blockers of the calcium channel, and as selective cardiovascular drugs without effects on neural excitability or on contraction of skeletal muscle.

The molecular mechanism of action of the calcium antagonists was however unknown. It was first assumed that they physically blocked the calcium channels in the manner proposed for the block of sodium channels by tetrodotoxin. This was however unlikely, as the prototype calcium antagonists verapamil, nifedipine and diltiazem had completely different chemical structures (see Fig. 3), and neither nifedipine nor diltiazem were charge-carrying molecules. The extreme potency of some of the drugs and the demonstration of marked stereoselectivity (e.g. TOWART et al. 1981) had suggested to some workers that some sort of "calcium antagonist receptor" might exist. The introduction of high specific activity tritiated nitrendipine and nimodipine in 1981 produced a wave of studies, which quickly established that specific binding sites for the calcium antagonists *did* exist on many cell membranes (see Sect. G). Although there are some discrepancies, it is now clear that the calcium channel contains binding sites for the main classes of calcium antagonists, and some workers feel that these "receptors" are the targets of at least one endogenous modulator of calcium channel function. This hypothesis has been supported by the recent discovery of calcium "agonists" such as Bay K 8644, which also bind to the dihydropyridine-binding site, but which increase rather than decrease the influx of calcium ions. The high affinity binding of labelled drugs to the calcium channel also provides a useful marker for purification studies, and we are convinced that the primary structure of the calcium channel will soon be determined, using methods comparable to similar studies on the voltage-dependent and receptor-operated sodium channels (e.g. NODA et al. 1984).

B. Calcium as a Biological Signalling Mechanism and the Role of Calcium Channels in Maintaining Its Homeostasis

WILLIAMS (1970) has elegantly summarized the differing roles of various cations in biological processes. Some cations, such as zinc, play a fairly static role as catalytic ions at the active centres of enzymes, whereas sodium and potassium are very much more mobile and act as charge carriers for electrical events. The calcium ion plays a unique "triggering" role as a second messenger responsible for much of the transfer of extracellular signals (e.g. from neurotransmitters and hormones) to the contractile, secretory or metabolic elements, sometimes deep inside the cell. Some authors have emphasized the antiquity of calcium as a second messenger, and have dated this from the first use by primitive organisms of energy-rich phosphates as a convenient energy source. To prevent calcium phosphate precipitation, the primitive cells developed active transport mechanisms to maintain very low intracellular free calcium concentrations (see CAMPBELL 1983). Thus, from early in evolution, a steep inwardly directed electrochemical gradient for calcium ions has existed. A controlled increase in cytoplasmic calcium concentration could then be utilized as a simple, but elegant, second messenger system. The antiquity (and importance) of this system can also be judged by the highly conserved nature of the main intracellular targets of increased calcium ion concentration, calcium-binding proteins such as calmodulin. Of the 148 amino acids which make up the calmodulin molecule, only 7 have been found to vary in the wide range of organisms so far tested (see MEANS et al. 1982). Given that the intracellular cytoplasmic free calcium concentration is low, and that "first messengers" of many types – depolarization, neurotransmitters, hormones, antigenic stimuli, etc. – may act on the cell to produce or modify contraction, secretion, degranulation, etc., how does the "second messenger" increase in free calcium concentration, which triggers these events, occur?

As outlined in other chapters of this book, intracellular free calcium concentrations may be increased by: (a) increased "leak" of calcium into the cell; (b) decreased calcium extrusion; (c) release of intracellularly sequestered calcium; or (d) controlled transmembrane influx ("calcium channels"). The first two events undoubtedly occur in poisoned or damaged tissue, and can lead to cell death (FARBER 1981). This topic has been much reviewed, especially in the context of myocardial or cerebral infarction (e.g. SIESJO 1981), and is outwith the topic discussed here. The release of intracellular sequestered calcium has been much studied. It is now known that many stimuli cause the breakdown of membrane phospholipids to inositol trisphosphate, which play a key role in the release of intracellular calcium. This topic is fully discussed elsewhere (see Chap. 12).

The controlled entry of calcium ions, down their electrochemical gradient, via "calcium channels", is the other major physiological mechanism for increasing the free calcium concentration, and in some tissues, especially heart, may precede and initiate release of intracellular calcium ions. It is however the mechanism most easily studied by modern electrophysiological techniques, and is the event most susceptible to drug action. The effects of drugs on calcium channel function, and their physiological and therapeutic implications, are the subject of this chapter.

C. How Many Types of Calcium Channels Exist?

Do different calcium channels exist? Although there have been many attempts to answer this question it remains unsolved. Functional studies have suggested at least four different sorts of calcium channels: leak channels, voltage-sensitive channels, receptor-operated channels, and stretch-sensitive channels.

The term "leak channel" denotes only the fact that even under resting conditions there is a basic calcium influx into the cell owing to the high concentration gradient across the cell membrane (FLAIM et al. 1984a). Usually this calcium influx is measured with radioactive ^{45}Ca and it is normally unaffected by drugs (GODFRAIND 1983; VAN BREEMEN et al. 1982; SCHRAMM et al. 1985; but see FLAIM et al. 1984b). So far, there is no experimental evidence that diffusion through water-filled (perhaps even gated) pores in the lipid membranes is the underlying mechanism, which would justify the use of the term "channel".

The same can be said for the "stretch-sensitive calcium channel", a term which is used to describe the fact that mechanical stress of blood vessels for example can result in a calcium-dependent contraction (HWA and BEVAN 1986). Again, calcium leaks through mechanically disturbed cell membranes could be responsible for such mechanical responses.

We have much more difficulty in assessing the role of the "receptor-operated channel". This expression is used to name the calcium influx pathway activated by neurotransmitters in the absence of a preceding membrane depolarization (Bolton 1979; Droogmans et al. 1977; Meisheri et al. 1981). Depending on the organ, this calcium influx seems to be insensitive to or even unaffected by organic calcium antagonists or agonists. However, as the voltage-sensitive calcium channel can be modulated by neurotransmitters (see Chap. 6) and calcium influx through receptor-operated channels may depolarize the cell membrane (Hermsmeyer et al. 1981; Siegel and Adler 1985), the question whether receptor-operated and voltage-sensitive calcium channels are two different channels or only two different activation mechanisms of one and the same calcium channel can not be answered definitively. Independently from this, neurotransmitters do not seem to bind directly to Ca channels. For this reason the receptor-operated calcium channel is also outwith the scope of this chapter.

Electrophysiological methods have allowed a direct, intensive study of the "voltage-dependent calcium channel". This term denotes the existence of a very strong depolarization-dependent current which must be carried by calcium ions through gated aqueous pores. These Ca channels are widely distributed both in different tissues in the body and across the animal kingdom. Pharmacologically they are very sensitive to many agents, especially organic calcium modulators.

Owing to the difficulties in separating calcium currents from potassium currents in classical electrophysiological studies, interpretation of experiments on Ca channels had remained difficult. So, in contrast to K channels, only one voltage-sensitive calcium channel was believed to exist until recently. Since the development of the patch-clamp method by Sakman and Neher (1983) and the technique of cell dialysis, allowing the study of whole-cell and single-cell channel calcium currents, the possibility of distinguishing calcium currents by their kinetics and their single-channel conductance and gating behaviour has increased dramatically. So, beginning in 1983, an increasing number of different voltage-sensitive calcium channels have been described: in neurons (e.g. Carbone and Lux 1984), heart muscle (e.g. Nilius et al. 1985), smooth muscle (e.g. Bean et al. 1985) and in endocrine cells (Armstrong and Matteson 1985). However, although these papers have demonstrated that the channels described were able to carry calcium ions under experimental (generally nonphysiological) conditions, they have not shown the selectivity of these channels for calcium in comparison with other physiological ions. Nor could the physiological or pathophysiological role of these channels be shown. It is therefore unclear whether all these channels can be demonstrated in the future to be proper calcium channels.

For these reasons and owing to the fact that so far these new calcium channels have not been shown to be influenced by drugs, we will limit ourselves throughout this chapter to the discussion of the best investigated calcium channel, the one which is sensitive to organic calcium antagonists and calcium agonists. This is the "calcium channel", referred to in the title of the chapter.

D. Electrophysiological Properties of the Calcium Channel

This subject has been reviewed several times during the last few years (HAGIWARA and BYERLY 1981; TSIEN 1983; REUTER 1984; NOBLE 1984). In addition, some very recent papers have extensively and carefully described and analysed both the whole-cell Ca current and the gating behaviour of the single calcium channel (CAVALIE et al. 1986; McDONALD et al. 1986). As some knowledge of the Ca current is necessary for understanding the mode of action of calcium-modulating drugs, we here briefly summarize the basic properties of the calcium current and the calcium channel, mainly for heart cells, whose electrophysiology is better known than that of other tissues.

The main characteristics of single-channel calcium currents can be derived from Fig. 1, in which barium currents are shown. Ba ions are often used as charge carriers in single Ca channel measurements, as these currents are rather similar to calcium currents, but bigger in amplitude and therefore easier to resolve (see FENWICK et al. 1982; CAVALIE et al. 1986). At the resting potential of −65 mV in Fig. 1 a the Ca current is 0. On a depolarization step of 75 mV, current fluctuations appear (inward currents are registered in a downward direction by convention), which are often followed by periods without any current. In addition, a remarkable number of records show no current fluctuations at all. Summarizing all tracings results in the mean Ca current shown in Fig. 1 b which resembles the macroscopic Ca current measured from a single cell.

The current fluctuations are interpreted as signs of statistically distributed openings and closure of the Ca channel. Only the open channel is able to conduct Ca ions, resulting in a uniform single-channel current amplitude. This current is determined by the driving voltage (actual membrane potential V minus the reversal potential V_{rev}, at which single Ca channel current is 0), and the single-channel conductance g.

As opening and closing events are randomly distributed, calcium channel kinetics can only be described in terms of statistics. Analysing the Ca channel kinetics within the periods of channel activity (the so-called clusters, Fig. 1 c) shows that dwell times can be well fitted (Fig. 1 d) by a monoexponential function for the open times, while two exponentials are necessary to describe the distribution of the shut times, suggesting that there is only one open state (lifetime under these experimental conditions 0.8 ms), but two closed states (lifetimes 0.4 and 2.1 ms). The long-lasting periods of rest between two clusters, shown in Fig. 1 c, suggest that there is at least one additional long-lived nonconducting state, the inactivated state.

Therefore single-channel current can be well characterized by the lifetimes of the different states, the single-channel conductance g, and the reversal potential V_{rev}. Instead of lifetimes of the channel states normally only the probability that the channel is open is given, i.e. the open probability P_o. In single-channel patch-clamp experiments it is usually defined as the quotient (total open time)/(duration of clamp pulse). On the basis of this definition P_o strongly depends on the experimental conditions, e.g. in Fig. 1 a a prolongation of the clamp pulse duration would reduce the open probability. Therefore, in this chapter we prefer to use the term P_o only for the open probability within one cluster of channel activity and

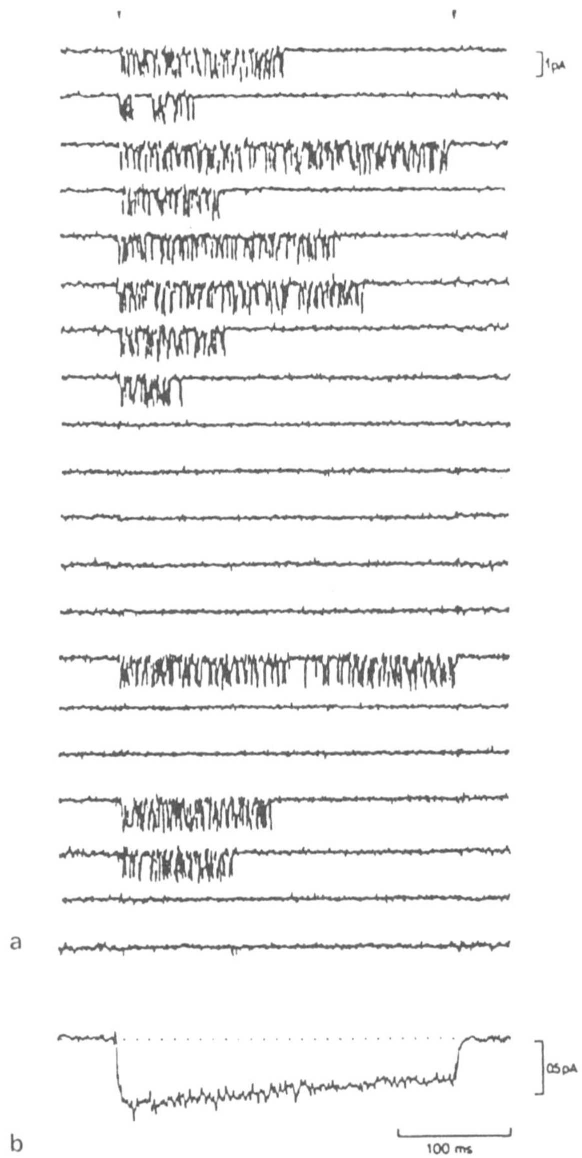

Fig. 1. a, b

Fig. 1. a Inward currents through a single Ca channel during step polarizations (marked by *arrows*) from −65 mV to +10 mV. Charge was carried by Ba ions (90 m*M* in the pipette, cell-attached configuration). **b** Mean current obtained by averaging 205 traces as shown in **a. c** Kinetic gating behaviour of the single Ca channel during steady depolarization of about 0 mV. Ca channel activity occurs in well-separated clusters. **d** Histograms of open and shut times. The open time histogram (*left*) was fitted by a monoexponential probability density function. Two exponentials were necessary to describe the shut times. The fact that channel activity is clustered is expressed by the long-lasting shut times summarized in the last bin of the shut times histogram. (CAVALIE et al. 1986)

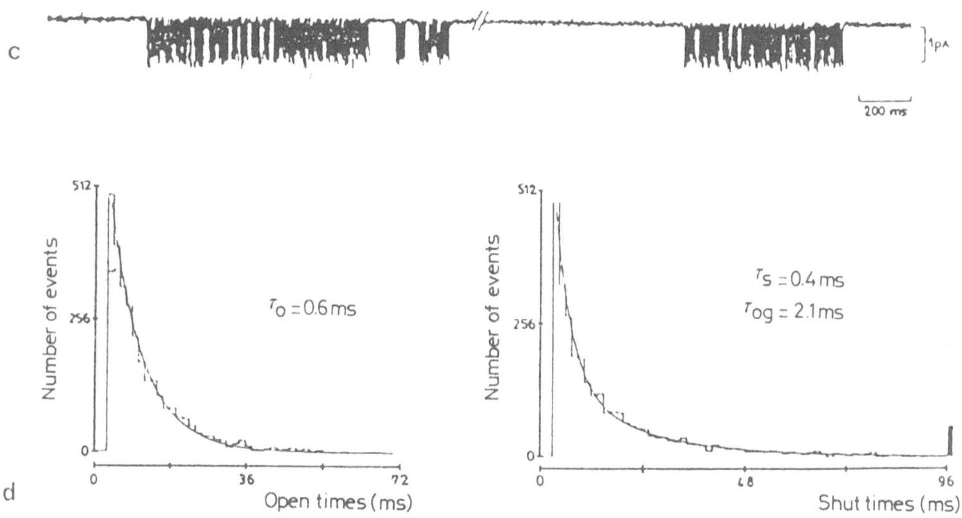

Fig. 1 c, d

not during the whole clamp pulse, as we are convinced that this definition can better describe the gating of the channel.

If n is the number of available (non-inactivated) Ca channels, the macroscopic Ca current I_{Ca} can be described as

$$I_{Ca}(V) = nP_o(V)g(V - V_{rev})$$

The advantage of this mathematical description is that its parameters can be obtained both from single-channel measurements, as described already, and from whole-cell measurements by means of fluctuation analysis and I–V analysis. In single-channel measurements the parameters can be determined much more directly, but whole-cell measurements can use a broader range of membrane potentials and more physiological conditions. A drug influencing the Ca channel directly should modify at least one of the parameters g, n, P_o, or V_{rev}.

E. Physiological Modulation of Calcium Channel Function

It is quite clear that current through voltage-dependent Ca channels is strongly dependent on extracellular Ca concentrations. Therefore, Ca itself could be considered as a modulator of the single-channel conductance g. However, owing to its universal role as second messenger in a variety of different tissues with quite different functions, the extracellular Ca concentration is kept quite constant in most organisms, and changing it is not a suitable way to modulate Ca channel function.

Physiologically, functions dependent on intracellular Ca concentration such as muscle contraction or hormone secretion are modulated by neurotransmitters, especially by catecholamines. The question whether voltage-dependent Ca channels are involved in these processes has not so far been answered. On one hand,

A+R
↓
AR
↓
Adenylate cyclase activation
↓
cAMP production
↓
Activation of a catalytic subunit of a cAMP dependent protein kinase
↓
Phosphorylation of the Ca-channel protein
↓
Increase in Ca-current

Fig. 2. Schematic view of the cascade induced by binding of a neurotransmitter A to its receptor R, ultimately resulting in an increased Ca current

some kinds of catecholamine-induced contractions are not influenced by organic Ca antagonists, which are known to block Ca currents through voltage-dependent Ca channels. For example, the noradrenaline-induced contraction of rabbit aortic rings is not influenced by Ca-antagonistic or Ca-agonistic dihydropyridines (see Sect. F). On the other hand, at least in the heart, voltage-dependent Ca channels are indirectly modulated by catecholamines: β-adrenergic stimulation by an agonist A is believed to induce the cascade reaction shown in Fig. 2. Binding of the agonist A to the receptor R activates the adenylate cyclase, resulting in an increased intracellular cAMP level. cAMP binds to a regulatory subunit of a protein kinase, which is inactive. However, this binding liberates the active catalytic C-subunit of the enzyme. This C-subunit catalyses the phosphorylation of the Ca channel or of a protein closely related to the channel, resulting in an increased Ca current.

Whole-cell and single Ca channel recordings in myocytes have demonstrated the effects of the single steps of this cascade: β-adrenergic stimulation by adrenaline or isoprenaline, intracellular applications of cAMP or of the catalytic subunit of a cAMP-dependent protein kinase, or extracellular 8-bromo-cAMP have very similar effects. Therefore, we summarize them here as effects of β-adrenergic stimulation by catecholamines (BRUM et al. 1983, 1984; CACHELIN et al. 1984; IRISAWA and KOKOBUN 1983; REUTER 1983; BEAN et al. 1984; KAMEYAMA et al. 1985):

1. β-Stimulation has no effect on single Ca channel conductance g.
2. It does not affect the reversal potential V_{rev}.
3. It increases the open probability P_o of the available Ca channel by increasing the mean open time and reducing the shut times.

4. The dominant effect seems to be an increase in the number *n* of *available* channels. As there is no hint of a change in the *total* number of Ca channels, this increase is caused by a reduction in the probability that the channel is inactivated.

Although these experiments show that, at least in myocytes, catecholamines are able to modulate Ca channel function, it has not so far emerged whether or not receptor-operated channels and voltage-dependent Ca channels are only different stimulation modes of the same channel (see Sect. C). At least it can be excluded that the Ca channels themselves are catecholamine receptors. Therefore, these questions are not discussed in more detail. Histamine also increases Ca currents via H_2-receptors, possibly by stimulation of adenylate cyclase and using the cascade of Fig. 2 (SANCHEZ-CHAPULA 1981; ECKEL et al. 1982). In contrast, acetylcholine inhibits Ca currents in cardiac tissue. However, this effect is only pronounced after β-stimulation and is accompanied by a decrease in intracellular cAMP concentration, suggesting that inhibition of adenylate cyclase is the mechanism of action (HESCHELER et al. 1985).

Besides Ca itself there is no physiological compound known which modulates Ca channel function *directly* by binding to the Ca channel. Nevertheless, some authors believe that an endogenous ligand may exist which is able to modulate the Ca channel directly by binding to the dihydropyridine receptor.

F. Calcium Antagonists

The first calcium channel blockers to be recognized were the "inorganic" calcium antagonists, di- or trivalent cations such as Ni^{2+}, Co^{2+} or Cd^{2+}, of ionic radius similar to calcium ions, which could be shown in electrophysiological experiments to block the inward calcium currents (e.g. KAUFMANN and FLECKENSTEIN 1965). Only very little information is available on the mode of action of inorganic Ca antagonists from a single-channel point of view. They seem to block the open Ca channel, resulting in a "bursting" behaviour of the single-channel Ca current and an appropriate reduction in the open probability P_o (HESS et al. 1985a). Owing to their chemical similarity to Ca they probably displace Ca^{2+} from its binding sites within or near the Ca channel, whose occupation with Ca^{2+} is necessary for Ca^{2+} to pass through the Ca channel (ALMERS et al. 1984; HESS and TSIEN 1984). Owing to their high toxicity, inorganic Ca channel blockers could not be used clinically nor in experiments in vivo, but they were useful as pharmacological tools in investigating the properties of the Ca channel.

As outlined already, during the last 20 years more and more organic calcium antagonists have become available. The most important chemical groups are: phenylalkylamines, e.g. verapamil, gallopamil (D600), tiapamil, anipamil; 1,4-dihydropyridines, e.g. nifedipine, nimodipine, nicardipine, nitrendipine, PN 200110; and benzothiazepines, e.g. diltiazem. In addition, some other structures like cinnarizine, flunarizine or bepridil are reported to have Ca-antagonistic properties. The chemical structures of some of these compounds are shown in Fig. 3. They are proving useful both as pharmacological tools and as therapeutic agents.

1,4-Dihydropyridines

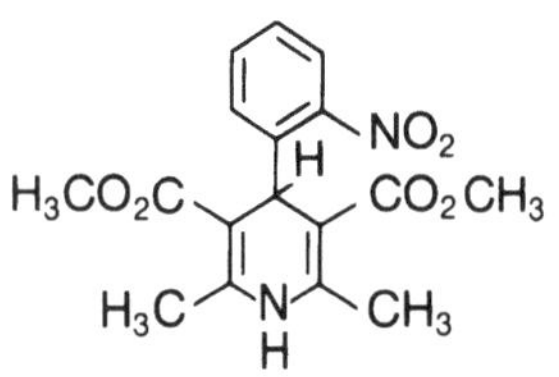

Nifedipine 10^{-8} M

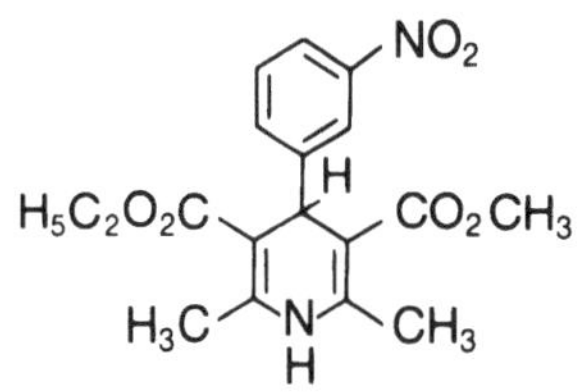

Nitrendipine 10^{-8} M

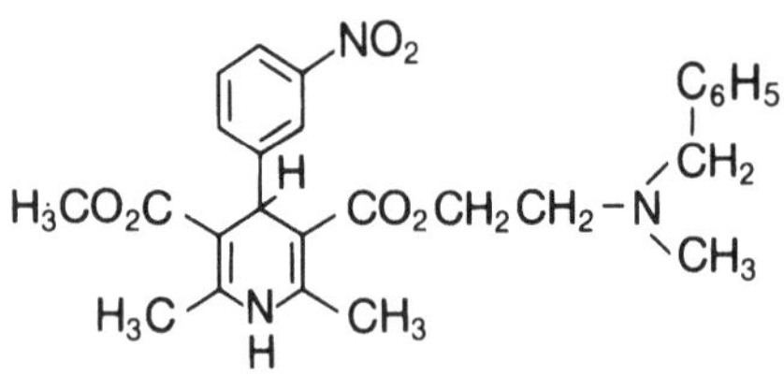

Nicardipine 3×10^{-9} M

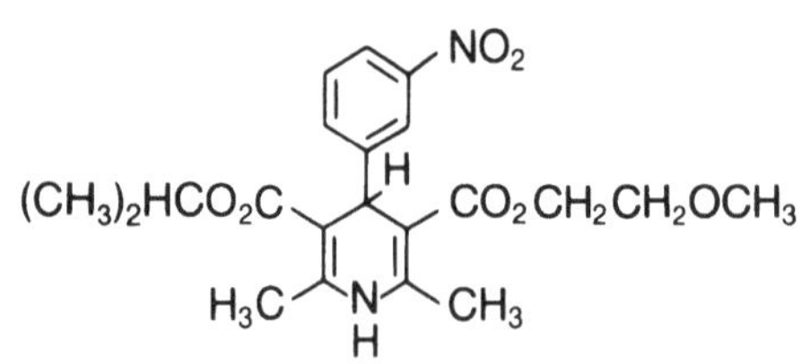

Nimodipine 10^{-8} M

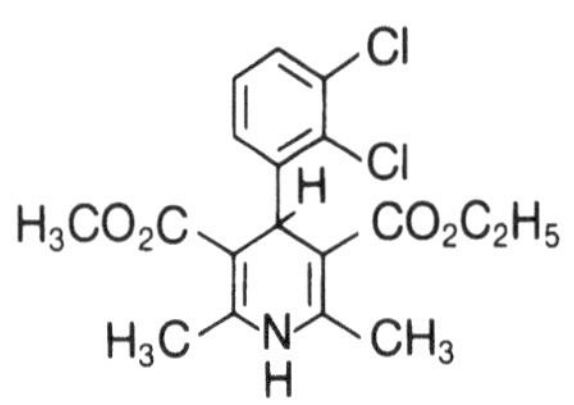

Felodipine 3×10^{-10} M

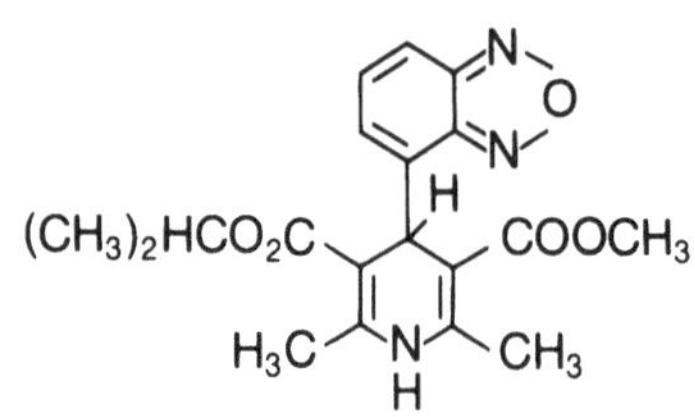

PN 200–110 3×10^{-9} M

a

Fig. 3a, b. Chemical structures of some Ca antagonists. The numbers give estimated concentrations for a 50% reduction of K^+-induced rabbit aorta contractions

The accumulated literature on their experimental and clinical use is now so vast that we can not attempt to review it here. Table 1, however, lists some topics of importance, and gives a selection of key publications or reviews in these areas. Although very different in chemical structure, these compounds all reduce rather specifically the Ca influx through the voltage-dependent Ca channels, which results in vasodilating, negative inotropic, and negative chronotropic and dromotropic properties of the compounds. Owing to their therapeutically most important effect, namely vasodilation in angina pectoris and hypertension, the compounds are well characterized in smooth muscle preparations, in which, however, electrophysiological studies are difficult to carry out. Therefore, instead of current–voltage dependence, the K^+-induced interaction is normally investigated.

Phenylalkylamines

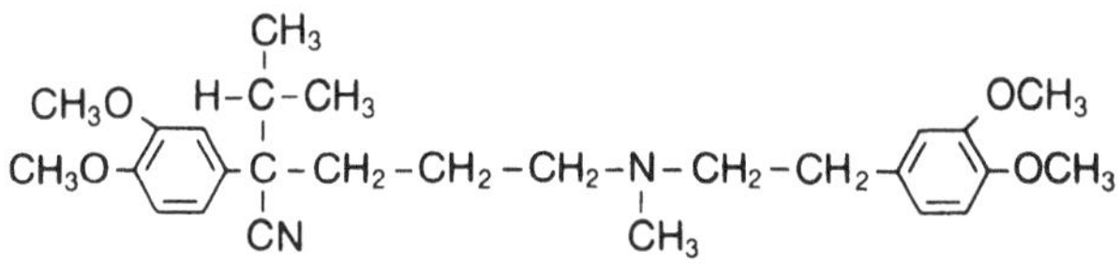

Verapamil 3×10^{-7} M

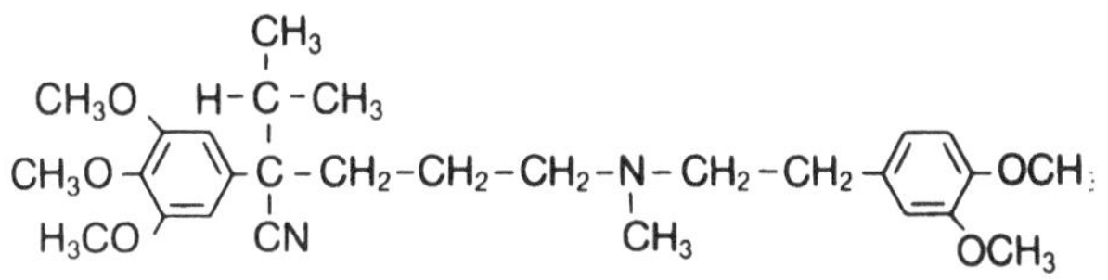

Gallopamil (D600) 10^{-7} M

Different Structures

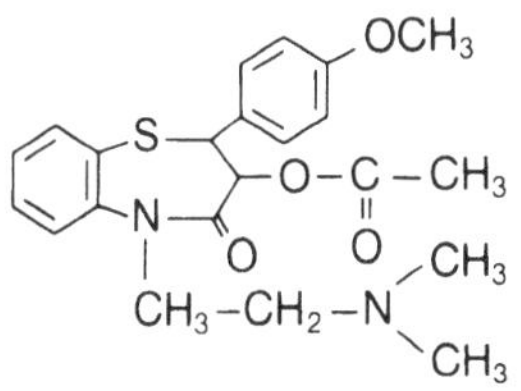

Diltiazem 3×10^{-7} M

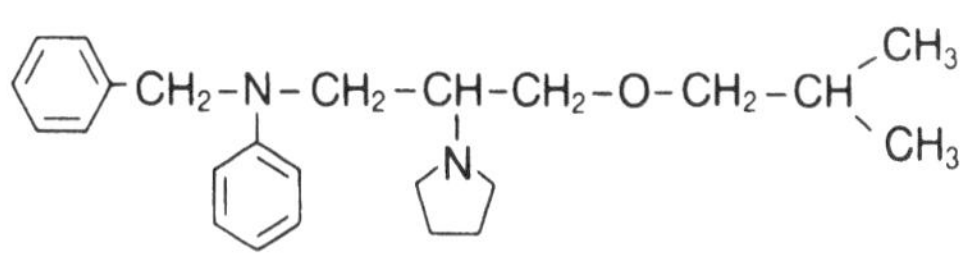

Bepridil 10^{-5} M

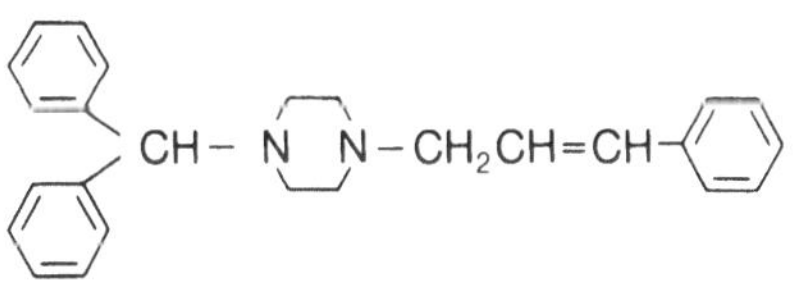

Cinnarizine 10^{-6} M

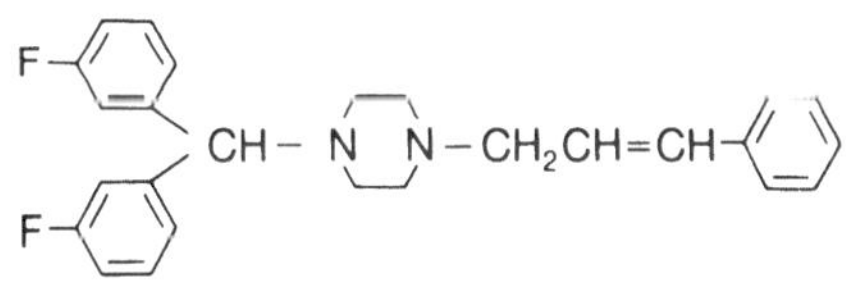

Flunarizine 10^{-6} M

Fig. 3b

The corresponding changes in membrane potential can be estimated by means of the Nernst equation. Such experiments show that the K^+-induced contraction of smooth muscle preparations is reduced dose-dependently to zero by Ca antagonists (Towart 1982; Kazda et al. 1983). Estimated ED_{50} values for inhibition of K^+-induced contractions of rabbit aortic rings are given in Fig. 3.

This reduction in tension is caused (at least for 1,4-dihydropyridines) only by a reduction in Ca influx, which has been shown by ^{45}Ca influx measurements

Table 1. Some selected topics of calcium antagonist application

Topic	References
Chemistry	BOSSERT et al. (1981); MEYER et al. (1984)
Structure-activity relations	SU et al. (1985)
Mechanism of action	JANIS and TRIGGLE (1983); SPEDDING (1985b)
Tissue selectivity	RATZ and FLAIM (1982)
Cardiac effects	HENRY (1983)
Genitourinary effects	ANDERSSON and FORMAN (1986); JANIS and TRIGGLE (1986)
Renal actions	LOUTZENHISER and EPSTEIN (1985); ZANCHETTI and LEONETTI (1985)
Secretory tissue and endocrines	MILLAR and STRUTHERS (1984)
Skeletal muscle	GALLANT and GOETL (1985)
Mast cells	KIM et al. (1985)
Platelets	METHA (1985)
Atherosclerosis	HENRY (1985)
Cardioprotection, cardioplegia	CLARK et al. (1985)
Cerebral blood flow	ALLEN (1985)
Migraine	MEYER (1985)
Asthma	LÖFDAHL and BARNES (1986); TOWART and ROUNDING (1986)

(GODFRAIND 1983; SCHRAMM et al. 1985) and the demonstration that the compounds had no effect in skinned vascular fibres at pharmacologically relevant concentrations (KANMURA et al. 1981; SAIDA and VAN BREEMEN 1983). Interestingly, in contrast to some other vascular tissues, in the rabbit aorta Ca antagonists do not affect noradrenaline-induced contractions (KAZDA et al. 1983), a finding which is often used as an argument for the differentiation between receptor-operated and voltage-dependent Ca channels (e.g. MEISHERI et al. 1981). In heart preparations negative inotropic and chronotropic effects of the compounds can also be attributed to their Ca channel-blocking effects as hinted at by the drug-induced reduction in action potential duration (see FLECKENSTEIN 1983).

Classical electrophysiological studies show nearly identical behaviour of calcium antagonists of the different chemical groups, the main distinction being differences in frequency dependence of the calcium channel-blocking activity (LEE and TSIEN 1983). While verapamil shows a strong use dependence, this is negligible with the 1,4-dihydropyridines. These differences may explain the lack of effect of 1,4-dihydropyridines in suppressing arrhythmias while verapamil is very effective in supraventricular tachycardias.

Not enough investigations have yet been published on the effects of Ca-antagonistic drugs on single Ca channels to decide which single-channel parameters are influenced by Ca antagonists. Generally, the number of empty sweeps is reduced in single-channel measurements (HESS et al. 1985 b; PELZER et al. 1984), which fits very well with the general belief that Ca-antagonistic dihydropyridines bind to the inactivated state or encourage inactivation (BEAN 1984; SANGUINETTI and KASS 1984 b). However, there is one major difference between phenylalkylamines and dihydropyridines: whereas D600 reduces open times and enhances shut times of the available Ca channel (PELZER et al. 1984), suggesting that the compound is an open channel blocker, 1,4-dihydropyridines seem to keep these time constants unchanged (HESS et al. 1985b).

Besides the general aspects of Ca-antagonistic drugs, mainly investigated in vessel or heart preparations, questions of tissue selectivity become more and more important. So far it is not clear whether the reported specificity of some drugs is caused by a tissue specificity of the chemical structure of the drug receptor (see Chap. 6), or perhaps by a different voltage dependence of Ca channel modulation, or even only by different pharmacokinetics. For example, nimodipine and flunarizine have been reported to have therapeutic effects in different kinds of CNS diseases (see GODFRAIND et al. 1985; BETZ et al. 1985). Whether these are special effects of these two compounds or general actions of Ca-antagonistic drugs crossing the blood–brain barrier is so far not known.

Another effect of Ca antagonists should be mentioned, which may be of importance in the future, whose connection to the Ca channel, however, is difficult to understand. During aging, rats (and humans) develop calcinosis (intra- or extracellularly?) of vessels, effects which can be prevented by adding Ca antagonists to the food (see FLECKENSTEIN 1983).

G. The 1,4-Dihydropyridine Receptor

The existence of very specific and selective calcium antagonists suggested specific binding sites of the compounds within or near the calcium channel with receptor properties, i.e. the existence of binding sites which are able to bind the drugs specifically and whose occupation is essential for drug action. Even before radioactive calcium antagonists had become available some strong evidence had been reported for the existence of such a receptor. As early as 1971 the stereospecificity of the diltiazem-binding site had been described: D-*cis*-diltiazem was shown to be the most potent isomer (SATO et al. 1971). Similar selectivity had been demonstrated for the phenylalkylamines (BAYER et al. 1975) and for 1,4-dihydropyridines (TOWART et al. 1981). In this latter paper the enantiomers of a nimodipine analogue showed large differences in potency. One relaxed rabbit aortic rings with $ED_{50} = 5 \times 10^{-10}$ *M*, while its antipode did so with 7×10^{-8} *M*.

At the end of 1981 tritiated nimodipine became available. This initiated an immense amount of experimental work with this and more and more newly tritiated 1,4-dihydropyridines (reviewed in JANIS and TRIGGLE 1984; JANIS et al. 1985; GLOSSMANN et al. 1985a, b). All these ligands bind with high affinity to one and the same binding site and can be displaced by other calcium-antagonistic dihydropyridines. The K_d value, the drug concentration of half-maximal receptor occupancy, coincides well with the concentrations for half-maximal reduction in K^+-induced smooth muscle contraction, indicating that binding is a *sine qua non* for action. Binding occurs with a rather similar affinity in different kinds of excitable tissues such as brain, heart, smooth and skeletal muscle, but also in nonexcitable cells like liver. In contrast, the number of receptor sites differs markedly, as shown in Table 2. Surprisingly, the receptor density does not go in parallel with the known pharmacological effect of the compound. For example, the density of ni-

Table 2. Tissue distribution of the 1,4-dihydropyridine receptor in the guinea pig. (Adapted from GLOSSMANN et al. 1985a)

Tissue	Specifically bound [^{3}H] nimodipine (fmol per milligram protein)
Duodenum	124
Heart	152
Lung	106
Kidney	31
Liver	48
Cerebral cortex	153
Medulla/pons	42
Thalamus/midbrain	68
Skeletal muscle	1770

modipine-binding sites in brain and smooth muscle was of the same order of magnitude, while calcium antagonists were believed to have no direct CNS effects. Even more strangely, the highest receptor density has been found in the t-tubuli of skeletal muscle, whose tone is hardly influenced by extracellular calcium concentration or 1,4-dihydropyridines. Nevertheless, it may be that, in these tissues also, the dihydropyridine receptors have some pharmacological, physiological or pathophysiological relevance which is unknown till now. For example, for the brain it has been demonstrated in some recent clinical reports that calcium-antagonistic dihydropyridines markedly potentiate the analgesic effect of fentanyl (VAN BORMANN et al. 1985). The additional possibility that the dihydropyridine receptors in the skeletal muscles are linked to calcium channels which are functional under certain conditions is now under discussion, and there are some indications for a role of these receptors in excitation–contraction coupling.

Are there any cells without dihydropyridine receptors? Only fat cells and platelets have been reported to have no high affinity dihydropyridine-binding site (see GLOSSMANN et al. 1985b). However, both calcium-antagonistic and -agonistic dihydropyridines modulate calcium influx into platelets, suggesting a calcium-carrying mechanism influenced by dihydropyridines (ERNE et al. 1984). One possible explanation could be that dihydropyridine receptors appear in these cells only after their stimulation.

Two very important questions arose from this work. What is the chemical structure of the dihydropyridine receptor and what is its connection with the calcium channel? During the last few years both questions could be answered only partly by rather indirect studies: target size analyses showed the dihydropyridine receptor to be a glycoprotein of approximate molecular weight 180 00–210 000 (FERRY et al. 1985; NORMAN et al. 1983). But what has this protein to do with the calcium channel?

One of the major arguments for a direct connection of the calcium channel and the dihydropyridine-binding site came from the comparison of the binding affinity and the efficacy of different calcium-antagonistic dihydropyridines. As mentioned already, an excellent 1 : 1 correlation exists between smooth muscle relaxing effects and concentrations of half-maximal receptor occupancy. In contrast, there is no 1 : 1 correlation between binding constants and the negative inotropic effect on the heart, reflecting the lower potency of dihydropyridines in reducing contractility of the heart (Su et al. 1985). Nevertheless, in heart preparations the rank order of potency of different dihydropyridines is the same as it is in vascular and intestinal smooth muscle.

The hypothesis that the dihydropyridine-binding site is an essential part of the calcium channel was also supported by investigating the effects of structurally unrelated compounds on the binding properties of dihydropyridines. The initial binding studies had shown that drugs like verapamil and diltiazem do not bind to the dihydropyridine receptor, although all these compounds seem to block one and the same calcium channel. However, all the binding sites are linked allosterically, suggesting that the different organic calcium antagonists bind to different sites of one macromolecule. Although the binding sites for benzothiazepines and phenylalkylamines have been characterized by using tritiated diltiazem and verapamil or D600, it is not totally clear whether these two binding sites are different (compare Glossmann et al. 1985b; Galizzi et al. 1984). Nevertheless, most workers believe that they are different, i.e. that the calcium channel contains at

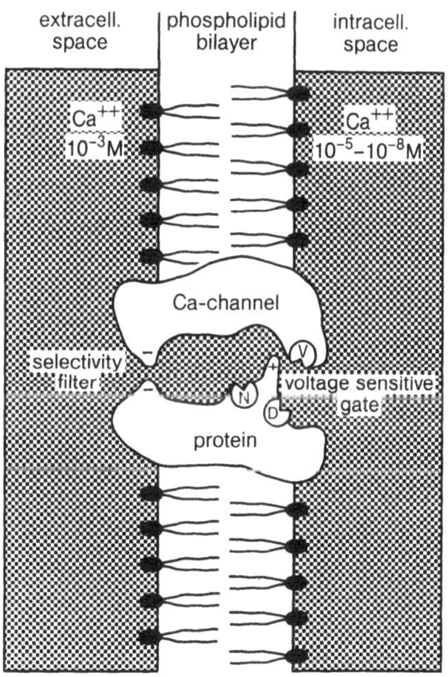

Fig. 4. Schematic picture of the proteinaceous calcium channel within a cell membrane. The negative charges on the left serve as a selectivity filter. Inorganic calcium antagonists possibly block the channel by interacting with this site. Whether organic calcium antagonists like nifedipine (*N*), verapamil (*V*) or diltiazem (*D*), which have different binding sites within the channel, modulate the function of the voltage-sensitive gate or of a (possibly different) inactivation mechanism is not clear. However dihydropyridine Ca agonists modulate the gating by interacting with the same binding site as nifedipine. (Schramm and Towart 1985)

least three different, but allosterically linked binding sites for 1,4-dihydropyridines, benzothiazepines, and phenylalkylamines, respectively (Fig. 4). The allosteric linkage of the three binding sites has also been shown in some functional studies (SPEDDING 1985 b).

The question of whether the dihydropyridine receptor is identical with the calcium channel may be answered much more directly. Using different tritiated dihydropyridines as probes different groups have started to solubilize and purify the calcium channels (CURTIS and CATTERALL 1984; BORSOTTO et al. 1985; FLOCKERZI et al. 1986). Depending on the purifying method and on the ligand used results are somewhat different. Normally two or three subunits can be found, one of molecular weight about 150000, others of 30000 or 50000. The purified large subunit seems to have the binding properties known for the calcium channel: it has a binding site for the three classes of calcium antagonists, which are also allosterically linked (STRIESSNIG et al. 1986). In addition, these binding sites can be phosphorylated, as is known from the intact channel (FLOCKERZI et al. 1986). Probably the amino acid sequence of this protein will soon be elucidated.

The first attempts to demonstrate the calcium channel properties of these purified proteins have now been made. When incorporated in artificial bilayers and studied electrophysiologically with the single-channel patch-clamp technique, these proteins have been demonstrated to carry calcium ions during statistically distributed opening events. These calcium currents are modulated by calcium channel modulating drugs in a way known for the intact channel. In addition, it could be shown that phosphorylation of these proteins results in an increased open probability as already described for the calcium channel (FLOCKERZI et al. 1986). Although many questions remain open (e.g. the voltage dependence of gating has not yet been demonstrated) it seems to be clear that the purified proteins are essential parts of the calcium channel and show, when incorporated into bilayers, properties similar to the whole calcium channel.

H. Calcium Agonists

Some years ago, independently of each other, three nifedipine-like dihydropyridines were reported which *increased* rather than decreased the force of contraction of smooth and cardiac muscle: Bay K 8644 (SCHRAMM et al. 1983 a), CGP 28392 (ERNE et al. 1984), and YC 170 (TAKENAKA and MAENO 1982). The chemical structures of these compounds, together with the structures of two recently published vasoconstricting dihydropyridines, are shown in Fig. 5.

Little is known about YC 170, H 160/51 (BEYER et al. 1985), and 202-791 (HOF et al. 1985), but more has been published about CGP 28392. The best investigated compound in this series is Bay K 8644. All these compounds have an asymmetric H atom at position 4 of the dihydropyridine ring. Therefore, two enantiomers of each compound exist. As far as these enantiomers have been investigated separately, the agonistic effects have been shown to be related to only one isomer, while the other has been shown to be an antagonist (shown for Bay K 8644 FRANCKOWIAK et al. 1985; 202-791 HOF et al. 1985; and H 160-51 GJÖRSTRUP et

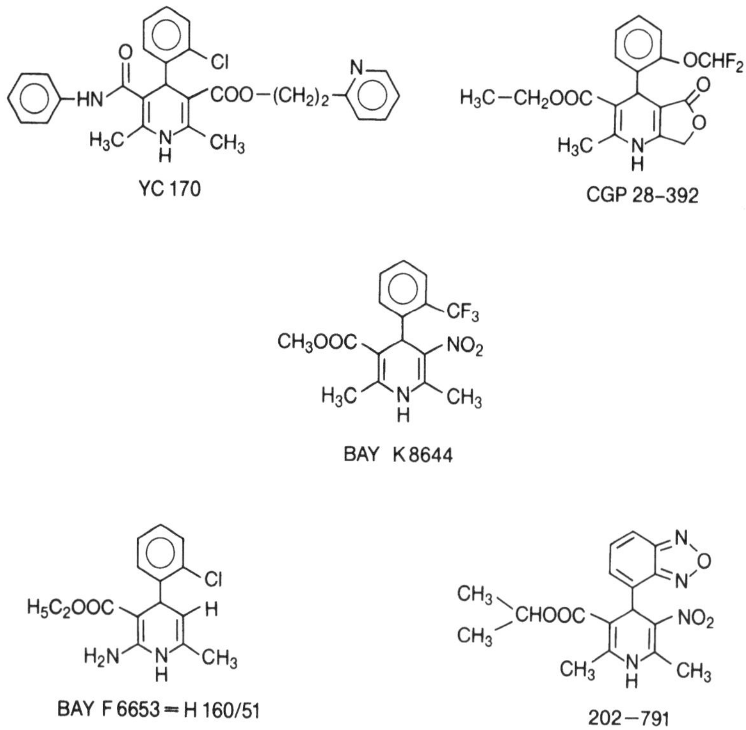

Fig. 5. Chemical structures of some Ca agonists

al. 1986). Nevertheless, results published with racemic Bay K 8644 (and also with racemic CGP 28 392) seem to be caused only by the contribution of the calcium-agonistic enantiomer, as this is much more potent than its antipode (BECHEM and SCHRAMM 1987).

The positive inotropic and vasoconstricting properties of Bay K 8644 were first observed in the anaesthetized dog. They were not influenced by sympatholytic drugs, but could be antagonized by nifedipine (SCHRAMM et al. 1983b). These findings suggested that this compound also bound to the dihydropyridine receptor, but increased rather than decreased the transmembrane calcium influx. For Bay K 8644 and other dihydropyridines this has now been proved by different authors using in vitro smooth muscle investigations (SCHRAMM et al. 1983b; KANMURA et al. 1984; SU et al. 1984). In different kinds of vascular or other smooth muscle preparations, the compounds have been shown to potentiate K^+-induced contractions. These effects could be competitively antagonized by nifedipine-like calcium antagonists, but only noncompetitively by non-dihydropyridine calcium antagonists like verapamil and diltiazem, suggesting that the Ca agonists also bind to the dihydropyridine receptor. This has now been proved in receptor binding studies by investigating the binding, either indirectly by displacement experiments, or directly using [^{3}H] Bay K 8644 (BELLEMANN 1984; JANIS et al. 1984).

It is interesting to note that in some tissues, e.g. rabbit aorta, the compounds have no effect under normal K^+ concentrations, but the tone increases dose-dependently after small depolarizations by a subthreshold increase in the extracellular K^+ concentration, suggesting a voltage dependence of drug action. For Bay K 8644, ^{45}Ca measurements showed a parallelism between drug-induced contractions and ^{45}Ca influx, suggesting that in smooth muscle preparations Bay K 8644 acts only by increase in transmembrane calcium influx without direct intracellular effects (SCHRAMM et al. 1985; HWANG and VAN BREEMEN 1985). The latter facts have been demonstrated in skinned smooth muscle (KANMURA et al. 1984) and cardiac muscle (THOMAS et al. 1985b) preparations, in which Bay K 8644 had no effect on contractions. From these findings the name "calcium agonist" has been derived for these dihydropyridines.

In heart preparations positive inotropic, chronotropic and dromotropic effects of the compound have been demonstrated. These effects occur in the same concentration range in which vasoconstriction is observed (for Bay K 8644 $ED_{50}=3\times10^{-8}$ *M*, SCHRAMM et al. 1983a; SCHWARTZ et al. 1984; WADA et al. 1985). The increase in contractility is accompanied (or caused) by increases in the duration of the action potential and in the maximum upstroke velocity of the slow action potential (SADA et al. 1986; THOMAS et al. 1985a). Conventional voltage-clamp investigations showed that Bay K 8644 enhances calcium current, especially at negative membrane potentials (THOMAS et al. 1985a). From patch-clamp studies it has been derived that the calcium current increase is due to a shift of the "open probability curve" $P_0(V)$ of the calcium channel to more negative potentials (BECHEM and SCHRAMM 1987, in press). This effect has been explained by single calcium channel recordings which showed a Bay K 8644-induced prolongation of the open times (OCHI et al. 1984; HESS et al. 1984; KOKUBUN and REUTER 1984; but see BROWN et al. 1984, who did not find a prolongation). Reports of the effects of Bay K 8644 on the shut times of the calcium channel are contradictory.

These single calcium channel investigations have been used to propose an interesting model of the calcium channel with three gating modes which have been differentiated by their different open probabilities in the current recordings (HESS et al. 1984). Mode 0 is characterized by tracings without calcium currents, mode 1 by tracings with short open events, and mode 2 by long-lasting openings. All three modes are reported to occur occasionally under control conditions, in which mode 1 is favoured. Bay K 8644 stabilized the channel in mode 2, whereas calcium-*antagonistic* dihydropyridines do so in mode 0.

Although this is a rather fascinating model which can explain many of the effects of dihydropyridines, there are some problems and findings which can not easily be explained:

1. CAVALIE et al. (1986) carefully analysed the open times under control conditions, whose distribution curve could be very well fitted by only one exponential; they did not find any evidence for the existence of mode 2 under control conditions.
2. The effects of Bay K 8644 are voltage dependent. Whereas it usually increases calcium currents, it behaves like a calcium antagonist under partly depolarized conditions (SANGUINETTI and KASS 1984a). This has now been shown also for the pure calcium-agonistic enantiomer of Bay K 8644 (KASS 1987).

3. The positive inotropic action of calcium agonists seems to be dose dependent: both in heart as well as in vessel preparations the dose–response curves are bell-shaped (SCHRAMM et al. 1983a). At concentrations higher than 10^{-7}–10^{-6} *M* its effect declined. In the isolated guinea pig and canine heart it shows, at about 10^{-5} *M*, a negative inotropic effect which can be overcome by additional calcium (THOMAS et al. 1984; VAGHY et al. 1984).

An alternative explanation for the mode of action of Bay K 8644 has been developed from electrophysiological whole-cell current measurements which can explain these experimental findings in a rather simple way (BECHEM and SCHRAMM 1987, in press). The compound is believed to bind only to the open state of the calcium channel, keeping it open as long as the drug is bound, resulting in a shift of the open probability curve $P_0(V)$ to more negative potentials, owing to an increased open time and an unchanged closed time of available calcium channels. The reduction in calcium current under partly depolarized conditions or at high drug concentrations is explained by an increased calcium channel inactivation from its open state.

The predictions of these two models are very different. While the three-mode model predicts identical effects for different calcium agonists, as they only stabilize a *physiological* mode of the calcium channel, the second model predicts different open times for compounds with different binding kinetics. Future experiments will demonstrate the relevance of these differing models.

For Bay K 8644, many very different effects have also been reported. As it is believed that the general mode of action of this compound is now well understood, these effects have mostly been used to show the involvement of voltage-dependent calcium channels in these processes. For example, Bay K 8644 stimulates neurotransmitter release in cells which have been stimulated. In partly depolarized (by addition of potassium) chromaffin cells and PC12 phaeochromocytoma cells it stimulated catecholamine release dose-dependently (GARCIA et al. 1984; SHALABY et al. 1984). In rat cortex slices it enhanced the potassium-stimulated serotonin release, an effect which could be antagonized by calcium antagonists (MIDDLEMISS and SPEDDING 1985). This may be one experimental explanation of the therapeutic effect of nimodipine or other calcium antagonists in cerebrovascular disorders like migraine, in which neurotransmitters are said to play a dominant role. The importance of calcium channels in hormone secretion has been demonstrated in studies with pancreatic beta cells and with pituitary cell lines. Bay K 8644 enhances glucose-stimulated calcium influx and insulin secretion of pancreatic islets, effects which are inhibited by calcium antagonists (MALAISSE-LAGAE et al. 1984). In other cell lines the compound increased prolactin (ENYEART et al. 1986) and aldosterone (KOJIMA et al. 1984) secretion. These findings may suggest future applications of new calcium-agonistic compounds.

If concentrations in the region of 10^{-5} *M* are necessary to demonstrate effects of dihydropyridines, data should be interpreted very carefully. In this concentration range calcium-antagonistic as well as calcium-agonistic dihydropyridines have been reported to block activity of calcium- or calmodulin-dependent enzymes (nonspecifically?) (BOSTRÖM et al. 1981; MOVSESIAN and ADELSTEIN 1984). Therefore, some effects of Bay K 8644 may not be related to its calcium channel modulating properties.

J. Summary and Conclusions

We have reviewed the present state of knowledge of the voltage-sensitive calcium channel, with special reference to the drugs which modulate its functioning. The way in which they affect single-channel parameters is summarized in Table 3. The most interesting of these drugs are the dihydropyridines, which bind directly to the channel protein. The use of these and other calcium channel modulators as probes and tools has now allowed detailed, multidisciplinary studies of the distribution, role and structure of the calcium channel. This increasing information about the molecular structure of the binding site, and probably of the channel itself, should allow the development of even better probes, and more specific drugs. There are also hints emerging from binding studies that calcium channel density or function may be altered in some disease states characterized by over- or underactivity of contractile or secretory tissue. With the new techniques of binding studies, immunochemistry and modern electrophysiological methods, the possible physiological role of potential endogenous calcium channel modulators may soon be resolved. Whatever the results of these studies, the biologist and the physician now possess both the means to investigate the role of calcium channels in disease, and potent drugs with which to normalize their activity.

Table 3. Effects of different calcium channel modulators on some single-channel parameters

Addition of	Effect on			
	n	g	P_o	V_{rev}
Calcium		↑		
Catecholamines	↑		↑	
Phenylalkylamines	↓		↓	
Ca-Antagonistic dihydropyridines	↓			
Ca-agonistic dihydropyridines	[a]		↑	

Abbreviations: n, number of functionally available channels; g, single-channel conductance; P_o, open probability; V_{rev}, reversal potential.

[a] Decreases at high concentrations or under depolarized conditions.

Note Added in Proof

The DHP receptor from rabbit skeletal muscle has now been purified and the complete amino acid sequence determined using cDNA techniques [TANABE et al., Nature 328:313–318 (1987)]. This DHP receptor has marked homologies with the voltage-dependent sodium channel, and the authors have suggested that it may function both as a voltage sensor and as a calcium channel.

References

Allen GS (1985) Role of calcium antagonists in cerebral arterial spasm. Am J Cardiol 55:149B–153B

Almers W, McCleskey EW, Palade PT (1984) A non-selective cation conductance in frog muscle membrane blocked by micromolar external calcium ions. J Physiol (Lond) 353:565–583

Andersson KE, Forman A (1986) Effects of calcium channel blockers on urinary tract smooth muscle. Acta Pharmacol Toxicol (Copenh) [Suppl 2] 58:193–200

Armstrong CM, Matteson DR (1985) Two distinct populations of calcium channels in a clonal line of pituitary cells. Science 227:65–67

Bayer R, Kalusche D, Kaufmann R, Mannhold R (1975) Inotropic and electrophysiological actions of verapamil and D 600 in mammalian myocardium. Naunyn Schmiedebergs Arch Pharmacol 290:81–97

Bean BP (1984) Nitrendipine block of cardiac calcium channels: high-affinity binding to the inactivated state. Proc Natl Acad Sci USA 81:6388–6392

Bean BP, Nowycky MC, Tsien RW (1984) β-Adrenergic modulation of calcium channels in frog ventricular heart cells. Nature 307:371–375

Bean BP, Sturek M, Puga A, Hermsmeyer K (1985) Calcium channels in smooth muscle cells from mesenteric arteries. J Gen Physiol 86:23a

Bechem M, Schramm M (1987) Calcium-agonists. J Mol Cell Cardiol 19:63–76

Bellemann P (1984) Binding properties of a novel calcium channel activating dihydropyridine in monolayer cultures of beating myocytes. FEBS Lett 167:88–92

Betz E, Deck K, Hoffmeister F (eds) (1985) Nimodipine, pharmacological and clinical properties. Schattauer, Stuttgart

Beyer T, Gjörstrup R, Ravens U (1985) Comparison of the cardiac effects of the dihydropyridine-derivative H160/51 with those of the "Ca-agonist" Bay K 8644. Naunyn Schmiedebergs Arch Pharmacol 330:142

Bolton TB (1979) Mechanisms of action of transmitters and other substances on smooth muscle. Physiol Rev 3:606–718

Bormann B von, Boldt J, Sturm G, Kling D, Weidler B, Lohmann E, Hempelmann G (1985) Ca-Agonisten in der Analgesie. Additive Analgesie durch Nimodipin während kardiochirurgischer Eingriffe. Anaesthesist 34:429–434

Borsotto M, Barhanin J, Fosset M, Lazdunski M (1985) The 1,4-dihydropyridine receptor associated with the skeletal muscle voltage-dependent Ca^{2+} channel. J Biol Chem 260:14255–14263

Bossert F, Meyer H, Wehinger E (1981) 4-Aryldihydropyridines, a new class of highly active calcium antagonists. Angew Chem Int Ed Engl 20:762–769

Boström SL, Ljung B, Mardh S, Forsén S, Thulin E (1981) Interaction of the antihypertensive drug felodipine with calmodulin. Nature 292:777–778

Breemen C van, Hwang O, Cauvin C (1982) Ca-antagonist inhibition of norepinephrine stimulated Ca-influx in vascular smooth muscle. In: Godfraind T, Albertini A, Paoletti R (eds) Calcium modulators. Elsevier, Amsterdam, pp 185–198

Brown AM, Kunze DL, Yatani A (1984) The agonist effect of dihydropyridines on Ca channels. Nature 311:570–572

Brum G, Flockerzi V, Hoffmann F, Osterrieder W, Trautwein W (1983) Injection of catalytic subunit of cAMP-dependent protein kinase into isolated cardiac myocytes. Pflugers Arch 398:147–154

Brum G, Osterrieder W, Trautwein W (1984) β-adrenergic increase in the calcium conductance of cardiac myocytes studied with the patch clamp. Pflugers Arch 401:111–118

Cachelin AB, de Peyer JE, Kokubun S, Reuter H (1984) Ca^{2+}-channel modulation by 8-bromo-cyclic AMP in cultured heart cells. Nature 304:462–464

Campbell AK (1983) Intracellular calcium: its universal role as regulator. Wiley, New York

Carbone E, Lux HD (1984) A low voltage-activated, fully inactivating Ca channel in vertebrate sensory neurones. Nature 310:501–502

Cavalie A, Pelzer D, Trautwein W (1986) Fast and slow gating behaviour of single calcium channels in cardiac cells. Pflugers Arch 406:241–258
Clark RE, Christlieb IY, Magovern GJ (1985) Use of nifedipine during cardiac surgery for improved myocardial protection. Am J Med [Suppl 2B] 78:6–8
Curtis BM, Catterall WA (1984) Purification of the calcium antagonist receptor of the voltage-sensitive calcium channel from skeletal muscle transverse tubules. Biochemistry 23:2113–2118
Droogsmans G, Raymaekers L, Casteels R (1977) Electro- and pharmacomechanical coupling in the smooth muscle cells of rabbit ear artery. J Gen Physiol 70:129–148
Eckel L, Gristwood RW, Nawrath H, Owen DAA, Satter P (1982) Inotropic and electrophysiological effects of histamine on human ventricular heart muscle. J Physiol (Lond) 330:111–123
Enyeart JJ, Aizawa T, Hinkle PM (1986) Interaction of dihydropyridine Ca^{2+} agonist Bay K 8644 with normal and transformed pituitary cells. Am J Physiol 250:C95–C102
Erne P, Bürgisser E, Bühler FR, Dubach B, Kühnis H, Meier M, Rogg H (1984) Enhancement of calcium influx in human platelets by CGP 28392, a novel dihydropyridine. Biochem Biophys Res Commun 118:842–847
Farber JL (1981) The role of calcium in cell death. Life Sci 29:1289–1295
Fatt P, Katz B (1953) The electrical properties of crustacean muscle fibers. J Physiol (Lond) 120:171–204
Fenwick EM, Marty A, Neher E (1982) Sodium and calcium channels in bovine chromaffin cells. J Physiol (Lond) 331:599–635
Ferry DR, Goll A, Rombusch M, Glossmann H (1985) The molecular pharmacology and structural features of calcium channels. Br J Clin Pharmacol 20:233S–246S
Flaim SF, Ratz PH, Ress RJ (1984a) Calcium channel blockers as a new therapeutic concept: cardiovascular physiology and clinical implications. In: Ong HH, Lewis JC (eds) Hypertension: physiological basis and treatment. Academic, New York, pp 269–316
Flaim SF, Wright DE, Gleason MM, Swigart SC (1984b) Neuromedin-K (tachykinin neuropeptide) has potent cardiotonic effects and stimulates leak channel calcium influx in rabbit aorta without altering vascular tone. Circulation 70:II–167
Fleckenstein A (1977) Specific pharmacology of calcium in myocardium, cardiac pacemakers, and vascular smooth muscle. Annu Rev Pharmacol Toxicol 17:149–166
Fleckenstein A (1983) Calcium antagonism in heart and smooth muscle. Wiley, New York
Fleckenstein A, Tritthart H, Fleckenstein B, Herbst A, Grün G (1969) A new group of competitive Ca-antagonists (iproveratil, D600, prenylamine) with highly potent inhibitory effects on excitation-contraction coupling in mammalian myocardium. Pflugers Arch 307:R25
Flockerzi V, Oeken HJ, Hofmann F, Pelzer D, Cavalié A, Trautwein W (1986) The purified dihydropyridine binding site from skeletal muscle T-tubules is a functional calcium channel. Nature 323:66–68
Franckowiak G, Bechem M, Schramm M, Thomas G (1985) The optical isomers of the 1,4-dihydropyridine Bay K 8644 show opposite effects on the Ca-channel. Eur J Pharmacol 114:223–226
Galizzi JP, Fosset M, Lazdunski M (1984) Properties of receptors for the Ca^{2+}-channel blocker verapamil in transverse-tubule membranes of skeletal muscle. Eur J Biochem 144:211–215
Gallant EM, Goetl VM (1985) Effects of calcium antagonists on mechanical responses of mammalian skeletal muscle. Eur J Pharmacol 117:259–266
Garcia AG, Sala F, Reig JA, Viniegra S, Frias J, Fonteriz R, Gandia L (1984) Dihydropyridine Bay K 8644 activates chromaffin cell calcium channels. Nature 309:69–71
Gjörstrup P, Hardin H, Isaksson R, Westerlund C (1986) The enantiomers of the dihydropyridine derivative H 160/51 show opposite effects of stimulation and inhibition. Eur J Pharmacol 122:357–361
Glossmann H, Ferry DR, Goll A, Striessnig J, Schober M (1985a) Calcium channels: basic properties as revealed by radioligand binding studies. J Cardiovasc Pharmacol 7 [Suppl 6]:S20–S30

Glossmann H, Ferry DR, Goll A, Striessnig J, Zernig G (1985b) Calcium channels and calcium channel drugs: recent biochemical and biophysical findings. Arzneimittelforsch 35:1917–1935
Godfraind T (1983) Actions of nifedipine on Ca-fluxes and contractions in isolated rat arteries. J Pharmacol Exp Ther 224:443–450
Godfraind T, Kaba A (1969) Blockade or reversal of the contraction induced by calcium and adrenaline in depolarized arterial smooth muscle. Br J Pharmacol 36:549–560
Godfraind T, Vanhoutte PM, Govoni S, Paoletti R (eds) (1985) Calcium entry blockers and tissue protection. Raven, New York
Hagiwara S, Byerly L (1981) Calcium channel. Annu Rev Neurosci 4:69–125
Henry PD (1983) Mechanisms of action of calcium antagonists in cardiac and smooth muscle. In: Stone PH, Antman EM (eds) Calcium channel blocking agents in the treatment of cardiovascular disorders. Futura, New York, pp 107–154
Henry PD (1985) Atherosclerosis, calcium and calcium antagonists. Circulation 72:456–459
Hermsmeyer K, Trapani A, Abel PW (1981) Membrane-potential dependent tension in vascular muscle. In: Vanhoutte PM, Leusen I (eds) Vasodilation. Raven, New York, pp 273–284
Hescheler J, Kameyama M, Trautwein W (1985) Mechanism of inhibition of the calcium current by acetylcholine in isolated ventricular cells of guinea pig heart. Pflugers Arch 405:R12
Hess P, Tsien RW (1984) Mechanism of ion permeation through calcium channels. Nature 309:453–456
Hess P, Lansmann JB, Tsien RW (1984) Different mode of calcium channel gating behaviour favoured by dihydropyridine Ca agonists and antagonists. Nature 311:538–544
Hess P, Lansman JB, Tsien RW (1985a) Blockade of single cardiac calcium channels by Mg^{2+}, Cd^{2+} and Ca^{2+} in guinea-pig ventricular cells. J Physiol (Lond) 358:61P
Hess P, Lansman JB, Tsien RW (1985b) Mechanism of calcium channel modulation by dihydropyridine agonists and antagonists. In: Fleckenstein A, van Breemen C, Groß R, Hoffmeister F (eds) Cardiovascular effects of dihydropyridine-type calcium antagonists and agonists. Springer, Berlin Heidelberg New York Tokyo, pp 34–55
Hodgkin AL, Huxley AF (1952) Currents carried by sodium and potassium ions through the membrane of the giant axon of LOLIGO. J Physiol 116:449–472
Hof RP, Rüegg UT, Hof A, Vogel A (1985) Stereoselectivity at the calcium channel: opposite action of the enantiomers of a 1,4-dihydropyridine. J Cardiovasc Pharmacol 7:689–693
Hwa JJ, Bevan JA (1986) Stretch-dependent (myogenic) tone in rabbit ear resistance arteries. Am J Physiol 250:H87–H95
Hwang KS, van Breemen C (1985) Effects of the agonist Bay K 8644 on ^{45}Ca-influx and net Ca uptake into rabbit smooth muscle. Eur J Pharmacol 116:299–305
Irisawa H, Kokubun S (1983) Modulation by intracellular ATP and cyclic AMP of the slow inward current in isolated single ventricular cells of the guinea-pig. J Physiol (Lond) 338:321–337
Janis RA, Triggle DJ (1983) Sites of action of Ca-channel inhibitors. Biochem Pharmacol 32:3499–3507
Janis RA, Triggle DJ (1984) 1,4-dihydropyridine Ca^{2+} channel antagonists and agonists. A comparison of binding characteristic with pharmacology. Drug Dev Res 4:257–274
Janis RA, Triggle DJ (1986) Effects of calcium channel antagonists on the myometrium. In: Huscar G (ed) Physiology and biochemistry of the uterus in pregnancy and labor. CRC, Boca Raton, pp 201–223
Janis RA, Rampe D, Sarmiento JG, Triggle DJ (1984) Specific binding of a Ca channel activator Bay K 8644 to membranes from cardiac muscle and brain. Biochem Biophys Res Commun 121:317–323
Janis RA, Bellemann P, Sarmiento JG, Triggle DJ (1985) The dihydropyridine receptor. In: Fleckenstein A, van Breemen C, Groß R, Hoffmeister F (eds) Cardiovascular effects of dihydropyridine type calcium antagonists and agonists. Springer, Berlin Heidelberg New York Tokyo, pp 140–155

Kameyama M, Hofmann F, Trautwein W (1985) On the mechanism of β-adrenergic regulation of the Ca channel in the guinea pig heart. Pflugers Arch 405:285–293
Kanmura Y, Itoh T, Suzuki H, Ito Y, Kuriyama H (1981) Nifedipine actions on smooth muscle cells of pig and rabbit skinned and intact coronary arteries. In: Hashimoto K, Kwai C (eds) Asian Pacific Adalat-Symposium. Medical Tribune, Tokyo, pp 3–30
Kanmura Y, Itoh T, Kuriyama H (1984) Agonist actions of Bay K 8644, a dihydropyridine derivative on the voltage dependent calcium influx in smooth muscle cells of the rabbit mesenteric artery. J Pharmacol Exp Ther 231:717–723
Kass RS (1987) Voltage-dependent modulation of cardiac Ca channel current by the optical isomers of Bay K 8644: implications for channel gating. Circ Res (in press)
Kaufmann R, Fleckenstein A (1965) Ca^{++}-kompetitive elektro-mechanische Entkopplung durch Ni^{++}- und Co^{++}-Ionen am Warmblütermyokard. Pflugers Arch 282:290–297
Kazda S, Knorr A, Towart R (1983) Common properties and differences between various calcium antagonists. Prog Pharmacol 5:83–116
Kim YY, Holgate ST, Church MK (1985) Inhibition of histamine release from dispersed human lung and tonsillar mast cells by nicardipine and nifedipine. Br J Clin Pharmacol 19:631–638
Kojima K, Kojima I, Rasmussen H (1984) Dihydropyridine calcium agonist and antagonist effects on aldosterone secretion. Am J Physiol 247:E645–E650
Kokubun S, Reuter H (1984) Dihydropyridine derivatives prolong the open state of Ca channels in cultured cardiac cells. Proc Natl Acad Sci USA 81:842–527
Lee KS, Tsien RW (1983) Mechanism of calcium channel blockade by verapamil, D600, diltiazem and nitrendipine in single dialysed heart cells. Nature 302:790–794
Löfdahl CG, Barnes PJ (1986) Calcium, calcium channel blockade and airways function. Acta Pharmacol Toxicol (Copenh) [Suppl 2] 58:91–112
Loutzenhiser R, Epstein M (1985) Effects of calcium antagonists on renal hemodynamics. Am J Physiol 249:F619–F629
Malaisse-Lagae F, Mathias PCF, Malaisse WJ (1984) Gating and blocking of calcium channels by dihydropyridines in the pancreatic B-cell. Biochem Biophys Res Commun 123:1062–1068
McDonald TF, Cavalie A, Trautwein W, Pelzer D (1986) Voltage-dependent properties of macroscopic and elementary calcium channel currents in guinea pig ventricular myocytes. Pflugers Arch 406:437–448
Means AR, Tash JS, Chafouleas JG (1982) Physiological implications of the presence, distribution, and regulation of calmodulin in eucaryotic cells. Physiol Rev 62:1–39
Mehta JL (1985) Influence of calcium-channel blockers on platelet function and arachidonic acid metabolism. Am J Cardiol 55:158B–164B
Meisheri K, Hwang O, van Breemen C (1981) Evidence for two separate Ca^{2+} influx pathways in smooth muscle plasmalemma. J Membr Biol 59:19–25
Meyer H, Wehinger E, Bossert F, Boeshagen H, Franckowiak G, Goldmann S, Seidel W, Stoltefuss J (1984) Chemistry of dihydropyridines. In: Fleckenstein A, van Breemen C, Groß R, Hoffmeister F (eds) Cardiovascular effects of dihydropyridine type calcium antagonists and agonists. Springer, Berlin Heidelberg New York, pp 90–103
Meyer JS (1985) Calcium channel blockers in the prophylactic treatment of vascular headache. Ann Intern Med 102:395–397
Middlemiss DN, Spedding M (1985) A functional correlate for the dihydropyridine binding site in rat brain. Nature 314:94–96
Millar JA, Struthers AD (1984) Calcium antagonists and hormone release. Clin Sci 66:249–255
Movsesian MA, Adelstein RS (1984) Inhibition of turkey gizzard myosin light chain kinase activity by Bay K 8644. Eur J Pharmacol 103:161–163
Nilius B, Hess P, Lansman JB, Tsien RW (1985) A novel type of calcium channel in ventricular cells. Nature 316:443–446
Noble D (1984) The surprising heart: a view on recent progress in cardiac electrophysiology. J Physiol (Lond) 353:1–50
Noda M, Shimizu S (1984) Primary structure of Electrophorus electricus sodium channel deduced from cDNA sequence. Nature 312:121–127

Norman RI, Borsotto M, Fosset M, Lazdunski M, Ellory JC (1983) Determination of the molecular size of the nitrendipine-sensitive Ca^{2+} channel by radiation inactivation. Biochem Biophys Res Commun 111:878–883
Ochi R, Hino N, Niimi Y (1984) Prolongation of calcium channel open time by the dihydropyridine derivative Bay K 8644 in cardiac myocytes. Proc Jpn Acad [B] 60:153–156
Pelzer D, Cavalié A, Trautwein W (1984) Guinea-pig ventricular myocytes treated with D 600: mechanism of calcium-channel blockade at the level of single channels. In: Lichten PR (ed) Recent aspects in calcium antagonism. Schattauer, Stuttgart, pp 3–26
Ratz PH, Flaim SF (1982) Species and blood vessel specificity in the use of calcium for contraction. In: Flaim SF, Zelis R (eds) Calcium blockers: mechanisms of action and clinical applications. Urban and Schwarzenberg, Munich, pp 77–98
Reuter H (1973) Divalent cations as charge carriers in excitable membranes. Prog Biophys Mol Biol 26:1–43
Reuter H (1983) Calcium channel modulation by neurotransmitters enzymes and drugs. Nature 301:569–574
Reuter H (1984) Ion channels in cardiac cell membranes. Annu Rev Physiol 46:473–484
Ringer S (1883) A further contribution regarding the influence of the different constituents of the blood on the contraction of the heart. J Physiol 4:29–42
Sada H, Sada S, Sperelakis N (1986) Recovery of the slow action potential is hastened by the calcium slow channel agonist, Bay K 8644. Eur J Pharmacol 120:17–24
Saida K, van Breemen C (1983) Mechanism of Ca^{++} antagonist-induced vasodilation, intracellular actions. Circ Res 52:137–142
Sakmann B, Neher E (eds) (1983) Single channel recording. Plenum, New York
Sanchez-Chapula J (1981) Differential effects of histamine on left and right guinea-pig atria. Eur J Pharmacol 74:253–256
Sanguinetti MC, Kass RS (1984a) Regulation of cardiac calcium channel current and contractile activity by the dihydropyridine Bay K 8644 is voltage-dependent. J Mol Cell Cardiol 16:667–670
Sanguinetti MC, Kass RC (1984b) Voltage-dependent block of calcium channel current in the calf purkinje fiber by dihydropyridine calcium channel antagonists. Circ Res 55:336–348
Sato M, Nagao T, Yamaguchi I, Nakajima H, Kiyomoto A (1971) Pharmacological studies on a new 1,5-benzothiazepine derivative (CRD-401). Arzneimittelforsch 21:1338–1348
Schramm M, Towart R (1985) Modulation of calcium channel by drugs. Life Sci 37:1843–1860
Schramm M, Thomas G, Towart R, Franckowiak G (1983a) Novel dihydropyridines with positive inotropic action through activation of Ca channels. Nature 303:535–537
Schramm M, Thomas G, Towart R, Franckowiak G (1983b) Activation of calcium channels by novel 1,4-dihydropyridines. A new mechanism for positive inotropics or smooth muscle stimulants. Arzneimittelforsch 33:1268–1272
Schramm M, Towart R, Lamp B, Thomas G (1985) Modulation of calcium ion influx by the 1,4-dihydropyridines nifedipine and Bay K 8644. J Cardiovasc Pharmacol 7:493–496
Schwartz A, Grupp IL, Grupp G, Williams JS, Vaghy PL (1984) Effects of dihydropyridine calcium channel modulators in the heart: pharmacological and radioligand binding correlations. Biochem Biophys Res Commun 125:387–394
Shalaby IA, Kongsamut S, Freedman SB, Miller RJ (1984) The effects of dihydropyridines on neurotransmitter release from cultured neuronal cells. Life Sci 35:1289–1295
Siegel G, Adler A (1985) The effect of noradrenaline on membrane potential and tension in vascular smooth muscle. Pflugers Arch 403:R58
Siesjo BK (1981) Cell damage in the brain: a speculative synthesis. J Cereb Blood Flow Metab 1:155–185
Spedding M (1985a) Competitive interactions between Bay K 8644 and nifedipine in K^+ depolarized smooth muscle: a passive role for Ca^{2+}? Naunyn Schmiedebergs Arch Pharmacol 328:464–466

Spedding M (1985b) Activators and inactivators of Ca^{++}-channels: new perspectives. J Pharmacol 16:319–343
Striessnig J, Goll A, Moosburger K, Glossmann H (1986) Purified calcium channels have three allostericly coupled drug receptors. FEBS Lett 197:204–210
Su CM, Swamy VC, Triggle DJ (1984) Calcium channel activation in vascular smooth muscle by Bay K 8644. Can J Physiol Pharmacol 62:1401–1410
Su CM, Yousif FB, Triggle DJ, Janis RA (1985) Structure-function relationships of 1,4-dihydropyridines: ligand and receptor perspectives. In: Fleckenstein A, van Breemen C, Groß R, Hoffmeister F (eds) Cardiovascular effects of dihydropyridine type calcium antagonists and agonists. Springer, Berlin Heidelberg New York Tokyo, pp 104–110
Takenaka T, Maeno H (1982) A new vasoconstrictor 1,4-dihydropyridine derivate, YC-170. Jpn J Pharmacol 32:139P
Thomas G, Groß R, Schramm M (1984) Calcium channel modulation: ability to inhibit or promote calcium influx resides in the same dihydropyridine molecule. J Cardiovasc Pharmacol 6:1170–1176
Thomas G, Chung M, Cohen CJ (1985a) A dihydropyridine (Bay K 8644) that enhances calcium currents in guinea pig and calf myocardial cells. A new type of positive inotropic agent. Circ Res 56:87–96
Thomas G, Gross R, Pfitzer G, Rüegg JC (1985b) The positive inotropic dihydropyridine Bay K 8644 does not affect calcium sensitivity or calcium release of skinned cardiac fibres. Naunyn Schmiedebergs Arch Pharmacol 328:378–381
Towart R (1982) Effects of nitrendipine (Bay E 5009), nifedipine, verapamil, phentolamine, papaverine and minoxidil on contractions of aortic smooth muscle. J Cardiovasc Pharmacol 4:895–902
Towart R, Rounding HP (1986) Calcium antagonists and airways smooth muscle. In: Kay AB (ed) Asthma. Clinical pharmacology and therapeutic progress. Blackwell, Oxford, pp 128–145
Towart R, Wehinger E, Meyer H (1981) Effects of unsymmetrical ester substituted 1,4-dihydropyridine derivatives and their optical isomers on contraction of smooth muscle. Naunyn Schmiedebergs Arch Pharmacol 317:183–185
Tsien RW (1983) Calcium channels in excitable cell membranes. Annu Rev Physiol 45:341–358
Vaghy PL, Grupp IL, Balwierczak JL, Williams JS, Schwartz A (1984) Correlation of nitrendipine and Bay K 8644 binding to isolated canine heart sarcolemma with their pharmacological effects of the canine heart. Eur J Pharmacol 102:373–374
Vater W, Kroneberg G, Hoffmeister F, Kaller H, Meng K, Oberdorf A, Puls W, Schloßmann K, Stoepel K (1972) Zur Pharmakologie von 4-(2-Nitrophenyl)-2,6-dimethyl-1,4-dihydropyridine-3,5-dicarbonsäuredimethylester (Nifedipine, Bay A 1040). Arzneimittelforsch 22:1–14
Wada Y, Satoh K, Taira N (1985) Cardiovascular profile of Bay K 8644, a presumed calcium channel activator, in the dog. Naunyn Schmiedebergs Arch Pharmacol 328:382–387
Williams RJP (1970) The biochemistry of sodium, potassium, magnesium and calcium. Q Rev Chem Soc 24:331–365
Zanchetti A, Leonetti G (1985) Natriuretic effect of calcium antagonists. J Cardiovasc Pharmacol [Suppl 4] 7:S33–S37

CHAPTER 6

The Chemistry of Calcium Channel Agonists and Antagonists

D. J. TRIGGLE

A. Introduction

The Ca^{2+} channel antagonists, including the clinically available verapamil, gallopamil (D600), diltiazem, and nifedipine (Fig. 1) enjoy a substantial therapeutic reputation in a variety of cardiovascular disorders ranging from some cardiac arrhythmias to angina, to hypertension and peripheral vascular disorders, cardioprotection and cardiomyopathy. However, the potential list of applications is far greater and includes virtually all smooth muscle disorders in which elevated tone is present or suspected (for reviews of therapeutic applications see BÜHLER et al. 1984; FLAIM and ZELAS 1982; FLECKENSTEIN 1983; GODFRAIND et al. 1985; OPIE

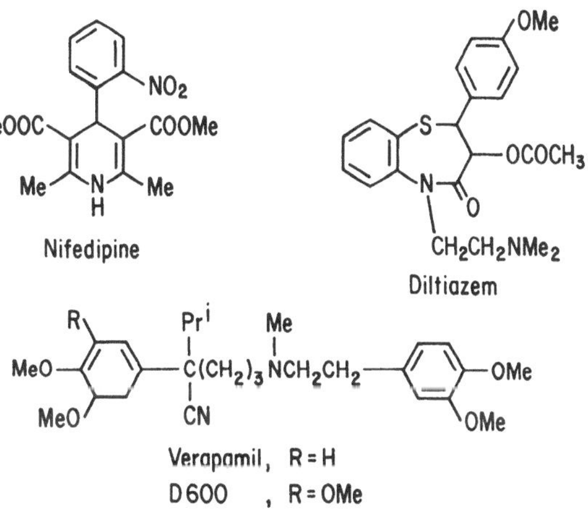

Fig. 1. Structural formulae of some Ca^{2+} channel antagonists representative of the three major structural classes

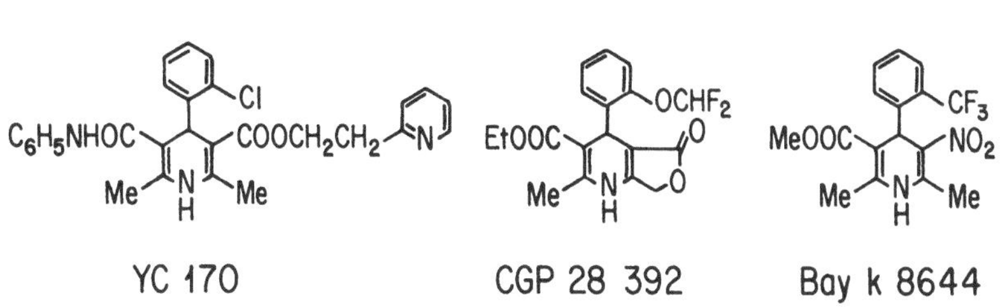

Fig. 2. Structural formulae and code numbers of three 1,4-dihydropyridine Ca^{2+} channel activators

1984; SCHWARTZ and TAIRA 1983; SINGH et al. 1983; STONE and ANTMAN 1983). The therapeutic and pharmacologic utility of the Ca^{2+} channel antagonists has served to focus attention on the roles of Ca^{2+} in cell regulation and of the several drug classes that may interact at critical loci of Ca^{2+} regulation. Additionally, the recent discovery of compounds closely related in structure to the 1,4-dihydropyridine, nifedipine, but which serve as Ca^{2+} channel activators (Fig. 2) serves to suggest that major new categories of uses ranging from inotropy to secretory stimulation may become available in the future.

The major clinically used agents are clearly a chemically heterogeneous group and, consistent with their therapeutic and pharmacologic heterogeneity, it is improbable that a single all-encompassing structure–function relationship describes this class of compounds (ROSENBERGER and TRIGGLE 1978; TRIGGLE and SWAMY 1983; TRIGGLE 1981; JANIS et al. 1987). Despite such heterogeneity these compounds do have as a common component of their action the ability to inhibit (or stimulate in the case of activators) Ca^{2+} current through voltage-dependent Ca^{2+} channels. It is to be anticipated, therefore, that the several structural classes of Ca^{2+} channel ligands will modulate Ca^{2+} current through voltage-dependent channels at different sites and by different mechanisms.

The preceding considerations raise a number of important questions concerning the molecular and structural definition of Ca^{2+} channel ligand action. These include:

1. Structure–function relationships
   a) Structural demands for each ligand class
   b) Differentiation of antagonist and activator ligands
2. Specific binding sites
   a) Detection of specific binding sites
   b) Pharmacologic significance of binding sites
3. Relationship of specific sites to channel function
   a) Ca^{2+} dependence
   b) Allosteric links between sites
   c) Stoichiometry of binding
4. Relationship of structural studies to tissue selectivity and function

B. Structure–Function Studies

Although primary attention has focused on Ca^{2+} channel ligands represented by diltiazem, nifedipine, and verapamil (see Fig. 1), it is clear that there is an abundance of structures possessing Ca^{2+} channel-blocking properties, although in most instances these are secondary to other and (often) more potent pharmacologic activities (ROSENBERGER et al. 1978; TRIGGLE and SWAMY 1980, 1983; JANIS et al. 1987). Among these agents (Fig. 3) are α-adrenoceptor antagonists, including phenoxybenzamine and benextramine, antischizophrenic drugs, including diphenylbutylpiperidines (GOULD et al. 1983; SNYDER and REYNOLDS 1985), barbiturates, including phenobarbital (KLEINHAUS and PRICHARD 1977), benzodiazepines, including diazepam (BOLGER et al. 1983; CANTOR et al. 1984), histamine antagonists, including cyproheptadine (LOWE et al. 1981), the antidiarrheal opiate

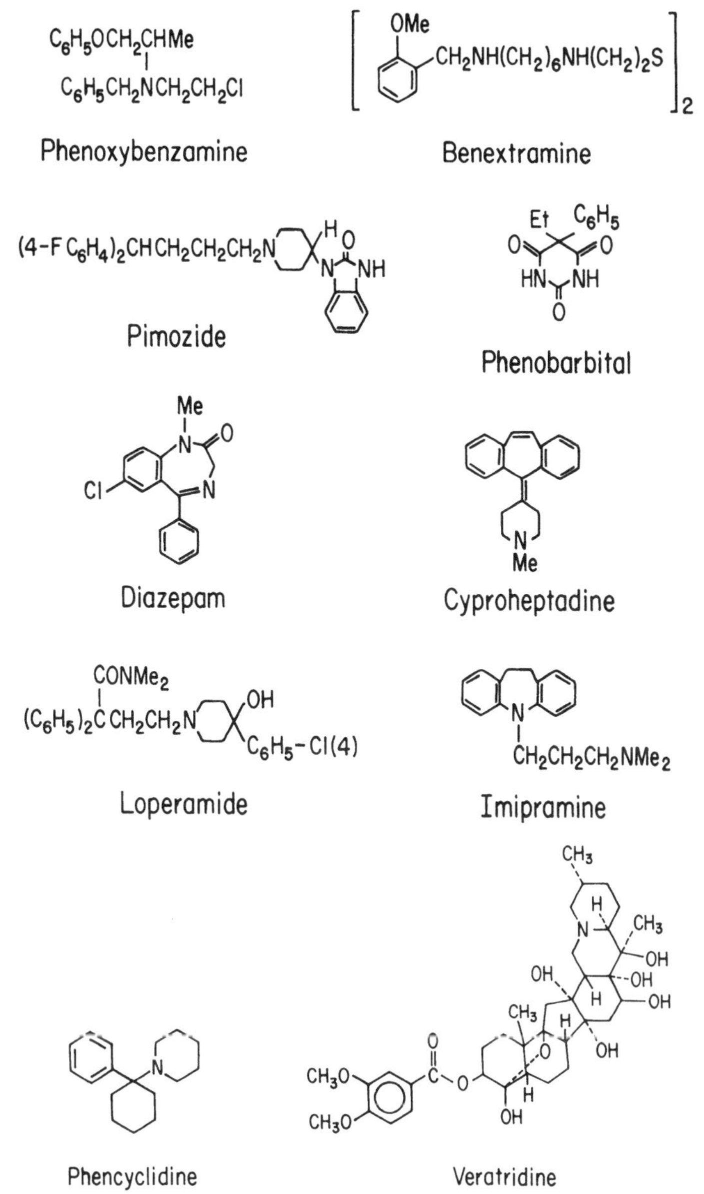

Fig. 3. Representative selection of molecular structures possessing Ca^{2+} channel antagonist activity

loperamide (REYNOLDS et al. 1984), the antidepressant imipramine (ISENBERG and TAMARGO 1985), the hallucinogen phencyclidine (EL-FAKAHAY et al. 1984), and the sodium channel toxin veratridine (BOLGER et al. 1983). Structural pursuit of these and other agents may well lead to other specific classes of Ca^{2+} channel ligands with properties distinct from those that currently exist. The major focus on structure–function relationships has, however, centered around the clinically used drug structures.

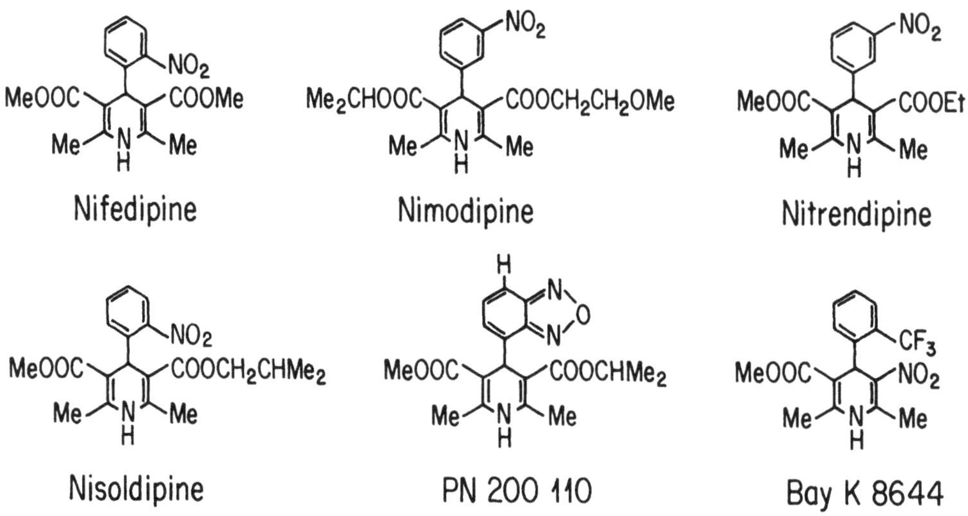

Fig. 4. Structural formulae of nifedipine and related 1,4-dihydropyridine Ca^{2+} channel ligands

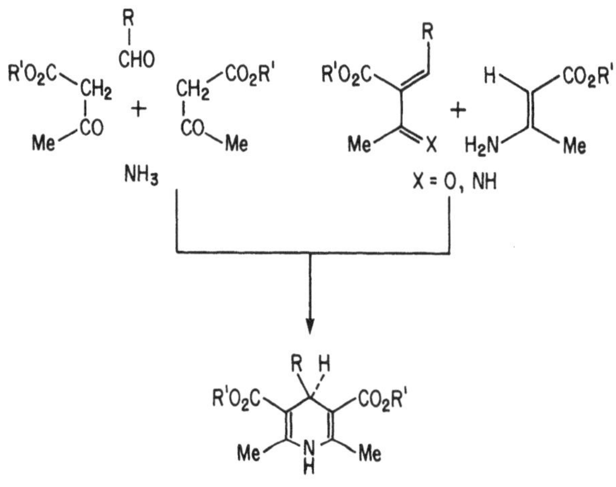

Fig. 5. Hantzsch 1,4-dihydropyridine synthesis

The largest number of analogs are available in the 1,4-dihydropyridine series (Fig. 4) in part because of the ease of synthesis. The typical synthetic methodology employs the Hantzsch dihydropyridine synthesis (EISNER and KUTHAN 1972; KUTHAN and KURFÜRST 1982; JANIS et al. 1987). The synthetic methodology is outlined in Fig. 5, depicting several variations on a theme by Hantzsch. The more complex synthetic procedures for diltiazem (Fig. 6) and verapamil (Fig. 7) have resulted in the generation of far fewer analogs and consequently more limited structure–function relationships.

From several studies of 1,4-dihydropyridine actions in whole animal systems, in isolated tissues, and in radioligand binding studies, a number of general conclusions may be drawn concerning the structural requirements contributing to an-

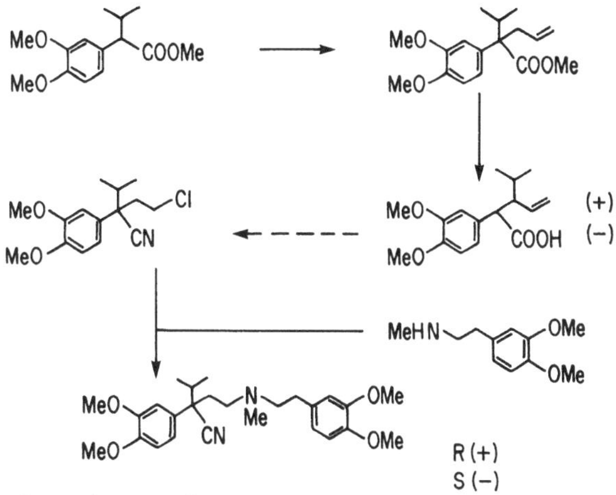

Fig. 6. Synthesis of enantiomers of verapamil

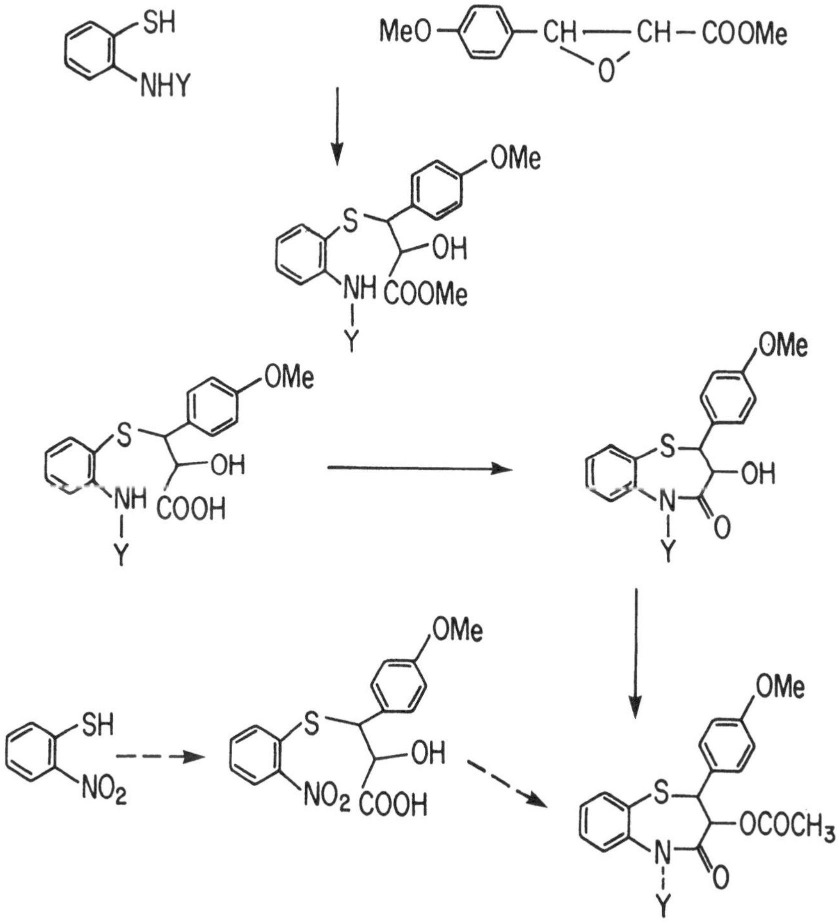

Fig. 7. Synthetic route to diltiazem

Y = $CH_2CH_2NMe_2$

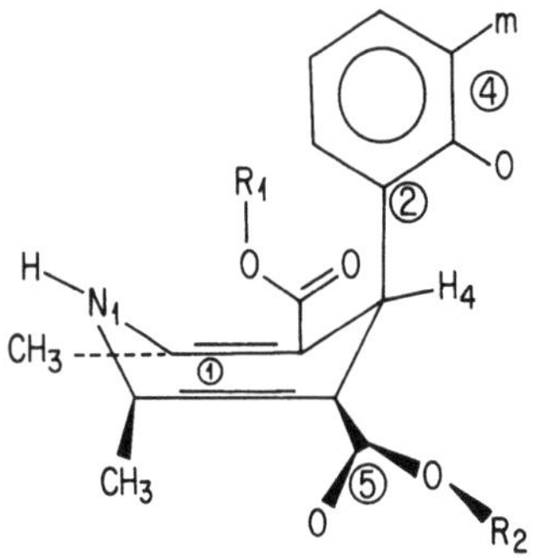

Fig. 8. Summary of structure–activity relationship for 1,4-dihydropyridine Ca^{2+} channel antagonists, indicating the basic conformation as demonstrated by solid state and solution studies. For further discussion see text

tagonist activity (Fig. 8; for reviews see ROSENBERGER and TRIGGLE 1978; TRIGGLE 1981, 1982; TRIGGLE and JANIS 1984a; MANNHOLD et al. 1982; LANGS and TRIGGLE 1984; JANIS et al. 1987):

1. The 1,4-dihydropyridine ring is essential: oxidation (to pyridine metabolites) or reduction leads to abolition of actions.
2. A 4-phenyl substituent is optimum, activity increasing with 4-substitution in the sequence, H < Me < cycloalkyl < heterocyclic < phenyl.
3. In the 4-phenyl substituent, incorporation of electron-withdrawing groups in the *ortho* and *meta* positions is optimum. Any substitution in the *para* position is detrimental to activity.
4. At least one ester function is required in the 3 or 5 position of the 1,4-dihydropyridine ring and replacement of ester functions with COMe, or CN groups leads to loss of activity. There is structural tolerance to the size of the ester groups that confer and maintain activity and a certain amount of evidence exists to suggest that differential recognition of ester groups may contribute to tissue selectivity.
5. When the substituents (usually esters) at the C-3 and C-5 positions of the 1,4-dihydropyridine ring are different, then C-4 becomes chiral and stereoselectivity of action results.

That a conformation in which the substituted phenyl ring is approximately orthogonal to the 1,4-dihydropyridine ring (Fig. 8) is both energetically and biologically important is indicated by structural studies of the solid state and solution conformations (TRIGGLE et al. 1980; FOSSHEIM et al. 1982; GOLDMANN and GEIGER 1984) and by the synthesis of rigid analogs bridged between the phenyl and 1,4-dihydropyridine rings where activity increases with increasing inter-ring angle (Fig. 9; SEIDEL et al. 1984). This conformation is shared by inactive analogs also, so that it is a necessary, but not a sufficient criterion for activity. Furthermore, the same general conformation is shared by activator 1,4-dihydropyridines, including CGP 28392 and Bay K 8644 (LANGS and TRIGGLE 1985).

Only minor structural variations differentiate activator from antagonist 1,4-dihydropyridines (TAKENAKA and MAENO 1982; SCHRAMM et al. 1983; LANGS and TRIGGLE 1985) and there is insufficient data available from which to draw structure–activity conclusions. Solid state studies of two activator molecules, CGP 28392 and Bay K 8644 (LANGS and TRIGGLE 1984, 1985), indicate that an en-

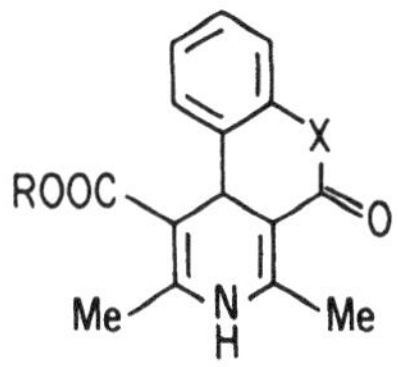

X = O, O–$(CH_2)_{1-2}$, O–$(CH_2)_{2-5}$–O

Fig. 9. Structural formulae of fused ring analogs of nifedipine

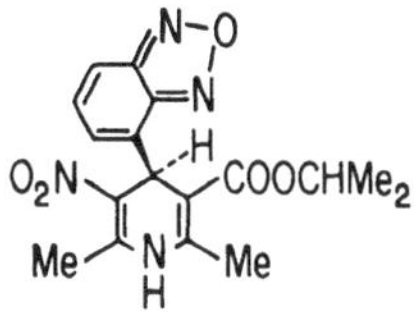

(+) S 202-791

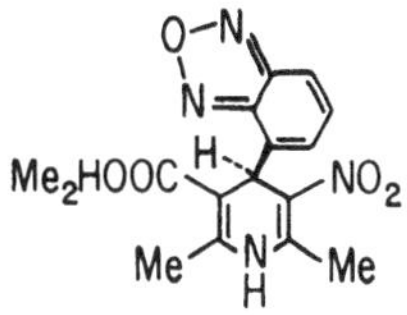

(−) R 202-791

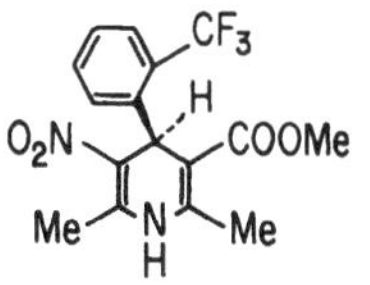

(−) S Bay k 8644

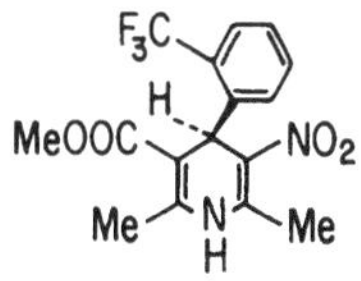

(+) R Bay k 8644

Fig. 10. Structural formulae of enantiomeric pairs of 1,4-dihydropyridines showing activator–antagonist properties

hanced planarity and acidity of the 1,4-dihydropyridine ring relative to antagonists might be important features. However, the dramatic observation that enantiomeric pairs of 1,4-dihydropyridines have opposing pharmacologic activities (Fig. 10) whereby the S- and R-forms of isopropyl-4-(2,3-benzoxadiazol-4-yl)-1,4-dihydro-2,6-dimethyl-5-nitro-3-pyridine carboxylate (202 791; HOF et al. 1985) and methyl-4-(2-trifluoromethylphenyl)-1,4-dihydro-2,6-dimethyl-5-nitro-3-pyridine carboxylate (Bay K 8644; FRANCKOWIAK et al. 1985) are activator and antagonist, respectively, indicates that other features must be important.

Much less information is available for structure–activity analyses in the verapamil, and particularly in the diltiazem series. MANNHOLD and his colleagues (MANNHOLD et al. 1978, 1981, 1982) have analyzed verapamil analogs for their negative inotropic effects in cardiac muscle and have shown that activity increases with increasing electron-withdrawing capacity (inductive substituent constant *F*) of substituents in the aromatic ring. However, the activity range of the series of compounds investigated was very narrow and there is clearly scope for more extensive studies in both the verapamil and diltiazem series.

C. Binding Sites for Ca^{2+} Channel Ligands

The discovery that specific binding sites existed for Ca^{2+} channel ligands in a variety of excitable cells including smooth, cardiac, and skeletal muscles and neurons (for reviews see GLOSSMANN et al. 1982; JANIS and TRIGGLE 1983; TRIGGLE and JANIS 1984a; GLOSSMANN and FERRY 1985) confirmed conclusions drawn from the original pharmacologic studies that specific, rather than nonspecific, sites mediated the actions of these ligands. A very extensive literature is now available detailing these radioligand binding studies (for review see JANIS et al. 1987).

The majority of studies have involved [^{3}H] 1,4-dihydropyridine antagonists, including nitrendipine, nimodipine, nisoldipine, and PN 200110 (see Fig. 4), and various membrane preparations from excitable cells. Limited studies only are available with the Ca^{2+} channel activator Bay K 8644. Quite generally, high affinity binding sites have been found in smooth, cardiac, and neuronal preparations with affinities of approximately 10^{-10}–10^{-11} M and with binding densities of 50–1000 fmol per milligram protein. Much higher concentrations of binding sites are found in skeletal muscle and these binding sites are of lower affinity by 1–2 orders of magnitude. A representative collection of data is presented in Table 1. There is general agreement that these binding sites are pharmacologically appropriate since they are sensitive only to drugs affecting Ca^{2+} channel function or expression and the sensitivity of the binding sites to series of Ca^{2+} channel ligands correlates with their pharmacologic activities (for reviews see TRIGGLE and JANIS 1984a; JANIS and TRIGGLE 1984; JANIS et al. 1987). Furthermore, the location of the binding sites accords with their anticipated function in the corresponding tissue. In smooth muscle they are associated with the plasma membrane (GROVER et al. 1984, 1985; GODFRAIND and WIBO 1985), in cardiac tissue with plasma membranes probably associated with junctional and cisternal complexes (BRANDT 1985), and in skeletal muscle with t-tubules (FOSSEL et al. 1983; GALIZZI et al. 1984).

A number of studies have described quantitative correlations between pharmacologic and radioligand binding activities in series of Ca^{2+} channel antagonists, particularly 1,4-dihydropyridines. Typical correlations are indicated in Fig. 11 and demonstrate important general observations. For smooth muscle there exists an essentially 1:1 correlation between the abilities of a series of 1,4-dihydropyridines to displace high affinity [^{3}H] 1,4-dihydropyridine binding and to inhibit depolarization-induced tension response. For cardiac muscle the same rank order of potency holds, but the correlation is less than 1:1 (Fig. 11 b), consistent with the pharmacologic and clinical observations that nifedipine and other 1,4-dihydropyridines are smooth muscle- rather than cardiac-selective drugs. It is noteworthy, however, that the same high affinity binding sites are found in smooth and cardiac muscle (Fig. 11 c). An even more striking discrepancy is found in neuronal preparations where high affinity binding sites are found, the properties of which are identical to those in a functional tissue (Fig. 11 d), despite the many observations that neuronal tissues are in general insensitive or weakly sensitive to Ca^{2+} channel antagonists (ROSENBERGER and TRIGGLE 1978; TRIGGLE and SWAMY 1980, 1983; TRIGGLE 1984; MILLER and FREEDMAN 1984; JANIS et al.

Table 1. Ca^{2+} channel radioligand binding

Tissue	Species	Radioligand	Temperature (°C)	K_d $\times 10^{-9}$ (*M*)	B_{max} (Fmol mg^{-1} protein)	Reference
Heart						
Ventricle	Rat	[^{3}H] Nitrendipine	25	0.18	400	JANIS et al. (1982)
Ventricle	Chick	[^{3}H] Desmethoxyverapamil	25	0.17	100	PTASIENSKI et al. (1985)
				39	2200	
Ventricle	Rabbit	[^{3}H] Nitrendipine	25	0.16	300	JANIS et al. (1984a)
Ventricle	Dog	[^{3}H] Nitrendipine	25	0.13	190	SARMIENTO et al. (1982)
Ventricle	Pig	[^{3}H] Verapamil	25	55	820	GARCIA et al. (1984)
Smooth muscle						
Ileal	Guinea-pig	[^{3}H] Nitrendipine	25	0.16	1130	BOLGER et al. (1983)
Mesenteric	Rat	[^{3}H] Nitrendipine	25	0.10	18	TRIGGLE et al. (1982)
Brain						
Whole	Rat	[^{3}H] Nitrendipine	25	0.17	114	YAMAMURA et al. (1982)
Whole	Rat	[^{3}H] Nimodipine	25	1.1	500	BELLEMANN et al. (1983)
Cerebral cortex	Guinea-pig	[^{3}H] Bay K 8644	15	3	440	JANIS et al. (1984b)
Skeletal muscle						
	Guinea-pig	[^{3}H] Diltiazem	2	39	11000	GLOSSMANN et al. (1983)
	Rabbit	[^{3}H] PN 200110	37	1.4	20000	FERRY et al. (1983)
	Rabbit	[^{3}H] Nitrendipine	10	1.8	50000	FOSSETT et al. (1983)
	Rabbit	[^{3}H] Nisoldipine	25	2.2	8250	TRIGGLE and JANIS (1984a)
	Guinea-pig	[^{3}H] Desmethoxyverapamil	25	2.2	18000	GOLL et al. (1984)

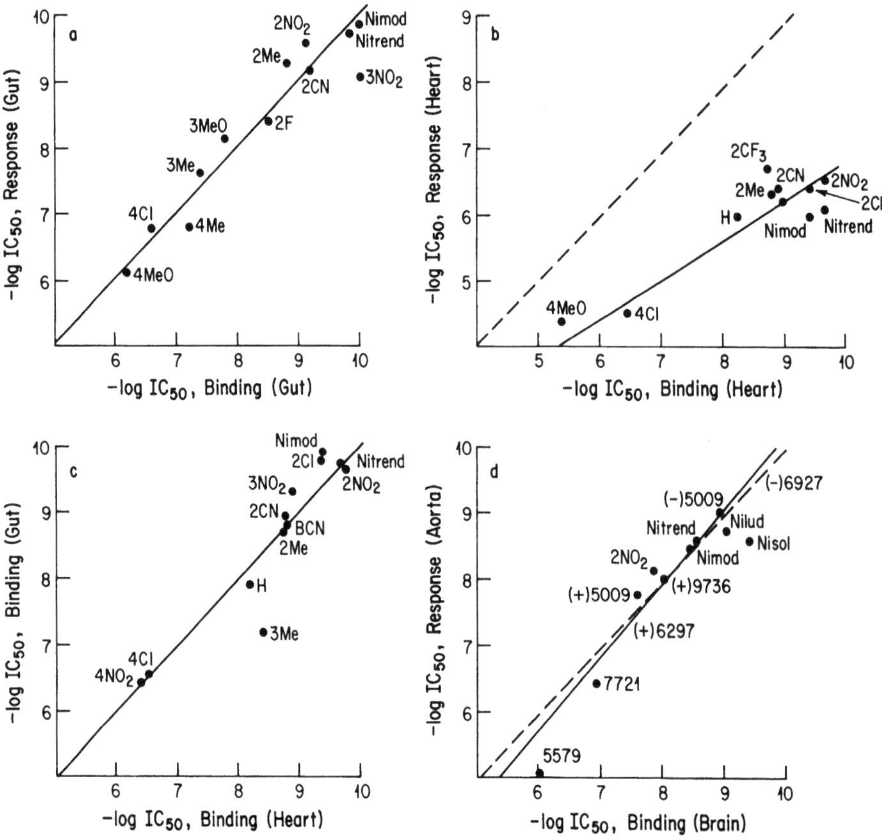

Fig. 11 a–d. Correlations between pharmacologic and receptor binding properties of a series of 1,4-dihydropyridines (including the nifedipine series and niludipine, nimodipine, nisoldipine, and nitrendipine). **a** Intestinal smooth muscle (guinea pig): binding, displacement of [^{3}H] nitrendipine; pharmacology, inhibition of K^+ depolarization-induced response (tonic phase), **b** cardiac muscle: binding, displacement of [^{3}H] nitrendipine (rat ventricle); pharmacology, inhibition of cat papillary muscle, **c** rat heart and guinea pig intestine; binding, displacement of [^{3}H] nitrendipine, **d** rat brain and rabbit aorta; binding, displacement of [^{3}H] nimodipine; pharmacology, inhibition of K^+ depolarization-induced response in rabbit aorta. Data from BOLGER et al. (1983) for intestinal smooth musle (**a, c**); from JANIS et al. (1984a) and JANIS and TRIGGLE (1984) for rat heart (**b, c**); and BELLEMANN et al. (1983) for rat brain and rabbit aorta (**d**). *Full lines* are regression, *broken lines* are 1:1 equivalency

1987). Similarly, the physiologic function of the extremely large number of binding sites associated with skeletal muscle remains apparently enigmatic (SCHWARTZ et al. 1985).

Observations that the quantitative expression of structure–function relationships can depend upon the tissue, its state, and the nature of the measured signal raise questions of fundamental concern both to the interpretation of structure–activity data and to the relationship of binding sites to Ca^{2+} channel function.

D. Ca^{2+} Channel Binding Sites: Relationship to Ca^{2+} Channel Function

The availability of radioligands for the Ca^{2+} channel, first nitrendipine and the 1,4-dihydropyridines and subsequently diltiazem, verapamil, and phenylalkylamines, has made possible attempts to define the relationship of these binding sites one to the other and to the functional machinery of the Ca^{2+} channel.

The high affinity binding site for nitrendipine and other 1,4-dihydropyridines is sensitive to low concentrations of divalent (and trivalent) cations including the group IIa cations, presumably reflecting an interconversion from low to high affinity states (Fig. 12; GLOSSMANN et al. 1982; GOULD et al. 1982; LUCHOWSKI et al. 1984). Additionally, binding of [^{3}H] 1,4-dihydropyridines is inhibited by calmodulin antagonists with potency paralleling their activity in calmodulin-dependent systems (LUCHOWSKI et al. 1984). Collectively, these data suggest a role for

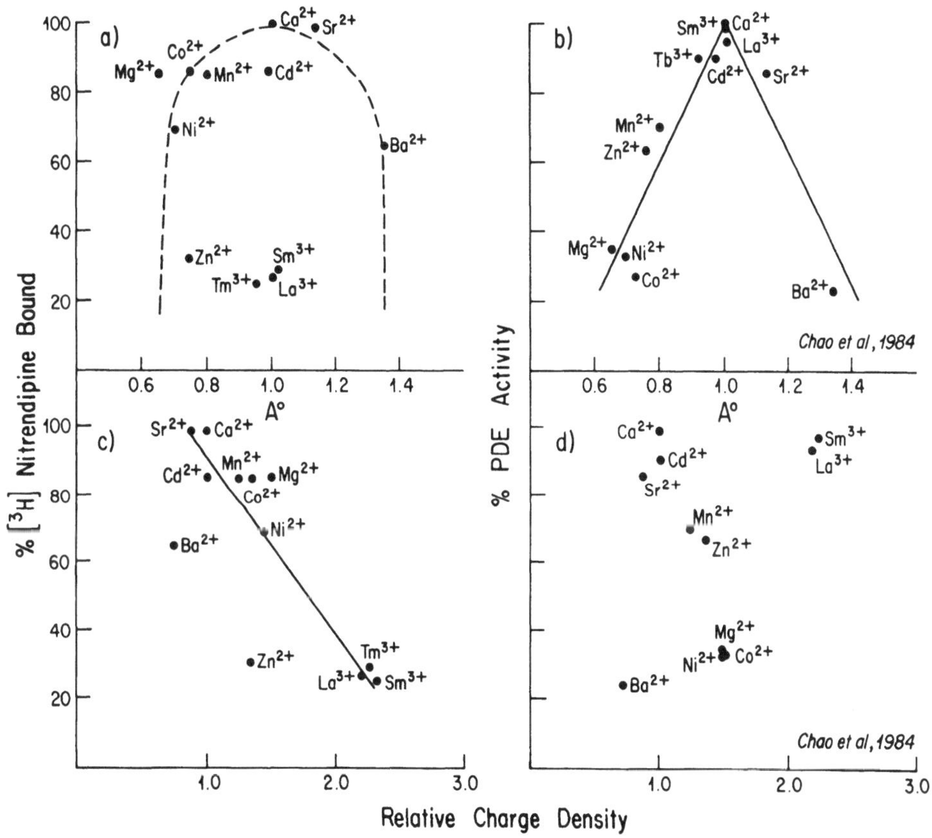

Fig. 12 a–d. Dependence of maximum [^{3}H] nitrendipine binding to microsomal preparations (cation-depleted) of guinea pig ileal longitudinal smooth muscle on ionic radius (**a**) and relative charge density (**c**, $Ca^{2+} = 1.0$) of added cation. Dependence of calmodulin-dependent phosphodiesterase activity on ionic radius (**b**) and relative charge density (**d**, $Ca^{2+} = 1.0$) of added cation. Data in **a, c** from LUCHOWSKI et al. (1984) and in **b, d** from CHAO et al. (1984)

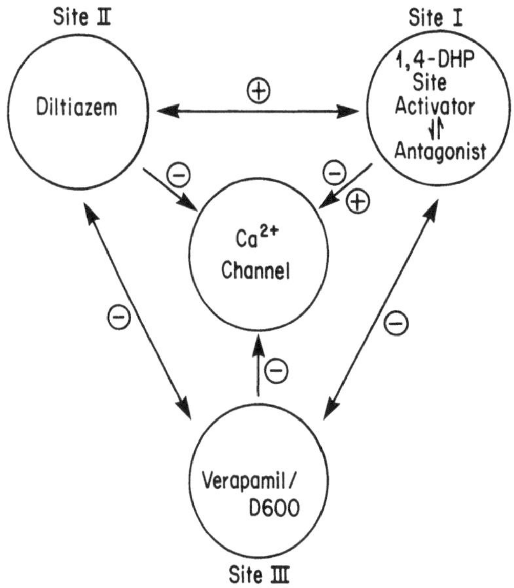

Fig. 13. Schematic representation of ligand binding sites (I, II, III) associated with the Ca^{2+} channel. The − and + signs represent inhibitory and stimulatory relationships, respectively

a high affinity–low affinity 1,4-dihydropyridine-binding site, modulated by divalent cation binding to a high affinity cation-binding protein, in the control of channel permeation. It is of interest that recent models of the Ca^{2+} channel (HESS and TSIEN 1984; ALMERS and MCCLESKEY 1984; ALMERS et al. 1984) contain high affinity binding sites for permeant cations.

The interactions between 1,4-dihydropyridines, diltiazem, and verapamil may best be described as a set of linked allosteric interactions. Verapamil and other phenylalkylamines inhibit 1,4-dihydropyridine binding by lowering affinity while diltiazem potentiates binding, in temperature-dependent fashion, by increasing B_{max} or affinity (for general discussion see GLOSSMANN et al. 1982; GLOSSMANN and FERRY 1985; BOLGER et al. 1983; JANIS et al. 1984a; TRIGGLE and JANIS 1984a). Parallels to these binding interactions have been demonstrated in tissue experiments (DEPOVER et al. 1983; SPEDDING 1983; YOUSIF and TRIGGLE 1985). Although the schematic representation of interactions shown in Fig. 13 has been deduced very largely from studies of 1,4-dihydropyridine binding, more limited studies with [^{3}H] diltiazem, [^{3}H] verapamil, and [^{3}H] desmethoxyverapamil (GLOSSMANN et al. 1983; GOLL et al. 1984; FERRY et al. 1984; SCHOEMAKER and LANGER 1985) have, in general, confirmed this model although the stoichiometry of the three recognition sites is in some dispute.

It is to be emphasized that such binding studies, the great majority of which have been performed in isolated plasma membranes, frequently uncharacterized as to purity, offer very limited molecular information concerning the topography of the binding sites one to the other and the relationship of these binding sites to the gating, voltage detection, and permeation machinery of the Ca^{2+} channel. Additionally, efforts to employ Ca^{2+} channel drugs as ligand probes for the channel are complicated by the state dependence of interaction of these ligands whereby Ca^{2+} channel activators may bind preferentially to an open state, and antagonists to open or inactivated states. Accordingly, such ligands will shift the voltage- and time-dependent equilibria existing between channel states and the apparent affinity of a ligand will be dependent upon the dominant channel state in the preparation. Thus, electrophysiologic data suggest that the affinity of nitrendipine is lower by some 2–3 orders of magnitude for the resting state than for the inactivated state of the cardiac Ca^{2+} (BEAN 1984; HESS et al. 1984; SANGUINETTI and KASS 1984). The expression of structure–activity relationships may thus be qualitatively and quantitatively different according to the channel state under consideration. These considerations may underlie the opposing stereochemical preferences exhibited by activator–antagonist pairs of 1,4-dihydropyridines. One further complication to the interpretation of structure–function relationships is the realization that there exist distinct classes of voltage-dependent Ca^{2+} channels with markedly different pharmacologic profiles (BEAN 1985; MILLER 1985; NILIUS et al. 1985; NOWYCKY et al. 1985). Since many pharmacologic preparations undoubtedly employ different Ca^{2+} channels, each capable of existing in different states, the interpretation of structure–function relationships must be viewed as potentially ambiguous.

E. Relationship of Structural and Functional Studies: A Prospective

Although the Ca^{2+} channel ligands have assumed great prominence in the past few years as molecular probes of Ca^{2} channel structure and function, their role as therapeutic agents is not to be ignored. Although these agents are used currently in the control of cardiovascular disorders, they project a very wide potential for use, including the control of neuronal function and, through activator action, cardiotonic and secretagogue activities. Questions of tissue and cellular selectivity are therefore of paramount concern. Knowledge of functional properties of channel behavior that underlie the expression of structure–activity relationships and the generation of tissue-selective ligands is therefore of substantial importance. The previous discussion indicates that a number of factors will contribute to the determination of tissue selectivity.

I. Different Categories of Ca^{2+} Channels

Electrophysiologic data have already indicated the existence of two, and possibly three, distinct categories of voltage-dependent Ca^{2+} channel whose sensitivitiy to the known Ca^{2+} channel ligands differ significantly. Since these channel categories may also enjoy a differential tissue distribution, for example, somatic versus dendritic in neuronal tissue (LLINÀS and YAROM 1981; JAHNSEN and LLINÀS 1984), the extent to which they are activated likely depends upon the signal, the tissue, and the cellular location.

It is quite clear that new categories of ligands will arise with different selectivities and that the search for such ligands will be greatly facilitated by the electrophysiologic distinctions now being made. Particular interest attaches to several toxins that appear to exert a spectrum of activities against different Ca^{2+} channels (MILLER 1984). Maitotoxin, isolated from the dinoflagellate *Gambierdiscus toxicus*, is of unknown composition and appears to serve as a Ca^{2+} channel activator in smooth and cardiac muscle and neuronal systems (TAKAHASHI et al. 1982, 1983; OHIZUMI and YASUMOTO 1983; FREEDMAN et al. 1984), an action which is sensitive to known inorganic or organic Ca^{2+} channel antagonists. Atrotoxin, a fraction of *Crotalus atrox* toxin, is similarly of unknown composition, but is also an apparent, but high molecular weight (<15000) Ca^{2+} channel activator in myocardial tissue (HAMILTON et al. 1985). Leptinotarsin-d is a 45000 dalton protein isolated from the Colorado potato beetle which appears to block specifically a subclass of nifedipine-sensitive Ca^{2+} channels in the mammalian CNS (KOENIG et al. 1985). A group of low molecular weight polypeptide toxins isolated from the fish-eating mollusk *Conus geographus* (OLIVERA et al. 1985) includes one group, the ω-toxins, basic peptides with 25–29 amino acids and three disulfide bridges, which appear to block selectively neuronal presynaptic voltage-dependent Ca^{2+} channels (KERR and YOSHIKAMI 1984). In this property they appear to exert a different selectivity from existing Ca^{2+} channel antagonists.

II. State-Dependent Interactions with Ca^{2+} Channels

It is now quite clear that, contrary to earlier views, the interaction of Ca^{2+} channel ligands is state dependent. Accordingly, the apparent affinity of a ligand, as well as the relative expression of activator and antagonist properties, will depend upon the state equilibrium (resting, open, and inactivated) of the channel. Presently available ligands have preferential affinity for the open or inactivated states, and their access pathways may depend upon the physicochemical properties of the ligand (HILLE 1977; HONDEGHEM and KATZUNG 1984). Thus, the neutral, nonpolar 1,4-dihydropyridines may access the channel via the lipid bilayer, whereas charged molecules, including the phenylalkylamines and diltiazem, may access the channel via the open state. More subtle considerations of molecular structure should make it possible to design agents that differ not only in their state dependence of binding, but also in their relative use of access pathway. Molecules with preferential affinity for the closed state of the channel should exist and these would be expected to possess a selectivity of action different from existing compounds and perhaps also to serve as "neutral" antagonists of existing Ca^{2+} channel ligands.

III. Pathologic State of Tissue

By analogy to a number of receptors for neurotransmitters and hormones, it is likely that channel expression and function is altered in disease. An altered dominance of one or other channel types, an altered state equilibrium, the presence of circulating antibodies, or the alteration in availability of an endogenous ligand for the channel or channels are all possibilities, hypothetical at the present, that will change ligand sensitivity between normal and pathologic states.

References

Almers W, McCleskey EW (1984) Non-selective conductance in calcium channels of frog muscle calcium selectivity in a single-file pore. J Physiol 353:586–608

Almers W, McCleskey EW, Palade PT (1984) A non-selective cation conductance in frog muscle membrane blocked by micromolar external calcium ions. J Physiol 353:565–583

Bean BP (1984) Nitrendipine block of cardiac calcium channels: high affinity binding to the inactivated state. Proc Natl Acad Sci USA 81:6388–6392

Bean BP (1985) Two kinds of calcium channels in canine atrial cells. Differences in kinetics, selectivity and pharmacology. J Gen Physiol 86:1–30

Bellemann P, Schade A, Towart R (1983) Dihydropyridine receptor in rat brain labelled with [^{3}H]nimodipine. Proc Natl Acad Sci USA 80:2356–2360

Bolger GT, Gengo P, Klockowski R, Siegel H, Janis RA, Triggle AM, Triggle DJ (1983) Characterization of binding of the Ca^{++} channel antagonist, [^{3}H]nitrendipine, to guinea pig ileal smooth muscle. J Pharmacol Exp Ther 225:291–309

Brandt N (1985) Identification of two populations of cardiac microsomes with nitrendipine receptors: correlation of the distribution of dihydropyridine receptors with organelle specific markers. Arch Biochem Biophys 242:306–319

Bühler FR, Distler A, Gross F (1984) Calcium antagonists for antihypertensive baseline therapy. 1st International nitrendipine symposium. J Cardiovasc Pharmacol 6 [Suppl 7]:S929–1113

Cantor EH, Kenessey A, Semenuk G, Spector S (1984) Interaction of calcium channel blockers with non-neuronal benzodiazepine binding sites. Proc Natl Acad Sci USA 81:1549–1552

Chao SH, Suzuki Y, Zysk J, Cheung WY (1984) Activation of calmodulin by various metal cations as a function of ionic radius. Mol Pharmacol 26:75–82

DePover A, Grupp IL, Grupp G, Schwartz A (1983) Diltiazem potentiates the negative inotropic action of nimodipine in heart. Biochem Biophys Res Commun 108:110–117

Eisner U, Kuthan J (1972) The chemistry of dihydropyridines. Chem Rev 72:1–42

El-Fakahany EE, Eldefrawi AT, Murphy DL, Aguayo LG, Triggle DJ, Albuquerque EX, Eldefrawi MW (1984) Interaction of phenycyclidine with crayfish muscle membranes. Sensitivity to calcium channel antagonists and other drugs. Mol Pharmacol 25:369–378

Ferry DR, Goll A, Glossmann H (1983) Differential labelling of putative skeletal muscle calcium channels by [^{3}H]nifedipine, [^{3}H]nitrendipine, [^{3}H]nimodipine and [^{3}H]PN 200110. Naunyn Schmiedebergs Arch Pharmacol 323:276–281

Ferry DR, Goll A, Gadow C, Glossmann H (1984) (−)-^{3}H-Desmethoxyverapamil labelling of putative calcium channels in brain: autoradiographic distribution and allosteric coupling to 1:4-dihydropyridine and diltiazem binding sites. Naunyn Schmiedebergs Arch Pharmacol 327:183–187

Flaim SF, Zelis R (1982) Calcium blockers. Mechanism of action and clinical applications. Urban and Schwarzenberg, Baltimore

Fleckenstein A (1983) Calcium antagonism in heart and smooth muscle. Experimental facts and therapeutic prospects. Wiley, New York

Fosset M, Jaimovich E, Delpont E, Lazdunski M (1983) [^{3}H]Nitrendipine receptors in skeletal muscle. Properties and preferential localization in transverse tubules. J Biol Chem 258:6086–6092

Fossheim R, Svarteng K, Mostad A, Rømming C, Shefter E, Triggle DJ (1982) Crystal structures and pharmacological activity of calcium channel antagonists: 2,6-dimethyl-3,5-dicarbomethoxy-4-(unsubstituted, 3-methyl-,4-methyl-,3-nitro-,4-nitro- and 2,4-dinitro phenyl) 1,4-dihydropyridine. J Med Chem 25:126–131

Franckowiak G, Bechem M, Schramm M, Thomas G (1985) The optical isomers of the 1,4-dihydropyridine Bay K 8644 show opposite effects on the Ca-channel. Eur J Pharmacol 114:223–226

Freedman SB, Miller RJ, Miller DM, Tindall DR (1984) Interactions of maitotoxin with voltage-sensitive calcium channels in cultured neuronal cells. Proc Natl Sci USA 81:4582–4585
Galizzi J-P, Fosset M, Lazdunski M (1984) Properties of receptors for the Ca^{2+} channel blocker verapamil in transverse tubule membranes of skeletal muscle. Eur J Biochem 144:211–215
Garcia ML, Trumble MJ, Reuben JP, Kaczorowski GJ (1984) Characterization of verapamil binding sites in cardiac membrane vesicles. J Biol Chem 259:15013–15016
Glossmann H, Ferry DR (1985) Assay for calcium channels. Methods Enzymol 109:513–551
Glossmann H, Ferry DR, Lübbecke F, Mewes R, Hofman F (1982) Calcium channels: direct identification with radioligand binding studies. Trends Pharmacol Sci 2:431–437
Glossmann H, Linn T, Rombusch M, Ferry DR (1983) Temperature-dependent regulation of d-cis [^{3}H]diltiazem binding to Ca^{2+} channels by 1,4-dihydropyridine channel agonists and antagonists. FEBS Lett 160:226–232
Godfraind T, Wibo M (1985) Subcellular localization of [^{3}H]nitrendipine binding sites in guinea-pig ileal smooth muscle. Br J Pharmacol 85:335–340
Godfraind T, Vanhoutte PM, Govoni S, Paoletti R (1985) Calcium entry blockers and tissue protection. Raven, New York
Goldman S, Geiger W (1984) Rotational barriers of 4-aryl-1,4-dihydropyridines (Ca antagonists). Angew Chem 23:301–302
Goll A, Ferry DR, Glossmann H (1984) Target size and molecular properties of Ca^{2+} channels labelled with [^{3}H]verapamil. Naunyn Schmiedebergs Arch Pharmacol 141:177–186
Gould RJ, Murphy KMM, Snyder SH (1982) [^{3}H]Nitrendipine-labelled calcium channels discriminate inorganic calcium agonists and antagonists. Proc Natl Acad Sci USA 79:3856–3660
Gould RJ, Murphy KMM, Reynolds IJ, Snyder SH (1983) Antischizophrenic drugs of the diphenylbutylpiperidine type as calcium channel antagonists. Proc Natl Acad Sci USA 80:5122–5125
Grover AK, Kwan C-Y, Luchowski E, Daniel EE, Triggle DJ (1984) Subcellular distribution of [^{3}H]nitrendipine binding in smooth muscle. J Biol Chem 259:2223–2226
Grover AK, Kwan C-Y, Daniel EE, Ahmad S, Ramlal T, Oakes P, Triggle DJ (1985) Subcellular distribution of dihydropyridine isothiocyanate binding in guinea-pig ileal smooth muscle. Arch Int Pharmacol 273:74–82
Hamilton SL, Yatani A, Hawkes MJ, Redding K, Brown AM (1985) Atrotoxin: a specific agonist for calcium currents in heart. Science 229:182–184
Hess P, Tsien RW (1984) Mechanism of ion permeation through calcium channels. Nature 309:453–456
Hess P, Lansman JB, Tsien RW (1984) Different modes of Ca channel gating behaviour favored by dihydropyridine Ca agonists and antagonists. Nature 311:538–544
Hille B (1977) Local anesthetics: hydrophilic and hydrophobic pathways for the drug-receptor reaction. J Gen Physiol 69:497–515
Hof RP, Rüegg UT, Hof A, Vogel A (1985) Stereo selectivity at the calcium channel: opposite action of the enantiomers of a 1,4-dihydropyridine. J Cardiovasc Pharmacol 7:689–693
Hondeghem LM, Katzung BG (1984) Antiarrhythmic agents: the modulated receptor mechanism of action of sodium and calcium channel-blocking drugs. Annu Rev Pharmacol Toxicol 24:387–423
Isenberg G, Tamargo J (1985) Effect of imipramine on calcium and potassium currents in isolated bovine ventricular myocytes. Eur J Pharmacol 108:121–131
Jahnsen H, Llinàs R (1984) Ionic basis for the electro responsiveness and oscillatory properties of the guinea-pig thalamic neurones in vitro. J Physiol 349:227–247
Janis RA, Triggle DJ (1983) New developments in Ca^{++} channel antagonists. J Med Chem 26:775–785
Janis RA, Triggle DJ (1984) 1,4-Dihydropyridine Ca^{2+} channel antagonists and activators: a comparison of binding characteristics with pharmacology. Drug Dev Res 4:257–274

Janis RA, Maurer SC, Sarmiento JG, Bolger GT, Triggle DJ (1982) Binding of [^{3}H]nimodipine to cardiac and smooth muscle membranes. Eur J Pharmacol 82:191–194
Janis RA, Sarmiento JG, Maurer SC, Bolger GT, Triggle DJ (1984a) Characteristics of the binding of [^{3}H]nitrendipine to rabbit ventricular membranes: modification by other Ca^{++} channel antagonists and by the Ca^{++} channel agonist Bay K 8644. J Pharmacol Exp Ther 231:8–15
Janis RA, Rampe D, Sarmiento JG, Triggle DJ (1984b) Specific binding of a calcium channel activator, [^{3}H]Bay K 8644, to membranes from cardiac muscle and brain. Biochem Biophys Res Commun 121:317–323
Janis RA, Silver P, Triggle DJ (1987) Adv Drug Res 17, in press
Kerr LM, Yoshikami D (1984) A venom peptide with a novel presynaptic blocking action. Nature 308:282–284
Kleinhaus AL, Prichard JW (1977) A calcium reversible action of barbiturates on the leech Retzius cell. J Pharmacol Exp Ther 201:332–339
Koenig ML, Connor J, Hsaio T, McClure WO (1985) Leptinotarsin-d opens rat brain calcium currents. Soc Neurosci Abs 11:794
Kuthan J, Kurfürst A (1982) Developments in dihydropyridine chemistry. Ind Eng Prod Dev Res 21:191–261
Langs DA, Triggle DJ (1984) Chemical structure and pharmacological activities of Ca^{++} channel antagonists. In: Paton W, Mitchell J, Turner P (eds) Proceedings of the 9th congress of the international union of pharmacology, vol 2. Macmillan, Basingstoke, pp 323–328
Langs DA, Triggle DJ (1985) Conformational features of calcium channel agonist and antagonist analogs of nifedipine. Mol Pharmacol 27:544–548
Llinàs R, Yarom Y (1981) Properties and distribution of ionic conductances generating electroresponsiveness of mammalian inferior olivary neurons in vitro. J Physiol 315:569–584
Lowe DA, Matthews EK, Richardson BP (1981) The calcium antagonistic effects of cyproheptadine on contraction, membrane electrical events and calcium influx in the guinea-pig taenia coli. Br J Pharmacol 74:651–663
Luchowski EM, Yousif F, Triggle DJ, Maurer SC, Sarmiento JG, Janis RA (1984) Effects of metal cations and calmodulin antagonists on [^{3}H]-nitrendipine binding in smooth and cardiac muscle. J Pharmacol Exp Ther 230:607–613
Mannhold R, Steiner R, Haas W, Kaufmann R (1978) Investigations on the structure-activity relationship of verapamil. Naunyn Schmiedebergs Arch Pharmacol 302:217–226
Mannhold R, Zierden P, Bayer R, Rodenkirchen R, Steiner R (1981) The influence of aromatic substitution on the negative inotropic action of verapamil in the isolated cat papillary muscle. Arzneimittelforsch 31:773–780
Mannhold R, Rodenkirchen R, Bayer R (1982) Qualitative and quantitative structure-activity relationships of specific Ca antagonists. Prog Pharmacol 5:25–52
Miller RJ (1984) Toxin probes for voltage sensitive calcium channels. Trends Neurosci 7:309
Miller RJ (1985) How many types of calcium channels exist in neurones? Trends Neurosci 8:45–47
Miller RJ, Freedman SB (1984) Are 1,4-dihydropyridine binding sites functional Ca^{2+} channels? Life Sci 34:1205–1221
Nilius B, Hess P, Lansman JB, Tsien RW (1985) A novel type of cardiac calcium channel in ventricular cells. Nature 316:443–446
Nowycky MC, Fox AP, Tsien RW (1985) Three types of neuronal calcium channel with different calcium agonist sensitivity. Nature 316:440–443
Ohizumi Y, Yasumoto T (1983) Contractile response of the rabbit aorta to maitotoxin, the most potent marine toxin. J Physiol 337:711–721
Olivera BM, Gray WR, Zeikus R, McIntosh JM, Varga J, Rivier J, deSantos V, Crus LJ (1985) Peptide neurotoxins from fish-eating snails. Science 230:1338–1343
Opie LH (1984) Calcium antagonists and cardiovascular disease. Raven, New York

Ptasienski J, McMahon KK, Hosey MM (1985) High and low affinity states of the dihydropyridine and phenylalkylamine receptors on the cardiac calcium channel and their interconversion by divalent cations. Biochem Biophys Res Commun 129:910–917
Reynolds IJ, Gould RJ, Snyder SH (1984) Loperamide: blockade of calcium channels as a mechanism for antidiarrheal effects. J Pharmacol Exp Ther 231:628–632
Rosenberger LB, Triggle DJ (1978) Calcium, calcium translocation and specific calcium antagonists. In: Weiss GB (ed) Calcium in drug action. Plenum, New York, pp 3–32
Sanguinetti MC, Kass RS (1984) Voltage-dependent block of calcium channel current in the calf cardiac Purkinge fiber by dihydropyridine calcium channel antagonists. Circ Res 55:336–348
Sarmiento JG, Janis RA, Colvin RA, Maurer SC, Triggle DJ (1982) Comparison of calcium channels in canine arterial and ventricular myocardium using [^{3}H]nitrendipine binding. Fed Proc 41:1707 abs
Schoemaker H, Langer SZ (1985) [^{3}H]Diltiazem binding to calcium channel antagonist recognition sites in rat cerebral cortex. Eur J Pharmacol 111:273–277
Schramm M, Thomas G, Towart R, Franckowiak G (1983) Novel dihydropyridines with positive inotropic action through activation of Ca^{2+} channels. Nature 303:535–538
Schwartz A, Taira N (1983) Calcium channel blocking drugs: a novel intervention for the treatment of cardiovascular disease. Circ Res 52:No. 2
Schwartz LM, McCleskey EW, Almers W (1985) Dihydropyridine receptors in muscle are voltage-dependent but most are not functional calcium channels. Nature 314:747–751
Seidel W, Meyer H, Born L, Kazda S, Dompert W (1984) Rigid calcium antagonists of the nifedipine type: geometrical requirements for the dihydropyridine receptor. Am Chem Soc 187th national meeting, St. Louis, April 8–14 (Medicinal chem abs 14)
Singh BN, Nademanee K, Baky SH (1983) Calcium antagonists: clinical use in the treatment of arrhythmias. Drugs 25:125–153
Snyder SH, Reynolds IJ (1985) Calcium antagonist drugs. Receptor interactions that clarify therapeutic effects. N Engl J Med 313:995–1002
Spedding MM (1983) Functional interactions of calcium antagonists in K^+-depolarized smooth muscle. Br J Pharmacol 80:485–488
Stone PH, Antman EM (1983) Calcium channel blocking agents in the treatment of cardiovascular disorders. Futura, Kiscoe
Takahashi M, Ohizumi Y, Yasumoto T (1982) Maitotoxin, a Ca^{2+} channel activator candidate. J Biol Chem 257:7287–7289
Takahashi M, Tatsumi M, Ohizumi Y, Yasumoto T (1983) Ca^{2+} channel activating function of maitotoxin, the most potent marine toxin known, in clonal rat pheochromocytoma cells. J Biol Chem 258:10944–10949
Takenaka T, Maeno T (1982) A new vasoconstrictor 1,4-dihydropyridine derivative. Jpn J Pharmacol 32:139P
Triggle DJ (1981) Calcium antagonists: basic chemical and pharmacological aspects. In: Weiss GB (ed) New perspectives on calcium antagonists. Am Physiol Soc, Bethesda, pp 1–18
Triggle DJ (1982) Chemical pharmacology of calcium antagonists. In: Rahwan RG, Witiak DT (eds) Chemical regulation by calcium antagonists. Am Chem Soc, Washington DC, pp 17–137
Triggle DJ (1984) Calcium channels revisited: problems and promises. Trends Pharmacol Sci 5:4–5
Triggle DJ, Janis RA (1984a) Calcium channel antagonists: new perspectives from the radioligand binding assay. In: Back N, Spector S (eds) Modern methods in pharmacology, vol 11. Liss, New York, pp 1–28
Triggle DJ, Janis RA (1984b) The 1,4-dihydropyridine receptor: a regulatory component of the Ca^{2+} channel. J Cardiovasc Pharmacol 6:S949–955
Triggle DJ, Swamy VC (1980) Pharmacology of agents that affect calcium: agonists and antagonists. Chest 78 [Suppl]:174–185
Triggle DJ, Swamy VC (1983) Calcium antagonists. Some chemical pharmacologic aspects. Circ Res 52(I):17–28

Triggle AM, Shefter E, Triggle DJ (1980) Crystal structures of calcium channel antagonists: 2,6-dimethyl-3,5-dicarbomethoxy-4[2-nitro,3-cyano-,4-(dimethylamino)- and 2,3,4,5,6-pentafluorophenyl]-1,4-dihydropyridine. J Med Chem 23:1442–1445
Triggle CR, Agrawal DK, Bolger GT, Daniel EE, Kwan CY, Luchowski EM, Triggle DJ (1982) Calcium channel antagonist binding to isolated vascular smooth muscle membranes. Can J Physiol Pharmacol 60:1738–1741
Yamamura HI, Schoemaker H, Boles RG, Roeske WR (1982) Diltiazem enhancement of [^{3}H]nitrendipine binding to calcium channel associated drug receptor sites in rat brain synaptosomes. Biochem Biophys Res Commun 108:640–646
Yousif FB, Triggle DJ (1985) Functional interactions between organic calcium channel antagonists in smooth muscle. Can J Physiol Pharmacol 63:193–195

CHAPTER 7

The Apamin-Sensitive Ca^{2+}-Dependent K^+ Channel: Molecular Properties, Differentiation, Involvement in Muscle Disease, and Endogeneous Ligands in Mammalian Brain

M. LAZDUNSKI, G. ROMEY, H. SCHMID-ANTOMARCHI, J. F. RENAUD, C. MOURRE, M. HUGUES, and M. FOSSET

A. Apamin, its Structure and its Active Site

Apamin is a neurotoxin extracted from bee venom (HABERMANN 1972). It is a polypeptide of 18 amino acids with two disulfide bridges (Fig. 1). It is the only polypeptide neurotoxin, as far as we know, that crosses the blood–brain barrier. Analysis of the structure–function relationships of this toxin has shown that two of the 18 amino acids in the sequence have particular importance for the action of the toxin; they are Arg-13 and Arg-14 (VINCENT et al. 1975). These two residues seem to be essential elements of the active site of the toxin. Chemical modifications elsewhere in the sequence may decrease the toxicity of the polypeptide, but do not suppress its biologic activity. Cumulative chemical modifications – of the amino group and of the imidazole of His-18 for example – may, however, abolish the activity of the toxin (VINCENT et al. 1975). Solid phase synthesis of apamin and analogs has been carried out (COSAND and MERRIFIELD 1977; GRANIER et al. 1978). This approach has confirmed that the active site of apamin comprises the two residues Arg-13 and Arg-14. The exact three-dimensional structure of the toxin remains unknown. However, recent solution analysis of apamin by NMR techniques has suggested that the toxin is highly ordered with an α-helical core and regions of β-type turns (BYSTROV et al. 1980; WEMMER and KALLENBACH 1983).

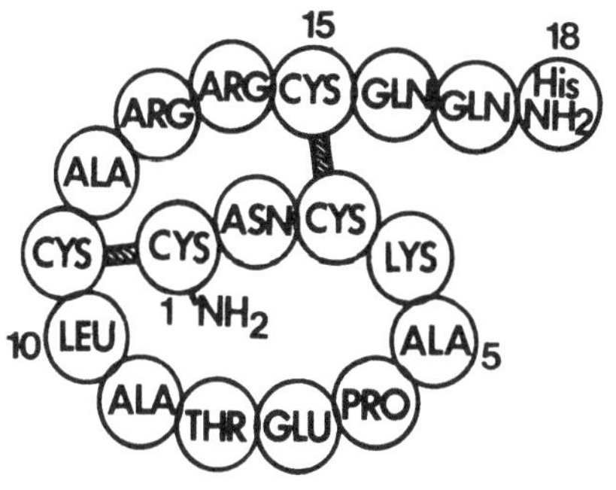

Fig. 1. The structure of apamin from bee venom

B. Apamin Blocks Ca^{2+}-Dependent K^+ Channels

Apamin does not seem to interact with receptors of the classical neurotransmitters (VINCENT et al. 1975). The toxin became a good candidate as a specific blocker of Ca^{2+}-dependent K^+ conductance when K^+ flux studies showed that apamin prevented the rise in potassium permeability in guinea pig hepatocytes and teniae coli (BANKS et al. 1979; MAAS and DEN HERTOG 1979; MAAS et al. 1980) produced by ATP and norepinephrine. A direct demonstration of the specific action of the toxin was obtained using voltage-clamp techniques with both neuroblastoma cells and rat muscle cells in culture (HUGUES et al. 1982a, b). The bee venom toxin suppresses the afterhyperpolarization potential (a.h.p.) and it was shown by voltage clamp that this blockade is due to an inhibition of the Ca^{2+}-dependent K^+ conductance.

C. The Apamin-Sensitive Ca^{2+}-Dependent K^+ Channel is Only One of Several Types of Ca^{2+}-Dependent K^+ Channels

Since the development of patch-clamp techniques (NEHER et al. 1978) single-channel recordings from Ca^{2+}-dependent K^+ channels have been reported from various preparations such as bovine chromaffin cells (MARTY 1981), rat myotubules (BARRETT et al. 1982; METHFESSEL and BOHEIM 1982), bullfrog ganglion cells (ADAMS et al. 1982), and rabbit muscle t-tubule membrane fragments reconstituted into planar lipid bilayers (LATORRE et al. 1982). In each case, the channel shows a large conductance (100–250 pS) and is highly selective for K^+ ions. One of the important problems to solve is whether the a.h.p. which is inhibited by apamin in several cellular systems is due to the same type of Ca^{2+}-dependent K^+ channel that has been identified so successfully using patch-clamp techniques. To solve this problem we have therefore carried out detailed work on rat skeletal muscle cells in culture. These cells have the following properties: (a) they have an a.h.p. which is inhibitable by apamin and which is due to a Ca^{2+}-dependent K^+ channel (HUGUES et al. 1982b); (b) they have ^{125}I-labeled apamin receptors; and (c) they have Ca^{2+}-dependent K^+ channels which can be identified by patch-clamp techniques (BARRETT et al. 1982). Our studies have shown that Ca^{2+}-dependent K^+ channels with a large conductance which are identified by patch-clamp techniques are inhibitable by tetraethylammonium while they are absolutely insensitive to apamin (ROMEY and LAZDUNSKI 1984). Conversely, Ca^{2+}-dependent K^+ channels, which generate the a.h.p., and which can be identified by voltage clamp, are very sensitive to apamin and insensitive to tetraethylammonium (ROMEY and LAZDUNSKI 1984). Apamin-sensitive Ca^{2+}-activated K^+ channels have a low conductance, near 10 pS. These results, which have also been obtained with neuroblastoma cells (HUGUES et al. 1982a) and bullfrog sympathetic ganglion cells (PENNEFATHER et al. 1985), clearly indicate the existence of different types of electrically expressed Ca^{2+}-dependent K^+ channels with a different pharmacology and a different physiologic function. The function of the apamin-sensitive Ca^{2+}-dependent K^+ channel is to generate a.h.p.s; the function of the Ca^{2+}-dependent K^+ channels identified as large conductance channels by patch-clamp techniques may be to prevent a prolonged depolarization of the cell which would favor intracellular Ca^{2+} accumulation. Measurements of single-channel activity of apamin-sensitive, Ca^{2+}-activated K^+ channels seem to indicate that the conductance of this channel is more than one order of magnitude lower than that of the tetraethylammonium-sensitive channel.

D. Biochemical Properties of the Apamin-Binding Component of the Ca^{2+}-Dependent K^+ Conductance

Radiolabeled monoiodoapamin can be prepared at a very high specific radioactivity ($\sim$2000 Ci/mmol) by incorporating iodine on His-18, the COOH terminal residue of the toxin (HABERMANN and FISCHER 1979; HUGUES et al. 1982c). The mean lethal dose LD_{50} of the monoiodo derivative of apamin measured by intracisternal injection in mice is 4.2 ± 1 µg/kg as compared with 2 ± 0.5 µg/kg for the native toxin (HUGUES et al. 1982c). This decrease of effectiveness by a factor of about two after iodination is also observed in a bioassay in which apamin inhibited (and even converted to a contraction) the neurotensin-induced relaxation in segments of guinea pig proximal colon (HUGUES et al. 1982c).

The main properties of the association of ^{125}I-labeled apamin binding with the Ca^{2+}-dependent K^+ channel are the following: (a) the specific binding component is much higher than the nonspecific binding component; (b) there is only one family of binding sites (the Scatchard plot is linear); (c) the affinity of apamin for its receptor is high, 15–25 p*M* for the monoiodo derivative and 10 p*M* for the native toxin; and (d) the binding of apamin to synaptosomes is characterized by a very low binding capacity; synaptosomes only bind 12–13 fmol ^{125}I-labeled apamin per milligram protein and therefore contain 150–300 times less apamin-sensitive channels than tetrodotoxin-sensitive Na^+ channels (HUGUES et al. 1982a).

^{125}I-labeled apamin has now been used to identify apamin-sensitive, Ca^{2+}-dependent K^+ channels in a variety of cell types, including: neuroblastoma cells (HUGUES et al. 1982a), smooth muscle (HUGUES et al. 1982d), skeletal muscle cells in culture (HUGUES et al. 1982b), hepatocytes (COOK et al. 1983), and primary neuronal cells in culture (SEAGAR et al. 1985). The high affinity of the toxin for its receptor and the low number of binding sites have been found in all systems. In neuroblastoma or muscle cells in culture the number of apamin-sensitive, Ca^{2+}-dependent K^+ channels is 5–7 times lower than the number of tetrodotoxin-sensitive Na^+ channels (HUGUES et al. 1982a, b).

The other properties of the apamin–receptor complex which are of interest are the following: (a) the toxin dissociates very slowly from its receptor. Dissociation rate constants of $1.5\text{–}4 \times 10^{-4}$ s^{-1} have been found in the different nerve and muscle systems investigated (HUGUES et al. 1982a, c), corresponding to dissociation half-lives which can be as low as about 60 min; (b) nontoxic derivatives of apamin do not bind to the apamin receptor; (c) the binding of monoiodoapamin to its receptor is sensitive to cations (HUGUES et al. 1982c; HABERMANN and FISCHER 1979). K^+ and Rb^+ at concentrations between 10 µ*M* and 5 m*M* are able to increase the binding of ^{125}I-labeled apamin to its receptor by a factor of about two while each of the many other cations tested (Li^+, Na^+, guanidinium, etc.) completely inhibits ^{125}I-labeled apamin binding in the concentration range 1–100 mM. These results have been interpreted (HUGUES et al. 1982c) as suggesting the existence of two different binding sites for cations. In this hypothesis, site 1 is specific for K^+ and Rb^+ and binds these cations with an affinity corresponding to a dissociation constant of about 500 μM. This site is distinct from the binding site of apamin; its occupancy by K^+ or Rb^+ results in an increased affinity of the receptor for the toxin. Site 2 most probably belongs to the apamin-binding site itself; it recognizes every cation tested with a low affinity. This anionic site which serves as a cation-binding site is probably the site to which Arg-13 and Arg-14, the two crucial residues of the toxin, bind. It is therefore not very surprising that molecules which contain guanidinium groups, like guanidinium itself, amiloride, or neurotensin (which has two contiguous arginine residues like apamin), possess a higher affinity for the apamin receptor than other inorganic cations or positively charged molecules (HUGUES et al. 1982a, c). Among this last class of molecules, quinine and quinidine prevent ^{125}I-labeled apamin binding to the Ca^{2+}-dependent K^+ channel with $K_{0.5}$ values of 100–200 μM (HUGUES et al. 1982b).

E. Apamin as a Tool to Purify the Apamin-Sensitive Ca^{2+}-Dependent K^+ Channel and to Determine its Molecular Weight and its Polypeptide Composition

The apamin receptor is of a polypeptide nature (HUGUES et al. 1982a, c, e) like the receptor of the many neurotoxins which are known to be specific for the fast Na^+ channel (LAZDUNSKI and RENAUD 1982). The problem is to know whether apamin can be used to purify the Ca^{2+}-dependent K^+ channel in the same way that tetrodotoxin, saxitoxin, or scorpion toxins have been used for the purification of the Na^+ channel (BARCHI et al. 1980; HARTSHORNE and CATTERALL 1981; MOORE et al. 1982; BARHANIN et al. 1983a; NORMAN et al. 1983). Properties of apamin in favor of its successful use to purify the Ca^{2+}-dependent K^+ channels are its high affinity for its receptor and the slow rate of dissociation of the complex. However, a clear difficulty for visualizing this purification in an optimistic way is the very small number of Ca^{2+}-dependent K^+ channels in all preparations which have been assayed up till now. The tetrodotoxin-sensitive Na^+ channel has been purified from brain synaptosomes (HARTSHORNE and CATTERALL 1981; BARHANIN et al. 1983a), but, as has been seen before, Na^+ channels are 150–300 times more numerous in this preparation than apamin-sensitive, Ca^{2+}-dependent K^+ channels.

Available data concerning the structure of the apamin-sensitive channel have been obtained directly on membranes. The molecular weight M_r of the apamin receptor (250000) was determined by the radiation inactivation technique which was so successful in establishing M_r of the Na^+ channel (LEVINSON and ELLORY 1973; BARHANIN et al. 1983b).

Affinity labeling of the apamin-sensitive Ca^{2+}-dependent K^+ channel was realized successfully by cross-linking the toxin to its receptor on the channel structure by disuccinimidyl suberate (DSS). To increase the chances of success, the experiment was carried out on synaptic membranes which contain about twice as much apamin-sensitive Ca^{2+} channel as synaptosomes, and at pH 9 where the binding capacity of the receptor is a maximum. The covalent labeling indicates that the apamin receptor in the Ca^{2+}-dependent K^+ channel of synaptic membranes is a single polypeptide chain with $M_r \sim 30\,000$ (HUGUES et al. 1982e; SCHMID-ANTOMARCHI et al. 1984).

It is not known at the present time whether the Ca^{2+}-dependent K^+ channel is made of only one type of polypeptide chain ($M_r = 30\,000$) or whether there are other polypeptide chains which have not been labeled by apamin. If there is only one type of polypeptide chain, the channel is then an oligomeric structure containing 8 ± 2 chains of $M_r = 30\,000$. Other authors have recently identified polypeptide chains of $M_r = 86\,000$ and $59\,000$ on synaptic membranes using other cross-linking derivatives of apamin (SEAGAR et al. 1985).

Knowing that the M_r of the apamin receptor is near 250000 (SCHMID-ANTOMARCHI et al. 1984), one can easily calculate that the specific activity of the pure preparation of apamin-sensitive, Ca^{2+}-dependent K^+ channel will have to be near 4000 pmol per milligram protein. The amount of apamin receptor in synaptosomes being about 12 fmol/mg (HUGUES et al. 1982c), the isolation of the pure neuronal channel will require purification by a factor of 300000–400000. We have successfully solubilized the apamin receptor with complete preservation of the same binding properties it had in the membrane. However, while classical purification steps have been sufficient to purify the tetrodotoxin-sensitive Na^+ channel (HARTSHORNE and CATTERALL 1981; BARHANIN et al. 1983a), it will certainly be necessary to use affinity columns containing either apamin or anti-Ca^{2+}-dependent K^+ channel antibodies to isolate this channel successfully in the pure form.

F. PC12 Pheochromocytoma Cells Hyperproduce the Apamin Receptor and Permit an Analysis of the Internal Ca^{2+} Concentration Dependence of the Apamin-Sensitive Ca^{2+} Channel

Undifferentiated PC12 cells produce high levels of apamin receptors (measured with ^{125}I-labeled apamin) after 7 days in culture. These levels are at least 50 times higher than those found in other cellular types which are also known to have apamin receptors and apamin-sensitive, Ca^{2+}-activated K^+ channels in their membranes. Treatment of undifferentiated PC12 cells with nerve growth factor maintains these cells in a state with a low level of apamin receptors (tenfold less after 7 days of culture). Ca^{2+} injection into PC12 cells with the calcium ionophore A23187 has been used to monitor the activity of the Ca^{2+}-activated K^+ channel following $^{86}Rb^+$ efflux. A large component (60%) of this Ca^{2+}-activated $^{86}Rb^+$ efflux is inhibited by apamin. Half-maximum inhibition by apamin of both $^{86}Rb^+$ efflux and ^{125}I-labeled apamin binding was observed at 240 pM apamin. Another component (40%) of $^{86}Rb^+$ efflux is due to another type of Ca^{2+}-activated K^+ channel which is resistant to apamin and sensitive to tetraethylammonium.

In normal physiologic conditions, the activity of Ca^{2+}-dependent K^+ channels is triggered by an increase of the internal Ca^{2+} concentration owing to the opening of Ca^{2+} channels which are present in the plasma membrane and/or to the release of Ca^{2+} from internal stores. It has been shown in this work that the voltage-dependent Ca^{2+} channel activator Bay K 8644 can be used instead of the Ca^{2+} ionophore A23187 to activate an apamin-sensitive $^{86}Rb^+$ efflux. This activation is linked to a measurable increase of $[Ca^{2+}]_i$ and is abolished by the Ca^{2+} channel inhibitor nitrendipine. The internal Ca^{2+} concentration can be modulated in the presence of Bay K 8644 by changing the external Ca^{2+} concentration. This procedure has offered a mean to determine for the first time the internal Ca^{2+} concentration dependence of the apamin-sensitive, Ca^{2+}-activated K^+ channel. This channel is essentially inactive, or has an activity which is too low to be detected, in normal concentrations of internal Ca^{2+} in standard culture conditions, i.e., 109 ± 17 n*M*. Full activation of the apamin-sensitive K^+ channel is observed near 320 n*M*. The resting membrane potential under these conditions is near -55 ± 5 mV. These results indicate that the apamin-sensitive channel is activated at very low $[Ca^{2+}]_i$ and that the response of the channel activity to variations of $[Ca^{2+}]_i$ is very cooperative. Whereas very cooperative responses with respect to $[Ca^{2+}]_i$ have also been previously observed for the tetraethylammonium-sensitive, Ca^{2+}-activated K^+ channel (BARRETT et al. 1982), this pharmacologically distinct type of channel responds to variations of internal Ca^{2+} at much higher values of $[Ca^{2+}]_i$. For example, at a membrane potential of -50 mV in skeletal muscle cells the tetraethylammonium-sensitive channels are half-activated at $[Ca^{2+}]_i > 100$ μ*M* (BARRETT et al. 1982). This different sensitivity to $[Ca^{2+}]_i$ of apamin-sensitive and tetraethylammonium-sensitive channels had already been suggested for skeletal muscle cells in culture (ROMEY and LAZDUNSKI 1984).

G. Autoradiographic Localization of Apamin-Sensitive Ca^{2+}-Dependent K^+ Channels in Rat Brain

The localization of the apamin receptor was studied in rat brain by an in vitro autoradiographic technique (MOURRE et al. 1984, 1986). Radiolabeled monoiodoapamin binds specifically to rat brain sections with a high affinity ($K_d = 25$ p*M*) to a single class of sites. Autoradiograms demonstrated a very heterogeneous distribution of the apamin receptor throughout the brain. Very high grain densities were localized on the habenula, lateral septum, supraoptic and supraschiasmatic nuclei. Areas containing high levels of apamin-binding sites included anterior olfactory nucleus, stratum oriens of hippocampus, pontine nuclei, and granular layer of the cerebellar cortex and inferior olive. The thalamus, some nuclei of the hypothalamus, hippocampus, tegmental area, red and oculomotor nuclei, vestibular nuclei, and superior olive, among others, presented intermediate grain densities. In the other main areas, in particular basal ganglia and raphe, low to very low levels of apamin-binding sites have been observed.

H. Developmental Properties of the Ca^{2+}-Dependent K^+ Channel in Mammalian Skeletal Muscle and the All-or-None Role of Innervation

The long-lasting a.h.p. which follows the action potential in rat myotubules differentiated in culture is due to Ca^{2+}-activated K^+ channels. These channels have the property of being specifically blocked by apamin at low concentrations. Apamin has been used (SCHMID-ANTOMARCHI et al. 1985) to analyze by electrophysiologic and biochemical techniques the role of innervation in the expression of these important channels. The main results are as follows:

1. Long-lasting a.h.p.s which follow the action potential in rat myotubules in culture disappear when myotubes are cocultured with nerve cells from the spinal cord under the conditions of in vitro innervation.
2. Extensor digitorum longus muscles from adult rats have action potentials which are not followed by a.h.p.s, but a.h.p.s are systematically recorded after muscle denervation and they are blocked by apamin.
3. Specific ^{125}I-labeled apamin binding is undetectable in innervated muscle fibers, but it becomes detectable 2–4 days after muscle denervation and is maximal 10 days after denervation.
4. Apamin receptors detected with ^{125}I-labeled apamin are present at fetal stages with biochemical characteristics identical to those found in myotubes in culture. The receptor number decreases as maturation proceeds and ^{125}I-labeled apamin receptors completely disappear after the first week of postnatal life in parallel with the disappearance of multi-innervation.

All these results taken together strongly suggest an all-or-none effect of innervation of the expression of apamin-sensitive, Ca^{2+}-activated K^+ channels. Apamin-sensitive Ca^{2+} channels and apamin receptors are present in noninnervated mammalian muscle and disappear after final innervation.

I. Expression of the Apamin Receptor in Muscles of Patients with Myotonic Muscular Dystrophy

Myotonic muscular dystrophy, or Steinert's disease, is a dominantly inherited disease of muscle which occurs with a frequency of between 1 in 18000 and 1 in 7500 people (GRIMM 1975; TODOROV et al. 1970). One of the prominent clinical manifestations is mucle stiffness and difficulty in relaxation of muscles after voluntary contractions. Electrophysiologic signs of myotonia include increased excitability with a tendency of fire trains of repetitive action potentials in response to direct electrical and mechanical stimulation. Most experimental and clinical data suggest that myotonic muscular dystrophy arises from genetically induced alterations of the muscle membrane (RUDEL and LEHMANN-HORN 1985). It has been shown (RENAUD et al. 1986) that muscle membranes of patients with myotonic muscular dystrophy contain the receptor for apamin. The apamin receptor is completely absent from normal human muscle and muscles of patients with spinal anterior horn disorders.

The most important property of myotonic dystrophic muscle fibers is their tendency to fire repetitive action potentials (RUDEL and LEHMANN-HORN 1985; LIPICKY 1977; TASMOUSH et al. 1983) which is related to electromyographic observations indicating that myotonic muscular dystrophy is characterized by myotonic runs (RUDEL and LEHMANN-HORN 1985), i.e., by repetitive bursts of activity. Another electrophysiologic feature of myotonic dystrophic muscle membranes is that they have less negative resting potential (MCCOMAS and MROZEK 1968; GRUENER 1977; GRUENER et al. 1975, 1979; MERICKEL et al. 1981). One way to generate repetitive action potentials is the combination of a decreased resting potential with a value close to the threshold for the activation of Na^+ channel and the presence of apamin-sensitive, Ca^{2+}-activated K^+ channels, creating an a.h.p. following the action potential which permits the transient reactivation of the Na^+ channel after activation and inactivation during the action potential. It is probably the combination of such effects that creates repetitive bursts of activity responsible for spontaneous contractions in noninnervated rat muscle cells in culture (BARRETT et al. 1981). The same combination seems to be present in myotonic muscular dystrophy.

J. An Endogenous Apamin-Like Factor Modulating Ca^{2+}-Dependent K^+ Channel Activity Exists in Mammalian Brain

An apamin-like factor has been purified (FOSSET et al. 1984) from pig brain after acidic extraction of the tissue and several purification steps on sulfopropyl Sephadex C-25 and C-18 reverse-phase high pressure liquid chromatography (FOSSET et al. 1984). The apamin-like activity was followed during the purification procedure by two biochemical assays: (a) a radioreceptor assay (HUGUES et al. 1982c) and a radioimmunoassay with anti-apamin antibodies (SCHWEITZ and LAZDUNSKI 1984) and two physiologic assays; and (b) measurement of the contractile activity of guinea pig teniae coli (HUGUES et al. 1982d) and electrophysiologic measurement of the ability of fractions to block the a.h.p. in rat skeletal muscle cells (HUGUES et al. 1982b).

Purification of an apamin-like factor has been achieved. Its properties are the following:

1. The factor is able to prevent ^{125}I-labeled apamin attaching to its binding site on rat brain synaptosomes.
2. It antagonizes ^{125}I-labeled apamin recognition by anti-apamin antibodies.
3. It contracts the intestinal smooth muscle previously relaxed by epinephrine.
4. It selectively blocks the hyperpolarization that follows the action potential on rat skeletal myotube in culture.
5. It is sensitive to the action of trypsin and insensitive to chymotrypsin digestion.

All these properties are those of apamin itself. These results strongly suggest the presence in pig brain of an endogenous equivalent of apamin. Knowing that a purification procedure gives an activity equivalent to 1.5 ± 0.5 pmol apamin per pig brain, large-scale purification will be necessary to obtain enough apamin-like factor to determine its sequence. Inactive precursors of the apamin-like factor could also be present in the pig brain; the structure of the native precursor will be known once the cloning of the gene encoding for this precursor is achieved.

Acknowledgments. This work was supported by the Associations Françaises de Lutte contre les Myopathies, the Centre National de la Recherche Scientifique [ATP 381, AIP 06931(1075)], the Fondation pour la Recherche Médicale, the Ministère de l'Industrie et de la Recherche (grant 83.C.0918), and the Mutuelle Générale de l'Education Nationale. We thank C. Roulinat-Bettelheim for expert technical assistance.

References

Adams PR, Constanti A, Brown DA, Clark RB (1982) Fast voltage-sensitive potassium current in vertebrate sympathetic neurons. Nature 296:746–749

Banks BEC, Brown C, Burgess GM, Burnstock G, Claret M, Cocks TM, Jenkinson DH (1979) Apamin blocks certain neurotransmitter-induced increases in potassium permeability. Nature 282:415–417

Barchi RL, Cohen SA, Murphy LE (1980) Purification from rat sarcolemma of the saxitoxin binding component of the excitable membrane sodium channel. Proc Natl Acad Sci USA 77:1306–1310

Barhanin J, Pauron D, Lombet A, Norman RI, Vijverberg HPM, Giglio JR, Lazdunski M (1983a) Electrophysiological characterization, solubilization and purification of the *Tityus* γ toxin receptor associated with the gating component of the Na^+ channel from rat brain. EMBO J 2:915–920

Barhanin J, Schmid A, Lombet A, Wheeler KP, Lazdunski M (1983b) Molecular size of different neurotoxin receptors on the voltage-sensitive Na^+ channel. J Biol Chem 258:700–702

Barrett JN, Barrett EF, Dribbin LB (1981) Calcium-dependent slow potassium conductance in rat skeletal myotubules. Dev Biol 82:258–266

Barrett JN, Magleby KL, Pallotta BS (1982) Properties of single calcium-activated potassium channels in cultured rat muscle. J Physiol (London) 331:211–230

Bystrov VF, Okhanov VV, Miroshnikov AI, Ovchinnikov YA (1980) Solution spatial structure of apamin as derived from NMR study. FEBS Lett 119:113–117

Cook NS, Haylett DN, Strong P (1983) High affinity binding of ^{125}I-monoiodo apamin to isolated guinea pig hepatocytes. FEBS Lett 152:265–269

Cosand WL, Merrifield RB (1977) Concept of internal structural controls for evaluation of inactive synthetic peptide analog: synthesis of (ORN 13, 14) apamin and its guanidination to an apamin derivative with full neurotoxic activity. Proc Natl Acad Sci USA 74:2771–2775

Fosset M, Schmid-Antomarchi H, Hugues M, Romey G, Lazdunski M (1984) The presence in pig brain of an endogenous equivalent of apamin, the bee venom peptide which specifically blocks Ca^{2+}-dependent K^+ channels. Proc Natl Acad Sci USA 81:7228–7232

Granier C, Pedroso Muller E, Van Rietschoten J (1978) Use of synthetic analogs for a study on the structure-activity relationship of apamin. Eur J Biochem 82:293–299

Grimm T (1975) The ages of onset and the age of death in patients with dystrophia myotonia. J Hum Genet 23:301–308

Gruener R (1977) In vitro membrane excitability of diseased human muscles. In: Rowland LP (ed) Pathogenesis of human muscular dystrophy. Excerpta Medica, Amsterdam, p 242

Gruener R, Stern LZ, Payne C, Hannapel L (1975) Hyperthyroid myopathy intracellular electrophysiological measurements in biopsied human intercostal muscle. J Neuro Sci 24:339–349

Gruener R, Stern LZ, Markovitz D, Gerdes C (1979) Electrophysiologic properties of intercostal muscle fibers in human neuromuscular diseases. Muscle Nerve 2:165–172
Habermann E (1972) Bee and wasp venom. Science 177:314–322
Habermann E, Fischer K (1979) Bee venom neurotoxin (apamin): iodine labelling and characterization of binding sites. Eur J Biochem 94:355–364
Hartshorne RP, Catterall WA (1981) Purification of the saxitoxin receptor of the sodium channel from rat brain. Proc Natl Acad Sci USA 78:4620–4624
Hugues M, Romey G, Duval D, Vincent JP, Lazdunski M (1982a) Apamin as a selective blocker of the calcium-dependent potassium channel in neuroblastoma cells: voltage-clamp and biochemical characterization of the toxin receptor. Proc Natl Acad Sci USA 79:1308–1312
Hugues M, Schmid H, Romey G, Duval D, Frelin C, Lazdunski M (1982b) The calcium-dependent slow potassium conductance in cultured rat muscle cells: characterization with apamin. EMBO J 9:1039–1042
Hugues M, Duval D, Kitabgi P, Lazdunski M, Vincent JP (1982c) Preparation of a pure monoiodo derivative of the bee venom neurotoxin apamin and its binding properties to rat brain synaptosomes. J Biol Chem 257:2762–2769
Hugues M, Duval D, Schmid H, Kitabgi P, Lazdunski M (1982d) Specific binding and pharmacological interactions of apamin, the neurotoxin from bee venom, with guinea-pig colon. Life Sci 31:437–443
Hugues M, Schmid H, Lazdunski M (1982e) Identification of a protein component of the calcium-dependent potassium channel by affinity labelling with apamin. Biochem Biophys Res Commun 107:1577–1582
Latorre R, Vergara C, Hidalgo C (1982) Reconstitution in planar lipid bilayers of a calcium-dependent potassium channel from transverse tubule membranes isolated from rabbit skeletal muscle. Proc Natl Acad Sci USA 79:805–809
Lazdunski M, Renaud JF (1982) The action of cardiotoxins on cardiac plasma membranes. Annu Rev Physiol 44:463–473
Levinson SR, Ellory JC (1973) Molecular size of the tetrodotoxin binding site estimated by irradiation inactivation. Nature 245:122–123
Lipicky RJ (1977) Studies in human myotonic dystrophy. In: Rowland LP (ed) Pathogenesis of human muscular dystrophy. Excerpta Medica, Amsterdam, p 729
Maas AD, Den Hertog A (1979) The effect of apamin on the smooth muscle cells of the guinea-pig taenia coli. Eur J Pharmacol 58:151–156
Maas AD, Den Hertog A, Ras R, Van Den Akker J (1980) The action of apamin on guinea-pig taenia caeci. Eur J Pharmacol 67:265–274
Marty A (1981) Calcium-dependent potassium channels with large unitary conductance in chromaffin cell membranes. Nature 291:497–500
McComas AJ, Mrozek K (1968) The electrical properties of muscle fiber membranes in dystrophic myotonia and myotonia congenita. J Neurol Neurosurg Psychiatry 31:441–447
Merickel M, Gray R, Chauvin P, Appel S (1981) Cultured muscle from myotonic muscular dystrophy patients: altered membrane electrical properties. Proc Natl Acad Sci USA 78:648–652
Methfessel C, Boheim G (1982) The gating of single calcium-dependent potassium channel is described by an activation blockade mechanism. Biophys Struct Mech 9:35–60
Moore HPM, Fritz LC, Raftery MA, Brokes JP (1982) Isolation and characterization of a monoclonal antibody against the saxitoxin-binding component from the electric organ of the eel *Electrophorus electricus*. Proc Natl Acad Sci USA 79:1673–1677
Mourre C, Schmid-Antomarchi H, Hugues M, Lazdunski M (1984) Autoradiographic localization of apamin-sensitive Ca^{2+}-dependent K^+ channels in rat brain. Eur J Pharmacol 100:135–136
Mourre C, Hugues M, Lazdunski M (1986) Quantitative autoradiographic mapping in rat brain of the receptor of apamin, a polypeptide toxin specific for one class of Ca^{2+}-dependent K^+ channel. Brain Res 382:239–249
Neher E, Sakmann B, Steinbach JH (1978) The extracellular patch-clamp: a method for resolving current through individual open channels in biological membranes. Pflugers Arch 375:219–228

Norman RI, Schmid A, Lombet A, Barhanin J, Lazdunski M (1983) Purification of binding protein for *Tityus* γ toxin identified with the gating component of the voltage-sensitive Na^+ channel. Proc Natl Acad Sci USA 80:4164–4168
Pennefather P, Lancaster B, Adams PR, Nicoll RA (1985) Two distinct Ca-dependent K currents in bullfrog sympathetic ganglion cells. Proc Natl Acad Sci USA 82:3040–3044
Renaud JF, Desnuelle C, Schmid-Antomarchi H, Hugues M, Serratrice G, Lazdunski M (1986) Expression of the apamin receptor in muscles of patients with myotonic muscular dystrophy. Nature 319:678–680
Romey G, Lazdunski M (1984) The coexistence in rat muscle cells of two distinct classes of Ca^{2+}-dependent K^+ channels with different pharmacological properties and different physiological functions. Biochem Biophys Res Commun 118:669–674
Rudel R, Lehmann-Horn F (1985) Membrane changes in cells from myotonia patients. Physiol Rev 65:310–356
Schmid-Antomarchi H, Hugues M, Norman RI, Ellory JC, Borsotto M, Lazdunski M (1984) Molecular properties of the apamin-sensitive Ca^{2+}-dependent K^+ channel: radiation-inactivation, affinity labelling and solubilization. Eur J Biochem 142:1–6
Schmid-Antomarchi H, Renaud JF, Romey G, Hugues M, Schmid A, Lazdunski M (1985) The all-or-none role of innervation in the expression of the apamin-sensitive Ca^{2+}-activated K^+ channel in mammalian skeletal muscle. Proc Natl Acad Sci USA 82:2188–2195
Schweitz H, Lazdunski M (1984) A microradioimmunoassay for apamin. Toxicon 22:985–988
Seagar MJ, Labbé-Julié C, Granier C, Van Rietshoten J, Couraud F (1985) Photoaffinity labeling of components of the apamin-sensitive K^+ channel in neuronal membranes. J Biol Chem 260:3895–3898
Tashmoush AJ, Askanas V, Nelson PG, Engel WK (1983) Electrophysiological properties of aneurally cultured muscle from patients with myotonic muscular atrophy. Neurology 33:311–316
Todorov A, Jequier M, Klein D, Morton NE (1970) Analyse de la ségrégation dans la dystrophie myotonique. J Hum Genet 18:387–406
Vincent JP, Schweitz H, Lazdunski M (1975) Structure-function relationships and site of action of apamin. A neurotoxic polypeptide of bee venom with an action on the central nervous system. Biochemistry 14:2521–2525
Wemmer D, Kallenbach NR (1983) Structure of apamin in solution, 2 dimensional NMR study. Biochemistry 22:1901–1906

CHAPTER 8

Drug Effects on Plasma Membrane Calcium Transport

F. F. VINCENZI and T. R. HINDS

A. Introduction

In this chapter we have concentrated mainly on the question of the effects of drugs on the plasma membrane Ca^{2+} pump. This is not an exhaustive review. We have reviewed mainly the literature files in our laboratory, concentrating on papers published in 1983, 1984, and 1985. There are a number of reviews in related areas, including calmodulin (CaM) (TOMLINSON et al. 1984) and its drug interaction (VINCENZI 1981; WEISS et al. 1985; ROUFOGALIS et al. 1983; ROUFOGALIS 1985), regulation of calcium in cells (CARAFOLI 1985), the effects of drugs on membrane fluidity (GOLDSTEIN 1984), and reviews on plasma membrane (PM) Ca^{2+} pump ATPase (ITANO and PENNISTON 1985) and Ca^{2+} transport (CARAFOLI 1984a, b, c, d, 1985; SCHATZMANN 1983, 1985). Because relatively more is known about the Ca^{2+} pump in human red blood cells (RBCs) than in other types and species of cells, most of our concern has been with this type of cell. Ca^{2+} transport in the heart has been reviewed by LANGER (1984) and CARAFOLI (1985).

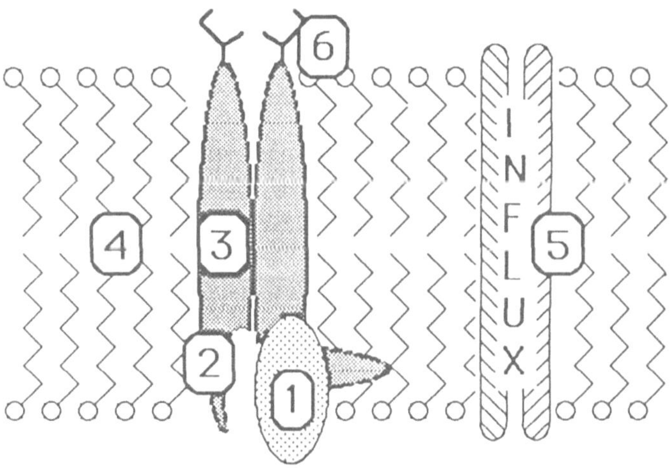

Fig. 1. Schematic diagram of the Ca^{2+} pump ATPase of the human RBC membrane. Six different, and somewhat arbitrarily designated, potential sites of action are indicated by the numbers. These sites or regions include: (1) calmodulin (CaM); (2) the CaM-binding site on the ATPase; (3) the ATPase itself, in particular its catalytic and other (non-CaM) regulatory regions; (4) the phospholipid environment of the membrane; (5) influx pathways for calcium; and (6) surface receptors which couple directly or indirectly to the pump ATPase

As a starting point the reader is referred to Fig. 1. In this figure we have designated six sites at which drug modification of the plasma membrane (PM) Ca^{2+} pump ATPase could occur. The figure is not meant to be a definitive representation of the state of affairs of the PM Ca^{2+} pump, but rather a way to organize both the writers' and readers' thoughts concerning the drugs to be discussed and their potential sites and mechanisms of action. Figure 1 is a schematic diagram of the Ca^{2+} pump ATPase of the human RBC membrane. Five different potential sites of action are indicated by the numbers. These somewhat arbitrarily designated sites include CaM, the CaM-binding site on the ATPase, the ATPase including its catalytic and non-CaM regulatory regions, the phospholipid environment of the membrane, influx pathway or pathways for calcium and surface receptors. The pump ATPase is portrayed as a dimer with some glycosylation of the catalytic portions and with one of two CaM-binding sites of the dimer occupied with CaM, or CaM(Ca^{2+}). Although a lack of glycosylation of the isolated Ca^{2+} pump ATPase has been claimed (GRAF et al. 1982), the nature of the pump as a transmembrane protein as well as some other considerations have led us to assume that the ATPase is glycosylated until definitely shown otherwise. Contamination by glycophorin was suggested to account for sugar residues in the isolated Ca pump ATPase, but glycophorin contains 12% by weight of galactosamine (WATKINS 1974) and galactosamine was not found when the isolated Ca pump ATPase was analyzed for sugars (PENNISTON et al. 1980). The dimeric structure is inferred from the azidocalmodulin cross-linking experiments of HINDS and ANDREASEN (1981), and the inactivation studies of MINOCHERHOMJEE et al. (1983) and CAVIERES (1984). Whether the Ca^{2+} pump contains CaM as a subunit in the intact RBC in vivo (i.e., at low internal Ca^{2+}) remains unresolved (VINCENZI and HINDS 1980). Evidence with anti-CaM drugs applied to intact, Ca^{2+}-loaded RBCs is in agreement, but does not prove this view (FODER et al. 1984; HINDS and VINCENZI 1986).

We will consider selected examples of agents which influence the Ca^{2+} pump ATPase by interactions at various sites. Of course, most agents are not selective and affect the ATPase via multiple interactions. Not all of these interactions may be mentioned even if known. Some agents are discussed under a certain site without any justification other than our intuition of where these agents probably act. At the outset one major limitation must be emphasized. In spite of several inhibitors such as vanadate (VARECKA and CARAFOLI 1982), it must be noted that no selective and potent inhibitor of the PM Ca^{2+} pump which works on intact cells has yet been found. Much of the tremendous progress in the field of Na^+–K^+ transport was fostered by the truly specific and potent inhibitors of the digitalis type. If and when a truly potent and specific inhibitor of the Ca^{2+} pump is found then we anticipate significant advancement in our understanding of the pump, and especially its relationships to cellular physiology and pathology.

B. Calmodulin

A wide variety of drugs bind to CaM, in the presence of Ca^{2+}. Beginning with the classical works of WOLFF and BROSTROM (1976) and LEVIN and WEISS (1977) it became apparent that phenothiazines, such as trifluoperazine (TFP), selectively antagonized certain CaM-dependent enzymes. Now these preliminary insights have been expanded to show that a wide range of amphipathic cations (PROZIALECK 1984; RAINTEAU et al. 1984), including calcium entry blockers (BOSTROM et al. 1981; AGRE et al. 1984; SCHLONDORFF and SATRIANO 1985), antiarrhythmics (LEVINE and HOLLIER 1983), beta-blockers (MELTZER and KASSER 1983b), tetracyclines (SCHLONDORFF and SATRIANO 1985), and peptides (COMTE et al. 1983; BARNETTE et al. 1983; COX et al. 1985; STEINER et al. 1986) can bind to and antagonize the effects of CaM in many systems, including the RBC Ca^{2+} pump ATPase (VINCENZI 1981; WEISS et al. 1985). So-called anti-CaM agents exert a variety of effects including inhibition of Ca^{2+} transport in the endoplasmic reticulum (WULFROTH and PETZELT 1985) and of the CaM-free PM Ca^{2+} pump ATPase (VINCENZI et al. 1982). Most anti-CaM agents are thought to prevent the binding of the CaM–Ca_n^{2+} complex to its various effectors. ROUFOGALIS (1982) noted that anti-CaM drugs do not antagonize Ca^{2+} binding to CaM. They could bind to the CaM–Ca_n^{2+} complex and/or to the enzyme in question. It was suggested that evidence favors the interpretation that most such agents interact with CaM and not the enzyme. The findings of ROUFOGALIS et al. (1983) and others fit with the notion that phenothiazines and antipsychotics bind to CaM, probably at two hydrophobic sites which are Ca^{2+} dependent (LAPORTE et al. 1980). While that view is valid, direct effects of anti-CaM agents have also been reported (VINCENZI et al. 1982). And while recognizing explicitly the severe limitations of an agent with so many nonspecific effects, one can nevertheless obtain evidence, using TFP, which is consistent with inhibition of the PM Ca^{2+} pump in intact RBCs loaded with Ca^{2+} (FODER et al. 1984; PLISHKER 1984; HINDS and VINCENZI 1986).

GIETZEN et al. (1983) found that compound 48/80 (a mixture of polycationic compounds) exhibited very selective anti-CaM activity against activation of Ca^{2+} pump ATPase in isolated RBC membranes. The lack of inhibition of other RBC membrane enzymes is presumably because compound 48/80 does not enter the lipid part of the membrane. This makes the agent much more selectively anti-CaM and useful in vitro, but precludes its usefulness as an inhibitor of CaM in intact cells. We have used compound 48/80 as an antagonist of CaM in saponin lysates of RBCs and isolated RBC membranes in order to define the basal Ca^{2+} pump ATPase activity. Compound 48/80 was less potent and somewhat less specific than reported by Gietzen et al. (DIJULIO and VINCENZI 1986).

The suggestion that one might be able to distinguish between CaM receptors (VINCENZI 1981) was furthered by INAGAKI and HIDAKA (1984), who divided CaM antagonists into CaM1 antagonists and CaM2 antagonists. CaM1 antagonists included W7, TFP, chlorpromazine, fluphenaxazine, dibucaine, and lidocaine. CaM2 antagonists included prenylamine and butaclamol, and possibly amitriptyline. The CaM1 antagonists suppress the increase in fluorescence of the hydrophobic probe *N*-phenyl-1-naphthylamine (NPN). The interpretation was that CaM1 antagonists occupy the hydrophobic region of CaM. CaM2 antagonists increase the fluorescence. CaM2 antagonists produce conformational changes which weaken the active hydrophobic domain by exposing the inactive hydrophobic domain. So CaM1 and CaM2 antagonists appear to interact with CaM at a different site and/or by a different means. INAGAKI et al. (1985) found Ca^{2+}-dependent interaction of the ionophore, A23187, with CaM. Curiously, A23187 did not suppress CaM-dependent myosin light chain kinase (MLCK) or phosphodiesterase (PDE) activity, perhaps because A23187 is not an amphipathic cation, but rather an amphipathic anion.

MINOCHERHOMJEE and ROUFOGALIS (1984) found PDE inhibition by nifedipine and related compounds. Inhibitory potencies did not correlate with potencies to inhibit Ca^{2+} channels. It was concluded that pharmacologic actions of nifedipine-related calcium entry blockers are unlikely to be due to intracellular inhibition of CaM at concentrations normally used to block Ca^{2+} entry. These and other data support the notion that, in addition to binding CaM on Ca^{2+} pump ATPase, such drugs bind to other sites. Cyclosporin was found to bind to CaM and to antagonize CaM activation of PDE (COLOMBANI et al. 1985). In our hands, by contrast, cyclosporin was ineffective in antagonizing CaM activation of Ca^{2+} pump ATPase, even at 10 μM (T. R. HINDS and F. F. VINCENZI 1986, unpublished work). Whether this apparent difference demonstrates a difference in CaM activation of these two enzymes remains to be determined. A difference may also be inferred from the work of GUERINI et al. (1984).

NELSON et al. (1983) advanced the idea that CaM is a potential regulator of RBC shape. The idea is based on the observation that, almost without exception, anti-CaM drugs (which are amphipathic cations) are all "cup formers," whereas "crenators" of RBCs are neutral or anionic compounds (and not CaM antagonists). It was argued that CaM antagonists favor cup morphology by preserving ATP (possibly by inhibiting an ATP-utilizing enzyme). And it was argued that drugs bind to CaM–Ca_n^{2+} and prevent Ca^{2+}-release, keeping Ca^{2+} low. The fact that the affinity of CaM for spectrin is some 1000-fold less than for the Ca^{2+} pump ATPase was not emphasized.

ROUFOGALIS et al. (1984) demonstrated that an EDTA extract of RBC membranes contains CaM and a protein to which CaM binds in a Ca^{2+}-dependent manner. This is a 56 kdalton protein. Its binding probably tends to obscure the negative charges on CaM. Thus, the complex sticks poorly to DEAE. On the other hand, CaM, even when bound to the complex, can still activate the Ca^{2+} pump ATPase. The fact that CaM is retained on a phenothiazine column in the presence of Ca^{2+} suggests that it still exhibits the Ca^{2+}-dependent hydrophobicity (LAPORTE et al. 1980; RAINTEAU et al. 1984) which presumably drives its interaction with the Ca^{2+} pump ATPase. ROUFOGALIS et al. (1984) suggested that this CaM-binding protein may provide CaM in proximity to certain other CaM-dependent proteins.

CaM did not stimulate the Ca^{2+}-dependent ATPase in isolated platelet membranes (STEINER and LUSCHER 1985) and there is doubt whether the activity in question represents the Ca^{2+} pump ATPase. Naturally, our bias is toward the view that all cells contain a PM Ca^{2+} pump. Failure to find a Ca^{2+} pump ATPase in isolated PMs, may be more methodological than biological; similar to the problem of finding an active Ca^{2+} pump ATPase in dog RBC membranes (SCHMIDT et al. 1985; HINDS and VINCENZI 1986). It may be noted that work from other laboratories suggests that platelets do contain an active (and important) Ca^{2+} pump ATPase, which may be relatively defective in hypertension (RESINK et al. 1986).

C. Calmodulin-Binding Site

Blumenthal et al. (1985) isolated a peptide fragment from MLCK that appeared to be from the CaM-binding region of the enzyme. A cyanogen bromide peptide from this fragment binds CaM in a 1 : 1 ratio. This fragment can also inhibit CaM-activated MLCK activity by binding with CaM. This fragment was tested for its general usefulness as an anti-CaM peptide using the Ca^{2+},Mg^{2+}-ATPase of the RBC and other CaM-dependent enzymes. The peptide was capable of inhibiting CaM activation of the Ca^{2+},Mg^{2+}-ATPase without affecting the basal activity (D.K. Blumenthal and T.R. Hinds 1985, unpublished work). It appears that other peptides that bind to and therefore inhibit CaM do so in part because of their structure which consists of basic residue clusters with adjacent hydrophobic residues that have α-helical structure (Steiner et al. 1986).

In addition to binding to CaM, some, or even most, anti-CaM drugs probably act at the CaM-binding site and/or directly on the catalytic part of the ATPase. Vincenzi et al. (1982) found that a variety of presumably anti-CaM drugs inhibited the activity of the isolated, CaM-free, Ca^{2+} pump ATPase in concentrations similar to those which antagonized activation of the isolated enzyme by added CaM. The simplest interpretation is that the site on the Ca^{2+} pump ATPase to which CaM binds is complementary and hydrophobic, like the site on CaM to which the drugs are known to bind (LaPorte et al. 1980). Binding of drugs to the CaM-binding site on the ATPase would prevent activation of the enzyme by CaM and may also result in direct inhibition of the enzyme.

Guerini et al. (1984) tested certain tryptic fragments of CaM. Some activate the Ca^{2+} pump ATPase, but do not activate PDE. The COOH terminal half of CaM, especially the Ca^{2+}-binding region III was necessary for interaction with the ATPase. Oxidation of methionines 78–148 blocked the ability of the CaM fragment to bind to Ca^{2+} pump ATPase. These results, and those from other laboratories, may be interpreted to suggest that CaM interacts differently with different effectors. This could potentially be exploited, as suggested previously (Vincenzi 1981). Different interactions of CaM with its effectors, including the Ca^{2+} pump ATPase may be the other side of the coin for the different types of CaM antagonists put forth by Hidaka and his co-workers (Inagaki and Hidaka 1984).

There is evidence that the CaM-binding region of the ATPase is inhibitory to the enzyme. CaM binding "disinhibits" the enzyme in this view. So also may partial proteolysis, which results in release of a CaM-binding fragment (Emelyaneko et al. 1985) and in an activated and CaM-independent enzyme. Benaim et al. (1984) produced different states of the Ca^{2+} pump ATPase by trypsinization. Various effectors of the ATPase changed the sensitivity of the enzyme to proteolysis, which would be expected if such effectors modified its conformation. Proteolysis of a variety of substrates may occur in intact RBCs, especially when they are "stressed," or otherwise loaded with Ca^{2+}. Lorand and Michalska (1985) observed that the proteolytic response to Ca^{2+} loading of RBCs was found only in fresh, not stored, blood. The activity of the Ca^{2+} pump ATPase was not tested.

D. ATPase Catalytic Unit

RAESS et al. (1985) investigated phenylglyoxal, an inhibitor which probably acts directly on the enzyme, presumably at a low affinity ATP site. A low affinity ATP site has been postulated by MUALEM and KARLISH (1983) to regulate turnover. Phenylglyoxal decreased the V_{max} and K_m of the Ca^{2+} pump ATPase with no change in Ca^{2+} affinity. Phenylglyoxal reacts with arginyl residues and RAESS et al. concluded that the binding of ATP to the low affinity site involves an arginyl residue.

Divicine (2,6-diamino-4,5-dihydroxypyrimidine) which is an active ingredient in fava beans, may be the cause of the hemolytic anemia associated with favism (BENATTI et al. 1985). In favic patients as well as in RBCs treated with divicine in vitro, the membrane Ca^{2+} pump ATPase was found to be inhibited and the calcium content was dramatically increased. It seems unlikely that the elevated Ca^{2+} content is due to an ionophoric effect of divicine, based on its structure and the fact that the ATP content of RBCs exposed to the compound remains normal. In RBCs exposed to the divalent cationophore, A23187, the half-life of ATP is about 5 min (HINDS and VINCENZI 1986). While some features of the hemolytic anemia of favism appear similar to those of glucose-6-phosphate dehydrogenase deficiency, glutathione depletion is probably not the cause of the inhibition of the Ca^{2+} pump induced by divicine. Divicine is not entirely specific. The Na^+,K^+ pump ATPase is also inhibited, if less reliably. The inhibitory effects may be mediated by reactive metabolites of divicine. Isolated membranes are much more sensitive than intact RBCs. BENATTI et al. (1985) suggested that this is due to more quenching of reactive species in intact cells. Another possible explanation is that divicine (and/or its products) acts at the inner membrane surface.

The Ca^{2+} pump ATPase may be chemically altered by the reactive metabolite of acetaminophen, *N*-acetyl-*p*-benzoquinone imine (NAPQI) (MOORE et al. 1985). NAPQI, produced by the microsomal cytochrome P-450 system, is a highly reactive electrophile which, if not detoxified by interaction with glutathione, results in oxidation and/or arylation of cellular proteins, primarily cysteinyl residues. Microsomal and/or Ca^{2+} pump ATPase may be affected by NAPQI. Results with NAPQI fit well with the idea that Ca^{2+} accumulation is a final common pathway in cell injury and death (SCHANNE et al. 1979; HUNT and WILLIS 1985).

E. Phospholipid Environment

The Ca^{2+} pump ATPase is exquisitely sensitive to the phospholipid environment of the RBC membrane (RONNER et al. 1977; NELSON and HANAHAN 1985). Certain amphipathic anions mimic the action of CaM in activating the Ca^{2+} pump ATPase (KOTAGAL et al. 1983). A simple example is oleic acid which we found increases both the ATPase (VINCENZI 1982) and transport (PINE et al. 1983) functions of the enzyme. TOKUMERA et al. (1985) found activation by synthetic lysophosphatidic acids and lysophosphatidylcholines. Amphipathic anions such as oleic acid or phosphatidyl serine (which are anionic and amphipathic, as is CaM in the presence of Ca^{2+}) may act either at the CaM-binding site and/or directly on the catalytic part of the enzyme. The activity of the Ca^{2+} pump ATPase appears to be a kind of algebraic sum of its inherent basal activity, plus phospholipid-dependent activation, plus CaM activation up to a characteristic maximum (which can be attained by either phospholipid or CaM alone; VINCENZI 1982). The implications of this have yet to be fully understood. It was suggested that phenobarbital alters Na^+, K^+ and Ca^{2+} pump ATPase activities by altering lipid fluidity of synaptosomal membranes (DELICONSTANTINOS 1983). Ethanol administration in vivo also alters Ca^{2+} transport in subsequently derived synaptosomal membranes (GARRETT and ROSS 1983).

FOURIE et al. (1983) described effects on the Ca^{2+} pump ATPase of cardiotoxin, a group of polypeptide toxins from cobra venom. Because of their amphipathic and cationic nature, we assume that they work in the phospholipid phase of the PM. Cardiotoxin causes depolarization and contraction of smooth and skeletal muscle, and leads to hemolysis. Thus, it is also called direct lytic factor (CONDREA 1984). FOURIE et al. found that cardiotoxin decreased the Ca^{2+} pump ATPase of RBCs. These effects of cardiotoxin, according to the authors, were not due to phospholipase activity. Other PM ATPases were not monitored, so the specificity of the effect is not yet clear.

LOCHER et al. (1984) altered the cholesterol content of RBC membranes in vitro. Then they measured Ca^{2+} flux in vanadate-containing medium (to inhibit the pump, thus to measure passive influx; VARECKA and CARAFOLI 1982). Increased cholesterol led to increased Ca^{2+} flux. Ca^{2+} flux was divided into nitrendipine-sensitive and nitrendipine-insensitive fractions. The nitrendipine-sensitive frac-

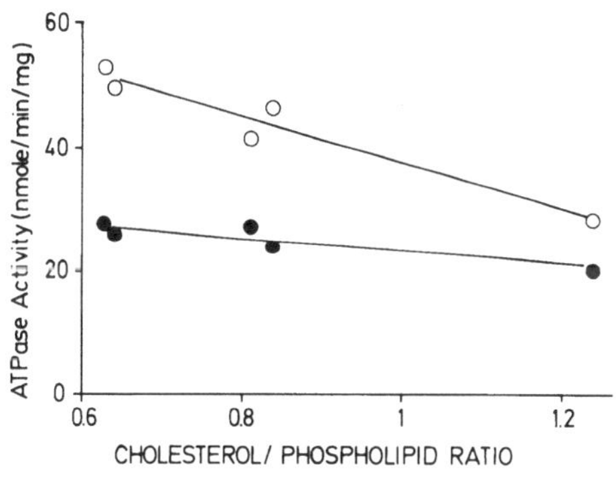

Fig. 2. Influence of cholesterol:phospholipid ratio of RBC membranes on Ca^{2+} pump ATPase activity. Intact fresh human RBCs were preincubated for 16 h at 37 °C in media designed to increase or decrease the cholesterol content of the membranes (SHINITZKY 1978). RBC membranes were then isolated according to FARRANCE and VINCENZI (1977) and assayed in the absence (*full symbols*) and presence (*open symbols*) of CaM (10 n*M*) according to RAESS and VINCENZI (1980) with the modification that all assays were performed in the presence of saponin 20 µg/ml. Saponin was found to be necessary to overcome differential resealing of the RBC membranes. As the cholesterol content of the membranes increased, the cholesterol:phospholipid ratio increased and Ca^{2+} pump ATPase activity decreased markedly in the presence of CaM. Much less effect was exerted on the ATPase in the absence of added CaM. Although not shown, the Na^+,K^+ pump ATPase and Mg^{2+} ATPase were affected similarly to the basal Ca^{2+} pump ATPase

tion of the leak was increased by cholesterol. If one assumes the existence of and participation of a Ca^{2+} channel protein in Ca^{2+} flux, then the cholesterol effect seems curious. Of course, one could assume that cholesterol exerts an effect on a "Ca^{2+} channel protein." In any event, altered cholesterol content of a plasma membrane alters its passive ion transport properties and, in the case of increased cholesterol content, may increase the load on the PM Ca^{2+} pump. We consider this result to be extremely important.

In addition to increasing the load on the PM Ca^{2+} pump, increased cholesterol in the PM may decrease its ability to respond to a load. Figure 2 presents data obtained some time ago in our laboratory by Charles W. Luetje. The data demonstrate a significant effect of the cholesterol: phospholipid ratio on the maximal activity of the Ca^{2+} pump ATPase of isolated RBC membranes. The influence of the cholesterol: phospholipid ratio on the basal activity of the ATPase (and on the Na^+,K^+ pump ATPase and Mg^{2+} ATPase, data not shown) was much less. In any event, it appears that the CaM-dependent state of the Ca^{2+} pump ATPase is exquisitely sensitive to the cholesterol: phospholipid ratio of the PM. Taken together, these results may mean that increased membrane cholesterol increases passive Ca^{2+} permeability and decreases the maximal capacity of the Ca^{2+} pump. Bearing in mind that lipids of the PM of cells in the vascular system exchange with plasma lipids (LANGE 1984), these facts may have much to do with cellular pathology of altered lipid states.

F. Influx Pathways

For reasons which should become clear, we wish to emphasize the influx pathways as the "true regulators" of the Ca^{2+} pump in vivo. The capacity of the RBC (and by inference other cell) membrane Ca^{2+} pump ATPase is considerably greater than the passive leak of Ca^{2+}, at least as has been measured in vitro (SCHATZMANN and VINCENZI 1969; LEW et al. 1982). Thus, in vivo the dominant "regulator" of the activity of the pump is the leak. The capacity of the PM Ca^{2+} pump has been compared with the leak in a previous review (VINCENZI and HINDS 1980).

VARECKA and CARAFOLI (1982) found that the influx of Ca^{2+} into RBCs is sensitive to verapamil, Co^{2+}, or quinidine. They suggested that since verapamil decreased Ca^{2+} uptake in RBCs "... it seems probable that Ca^{2+} uptake in RBCs and during activation of excitable cells is mediated by the same structural components of the plasma membrane ..." Concentrations of verapamil in the Varecka and Carafoli study were rather high. Approximately 70 μM verapamil was needed to decrease Ca^{2+} influx half-maximally. While so-called Ca^{2+} channels have been more or less convincingly demonstrated in other cell types, we remain somewhat skeptical about their (functional) existence and/or importance in RBCs. First, the membrane potential of RBCs is low. Thus, voltage-operated Ca^{2+} channels are not likely to be very important. Second, surface receptors which might couple to receptor-operated Ca^{2+} channels are also usually thought of as quite limited in RBCs. From the previous section one might suggest, somewhat tongue in cheek, that cholesterol is a ligand for such receptors.

With the same tongue-in-cheek attitude, we have suggested that physiologic Ca^{2+} influx in RBCs can occur via "mechanically operated channels" (VINCENZI and CAMBARERI 1985). A similar notion was advanced by LARSEN et al. (1981). In our laboratory, shear stress increased the total ATP consumption from approximately 420 to 2400 μmol per liter RBC per hour. Most of this increase was prevented by the simultaneous addition of ouabain and EGTA (VINCENZI and CAMBARERI 1985). ATP utilization is increased by shear stress, but the capacity of the Ca^{2+} pump to utilize ATP is considerably greater. In RBCs exposed to A23187 in the presence of Ca^{2+} the initial rate of ATP consumption is approximately 18000 μmol per liter RBC per hour; most of which is contributed by the Ca^{2+} pump ATPase. VARECKA and CARAFOLI (1982) reported greater turnover of Ca^{2+} (approximately 40–200 μmol per liter RBC per hour) than that reported by LEW et al. (1982) (45–48 μmol per liter RBC per hour). Still, the Ca^{2+} pump is utilized for less than 1% of its capacity, at least in vitro. Thus, to observe significant Ca^{2+} accumulation in RBCs in vitro by inhibiting the pump, one must inhibit the pump rather completely. This serves to reemphasize the need for a potent and specific inhibitor of the PM Ca^{2+} pump. Of course, one way to "inhibit" the Ca^{2+} pump is to deplete ATP. For example, in ATP-depleted cells, YINGST et al. (1985) used 0.76–2.4 m*M* ethanol and found increased Ca^{2+} in RBCs, as monitored by arsenazo III. These effects were (somewhat incredibly) reversible. Whether such immense concentrations of lipid-soluble agents alter Ca^{2+} transport in any way other than nonspecifically is not clear to us.

MIKKELSEN et al. (1984) found that plasmodium-infected RBCs exhibit altered Ca^{2+} levels as a consequence of modification of host membrane permeability to Ca^{2+}. This is reminiscent of increased Ca^{2+} permeability (LEIDA et al. 1981) reported in plasmodium-containing cells. Levels of Ca^{2+} increase 10- to 20-fold and 90% of cell Ca^{2+} is localized in the parasite compartment. Plasmodium infection leads to a decrease in Ca^{2+} pump ATPase enzyme activity and no change in affinity. CaM stimulation of the enzyme is normal (even if the level of CaM is decreased by 30%). ATP in plasmodium-infected cells is low; 0.4 m*M* compared with approximately 1 m*M* in normal cells. Partial uncoupling of the Ca^{2+} pump was suggested.

BARZILAI and RAHAMIMOFF (1983) examined the effects of certain flavonoids on synaptosomal ATPase and transport activities. They compared quercetin, morin, and rutin. Based on their potencies and lipid partitioning, it is reasonable to assume these agents work in the lipid phase of the membrane, but what they do is not clear.

G. Surface Receptors

It has recently been shown that RBCs respond to parathyroid hormone (PTH). PTH incudes an influx of Ca^{2+} into human RBCs. Presumably, this places an added load on the RBC Ca^{2+} pump. PTH induces increased osmotic fragility in vitro (BOGIN et al. 1982) and it has more recently been shown that RBCs in hyperparathyroid patients exhibit a shortened life span (SALTISSI and CARTER 1985). The mechanism by which PTH increases Ca^{2+} influx is not yet clarified. BRAUT-

BAR et al. (1985) showed that PTH produces decrements in ^{32}P incorporation into phosphatidylinositol, di- and triphosphatidylinositol, and an increase in phosphatidic acid. In addition, there was a significant increase in RBC phosphatidylserine. It was suggested that increased phosphatidylserine may increase rigidity of the RBC membrane and enhance osmotic fragility. Because phosphatidic acid has ionophoric properties, one mechanism of PTH-induced Ca^{2+} entry in RBCs is via altered phospholipid metabolism and increased phosphatidic acid-mediated flux. Alternatively, or additionally, PTH may increase Ca^{2+} influx by other mechanisms and altered phospholipid metabolism may be an effect, not a cause, of increased Ca^{2+} influx. These data demonstrate that a cell not usually thought of as a PTH effector responds in ways which impact on the activity of the PM Ca^{2+} pump. We suggest that PTH acts via surface receptors to bring about increased Ca^{2+} influx and that, while the Ca^{2+} pump presumably responds to such a "physiologic" signal, the consequences on the cell are not insignificant. These data suggest that the RBC may be a useful model for elucidating effects of certain cell surface active agents.

DAVIS and DAVIS (1985) recently reviewed a series of their papers, on the influence of thyroid hormones, triiodothyronine (T_3) and tetraiodothyronine (T_4) on the Ca^{2+} pump ATPase. Their work may demonstrate that thyroid hormones act, at least in part, via a non-nuclear mechanism. The significance of thyroid actions on the RBC is not clear, but may be useful as a model of effects on other cells. It was concluded that thyroid hormone increases the activity of the Ca^{2+} pump ATPase in a CaM-dependent manner. Since thyroid hormone neither interacted with CaM, nor altered CaM binding to RBC membranes, the authors suggested that thyroid hormone stimulation depends upon the intact CaM–enzyme complex (DAVIS and DAVIS 1985). DAVIS et al. (1983b) found that quercetin blocked the T_4 effect (prevented the increase in Ca^{2+} pump ATPase activity). Curiously, in the absence of T_1, quercetin stimulated enzyme activity. There are some structural characteristics of quercetin which resemble T_4 and this might be a matter of comparing two agonists of different intrinsic activity. Based on the particular assay system used and the results of others who have examined flavonoids (BARZILAI and RAHAMIMOFF 1983), we suspect that the T_3 and T_4 as well as flavonoid effects actually occur on the lipid phase of the membrane. While the results of the Davis group are intriguing and apparently internally consistent, we wonder if they are in some way peculiar to their enzyme assay system. These authors typically preincubate isolated RBC membranes in vitro with T_3 or T_4 for 60 min at 37 °C before the ATPase assay (DAVIS et al. 1983a). The fact that CaM gives a relatively modest stimulatory effect and that specific activities of the Ca^{2+} pump ATPase are rather low (approximately 3 and 4 nmol phosphate per milligram protein per minute for basal and CaM-activated, respectively; see Fig. 2 for comparison) suggests to us that the membranes may have resealed to a significant extent; or in some other manner acquired low activity. It may be that the effect of thyroid hormone under these conditions is related to altered resealing and/or low activity. It would be of interest to examine thyroid hormone effects in a system with higher specific activity and especially on the isolated enzyme (NIGGLI et al. 1979). It was stated by DAVIS et al. (1983a) that the presence of imidazole, such as used in assay buffers in our laboratory (RAESS and VINCENZI 1980), and a number of others, causes loss of the effect of T_3 or T_4. In our view, the concomitant presence of a detergent in the assay system would help to demonstrate that the data are not simply due to differential resealing of the membranes during preincubation.

YAMASAKI and WAY (1983) reported that nanomolar concentrations of the opioid kappa agonists, ethylketocyclazocine (EKC) and dynorphin 1–13 decreased La^{3+}-sensitive ^{45}Ca efflux from rat RBCs. This inhibition was reversible by naloxone. Morphine and β-endorphin were effective at higher concentrations. Levorphanol and Leu-enkephalin, relatively selective mu and delta receptor agonists, respectively, were ineffective. It was concluded that a kappa-type opioid receptor present on the RBC, when activated, leads to inhibition of the plasma membrane Ca^{2+} pump. Kappa agonist-mediated inhibition of the Ca^{2+} pump ATPase was also extended (YAMASAKI and WAY 1985). EKC led to decreased ATPase activity of isolated rat RBC membranes. This effect was blocked by the opioid antagonist, Win 44441. Other opioid agonists mimicked the effects of EKC. The kappa receptor–Ca^{2+} pump ATPase coupling effect is apparently not related to the β-endorphin and endorphin effects on CaM; a presumably separate phenomenon (BARNETTE and WEISS 1982). To our knowledge, the observations of Way and his co-workers have not been confirmed. If kappa agonists work centrally by inhibiting active efflux of Ca^{2+} from neurons, then the observations of this group could open up avenues to new understanding of opiate drugs and their mechanisms.

Other possible examples of surface ligand-induced inhibition of Ca^{2+} have appeared. ROSS et al. (1985) reported that activation of muscarinic receptors inhibits Ca^{2+} pump ATPase and transport in brain membranes. Likewise, LOTERSZTAJN et al. (1984) reported that glucagon inhibits the calcium pump in liver PM. The idea that increased intracellular signaling might arise from inhibition of outward Ca^{2+} transport (as well as the presently fashionable Ca^{2+} channel mechanisms) has been put forth in other contexts. SOLOFF and SWEET (1982) found that oxytocin inhibited a portion of the Ca^{2+}-dependent ATPase activity of rat myometrial plasma membranes with half-maximal inhibition occurring at about 1 nM. The oxytocin-sensitive fraction of ATPase activity was inhibited by TFP, and the interference is that oxytocin inhibits the CaM-dependent fraction of Ca^{2+} pump ATPase in the myometrial PM. These results are an extension of an observation originally made by AKERMAN and WIKSTROM (1979) using uterine membranes. We failed to find inhibition of the RBC Ca^{2+} pump ATPase by oxytocin (F.F. VINCENZI 1982, unpublished work), presumably because RBCs lack oxytocin receptors.

H. Summary

A number of agents and conditions impinge on the activity of the PM Ca^{2+} pump, but few specific modifiers of the activity of the pump or its associated ATPase have been identified. No specific and widely useful inhibitor of the ATPase and pump has been identified. From our point of view, the simplest interpretation of the available literature is that a number of drugs exert indirect effects on the Ca^{2+} pump ATPase and that many of such effects are mediated via alterations in the phospholipid environment of the membrane. Certain relatively selective (for example anti-CaM) drug alterations in the pump ATPase activity can be demonstrated in vitro. Agents which change the passive entry of Ca^{2+} into cells alter the load presented to the pump. Most available evidence suggests that the capacity of the pump is large compared with the load normally imposed on it and/or that, in cells which must transport relatively large amounts of Ca^{2+}, additional mechanisms, such as Na^{+}/Ca^{2+} exchange are also present. There are a number of reports which may be interpreted to suggest that certain surface receptors are linked functionally to the Ca^{2+} pump ATPase, although these links may be somewhat indirect. Thus, endogenous ligands acting on the surfaces of certain cells may be the most significant drugs affecting PM Ca^{2+} transport in vivo.

Acknowledgments. David Godin performed the assays of cholesterol : phospholipid ratio in certain experiments presented here. Work in our laboratory has been supported in part by DHHS grant AM-16436 and by grants from the Cystic Fibrosis Foundation and the American Diabetes Association.

References

Agre P, Virshup D, Bennett V (1984) Bepridil and cetiedil. Vasodilators which inhibit Ca^{2+}-dependent calmodulin interactions with erythrocyte membranes. J Clin Invest 74:812–820

Akerman KEO, Wikstrom MKF (1979) (Ca^{2+} + Mg^{2+})-stimulated ATPase activity of rabbit myometrium plasma membrane is blocked by oxytocin. FEBS Lett 97:283–287

Barnette SM, Weiss B (1982) Interaction of beta-endorphin and other opioid peptides with calmodulin. Mol Pharmacol 21:86–91

Barnette MS, Daly R, Weiss B (1983) Inhibition of calmodulin activity by insect venom peptides. Biochem Pharmacol 32:2929–2933

Barzilai A, Rahamimoff H (1983) Inhibition of Ca^{2+}-transport ATPase from synaptosomal vesicles by flavonoids. Biochim Biophys Acta 730:245–254

Benaim G, Zurini M, Carafoli E (1984) Different conformational states of the purified Ca^{2+}-ATPase of the erythrocyte plasma membrane revealed by controlled trypsin proteolysis. J Biol Chem 259:8471–8477

Benatti U, Guida L, Forteleoni G, Meloni T, De Flora A (1985) Impairment of the calcium pump of human erythrocytes by divicine. Arch Biochem Biophys 239:334–341

Blumenthal DK, Takio K, Edelman AM, Charbonneau H, Titani K, Walsh KA, Krebs EG (1985) Identification of the calmodulin-binding domain of skeletal muscle myosin light chain kinase. Proc Natl Acad Sci USA 82:3187–3191

Bogin E, Marrsy SG, Levi J, Djaldeti M, Bristol G, Smith J (1982) Effect of parathyroid hormone on osmotic fragility of human erythrocyte. J Clin Invest 69:1017–1025

Bostrom S-L, Ljung B, Mardh S, Forsen S, Thulin E (1981) Interaction of the antihypertensive drug felodipine with calmodulin. Nature 292:777–778

Brautbar N, Chakrabarty J, Coats J, Massry SG (1985) Calcium, parathyroid hormone and phospholipid turnover of human red blood cells. Mineral Electrolyte Metab 11:111–116

Carafoli E (1984a) Molecular, mechanistic, and functional aspects of the plasma membrane calcium pump. In: Bronner F, Peterlik M (eds) Epithelial calcium and phosphate transport: molecular and cellular aspects, vol 168. Liss, New York, p 13

Carafoli E (1984b) Plasma membrane Ca^{2+} transport, and Ca^{2+} handling by intracellular stores: an integrated picture with emphasis on regulation. In: Donowitz M, Sharp GWG (eds) Mechanisms of intestinal electrolyte transport and regulation by calcium. Liss, New York, p 121

Carafoli E (1984c) Calmodulin-sensitive calcium-pumping ATPase of plasma membranes: isolation, reconstitution, and regulation. Fed Proc 43:3005–3010

Carafoli E (1984d) Membrane transport in the messenger function of calcium. In: Ovchinnikov YA (ed) Progress in bioorganic chemistry and molecular biology. Elsevier, New York, p 233

Carafoli E (1985) The homeostasis of calcium in heart cells. J Mol Cell Cardiol 17:203–212

Cavieres JD (1984) Calmodulin and the target size of the $(Ca^{2+} + Mg^{2+})$-ATPase of human red-cell ghosts. Biochim Biophys Acta 771:241–244

Colombani PM, Robb A, Hess AD (1985) Cyclosporin A binding to calmodulin: a possible site of action on T lymphocytes. Science 228:337–339

Comte M, Maulet Y, Cox JA (1983) Ca^{2+}-dependent high-affinity complex formation between calmodulin and melittin. Biochem J 209:269–272

Condrea E (1984) Membrane-active polypeptides from snake venom: cardiotoxins and haemocytotoxins. Experientia 30:121–129

Cox JA, Comte M, Fitton JE, DeGrado WF (1985) The interaction of calmodulin with amphiphilic peptides. J Biol Chem 260:2527–2534

Davis FB, Davis PJ, Blas SD (1983a) Role of calmodulin in thyroid hormone stimulation in vitro of human erythrocyte Ca^{2+}-ATPase activity. J Clin Invest 71:579–586

Davis FB, Middleton E, Davis PJ, Blas SD (1983b) Inhibition by quercetin of thyroid hormone stimulation in vitro of human red blood cell Ca^{2+}-ATPase activity. Cell Calcium 4:71–81

Davis PJ, Davis FB (1985) Thyroid hormone and calmodulin. In: Hidaka H, Hartshorne DJ (eds) Calmodulin antagonists and cellular physiology. Academic, New York, p 185

Deliconstantinos G (1983) Phenobarbital modulates the (Na^+,K^+)-stimulated ATPase and Ca^{2+}-stimulated ATPase activities by increasing the bilayer fluidity of dog brain synaptosomal plasma membranes. Neurochem Res 8:1143–1152

DiJulio D, Vincenzi FF (1986) Evaluation of trifluoperazine and compound 48/80 as selective antagonists of calmodulin activation of the Ca^{2+} pump ATPase. Proc West Pharmacol Soc 29:445–446

Emelyanenko EI, Shakhparonov MI, Modyanov NM (1985) Limited proteolysis of human erythrocyte Ca^{2+}-ATPase in membrane-bound form. Identification of calmodulin-binding fragments. Biochem Biophys Res Commun 126:214–219

Farrance ML, Vincenzi FF (1977) Enhancement of $(Ca^{2+} + Mg^{2+})$-ATPase activity of human erythrocyte membranes by hemolysis in isosmotic imidazole buffer. I. General properties of variously prepared membranes and the mechanism of the isosmotic imidazole effect. Biochim Biophys Acta 471:49–58

Foder B, Skibsted U, Scharff O (1984) Effect of trifluoperazine, compound 48/80, TMB-8 and verapamil on ionophore A23187 mediated calcium uptake in ATP depleted human red cells. Cell Calcium 5:441–450

Fourie AM, Meltzer S, Berman MC, Louw AI (1983) The effect of cardiotoxin on $(Ca^{2+} + Mg^{2+})$-ATPase of the erythrocyte and sarcoplasmic reticulum. Biochem Int 6:581–591

Garrett KM, Ross DH (1983) Effects of in vivo ethanol administration on Ca^{2+}/Mg^{2+} ATPase and ATP-dependent Ca^{2+} uptake activity in synaptosomal membranes. Neurochem Res 8:1013–1028

Gietzen K, Adamczyk-Engelmann P, Wuthrich A, Konstantinova A, Bader H (1983) Compound 48/80 is a selective and powerful inhibitor of calmodulin-regulated functions. Biochim Biophys Acta 736:109–118

Goldstein DB (1984) The effects of drugs on membrane fluidity. Annu Rev Pharmacol Toxicol 24:43–64

Graf E, Verma AK, Gorski JP, Lopaschuk G, Niggli V, Zurini M, Carafoli E, Penniston JT (1982) Molecular properties of calcium-pumping ATPase from human erythrocytes. Biochemistry 21:4511–4516

Guerini D, Krebs J, Carafoli E (1984) Stimulation of the purified erythrocyte Ca^{2+}-ATPase by tryptic fragments of calmodulin. J Biol Chem 259:15172–15177

Hinds TR, Andreasen TJ (1981) Photochemical cross-linking of azidocalmodulin to the (Ca^{2+} + Mg^{2+})-ATPase of the erythrocyte membrane. J Biol Chem 256:7877–7882
Hinds TR, Vincenzi FF (1986) Evidence for a calmodulin-activated Ca^{2+} pump ATPase in dog erythrocytes (42290). Proc Soc Exp Biol Med 181:542–549
Hunt WG, Willis RJ (1985) Calcium exposure required for full expression of injury in the calcium paradox. Biochem Biophys Res Commun 126:901–904
Inagaki M, Hidaka H (1984) Two types of calmodulin antagonists: a structurally related interaction. Pharmacology 29:75–84
Inagaki M, Tanaka T, Sasaki Y, Hidaka H (1985) Calcium-dependent interactions of an ionophore A23187 with calmodulin. Biochem Biophys Res Commun 130:200–206
Itano T, Penniston JT (1985) Ca^{2+}-pumping ATPase of plasma membranes. In: Hidaka H, Hartshorne DJ (eds) Calmodulin antagonists and cellular physiology. Academic, New York, p 335
Kotagal N, Colca JR, McDaniel ML (1983) Activation of an islet cell plasma membrane (Ca^{2+} + Mg^{2+})-ATPase by calmodulin and Ca-EGTA. J Biol Chem 258:4808–4813
Lange Y (1984) The dynamics of erythrocyte membrane cholesterol. In: Kruckeberg WC, Eaton JW, Brewer GJ (eds) Erythrocyte membranes: recent clinical and experimental advances, vol 3. Liss, New York, p 137
Langer GA (1984) Calcium at the sarcolemma. J Mol Cell Cardiol 16:147–153
LaPorte DC, Wierman BM, Storm DR (1980) Calcium-induced exposure of a hydrophobic surface on calmodulin. Biochemistry 19:3814–3819
Larsen FL, Katz S, Roufogalis BD, Brooks DE (1981) Physiological shear stresses enhance the Ca^{2+}-permeability of human erythrocytes. Nature 294:667–668
Leida MN, Mahoney JR, Eaton JR (1981) Intraerythrocytic plasmodial calcium metabolism. Biochem Biophys Res Commun 103:402–406
Levin RM, Weiss B (1977) Binding of trifluoperazine to the calcium-dependent activator of cyclic nucleotide phosphodiesterase. Mol Pharmacol 13:690–697
Levine SN, Hollier B (1983) Aprindine inhibits calmodulin-stimulated phosphodiesterase and Ca-ATPase activities. J Cardiovasc Pharmacol 5:151–156
Lew VL, Tsien RY, Miner C, Bookchin RM (1982) Physiological $[Ca^{2+}]_i$ level and pump-leak turnover in intact red cells measured using an incorporated Ca chelator. Nature 298:478–481
Locher R, Neyses L, Stimpel M, Kuffer B, Vetter W (1984) The cholesterol content of the human erythrocyte influences calcium influx through the channel. Biochem Biophys Res Commun 124:822–828
Lorand L, Michalska M (1985) Altered response of stored red blood cells to Ca^{2+} stress. Blood 65:1025–1027
Lotersztajn S, Epand RM, Mallat A, Pecker F (1984) Inhibition by glucagon of the calcium pump in liver plasma membranes. J Biol Chem 259:8195–8201
Meltzer HL, Kassir S (1983b) Inhibition of calmodulin-activated Ca^{2+}-ATPase by propanolol and nadolol. Biochim Biophys Acta 755:452–456
Mikkelsen RB, Geller E, Van Doren E, Asher CR (1984) Ca^{2+} metabolism of *plasmodia*-infected erythrocytes. In: Eaton JW, Brewer GJ (eds) Malaria and the red cell, vol 155. Liss, New York, p 25
Minocherhomjee A-E-VM, Roufogalis BD (1984) Antagonism of calmodulin and phosphodiesterase by nifedipine and related calcium entry blockers. Cell Calcium 5:57–63
Minocherhomjee AM, Beauregard B, Potier M, Roufogalis BD (1983) The molecular weight of the calcium-transport-ATPase of the human red blood cell determined by radiation inactivation. Biochem Biophys Res Commun 116:895–900
Moore M, Thor H, Moore G, Nelson S, Moldeus P, Orrenius S (1985) The toxicity of acetaminophen and *N*-acetyl-*p*-benzoquinone imine in isolated hepatocytes is associated with thiol depletion and increased cytosolic Ca^{2+}. J Biol Chem 260:13035–13040
Mualem S, Karlish SJD (1983) Catalytic and regulatory ATP-binding sites of the red cell Ca^{2+} pump studied by irreversible modification with fluorescein isothiocyanate. J Biol Chem 258:169–175
Nelson DR, Hanahan DJ (1985) Phospholipid and detergent effects on (Ca^{2+} + Mg^{2+}) ATPase purified from human erythrocytes. Arch Biochem Biophys 236:720–730

Nelson GA, Andrews ML, Karnovsky MJ (1983) Control of erythrocyte shape by calmodulin. J Cell Biol 96:730–735
Niggli V, Penniston JT, Carafoli E (1979) Purification of the ($Ca^{2+}+Mg^{2+}$)ATPase from human erythrocyte membranes using a calmodulin affinity column. J Biol Chem 254:9955–9958
Penniston JT, Graf E, Niggli V, Verma AK, Carafoli E (1980) The plasma membrane calcium ATPase. In: Siegel FL, Carafoli E, Kretsinger DH, Maclennan DH, Wasserman RH (eds) Calcium-binding proteins: structure and function. Elsevier, New York, pp 23–30
Pine RW, Vincenzi FF, Carrico CJ (1983) Apparent inhibition of the plasma membrane Ca^{2+} pump by oleic acid. J Trauma 23:366–370
Plishker GA (1984) Phenothiazine inhibition of calmodulin stimulates calcium-dependent potassium efflux in human red blood cells. Cell Calcium 5:177–185
Prozialeck WC (1984) Interaction of quaternary phenothiazine salts with calmodulin. J Pharmacol Exp Ther 231:473–479
Raess BU, Vincenzi FF (1980) A semi-automated method for determination of multiple membrane ATPase activities. J Pharmacol Meth 4:273–283
Raess BU, Record DM, Tunnicliff G (1985) Interaction of phenylglyoxal with the human erythrocyte ($Ca^{2+}+Mg^{2+}$)-ATPase. Mol Pharmacol 27:444–450
Rainteau D, Wolf C, Bereziat G, Polonovski J (1984) Binding of a spin-labelled chlorpromazine analogue to calmodulin. Biochem J 221:659–663
Resink TJ, Tkachuk VA, Erne P, Buhler FR (1986) Platelet membrane calmodulin-stimulated calcium-adenosine triphosphatase. Altered activity in essential hypertension. Hypertension 8:159–166
Ronner P, Gazzotti P, Carafoli E (1977) A lipid requirement for the ($Ca^{2+}+Mg^{2+}$)-activated ATPase of erythrocyte membranes. Arch Biochem Biophys 179:578–583
Ross DH, Shreeve SM, Hamilton MG (1985) Activation of central muscarinic receptors inhibit Ca^{2+}/Mg^{2+} ATPase and ATP-dependent Ca^{2+} transport in synaptic membranes. Brain Res 329:39–47
Roufogalis BD (1982) Specificity of trifluoperazine and related phenothiazines for calcium-binding proteins. In: Cheung WY (ed) Calcium and cell function, vol III. Academic, New York, p 129
Roufogalis BD (1985) Calmodulin antagonism. In: Marme D (ed) Calcium and cell physiology. Springer, Berlin Heidelberg New York Tokyo, p 148
Roufogalis BD, Minocherhomjee A-E-VM, Al-Jobore A (1983) Pharmacological antagonism of calmodulin. Can J Biochem Cell Biol 61:927–933
Roufogalis BD, Elliott CT, Ralston GB (1984) Characterization of a ($Ca^{2+}+Mg^{2+}$)ATPase activator bound to human erythrocyte membranes. Cell Calcium 5:77–88
Saltissi D, Carter GD (1985) Association of secondary hyperparathyroidism with red cell survival in chronic haemodialysis patients. Clin Sci 68:29–33
Schanne FAX, Kane AB, Young EE, Farber JL (1979) Calcium dependence of toxic cell death: a final common pathway. Science 206:700–702
Schatzmann HJ (1983) The red calcium pump. Annu Rev Physiol 45:303–312
Schatzmann HJ (1985) Calcium extrusion across the plasma membrane by the calcium-pump and the Ca^{2+}-Na^{+} exchange system. In: Marme D (ed) Calcium and cell physiology. Springer, Berlin Heidelberg New York Tokyo, p 18
Schatzmann HJ, Vincenzi FF (1969) Calcium movements across the membrane of human red cells. J Physiol 201:369–395
Schlondorff D, Satriano J (1985) Interactions with calmodulin: potential mechanism for some inhibitory actions of tetracyclines and calcium channel blockers. Biochem Pharmacol 34:3391–3393
Schmidt JW, Hinds TR, Vincenzi FF (1985) On the failure of calmodulin to activate Ca^{2+} pump ATPase of dog red blood cells. Comp Biochem Physiol 82A:601–607
Shinitzky M (1978) An efficient method for modulation of cholesterol level in cell membranes. FEBS Lett 85:317–320
Soloff MS, Sweet P (1982) Oxytocin inhibition of ($Ca^{2+}+Mg^{2+}$)-ATPase activity in rat myometrial plasma membranes. J Biol Chem 257:10687–10693

Steiner B, Luscher EF (1985) Evidence that the platelet plasma membrane does not contain a (Ca^{2+} + Mg^{2+})-dependent ATPase. Biochim Biophys Acta 818:299–309
Steiner RF, Marshall L, Needleman D (1986) The interaction of melittin with calmodulin and its tryptic fragments. Arch Biochem Biophys 246:286–300
Tokumura A, Mostafa MH, Nelson DR, Hanahan DJ (1985) Stimulation of (Ca^{2+} + Mg^{2+})-ATPase activity in human erythrocyte membranes by synthetic lysophosphatidic acids and lysophosphatidylcholines. Effects of chain length and degree of unsaturation of the fatty acid groups. Biochim Biophys Acta 812:568–574
Tomlinson S, Macneil S, Walker SW, Ollis CA, Merritt JE, Brown BL (1984) Calmodulin and cell function. Clin Sci 66:497–508
Varecka L, Carafoli E (1982) Vanadate-induced movements of Ca^{2+} and K^{+} in human red blood cells. J Biol Chem 257:7414–7421
Vincenzi FF (1981) Calmodulin pharmacology. Cell Calcium 2:387–409
Vincenzi FF (1982) Pharmacological modification of the Ca^{2+}-pump ATPase activity of human erythrocytes. Ann NY Acad Sci 402:368–380
Vincenzi FF, Cambareri JJ (1985) Apparent ionophoric effects of red blood cell deformation. In: Eaton JW, Konzen DL, White JG (eds) Cellular and molecular aspects of aging. The red cell as a model. Liss, New York, p 213
Vincenzi FF, Hinds TR (1980) Calmodulin and plasma membrane calcium transport. In: Cheung WY (ed) Calcium and cell function, vol I. Academic, New York, p 128
Vincenzi FF, Adunyah ES, Niggli V, Carafoli E (1982) Purified red blood cell Ca^{2+}-pump ATPase: evidence for direct inhibition by presumed anti-calmodulin drugs in the absence of calmodulin. Cell Calcium 3:545–559
Watkins WM (1974) Blood-group substances: their nature and genetics. In: Surgenor DM (ed) The red blood cell, 2nd ed. Academic, New York, pp 293–260
Weiss B, Sellinger-Barnette M, Winkler JD, Schechter LE, Prozialeck WC (1985) Calmodulin antagonists: structure-activity relationships. In: Hidaka H, Hartshorne DJ (eds) Calmodulin antagonists and cellular physiology. Academic, New York, p 45
Wolff DJ, Brostrom CO (1976) Calcium-dependent cyclic nucleotide phosphodiesterase from brain identification of phospholipids as calcium-independent activators. Arch Biochem Biophys 173:720–731
Wulfroth P, Petzelt C (1985) The so-called anticalmodulins fluphenazine, calmidazolium, and compound 48/80 inhibit the Ca^{2+}-transport system of the endoplasmic reticulum. Cell Calcium 6:295–310
Yamasaki Y, Way EL (1983) Possible inhibition of Ca^{++} pump of rat erythrocyte ghosts by opioid K agonists. Life Sci 33:723–726
Yamasaki Y, Way EL (1985) Inhibition of Ca^{++}-ATPase of rat erythrocyte membranes by K-opioid agonists. Neuropeptides 5:359–362
Yingst DR, Polasek PM, Kilgore P (1985) The effect of ethanol on the passive Ca permeability of human red cell ghosts measured by means of arsenazo III. Biochim Biophys Acta 813:277–281

CHAPTER 9

Development of Inhibitors of Sodium, Calcium Exchange

G.J. KACZOROWSKI, M.L. GARCIA, V.F. KING, and R.S. SLAUGHTER

A. Introduction

Electrically excitable cells process the Ca^{2+} which triggers excitation–response coupling in an efficient manner. Several Ca^{2+} transport systems have evolved by which to accomplish this task. These mechanisms typically act together to maintain Ca^{2+} homeostasis, thereby preventing buildup of cytoplasmic Ca^{2+} which would be cytotoxic. There are two plasmalemmal active transport systems which directly regulate Ca^{2+} fluxes in many types of cells. The first, a Ca^{2+},Mg^{2+}-ATPase, is an ATP-dependent pump of low capacity, but high affinity that extrudes Ca^{2+} unidirectionally from the cell (Chap. 8). The other, an Na,Ca antiporter, moves Ca^{2+} via a carrier mechanism controlled by transmembrane electrical and Na^+ concentration gradients. Since Na,Ca exchange is completely reversible, transmembrane Ca^{2+} movement can occur in either direction, depending on cellular conditions which regulate carrier activity. These are the only two Ca^{2+} transport processes known to remove Ca^{2+} from the cell.

In order to assess the physiologic role of Na,Ca exchange and determine its overall contribution to Ca^{2+} homeostasis in various cells, specific inhibitors must be used to modulate this transport reaction in vivo. This chapter will focus on recent studies directed at developing mechanism-based exchange inhibitors. For more complete treatment of the characteristics, regulation, and possible involvement of this transport reaction in the physiology and pathology of different excitable tissues, the reader is directed to several other reviews (MULLINS 1981; REUTER 1982; LANGER 1982; CHAPMAN 1983; BAKER and DIPOLO 1984; PHILIPSON 1985a; KACZOROWSKI 1985; REEVES 1985) because only sufficient details will be provided to form a conceptual framework for the inhibitor studies.

B. Characteristics and Physiologic Properties of Na,Ca Exchange

The Na,Ca exchange reaction is driven by transmembrane ion gradients and electrical potentials without any direct involvement of ATP. The energy derived from an ion moving down its electrochemical gradient (i.e., Na^+) is coupled to the movement of another ion (i.e., Ca^{2+}) in the opposite direction in an antiport reaction. Since the plasmalemmal exchange system has been shown to be electrogenic in every case examined (i.e., net charge translocation occurs), the transmembrane electrical potential also influences carrier activity. Therefore, Ca^{2+} movement can occur in either direction, depending on the magnitude and polarity of the Na^+ gradient and membrane potential.

The concept of Na,Ca exchange has its origin in early studies which assessed the effects of Na^+ and Ca^{2+} on myocardial contractility and revealed a correlation between the extracellular concentration of these ions and tension development. However, the first direct demonstration that a unique transporter promotes this reaction came from experiments monitoring $^{45}Ca^{2+}$ fluxes in response to changing transmembrane Na^+ concentration gradients in squid axon (BAKER et al. 1967) and atria (REUTER and SEITZ 1968; GLITSCH et al. 1970). Since these pioneering studies, Na,Ca exchange has been described in several types of excitable cells, including those derived from neural, muscle (e.g., cardiac, skeletal, crustacean, and smooth muscle), endocrine, and visual transducing tissues. This reaction is also found in nonexcitable cells (e.g., kidney, bladder, intestine, sperm, and liver). A different nonelectrogenic Na,Ca exchange system is present in mitochrondria isolated from brain and heart. The kinetic and thermodynamic properties of Na,Ca exchange have been characterized by using either isotopic flux measurements in intact tissue preparations, electrophysiologic recording techniques, or transport measurements in isolated membrane vesicles.

REEVES and SUTKO (1979) first described Na,Ca exchange in membrane vesicles by showing that a crude cardiac sarcolemmal membrane preparation would accumulate Ca^{2+} in response to an outwardly directed Na gradient with an absolute requirement for Na^+ as the monovalent cation substrate. With more highly purified sarcolemmal vesicle preparations, it has been demonstrated that the reaction is electrogenic (REEVES and SUTKO 1980), completely reversible (PHILIPSON and NISHIMOTO 1982a), and has a stoichiometry of $3Na^+:Ca^{2+}$ (REEVES and HALE 1984). Moreover, the kinetics of Na,Ca exchange have been studied in detail in cardiac vesicles (KADOMA et al. 1982; REEVES and SUTKO 1983; PHILIPSON 1985b). This work revealed several interesting features of the transport reaction. Na,Ca exchange can operate at a very high capacity. Typical V_{max} values from these vesicle studies translate into a Ca^{2+} flux of 10 pmol s^{-1} cm^{-2} in cardiac tissue (REEVES 1985), which compares favorably with a value for exchange activity of 30 pmol s^{-1} cm^{-2} measured in guinea pig atria (CHAPMAN et al. 1983). Multiple classes of ion-binding sites have been identified and the interaction of Na^+ is highly cooperative. Studies of transport in right-side-out and inside-out vesicles indicate that most properties of the reaction are symmetric. Investigation of nonproductive modes of carrier action (i.e., Ca,Ca and Na,Na exchange) have also provided insight into mechanism. For example, measurement of transstimulated Ca^{2+} fluxes in cardiac vesicles indicate that V_{max} of Ca,Ca exchange can be increased by various alkali metal ions without subsequent transport of the stimulatory cation (SLAUGHTER et al. 1983), as has previously been shown in squid axon (BLAUSTEIN 1977).

Together, these flux studies have led to the proposal of a mechanistic model for Na,Ca exchange in heart (REEVES 1985). In this scheme (Fig. 1), two classes of ion-binding sites exist; a site to which either 1–$2Na^+$ or Ca^{2+} bind (A-site) and a separate site where Na^+ can bind (B-site). Occupation of the A-site by $2Na^+$ places the carrier in a conformation which allows the third Na^+ to bind at the B-site. Binding of $3Na^+$ promotes transport of Ca^{2+} bound to an A-site on the

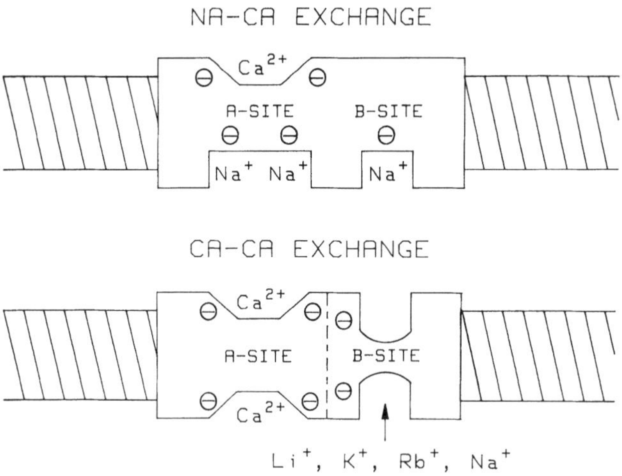

Fig. 1. Model of the Na,Ca exchange reaction in heart. This scheme illustrates proposed ion-binding sites on the cardiac Na,Ca exchange carrier as it functions in either the Na,Ca or Ca,Ca exchange mode. Anionic charges exist at both the A- and B-sites which bind Ca^{2+}, Na^{+}, and various monovalent cations which stimulate Ca,Ca exchange. The *broken line* indicates that monovalent cations are not translocated during Ca,Ca exchange. See text for further details

transmembrane surface in a simultaneous reaction. Occupation of an A-site by Ca^{2+} allows broader substrate specificity at the B-site and it was postulated that this is the site of interaction of alkali metal ions which stimulate Ca,Ca exchange. A conceptually similar model has been proposed for Na,Ca exchange in squid axon (BAKER et al. 1969; BLAUSTEIN 1977). Although most vesicular studies have been performed with cardiac sarcolemmal membranes, similar properties for Na,Ca exchange have been noted with membranes isolated from other sources (e.g., brain, pituitary, and smooth muscle). There is also general agreement between data obtained with vesicles and intact tissues in terms of the kinetic and thermodynamic properties of this reaction, although some systems (e.g., squid axon) display unique properties.

While the existence of Na,Ca exchange has been demonstrated in a number of tissues, its physiologic role is the subject of considerable controversy. The most extensive investigations have been made in cardiac tissue. The relative contributions of this system, other Ca^{2+} transport reactions, sequestration organelles, and intracellular Ca^{2+}-binding sites in maintaining cardiac Ca^{2+} homeostasis have yet to be precisely determined, and available data are very dependent on experimental conditions. Given recent measurements of intracellular Na^{+} and Ca^{2+} activities and the stoichiometry of the exchange reaction, it has been argued on thermodynamic grounds that Na,Ca exchange functions in net Ca^{2+} efflux in heart cells at resting membrane potentials (e.g., −80 mV: CHAPMAN 1983; REEVES 1985). However, because of reversibility and electrogenicity, it is possible to change the direction of Ca^{2+} flux by depolarizing the membrane to the reversal potential of Na,Ca exchange and drive net Ca^{2+} uptake (MULLINS 1979). This might occur during the plateau phase of the cardiac action potential. Therefore, conceptually it is possible for Na,Ca exchange to contribute to Ca^{2+} uptake or efflux, depending on the cell's ionic and electrical gradients. Unfortunately, complications arise from the application of equilibrium thermodynamics to in vivo situations where the intracellular ionic environment is under steady state conditions. There is also uncertainty as to the concentration of ionic species at the sarcolemmal membrane surface during cycles of excitation–contraction coupling (Langer 1984). Thus, one cannot predict at present the extent to which Na,Ca exchange controls intracellular Ca^{2+} in heart, either in a direct fashion or by modulating total Ca^{2+} stores.

Nonetheless, many situations have been described which are consistent with the functional expression of Na,Ca exchange activity in intact cardiac tissue (Mullins 1981; Chapman 1983; Reeves 1985). For instance, measurements of intracellular ion activities indicate that Na^+ concentration rapidly decreases in heart cells when extracellular Na^+ is lowered, while increasing intracellular Na^+ promotes a corresponding increase in intracellular Ca^{2+} activity. These ionic fluctuations are thought to be mediated directly by the exchange protein. Amphibian cardiac cells have elevated intracellular Na^+ levels and lack a significant sarcoplasmic reticulum Ca^{2+}-buffering capacity. In this tissue, tension measurements indicate that Na,Ca exchange is in equilibrium with the transmembrane Ca^{2+} gradient, and that it can directly control intracellular Ca^{2+}. Furthermore, conditions which elevate intracellular Na^+ in mammalian heart produce a positive inotropic effect, suggesting that Na,Ca exchange can affect contractility by controlling Ca^{2+} available for loading intracellular Ca^{2+} stores. However, until the directionality of Na,Ca exchange-mediated Ca^{2+} flux is resolved in heart, it will be difficult to assess the relative importance of this system in maintaining Ca^{2+} homeostasis.

Na,Ca exchange has also been implicated in certain pathophysiologic situations. These conditions have in common the elevation of intracellular Na^+. Baker et al. (1969) originally suggested that the cardiac inotropic response to digitalis, which raises cell Na^+ by blocking the Na^+,K^+-ATPase, is a consequence of Na,Ca exchange activity. In support of this suggestion, treatment with Na pump inhibitors has clearly been shown to increase intracellular Ca^{2+} of heart cells via Na,Ca exchange (Sheu and Fozzard 1982; Lee and Dagostino 1982; Eisner et al. 1983). Ca^{2+} overloading of the sarcoplasmic reticulum by this mechanism could lead to arrhythmogenic conditions as induced by toxic levels of ouabain. Spontaneous fluctuations in intracellular Ca^{2+} levels occur in heart owing to asynchronous release of Ca^{2+} from the sarcoplasmic reticulum, and this is accentuated when overloading occurs (Orchard et al. 1983; Wier et al. 1983). Since intracellular Ca^{2+} is known to regulate voltage-dependent Ca^{2+} channels and other channels responsible for a transient inward current (Noble 1984), spontaneous fluctuations in Ca^{2+} levels could interfere with normal rhythmicity and lead to the onset of arrhythmias. In ischemic tissue, intracellular Na^+ has also been shown to increase. Some of the cardiac damage and subsequent arrhythmias which occur during ischemia and reperfusion injury (Braunwald and Kloner 1985) are likely to result from exchange-mediated Ca^{2+} influx (Daly et al. 1984). One hypothesis suggests that the development of essential hypertension is linked to the action of Na,Ca exchange in vascular smooth muscle (Blaustein and Ham-

LYN 1984). In a hypothetical mechanism, the production of an endogenous Na pump inhibitor in response to excessive Na^+ and water retention by the body could block Na pump activity in smooth muscle and lead to an elevation in intracellular Na^+. As a result, Ca^{2+} would enter the cell via Na,Ca exchange, promote muscle contractility, and thus raise vascular tone and elevate peripheral resistance. Although these ideas provide a basis for blood pressure elevation, the endogenous Na pump inhibitor has yet to be purified, and the pathophysiologic condition may have other contributing factors.

Studies on the role of Na,Ca exchange in intact systems have been hindered by lack of potent inhibitors which can selectively modulate exchange activity in vivo. Unquestionably, such inhibitors would be very useful in a wide variety of tissues for dissecting the involvement of Na,Ca exchange in excitation–response mechanisms, just as specific inhibitors of voltage-dependent Ca^{2+} channels have been successfully employed in studies of this type. However, some compounds which block Na,Ca exchange have been identified and these provide a basis for future development.

C. Identification of Na,Ca Exchange Inhibitors

The search for inhibitors of Na,Ca exchange has proven problematic. It is known that various inorganic ions block exchange-mediated Ca^{2+} fluxes. These include Mg^{2+}, Mn^{2+}, Cd^{2+}, La^{3+}, Ba^{2+}, and Sr^{2+} (TROSPER and PHILIPSON 1983). The latter two ions also serve as substrates for the transporter (PHILIPSON 1985a). La^{3+} is the most potent inorganic inhibitor of exchange activity and functions in vesicles (REEVES and SUTKO 1979), as well as in intact tissues such as heart (BARRY and SMITH 1982) and squid axon (BAKER et al. 1969). Since La^{3+} interferes with most Ca^{2+} transport reactions, its utility is limited, although some progress has been reported using this inhibitor to study Na,Ca exchange in isolated myocytes (BARRY and SMITH 1982).

Several organic molecules have been described as inhibitors of Na,Ca exchange. Chlorpromazine (CARONI et al. 1980), dibucaine, tetracaine, and ethanol (MICHAELIS and MICHAELIS 1983), verapamil (ERDREICH et al. 1983), and amrinone (MALLOV 1983) inhibit Na,Ca exchange in vesicles, but all function at very high concentrations. The antibiotic polymyxin B (PHILIPSON and NISHIMOTO 1982b) and quinidine (MENTRARD et al. 1984) inhibit this transport system in cardiac preparations. Adriamycin, an anthracycline antibiotic which is cardiotoxic, has been reported to be a potent blocker of exchange activity in cardiac sarcolemmal membranes (CARONI et al. 1981). However, its ability to inhibit Na,Ca exchange in cardiac (REEVES 1985) or pituitary (KACZOROWSKI et al. 1984) vesicles has not been confirmed by other investigators, and adriamycin was only partially effective in blocking a current related to Na,Ca exchange in heart cells (MENTRARD et al. 1984). The reason for this discrepancy is not clear at present. Quinacrine inhibits Na,Ca exchange in sarcolemmal vesicles ($K_i = 20\ \mu M$; REEVES 1985), but its utility for physiologic investigations is limited because of the many actions of this compound. Various cationic amphiphilic molecules such as dodecylamine are relatively potent inhibitors of exchange in sarcolemmal membranes

(IC_{50} of dodecylamine is approximately 20 μM; Philipson 1984) and have been shown to modulate cardiac contractility in isolated tissues (Philipson et al. 1985). It is uncertain whether these agents act directly at the level of the carrier or alter the membrane environment owing to their detergent-like properties, thus affecting carrier activity in a secondary fashion. The Ca^{2+} indicator dye, quin2, inhibits Na,Ca exchange in squid axon, blocking Na-dependent Ca^{2+} influx without affecting the efflux reaction (Allen and Baker 1985). These findings question the validity of using quin2 to study Ca^{2+} homeostasis in tissues where Na,Ca exchange is involved. Harmaline, an inhibitor of many different Na^+ transport reactions, has also been shown to block Na,Ca exchange in ileum smooth muscle (Suleiman and Hider 1985).

Organic Ca^{2+} entry blockers, in general, are not inhibitors of Na,Ca exchange. However, some inhibitory activity with certain of these agents has been reported. For example, diltiazem (Vaghy et al. 1982) and other benzothiazepines and benzodiazepines (Matlib et al. 1983) block Na,Ca exchange in cardiac mitochondria at micromolar concentrations without affecting the sarcolemmal transport system. One such agent, clonezapam, inhibits the mitochondrial exchanger, but does not possess significant Ca^{2+} channel blocker activity (Matlib et al. 1985), making it perhaps a relatively specific probe of mitochondrial Na,Ca exchange activity in heart. Nicardipine, a dihydropyridine, is reported to block Na,Ca exchange in sarcolemmal vesicles, however, only Na-dependent Ca^{2+} uptake, not Na-dependent Ca^{2+} efflux was affected (Takeo et al. 1985). Since the exchange reaction has been shown to be symmetric in cardiac membranes (Philipson 1985b), the specificity of this interaction is uncertain. The aralkylamine Ca^{2+} channel inhibitor bepridil also inhibits sarcolemmal Na,Ca exchange (Garcia et al. 1985), and it has been a useful agent for mechanistic studies (see Sect. D).

Several laboratories have been attempting to raise antibodies that would be specific inhibitors of Na,Ca exchange. A monoclonal antibody, 4F2, has recently been reported to increase intracellular Ca^{2+} and concomitantly inhibit parathyroid hormone secretion in human adenomatous parathyroid cells (Posillico et al. 1985), suggesting that it is directed against an antigen responsible for clearing Ca^{2+} from these cells. Another monoclonal antibody, 44D7, produced against the non-T,non-B acute lymphoblastic leukemia cell line HOON reacts with similar plasmalemmal antigenic determinants as does 4F2 (Quackenbush et al. 1986). Interestingly, when 44D7 was tested against Na-dependent Ca^{2+} fluxes in cardiac and skeletal muscle sarcolemmal vesicles, Na,Ca exchange was specifically inhibited (Michalak et al. 1986). These results imply that the antibody reacts either directly with the Na,Ca exchange carrier protein or with a regulatory protein associated with the transporter. Production of such antibodies raises the exciting possibilities that they can be used in the isolation of the Na,Ca exchange carrier, as well as in experiments directed at determining the physiologic role of Na,Ca exchange in intact tissue.

Although this chapter is concerned with inhibitor development, it is noteworthy that different classes of Na,Ca exchange stimulatory agents have also been identified. One class consists of hydrophobic anions which presumably perturb the membrane environment in which the carrier is located and thereby enhance transport activity. Anionic phospholipids (PHILIPSON et al. 1983; PHILIPSON and NISHIMOTO 1984), negatively charged amphiphilic molecules (PHILIPSON 1984), and fatty acids (PHILIPSON and WARD 1985) have this effect in cardiac sarcolemmal vesicles. Since fatty acids are produced in ischemic myocardium, stimulation of exchange activity by these agents may have important physiologic consequences in heart. It should be noted that rather high concentrations of amphiphile are required relative to membrane phospholipid to elicit these effects in vesicles (ASHAVAID et al. 1985). The other class of stimulatory agents is believed to act through a direct redox modification of the carrier protein (REEVES et al. 1986). Both an oxidizing agent (e.g., H_2O_2, O_2, oxidized glutathione) and a reducing agent (e.g., dithiothreitol, glutathione, Fe^{2+}, $O_2^{\bar{\cdot}}$) are required to elicit the effect, and stimulation is substrate dependent (i.e., Na^+ is required, but Ca^{2+} blocks the effect). Stimulation of Na,Ca exchange in vivo by a redox modification may be important in conditions of ischemia, but could also be a means for regulating exchange activity in normal tissue. The physiologic consequences of stimulating Na,Ca exchange remain to be elucidated.

The diuretic, amiloride (Fig. 2) is an inhibitor of a number of different transport systems in which Na^+ is a substrate, such as epithelial Na^+ channels and Na,H exchange (BENOS 1982). It is also a weak inhibitor of Na,Ca exchange in murine erythroleukemia cells (SMITH et al. 1982), and in vesicles from brain (SCHELLENBERG et al. 1983; GILL et al. 1984), pituitary (KACZOROWSKI et al. 1984), and heart (SIEGL et al. 1984), where if functions at millimolar concentrations. However, because of poor potency and selectivity, amiloride is an unattractive agent for studying the exchange reaction. A wide variety of amiloride derivatives have been prepared, and some of these possess significantly greater potency and selectivity than amiloride as transport inhibitors in different systems. Two major classes of compounds have been synthesized: amiloride substituted on either (a) the terminal guanidino nitrogen or (b) the 5-pyrazine ring nitrogen.

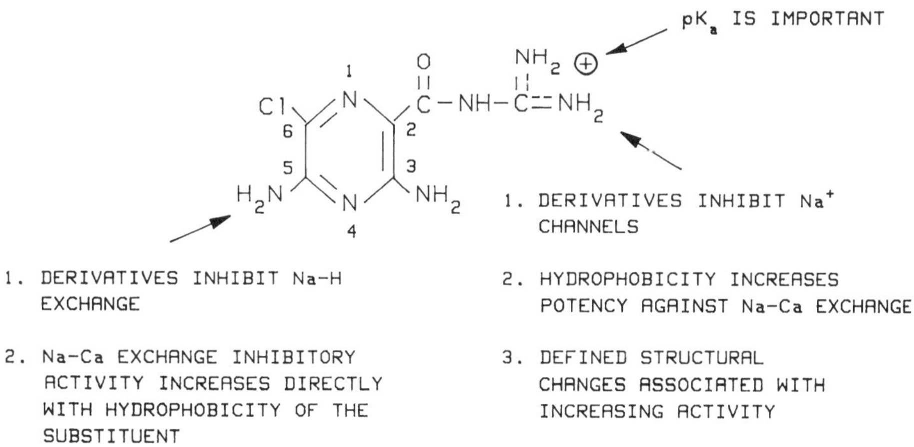

Fig. 2. Structure–activity relationship of amiloride and amiloride derivatives as transport inhibitors. The structure of amiloride, a pyrazinoyl acylguanidine, is shown in its protonated form. Also shown are positions at which substitution will increase the potency of amiloride as an inhibitor of epithelial Na^+ channels, Na,H exchange, or Na,Ca exchange

Table 1. The effect of various guanidino-nitrogen and 5-pyrazine-nitrogen-substituted amiloride derivatives on Na, Ca exchange activity in pituitary, brain, and heart[a]

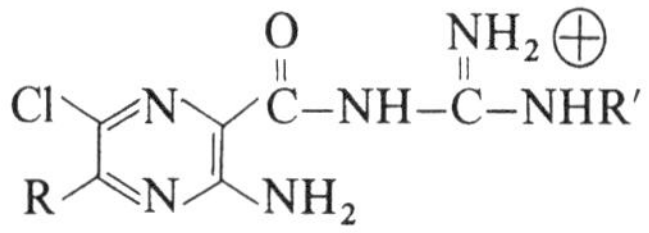

Compound	Inhibition (%) at 100 μ*M* (IC_{50})		
	Pituitary	Brain	Heart
R = $-NH_2$ R′ = H (Amiloride)	5	20	10
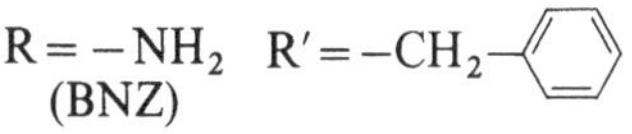 (BNZ)	50	47	45
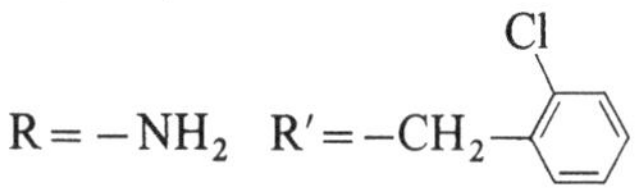	100 (20 μ*M*)	00	100 (29 μ*M*)
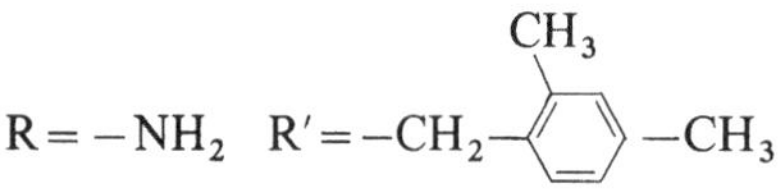	100 (10 μ*M*)	90 (40 μ*M*)	100 (14 μ*M*)
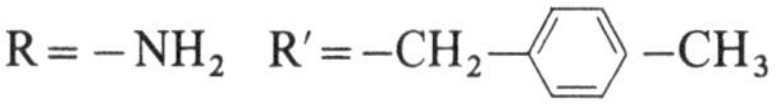	62	58	
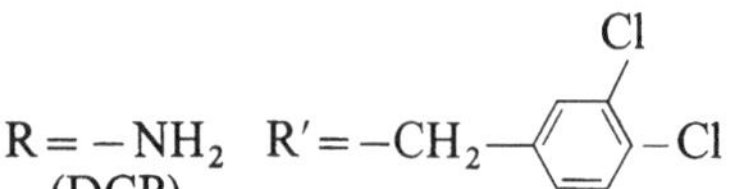 (DCB)	100 (30 μ*M*)	100 (35 μ*M*)	100 (17 μ*M*)
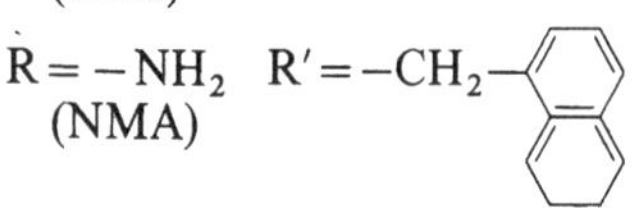 (NMA)	100 (35 μ*M*)	90 (45 μ*M*)	100 (10 μ*M*)
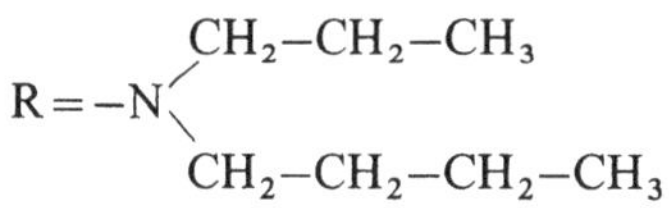 R′ = H	100 (20 μ*M*)	66	90 (40 μ*M*)
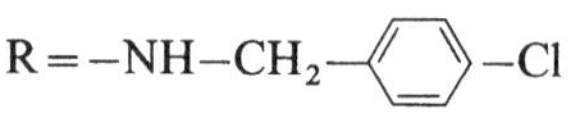 R′ = H	100 (20 μ*M*)	57	90 (45 μ*M*)

[a] Initial rates of Na, Ca exchange activity were measured in purified plasma membrane vesicles derived from rat pituitary, bovine brain, or porcine cardiac tissue. The experiment was then repeated in the presence of various amiloride derivatives at a concentration of 100 μ*M*. In some cases, IC_{50} values for inhibition were determined graphically from titrations of inhibitor activity and these values are listed in parentheses.

Members of the first class of derivatives are specific, high affinity blockers of passive Na^+ flux through non-tetrodotoxin-sensitive Na^+ channels (CUTHBERT and FANELLI 1978) while certain pyrazine-nitrogen-substituted analogs selectively block Na,H exchange at low concentrations (LAZDUNSKI et al. 1985). If certain amiloride derivatives are more effective than the parent compound in inhibiting Na,Ca exchange, one might use these molecules as specific probes of carrier action. Because of amiloride's known ability to interact at Na^+-binding sites on other transporters (i.e., function as an Na^+ mimic), many different analogs were tested for activity using heart, brain, and pituitary plasma membrane vesicles (KACZOROWSKI et al. 1985). Both terminal guanidino-nitrogen-substituted analogs and 5-amino-nitrogen-substituted derivatives were found to possess increased potency over that of amiloride as Na,Ca exchange inhibitors (Table 1). The structure–activity relationship determined for this interaction is different from that described for Na^+ channels or Na,H exchange inhibitors in two respects: first, inhibitory activities are restricted to only one type of amiloride derivative in these other systems, whereas both classes inhibit Na,Ca exchange; and second, within each class of analogs, the chemical substitution which gives rise to increased potency differs between transport systems. In the case of Na,Ca exchange inhibitors, defined structural modification of guanidino substituents are associated with increasing inhibitory activity while analogs bearing 5-amino substituents generally increase in potency with increasing hydrophobicity. The most efficacious inhibitors discovered are approximately 100-fold more potent than amiloride ($IC_{50} \sim 10\ \mu M$), and only minor differences were noted in the rank order of potency of these compounds in pituitary, heart, and brain vesicles.

Given that certain amiloride derivatives function as effective inhibitors of Na,Ca exchange in vesicles from a variety of tissues, the specificity of this interaction was assessed. Most of these studies were performed with potent inhibitors and either GH_3 pituitary cells or purified plasma membrane vesicles derived from that cell line (KACZOROWSKI et al. 1985). Selected members of either class of analog did not disrupt the osmotic integrity of the membrane, did not significantly interfere with Ca^{2+}-ATPase-dependent transport activity or Na pump enzymatic activity, and did not block ligand–receptor interactions (e.g., ouabain or thyrotropin-releasing hormone binding) at concentrations where they inhibited Na,Ca exchange. Whole cell voltage-clamp recordings of GH_3 cells revealed that voltage-dependent Ca^{2+} and K^+ channel and Ca^{2+}-activated K^+ channel activities were not significantly affected. However, some Ca^{2+} channel blocker activity was noted at high concentrations of 2,4-dimethylbenzylamiloride (40% block at $100\ \mu M$), the most potent guanidino-substituted amiloride inhibitor of GH_3 Na,Ca exchange ($IC_{50} = 10\ \mu M$). Other studies in heart indicate that a small window of specificity exists where Na,Ca exchange is blocked before other transport reactions (e.g., sarcolemmal Ca^{2+}-ATPase, Na^+ channels, Ca^{2+} channels) are inhibited (SIEGL et al. 1984).

It is important, however, that caution be exercised in the interpretation of physiologic results derived from use of these inhibitors, because they are hydrophobic cations and are likely to interfere with other cellular systems. Amiloride is an inhibitor of protein synthesis (LUBIN et al. 1982), Na pump activity (SOLTOFF and MANDEL 1983), and three different protein kinase activities (RALPH et al.

1982; HOLLAND et al. 1983; DAVIS and CZECH 1985); it also collapses transmembrane pH gradients (DUBINSKY and FRIZZELL 1983). Moreover, the interference of amiloride derivatives in various cell functions has been reported (ZHUANG et al. 1984; SCHELLENBERG et al. 1985b). We have found that some amiloride analogs interfere with the binding of Ca^{2+} channel inhibitors (e.g., nitrendipine, verapamil, diltiazem) to sarcolemmal vesicles, indicating that they may possess Ca^{2+} channel blocker activity, and that they also block β-receptor binding in cardiac membranes (ML GARCIA, VF KING, EJ CRAGOE and GJ KACZOROWSKI 1986, unpublished work). Therefore, if this group of compounds is to be used in intact tissue studies, their specificity must be assessed under all experimental conditions to ensure that the responses elicited are due only to Na,Ca exchange inhibition and not to interference with other systems.

The specificity of amiloride analogs as inhibitors of Na,Ca exchange in GH_3 vesicles was demonstrated in several ways (KACZOROWSKI et al. 1985). It was shown that protonation of the guanidino moiety (i.e., a positive charge) is essential for activity, that inhibition is freely reversible since block can be removed by washing membranes, that Na-dependent Ca^{2+} efflux as well as uptake is inhibited, and that there is no involvement of ΔpH in the Ca^{2+} transport reaction, excluding inhibition due to block of a contributing Na,H exchange activity. In addition, a kinetic analysis revealed that the two different classes of derivatives are noncompetitive inhibitors with respect to Ca^{2+} (Fig. 3), but are competitive with Na^+. This suggests that both inhibitor classes function in an identical fashion to block exchange. Furthermore, it was postulated that they do so by acting as Na^+ analogs, reversibly tying up the transporter in an inactive complex by interacting at an Na^+-binding site (or sites). This idea has been confirmed and extended by mechanistic studies of Na,Ca exchange inhibition in cardiac sarcolemmal vesicles (see Sect. D). Taken together with the defined structure–activity relationship which exists for amiloride derivatives, these results are a strong indication that amiloride inhibition of Na,Ca exchange is a specific phenomenon which occurs by direct interaction of inhibitors with the antiporter.

The efficacy of various other substituted amiloride molecules as inhibitors of Na,Ca exchange has been explored in brain synaptosomal membrane vesicles (SCHELLENBERG et al. 1985a). Most of the 5- and 6-pyrazine ring substituted analogs investigated were as weak as amiloride, or completely ineffective, in inhibiting exchange activity, but substitution of the terminal guanidino nitrogen with a benzyl group (benzamil, BNZ) significantly increased potency. This is consistent with previous observations that BNZ is tenfold more potent than amiloride in inhibiting the exchange reaction in vesicles from different tissues (Table 1). However, upon investigating the kinetics of inhibition, amiloride analogs were noted to be competitive inhibitors versus Ca^{2+} (SCHELLENBERG et al. 1983, 1985a). These observations are in contrast to other findings that all amiloride analogs are noncompetitive inhibitors versus Ca^{2+} in pituitary (see Fig. 3), heart, and brain membranes (KACZOROWSKI et al. 1985). Possible reasons for this discrepancy are discussed in Sect. D. In addition to inhibiting the plasmalemmal exchange reaction, amiloride (SORDAHL et al. 1984) and various guanidino-nitrogen-substituted analogs (JURKOWITZ et al. 1983) block Na,Ca exchange in cardiac mitochondria. For comparison, the K_i values of BNZ as an inhibitor are ap-

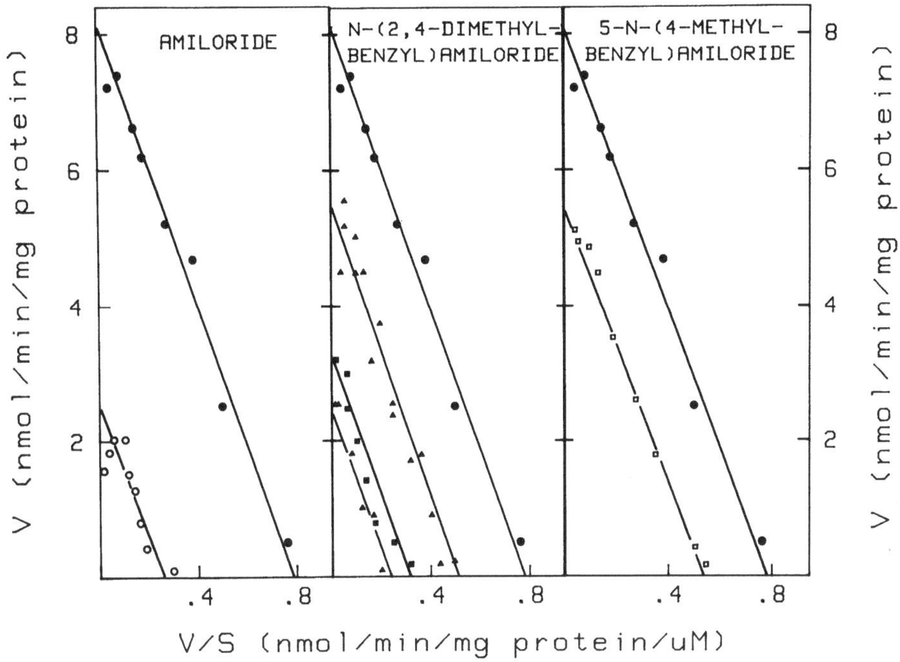

Fig. 3. Kinetic analysis of Na,Ca exchange inhibition in pituitary plasma membrane vesicles by amiloride derivatives. Initial rates of Na,Ca exchange were monitored in plasma membrane vesicles derived from GH_3 pituitary cells as a function of increasing Ca^{2+} concentration (1–100 μ*M*). Na-loaded vesicles (100 m*M* NaCl) were diluted into a medium of 100 m*M* KCl, 20 m*M* TRIS–Hepes, pH 7.4, containing $^{45}CaCl_2$, and uptake of Ca^{2+} was measured by filtration techniques (*full circles*). The experiment was then repeated with amiloride (*open circles* 1.5 m*M*), with a guanidino-nitrogen-substituted derivative, *N*-(2,4-dimethylbenzyl)amiloride (*full triangles* 10 μ*M*; *full squares* 20 μ*M*; *open triangles* 40 μ*M*), or with a 5-pyrazine-nitrogen-substituted analog, 5-*N*-(4-methylbenzyl)amiloride (*open squares* 20 μ*M*) present. The data are presented in the form of an Eadie–Hofstee plot

proximately equal in the mitochondrial and sarcolemmal systems, but the mechanism of block appears different for the two transport reactions. Amiloride analogs have also been shown to block Na,Ca exchange in brain mitochondria, but there is a marked difference in the structure–activity relationship for transport inhibition in mitochondria from the two tissues (SCHELLENBERG et al. 1985 b). Since some amiloride analogs may interfere with mitochondrial as well as with plasmalemmal Na,Ca exchange, use of these agents in vivo must take into account possible contributions of both transport processes in the control of cellular Ca^{2+} metabolism.

D. Mechanism of Na,Ca Exchange Inhibition by Amiloride and Bepridil

Three different guanidino-substituted amiloride analogs, benzamil (BNZ, K_i = 100 μ*M*), 3,4-dichlorobenzamil (DCB, K_i = 20 μ*M*) and *N*-1-naphthylmethylamiloride (NMA, $K_i = 10\ \mu M$) have been used to study the mechanism of Na,Ca exchange inhibition in purified cardiac sarcolemmal vesicles (CRAGOE et al. 1984; SLAUGHTER et al. 1984; REEVES 1985; GARCIA et al. 1986). Most of the data were obtained from kinetic studies of $^{45}Ca^{2+}$ fluxes performed under initial rate conditions (i.e., filtration measurements were made at 1 s). If initial rate conditions are not employed, the kinetic data can be equivocal, which would account for some conflicting reports in the literature. Inhibition by these three agents is concentration dependent, reversible, and complete. In addition, both Na-dependent Ca^{2+} uptake and Na-dependent Ca^{2+} efflux reactions are blocked in parallel, indicating that inhibitors act directly at the level of the carrier.

Inhibition patterns were assessed as a function of either Ca^{2+} or Na^+ concentration. Under the standard conditions of the assay (i.e., 160 mM Na^+_{in}, 160 mM K^+_{out}) an Eadie–Hofstee plot of initial transport rates versus Ca^{2+} concentration yielded a series of parallel lines as inhibitor concentration increased, indicating noncompetitive inhibition with respect to this ion. There was some indication of mixed kinetics (i.e., both competitive and noncompetitive components of inhibition) with weak inhibitors such as BNZ, although the inhibitory pattern was primarily noncompetitive. However, when sucrose was substituted for K^+ in the dilution buffer, inhibition clearly displayed mixed kinetics with all three compounds. These results demonstrate that under certain conditions, amiloride analogs display some competitive interaction with Ca^{2+}, but under physiologic ionic conditions they are primarily noncompetitive inhibitors. In contrast, when inhibition was studied as a function of Na^+ concentration, a strictly competitive interaction was observed. This was apparent in studies of Na-dependent Ca^{2+} efflux where increasing extravesicular Na^+ reverses exchange inhibition, and in measurements of Na-dependent Ca^{2+} uptake where increasing intravesicular Na^+ has the same effect.

Amiloride analogs block Na,Na and Ca,Ca exchange reactions, indicating that they interfere with all modes of carrier action (SLAUGHTER et al. 1984). Inhibition of Na,Na exchange was completly reversed by increasing Na^+ concentration, consistent with the competitive interaction between amiloride and Na^+ noted previously. Inhibition of Ca,Ca exchange was studied under two conditions; in the absence or presence of the monovalent stimulatory cation K^+. Greater inhibition was observed in sucrose than in K^+-containing media. For example, 160 mM K^+ shifts the K_i of DCB to a tenfold higher value. Indeed, increasing K^+ concentration reverses the inhibition of Ca,Ca exchange by BNZ, whereas substituting choline (which does not stimulate exchange) for K^+ does not. This suggests a competitive interaction between inhibitor and monovalent cations which stimulate exchange rates, but are not themselves transported. However, this interpretation is complicated by the observations that K^+ also competes with Ca^{2+} at the carrier's A-site and that there is little correlation between the ability of different monovalent cations to reverse inhibition and stimulate Ca,Ca exchange (J REEVES 1986, personal communication). Kinetic experiments indicate that Ca,Ca exchange inhibition is mixed versus Ca^{2+} in both sucrose and K^+-containing media, but it appears to be of a more competitive nature in sucrose. These data are consistent with the idea that amiloride analogs bind to different classes of ion-binding sites on the carrier, but that the interaction is primarily directed to Na^+-binding sites with a preference for the B-site. Since $2Na^+$ or Ca^{2+} have been hypothesized to share the common A-site (REEVES and SUTKO 1983), mixed patterns of inhibition may be expected under certain conditions. Monovalent cations which stimulate Ca,Ca exchange may do so by interacting at the B-site, although recent data suggest that they bind to a distinct site on the carrier (see discussion later in this section).

Two approaches were used to explore the action of these inhibitors in greater detail. If all amiloride analogs interact at one unique site, measurements of inhibitory activity using mixtures of different amiloride derivatives should yield parallel lines in a Dixon plot (i.e., behavior expected of mutually exclusive inhibitors). When the experiment was performed with mixtures of DCB and NMA, these results were not obtained. The data with DCB alone were not linear, but curved upward as DCB concentration was increased. Furthermore, repeating the experiment with varying DCB and increasing amounts of NMA generated a family of lines curving upward. All data became linear, however, when they were replotted as the square of DCB concentration. Thus, the interaction between amiloride analogs and the Na,Ca exchange protein appears to occur at two different sites. This was confirmed by monitoring the ability of different amiloride analogs to inhibit exchange activity and then subjecting the data to analysis in a Hill plot. Data thus obtained with DCB or NMA yielded a biphasic Hill plot with Hill coefficients of $n=1$ and $n=2$, at low and high inhibitor concentration, respectively. This is reminiscent of the relationship noted with Na^+ in its ability to stimulate Na,Ca exchange activity in cardiac vesicles (REEVES and SUTKO 1983). If the experiment was repeated with varying DCB in the presence of a concentration of NMA equal to its K_i, the Hill plot was monophasic with $n=2$. Taken together, these data suggest that amiloride analogs can interact at both the A- and B-sites on the carrier, but that the interaction is directed to only one site at low inhibitor concentration. That the interaction first occurs at the B-site is inferred from the noncompetitive kinetics of inhibition with respect to Ca^{2+}.

In order to test this postulate more completely, we sought to compare the action of other inhibitors with that of amiloride analogs. Bepridil (BPD) has recently been shown to be a relatively potent inhibitor of Na,Ca exchange in sarco lemmal vesicles ($K_i=30$ μM), and it appears to possess several interesting inhibitory properties which are different from those of amiloride (GARCIA et al. 1985, 1986). When Na-dependent Ca^{2+} uptake is measured in the presence of BPD, inhibition is concentration dependent, but only partial (i.e., maximum block is 50%). A different pattern was noted in studies of Na-dependent Ca^{2+} efflux. In this case, the concentration dependence of inhibition was similar to that obtained for influx (e.g., $K_i \sim 30$ μM), but inhibition was complete. One major difference in these two transport protocols is that in the former situation, Na^+ is already bound to the carrier before inhibitor is added, while in the latter case, inhibitor and Na^+ are presented to the carrier at the same time. These results are different from those obtained with amiloride where complete inhibition is observed under all conditions. A kinetic analysis of Na,Ca exchange inhibition versus Ca^{2+} indicates purely noncompetitive kinetics in both K^+ and sucrose media. In no instance were mixed or competitive kinetics observed. However, BPD inhibition was competitive with Na^+ since block was removed by increasing extravesicular

Na^+ in Na-dependent Ca^{2+} efflux experiments, and enhanced by lowering intravesicular Na^+ in Na-dependent Ca^{2+} uptake studies. Thus, although BPD and amiloride are similar in that both inhibitors are competitively antagonized by Na^+, differences in the kinetic profiles versus Ca^{2+} suggest a difference in the interactions of these compounds with the transporter.

Ca,Ca exchange is completely inhibited by BPD when the reaction is carried out in sucrose medium and inhibition is noncompetitive versus Ca^{2+}. Increasing K^+ concentration in the reaction buffer produces two effects: first, it raises the K_i of BPD and second, the extent of block becomes incomplete. However, the kinetics of inhibition remain noncompetitive as a function of Ca^{2+} in K^+ medium. These results confirm that differences exist in the mechanism by which BPD and amiloride analogs inhibit exchange and, together with the other studies, are a preliminary indication that BPD interacts with only one type of Na^+-binding site (i.e., the B-site). Moreover, they suggest that the action of monovalent stimula-

Table 2. Comparison of inhibitory activities of amiloride derivatives and bepridil[a]

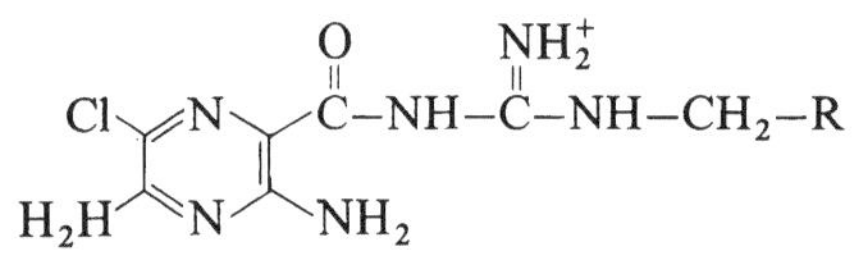

Amiloride derivatives

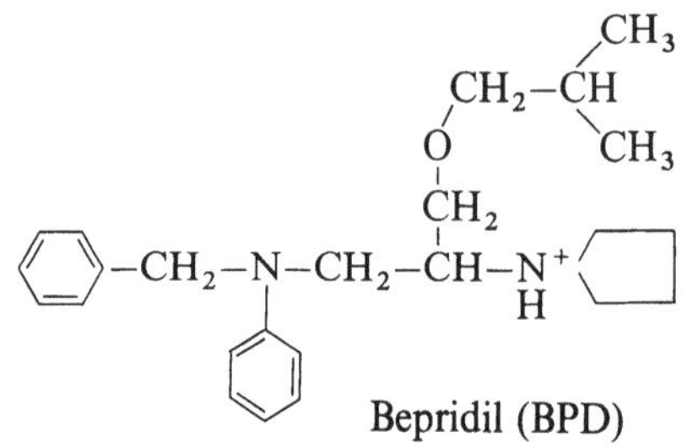

Bepridil (BPD)

1. Complete block of Na, Ca exchange under all conditions	Block of Na, Ca exchange is partial or complete depending on conditions
2. Inhibition noncompetitive vs Ca^{2+} in K^+, but mixed in sucrose medium	Inhibition noncompetitive vs Ca^{2+} in both K^+ and sucrose medium
3. Inhibition competitive with Na^+	Inhibition competitive with Na^+
4. Dixon plots of inhibition with mixtures of analogs are not linear	Dixon plots only yield parallel lines with varying BPD and low amiloride analog concentrations
5. Hill plots of inhibition are biphasic with $n=1$ and $n=2$, at low and high inhibitor concentration	Hill plots of inhibition are monophasic with $n=1$
6. Ca, Ca exchange inhibited with mixed kinetics vs Ca^{2+}	Ca, Ca exchange inhibited with non competitive kinetics vs Ca^{2+}
7. K^+ competitively antagonizes Ca, Ca exchange inhibition	K^+ competitively antagonizes Ca, Ca exchange inhibition and reduces the extent of block
8. Na, Na exchange inhibited with competitive kinetics vs Na^+	

[a] The inhibition profiles of three guanidino-nitrogen-substituted amiloride analogs (R = phenyl, BNZ; R = 3,4-dichlorophenyl, DCB; R = naphthyl, NMA) are compared with bepridil for their ability to block cardiac Na, Ca and Ca, Ca exchange.

tory cations on Ca,Ca exchange is complex and that their interaction may occur at a unique, previously undescribed site on the cardiac transport protein.

The relationship between BPD and amiloride analogs was investigated further by monitoring exchange activity with mixtures of these blockers present. At low fixed DCB or NMA concentrations and varying BPD, Dixon plots of inhibition display parallel lines, suggesting a competitive interaction between the different inhibitors. As the concentration of either DCB or NMA is increased, however, parallel line behavior degenerates. This kinetic profile suggests that as concentration increases, amiloride analogs interact at another site on the carrier where BPD does not interact. These results predict that analysis of BPD inhibition in a Hill plot should yield monophasic data with a Hill coefficient of 1. Consistent with this prediction, Hill plots of BPD inhibition of Na-dependent Ca^{2+} uptake, Na-dependent Ca^{2+} efflux, and Ca,Ca exchange in either sucrose or K^+ medium are all monophasic with $n=1$ in each case. Thus, BPD binding is restricted to only one class of Na^+-binding sites on the carrier. The data obtained with amiloride analogs and BPD are summarized in Table 2.

A proposed mechanism for Na,Ca exchange inhibition by amiloride and BPD is shown schematically in Fig. 4. Both chemically distinct classes of inhibitors interact with specific substrate-binding sites on the carrier. Amiloride derivatives inhibit Na,Ca exchange by binding to both the A- and B-sites, but they have a preference for the B-site. When an interaction occurs at the A-site, a competitive component of inhibition is observed in the kinetics measured with varying Ca^{2+}. BPD, on the other hand, binds to the B-site only, and this is sufficient to inhibit

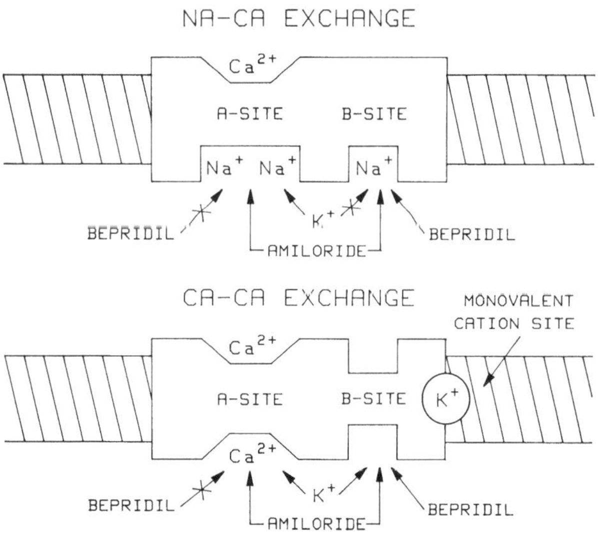

Fig. 4. Mechanism of Na,Ca exchange inhibition by amiloride and bepridil. This scheme illustrates the sites of interaction of amiloride and bepridil with the cardiac Na,Ca exchanger as it functions in either the Na,Ca or Ca,Ca exchange mode. Amiloride derivatives interact at both A- and B-sites while bepridil binds to only the B-site of the transporter. The binding site for monovalent cations which stimulate Ca,Ca exchange is represented as being distinct from the B-site. See text for further details

carrier activity. The partial inhibition by BPD under certain conditions could be related to its interaction at only one Na^+-binding site and/or to the order of substrate binding and release during carrier turnover. Because K^+ interacts at the A-site, it can alter the potency of block by amiloride, but has no effect on BPD inhibition of Na,Ca exchange. In the Ca,Ca exchange mode, the B-site must be accessible to amiloride, BPD, and K^+. Occupation of the B-site by either inhibitor class is sufficient to block carrier turnover. In addition, amiloride inhibitors interact competitively with Ca^{2+} at the A-site, causing mixed inhibition kinetics versus Ca^{2+}, a pattern which is not observed with BPD. The effects of K^+ on the Ca,Ca exchange reaction are complex. It can interact at both the A- and B-sites, but also appears to bind to another site which may be involved in stimulating Ca,Ca exchange rates. The fact that, during Ca,Ca exchange, K^+ shifts the BPD concentration dependence, and also causes BPD inhibition to become incomplete, is consistent with this idea. More kinetic experiments are in progress with other inhibitors to verify this proposed mechanism for the Na,Ca exchange reaction in heart.

E. Pharmacology of Na,Ca Exchange Inhibitors

Our present understanding of the pharmacology of Na,Ca exchange inhibitors is complicated because studies have been performed without potent selective inhibitors of this transport reaction. However, several investigations made with amiloride and various amiloride analogs are worth noting. Amiloride, at high concentration, has been shown to produce a positive inotropic and negative chronotropic effect in isolated atrial muscle from guinea pig (POUSTI and KHOYI 1979). These effects have been confirmed by others who postulated that they are due to a direct non-catechol-mediated mechanism which involves inhibition of Na,Ca exchange activity (FLOREANI and LUCIANI 1984). However, amiloride may inhibit other ion flux pathways in heart, such as K^+ channels (YAMASHITA et al. 1981), which could account for some of this pharmacologic activity. DCB was found to produce a small positive inotropic effect in guinea pig atria (SIEGL et al. 1983), but to cause a negative inotropic effect in guinea pig papillary muscle (SIEGL et al. 1984). Interestingly, amphiphilic cation inhibitors of Na,Ca exchange also produce negative inotropism in rabbit papillary while other agents which stimulate Na,Ca exchange are positive inotropic agents in this muscle (PHILIPSON et al. 1985). Whether the negative inotropic effect of DCB in papillary is related to blocking Ca^{2+} influx via the exchanger, or is due to Ca^{2+} channel blocker activity of this compound, is not clear. However, it is possible that Na,Ca exchange may be responsible for net Ca^{2+} influx in some types of cardiac tissue (e.g., papillary), while it promotes Ca^{2+} efflux in other tissue (e.g., atria). This would explain the ability of Na,Ca exchange inhibitors to produce both positive and negative inotropic effects. Based on data such as these, it has been suggested that Na,Ca exchange inhibitors would be a useful new class of inotropic agents (LUCIANI and FLOREANI 1985).

The best evidence that amiloride analogs inhibit Na,Ca exchange in intact tissue comes from studies with Na-loaded preparations. Treatment of papillary muscle with either ouabain or veratridine is known to produce a positive inotropic effect due to Na^+ loading and subsequent Ca^{2+} influx via the exchanger. DCB was shown to block the positive inotropic effect elicited by either treatment over a concentration range in which Na,Ca exchange is inhibited in vesicles (SIEGL et al. 1984). DCB also prevents arrhythmias produced by toxic levels of ouabain in isolated muscle preparations, as well as in anesthetized dogs (BUSH et al. 1985). Moreover, the positive inotropic effect due to elevating intracellular Na^+ by increasing electrical stimulation frequency (i.e., treppe response) is blocked by DCB (PKS SIEGL 1985, personal communication). Since DCB has been shown to block a current generated by Na,Ca exchange in isolated myocytes (HADLEY et al. 1985; HUME et al. 1985), it is clear that this compound functions as a transport inhibitor in intact cells with similar properties to those observed in vesicle experiments. These results, then, are completely consistent with the hypothesis that elevating intracellular Na^+ results in Na,Ca exchange-mediated Ca^{2+} influx which, in turn, can be blocked by amiloride analog inhibitors of the exchange reaction.

Studies of the Ca^{2+} paradox in cardiac muscle are a further test of this hypothesis. It has been observed that when cardiac tissue is perfused with Ca^{2+}-free buffer and then reexposed to Ca^{2+}, the cells accumulate large quantities of Ca^{2+}, go into contracture, lose excitability, and show significant intracellular damage (BONVALLET et al. 1984; NAYLER et al. 1984). Ca^{2+} influx via Na,Ca exchange has been postulated to be involved in these events (CHAPMAN 1983; NAYLER et al. 1984). When extracellular Ca^{2+} is removed, there is a large rise in intracellular Na^+ owing to Na^+ entry through Ca^{2+} channels (CHAPMAN et al. 1984). Upon readmission of Ca^{2+}, Na,Ca exchange is thought to mediate a large influx of Ca^{2+}, which can rise to cytotoxic levels. Recent studies of the Ca^{2+} paradox in frog atrial muscle have shown that if DCB is added during a 5 min Ca^{2+}-free perfusion, there is a concentration-dependent inhibition of the contracture which normally occurs upon Ca^{2+} readmission and the cells regain excitability (i.e., the ability to elicit tension is completely restored; G SUAREZ-KURTZ 1986, personal communication). Since DCB does not produce a negative inotropic effect in frog atria, these effects are presumably not due to Ca^{2+} channel blocker activity of the compound. Taken together with other studies on Na loaded preparations, the data suggest that amiloride analogs are effective inhibitors of Na,Ca exchange in intact tissue.

In order to have a better understanding of the physiologic role of Na,Ca exchange in excitation–response coupling, more specific inhibitors must be developed. Inhibitors useful for such studies could be small organic molecules, more complex structures isolated from natural product sources which are specific toxins, or antibodies directed against the Na,Ca exchange transport protein. Until these specific inhibitors are found, the function of Na,Ca exchange in Ca^{2+} homeostasis of various cells will remain the subject of debate. Nonetheless, it is exciting to speculate that once specific inhibitors are developed, they could provide a new class (or classes) of therapeutically effective agents.

References

Allen TJ, Baker PF (1985) Intracellular Ca indicator Quin-2 inhibits Ca influx via Na-Ca exchange in squid axon. Nature 315:755–756

Ashavaid TF, Colvin RA, Messineo FC, Macalister T, Katz AM (1985) Effects of fatty-acids on sodium-calcium exchange in cardiac sarcolemmal membranes. J Mol Cell Cardiol 17:851–862

Baker PF, DiPolo R (1984) Axonal calcium and magnesium homeostasis. Curr Top Membr Trans 22:195–247

Baker PF, Blaustein MP, Hodgkin AL, Steinhardt RA (1967) The effect of sodium concentration on calcium movements in giant axons of Loligo forbesi. J Physiol 192:43P–44P

Baker PB, Blaustein MP, Hodgkin AL, Steinhardt RA (1969) The influence of calcium on sodium efflux in squid axons. J Physiol 200:431–458

Barry WH, Smith TW (1982) Mechanisms of transmembrane calcium movement in cultured chick embryo ventricular cells. J Physiol 325:243–260

Benos DJ (1982) Amiloride: a molecular probe of sodium transport in tissues and cells. Am J Physiol 242:C131–C145

Blaustein MP (1977) Effects of internal and external cations and of ATP on sodium-calcium and calcium-calcium exchange in squid axons. Biophys J 20:79–111

Blaustein MP, Hamlyn JM (1984) Sodium transport inhibition, cell calcium, and hypertension. Am J Med 77:45–59

Bonvallet R, Rougier O, Tourneur Y (1984) Role of Na-Ca exchange in the calcium paradox in frog auricular trabeculae. J Mol Cell Cardiol 16:623–632

Braunwald E, Kloner RA (1985) Myocardial reperfusion: a double-edged sword? J Clin Invest 76:1713–1719

Bush LR, Kaczorowski GJ, Siegl PKS (1985) Antiarrhythmic properties of dichlorobenzamil, a sodium-calcium exchange inhibitor. Circulation 72-III:313

Caroni P, Reinlib L, Carafoli E (1980) Charge movements during the Na-Ca exchange in heart sarcolemmal vesicles. Proc Natl Acad Sci USA 77:6354–6358

Caroni P, Villani F, Carafoli E (1981) The cardiotoxic antibiotic doxorubicin inhibits the Na-Ca exchange of dog heart sarcolemmal vesicles. FEBS Lett 130:184–186

Chapman RA (1983) Control of cardiac contractility at the cellular level. Am J Physiol 245:H535–H552

Chapman RA, Coray A, McGuigan JAS (1983) Sodium-calcium exchange in mammalian heart: the maintenance of low intracellular calcium concentration. In: Drake-Holland AJ, Noble MIM (eds) Cardiac metabolism. Wiley, New York, p 117

Chapman RA, Rodrigo GC, Tunstall J, Yates RJ, Busselen P (1984) Calcium paradox of the heart: a role for intracellular sodium ions. Am J Physiol 247:H874–H879

Cragoe E, Kaczorowski GJ, Reeves JP, Slaughter RS (1984) Amiloride analogs interact with the monovalent cation binding site of the bovine heart sodium-calcium exchange carrier. J Physiol 353:74p

Cuthbert AW, Fanelli GM (1978) Effect of some pyrazinecarboxamides on sodium transport in frog skin. Br J Pharmacol 63:139–149

Daly MJ, Elz JS, Nayler W (1984) Sarcolemmal enzymes and the Na-Ca exchange in hypoxic, ischemic, and reperfused rat hearts. Am J Physiol 247:H237–H243

Davis R, Czech M (1985) Amiloride directly inhibits growth factor receptor tyrosine kinase activity. J Biol Chem 260:2543–2551

Dubinsky WP, Frizzell RA (1983) A novel effect of amiloride on H-dependent Na transport. Am J Physiol 245:C157–C159

Eisner DA, Lederer WJ, Vaughan-Jones RD (1983) The control of tonic tension by membrane potential and intracellular sodium activity in the sheep cardiac Purkinje fiber. J Physiol 335:723–743

Erdreich A, Spanier R, Rahamimoff H (1983) The inhibition of Na-dependent Ca uptake by verapamil in synaptic plasma membrane vesicles. Eur J Pharmacol 90:193–202

Floreani M, Luciani S (1984) Amiloride: relationship between cardiac effects and inhibition of Na-Ca exchange. Eur J Pharmacol 105:317–322

Garcia ML, King VF, Kaczorowski GJ (1985) Inhibition of Na-Ca exchange in cardiac sarcolemmal membrane vesicles by bepridil. Circulation 72-III:313

Garcia ML, King VF, Kaczorowski GJ (1986) Interaction of bepridil and amiloride with the Na-Ca antiporter in cardiac sarcolemmal vesicles. Biophys J 49:545a

Gill DL, Chueh S-H, Whitlow CL (1984) Functional importance of the synaptic plasma membrane calcium pump and sodium-calcium exchanger. J Biol Chem 259:10807–10813

Glitsch HG, Reuter H, Scholz H (1970) The effect of the internal sodium concentration on calcium fluxes in isolated guinea-pig auricles. J Physiol 209:25–43

Hadley RW, Hume JR, Kaczorowski GJ, Siegl PKS, Vassilev PM (1985) Block of "creep currents" in single frog atrial cells by vesicular Na-Ca exchange inhibitors. J Physiol 369:89

Holland R, Woodgett J, Hardie D (1983) Evidence that amiloride antagonises insulin-stimulated protein phosphorylation by inhibiting protein kinase activity. Fed Eur Biochem Soc 154:269–273

Hume JR, Kaczorowski GJ, Siegl PKS (1985) Lanthanum and 3′,4′-dichlorobenzamil block "creep currents" in single atrial myocytes. Circulation 72-III:230

Jurkowitz MS, Altschuld RA, Brierley GP, Cragoe EJ Jr (1983) Inhibition of Na-dependent Ca efflux from heart mitochondria by amiloride analogs. FEBS Lett 162:262–265

Kaczorowski GJ (1985) Sodium-calcium exchange and calcium homeostasis in excitable tissue. Annu Rep Med Chem 20:215–226

Kaczorowski GJ, Costello L, Dethmers J, Trumble MJ, Vandlen RL (1984) Mechanism of Ca transport in plasma membrane vesicles prepared from cultured pituitary cells. J Biol Chem 259:9395–9403

Kaczorowski GJ, Barros F, Dethmers JK, Trumble MJ (1985) Inhibition of Na-Ca exchange in pituitary plasma membrane vesicles by analogues of amiloride. Biochemistry 24:1394–1403

Kadoma M, Froehlich J, Reeves JP, Sutko J (1982) Kinetics of sodium ion induced calcium ion release in calcium ion loaded cardiac sarcolemmal vesicles: determination of initial velocities by stopped flow spectrophotometry. Biochemistry 21:1914–1918

Langer GA (1982) Sodium-calcium exchange in the heart. Annu Rev Physiol 44:435–449

Langer GA (1984) Calcium at the sarcolemma. J Mol Cell Cardiol 16:147–153

Lazdunski M, Frelin C, Vigne P (1985) The sodium-hydrogen exchange system in cardiac cells: its biochemical and pharmacological properties and its role in regulating internal concentrations of sodium and internal pH. J Mol Cell Cardiol 17:1029–1042

Lee CO, Dagostino M (1982) Effect of strophanthidin on intracellular Na ion activity in twitch tension of constantly driven canine cardiac Purkinje fibers. Biophys J 40:185–198

Lubin M, Cahn F, Coutermarsh B (1982) Amiloride, protein synthesis, and activation of quiescent cells. J Cell Physiol 113:247–251

Luciani S, Floreani M (1985) Na-Ca exchange as a target for inotropic drugs. Trends Pharmacol Sci 6:316

Mallov S (1983) Effect of amrinone on sodium-calcium exchange in cardiac sarcolemmal vesicles. Res Commun Chem Pathol Pharmacol 41:197–210

Matlib MA, Lee S-W, Depover A, Schwartz A (1983) A specific inhibitory action of certain benzothiazepines and benzodiazepines on the sodium-calcium exchange process of heart and brain mitochondria. Eur J Pharmacol 89:327–328

Matlib MA, Doane JD, Sperelakis N, Riccippo-Neto F (1985) Clonazepam and diltiazem both inhibit sodium-calcium exchange of mitochondria but only diltiazem inhibits the slow action potentials of cardiac muscles. Biochem Biophys Res Commun 128:290–296

Mentrard D, Vassort G, Fischmeister R (1984) Changes in external Na induce a membrane current related to the Na-Ca exchange in cesium-loaded frog heart cells. J Gen Physiol 84:201–220

Michaelis ML, Michaelis EK (1983) Alcohol and local anesthetic effects on Na-dependent Ca fluxes in brain synaptic membrane vesicles. Biochem Pharmacol 32:963–969

Michalak M, Quackenbush EJ, Letarte M (1986) Inhibition of Na-Ca exchanger activity in cardiac and skeletal muscle sarcolemmal vesicles by monoclonal antibody 44D7. J Biol Chem 261:92–95
Mullins LJ (1979) The generation of electric currents in cardiac fibers by Na-Ca exchange. Am J Physiol 236:C103–C110
Mullins LJ (1981) Ion transport in the heart. Raven, New York
Nayler WG, Perry SE, Elz JS, Daly MJ (1984) Calcium, sodium, and the calcium paradox. Circ Res 55:227–237
Noble D (1984) The surprising heart: a review of recent progress in cardiac electrophysiology. J Physiol 353:1–50
Orchard CH, Eisner DA, Allen DG (1983) Oscillations of intracellular Ca in mammalian cardiac muscle. Nature 304:735–738
Philipson KD (1984) Interaction of charged amphiphiles with Na-Ca exchange in cardiac sarcolemmal vesicles. J Biol Chem 259:13999–14002
Philipson KD (1985a) Sodium-calcium exchange in plasma membrane vesicles. Annu Rev Physiol 47:561–571
Philipson KD (1985b) Symmetry properties of the Na-Ca exchange mechanism in cardiac sarcolemmal vesicles. Biochim Biophys Acta 821:367–376
Philipson KD, Nishimoto AY (1982a) Na-Ca exchange in inside-out cardiac sarcolemmal vesicles. J Biol Chem 257:5111–5117
Philipson KD, Nishimoto AY (1982b) Stimulation of Na-Ca exchange in cardiac sarcolemmal vesicles by proteinase pretreatment. Am J Physiol 243:C191–C195
Philipson KD, Nishimoto AY (1984) Stimulation of Na-Ca exchange in cardiac sarcolemmal vesicles by phospholipase D. J Biol Chem 259:16–19
Philipson KD, Ward R (1985) Effects of fatty-acids on sodium-calcium exchange and calcium permeability of cardiac sarcolemmal vesicles. J Biol Chem 260:9666–9671
Philipson KD, Frank JS, Nishimoto AY (1983) Effects of phospholipase C on the Na-Ca exchange and Ca permeability of cardiac sarcolemmal vesicles. J Biol Chem 258:5905–5910
Philipson KD, Langer GA, Rich TL (1985) Regulation of myocardial contractility and sarcolemmal Ca binding and transport by charged amphiphiles. Am J Physiol 248:H147–H150
Posillico JT, Srikant S, Brown EM, Eisanbarth GS (1985) The 4F2 cell surface protein modulates intracellular calcium. Clin Res 33:385A
Pousti A, Khoyi MA (1979) Effect of amiloride on isolated guinea-pig atrium. Arch Int Pharmacodyn 242:222–229
Quackenbush EJ, Gougos A, Baumal R, Letarte M (1986) Differential localization within human kidney of five membrane proteins expressed on acute lymphoblastic leukemia cells. J Immunol 136:118–124
Ralph RK, Smart J, Wojcik SJ, McQuillan J (1982) Inhibition of mouse mastocytoma protein kinases by amiloride. Biochem Biophys Res Commun 104:1054–1059
Reeves JP (1985) The sarcolemmal sodium-calcium exchange system. Curr Top Membr Trans 25:77–127
Reeves JP, Hale CC (1984) The stoichiometry of the cardiac sodium-calcium exchange system. J Biol Chem 259:7733–7739
Reeves JP, Sutko JL (1979) Sodium-calcium ion exchange in cardiac membrane vesicles. Proc Natl Acad Sci USA 76:590–594
Reeves JP, Sutko JL (1980) Sodium-calcium exchange activity generates a current in cardiac membrane vesicles. Science 208:1461–1464
Reeves JP, Sutko JL (1983) Competitive interactions of sodium and calcium with the sodium-calcium exchange system of cardiac sarcolemmal vesicles. J Biol Chem 258:3178–3182
Reeves JP, Bailey CA, Hale CC (1986) Redox modification of sodium-calcium exchange activity in cardiac sarcolemmal vesicles. J Biol Chem 261:4948–4955
Reuter H (1982) Na-Ca countertransport in cardiac muscle. In: Martonosi AN (ed) Membranes and transport. Plenum, New York, p 623
Reuter H, Seitz N (1968) The dependence of calcium efflux from cardiac muscle on temperature and external ion composition. J Physiol 195:451–470

Schellenberg GD, Anderson L, Swanson PD (1983) Inhibition of Na-Ca exchange in rat brain by amiloride. Mol Pharmacol 24:251–258
Schellenberg GD, Anderson L, Cragoe EJ Jr, Swanson PD (1985a) Inhibition of synaptosomal membrane Na-Ca exchange transport by amiloride and amiloride analogues. Mol Pharmacol 27:537–543
Schellenberg GD, Anderson L, Cragoe EJ Jr, Swanson PD (1985b) Inhibition of brain mitochondrial Ca transport by amiloride analogues. Cell Calcium 6:431–447
Sheu S-S, Fozzard HA (1982) Transmembrane Na and Ca electrochemical gradients in cardiac muscle and their relationship to force development. J Gen Physiol 80:325–351
Siegl PKS, Kaczorowski GJ, Trumble MJ, Cragoe EJ Jr (1983) Inhibition of Na-Ca exchange in guinea pig heart sarcolemmal vesicles and mechanical response by isolated atria and papillary muscle to 3,4-dichlorobenzamil (DCB). J Mol Cell Cardiol 14 [Suppl 1]:363 (abstract)
Siegl PKS, Cragoe EJ, Trumble MJ, Kaczorowski GJ (1984) Inhibition of Na-Ca exchange in membrane vesicle and papillary muscle preparations from guinea pig heart by analogs of amiloride. Proc Natl Acad Sci USA 81:3238–3242
Slaughter RS, Sutko JL, Reeves JP (1983) Equilibrium calcium-calcium exchange in cardiac sarcolemmal vesicles. J Biol Chem 258:3183–3190
Slaughter R, de la Pena P, Reeves JP, Cragoe E, Kaczorowski GJ (1984) Amiloride analogs, non-competitive inhibitors of sodium-calcium exchange in cardiac sarcolemmal vesicles. Biophys J 45:81a
Smith RL, Macara IG, Levenson R, Housman D, Cantley L (1982) Evidence that a Na-Ca antiport system regulates murine erythroleukemia cell differentiation. J Biol Chem 257:773–780
Soltoff SP, Mandel LJ (1983) Amiloride directly inhibits the Na,K-ATPase activity of rabbit kidney proximal tubules. Science 220:957–959
Sordahl LA, LaBelle EF, Rex KA (1984) Amiloride and diltiazem inhibition of microsomal and mitochondrial Na and Ca transport. Am J Physiol 246:C172–C176
Suleiman MS, Hider RC (1985) The influence of harmaline on the movements of sodium ions in smooth muscle of the guinea pig ileum. Mol Cell Biochem 67:145–150
Takeo S, Adachi K, Sakanashi M (1985) A possible action of nicardipine on the cardiac sarcolemmal Na-Ca exchange. Biochem Pharmacol 34:2303–2308
Trosper TL, Philipson KD (1983) Effects of divalent and trivalent cations on Na-Ca exchange in cardiac sarcolemmal vesicles. Biochim Biophys Acta 731:63–68
Vaghy PL, Johnson JD, Matlib MA, Wang T, Schwartz A (1982) Selective inhibition of Na-induced Ca release from heart mitochondria by diltiazem and certain other Ca antagonist drugs. J Biol Chem 257:6000–6002
Wier WG, Kort AA, Stern MD, Lakatta EG, Marban E (1983) Cellular calcium fluctuations in mammalian heart: direct evidence from noise analysis of aequorin signals in Purkinje fibers. Proc Natl Acad Sci USA 80:7367–7371
Yamashita S, Motomura S, Taira N (1981) Cardiac effects of amiloride in the dog. J Cardiovasc Pharmacol 3:704–715
Zhuang Y-X, Cragoe EJ Jr, Shaikewitz T, Glaser L, Cassel D (1984) Characterization of potent Na-H exchange inhibitors from the amiloride series in A431 cells. Biochemistry 23:4481–4488

CHAPTER 10

The Effect of Ruthenium Red and Other Agents on Mitochondrial Calcium Metabolism

M. CROMPTON

A. Introduction

The physiological consequences of pharmacological interventions that change mitochondrial Ca^{2+} will depend on the role of Ca^{2+} in the control of mitochondrial metabolism and on the adverse effects of increased Ca^{2+} load on mitochondrial function. Until quite recently, it was generally believed that the function of mitochondria in cellular Ca^{2+} metabolism was to provide a sink for Ca^{2+}. This view arose principally from the relatively low affinity of the Ca^{2+} uniporter (mediating Ca^{2+} influx) for cytosolic Ca^{2+} together with the massive capacity for Ca^{2+} accumulation displayed by isolated mitochondria when presented with supraphysiological levels of Ca^{2+}. Although mitochondria may act as an intracellular sink under pathological conditions, this capacity is most probably quite incidental to the natural role of the mitochondrial Ca^{2+} transport systems in controlling intramitochondrial free Ca^{2+} at low levels according to the regulatory requirements of oxidative metabolism. The capacity for Ca^{2+} accumulation is also constrained by the tolerance of mitochondrial function to increased Ca^{2+} load. It is important to stress that there are therefore two aspects to the behaviour of mitochondria with respect to Ca^{2+}. These two aspects are indicated in Fig. 1.

The key observations that led to the concept of the control of oxidative metabolism by intramitochondrial free Ca^{2+} are that several key enzymes are markedly activated by increase in free Ca^{2+} over the range 0.1–10 μM; these enzymes are α-oxoglutarate dehydrogenase, NAD-linked isocitrate dehydrogenase and pyruvate dehydrogenase (DENTON and MCCORMACK 1980). In isolated mitochondria, at least, about 0.1% of total intramitochondrial Ca^{2+} is free, so that 1 μM intramitochondrial free Ca^{2+} would correspond to about 1 nmol Ca^{2+} per milligram mitochondrial protein (COLL et al. 1982). In vitro studies with isolated mitochondria (DENTON and MCCORMACK 1980), in vivo studies in heart and liver (CROMPTON et al. 1983; MCCORMACK and DENTON 1984; MCCORMACK 1985) and computer simulations of heart mitochondrial behaviour (CROMPTON et al. 1986; CROMPTON 1986) all indicate that the Ca^{2+} transport cycle, described later in this section, is able to set intramitochondrial free Ca^{2+} within the range required for regulation of these enzymes under physiological conditions. In essence the cycle is envisaged as a means whereby increases in cytosolic free Ca^{2+} on cell activation are relayed into an increase in intramitochondrial free Ca^{2+} with consequent activation of the Ca^{2+}-sensitive dehydrogenases and increased ATP production.

In contrast to its metabolic effects, the adverse effects of intramitochondrial Ca^{2+} on mitochondrial function are poorly understood. What is known suggests

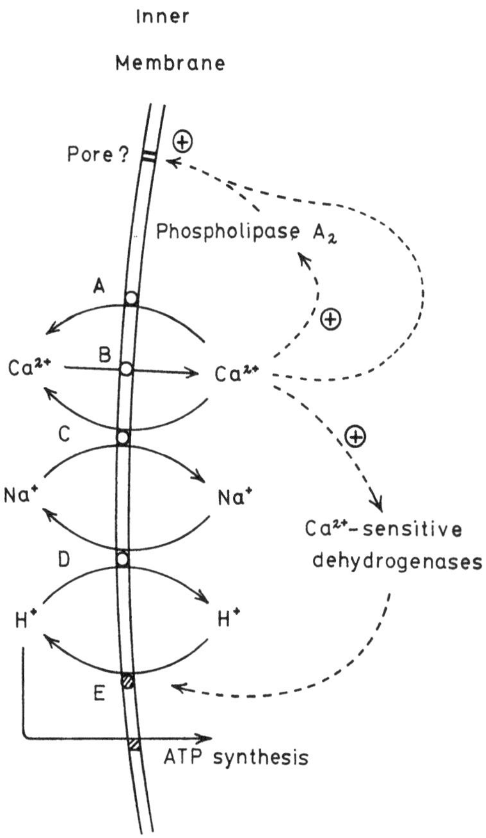

Fig. 1. The mitochondrial Ca^{2+} transport cycle and the effects of intramitochondrial Ca^{2+} on mitochondrial function. *A* the Na^+-independent Ca^{2+} transport system; *B* the Ca^{2+} uniporter; *C* the Na^+–Ca^{2+} carrier; *D* the Na^+–H^+ antiporter; *E* the respiratory chain

a complex series of events. Excessive Ca^{2+} accumulation by liver mitochondria (> 50 nmol per milligram mitochondrial protein) leads to a nonspecific permeability increase of the inner membrane. The mitochondria become uncoupled, and lose Mg^{2+},K^+, adenine nucleotides, preaccumulated Ca^{2+} and, presumably, low molecular weight matrix constituents generally (for review see CROMPTON 1985). The deleterious effects are promoted by inorganic phosphate, which acts intramitochondrially, and opposed by ATP, ADP and Mg^{2+}. The studies of HAWORTH and HUNTER (1979) indicated that Ca^{2+} induces an increase in inner membrane permeability to solutes with molecular weight < 1000, and soluble proteins of the mitochondrial matrix are retained (LÖTSCHER et al. 1980; I AL NASSER and M CROMPTON 1986, unpublished work). It appears then that the lesion is not caused by inner membrane rupture; rather, activation of a pore permeable to small solutes is indicated. Moreover, inner membrane impermeability in isolated mitochondria is restored when Ca^{2+} is removed, and the hypothetical pore appears to open in a fully reversible manner (CROMPTON et al. 1986; I AL NASSER and M CROMPTON 1986, unpublished work). For reasons that are not understood,

Ca^{2+}-induced permeabilization seems to depend on Ca^{2+}-induced activation of phospholipase A_2 and oxidation of the mitochondrial pyridine nucleotide and glutathione pools (BEATRICE et al. 1980, 1982). These effects are also observed on exposure of isolated mitochondria to H_2O_2 and organic hydroperoxides (LÖTSCHER et al. 1980; HOFSTETTER et al. 1981; BAUMHÜTTER and RICHTER 1982) and, under these conditions, ADP ribosylation of a mitochondrial protein has been detected. ATP, a protective agent, inhibits ADP ribosylation in submitochondrial particles (RICHTER et al. 1983).

Mitochondria contain separate transport systems for influx and efflux of Ca^{2+}. These operate unidirectionally and mediate continuous Ca^{2+} cycling across the inner membrane. The resultant Ca^{2+} distribution is far removed from electrochemical equilibrium (for reviews see CROMPTON 1985, 1986). Passive Ca^{2+} influx is opposed by two distinct active transport systems, the Na^+–Ca^{2+} carrier and a Na^+-independent system, the reaction of which has not been resolved. In many tissues, the activity of the Na^+-independent system is negligible in comparison with Na^+–Ca^{2+} carrier activity, e.g. cardiac and skeletal muscle, brain and adrenal cortex. In other tissues, however, approximately equal activities are found (e.g. liver, kidney). This chapter examines the agents that affect mitochondrial Ca^{2+} fluxes and how one of these in particular, ruthenium red, has been used to investigate mitochondrial dysfunction in ischaemic injury.

B. Inhibitors of the Mitochondrial Ca^{2+} Transport Systems

I. Ruthenium Red

Ruthenium red is a polynuclear ionic complex. FLETCHER et al. (1961) give the empirical formula $Ru_3O_2(NH_3)_{14}ICl_6 \cdot 2H_2O$, and propose the following structural formula: $[(NH_3)_5Ru\text{-}O\text{-}Ru(NH_3)_4\text{-}O\text{-}Ru(NH_3)_5]_6$. Commercially obtained samples are impure (e.g. 11% pure; REED and BYGRAVE 1974), the principal contaminants being ruthenium violet and the oxidation product ruthenium brown (LUFT 1971). Ruthenium red may be purified by crystallization (FLETCHER et al. 1961) and characterized by its visible absorption spectrum which shows a strong absorption peak at 533 nm. Ruthenium red sticks tenaciously to glassware, but is removed by dilute acid. Ruthenium red has been used in light and electron microscopy as an extracellular stain of pectins and acidic mucopolysaccharides, although in his extensive study of the use of ruthenium red as a stain, LUFT (1971) noted some entry of ruthenium red into certain types of apparently intact cell, e.g. muscle, where staining of the sarcoplasmic sacs of the triads was observed. This aspect is considered further in Sect. D.

The capacity of ruthenium red to inhibit mitochondrial Ca^{2+} accumulation was first reported by MOORE (1971). Subsequently, VASINGTON et al. (1972) showed that 3–6 nmol ruthenium red (unpurified) per milligram mitochondrial protein (liver) completely blocked the accumulation of Ca^{2+} when driven by respiration or by K^+ diffusion potentials generated with the ionophore, valinomycin. The latter observation locates the action of ruthenium red on the uniporter rather than on the generation of the membrane potential, the driving force for Ca^{2+} accumulation. Conversely, AKERMAN (1978) observed that Ca^{2+} diffusion potentials in respiration-inhibited mitochondria were abolished by ruthenium red. Other studies have confirmed that ruthenium red does not affect mitochondrial energy transduction at concentrations sufficient to block the uniporter although higher concentrations of unpurified ruthenium red (>10 nmol per milligram mitochondrial protein) do inhibit respiration and cause some decrease in inner membrane potential (VASINGTON et al. 1972; PUSKIN et al. 1976; NICHOLLS 1978).

In this laboratory, unpurified ruthenium red from Johnson Matthey has been used routinely at 2 nmol per milligram mitochondrial protein for >99.9% inhibition of the uniporter of isolated rat heart mitochondria. Rather higher concentrations (5–6 nmol/mg) are required for a similar inhibition with isolated liver mitochondria. The Na^+–Ca^{2+} carrier and the Na^+-independent efflux of Ca^{2+} are assayed typically in the presence of ruthenium red to prevent Ca^{2+} reuptake (e.g. CROMPTON et al. 1977) and the effects of ruthenium red on these processes have not been examined systematically. Nevertheless work in this laboratory has not detected any effect of increasing ruthenium red from 1 to 10 nmol/mg on either process.

A detailed study of the interaction of purified ruthenium red with the liver mitochondrial uniporter indicated that the ion inhibits noncompetitively with respect to Ca^{2+} with a K_i value of 30 nM (REED and BYGRAVE 1974). Only a very small fraction of the ruthenium red added to isolated mitochondria interacts with the uniporter. Reed and Bygrave determined a total of 15 nmol binding sites per milligram protein. The number of uniporter units per milligram protein is almost certainly several orders of magnitude less than this. Perhaps the best indications are provided by studies with lanthanides which are also highly effective uniporter inhibitors (K_i values, 20–30 nM; SCARPA and AZZONE 1970; REED and BYGRAVE 1974). Kinetic analyses of La^{3+} inhibition of the liver mitochondrial uniporter indicate the presence of <1 pmol La^{3+}-binding site per milligram mitochondrial protein (REED and BYGRAVE 1974). This is corroborated by the finding that the most potent lanthanide, Dy^{3+}, yields 50% inhibition in heart mitochondria at about 1 pmol per milligram mitochondrial protein (CROMPTON et al. 1979).

The sensitivity of the uniporter to ruthenium red is thought to reflect its interaction with a glycoprotein, although the role of the glycoprotein in Ca^{2+} transport has not been resolved. The glycoprotein of CARAFOLI and SOTTOCASA (1974) is highly acidic and H_2O soluble, and is partially released when the mitochondrial outer membrane is lysed osmotically, suggesting a loose attachment to the inner membrane. Such treatment leaves the inner membrane intact and the resultant particles (mitoplasts) accumulate Ca^{2+} at a decreased rate with respect to normal mitochondria. Mitosomal capacity for Ca^{2+} accumulation is improved by addition of glycoprotein (SANDRI et al. 1979). In addition, Ca^{2+} uptake by mitosomes is inhibited by antibodies raised against the glycoprotein (PANFILI et al. 1976). Prolonged incubation of the antibodies with normal mitochondria also leads to uniporter inhibition without affecting Na^+–Ca^{2+} carrier activity (PANFILI et al. 1981). MIRANOVA et al. (1982) have reported incorporation of a glycoprotein purified from heart mitochondria into black lipid membranes; these developed Nernst potentials with applied Ca^{2+} gradients in a manner sensitive to ruthenium red. The work of AMBUDKAR et al. (1984) has led to the isolation of a small protein (molecular weight 3000–6000 that binds Ca^{2+} with an affinity similar to that of the native uniporter ($K_d = 10$–$14\ \mu M$) and is sensitive to ruthenium red.

II. Benzothiazepines

The Na^+–Ca^{2+} carrier of heart and brain mitochondria is inhibited by *d-cis*-diltiazem ($IC_{50} = 7$–$10\ \mu M$; VAGHY et al. 1982; MATLIB et al. 1983). The action is stereospecific, *l-cis*-diltiazem being 50-fold less active. The same group reported a very low affinity of the Na^+–Ca^{2+} carrier for thiazesim ($IC_{50} = 190$–$250\ \mu M$).

III. Other Ca^{2+} Antagonists

In addition to diltiazem, other Ca^{2+} antagonists inhibit the Na^+–Ca^{2+} carrier of heart mitochondria with varying affinities. Of those tested the most potent were prenylamine and fendiline ($IC_{50} = 12$–$13\ \mu M$); much less effective were nifedipine and verapamil ($IC_{50} = 66$ and $150\ \mu M$, respectively; VAGHY et al. 1982). The low affinity for verapamil was confirmed for dog heart mitochondria by WOLKOWICZ et al. (1983). DEANNA et al. (1984) reported inhibition of the heart mitochondrial Na^+–Ca^{2+} carrier with the compound YS035 (*N,N*-bis-(3,4-dimethoxyphenethyl)-*N*-methylamine; $IC_{50} = 28\ \mu M$). This compound yielded half-maximal inhibition of Ca^{2+} influx into chick muscle cells at $20\ \mu M$. There are no reports that Ca^{2+} antagonists inhibit the uniporter. Studies in this laboratory have shown that prenylamine inhibits the Na^+–Ca^{2+} carrier of liver mitochondria ($IC_{50} = 6\ \mu M$) and that the Na^+-independent system is much less sensitive ($IC_{50} > 80\ \mu M$).

IV. Benzodiazepines

The Na^+–Ca^{2+} carriers of heart and brain mitochondria display a range of sensitivities to benzodiazepines (MATLIB et al. 1985). The most potent investigated was clonazepam ($IC_{50} = 5$–$6\ \mu M$), followed by diazepam ($IC_{50} = 10$–$12\ \mu M$) and flunitrazepam ($IC_{50} = 156$ and $390\ \mu M$ in heart and brain mitochondria, respectively). At concentrations that block the Na^+–Ca^{2+} carrier of mitochondria, it was reported that clonazepam had no effect on plasma membrane voltage-dependent Ca^{2+} channels, and Na^+,Ca^{2+} exchange or on Ca^{2+} uptake by sarcoplasmic reticulum (guinea pig papillary muscle).

V. Trifluoperazine

Trifluoperazine is a weak inhibitor of the Na^+–Ca^{2+} carrier of heart mitochondria ($IC_{50} = 70\ \mu M$). However, in mitoplasts (see Sect. B.I) prepared with digitonin, trifluoperazine is a potent inhibitor ($IC_{50} = 5\ \mu M$; HAYAT and CROMPTON 1985). Strong inhibition is acutely dependent on the presence of extramitochondrial Ca^{2+} ($2\ \mu M$), and only weak inhibition is observed in its absence ($IC_{50} = 70\ \mu M$). The molecular basis is obscure although it was suggested that access of trifluoperazine to its binding sites may be restricted in mitochondria, and that some perturbation of the inner membrane in digitonin-prepared mitoplasts may lead to exposure of these sites. In this connection HAYAT and CROMPTON (1982) reported the presence of external regulatory sites on the heart mitochondrial Na^+–Ca^{2+} carrier, that confer Ca^{2+} inhibition and are fully saturated with 2 μ*M* external free Ca^{2+}. The Ca^{2+} dependence of trifluoperazine sensitivity in mitoplasts may be related to occupation of the Ca^{2+}-regulatory sites. Trifluoperazine also inhibits the Ca^{2+}-induced permeabilization of the inner membrane of liver, heart and kidney mitochondria with low affinity (IC_{50} = 25–30 μ*M*; HARRIS and COOPER 1982). This may be related to inhibition of phospholipase A_2 by trifluoperazine (BROEKEMEIER et al. 1985).

VI. Gentamicin

Gentamicin is a cationic, aminoglycoside antibiotic. Nephrotoxicity is a serious side effect. SASTROSINH et al. (1982) reported that gentamicin is a competitive inhibitor of Ca^{2+} uptake by kidney mitochondria (K_i = 230 μ*M*). A specific interaction with the uniporter was indicated by the absence of an effect on electron transport.

VII. Amiloride Analogues

JURKOWITZ et al. (1983) reported inhibition of the heart mitochondrial Na^+–Ca^{2+} carrier by amiloride derivatives. The potency of amiloride (IC_{50} = 400 μ*M*) was increased by the introduction of a benzyl group on the terminal guanidino nitrogen (benzamil, IC_{50} = 160 μ*M*) and increased further with the *p*-fluoro derivative of benzamil (IC_{50} = 100 μ*M*). The introduction of a carbamoyl group on the terminal guanidino nitrogen of amiloride decreased its potency (IC_{50} > 600 μ*M*). These compounds inhibited noncompetitively with respect to Na^+. The effectiveness of these compounds was improved in the presence of inorganic phosphate, which dissipates the pH gradient across the inner membrane, and thereby could slow Na^+–Ca^{2+} exchange in principle. Although this might suggest that these compounds inhibit indirectly, by affecting Na^+,H^+ exchange, this was not borne out by measurements of Na^+,H^+ exchange and by the amiloride concentration dependence of the inhibition.

C. Effectors of Ca^{2+}-Induced Permeabilization

A number of local anaesthetics that inhibit mitochondrial phospholipase A_2 (WAITE and SISSON 1971) also inhibit Ca^{2+}-induced permeabilization of the inner membrane, e.g. dibucaine, Nupercaine (IC_{50} = 70–250 μ*M*; HARRIS and COOPER 1982; COCKRELL 1982; PFEIFFER et al. 1979). A correlation between the capacity of various agents to inhibit the Ca^{2+}-induced permeabilization and phospholipase A_2 has been made (BROEKEMEIER et al. 1985). The most potent inhibitors of both were dibucaine (250 μ*M*), quinacrine (250 μ*M*) and trifluoperazine (60 μ*M*); Ca^{2+} channel blockers inhibited both processes very weakly at high concentration (> 50 μ*M*; nifedipine, diltiazem, prenylamine, verapamil).

The effects of alloxan on mitochondrial Ca^{2+} may be considered in this category since certain phenomena are common to both. Alloxan causes necrosis of pancreatic beta-cells and has been used to produce experimental diabetes. BOQUIST (1984) observed that alloxan induced the release of Ca^{2+} from mouse liver mitochondria and suggested that disturbances of Ca^{2+} homeostasis underlie its cytotoxicity. The capacity of 1–5 m*M* alloxan to release mitochondrial Ca^{2+} was confirmed by FREI et al. (1985). The characteristics of the release, i.e. a lag phase, pronounced at low alloxan concentration, concomitant oxidation and hydrolysis of pyridine nucleotides, and inhibition by ATP are features associated with Ca^{2+} release induced by peroxides. In this connection it has been proposed that alloxan action may be caused by OH· radicals; these may be generated by autooxidation of dialuric acid, derived from alloxan, with the formation of O_2^- and H_2O_2, which generate OH· via the Haber–Weiss reaction (HEIKKILA et al. 1976; GRANKVIST et al. 1979). However, in contrast to peroxide-induced Ca^{2+} release, the release induced by alloxan was insensitive to inhibition of glutathione reductase so that its action was not apparently mediated via production of H_2O_2. FREI et al. (1984) suggested a nonenzymatic oxidation of pyridine nucleotides by alloxan. Exactly how pyridine nucleotide oxidation relates to permeabilization is not known (see Sect. A).

As noted already, a feature of permeabilization in liver mitochondria, at least, is that it exhibits a lag phase. This has been reported in the context of the capacity of isolated mitochondria to retain accumulated Ca^{2+} for prolonged periods (the so-called retention time). The retention time of mitochondria isolated from perfused liver (15–45 min) is improved about twofold by 0.35 μ*M* glucagon (HUGHES and BARRITT 1978). Similar improvements were observed after perfusing with phenylephrine (10 μ*M*) or adrenaline (1 μ*M*), and the effects of adrenaline were blocked by the α-adrenergic antagonist, phenoxybenzamine (10 μ*M*), but not by the β-antagonist, propranolol (10 μ*M*; TAYLOR et al. 1980). It appears therefore that glucagon and α-adrenergic agonists stabilize mitochondria against Ca^{2+}-induced permeabilization, but it is not known how this is brought about.

D. Mitochondrial Ca^{2+} Overload

There is increasing interest in the question of mitochondrial Ca^{2+} overload during ischaemia and in particular during postischaemic reoxygenation, and in the type of intervention that may be useful in protecting mitochondrial and tissue function under these conditions. This topic has received considerable attention in heart. During the early phase of hypoxia or ischaemia the rise in cardiac resting tension is not associated with changes in either Ca^{2+} uptake by the tissue, total tissue Ca^{2+} (NAYLER et al. 1979; LEWIS et al. 1979; SHINE 1981) or cytosolic free Ca^{2+} (ALLEN and ORCHARD 1983). On prolonged hypoxia, however, tissue Ca^{2+} uptake increases (NAYLER et al. 1979), HARDING and POOLE-WILSON 1980). The reintroduction of oxygen after prolonged hypoxia is associated with a large uptake of Ca^{2+} by the tissue and results in irreversible injury. The essential question regarding mitochondria is whether they become overloaded with Ca^{2+} under these conditions to such an extent that mitochondrial energy transduction is impaired via permeabilization of the inner membrane. If this is the case, tissue recovery may be compromised by ATP depletion and a decreased capacity for active Ca^{2+} extrusion from the cell, which is then in a vicious circle.

Mitochondria in vitro do exhibit a remarkable capacity for Ca^{2+} accumulation in the presence of ATP or ADP and Mg^{2+} when presented with a finite amount of Ca^{2+}, and may accumulate at least 200 nmol Ca^{2+} per milligram mitochondrial protein without affect on inner membrane potential development (e.g. NICHOLLS and BRAND 1980). Recent measurements indicate that heart mitochondria in situ may contain 1–2 nmol Ca^{2+} per milligram mitochondrial protein (CROMPTON et al. 1983), and it would seem therefore that accumulation of 100-fold excess may occur with impunity, at least with isolated mitochondria. Clearly, there may be factors in ischaemic/reperfused tissue that decrease mitochondrial tolerance to Ca^{2+} load. In addition, in one respect mitochondria behave in quite an uncompromising manner when faced with increased external Ca^{2+}. This refers to the fact that there is a limit of external free Ca^{2+} that is compatible with normal mitochondrial function. This aspect emerges from the relation between intramitochondrial and extramitochondrial free Ca^{2+} established by the transport cycle of the inner membrane. Examination of this critical relation is beyond the scope of this chapter, but the relation has been analysed in some detail (CROMPTON et al. 1986; CROMPTON 1986). The essential point is that the relation between time-averaged cytosolic free Ca^{2+} and intramitochondrial free Ca^{2+} in heart (and other tissues) is highly nonlinear, and an increase in the former produces a much larger increase in the latter. Moreover, there is a limit to time-averaged cytosolic free Ca^{2+} above which the (quasi) steady state distribution of Ca^{2+} across the inner membrane cannot be attained irrespective of mitochondrial Ca^{2+} content, which increases continually with time until, inevitably, energy transduction is impaired. This condition can only be avoided if either the time-averaged cytosolic free Ca^{2+} is decreased below the critical limit or the kinetic properties of the mitochondrial Ca^{2+} carriers are changed (inhibited influx, activated efflux) so that the critical limit is shifted to a higher value. Computer simulations suggest that the critical limit in time-averaged cytosolic free Ca^{2+} in rat heart might be close to 1 μM, but no precise figure can be given because it would depend on the profile of cytosolic free Ca^{2+} with time (CROMPTON 1986).

I. Mitochondrial Ca^{2+} and Oxidative Phosphorylation During Ischaemia/Reperfusion

Mitochondria isolated from ischaemic myocardium have a defective capacity for oxidative phosphorylation and this capacity is impaired further by reperfusion (LOCHNER et al. 1975; JENNINGS 1976; PENG et al. 1980). This defect has been correlated with increased mitochondrial Ca^{2+}. HENRY et al. (1977) reported that the endogenous Ca^{2+} content of rabbit heart mitochondria (7.6 nmol per milligram mitochondrial protein) was increased 3.7-fold by ischaemia and 7.2-fold by ischaemia/reperfusion. NAKANISHI et al. (1982) observed that the Ca^{2+} content of mitochondria isolated from septal preparations (8.6 nmol per milligram mitochondrial protein) increased 2.2-fold after hypoxia/reoxygenation. The work of FERRARI et al. (1982) showed that the capacity of rat heart mitochondria for ATP synthesis was decreased by 32% after 60 min anoxia, and by 60% after 30 min post-anoxic reperfusion. The mitochondrial Ca^{2+} content increased from 15 nmol Ca^{2+} per milligram mitochondrial protein to 22 nmol/mg (anoxia) and 40 nmol/mg (anoxia/reperfusion). The capacity of isolated pig heart mitochondria for ATP synthesis was decreased by 13% (30 min ischaemia/2 h reperfusion) and by 90% (2 h ischaemia/2 h reperfusion; PENG et al. 1980). The same study showed an increase in mitochondrial Ca^{2+} content from 10–14 nmol per milligram mitochondrial protein to 161 nmol/mg (30 min occlusion/reperfusion) and 55 nmol/mg (2 h occlusion/reperfusion). The last study is particularly interesting since the rise in endogenous Ca^{2+} content during ischaemia was actually reversed with prolonged ischaemia. Conceivably the mitochondria were becoming more permeabilized at this time so that accumulated Ca^{2+} was lost. In all these studies the normal Ca^{2+} content of the isolated mitochondria was considerably higher than that determined in rat heart mitochondria following isolation in the presence of EGTA (1.8 nmol per milligram mitochondrial protein; CROMPTON et al. 1983); EGTA does not appear to cause significant loss of mitochondrial Ca^{2+} (MCCORMACK 1985). In addition an intramitochondrial content close to 1 nmol/mg would be predicted if the normal function of mitochondrial Ca^{2+} is to control matrix dehydrogenases. It cannot be discounted therefore that some Ca^{2+} accumulation occurred during mitochondrial preparation. If such is the case, the relative changes in mitochondrial Ca^{2+} induced by ischaemia/reperfusion might be greater than the values indicate.

The most straightforward way of interpreting the increases in mitochondrial Ca^{2+} is that the accumulation occurred in response to increased cytosolic free Ca^{2+} as a result of increased sarcolemmal permeability to Ca^{2+}. NAKANISHI et al. (1982) observed a correlation between interventions that limit mitochondrial uptake of Ca^{2+} and Ba^{2+} (low temperature, CN^-, uncoupling agents) and also limit tissue accumulation of these following reoxygenation. This supports the concept that considerable tissue Ca^{2+} resides in mitochondria of reperfused heart, although in my view, one can not conclude (like NAKANISHI et al. 1982) that the primary lesion is mitochondrial.

The concept of a causative link between mitochondrial Ca^{2+} overload and cell necrosis has been applied more generally, e.g. to cardiomyopathic hamsters (WROGEMANN and NYLEN 1978), liver ischaemia (CHIEN et al. 1977) and ischaemia/reperfusion-induced renal failure (ARNOLD et al. 1985). In the latter study, defects of oxidative phosphorylation were correlated with a sixfold increase in mitochondrial Ca^{2+}. Interestingly, however, studies of ZYDOWA et al. (1985) concluded that mitochondrial Ca^{2+} overload did not occur during vitamin D-induced cardionecrosis.

II. The Effects of Ruthenium Red and Other Agents on Mitochondrial Ca^{2+}

Reperfusion of ischaemic heart with medium containing 1 μM ruthenium red largely prevented the increase in mitochondrial Ca^{2+} and abolished completely the deleterious effects on oxidative phosphorylation (PENG et al. 1980; FERRARI et al. 1982). SMITH and KENT (1980) observed that ruthenium red also improved myocardial contractility during reperfusion in anaesthetized dogs. The rationale of these experiments, that ruthenium red might protect mitochondrial function during tissue Ca^{2+} overload by inhibiting the mitochondrial Ca^{2+} uniporter, appears to be borne out. This begs the question of how ruthenium red might gain access to the intracellular compartment. Although alterations in sarcolemmal permeability properties may occur on reoxygenation, this does not extend to permeability to small solutes generally, e.g. Cr-EDTA (HARDING and POOLE-WILSON 1980). Moreover, there is evidence that perfusion with ruthenium red restricts mitochondrial Ca^{2+} uptake even in otherwise normal myocardium. As stated in Sect. A, a plausible case can now be made that increases in time-averaged cytosolic free Ca^{2+} are relayed by the Ca^{2+} cycle to the mitochondrial matrix with consequent activation of the Ca^{2+}-sensitive dehydrogenases. Thus, raised extracellular Ca^{2+}, isoprenaline and adrenaline, all of which increase cytosolic Ca^{2+}, also increase pyruvate dehydrogenase activity (HIRAOKA et al. 1980; MCCORMACK and DENTON 1981) and increase mitochondrial Ca^{2+} (CROMPTON et al. 1983). MCCORMACK and ENGLAND (1984) showed that perfusion with 3 μM ruthenium red completely abolished the activation of pyruvate dehydrogenase caused by these treatments. Ruthenium red may also act on plasma membrane Ca^{2+} fluxes. Thus, ruthenium red blocks transmitter release from synaptosomes and the neuromuscular junction and inhibits Ca^{2+} entry (TAPIA and MESA-RUIZ 1977; PERSON and KUHN 1979). STIMERS and BYERLY (1982) reported that 5 μM ruthenium red inhibited the Ca^{2+} channel of snail neurons by 50%. Indeed, in the studies of MCCORMACK and ENGLAND (1984), a negative inotropic effect of ruthenium red was noted with 1.5 mM extracellular Ca^{2+}. Nevertheless, with 6 mM extracellular Ca^{2+}, ruthenium red decreased neither contractile force nor the activity of glycogen phosphorylase *a*, an index of cytosolic free Ca^{2+}, yet did abolish pyruvate dehydrogenase activation, as noted. In addition, the selective action of ruthenium red on pyruvate dehydrogenase activity was not shared with the recognized Ca^{2+} channel blocker, verapamil (0.1 μM).

Other studies have examined the effectiveness of Ca^{2+} channel blockers in protecting the myocardium. NAYLOR et al. (1980) and PIACENZA et al. (1981) observed a protective effect of verapamil and nifedipine against hypoxia-induced defects of adenine nucleotide metabolism. PINSKY et al. (1981) reported that verapamil preserved mitochondrial electron transport activity during ischaemic arrest. In addition, 0.1 μM nifedipine largely abolished the sevenfold increase in mitochondrial Ca^{2+} induced by ischaemia/reperfusion in rabbit heart. BOURDILLON and POOLE-WILSON (1981) however concluded that administration of nifedipine late in anoxia does not prevent Ca^{2+} uptake by the tissue on reoxygenation.

References

Akerman KEO (1978) Changes in membrane potential during Ca^{2+} influx and efflux across the mitochondrial membrane. Biochim Biophys Acta 502:359–365

Allen DG, Orchard CH (1983) Intracellular calcium concentration during hypoxia and metabolic inhibition in mammalian ventricular muscle. J Physiol 339:107–122

Ambudkhar IS, Kima PE, Shamoo AE (1984) Characterisation of calciphorin, the low molecular weight Ca^{2+} ionophore from rat liver mitochondria. Biochim Biophys Acta 771:165–170

Arnold PE, Lumlertgul D, Burke TJ, Schrier RW (1985) In vitro versus in vivo mitochondrial Ca^{2+} loading in ischaemic acute renal failure. Am J Physiol 248:F845–F850
Baumhütter S, Richter C (1982) The hydroperoxide-induced release of mitochondrial Ca^{2+} occurs via a distinct pathway and leaves mitochondria intact. FEBS Lett 148:271–275
Beatrice MC, Palmer JW, Pfeiffer DR (1980) The relationship between mitochondrial membrane potential, permeability and the retention of Ca^{2+} by mitochondria. J Biol Chem 255:8663–8671
Beatrice MC, Stiers DL, Pfeiffer DR (1982) Increased permeability of mitochondria during Ca^{2+} release induced by t-butylhydroperoxide or oxaloacetate. J Biol Chem 257:7161–7171
Boquist L (1984) Alloxan effects on mitochondria in vitro: correlation between endogenous adenine nucleotides and efflux of Ca^{2+}. Biochem Int 9:637–641
Bourdillon PDV, Poole-Wilson PA (1981) Effects of ischaemia and reperfusion on Ca^{2+} exchange and mechanical function in isolated rabbit myocardium. Cardiovasc Res 15:121–130
Broekemeier KM, Schmid PC, Schmid HHO, Pfeiffer DR (1985) Effects of phospholipase A_2 inhibitors on ruthenium red-induced Ca^{2+} release from mitochondria. J Biol Chem 260:105–113
Carafoli E, Sottocasa GL (1974) The Ca^{2+} transport systems of the inner mitochondrial membrane and the problem of the Ca^{2+} carrier. In: Ernster L, Estabrook RW, Slater EC (eds) Dynamics of energy transducing membranes. Elsevier, Amsterdam, p 455
Chien ICR, Abrams J, Pfau RG, Farber JL (1977) Prevention by chlorpromazine of ischaemic liver cell death. Am J Pathol 88:539–558
Cockrell RS (1982) The influence of nupercaine on Ca^{2+} transport by rat liver and Ehrlich ascites tumour cell mitochondria. FEBS Lett 144:279–282
Coll KE, Joseph SK, Corkey BE, Williamson JR (1982) Determination of the matrix free Ca^{2+} concentration and kinetics of Ca^{2+} efflux in liver and heart mitochondria. J Biol Chem 257:8696–8704
Crompton M (1985) The calcium carriers of mitochondria. In: Martonosi AN (ed) The enzymes of biological membranes. 2nd edn, vol 3. Plenum, New York, p 249
Crompton M (1986) The regulation of mitochondrial calcium transport in heart. In: Shamoo A (ed) Current topics in membranes and transport, vol 25. Academic, New York, pp 231–276
Crompton M, Kunzi M, Carafoli E (1977) The calcium-induced and the sodium-induced effluxes of calcium from heart mitochondria; evidence for a sodium-calcium carrier. Eur J Biochem 79:549–558
Crompton M, Heid I, Baschera C, Carafoli E (1979) The resolution of calcium fluxes in heart and liver mitochondria using the lanthanide series. FEBS Lett 104:352–354
Crompton M, Kessar P, Al Nasser I (1983) The α-adrenergic mediated activation of the cardiac mitochondrial Ca^{2+} uniporter and its role in the control of intramitochondrial Ca^{2+} in vivo. Biochem J 216:333–342
Crompton M, Goldstone TP, Al Nasser I (1986) The regulation of mitochondrial calcium. In: Bader H, Gietzen K, Rosenthal J, Rudel R, Wolf HU (eds) Intracellular calcium regulation. Manchester University Press, Manchester, pp 67–78
Deanna R, Panata L, Cancellotti FM, Quadro G, Galzigna L (1984) Properties of a new Ca^{2+} antagonist on cellular uptake and mitochondrial efflux of Ca^{2+}. Biochem J 218:899–905
Denton RM, McCormack JG (1980) On the role of the calcium transport cycle in heart and other mammalian tissues. FEBS Lett 119:1–8
Ferrari R, diLisa F, Raddino R, Visioli O (1982) The effects of ruthenium red on mitochondrial function during post ischaemic reperfusion. J Mol Cell Cardiol 14:737–740
Fletcher JM, Greenfield BF, Hardy CJ, Scargill D, Woodhead JL (1961) Ruthenium red. J Chem Soc Lond 2000–2006
Frei B, Winterhalter KH, Richter C (1985) Mechanism of alloxan-induced Ca^{2+} efflux from rat liver mitochondria. J Biol Chem 260:7395–7401
Grankvist K, Marklund S, Sehlin J, Taljedahl I (1979) Superoxide dismutase, catalase and scavengers of hydroxyl radicals protect against the toxic action of alloxan on pancreatic islet cells in vitro. Biochem J 182:17–25

Harding DP, Poole-Wilson PA (1980) Ca^{2+} exchange in rabbit myocardium during and after hypoxia: effect of temperature and substrate. Cardiovasc Res 14:435–445
Harris EJ, Cooper MB (1982) Inhibition of Ca^{2+}-stimulated ion losses in mitochondria by inhibitors of calmodulin. Biochem Biophys Res Commun 108:1614–1618
Haworth RA, Hunter DR (1979) The Ca^{2+}-induced membrane transition in mitochondria. Arch Biochem Biophys 195:460–467
Hayat L, Crompton M (1982) Evidence for the existence of regulatory sites for Ca^{2+} on the Na^+–Ca^{2+} carrier of cardiac mitochondria. Biochem J 202:509–518
Hayat LH, Crompton M (1985) Ca^{2+}-dependent inhibition by trifluoperazine of the Na^+–Ca^{2+} carrier in mitoplasts derived from heart mitochondria. FEBS Lett 182: 281–285
Heikkila RE, Winston B, Cohen G, Barden H (1976) Alloxan-induced diabetes: evidence for hydroxyl radical as a cytotoxic intermediate. Biochem Pharmacol 25:1085–1092
Henry PD, Schuchleib R, Davis R, Weiss ES, Sobel BD (1977) Myocardial contracture and accumulation of mitochondrial Ca^{2+} in ischaemic rabbit heart. Am J Physiol 233:H677–H684
Hiraoka T, DeBuysere M, Olson MS (1980) Studies of the effect of β-adrenergic agonists on the regulation of pyruvate dehydrogenase in the perfused rat heart. J Biol Chem 255:7604–7609
Hofstetter W, Mühlebach T, Lötscher HR, Winterhalter K, Richter C (1981) ATP prevents both hydroperoxide induced hydrolysis of pyridine nucleotides and release of calcium in rat liver mitochondria. Eur J Biochem 117:361–367
Hughes BP, Barritt GJ (1978) Effects of glucagon and N^6,O^2-dibutyryladenosine 3′:5′-cyclic monophosphate on calcium transport in isolated rat liver mitochondria. Biochem J 176:295–304
Jennings RB (1976) Relationship of acute ischaemia to functional defects and irreversibility. Circulation [suppl 1] 53:26–29
Jurkowitz MS, Altschud RA, Brierley GP, Cragoe EJ (1983) Inhibition of Na^+-dependent Ca^{2+} efflux from heart mitochondria by amiloride analogues. FEBS Lett 162:262–265
Lewis MJ, Grey AC, Henderson AH (1979) Determinants of hypoxic contracture in isolated heart muscle preparations. Cardiovasc Res 13:86–94
Lochner A, Opie LH, Owen P, Kotze JCN (1975) Oxidative phosphorylation in infarcting baboon and dog myocardium. J Mol Cell Cardiol 7:203–217
Lötscher H, Winterhalter KH, Carafoli E, Richter C (1980) Hydroperoxide-induced loss of pyridine nucleotides and release of Ca^{2+} from rat liver mitochondria. J Biol Chem 255:9325–9330
Luft JH (1971) Ruthenium red and violet 1. Chemistry, purification, methods of use for electron microscopy and mechanism of action. Anat Rec 171:347–368
Matlib MA, Less S, Depover A, Schwartz A (1983) A specific inhibitory action of certain benzothiazepines and benzodiazepines on the sodium-calcium exchange process of heart and brain mitochondria. Eur J Pharm 89:377–385
Matlib MA, Doane JD, Sperelakis N, Riccippo-Neto F (1985) Clonazepam and diltiazem both inhibit the Na^+-Ca^{2+} exchange of mitochondria but only diltiazem inhibits the slow action potential of cardiac muscle. Biochem Biophys Res Commun 128:290–296
McCormack JG (1985) Studies on the activation of rat liver pyruvate dehydrogenase by adrenaline and glucagon. Role of increases in intramitochondrial Ca^{2+} concentration. Biochem J 231:597–608
McCormack JG, Denton RM (1981) The activation of pyruvate dehydrogenase in the perfused rat heart by adrenaline and other inotropic agents. Biochem J 194:639–643
McCormack JG, Denton RM (1984) Role of Ca^{2+} ions in the regulation of intramitochondrial metabolism in rat heart. Evidence from studies with isolated mitochondria that adrenaline activates the pyruvate and 2-oxoglutarate complexes by increasing the intramitochondrial concentration of Ca^{2+}. Biochem J 218:235–247
McCormack JG, England PJ (1984) Ruthenium red inhibits the activation of pyruvate dehydrogenase caused by positive inotropic agents in the perfused rat heart. Biochem J 214:581–585

Miranova GD, Tutjana VS, Pronevitch LA, Trofimenko NT, Miranov GP, Grigorjev PA, Kondrashova M (1982) Isolation and properties of a Ca^{2+} transporting glycoprotein and peptide from beef heart mitochondria. J Bioenerg Biomembr 14:213–219

Moore CL (1971) Specific inhibition of mitochondrial calcium transport by ruthenium red. Biochem Biophys Res Commun 42:298–305

Nakanishi T, Nishioka K, Jarmakana JM (1982) Mechanism of tissue Ca^{2+} gain during reoxygenation after hypoxia in rabbit myocardium. Am J Physiol 242:H437–H449

Nayler WG, Poole-Wilson PA, Williams A (1979) Hypoxia and calcium. J Mol Cell Cardiol 11:683–706

Nayler WF, Ferrari R, Williams A (1980) Protective effect of pretreatment with verapamil nifedipine, and propranolol on mitochondrial function in ischaemic and reperfused myocardium. Am J Cardiol 46:242–248

Nicholls DG (1978) Calcium transport and proton electrochemical gradient in mitochondria from guinea pig cerebral cortex and rat heart. Biochem J 170:511–522

Nicholls DG, Brand MD (1980) The nature of the calcium ion efflux induced in rat liver mitochondria by the oxidation of endogenous nicotinamide nucleotides. Biochem J 188:113–118

Panfili E, Sandri G, Sottocasa GL, Lunazzi G, Liut G (1976) Specific inhibition of mitochondrial Ca^{2+} transport by antibodies directed to the Ca^{2+} binding glycoprotein. Nature 264:185–186

Panfili E, Crompton M, Sottocasa GL (1981) Immunochemical evidence of the independence of the Na^{+}-Ca^{2+} antiporter and the electrophoretic Ca^{2+} uniporter in heart mitochondria. FEBS Lett 123:30–32

Peng C, Kane JJ, Staub KD, Murphy ML (1980) Improvement of myocardial energy production in ischaemic myocardium by in vivo infusion of ruthenium red. J Cardiovasc Pharm 2:45–54

Person RJ, Kuhn JA (1979) Depression of spontaneous and ionophore-induced transmitter release by ruthenium red at the neuromuscular junction. Brain Res Bull 4:669–674

Pfeiffer DR, Schmid PC, Beatrice MC, Schmid HHO (1979) Intramitochondrial phospholipase activity and the effects of Ca^{2+} plus N-ethylmaleimide on mitochondrial function. J Biol Chem 254:11485–11494

Piacenza A, Osella R, Borgoglio R (1981) Effect of nifedipine on mitochondrial function of isolated rabbit hearts perfused under hypoxic conditions. J Mol Cell Cardiol 13:709 (abstract)

Pinsky WW, Lewis RM, McMillin-Wood JM, Hara H, Hartley CJ, Gillette DC, Entman ML (1981) Myocardial protection from ischaemic arrest: potassium and verapamil cardioplagia. Am J Physiol 240:H326–H335

Puskin JS, Gunter TE, Gunter KK, Russell PR (1976) Evidence for more than one Ca^{2+} transport system in mitochondria. Biochemistry 15:3834–3842

Reed KC, Bygrave FL (1974) Inhibition of mitochondrial calcium transport by lanthanides and ruthenium red. Biochem J 140:143–150

Richter C, Winterhalter KH, Baumhutter S, Lotscher H, Moser R (1983) ADP-ribosylation in the inner membrane of rat liver mitochondria. Proc Natl Acad Sci USA 80:3188–3192

Sandri G, Sottocasa GL, Panfili E, Liut G (1979) The ability of the mitochondrial Ca^{2+}-binding glycoprotein to restore Ca^{2+} transport in glycoprotein depleted rat liver mitochondria. Biochim Biophys Acta 558:214–220

Sastrosinh M, Weinberg JM, Hulmes HD (1982) The effect of gentamicin on calcium uptake by renal mitochondria. Life Sci 30:2309–2315

Scarpa A, Azzone GF (1970) The mechanism of ion translocation in mitochondria. Eur J Biochem 12:328–335

Shine KI (1981) Ionic events in ischaemia and anoxia. Am J Pathol 102:256–261

Smith HJ, Kent KM (1980) Depressed contractile function in reperfused canine myocardium: metabolism and response to pharmacological agents. Cardiovasc Res 14:458–468

Stimers JR, Byerly L (1982) Slowing of sodium current inactivation by ruthenium red in snail neurons. J Gen Physiol 80:485–497

Tapia R, Mesa-Ruiz G (1977) Inhibition by ruthenium red of Ca^{2+}-dependent release of ^{3}H-GABA in synaptosomal fractions. Brain Res 126:160–166
Taylor WM, Prpic V, Exton JH, Bygrave FL (1980) Stable changes to Ca^{2+} fluxes in mitochondria isolated from rat livers perfused with α-adrenergic agonists and with glucagon. Biochem J 188:443–450
Vaghy P, Johnson JD, Matlib MA, Wang T, Schwartz A (1982) Selective inhibition of Na^{+}-induced Ca^{2+} release from heart mitochondria by diltiazem and other Ca^{2+} antagonist drugs. J Biol Chem 257:6000–6004
Vasington FD, Gazzotti P, Tiozzo R, Carafoli E (1972) The effect of ruthenium red on Ca^{2+} transport and respiration in rat liver mitochondria. Biochim Biophys Acta 256:43–54
Waite M, Sisson P (1971) Partial purification and characterisation of the phospholipase A_2 from rat liver mitochondria. Biochemistry 10:2377–2383
Wolkowicz PE, Michael LA, Lewis RM, McMillin-Wood J (1983) Sodium-calcium exchange in dog heart mitochondria: effects of ischaemia and verapamil. Am J Physiol 244:H644–H651
Wrogemann K, Nylen EG (1978) Mitochondrial calcium overloading in cardiomyopathic hamsters. J Mol Cell Cardiol 10:185–195
Zydowo MM, Swierczynski I, Nagel G, Wrzotkowa T (1985) The respiration and Ca^{2+} content of heart mitochondria with vitamin D-induced cardionecrosis. Biochem J 226:155–161

Pharmacology of Calcium Uptake and Release from the Sarcoplasmic Reticulum: Sensitivity to Methylxanthines and Ryanodine

R. A. CHAPMAN and J. TUNSTALL

A. Introduction

A number of plant alkaloids affect the contraction of striated muscles, amongst these the methylxanthines and ryanodine have been widely used as their action may throw light upon the role of the specialised endoplasmic reticulum of muscle in the excitation–contraction–relaxation sequence. The methylxanthines popularly used, caffeine, theobromine, theophylline and paraxanthine have a common chemical structure consisting of a double ring where a pyramidine and an imidazole combine to give the xanthine structure (Fig. 1 a). Ryanodine is a more complex molecule, incorporating a single nitrogen-containing pyrrole ring (Fig. 1 b).

A general picture of the action of these compounds would be that in intact skeletal muscle the application of the methylxanthines, caffeine in particular, potentiates the twitch and at high concentration produces a sustained contracture. In cardiac muscle the action may be complex, producing either a potentiation or an inhibition of the heart beat and if it produces a contracture this is usually phasic. Ryanodine, on the other hand, inhibits the twitch of skeletal muscle and with prolonged exposure will produce a rigor-like state. In the heart, although certain species seem to be insensitive to ryanodine, in mammalian preparations

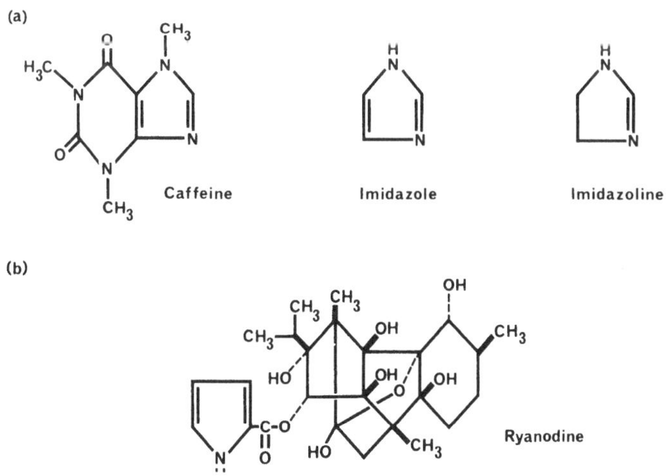

Fig. 1 a, b. The chemical structure of **a** caffeine (a typical methylxanthine), imidazole, imidazoline, and **b** ryanodine

the response is usually a negative inotropism without a rise in resting tension. In some preparations however a negative inotropism at low concentration of the drug may be followed by a positive inotropic phase as the drug concentration is raised.

In a review of the literature on the effects of these agents, it may be possible in the case of caffeine and the methylxanthines to come to conclusions about their effects upon subcellular fractions of striated muscles which may explain their action in intact preparations. A similar approach to the action of ryanodine is much more difficult. In this case it is necessary to rely much more on the effect of the alkaloid upon whole muscles in any attempt to understand the observable effects upon muscle fractions. Because of these differences of approach we have considered the methylxanthines and ryanodine separately in what follows.

B. Methylxanthines

The ability of caffeine to potentiate the strength of contraction, prolong the active state and produce contractures in intact skeletal muscles, independently of a depolarisation of the cell membrane and the integrity of the transverse tubular system, provided the motivation for the first studies of the action of this alkaloid on preparations of isolated sarcoplasmic reticulum (SR) (AXELSON and THESLEFF 1958; GAGE and EISENBERG 1967; WEBER 1968; WEBER and HERZ 1968). The experiments on isolated subcellular fragmented preparations of sarcoplasmic reticulum (FSR) from frog and rabbit skeletal muscle, established some of the important features of the several effects of caffeine, namely that it: (a) inhibited ATP-dependent Ca uptake; (b) activated a Ca,Mg-ATPase; (c) caused a release of accumulated Ca that was antagonised by procaine; and (d) that a heavy fraction was the more sensitive to caffeine.

Subsequently, the techniques employed in these studies, differential centrifugation to prepare the vesicles from muscle homogenates, ^{45}Ca Millipore filtration to measure the movements of Ca across the vesicular membrane and measurement of the rate of ATP hydrolysis from the production of inorganic phosphate, have been further developed and extended. Other important advances followed the development of rapid mixing techniques and the use of calcium indicators (OGAWA 1970). Further understanding came with the development of skinned muscle fibre preparations, in which Ca release from the SR is deduced either from the tension developed or by indicators of free Ca or from the efflux of ^{45}Ca (ENDO 1977; STEPHENSON 1981 a; FABIATO 1983; FABIATO and BAUMGARTEN 1984).

In one area of study, caffeine was used as a probe for SR function in intact muscles. This approach has been complicated by the other effects the methylxanthines may have upon such preparations, including an effect upon the phosphodiesterase activity of the cells, an interaction with the contractile proteins to alter their Ca sensitivity and an interaction with adenosine receptors, especially in cardiac and smooth muscles (BEAVO et al. 1970; ENDO and KITAZAWA 1978; FREDHOLM and PERSSON 1982; WENDT and STEPHENSON 1983).

A second area of study was developed by BIANCHI (1968) who suggested a possible structure–activity relationship for the methylxanthines. Apart from a comparative study of the Ca efflux from isolated FSR and several studies on intact muscle (JOHNSON and INESI 1969; CHAPMAN and MILLER 1974; CHAPMAN and LEOTY 1975), this has not proved a fruitful area so far. However, we wish to include in this chapter some unpublished data on the effects of various methylxanthines and related chemicals upon FSR isolated from rabbit skeletal muscle that may be relevant to a study of the structure–activity relationship of these compounds.

I. The Effect of Caffeine on Ca Uptake

In isolated vesicles of FSR, caffeine reduces the extent and rate of the ATP-dependent Ca uptake either in the presence or absence of binding anions such as phosphate or oxalate. This action, which is dose and drug dependent, appears to be a general property of methylxanthines, where caffeine and theophylline are less effective than aminophylline (JOHNSON and INESI 1969).

II. The Release of Ca from the Sarcoplasmic Reticulum by Caffeine

Although WEBER and HERZ (1968) showed that caffeine may induce a sustained release of Ca from FSR, subsequent work indicates some variability in this response which may arise from the different experimental conditions and fraction of FSR used. Several authors found that caffeine produced little or no release of Ca, even in preparations where it was seen to inhibit the ATP-dependent accumulation of Ca (THORPE and SEEMAN 1971; CHAPMAN et al. 1976; BLAYNEY et al. 1978).

This apparent confusion was resolved by experiments which showed that the Ca release may be phasic and hence too fast to detect with the convential Millipore technique (Fig. 2; CHAPMAN et al. 1976; SU and HASSELBACH 1984). The sustained release, produced by caffeine, is only seen in the presence of sufficient EGTA or EDTA in the suspending medium to prevent the reaccumulation of the ion (WEBER and HERZ 1968; JOHNSON and INESI 1969).

Evidence was also produced showing that the small caffeine-induced release of Ca from loaded vesicles, which is seen when the bathing [Ca] is below 0.2 nM, can be increased by raising [Ca] up to about 3 μM (KATZ et al. 1977 a; MIYAMOTO and RACKER 1981, 1982; SU and HASSELBACH 1984). This effect of the bathing Ca is still seen even if a large part of the intravesicular Ca is complexed by phosphate so that the vesicles become highly loaded with Ca (KATZ et al. 1977 b), suggesting that any augmentation of the release by Ca depends on the [Ca] outside rather than inside the vesicles.

Essentially a similar conclusion has been drawn from experiments on skinned muscle fibres. In these preparations the release of ^{45}Ca from the SR by caffeine is also acutely dependent on the free Ca in the surrounding matrix, although a small release of ^{45}Ca remains in the absence of Ca (STEPHENSON 1981 a). This work, which has been most thoroughly reviewed by ENDO (1977) and FABIATO (1983), suggests that the SR may accumulate Ca even if the free "sarcoplasmic" [Ca] is well below that which activates contraction, and that the stored Ca can be released from the SR to induce the development of tension, not only by caf-

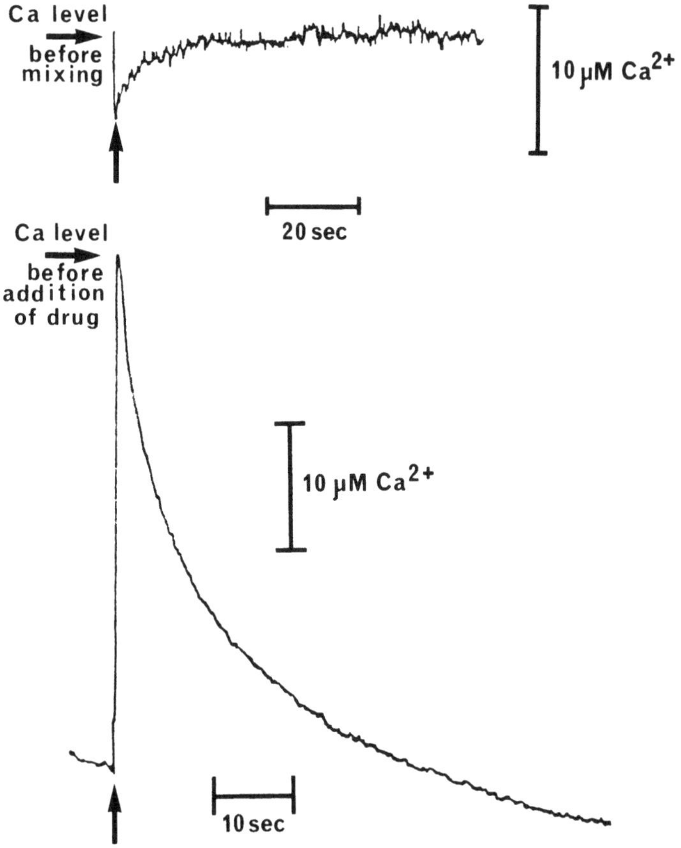

Fig. 2. The rise in extravesicular Ca caused by the addition of caffeine (*upper trace*) and antazoline (*lower trace*) to vesicles of sarcoplasmic reticulum isolated from rabbit skeletal muscles by the method of OGAWA (1970). The change in Ca was detected using ammonium purpurate. Two solutions were mixed in a stopped flow chamber in a differential dual beam spectrophotometer. One contained 20 m*M* Tris-maleate buffer (pH 6.5), 10 m*M* $MgCl_2$, 100 m*M* KCl, 0.4 m*M* ATP, 0.4 ml ammonium purpurate (1.76 g/l), 60 μ*M* $CaCl_2$ and 0.42–0.47 mg/ml microsomal protein; the other solution contained the drug (which had a final concentration in the mixture of 5 m*M*) with the same constituents except that there was no added Ca or SR vesicles. A downward deflection of the trace shows an increase in the extravesicular Ca. Caffeine causes a rapid, but transient release while antazoline causes a sustained release of virtually all the Ca accumulated by the SR. The *vertical arrow* indicates the moment when the mixing flow was stopped. The *horizontal arrow* indicates the level of free Ca before mixing commenced. Temperature 25 °C

feine, but also by a small but rapid increase in the free bathing [Ca]. It would appear that caffeine may trigger the Ca-induced Ca release mechanism, either by producing a small efflux of Ca as seen by STEPHENSON (1981 a, b) and SU and HASSELBACH (1984), which in turn activates its own release, or by lowering the threshold for the Ca-induced Ca release mechanism.

The Ca-induced Ca release, in skinned fibres, has so many properties in common with the caffeine-induced release that they appear to have a common origin. Indeed both Ca and caffeine act synergistically to trigger a Ca release which is blocked by ryanodine and by local anaesthetics, while no release can be evoked from preparations where the SR is destroyed by detergents (ENDO 1977; FABIATO 1983, 1985).

Previous reports noted that Mg would also inhibit the caffeine-induced Ca release from isolated FSR, and that a lowering of [Mg] would itself induce Ca release from the SR of skinned muscle fibres (FAIRHURST and HASSELBACH 1970; INESI and MALAN 1976; MIYAMOTO and RACKER 1981; STEPHENSON 1981 a, b; WEBER 1968). However, SU and HASSELBACH (1984) found that if the free ATP was maintained at a constant level, variation of [Mg] was without effect. By loading the vesicles with Ca, using acetylphosphate as the substrate, they demonstrated that free ATP was essential for caffeine-induced Ca release. They noted that Ca release occurred only when ATP and caffeine were added together, each was ineffective if added alone. Similarly KAKUTA (1984) noticed that in skinned skeletal muscle fibres, a Ca release followed the addition of either ATP, the non-hydrolysable ATP analogue β,γ-methylene adenosine triphosphate (AMPOPCP), ADP, AMP, and even adenosine. The effectiveness of these compounds decreased with the number of phosphate groups present, while UTP, CTP and ITP were much less effective. Interestingly, the release induced by ATP was facilitated by caffeine and raised free Ca and was inhibited by procaine, suggesting that free ATP facilitates the Ca-induced release mechanism in much the same way as does caffeine.

Very high rates of nucleotide-stimulated Ca release have now been reported (NAGASAKI and KASAI 1983) suggesting that Ca leaves the SR down its electrochemical gradient by a passive movement through an ionic channel. The existence of such an ion channel in SR membranes has been demonstrated by SMITH et al. (1985) who were able to fuse FSR membranes into planar lipid bilayers and to show step changes in conductance typical of the opening and closing of ionic channels. These channels which can carry either Ca or Ba ions, discriminate well against monovalent cations, are activated by adenine nucleotides and are inhibited by ruthenium red. An estimated channel conductance of 170 pS in 50 mM Ba is unusually high and sufficient to account for the high rates of Ca efflux from SR vesicles. STEPHENSON (1981 b) hypothesised the existence of such a channel in SR membranes. She suggested that the channel could be activated either by Ca at its opening into the sarcoplasm (R1 in Fig. 3) or by caffeine at another site within the lumen of the SR (R2 in Fig. 3). SU and HASSELBACH (1984) supported this idea, but noting that the effects of Mg depended upon the levels of free ATP, suggested that the effect may be due to the binding of free ATP to a site in the supposed Ca channel. If this site can bind the adenine moiety then the results of KAKUTA (1984) and SMITH et al. (1985) can be brought into the same scheme. Such a scheme would propose that the binding site (R1) can bind substances with an adenine moiety as well as caffeine, to have an effect on the Ca-binding site (R2) to facilitate the opening of the channel. This interpretation may go some way to explain the observed synergism of these compounds on Ca release.

The suggestion that both caffeine and adenine compounds may bind to a common site requires that they have a similar chemical conformation. Further, if the proposed binding site is within the lumen of the SR, the potency of any agent may depend upon the ease with which it permeates the SR membrane. We shall deal with this second point later, however, the obvious chemical similarity of the vari-

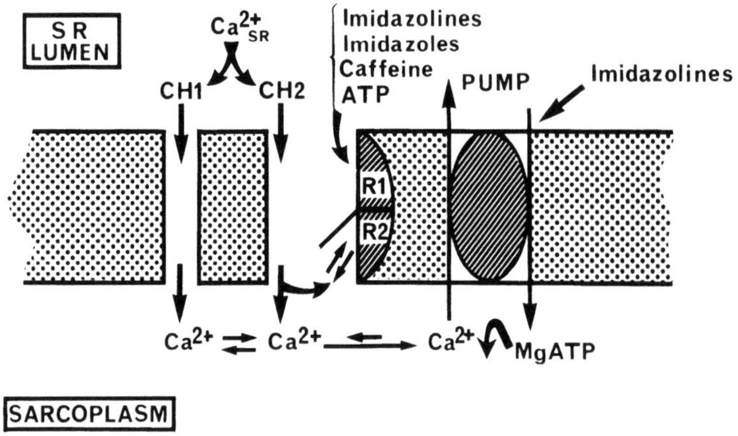

Fig. 3. A schematic representation of the pathways for Ca movement across the membrane of the SR. CH1 is the passive efflux pathway, CH2 is a gated high conductance Ca channel of the type described by SMITH et al. (1985). Two binding sites, R1 and R2 are indicated. R1 is supposed to contain an activating site that binds substances with an imidazole moiety such as caffeine, ATP, etc. R2 is the site where the binding of Ca opens the channel. The ATP-dependent pump is also represented. Ca activates the pump from the sarcoplasmic side, but the imidazolines may inhibit from the luminal side. (Modified from STEPHENSON 1981 a)

ous compounds is the presence of the five-membered imidazole ring. This structure is shared not only by adenine compounds and the methylxanthines, but is also a feature of the imidazoles and the imidazolines. The possible importance of this structure in binding to the Ca channel seems confirmed particularly as all of these compounds and adenosine give rise to contractures in cardiac muscle which are inhibited by local anaesthetics (CHAPMAN and MILLER 1974; CHAPMAN and LEOTY 1976).

Table 1. The release[a] of Ca accumulated in the presence of 6 m*M* MgATP, caused by the addition of 7 m*M* of a variety of methylxanthines and imidazoles

Substance	Ca released (%) after 30 s ± SD	*n*	*P*
1-Ethyl-2-methylbenzimidazole	19.1 ± 0.9	5	<0.01
Aminophilline	17.7 ± 2.4	6	<0.01
1,7-Dimethylxanthine	13.9 ± 1.7	6	<0.01
1,9-Dimethylxanthine	9.8 ± 2.7	8	<0.05
1,3,9-Trimethylxanthine	5.6 ± 1.0	3	>0.05
1-Methylimidazole	3.4 ± 1.4	5	>0.05
Caffeine	3.3 ± 1.3	6	>0.05
1,3,7,9-Tetramethylxanthine	2.2 ± 2.0	8	>0.05
Imidazole	1.0 ± 1.5	6	>0.05

[a] Expressed as a percentage of the total ATP-dependent Ca uptake in the absence of the drug. Rabbit skeletal muscle SR prepared by the method of OGAWA (1970). Ca release measured with the Millipore technique.
P = level of significance for a paired *t*-test of the drug against a no-drug control, *n* = number of experiments.

Table 2. The release[a] of Ca accumulated in the presence of 6 m*M* MgATP, caused by the addition of 7 m*M* of a variety of imidazolines

Substance	Ca released (%) after 2 mins ± SD	*n*	*P*
2-(O-Diphenyloxymethyl)imidazoline	85.5 ± 5.0	6	< 0.01
Oxymetazoline	80.0 ± 4.0	6	< 0.01
Antazoline	77.4 ± 8.0	24	< 0.01
2-(4-Methoxynaphthyl-1-methyl)imidazoline	75.1 ± 13.9	6	< 0.01
Naphazoline	45.2 ± 8.3	9	< 0.01
Xylometazoline	30.0 ± 6.1	3	< 0.5
2-(Phenylaminomethyl)imidazoline	15.5 ± 12.1	6	> 0.5
2-(3α-Dihydroxybenzyl)imidazoline	14.3 ± 9.8	9	> 0.5
Tolazoline	9.3 ± 4.2	9	> 0.5
Tetrahydrozoline	9.2 ± 6.1	9	> 0.5

[a] Expressed as a percentage of the total ATP-dependent Ca uptake in the absence of the drug. Rabbit skeletal muscle SR prepared by the method of OGAWA (1970). Ca release measured with the Millipore technique.
P = level of significance for a paired *t*-test of the drug against a no-drug control, *n* = number of experiments.

In our laboratory we have compared the actions of methylxanthines, imidazoles and imidazolines upon Ca release from FSR of rabbit skeletal muscle. Experiments using the stopped flow technique showed a phasic release in response to caffeine and aminophylline generally missed with the Millipore method. The Millipore method was sufficient to demonstrate that both the imidazoles and the methylxanthines can produce a significant release of Ca (Table 1). The imidazolines show a wide range of potency, but the most effective produce a sustained release of virtually all the Ca accumulated by the vesicles (Fig. 2, Table 2).

These observations may be particularly important as imidazole is often used as a pH buffer in experiments with isolated SR and with skinned fibres. Although imidazole has only a weak effect on the Ca metabolism of isolated SR (see Table 1), it does induce contractures in cardiac muscle (CHAPMAN and MILLER 1974; CHAPMAN and LEOTY 1976). Its presence at concentrations as high as 20 m*M* may have a facilitating effect on the Ca channel.

III. The Effect of Caffeine on the Passive Efflux of Ca

JOHNSON and INESI (1969), showed that if the Mg in the incubating medium was suddenly removed by the addition of EDTA, intravesicular Ca then falls exponentially (the CH1 pathway in Fig. 3). Caffeine, theophylline and aminophylline temporarily increase the rate of this Ca efflux. In our laboratory, a similar effect was produced by a range of methylxanthines and imidazoles. The effect was dose dependent (Fig. 4b) and presumably activates the Ca-induced Ca release mechanism which may be facilitated by the rise in free [ATP] (CH2 in Fig. 3). This release will be sustained because of the high EDTA concentration. In similar experiments, the imidazolines have a different effect on Ca efflux. Below 1 m*M* they

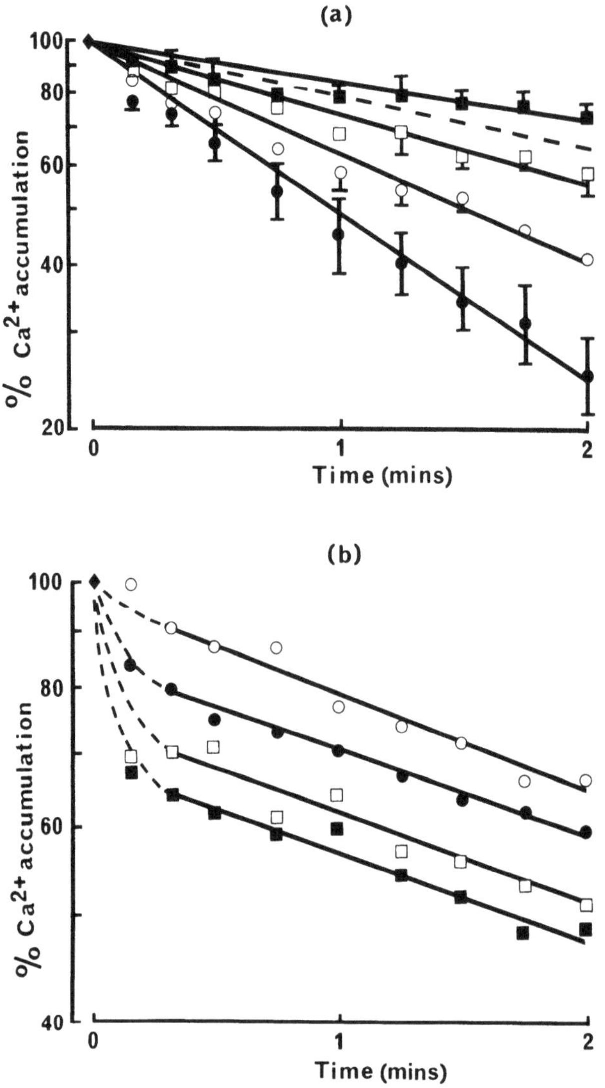

Fig. 4a, b. A semilogarithmic plot of the efflux of Ca from loaded vesicles of rabbit SR in the absence of MgATP and an excess of EDTA (the method of JOHNSON and INESI 1969). The Ca remaining in the vesicles as a percentage of that at time zero (*lozenge*) is plotted. **a** The effect of antazoline, the *broken line* is the buffer control; *full squares* 1.4 m*M*; *open squares* 2.7 m*M*; *open circles* 4.1 m*M*, and *full circles 5.5 mM*. The *bars* represent standard errors of five separate determinations, **b** the effect of 1-ethyl-2-methylbenzimidazole: *open circles* buffer control; *full circles* 1.4 m*M*; *open squares* 2.7 m*M*; *full squares* 5.5 m*M*. The *broken lines* show the phase of rapid efflux. A single experiment. Temperature 25 °C

slow the rate of efflux, but a higher concentration they cause a dose-dependent increase in the monotonic loss of Ca, indicating a more complex action (Fig. 4a).

IV. The Effect of Caffeine on Ca,Mg-ATPase Activity

In isolated FSR from both skeletal and cardiac muscle, caffeine has no effect on the basal ATPase activity, but induces an increase in the Ca-dependent Mg-ATPase activity. This activation shows a marked dependence upon [Ca] in the suspension medium (WEBER 1968; THORPE and SEEMAN 1971; BLAYNEY et al. 1978). In our studies on rabbit skeletal muscle FSR, all the methylxanthines tested showed this behaviour. This stimulation was greatest between 5 and 50 μ*M* Ca, i.e. over the ascending limb of a bell-shaped relationship between the ATPase activity and [Ca] (Fig. 5a).

It has been argued that one effect of caffeine may be to reduce the coupling between ATP hydrolysis and Ca uptake (WEBER 1968). As this effect is most pronounced when the incubating [Ca] is low and is not seen when it is high, it is possible that the increased ATPase activity is associated with an increased Ca cycling across the vesicular membrane. This idea seems consistent with a number of observations: that procaine, which blocks both caffeine-induced and Ca-induced release of Ca, also blocks the stimulating effects of caffeine on the ATPase activity (WEBER 1968); that ATP-dependent Ca release is not associated with the binding of ATP to the Ca,Mg pump ATPase (KUKATU 1984) and that the effect of the Ca ionophore A23187 to increase the hydrolysis of ATP is blocked by a high [EGTA] in the incubating medium (PALADE et al. 1983). It is therefore generally

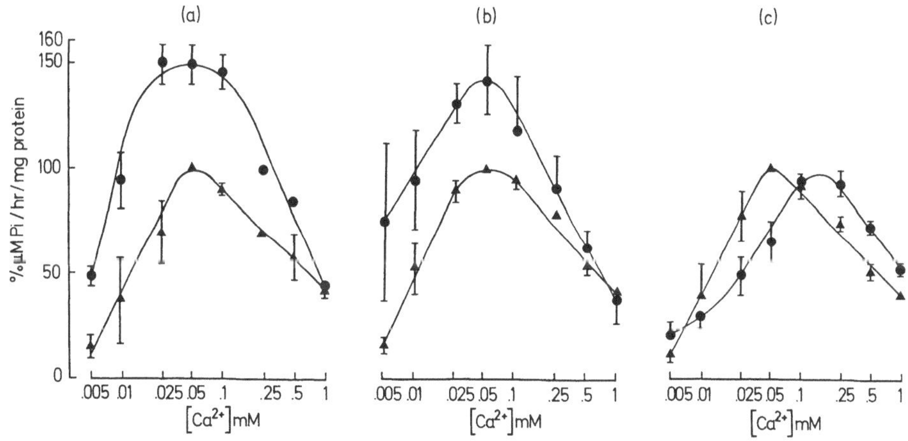

Fig. 5a–c. The effects of 5 m*M* (*full circles*) of **a** a methylxanthine (aminophylline), **b** an imidazole (1-ethyl-2-methylbenzimidazole) and **c** an imidazoline (antazoline) on the relationship between the Ca-activated ATPase of rabbit skeletal SR vesicles and [Ca] in the incubating medium. The rate of ATP hydrolysis is expressed as a percentage of that obtained at the peak of the controls (*full triangles*) which had a mean value of 17.3 ± 1.1 μ*M* inorganic phosphate liberated per hour per milligram protein. The symbols represent the mean and the *bars* standard errors of three determinations when this is larger than the symbol. Temperature 25 °C

Table 3. The effect[a] of various imidazolines on Ca-ATPase activity of rabbit skeletal muscle SR measured in the presence of 0.1 mM Ca

Substance	Ca-ATPase activity (%)	n	P
2-(O-Diphenyloxymethyl)imidazoline	95.0±4.5	5	<0.01
Xylometazoline	86.5±10.5	5	<0.01
2-(4-Methoxynaphthyl-1-methyl)imidazole	80.6±9.6	4	<0.01
Antazoline	65.0±10.8	5	<0.01
Oxymethazoline	59.7±14.7	5	<0.05
Naphazoline	37.5±6.5	5	<0.05
Tetrahydrozoline	9.6±3.8	5	>0.05
2-(3α-Dihydroxybenzyl)imidazoline	6.0±2.0	5	>0.05
Tolazoline	3.5±2.5	6	>0.05
2-(Phenylaminomethyl)imidazoline	2.0±2.5	5	>0.05

[a] Expressed as a percentage of the value observed in the absence of the drug (17.2 μM inorganic phosphate per hour per milligram protein)±SD.
P = level of significance for a paired t-test of the drug against a no-drug control; n = number of experiments.

agreed that caffeine does not have a direct effect on the ATP-driven Ca pump (INESI and MALAN 1976; PALADE et al. 1983).

The imidazoles also produce an activation of the Ca,Mg-ATPase of rabbit skeletal FSR. The most potent being the benzimidazoles (Fig. 5b). On the other hand the imidazolines inhibit the breakdown of ATP, especially on the rising phase of the relationship between phosphate production and the extravesicular [Ca] (Fig. 5c, Table 3). This suggests that they not only release Ca from the SR, but also inhibit its active accumulation which would account for the sustained release seen in Fig. 2.

V. Heavy and Light Fractions of Isolated Sarcoplasmic Reticulum

WEBER (1968) and WEBER and HERZ (1968) were the first to notice that the heavier fractions of FSR were more sensitive to caffeine than the lighter fractions. It has been argued that the heavier fraction originates from the terminal cisternae and the lighter from the longitudinal elements of the SR (WATRAS and KATZ 1984). The heavier fractions are able to release Ca more rapidly, suggesting that they possess more Ca channels. The phasic release described by SU and HASSELBACH (1984) shows a refractory period. This they note could be a property of a release mechanism or due to the presence of a heterogeneous population of vesicles. A differential sensitivity of the SR membrane to caffeine has been suggested by experiments on intact mammalian cardiac muscle and functional differences of the SR from the movements of radioactive Ca in skeletal muscle fibres (WINEGRAD 1968; CHAPMAN and LEOTY 1976).

VI. Structure–Activity Relations for the Methylxanthines

We noted earlier that substances which release Ca from the SR, and/or cause an increase in the activity of the Ca,Mg-ATPase have an imidazole moiety in common, and that the potency of the adenine compounds decreased with the number of phosphates attached to the molecule. Although the methylxanthines, imidazoles and imidazolines show a wide range of potency on isolated FSR and intact muscles. There is an approximate relationship between the water/octanol partition coefficient and the potency of the substance. The lipid solubility of these chemicals will clearly affect their potency in intact muscles and may be important even in isolated SR if the binding site is on the inner surface of the SR Ca channel as STEPHENSON (1981 b) suggests. The lipophilicity of these compounds may be another variable which must be taken into account in any study of the structure–activity relationship.

C. Ryanodine

The wide variety of the responses of striated muscles to ryanodine, e.g. the slowly developing rigor-like contracture in skeletal muscle, an absence of effect upon amphibian or foetal heart, the appearance of both positive and negative inotropic effects in adult mammalian hearts, the effect upon electrical parameters of the cells as well as the modification of the drug action by stimulus frequency and other physiological parameters are difficult to ascribe to a single action of the drug (NAYLER 1963; HAJDU 1969; JENDEN and FAIRHURST 1969; CIOFALO 1973; PENEFSKY 1974a, b; SUTKO et al. 1979; SUTKO and KENYON 1983). These effects, however, are seen at nanomolar concentrations, suggesting a highly specific action in the excitation–contraction sequence, and that the variation of the response may be secondary effects.

Of a number of possibilities, an action of the drug upon Ca movements in the SR seems to be its only direct action. The contractile proteins do not seem to be involved because the Ca sensitivity of skinned fibres, the binding of Ca to troponin, and the myofibrillar Ca-ATPase are all unaffected by the drug (ELISON and JENDEN 1967; NAYLER et al. 1970; SUTKO and KENYON 1983; FABIATO 1985). A number of other possible actions such as an effect upon the mitochondria, or an inhibition of Ca movements across cell membranes by way of either the Ca channels, the Na/Ca exchange or the Ca pump, are similarly unlikely as the Ca metabolism of these structures is insensitive to the alkaloid (JENDEN and FAIRHURST 1969; FAIRHURST 1974; ITO et al. 1984; MITCHELL et al. 1984a, b; WIER et al. 1985).

In the heart, many of the features of the effect of ryanodine upon the intact preparation, including its negative inotropic effect and changes in the electrical properties of the cells, such as a prolongation of the action potential, a slowing of the repolarisation (SLEATOR et al. 1964; MARBAN and WIER 1985), an inhibition of the transient inward current (SUTKO and KENYON 1983) and a prolongation of the slow inward current (MITCHELL et al. 1984b) give the appearance that the drug may reduce the availability of intracellular Ca. This notion is strengthened by the observation that many of these effects can be partially reversed by agents or procedures likely to raise intracellular [Ca], including the application of caffeine or quinidine, increased heart rate, raised extracellular [Ca], reduced extracellular [Na] or lowered extracellular [K], or the aftereffects of EDTA (NAYLER 1963; SLEATOR et al. 1964; PENEFSKY and KAHN 1969; KATZ et al. 1970; SUTKO et al. 1979).

A number of other effects of the drug may also be interpreted in this way and suggest a specific action on the mechanisms which lead to Ca release from the SR. The most marked is the abolition of the "post-rest" contracture and the potentiation which accompanies paired electrical stimulation, responses thought to involve the release of Ca from intracellular stores. These responses are particularly sensitive to ryanodine at concentrations which have only moderate effects on the regularly beating preparation (HAJDU 1969; SLEATOR et al. 1964; FRANK and SLEATOR 1975; RUMBERGER 1976; SUTKO et al. 1979; SUTKO and WILLERSON 1980). Furthermore, ryanodine inhibits the transient depolarisations and arrhythmias which arise from a "Ca overload" of the SR following sodium pump inhibition and are typical of glycoside poisoning (SUTKO and KENYON 1983; VALDEOLMILLOS and EISNER 1985).

An action of ryanodine to reduce the availability of calcium to activate the heartbeat could be due to either a loss of Ca from the SR or an inhibition of its release. Evidence for either possibility would seem to be available. Measurements of ^{45}Ca movements suggest that ryanodine may induce an efflux from a fast exchanging intracellular site and inhibit uptake into a procaine-sensitive store within the cells (BIANCHI 1963; HAJDU 1970 ;NAYLER et al. 1970; FRANK and SLEATOR 1975; HUNTER et al. 1983). A similar inhibition of uptake is recorded for cultured chick heart cells although in this preparation an efflux is not seen (RASMUSSEN et al. 1983).

Direct evidence of an effect of the drug upon the amount of Ca liberated during the heartbeat is provided by studies using the photoprotein aequorin (MARBAN and WIER 1985; WIER et al. 1985). These authors show that in canine Purkinje fibres both components of the aequorin light signal are reduced under the influence of ryanodine. The remaining small light signal is abolished by the Ca channel blocker nitrendipine, but potentiated by the Ca channel agonist Bay K 8644, suggesting that the ryanodine-insensitive signal represents Ca movements across the sarcolemma, but the ryanodine-sensitive response reflects Ca release from the SR. The two components of the ryanodine-sensitive signal are affected differentially by caffeine and the pattern of electrical stimulation in ways which suggest that the early light signal may arise from SR in intimate contact with the sarcolemma, while the later signal originates from stores deeper in the cells (WIER 1980; HESS and WIER 1984). In ferret ventricular muscle, which possesses t-tubules (SIMPSON and RAYNS 1968), ryanodine reduces both the twitch and the aequorin signal. The aequorin signal has not been separated in more than one component (WIER et al. 1985), however both the twitch and the light signal are prolonged (presumably as the duration of the action potential is prolonged).

The functional heterogeneity of the SR implicit in these observations is also a conclusion drawn from experiments on the effect of ryanodine upon isolated fractions of the SR (JONES et al. 1979). These show a wide species variability in the effect of the drug and within a single species its activity varies markedly with the particular microsomal fraction studied. Whilst skeletal muscle FSR may show very little sensitivity to the drug (FAIRHURST 1973), successively heavier fractions of the SR show an increasing inhibition of Ca uptake, which is accompanied by an increased ATPase activity (FAIRHURST and HASSELBACH 1970; FAIRHURST 1974). JONES and CALA (1981) investigated the suggestion that the effect of ryanodine, like caffeine, may depend upon the fraction of the FSR studied. In SR isolated from the heart they found that the lighter fractions had a low sensitivity to the alkaloid, but in the absence of the drug accumulated more Ca than the heavier fractions (i.e. they exhibited the most Ca pump sites). The heavier fractions, however, responded to the drug with a five- to tenfold increase in their ability to accumulate Ca without any change in the rate of ATP hydrolysis. Interestingly, these ryanodine-sensitive fractions were associated with a protein of around 55000 dalton. SEILER et al. (1984) found a similar ryanodine-sensitive fraction in both cardiac and skeletal muscle FSR. They identified this fraction with elements of the terminal cisternae of the SR and it may be that the drug acts here to promote a net Ca uptake by inhibition of the efflux. They further noted that these "junctional" SR vesicles, in contrast to the "free" SR vesicles, were enriched with calsequestrin.

Other evidence would support the notion that ryanodine acts at the junction between the SR and the t-tubules. PENEFSKY (1974a, b) noted that amphibian and foetal hearts which lack t-tubules are insensitive to the drug and that a sensitivity develops in the mammalian species as the t-system develops. She further noted that damage at the level of the Z-disc was associated with ryanodine treatment. A need for a much higher dose of the drug to produce effects in skinned fibres may also be relevant if the terminal portions of the SR are damaged during the skinning process (FABIATO 1985).

The interpretation of the action of ryanodine as an inhibitor of Ca release from the SR finds some support from the observations that:

1. In skinned fibres the caffeine-induced release of Ca is inhibited – albeit at a higher ryanodine concentration than that needed in intact heart (FABIATO 1985).

2. The species variability of the action of ryanodine on intact heart is paralleled by the sensitivity of skinned fibrers to Ca-induced Ca release (HAJDU 1969; CIOFALO 1973; RUMBERGER 1976; FABIATO and FABIATO 1978; BERS et al. 1980; SUTKO and WILLERSON 1980; BERS 1985).

3. Ryanodine inhibits Ca-induced Ca release in isolated cardiac FSR (CHAMBERLAIN et al. 1984).

4. Ruthenium red, which acts like ryanodine to increase Ca uptake in the heavy or junctional SR, also blocks the high conductance Ca channel isolated from SR membranes (SEILER et al. 1984; SMITH et al. 1985).

This evidence leads to an inevitable conclusion that ryanodine inhibits the release of Ca from the junctional SR to reduce the strength of the heartbeat or the skeletal muscle twitch. A number of other observations, e.g. the effect of ryanodine on the lighter fractions of FSR to increase Ca efflux and Ca-ATPase activity, the development of a rigor-like contracture in skeletal muscles, the increased ^{45}Ca efflux, the late positive inotropic phase and its apparent use dependence do not immediately and simply follow from that conclusion. This may reflect both agonist and antagonist actions on the SR (SUTKO et al. 1985), however, if the difference in the effect of the drug on the two major fractions of the SR are real then both will be operating in the intact muscle. The appearance of two effects may reflect that action of the drug on Ca movements between different elements of the SR, from the longitudinal tubules (the lighter fraction) which has the higher concentration of Ca pump sites, i.e. Ca uptake sites, to release sites, i.e. the Ca channels in the junctional SR (the heavier fraction). A model where Ca cycles in the SR has been a popular notion since it was first proposed by WINEGRAD (1968). A ryanodine-induced increase in the efflux of Ca from the longitudinal tubules may well be compensated by a rise in ATPase activity and Ca pumping, but the reduced efflux and the increased Ca accumulation will heavily load the junctional SR so that Ca may back up into the longitudinal SR. This will have a number of secondary effects. It could progressively reduce the ability of the SR to accumulate Ca, which may account for the increase in the slope of the relationship between intracellular [Na] and tonic tension which occurs in cardiac muscle (CANNELL et al. 1985). It may also explain the late positive inotropic effect of the drug because of a reduction in the intracellular Ca buffering. If the overload of the junctional SR eventually results in a release of Ca this effect may not be too severe in cardiac muscle where the Na/Ca exchange can still regulate intracellular [Ca]. In skeletal muscle, however, where the Na/Ca exchange is absent intracellular [Ca] may rise to cause a rigor-like contracture which may lead to structural damage as PENEFSKY (1974b) reports and to increased ^{45}Ca efflux from the muscle.

D. Conclusions

In response to the data available we have been obliged to review the literature on the methylxanthines and ryanodine in rather different ways. We come to the general conclusion, however, that both have their primary action upon the Ca release mechanism of the sarcoplasmic reticulum. In the case of the methylxanthines we have noted that a variety of structurally similar compounds have the same action. They would seem to facilitate the Ca-induced Ca release mechanism and we suggest that this is due to a binding of substances with an imidazole moiety to a site with a relatively low affinity in a Ca channel in the SR. This site may normally be regulated in the intact muscle by the free ATP concentration or even naturally occurring imidazoles like carnosine (HARRISON et al. 1986). The regulation of the Ca channel by caffeine, ATP, etc. shows some cooperativity because caffeine facilitates the effect of ATP and vice versa. The effect of binding of such substances to the site in the channel could be mediated either by the initiation of a small triggering efflux of Ca or by the lowering of the threshold for the Ca-induced Ca release mechanism by an allosteric influence. Ionic channels that are also affected by free ATP are found in the sarcolemma of both cardiac and skeletal muscles (NOMA 1983; SPRUCE et al. 1985). Part of the action of methylxanthines on the intact muscle, however, must be due to its other actions, particularly its effects on phosphodiesterase enzymes and the contractile proteins, as well as an interaction with other systems that regulate intracellular [Ca].

Ryanodine's primary action would also seem to be on the Ca channels of the SR to block the release of Ca. The potentiating effect of depolarisation upon the action of the drug suggests that it may bind more strongly to the activated Ca channel. The site to which ryanodine binds may be quite different from the ATP–methylxanthine site, especially as ruthenium red has a similar action to ryanodine. The presence of a pyrrole ring in ryanodine might be important in the binding, as it is somewhat similar to the imidazole moiety. We have been unable, however, to test this notion as few experiments have been reported where the effects of simple pyrroles have been tested upon either intact muscles or SR fractions.

Acknowledgments. We thank the British Heart Foundation for financial support, and A Boon, NG Rutherford, M Kashefiolasl and RJ Yates who performed some of the experiments.

References

Axelson J, Thesleff S (1958) Activation of the contractile mechanism in striated muscle. Acta Physiol Scand 44:55–66

Beavo JA, Rogers NL, Crofford OB, Hardman JG, Sutherland EW, Newman EV (1970) Effects of xanthine derivatives on lipolysis and adenosine 3′,5′-monophosphate phosphodiesterase activity. Mol Pharmacol 6:597–603

Bers DM (1985) Ca influx and sarcoplasmic reticulum Ca release in cardiac muscle activation during post rest recovery. Am J Physiol 248:H366–H381

Bers DM, Philipson KD, Langer GA (1980) Cardiac contractility and sarcolemmal calcium binding in several cardiac muscle preparations. Am J Physiol 240:H576–H583

Bianchi CP (1963) Action on calcium movements in frog sartorius muscles by drugs producing rigor. J Cell Comp Physiol 61:255–263

Bianchi CP (1968) Pharmacological actions on excitation-contraction coupling in striated muscle. Fed Proc 27:126–131

Blayney L, Thomas H, Muir J, Henderson A (1978) Action of caffeine on calcium transport by isolated fractions of myofibrils, mitochondria, and sarcoplasmic reticulum from rabbit heart. Circ Res 43:520–526

Cannell MB, Vaughan-Jones RD, Lederer WJ (1985) Ryanodine block of calcium oscillations in heart muscle, and the Na-tension relationship. Fed Proc 44:2964–2969

Chamberlain EK, Volpe P, Fleischer S (1984) Inhibition of Ca-induced Ca release from purified cardiac sarcoplasmic reticulum vesicles. J Biol Chem 259:7547–7553

Chapman RA, Leoty C (1975) Which of caffeine's chemical relatives are able to induce contractures in mammalian heart? In: Harris P, Bing RJ (eds) Recent advances in studies on cardiac structure and metabolism, vol 7. University Park Press, Baltimore, pp 425–430

Chapman RA, Leoty C (1976) Time-dependent and dose-dependent effects of caffeine on the contraction of the ferret heart. J Physiol (Lond) 256:287–314

Chapman RA, Miller DJ (1974) Structure-activity relations for caffeine: a comparative study of the inotropic effects of the methylxanthines, imidazoles and related compounds on the frog's heart. J Physiol (Lond) 242:589–613

Chapman RA, Rutherford NG, Wallace S (1976) The use of a dual-beam differential spectrophotometer with a stopped flow mixing chamber, in the study of the release of Ca^{2+} from preloaded vesicles of the sarcoplasmic reticulum. J Physiol (Lond) 258:2–3P

Ciofalo FR (1973) Relationship between ^{3}H-ryanodine uptake and myocardial contractility. Am J Physiol 225:324–327

Elison C, Jenden DJ (1967) The effect of ryanodine on model systems derived from muscle – II. Myofibrils and natural actomyosin. Biochem Pharmacol 16:1347–1354

Endo M (1977) Calcium release from sarcoplasmic reticulum. Physiol Rev 57:71–108

Endo M, Kitizawa T (1978) Excitation-contraction coupling in chemically skinned cardiac muscle. In: Hayane S, Murao S (eds) Proceedings of the VIII world congress of cardiology. Excerpta Medica, Amsterdam, pp 800–803

Fabiato A (1983) Calcium-induced release of calcium from cardiac sarcoplasmic reticulum. Am J Physiol 245:C1–C14
Fabiato A (1985) Effects of ryanodine on skinned cardiac cells. Fed Proc 44:2970–2976
Fabiato A, Baumgarten CM (1984) Methods for detecting calcium release from the sarcoplasmic reticulum of skinned cardiac cells and the relationships between calculated transsarcolemmal calcium movement and calcium release. In: Sperelakis N (ed) Physiology and pathophysiology of the heart. Martinus Nijhoff, Boston, pp 215–254
Fabiato A, Fabiato F (1978) Ca-induced release of calcium from the sarcoplasmic reticulum of skinned cells from adult human, dog, cat, rabbit, rat, and frog hearts and from fetal and new-born rat ventricles. Ann NY Acad Sci 307:491–522
Fairhurst AS (1973) Effect of ryanodine on skeletal muscle reticulum calcium adenosine triphosphatase (CaATPase). Biochem Pharmacol 22:2815–2827
Fairhurst AS (1974) A ryanodine caffeine sensitive membrane fraction of skeletal muscle. Am J Physiol 227:1124–1131
Fairhurst AS, Hasselbach W (1970) Calcium efflux from a heavy sarcotubular fraction: effects of ryanodine, caffeine and magnesium. Eur J Biochem 13:504–509
Frank M, Sleator WW (1975) Effect of ryanodine on myocardial calcium. N S Arch Pharmacol 290:35–47
Fredholm BB, Persson GGA (1982) Xanthine derivatives as adenosine receptor antagonists. Eur J Pharmacol 81:673–676
Gage PW, Eisenberg RS (1967) Action potentials without contraction in frog skeletal muscle fibers with disrupted transverse tubules. Science 158:1702–1703
Hajdu S (1969) Mechanism of the Woodworth staircase phenomenon in heart and skeletal muscle. Am J Physiol 216:206–214
Hajdu S (1970) Effect of drugs, temperature, and ions on Ca^{++} of coupling system of skeletal muscle. Am J Physiol 218:966–972
Harrison SM, Lamont C, Miller DJ (1986) Carnosine and other natural imidazoles enhance muscle Ca sensitivity and are mimicked by caffeine and AR-L 115BS. J Physiol (Lond) 371:197P
Hess P, Wier WG (1984) Excitation-contraction coupling in cardiac Purkinje fibres. J Gen Physiol 83:417–433
Hunter DR, Haworth RA, Berkoff HA (1983) Modulation of cellular calcium stores in the perfused rat heart by isoproterenol and ryanodine. Circ Res 53:703–712
Inesi G, Malan N (1976) Mechanism of calcium release in sarcoplasmic reticulum. Life Sci 18:773–779
Ito K, Kenyon JL, Isenberg G, Sutko JL (1984) The existence of two components of transient outward current in isolated cardiac ventricular myocytes. Biophys J 45:54a
Jenden DJ, Fairhurst AS (1969) The pharmacology of ryanodine. Pharmacol Rev 21:1–25
Johnson PN, Inesi G (1969) The effect of methylxanthines and local anaesthetics on fragmented sarcoplasmic reticulum. J Pharmacol Exp Ther 169:308–314
Jones LR, Cala SE (1981) Biochemical evidence for functional heterogeneity of cardiac sarcoplasmic reticulum vesicles. J Biol Chem 256:11809–11818
Jones LR, Besch HR, Sutko JL, Willerson JT (1979) Ryanodine-induced stimulation of net Ca^{++} uptake by cardiac sarcoplasmic reticulum vesicles. J Pharmacol Exp Ther 209:48–55
Kakuta Y (1984) Effects of ATP and related compounds on the Ca-induced Ca release mechanism of the Xenopus SR. Pflugers Arch 400:72–79
Katz AM, Repke DI, Fudyma G, Shingekawa M (1977a) Control of calcium efflux from sarcoplasmic reticulum vesicles by external calcium. J Biol Chem 252:4210–4212
Katz AM, Repke DI, Hasselbach W (1977b) Dependence of ionophore- and caffeine-induced calcium release from sarcoplasmic vesicles on external and internal calcium ion concentration. J Biol Chem 252:1938–1949
Katz N, Ingenito A, Procita L (1970) Ryanodine-induced contractile failure of skeletal muscle. J Pharmacol Exp Ther 171:242–248
Marban E, Wier WG (1985) Ryanodine as a tool to determine the contributions of calcium entry and calcium release to the calcium transient and contraction of cardiac Purkinje fibres. Circ Res 56:133–138

Mitchell MR, Powell T, Terrar DA, Twist VW (1984a) Ryanodine prolongs Ca-currents while suppressing contraction in rat ventricular muscle cells. Br J Pharmacol 81:13–15
Mitchell MR, Powell T, Terrar DA, Twist VW (1984b) The effects of ryanodine, EGTA and low-sodium on action potentials in rat and guinea-pig ventricular myocytes: evidence for two inward currents during the plateau. Br J Pharmacol 81:543–550
Miyamoto H, Racker E (1981) Calcium-induced release at terminal cisternae of skeletal sarcoplasmic reticulum. FEBS Lett 133:235–238
Miyamoto H, Racker E (1982) Mechanism of calcium release from skeletal sarcoplasmic reticulum. J Membr Biol 66:193–201
Nagasaki K, Kasai M (1983) Fast release of calcium from sarcoplasmic reticulum vesicles monitored by chlortetracycline fluorescence. J Biochem 94:1101–1109
Nayler WG (1963) Effect of ryanodine on cardiac muscle. Am J Physiol 204:975–978
Nayler WG, Daile P, Chipperfield D, Gan K (1970) Effect of ryanodine on calcium in cardiac muscle. Am J Physiol 219:1620–1626
Noma A (1983) ATP-regulated K channels in cardiac muscle. Nature 305:147–148
Ogawa Y (1970) Some properties of frog fragmented sarcoplasmic reticulum with particular reference to its response to caffeine. J Biochem 67:667–683
Palade P, Mitchell RD, Fleischer S (1983) Spontaneous calcium release from sarcoplasmic reticulum. J Biol Chem 258:8098–8107
Penefsky ZJ (1974a) Studies on mechanism of inhibition of cardiac muscle contractile tension by ryanodine. Pflugers Arch 347:173–184
Penefsky ZJ (1974b) Ultrastructural studies of the site of action of ryanodine on heart muscle. Pflugers Arch 347:185–198
Penefsky ZJ, Khan M (1970) Mechanical and electrical effects of ryanodine on mammalian heart muscle. Am J Physiol 218:1682–1686
Rasmussen CA, Sutko JL, Barry WH (1983) Effects of ryanodine on contractility, transmembrane voltage and calcium fluxes in cultured embryo ventricular cells. Fed Proc 42:573
Rumberger E (1976) The role of the sarcoplasmic reticulum in the pure frequency potentiation: the effect of ryanodine. Pflugers Arch 364:203–204
Seiler S, Wegener AD, Whang DD, Hathaway DR, Jones LR (1984) High molecular weight proteins in cardiac and skeletal muscle junctional sarcoplasmic reticulum vesicles bind calmodulin, are phosphorylated, and are degraded by Ca^{2+}-activated protease. J Biol Chem 259:8550–8557
Simpson F, Raynes D (1968) The relationship between the transverse tubular system and other tubules at the Z disc levels of myocardial cells in the ferret. Am J Anat 122:193–198
Sleator W, Furchgott RF, Gubareff T De, Krespl V (1964) Action potentials of guinea pig atria under conditions which alter contraction. Am J Physiol 206:270–282
Smith JS, Coronado R, Meissner G (1985) Sarcoplasmic reticulum contains adenine nucleotide-activated calcium channels. Nature 316:446–449
Spruce AE, Standen NB, Stanfield PR (1985) Voltage-dependent ATP-sensitive potassium channels of skeletal muscle membrane. Nature 316:736–738
Stephenson EW (1981a) Activation of fast skeletal muscle: contributions of studies on skinned fibers. Am J Physiol 240:C1–C19
Stephenson EW (1981b) Ca^{2+} dependence of stimulated ^{45}Ca efflux in skinned muscle fibres. J Gen Physiol 77:419–443
Su JY, Hasselbach W (1984) Caffeine-induced calcium release from isolated sarcoplasmic reticulum of rabbit skeletal muscle. Pflugers Arch 400:14–21
Sutko JL, Kenyon JL (1983) Ryanodine modification of cardiac muscle responses to potassium free solutions. Evidence for inhibition of sarcoplasmic reticulum calcium release. J Gen Physiol 82:385–404
Sutko JL, Willerson JT (1980) Ryanodine alteration of the contractile state of rat ventricular myocardium. Circ Res 46:332–343
Sutko JL, Willerson JT, Templeton GH, Jones LR, Besch HR (1979) Ryanodine: its alterations of cat papillary muscle contractile state and responsiveness to inotropic interventions and a suggested mode of action. J Pharmacol Exp Ther 209:37–47

Sutko JL, Ito K, Kenyon JL (1985) Ryanodine: a modifier of sarcoplasmic reticulum calcium release in striated muscle. Fed Proc 44:2984–2988
Thorpe WR, Seeman P (1971) The site of action of caffeine and procaine in skeletal muscle. J Pharmacol Exp Ther 179:324–330
Valdeolmillos M, Eisner DA (1985) The effects of ryanodine on calcium overloaded sheep heart Purkinje fibres. Circ Res 56:452–456
Watras J, Katz AM (1984) Calcium release from two fractions of sarcoplasmic reticulum from rabbit skeletal muscle. Biochem Biophys Acta 796:429–439
Weber A (1968) The mechanism of action of caffeine on sarcoplasmic reticulum. J Gen Physiol 52:760–772
Weber A, Herz R (1968) The relationship between caffeine contracture of intact muscle and the effect of caffeine on reticulum. J Gen Physiol 52:750–759
Wendt IR, Stephenson DG (1983) The effects of caffeine on Ca-activated force in skinned cardiac and skeletal fibres of the rat. Pflugers Arch 398:210–216
Wier WG (1980) Calcium transients during excitation-contraction coupling in mammalian heart: aequorin signals of canine Purkinje fibers. Science 207:1085–1087
Wier WG, Yue DT, Marban E (1985) Effects of ryanodine on intracellular Ca^{2+} transients in mammalian cardiac muscle. Fed Proc 44:2989–2993
Winegrad S (1968) Intracellular calcium movements of frog skeletal muscle during recovery from tetanus. J Gen Physiol 51:65–83

CHAPTER 12

Effect of Lithium in Stimulus – Response Coupling

I. SCHULZ

A. Introduction

Questions on the mode of intracellular action of Li^+ have attracted special interest because of its therapeutic effects in manic-depressive illness. The original report of control and prevention of mania by Li^+ salts (CADE 1949) initiated widespread use of Li^+ for this illness. Since Li^+ is most effective in reducing the mood swings of bipolar manic-depressive patients, patients suffering from this disease often take Li^+ for the rest of their lives (W. R. SHERMAN 1986, personal communication). The mode of Li^+ action in these patients is not yet clear, although many different functions of Li^+ on ion transport systems, enzymes, synaptic transmission, and receptor sensitivity have been described (EMRICH et al. 1982) from which a hypothesis has emerged that Li^+ modulates central nervous activity by influencing receptor-mediated phosphoinositide metabolism – and consequently also Ca^{2+} mobilization and control of Ca^{2+}-dependent neurosecretion (BERRIDGE et al. 1982).

Thus, what had been separate problems for decades of research have merged today into related areas of investigation. Lithium has become a useful tool in studying phosphoinositide metabolism and the latter has been shown to be closely connected to receptor-mediated Ca^{2+} mobilization. In the following, a brief review of the main findings related to research on these three topics, including the action of lithium, phosphoinositide metabolism, and the mobilization of Ca^{2+}, will be given. Finally, new concepts of cellular function which have emerged from understanding how these three topics are related will be discussed.

Receptor-mediated breakdown of phosphoinositides, subsequent rise of their hydrolysis products, inositol phosphates and diacylglycerol, and consequent inositol-1,4,5-triphosphate-induced Ca^{2+} release from the endoplasmic reticulum can be considered as the main pathways by which hormones and neurotransmitters stimulate physiologic functions of cells. However, other pathways, leading to rise of cyclic AMP or which involve diacylglycerol are no less important. Since interaction between these pathways occurs at different steps in the cascade of events between stimulation and cellular response, they will be mentioned in relation to the phosphatidylinositol Ca^{2+} pathway wherever it is relevant.

B. Lithium

Since Li^+ is able to replace other biologically important ions, it has often been used as a cation substitute mainly of Na^+ when investigating Na^+-dependent phenomena (OVERTON 1902). With its ionic radius of 0.60 Å it is the smallest of the alkali cations in group Ia of the periodic table, the others being Na^+ (0.95 Å), K^+ (1.33 Å), Rb^+ (1.48 Å), and Cs^+ (1.59 Å).

Li^+ can replace Na^+ in some cellular processes. For example, it can substitute for Na^+ in maintaining cell excitability (HILLE 1970; OVERTON 1902) and it has been shown that Li^+ can pass through potential-sensitive Na^+ channels in frog sciatic nerve node of Ranvier (HILLE 1970), in mouse neuroblastoma cells (RICHELSON 1977), in frog skeletal muscle (KEYNES and SWAN 1959b), and in rabbit cardiac muscle (NIELSEN-KUDSK and PEDERSEN 1978). Furthermore, Li^+ can pass through amiloride-sensitive channels of the apical membrane of frog skin (NAGEL 1977).

In nonexcitable cells such as red blood cells, Li^+ can be transported via Na^+–Li^+ countertransport (DUHM et al. 1976; HAAS et al. 1975), by the furosemide-sensitive Cl^--dependent Na^+–K^+ cotransport system (GECK et al. 1980), and by the SITS-sensitive Cl^-/HCO_3^- exchange system (DUHM and BECKER 1977). Li^+ can partly substitute for Na^+ in coupled amiloride-sensitive Na^+/H^+ exchange (HELLMESSEN et al. 1985) and Na^+/Ca^{2+} exchange systems (BAYERDÖRFFER et al. 1985). Furthermore, Li^+ can substitute for Na^+ in some coupled Na^+ cotransport and countertransport systems such as Na^+–phenylalanine cotransport in the kidney (EVERS et al. 1976).

In other processes, such as Na^+,K^+-ATPase activity, Li^+ is not handled effectively (ARAKI et al. 1965; KEYNES and SWAN 1959a, b). The charge density of Li^+ and Ca^{2+} and the ionic radii of Li^+ and Mg^{2+} (0.60 Å and 0.65 Å, respectively) are comparable (WILLIAMS 1973). Indeed, it has been suggested that Li^+ can mimic and interfere with Mg^{2+}- and Ca^{2+}-dependent functions in several systems such as platelet aggregation (HARGREAVES and HAYES 1978), Ca^{2+} transport across membranes (BAKER 1972; BAKER and MCNAUGHTON 1978; BAKER and SCHLAEPFER 1978; BLAUSTEIN 1974; RINK 1977; ULLRICH et al. 1976), and activation of enzymes, especially those involved in phosphoinositide metabolism, as will be discussed in Sect. D. With regard to its multifunctional properties in intact cells, one should distinguish between effects in which Li^+ can replace other cations and those in which Li^+ leads to changes in other cation concentrations. Thus, ALDENHOFF and LUX (1982) have observed that injection of Li^+ into *Helix* neurons decreased inward Ca^{2+} currents and changed Ca^{2+}-dependent K^+ currents, which were similar to those observed after intracellular Ca^{2+} injections. It was further shown that Li^+ induced elevation of cytosolic free Ca^{2+} concentration which was interpreted by the authors as due to impairment of Ca^{2+} outward transport in the presence of Li^+ (ALDENHOFF and LUX 1982).

C. Phosphoinositides and Calcium

The importance of phosphoinositide metabolism in receptor-mediated stimulation of cellular response has been recognized in a variety of systems, including exocrine and endocrine secretion, platelet aggregation, DNA synthesis, smooth muscle contraction, and phototransduction (BERRIDGE and IRVINE 1984). The initial discovery that hormones have effects on phosphoinositide metabolism was

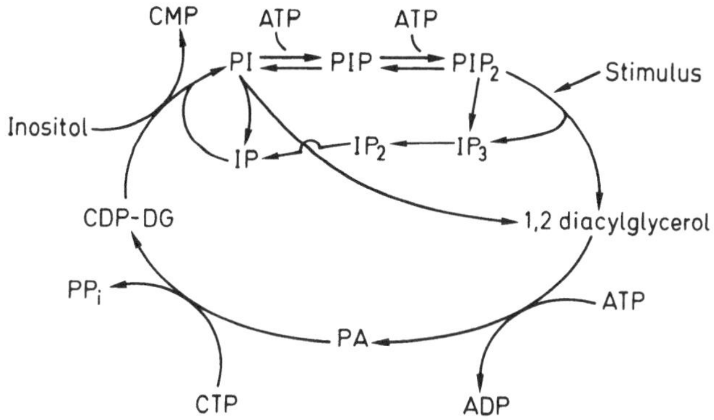

Fig. 1. Scheme for the phosphoinositide cycle. In the old version of this model (HOKIN and HOKIN 1964), it consisted of phosphatidylinositol (PI) breakdown and phosphorylation of the released 1,2-diacylglycerol (DG) to phosphatidic acid (PA). PI is resynthesized from cytidine diphosphodiglyceride (CDP-DG). We now know that phosphatidylinositol-4,5-bisphosphate (PIP_2) is the precursor for stimulus-induced activation of phospholipase C and breakdown to inositol-1,4,5-trisphosphate (IP_3) and DG. IP is inositol phosphate; IP_2 is inositol-1,4-bisphosphate; PP_i is pyrophosphate

made by HOKIN and HOKIN (1953, 1955, 1958 a, b). They showed that administration of acetylcholine to secretory cells of the pancreas increased the incorporation of $^{32}P_i$ into phosphatidylinositol (PI), one of the phospholipid constituents of the plasma membranes, and into phosphatidic acid (PA).

Other secretagogues of enzyme secretion also increased turnover of PI (HOKIN and HOKIN 1956). An increase in the incorporation of [^{3}H] inositol into PI on stimulation was also observed in pancreas (HOKIN and HOKIN 1958 a), in brain (HOKIN and HOKIN 1958 b), and many other tissues (R. H. MICHELL 1975). On the basis of these studies, a scheme was proposed, called the phosphatidylinositol–phosphatidate cycle (Fig. 1; HOKIN and HOKIN 1965, 1967). According to this model, receptor-mediated stimulation of PI phosphodiesterase leads to PI breakdown and consequent rise in both hydrolysis products, diacylglycerol (DG) and inositol phosphate (IP). DG is phosphorylated from ATP to form PA (HOKIN and HOKIN 1959). On removal of stimulus, PA is converted back to PI by the sequential actions of CDP-DG[1] cytidyl transferase and PI synthase (AGRANOFF et al. 1958; PAULUS and KENNEDY 1960). The authors proposed that this "phospholipid effect" had something to do with the mechanism of exocytosis (HOKIN and HOKIN 1953). Later, it was shown that, in addition to PI, phosphatidylinositol phosphate (PIP) and phosphatidylinositol bisphosphate (PIP_2) were also present. The rapid turnover of polyphosphoinositides was first described in brain (BROCKERHOFF and BALLOU 1962) and after that in non-neuronal tissue (SANTIAGO-CALVO et al. 1964). Stimulated catabolism of one of the polyphosphoinositides and formation of inositol mono- and bisphosphate was detected by DURELL et al. (1968, 1969; DURELL and GARLAND 1969). PIP_2 breakdown and inositol trisphosphate (IP_3) formation was documented in detail in iris smooth muscle (ABDEL-LATIF et al.

[1] Cytidine diphosphodiacylglycerol, also called cytidine monophosphorylphosphatidate, is variously abbreviated CDP-DG, CMPPA, and CMP-PA.

1977; AKHTAR and ABDEL-LATIF 1980). It was not yet clear, however, if this reaction was a consequence of the calcium signal or its cause. An important discovery that favored the concept of the primary receptor-stimulated event being hydrolysis of PIP_2 and generation of IP_3 was the observation that in hepatocytes stimulated by vasopressin, decrease in PIP_2 preceded changes in PI and were largely independent of external Ca^{2+} (MICHELL et al. 1981). BERRIDGE and co-workers then observed rapid increase in formation of IP_3 that preceded that of IP in cells of insect salivary gland stimulated with serotonin. Since IP_3 formation occurred without an apparent lag period, whereas the onset of calcium-dependent response was delayed by at least 1 s, the authors suggested that IP_3 could function as an intracellular messenger to mobilize calcium (BERRIDGE 1983; BERRIDGE et al. 1984a).

Another important finding on the action of hormones which lead to enzyme secretion from the exocrine pancreas was the observation that calcium was necessary in the secretory process. The importance of Ca^{2+} for intracellular mediation in stimulation of cell function, such as exocrine and endocrine secretion (DOUGLAS and POISNER 1964; KATZ and MILEDI 1967), muscle contraction (BLUMENTHAL and STULL 1980), cell proliferation (ALLAN and MICHELL 1977), fertilization (SILLERS and FORER 1985), and phototransduction (YOSHIKAMI and HAGINS 1970) has been demonstrated. The rise of cytosolic free Ca^{2+} concentration from its resting level of $\sim 10^{-7}$ to $\sim 10^{-6} M$ is considered to trigger the physiologic response. The questions which had intrigued investigators were mostly those on the source of Ca^{2+}: does it originate from the inside of the cell or does it come from the extracellular fluid? The other important problem concerned the mechanism by which hormonal signals transfer the message to Ca^{2+} release and to Ca^{2+} influx into the cell.

The source of Ca^{2+} that leads to rise of cytosolic free Ca^{2+} concentration differs in different cell types. Whereas in some cells, such as nerve terminals (KATZ and MILEDI 1967), neurohypophysis chromaffin and mast cells (DOUGLAS 1974), the physiologic response to hormonal stimulation is clearly dependent on extracellular Ca^{2+}, and receptor-mediated stimulation leads to increased net uptake of Ca^{2+} into the cell, in other tissues, such as the exocrine pancreas and smooth muscle, the source of Ca^{2+} is both intracellular stores and the extracellular milieu (SCHULZ 1980). In the pancreas, an immediate cessation of stimulated secretion in the absence of extracellular calcium could not be observed. Only prolonged preincubation or perfusion (30–60 min) of the pancreatic tissue in the absence of calcium and with the Ca^{2+} chelator EGTA was found to markedly diminish (PETERSEN and UEDA 1976; ROBBERECHT and CHRISTOPHE 1971) or abolish (ARGENT et al. 1973; CASE and CLAUSEN 1973; HEISLER et al. 1972; HOKIN 1966; KANNO 1972) enzyme secretion.

A study on superfused mouse pancreatic fragments (PETERSEN and UEDA 1976) showed that amylase secretion in response to short pulses of acetylcholine stimulation at half-hour intervals was little affected by prolonged exposure to Ca^{2+}-free solution, even when the Ca^{2+}-chelating agent EGTA was present. However, in Ca^{2+}-free solution containing EGTA, sustained exposure to acetylcholine (normally causing sustained enzyme secretion) or sustained exposure to other secretagogues, such as cholecystokinin, cerulein, or bombesin resulted in only a short burst of secretion (PETERSEN and IWATSUKI 1978; SCHREURS et al. 1976). Readmission of Ca^{2+} to the bathing fluid during continued maximal stimulation with any of these secretagogues resulted in an immediate and sustained secretory response. Removal of Ca^{2+} during sustained stimulation abolished the stimulant-evoked secretion immediately (PETERSEN and IWATSUKI 1978; SCHREURS et al. 1976). The authors concluded that secretion evoked by a short-pulse stimulation is triggered by Ca^{2+} ions released from intracellular Ca^{2+} stores, whereas for sustained secretion Ca^{2+} influx into the cell is necessary.

The argument whether Ca^{2+} is released from mitochondria, the plasma membrane, or endoplasmic reticulum during stimulation had been a matter of controversy for a long time (CLEMENTE and MELDOLESI 1975; SCHULZ 1980). Moreover, the question of how hormonal stimulation is translated to intracellular signals was not answered until recently. From the initial disovery by HOKIN and HOKIN (1953, 1955, 1958 a, b) that phosphatidylinositol metabolism is stimulated by the action of hormones, it was some 20 years before Robert Michell recognized a correlation between stimulation of phosphatidylinositol breakdown and mobilization of Ca^{2+}. He suggested a close connection between PI breakdown and intracellular calcium signals and proposed that PIP_2, and not PI, is the precursor that leads to the calcium signal (R. H. MICHELL 1975; MICHELL et al. 1981).

D. A Link Between Lithium, Phospholipids and Ca^{2+} Mobilization

The first concrete finding, how Li^+ might act in the metabolism of phosphoinositides, was the observation by Allison, in the St. Louis laboratory of Stewart and Sherman, that treatment of rats with doses of Li^+ similar to those used in the therapy of the manic-depressive syndrome, caused decrease in inositol levels in cerebral cortex and increase in the concentration of inositol-1-phosphate (IP_1) (ALLISON and STEWART 1971; ALLISON et al. 1976; SHERMAN et al. 1981).

The molecular basis of these effects did not become evident until it was found that lithium is a powerful inhibitor of the enzyme inositol-1-phosphatase (HALLCHER and SHERMAN 1980). This enzyme is almost equally active against the D and L enantiomers of IP_1. Whereas the L enantiomer is made during inositol synthesis from glucose-6-phosphate, the D enantiomer is formed during the breakdown of phosphoinositides. The principal enantiomer of IP_1 (D-IP_1) that is increased in brain is derived from phosphoinositide metabolism. This finding has focused attention on inositol lipids and inositols released by receptor activities in the central nervous system. Chronic administration of LiCl to adult male rats resulted in an increase in the cerebral cortex level of IP_1. Thus, the previously observed acute effect of lithium on IP_1 (ALLISON et al. 1976) is both prolonged and agumented by repeated doses of lithium. Some 90% of the increase was due to the D enantiomer, although lithium was shown to be an uncompetitive inhibitor of the hydrolysis of both D- and L-inositol-1-phosphate by inositol-1-phosphatase. The effect of lithium to decrease inositol could be entirely reversed by muscarinic–cholinergic antagonists such as atropine or scopolamine (ALLISON 1978). When administered by itself, atropine was found to diminish D-IP_1 levels in the cortex, suggesting that it is formed during agonist-stimulated inositol phospholipid breakdown (SHERMAN et al. 1981).

The ability of Li^+ to inhibit IP_1 phosphatase and hence to cause accumulation of IP_1 in stimulated cells was not confined to brain. Studies on parotid salivary gland fragments exposed to acetylcholine and on insect salivary gland stimulated by 5-hydroxytryptamine (BERRIDGE et al. 1982) indicated that this effect of Li^+ is a general phenomenon. Since Li^+ inhibition of IP_1 phosphatase occurs at therapeutically appropriate Li^+ concentrations ($K_i \sim 0.8$ mM, in the range of the plasma level of lithium sought in the treatment of manic illness; HALLCHER and SHERMAN 1980), the implications of these observations for future work on the understanding of the therapeutic actions of Li^+ are immediate.

There are some tissues, such as parotid (AUB and PUTNEY 1984), guinea pig ileum (SEKAR and ROUFOGALIS 1984a, b), and pancreatic acini (RUBIN 1984) where high concentrations of lithium (10 mM) in the presence of stimulants result in the enhanced accumulation of polyphosphates. From effects of varying lithium concentrations on the accumulation of inositol phosphates in hepatocytes, it was concluded that lithium inhibited the three phosphatases which degrade IP_3, IP_2, and IP_1 with half-maximal concentrations of 5, 1, and 0.5 mM, respectively (THOMAS et al. 1984).

Using refined methods such as high pressure liquid chromatography, it has recently been shown that the increase in IP_3 production in the presence of lithium is due to increase in inositol-1,3,4-triphosphate and not to inositol-1,4,5-trisphosphate (BURGESS et al. 1985). As will be discussed later, this increase in inositol-1,3,4-trisphosphate is most likely due to inositol-1,3,4,5-P_4 formation from D-inositol-1,4,5-P_3 and further breakdown to inositol-1,3,4-P_3. The conclusion that Li^+ does not inhibit inositoltrisphosphate breakdown was also drawn from studies in which the hydrolysis of ^{32}P-labeled IP_3 and IP_2 had been examined in subcellular fractions of rat liver. Whereas a plasma membrane-bound phosphatase degraded IP_3, but not IP_2, and was not inhibited by lithium, a cytosolic phosphatase degraded both IP_3 and IP_2 (JOSEPH and WILLIAMS 1985). The soluble enzyme dephosphorylated IP_2 at either the 1- or the 4-phosphate position and was, unlike the IP_3-5-phosphatase, Li^+ sensitive (STOREY et al. 1984). IP phosphatase is also soluble and inhibited by Li^+ (HALLACHER and SHERMAN 1980; STOREY et al. 1984). Similarly, CONNOLLY et al. (1985) have isolated a soluble enzyme that degrades inositol-1,4-5-trisphosphate from human platelets which is not inhibited by Li^+ up to 40 mM.

E. Inositol-1,4,5-Trisphosphate as Second Messenger in the Action of Ca^{2+}-Mobilizing Hormones

The hypothesis that hormonal stimulation and consequent phospholipid breakdown leads to Ca^{2+} release from intracellular stores received experimental support from the finding that direct application of inositol-1,4,5-trisphosphate to permeabilized acinar cells from the exocrine pancreas resulted in rapid release of Ca^{2+} (Fig. 2; STREB et al. 1983). This effect of inositol-1,4,5-trisphosphate was not additive with the Ca^{2+}-mobilizing secretagogue acetylcholine. Since inositol-

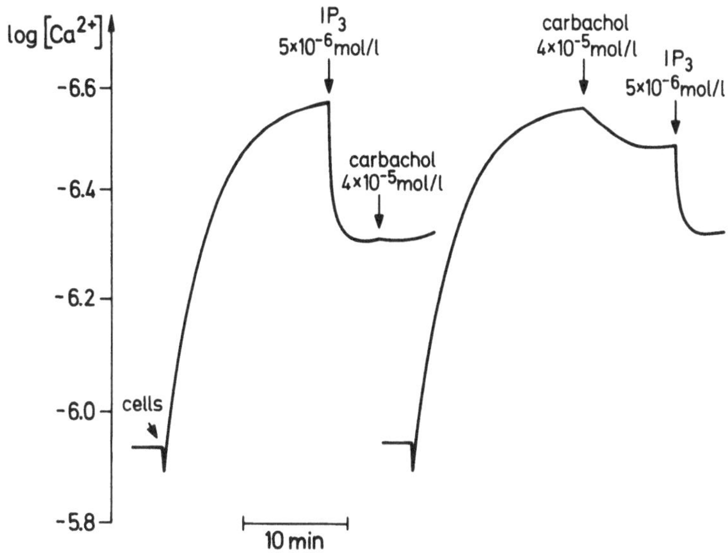

Fig. 2. Effect of sequential additions of inositol-1,4,5-trisphosphate (IP_3 5 μ*M*) and carbamylcholine (carbachol 4×10^{-5} *M*) to isolated permeabilized pancreatic acinar cells. (STREB et al. 1983)

1,4,5-trisphosphate is the degradation product of phosphatidylinositol-4,5-bisphosphate (PIP_2) following receptor-mediated activation of phospholipase C, these experiments provided strong evidence that inositol-1,4,5-trisphosphate is the second messenger for hormone action (STREB et al. 1983). Other degradation products of phosphoinositide hydrolysis such as inositol-1,4-bisphosphate, inositol-1-monophosphate, cyclic inositol-1-monophosphate, and inositol had no effect on Ca^{2+} release (STREB et al. 1983). Since inositol-1,4,5-trisphosphate-induced Ca^{2+} release was unchanged when mitochondrial inhibitors were added to the incubation medium, it was concluded that Ca^{2+} was being released from a nonmitochondrial store. Further studies on isolated subfractions of pancreatic acinar cells showed that the endoplasmic reticulum is the target for inositol-1,4,5-trisphosphate (STREB et al. 1984; PRENTKI et al. 1984; DAWSON and IRVINE 1984). These initial observations of the inositol-1,4,5-trisphosphate effect in pancreatic acinar cells were confirmed in a number of different cell types (for review see BERRIDGE and IRVINE 1984; BERRIDGE 1984), including platelets (AUTHI and CRAWFORD 1985; AUTHI et al. 1986; O'ROURKE et al. 1985), cardiac microsomes (HIRATA et al. 1984), smooth muscle cells (SUEMATSU et al. 1984), sarcoplasmic reticulum (VOLPE et al. 1985), and photoreceptors (BROWN et al. 1984).

F. Other Inositol Polyphosphates

In several tissues, inositoltrisphosphate levels in response to agonists are elevated over many minutes (AUB and PUTNEY 1984; BURGESS et al. 1984; DOWNES and WUSTEMAN 1983; THOMAS et al. 1984). Separation of inositoltrisphosphates by high pressure liquid chromatography showed formation of inositol-1,3,4-tris-

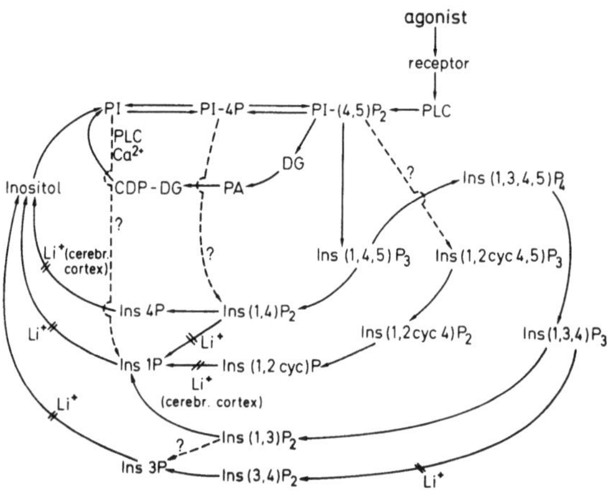

Fig. 3. Agonist-dependent phosphoinositide metabolism and sites of Li^+ action. Known (*full arrows*) and probable (*broken arrows with question marks*) interconversions of inositol lipids and inositol phosphates in stimulated cells

phosphate beside inositol-1,4-5-trisphosphate (BURGESS et al. 1985; IRVINE et al. 1984b). Inositol-1,3-4-trisphosphate releases Ca^{2+} from intracellular stores of Swiss-mouse 3T3 cells at a half-maximal dose ~30 times higher than for IP_3 (IRVINE et al. 1986). So far any specific function of this isomer is not known. It might be just a by-product in the process by which D-inositol-1,4,5-trisphosphate is inactivated. This occurs in a two-step mechanism. Inositol-1,4,5-trisphosphate can readily be phosphorylated in the 3 position (IRVINE et al. 1986) which yields inositol-1,3,4,5-tetrakisphosphate (IP_4) (Fig. 3). This compound does not mobilize Ca^{2+} from a permeabilized Swiss mouse 3T3 preparation but might control Ca^{2+} entry across the plasma membrane (IRVINE and MOOR 1986). IP_4 is then dephosphorylated in the 5 position, probably by the same phosphatase which hydrolyzes D-inositol-1,4,5-trisphosphate (BATTY et al. 1985), to form inositol-1,3,4-trisphosphate (IRVINE et al. 1984b, 1985).

Although the function of inositol-1,3,4-trisphosphate is not known, there are speculations that inositol-1,3,4-trisphosphate could have a different inositide messenger function (BERRIDGE and IRVINE 1984; IRVINE et al. 1984b). For example, hyperstimulation of the parotid gland by cholinergic agonists, which results in a large accumulation of inositol-1,3,4-trisphosphate (IRVINE et al. 1984b), also stimulates DNA synthesis (SCHNEYER 1974). Furthermore, treatment of cells with lithium, which enhances the formation of inositol-1,3,4-trisphosphate (BURGESS et al. 1985; IRVINE et al. 1984b), is a potent growth stimulus in lymphocytes (HART 1979), in mammary (TOMOOKA et al. 1983), and kidney epithelial cells (RYBACK and STOCKDALE 1981).

It has also been found that a phospholipase C from sheep seminal vesicles that produces inositol-1,4,5-trisphosphate from PIP_2 also produces inositol-1,2-(cyclic)-4,5-trisphosphate (cyclic IP_3) (Fig. 3; WILSON et al. 1985a). The latter substance is some five times more active than inositol-1,4,5-trisphosphate in stimulating depolarization in *Limulus* photoreceptors similar to that produced by light. Cyclic IP_3 mobilizes Ca^{2+} to the same extent as inositol-1,4,5 trisphosphate in Swiss-mouse 3T3 cells (IRVINE et al. 1986) but is less active in platelets (WILSON et al. 1985b).

Degradation of each of these inositol polyphosphates leads to inositol monophosphates and to inositol by the action of a inositol-1-phosphatase which can be inhibited by lithium and by a 3-phosphatase and a 4-phosphatase which are not inhibited by lithium ions. The three inositol monophosphates, i.e. Ins 1-P, Ins 3-P, and Ins 4-P, are converted to inositol by a Li^+ sensitive inositol monophosphate phosphatase (Fig. 3; ACKERMANN et al. 1987; BERRIDGE et al. 1982, BANSAL et al. 1987, HALLCHER and SHERMAN 1980, INHORN et al. 1987, STOREY et al. 1984). Recent studies have shown, however, that inositol-1,4-bisphosphate can be degraded to inositol via both inositol-1-phosphate and inositol-4-phosphate (B. MICHELL 1986; SHERMAN et al. 1985; STOREY et al. 1984). In liver, degradation via the 4-phosphate is Li^+ insensitive, whereas in brain it is Li^+ sensitive (INHORN et al. 1987). Both steps in degradation of inositol-1,4-bisphosphate to inositil-1-phosphate and further to free inositol are Li^+ sensitive (B. MICHELL 1986). In cerebral cortex, inositol-1-phosphate and inositol-4-phosphate levels increased in parallel over a 24-h period following administration of a single subcutaneous dose of LiCl to rats, whereas inositol-5-phosphate remained unchanged (SHERMAN et al. 1985). Inositol-1,2-cyclic phosphate did not change over a range of LiCl doses in rat cerebral cortex (W. R. SHERMAN 1986, personal communication). The authors concluded from these observations that inositol-1,4,5-trisphosphate is converted to inositol-1,4-bisphosphate, not to inositol-4,5-bisphosphate or inositol-1,5-bisphosphate, and that this conversion is lithium insensitive.

Muscarinic anticholinergic agents, atropine and scopolamine, brought about partial reversal of the Li^+ effects on inositol-1-phosphate (ALLISON 1978) and inositol-4-phosphate levels of rat cortex, when administered 2 h before killing (SHERMAN et al. 1985b). Chronic dietary administration of LiCl for 22 days showed that the effects of lithium on increased inositol-1-phosphate levels and decreased inositol levels persisted for that period and that levels of inositol-1-phosphate, inositol, and lithium in the cortex remained significantly correlated. This indicated to the authors that prolonged administration of LiCl did not result in compensatory changes in inositol-1-phosphate synthase or inositol-1-phosphatase (SHERMAN et al. 1985a, b). Furthermore, these investigators believe that increases in inositol phosphates result from endogenous phosphoinositide metabolism in cerebral cortex and that lithium is capable of modulating that metabolism by reducing cellular inositol levels. The only partial reversibility of the lithium effect by atropine was considered to be present because muscarinic activity was not the only source of the lithium effect.

In a study on effects of lithium on muscarinic receptor stimulation of individual inositol phosphate levels in rat cerebral cortex slices, lithium was found to potentiate carbachol-stimulated accumulation of inositol monophosphate and inositol bisphosphate. However, exposure to lithium in the presence of the muscarinic agonist produced a concentration- and time-dependent inhibition of inositol trisphosphate accumulation that was not related to receptor desensitization. Whereas, in the presence of lithium, inositol monophosphate continued to accumulate linearly for at least 20 min, inositol trisphosphate began to decline by 5–10 min (BATTY and NAHORSKI 1985). The authors concluded that these results were not entirely consistent with the production of inositol mono- and bisphosphate through inositol trisphosphate dephosphorylation, and could indicate the formation of the monophosphate directly from phosphatidylinositol in addition to production through polyphosphate dephosphorylation.

G. Lithium and Manic-Depressive Illness

Since, in the therapy of manic-depressive patients, lithium is given in doses which inhibit inositolphosphatase, metabolism of phosphoinositides which are finally degraded to inositol, will be reduced. It is not yet clear how lithium acts in the moderation of the symptoms of manic illness. The idea is that Li^+ inhibits inositolphosphatase, thus reducing the supply of free inositol to synthesis of phosphatidylinositol and polyphosphatidylinositols, thereby limiting a cell's signaling capacity, especially in those neurons that are being stimulated excessively (BERRIDGE and IRVINE 1984). As a result, receptors dependent on the phosphoinositides could be downregulated with consequent reduced responses of their functions in neural activity of the central nervous system in manic illness. This alteration in phosphatidylinositol metabolism may serve to reset the sensitivity of intracellular messengers and their targets such as inositol-1,4,5-trisphosphate, Ca^{2+}, and subsequent stimulatory steps (BERRIDGE et al. 1982).

Reports that manic symptoms could be successfully controlled by treatment of patients with Ca^{2+} antagonists (DOSE and EMRICH 1986; DUBOVSKY et al. 1982) might point to the importance of the phosphatidylinositol-4,5-bisphosphate–inositol-1,4,5-trisphosphate–Ca^{2+} messenger system in manic-depressive illness. In an attempt to elucidate the pathology of mania, the effects of treatments with lithium and antidepressant drugs in unipolar mania and also in patients suffering from the bipolar form of the disease, in which mania alternates with depression, have been described (WOOD 1985). The authors concluded that in order to maintain a euthymic state in a bipolar patient one should maintain a certain degree of sensitivity of 5-hydroxytryptamine (5-HT) receptors. This sensitivity may be related to the activity of calcium channels and to the polyphosphoinositide system. Furthermore, it is suggested that coadministration of a calcium channel antagonist and lithium may be a very effective treatment of mania.

In a study on the effects of the antidepressants iprindole and imipramine in rat cerebral cortex slices, it was found that chronic, but not acute treatment resulted in a loss of 5-HT-induced phosphatidylinositol hydrolysis. This effect of antidepressants was specific for 5-HT in that neither α_1-adrenoceptor nor muscarinic receptor stimulation were altered by acute or chronic treatment (KENDALL and NAHORSKI 1985). Taken together, these studies are strong indications for altered receptor-mediated phosphoinositide metabolism in manic-depressive illness.

H. Is Lithium a Secretagogue?

The discovery that lithium is a potent and specific inhibitor of the phosphatase that cleaves inositol-1-phosphate to inositol (HALLCHER and SHERMAN 1980) connects the action of lithium to the actions of several neurotransmitters and hormones that stimulate inositol phospholipid turnover and calcium mobilization.

Although Li^+ has been found to mimic and to potentiate Ca^{2+}-mobilizing hormones (HONCHAR et al. 1983), there are only a few reports that Li^+ has agonist-like effects on its own. In a preparation of isolated pancreatic acini prelabeled with [^{3}H] inositol, the peptidergic agonist cerulein elicited increases in [^{3}H] IP_3, [^{3}H] IP_2, and [^{3}H] IP. Lithium enhanced the action of cerulein on the accumulation of inositol phosphate, but failed to augment the response to cerulein on amylase secretion (RUBIN 1984).

In contrast, in a study on cultured anterior pituitary cells and on a primary culture of rat anterior pituitary cells, ZATZ andREISINE (1985) have found that lithium stimulates corticotropin (ACTH) secretion by mouse pituitary tumor cells (At T-20/D16-16). Effects were observed at less than 2 m*M* LiCl and ACTH secretion was comparable in magnitude to that induced by other secretagogues. Increases in [^{3}H] IP and [^{3}H] IP_2 which accompanied ACTH secretion were also observed, the most prominent effect was to increase [^{3}H] IP. Other secretagogues had no effect on [^{3}H] inositides, nor did these secretagogues alter the effects of lithium on IP. Similar to other secretagogues of ACTH secretion, lithium failed to stimulate ACTH secretion in the absence of extracellular calcium and the increase in IP was also markedly diminished. Somatostatin, which inhibits the stimulation of ACTH secretion by several hormones (HEISLER et al. 1982) also diminished the effects of lithium on ACTH secretion and IP. Further similarities between the action of other secretagogues and Li^+ included the ability to desensitize cells to subsequent stimulation. Pretreatment of cells with lithium for 3 h reduced ACTH secretion upon subsequent stimulation with lithium to $\sim$25% of the effect obtained without pretreatment. However, there was no desensitization by lithium pretreatment of the response to other stimulators such as corticotropin-releasing factor (CRF) or increase in potassium concentration, although these agents desensitized the At T-20/D16-16 cells to their own effects under similar conditions. There was also no desensitization by pretreatment with potassium to the effects of lithium, nor between lithium and isoproterenol in either direction. There was, however, one class of secretagogues whose action was strongly affected by lithium – the phorbol esters. Both lithium and phorbol myristate acetate (PMA) stimulated ACTH secretion; their effects were not additive. Pretreatment with LiCl markedly reduced the effect of subsequent exposure to either LiCl or PMA. Pretreatment with PMA desensitized the cells to the subsequent action of PMA, but not to the stimulatory effect of lithium. In former concepts, the potentiating action of Li^+ on the effect of receptor-mediated stimulation of phospholipase C had been explained by inhibition of IP_1 phosphatase (HALLCHER and SHERMAN 1980). Under these conditions, degradation of IP_1 and, at higher Li^+ concentrations, also recycling of other inositol phosphates is blocked. However, the data obtained on ACTH secretion from At T-20/D16-16 cells (ZATZ and REISINE 1985) require an additional postulate. The interaction between Li^+ and phorbol esters suggests the involvement of protein kinase C (BLUMBERG et al. 1984). The way that Li^+ leads to activation of this enzyme, however, is not yet clear.

Evidence that Li^+ might activate protein kinase C indirectly by increasing the level of 1,2-DG was obtained from studies on GH_3 pituitary tumor cells (DRUMMOND and RAEBURN 1984). Addition of thyrotropin-releasing hormone (TRH) to suspensions of GH_3 cells led to reduction in the cellular content of PI with no change in PIP and PIP_2. There was also an increase in cellular 1,2-DG levels and a decline in the level of triacylglycerol. When the cells were suspended in lithium-containing buffer in the absence of exogenous inositol, there was a 15% decrease in GH_3 cell inositol levels. This was associated with a small, but significant, increase in the cellular content of phosphatidylinositol-4,5-bisphosphate and 1,2-DG. Addition of TRH to cells suspended in lithium-containing medium depleted cellular inositol levels by around 65% within 30 min. By this time, there was also a 50% reduction in the cellular content of PI and 20% reduction in PIP. Control levels of PIP_2 were maintained in the combined presence of TRH and lithium. Under these conditions, TRH no longer depleted cellular triacylglycerol and there was a marked increase in the ability of TRH to elevate the GH_3 cell content of 1,2-DG. The effect of TRH on the cellular content of PA was not altered by the presence of lithium. These results show that when PI resynthesis is inhibited by lithium-induced inositol depletion, its glycerol backbone accumulates, at least in part, in 1,2-DG. The data further show that GH_3 cells preserve their cellular levels of PIP_2 in the face of a considerable reduction in the cellular content of PI. The reason for the absence of TRH-induced PA production could be activation of phosphatidate phosphohydrolase, as shown for Ca^{2+}-mobilizing agonists (POLLARD and BRINDLEY 1984), although it is not known whether TRH can activate this enzyme in GH_3 cells. The source of the 1,2-DG in the presence of lithium remained uncertain, since there was no accompanying change in the cellular levels of any other lipid. The authors assumed that the most likely source of the 1,2-DG was PI (presumably via PI hydrolysis). They further suggested that treatment of cells and tissues with lithium may, under certain conditions, lead to selective activation of protein kinase C in the absence of a change in cytoplasmic free calcium levels and that this manipulation may thus produce similar effects to the addition of phorbol esters which activate protein kinase C. The observation of enhanced accumulation of 1,2-DG, observed with TRH under inositol-depleting conditions (DRUMMOND and RAEBURN 1984), might have some significance for the question of how symptoms are altered by Li^+ treatment of manic-depressive patients.

Similar observations have been made by DOWNES and STONE (1986) using rat parotid gland slices and isolated acinar cells labeled with $^{32}P_i$. Cholinergic stimulation caused substantial breakdown of phosphatidylinositol-4,5-bisphosphate and enhanced labeling of PA and PI. Lithium alone had little effect upon $^{32}P_i$ incorporation, but in combination with carbachol it greatly reduced the PI-labeling response to the agonist carbachol. Instead, the label accumulated in a lipid identified as cytidine monophosphorylphosphatidate (CMP-PA; see Sect. C, footnote). There was also an enhancement of the PA labeling response to carbachol. Despite reduced PI synthesis, lithium had relatively little effect on polyphosphoinositide labeling in stimulated cells. Treatment with lithium during the carbachol-stimulated phase reduced the rate of phosphatidylinositol-4-phosphate synthesis, but had no significant effect on PIP_2. These results suggest than an active inositol phosphatase pathway is essential to maintain intracellular inositol levels, but that PIP_2 synthesis is not markedly reduced by a substantial fall in intracellular inositol, even when PI and PIP synthesis are seriously diminished, indicating a close control over the rates of PIP_2 resynthesis during agonist stimulation. It was not surprising therefore that lithium did not reduce the secretory response to muscarinic stimulation of parotid gland slices. Potentiation of PA accumulation by lithium observed in carbachol-stimulated cells suggested to the authors that some interconversion of CMP-PA and PA may occur. Since PA can be dephosphorylated to DG, this suggested that lithium may significantly raise DG levels in agonist-stimulated cells, as noted by DRUMMOND and RAEBURN (1984) for thyrotropin-stimulated GH_3 pituitary cells. Since DG formed during agonist-stimulated PI breakdown may serve to activate protein kinase C, this raises the possibility that lithium may potentiate responses in some cells which are influenced by activators of protein kinase C.

J. A Link Between Li^+ Phosphoinositides and Cell Proliferation

Phosphoinositides may play a crucial role in mediating the action of growth factors (R. H. MICHELL 1982). One of the earliest observations regarding the involvement of phosphoinositides in cell growth was the discovery by FISHER and MUELLER (1968, 1971) that phytohemagglutinin (PHA), a polyclonal mitogen for T-lymphocytes, stimulates the turnover of PI within minutes of its addition to human blood leukocytes. Since then, stimulation of PI turnover by growth-promoting agents has been observed in different cells and tissues (DIRINGER and FRIIS 1977; FISHER and MUELLER 1968, 1971; HABENICHT et al. 1981; HOFFMANN et al. 1974; RISTOW et al. 1973; SAWYER and COHEN 1981). These observations were of special interest when it became apparent that elevation of cytosolic Ca^{2+} in lymphocytes by the Ca^{2+} ionophore A23187 could mimic the effect of PHA (ALLAN and MICHELL 1977; CRUMPTON et al. 1975; GREENE et al. 1976) and that PHA produced a transient elevation of cytosolic Ca^{2+} concentration in lymphocytes (CRUMPTON et al. 1975; GREENE et al. 1976). An increase in the intracellular free $[Ca^{2+}]$ has been recognized as part of the mitogenic signal for many different cell types (BERRIDGE 1975; BOYNTON et al. 1974; METCALFE et al. 1980). More recently, the enhanced breakdown of PIP_2 was shown to be a very early event in proliferation (HASEGAWA-SASAKI and SASAKI 1983).

A further line of evidence for the involvement of phosphoinositide turnover in cell proliferation came from studies showing the importance of DG formation in cell growth. Stimulation of Swiss mouse 3T3 cells with platelet-derived growth factor (PDGF) caused an increase in DG, suggesting that the inositol lipid was being hydrolyzed by the same mechanism as used by Ca^{2+}-mobilizing hormones and neurotransmitters which act by hydrolyzing phosphatidylinositol-4,5-bisphosphate to give DG and inositol-1,4,5-trisphosphate (BERRIDGE et al. 1984b).

It has been known for some time that increase in pH is an intracellular mitogenic signal (EPEL 1978; MOOLENAAR et al. 1983). Whereas IP_3 mobilizes Ca^{2+}, DG is thought to control the change in pH. Phorbol esters which mimic the action of DG to stimulate protein kinase C (CASTAGNA et al. 1982) enhance plasma membrane-located Na^+,H^+ exchange, leading to cytosolic alkalinization (DICKER and ROZENGURT 1981; ROSOFF et al. 1984). Synergism in the response to phorbol ester and the Ca^{2+} ionophore ionomycin during proliferation of chicken heart mesenchymal cells (BALK et al. 1984) further supports the involvement of both branches of phosphoinositide turnover, i.e., DG formation and IP_3 production–Ca^{2+} release for optimal cell growth. Since signal transfer consists of a sequence of reactions controlled by specific proteins, the genetic material of the cell must include the genes responsible for the synthesis of these proteins. Genes which appear to be involved in the regulation of cell growth and whose inappropriate function has been linked to tumor growth, are collectively termed oncogenes. Oncogenes show very close homology with transforming genes of retroviruses. Thus, the protein products of the *ros* and *src* genes phosphorylate tyrosine on proteins. The protein product of the *ros* gene from avian sarcoma virus UR2 phosphorylates PI to form PIP (MACARA et al. 1984) and that from the *src* gene from Rous sarcoma virus phosphorylates both PI and PIP (SUGIMOTO et al. 1984). In cells transformed with avian sarcoma virus UR2, the concentrations of the hydrolysis products of PIP and PIP_2, i.e., IP_2 and IP_3, respectively, were increased. Further evidence that links oncogenes to phosphatidylinositide turnover was the discovery that the oncogene called *sis* controls the synthesis of PDGF (DOOLITTLE et al. 1983; WATERFIELD et al. 1983). Also the *erb b* gene proved to encode the structure of a protein almost identical with the epidermal growth factor (EGF) receptor (DOWNWARD et al. 1984) which also stimulated phosphoinositide metabolism (SAWYER and COHEN 1981). The product of the *erb b* gene, however, lacks the external part of the protein that binds EGF. This version of the protein initiates signals inside the cell, even in the absence of EGF.

The finding that 10 m*M* Li^+ enhances the formation of IP_3 induced by PDGF is interesting, because this concentration of Li^+ is mitogenic in lymphoid cells (HART 1979), lymphocytes (GELFAND et al. 1979), mammary epithelium (HORI and OKA 1979; PTASHNE et al. 1980; TOMOOKA et al. 1983), kidney epithelial cells (TOBACK 1980), and BALB/c 3T3 fibroblasts (RYBACK and STOCKDALE 1981). Li^+ had no effect when added alone, but appeared to act either additively or synergistically with various growth factors. HART (1979) suggested that growth factors may induce a mitogenic signal that is retarded by some rate-limiting enzyme whose activity is diminished by Li^+. The discovery that growth factors stimulate increase in the formation of inositol-1,4,5-trisphosphate which releases Ca^{2+} and that increase in cytosolic free Ca^{2+} concentration is part of the mitogenic signal, suggests that the mitogenic signal is IP_3 and IP_3-induced increase in cytosolic free Ca^{2+} concentration. In this case, the rate-limiting enzyme which is inhibited by Li^+ would be inositol-1-phosphatase. On the other hand, evidence that Li^+ increases the level of 1,2-DG in GH_3 pituitary tumor cells, as discussed in Sect. H (DRUMMOND and RAEBURN 1984), might indicate a role for the protein kinase C pathway in the mitogenic action of Li^+. It is not excluded, however, that a quite different pathway is involved in Li^+-induced cell growth. Lithium inhibits inositol-1-phosphatase (HALLCHER and SHERMAN 1980) and therefore also degradation of inositol phosphates other than IP_3. For instance, Li^+-induced increase in inositol-1,3,4-trisphosphate has led to speculation on the involvement of this inositoltrisphosphate in stimulation of cell growth (BERRIDGE and IRVINE 1984; BERRIDGE et al. 1985; IRVINE et al. 1984b).

HORI and OKA (1979) showed that Li^+ exerts insulin-like effects on mouse mammary explants in culture by eliciting a series of responses such as synthesis of RNA, DNA, and cell multiplication. In a study on induction of growth in kidney epithelial cells in culture, it was shown, however, that Na^+ leads to initiation of DNA synthesis in an increased number of cells and to enhanced cell growth, whereas Li^+ could only partly replace Na^+ in this effect (TOBACK 1980). In initiation of DNA synthesis in mammary epithelia and mammary tumors, different monovalent cations such as Li^+, Na^+, NH_4^+, Rb^+, Cs^+, but not choline, were found to stimulate growth (PTASHNE et al. 1980).

K. Effects of Li^+ on Neurotransmitter-cAMP-Stimulated Pathways

So far, this chapter has mainly discussed Li^+ effects in connection with those hormones and neurotransmitters acting via phosphatidylinositide breakdown and intracellular Ca^{2+} mobilization. It should be mentioned, however, that intracellular pathways involving cAMP are also affected by Li^+. Since cAMP can modulate Ca^{2+} signals in different steps of stimulus–response coupling (RASMUSSEN and BARRETT 1984) and has been suggested to release Ca^{2+} from the endoplasmic reticulum in rat hepatocytes (MAUGER and CLARET 1986), interaction of Li^+ with cAMP-dependent pathways could also be relevant for cellular responses dependent on Ca^{2+}-mediated pathways.

Li^+ has been shown to reduce the release of neurotransmitters such as norepinephrine induced by electrical stimulation from brain tissue and peripheral adrenergic neurons (BINDLER et al. 1971; KATZ and KOPIN 1969; KATZ et al. 1968) and increased calcium concentrations in the perfusion medium reversed these lithium effects (KATZ and KOPIN 1969). Furthermore, several studies demonstrated reduced brain norepinephrine levels and an increase in the turnover of norepinephrine after both acute and chronic lithium administration (CORRODI et al. 1967; GREENSPAN et al. 1970; STERN et al. 1969). Li^+ has also been shown to interfere with these neurotransmitters at the receptor level. Treatment with Li^+ reduced the number of β-adrenoceptors, but increased α-adrenoceptor binding sites in whole brain homogenate (ROSENBLATT et al. 1979), whereas in rat cerebral cortex the density of β-adrenoceptors was reduced and α-adrenoceptors were unchanged (TREISER and KELLAR 1979). If mania is associated with supersensitive β-adrenergic receptors then there should be some evidence of this in patients with an episode of mania. EXTEIN et al. (1979) found that binding of the β-receptor agonist [^3H] dihydroalprenolol to lymphocytes was decreased in both depressed and manic patients when compared with controls. Similarly, PANDEY et al. (1979) have reported that isoprenaline- and noradrenaline-induced cAMP responses in leukocytes were significantly lower in depressed and bipolar manic patients. Plasma levels of cAMP are raised in mania and low in depression (LYKOURAS et al. 1979). Elevated levels of cAMP occurring during mania fell during treatment with neuroleptic drugs (LYKOURAS et al. 1978) and lithium (ARATO et al. 1980). Lithium has been shown to influence receptor responses to norepinephrine that involve the adenylate cyclase–cAMP system, including receptor coupling to the enzyme by GTP-binding proteins, the catalytic unit of the adenylate cyclase, as well as subsequent cAMP-dependent reactions such as modulation of cAMP-dependent phosphodiesterase or protein kinase (GEISLER et al. 1985).

The stimulation of human platelet adenylate cyclase and cyclic AMP production by prostaglandin E_1 is markedly inhibited by lithium (MURPHY et al. 1973;

WANG et al. 1974). Li^+ has also been shown to inhibit adenylate cyclase activity in rat cortical slices as stimulated by norepinephrine or histamine (FORN and VALDECASAS 1971).

Forskolin-stimulated adenylate cyclase activity in vivo and in vitro is inhibited by Li^+ (GEISLER et al. 1985). Since Li^+ influenced neither basal nor fluoride-stimulated adenylate cyclase activity, the authors felt that these results indicate interference of Li^+ with conformational change of the enzyme during activation with forskolin rather than a direct effect of Li^+ on the catalytic subunit or on the coupling GTP-binding protein (GEISLER et al. 1985). Furthermore, studies on human adrenal medulla showed an inhibitory effect by Li^+ on cAMP-dependent protein kinase (DOUSA 1974), indicating that Li^+ may also interfere with cAMP-dependent processes.

L. Conclusion

Lithium has been widely used in the treatment of manic-depressive disorders, yet its precise mode of action remains obscure. Since the discovery that Li^+ interferes with the phosphoinositide turnover and that inositol-1,4,5-trisphosphate, a hydrolysis product of phosphatidylinositol-4,5-bisphosphate, is a second messenger for the release of intracellular Ca^{2+}, Li^+ has moved into the center of research interest. By inhibiting inositol P_1-phosphatase, lithium treatment increases the inositol phosphate level and reduces the intracellular inositol supply, when stimulated intensely, but in most cells has no effect on its own. Consequently, in stimulated cells, resynthesis of phosphatidylinositol (PI) from both inositol and diacylglycerol (DG) is reduced.

Since polyphosphoinositides are generated from PI by phosphorylation of PI to phosphatidylinositol-4-phosphate and further to phosphatidylinositol-4,5-bisphosphate (PIP_2), one would expect that, during agonist stimulation, PIP_2 synthesis is also reduced by a fall in intracellular inositol following lithium treatment. In some cells, this does not seem to be the case and implies a close control over the rates of PIP_2 breakdown and resynthesis. When PI resynthesis is inhibited by lithium-induced inositol depletion, DG is increased. In some cells, e.g., cultured anterior pituitary cells, lithium stimulates corticotropin secretion, most likely indirectly by increasing the level of DG and consequent activation of protein kinase C.

The widespread occurrence of inositol phospholipid-dependent hormone and neurotransmitter receptors in the brain and the sensitivity of inositol P_1-phosphatase to lithium may hold an explanation for the unique effectiveness of lithium in the treatment of manic-depressive psychoses.

There are indications that the cells most sensitive to lithium are those that are excessively stimulated, as may happen in a condition such as mania. Whereas lithium has little effect in the normal operational range, it becomes increasingly effective when the receptors are hyperactive.

The finding that lithium is mitogenic and enhances formation of inositoltrisphosphate induced by growth hormones is suggestive of the involvement of inositolphospholipids in cell growth. Both branches of PIP_2 breakdown, yielding inositol-1,4,5-trisphosphate and DG, seem to play a role in cell growth. Furthermore, increase in both cytosolic free Ca^{2+} concentration and pH are also mitogenic signals. Whereas cytosolic free Ca^{2+} concentration is increased by inositol-1,4,5-trisphosphate-induced Ca^{2+} release from intracellular stores, DG activates protein kinase C that phosphorylates and thereby stimulates the Na^+/H^+ exchange system in the plasma membrane. Since inhibition of inositol P_1-phosphatase by Li^+ leads mainly to increase in inositol-1,3,4-trisphosphate and not to inositol-1,4,5-trisphosphate, it is speculated that inositol-1,3,4-trisphosphate could have a messenger function in cell growth on its own. It appears that lithium is one of the most important tools in research on cellular function and dysfunction which might be related to symptoms of manic-depressive disease.

Acknowledgments. I wish to thank Professor Dr. K. J. Ullrich and Dr. Robin Irvine for reading the manuscript and for helpful discussions. I am extremely grateful to Dr. William Sherman for giving me his published and unpublished papers, for his expert criticism of this manuscript, and many helpful suggestions.

References

Abdel-Latif AA, Akhtar RA, Hawthorne JN (1977) Acetylcholine increases the breakdown of triphosphoinositide of rabbit iris muscle prelabelled with (^{32}P)phosphate. Biochem J 162:61–73

Ackermann KE; Gigh BG, Honchar MP, Sherman WR (1987) Evidence that inositol 1-phosphate in brain of lithium-treated rats results mainly from phosphatidyl-inositol metabolism. Biochem J 242:517–524

Agranoff BW, Bradley RM, Brady RO (1958) The enzymatic synthesis of inositol phosphatide. J Biol Chem 233:1077–1083

Akhtar RA, Abdel-Latif AA (1980) Requirement for calcium ions in acetylcholine-stimulated phosphodiesteratic cleavage of phosphatidyl-myoinositol 4,5-bisphosphate in rabbit iris smooth muscle. Biochem J 192:783–791

Aldenhoff JP, Lux HD (1982) Effects of lithium on calcium-dependent membrane properties and on intracellular calcium-concentration in helix neurons. In: Emrich HM, Aldenhoff JB, Lux HD (eds) Basic mechanisms in the action of lithium. Excerpta Medica, Amsterdam, pp 50–63

Allan D, Michell RH (1977) A comparison of the effects of phytohaemagglutinin and of calcium ionophore A23187 on the metabolism of glycerolipids in small lymphocytes. Biochem J 164:389–397

Allison JH (1978) Lithium and brain *myo*-inositol metabolism. In: Wells WW, Eisenberg F Jr (eds) Cyclitols and phosphoinositides. Academic, New York, pp 507–519

Allison JH, Stewart MA (1971) Reduced brain inositol in lithium-treated rats. Nature 233:267–268

Allison JH, Blisner ME, Holland WH, Hipps PP, Sherman WR (1976) Increased brain myo-inositol 1-phosphate in lithium-treated rats. Biochem Biophys Res Commun 71:664–670

Araki T, Ito M, Kostyuk PG, Oscarsson O, Oshima T (1965) The effects of alkaline cations on the responses of cat spinal motoneurons, and their removal from the cells. Proc R Soc Lond [Biol] 162:319–332

Arato M, Rihmer Z, Felszeghy K (1980) Reduced plasma cyclic AMP level during prophylactic lithium treatment in patients with affective disorder. Biol Psychiatry 15:319–322

Argent BE, Case RM, Scratcherd T (1973) Amylase secretion by the perfused cat pancreas in relation to the secretion of calcium and other electrolytes and as influenced by the external ionic environment. J Physiol (Lond) 230:575–593

Aub DL, Putney JW Jr (1984) Metabolism of inositol phosphates in parotid cells: implications for the pathway of the phosphoinositide effect and for the possible messenger role of inositol trisphosphate. Life Sci 34:1347–1355

Authi KS, Crawford N (1985) Inositol 1,4,5-trisphosphate-induced release of sequestered Ca^{2+} from highly purified human platelet intracellular membranes. Biochem J 230:247–253

Authi KS, Evenden BJ, Crawford N (1986) Metabolic and functional consequences of introducing inositol 1,4,5-trisphosphate into saponin-permeabilized human platelets. Biochem J 233:709–718

Baker PF (1972) Transport and metabolism of calcium ions in nerve. Prog Biophys Mol Biol 24:177–223

Baker PF, McNaughton PA (1978) The influence of extracellular calcium binding on the calcium efflux from squid axons. J Physiol (Lond) 276:127–150

Baker PF, Schlaepfer WW (1978) Uptake and binding of calcium by axoplasm isolated from giant axons of *loligo* and *myxicola*. J Physiol (Lond) 276:103–125

Balk SD, Morisi A, Gunther HS (1984) Phorbol 12-myristate 13-acetate, ionomycin or ouabain, and raised extracellular magnesium induce proliferation of chicken heart mesenchymal cells. Proc Natl Acad Sci USA 81:6418–6421

Bansal VS, Inhorn RC, Majerus PW (1987) The metabolism of inositol 1,3,4-trisphosphate to inositol 1,3-biphosphate. J Biol Chem 262:9444–9447

Batty I, Nahorski SR (1985) Differential effects of lithium on muscarinic receptor stimulation of inositol phosphates in rat cerebral cortex slices. J Neurochem 45:1514–1521

Batty IR, Nahorski SR, Irvine RF (1985) Rapid formation of inositol 1,3,4,5-tetrakisphosphate following muscarinic receptor stimulation of rat cerebral cortical slices. Biochem J 232:211–215

Bayerdörffer E, Haase W, Schulz I (1985) Na^{+}/Ca^{2+} countertransport in plasma membrane of rat pancreatic acinar cells. J Membr Biol 87:107–119

Berridge MJ (1975) Control of cell division: a unifying hypothesis. J Cyclic Nucleotide Res 1:305–320

Berridge MJ (1983) Rapid accumulation of inositol trisphosphate reveals that agonists hydrolyse polyphosphoinositides instead of phosphatidylinositol. Biochem J 212:849–858

Berridge MJ (1984) Inositol trisphosphate and diacylglycerol as second messengers. Biochem J 220:345–360

Berridge MJ, Irvine RF (1984) Inositol trisphosphate, a novel second messenger in cellular signal transduction. Nature 312:315–321

Berridge MJ, Downes CP, Hanley MR (1982) Lithium amplifies agonist-dependent phosphatidylinositol responses in brain and salivary glands. Biochem J 206:587–595

Berridge MJ, Buchan PB, Heslop JP (1984a) Relationship of polyphosphoinositide metabolism to the activation of the insect salivary gland by 5-hydroxytryptamine. Mol Cell Endocrinol 36:37–42

Berridge MJ, Heslop JP, Irvine RF, Brown KD (1984b) Inositol trisphosphate formation and calcium mobilization in Swiss 3T3 cells in response to platelet-derived growth factor. Biochem J 222:195–201

Berridge MJ, Heslop JP, Irvine RF, Brown KD (1985) Inositol lipids and cell proliferation. Biochem Soc Trans 13:67–71

Bindler EH, Wallach MB, Gershon S (1971) Effect of lithium on the release of ^{14}C-norepinephrine by nerve stimulation from the perfused cat spleen. Arch Int Pharmacodyn Ther 190:150–154

Blaustein MP (1974) The interrelationship between sodium and calcium fluxes across cell membranes. Rev Physiol Biochem Pharmacol 70:33–82

Blumberg PM, Jaken S, König B, Sharkey NA, Leach KL, Jeng AY, Yeh E (1984) Mechanism of action of the phorbol ester tumor promoters: specific receptors for lipophilic ligands. Biochem Pharmacol 33:933–940

Blumenthal DK, Stull JT (1980) Activation of skeletal muscle myosin light chain kinase by calcium and calmodulin. Biochemistry 19:5608–5614

Boynton AL, Whitfield JF, Isaacs RJ, Morton HJ (1974) Control of 3T3 cell proliferation by calcium. In Vitro 10:12–17

Brockerhoff H, Ballou CE (1962) Phosphate incorporation in brain phosphoinositides. J Biol Chem 237:49–52

Brown JE, Rubin LJ, Ghalayini AJ, Tarver AP, Irvine RF, Berridge MJ, Anderson RE (1984) Myo-inositol polyphosphate may be a messenger for visual excitation in Limulus photoreceptors. Nature 311:160–162
Burgess GM, Godfrey PP, McKinney JS, Berridge MJ, Irvine RF, Putney JW Jr (1984) The second messenger linking receptor activation to internal Ca^{2+} release in liver. Nature 309:63–66
Burgess GM, McKinney JS, Irvine RF, Putney JW Jr (1985) Inositol 1,4,5-trisphosphate and inositol 1,3,4-trisphosphate formation in Ca^{2+}-mobilizing-hormone-activated cells. Biochem J 232:237–243
Cade JFJ (1949) Lithium salts in the treatment of psychotic excitement. Med J Aust 2:349–352
Case RM, Clausen T (1973) The relationship between calcium exchange and enzyme secretion in the isolated rat pancreas. J Physiol (Lond) 235:75–102
Castagna M, Takai Y, Kaibuchi K, Sano K, Kikkawa U, Nishizuka Y (1982) Direct activation of calcium-activated, phospholipid-dependent protein kinase by tumor promoting phorbol esters. J Biol Chem 257:7847–7851
Clemente F, Meldolesi J (1975) Calcium and pancreatic secretion-dynamics of subcellular calcium pools in resting and stimulated acinar cells. Br J Pharmacol 55:369–379
Connolly TM, Bross TE, Majerus PW (1985) Isolation of a phosphomonoesterase from human platelets that specifically hydrolyzes the 5-phosphate of inositol 1,4,5-trisphosphate. J Biol Chem 260:7868–7874
Corrodi H, Fuxe K, Hökfelt T, Schou M (1967) The effect of lithium on cerebral monoamine neurons. Psychopharmacologia 11:345–353
Crumpton MJ, Allan D, Auger J, Green NM, Maino VC (1975) Recognition at cell surfaces: PHA-lymphocyte interaction. Philos Trans R Soc Lond [Biol] 272:123–180
Dawson AP, Irvine RF (1984) Inositol (1,4,5) trisphosphate-promoted Ca^{2+} release from microsomal fractions of rat liver. Biochem Biophys Res Commun 120:858–864
Dicker P, Rozengurt E (1981) Phorbol ester stimulation of Na influx and Na-K pump activity in Swiss 3T3 cells. Biochem Biophys Res Commun 100:433–441
Diringer H, Friis RR (1977) Changes in phosphatidylinositol metabolism correlated to growth state of normal and Rous sarcoma virus-transformed Japanese quail cells. Cancer Res 37:2979–2984
Doolittle RF, Hunkapiller MW, Hood LE, Devare SG, Robbins KC, Aaronson SA, Antoniades HN (1983) Simian sarcoma virus onc gene, v-sis, is derived from the gene (or genes) encoding a platelet-derived growth factor. Science 221:275–277
Dose M, Emrich HM (1986) Calcium antagonistic properties of antimanic compounds. In: Heinemann U, Klee M, Neher E, Singer W (eds) Calcium electrogenesis and neuronal function. Springer, Berlin Heidelberg New York (Experimental brain research, series 14)
Douglas WW (1974) Exocytosis and the exocytosis-vesiculation sequence: with special reference to neurohypophysis, chromaffin and mast cells, calcium and calcium ionophore. In: Thorn NA, Petersen OH (eds) Secretory mechanisms of exocrine glands. Academic, New York, pp 116–129
Douglas WW, Poisner AM (1964) Stimulus-secretion coupling in a neurosecretory organ: the role of calcium in the release of vasopressin from the neurohypophysis. J Physiol (Lond) 172:1–18
Dousa TP (1974) Interaction of lithium with vasopressin-sensitive cyclic AMP system of human renal medulla. Endocrinology 95:1359–1366
Downes CP, Stones MA (1986) Lithium-induced reduction in intracellular inositol supply in cholinergically stimulated parotid gland. Biochem J 234:199–204
Downes CP, Wusteman MM (1983) Breakdown of polyphosphoinositides and not phosphatidylinositol accounts for muscarinic agonist-stimulated inositol phospholipid metabolism in rat parotid glands. Biochem J 216:633–640
Downward J, Yarden Y, Mayes E, Scrace G, Totty N, Stockwell P, Ullrich A, Schlessinger J, Waterfield MD (1984) Close similarity of epidermal growth factor receptor and v-erb-B oncogene protein sequences. Nature 307:521–527
Drummond AH, Raeburn CA (1984) The interaction of lithium with thyrotropin-releasing hormone-stimulated lipid metabolism in GH_3 pituitary tumour cells. Enhancement of stimulated 1,2-diacylglycerol formation. Biochem J 224:129–136

Dubovsky SL, Franks RD, Lifschitz M, Coen P (1982) Effectiveness of verapamil in the treatment of a manic patient. Am J Psychiatry 139:502–504
Duhm J, Becker BF (1977) Studies in the lithium transport across the red cell membrane. II. Characterization of ouabain-sensitive and ouabain insensitive Li^+ transport. Effects of bicarbonate and dipyridamole. Pflügers Arch 367:211–219
Duhm J, Eisenried F, Becker BF, Greil W (1976) Studies on the lithium transport across the red cell membrane. I. Li^+ uphill transport by the Na^+-dependent Li^+ countertransport system of human erythrocytes. Pflügers Arch 364:147–155
Durell J, Garland JT (1969) Acetylcholine-stimulated phosphodiesteratic cleavage of phosphoinositides: hypothetical role in membrane depolarization. Ann NY Acad Sci 165:743–754
Durell J, Sodd MA, Friedel RO (1968) Acetylcholine stimulation of the phosphodiesteratic cleavage of guinea pig brain phosphoinositides. Life Sci 7:363–368
Durell J, Garland JT, Friedel RO (1969) Acetylcholine action: biochemical aspects. Science 165:862–866
Emrich HM, Aldenhoff JB, Lux HD (eds) (1982) Basic mechanisms in the action of lithium. Excerpta Medica, Amsterdam
Epel D (1978) Mechanisms of activation of sperm and egg during fertilization of sea urchin gametes. Topics Dev Biol 12:185–246
Evers J, Murer H, Kinne R (1976) Phenylalanine uptake in isolated renal brush border vesicles. Biochim Biophys Acta 426:598–615
Extein I, Tallman J, Smith CC, Goodwin FK (1979) Changes in lymphocyte beta-adrenergic receptors in depression and mania. Psychiatry Res 1:191–197
Fisher DB, Mueller GC (1968) An early alteration in the phospholipid metabolism of lymphocytes by phytohemagglutinin. Proc Natl Acad Sci USA 60:1396–1402
Fisher DB, Mueller GC (1971) Studies on the mechanism by which phytohemagglutin rapidly stimulates phospholipid metabolism of human lymphocytes. Biochim Biophys Acta 248:434–448
Forn J, Valdecasas FG (1971) Effects of lithium on brain adenylate cyclase activity. Biochem Pharmacol 20:2773–2778
Geck P, Pietrzyk C, Burckhardt BC, Pfeiffer B, Heinz E (1980) Electrically silent cotransport of Na^+, K^+ and Cl^- in Ehrlich cells. Biochim Biophys Acta 600:432–447
Geisler A, Klysner R, Andersen PH (1985) Influence of lithium in vitro and in vivo on the catecholamine-sensitive cerebral adenylate cyclase systems. Acta Pharmacol Toxicol 56 [Suppl 1]:80–97
Gelfand EW, Dosch HM, Hastings D, Shore A (1979) Lithium: a modulator of cyclic AMP-dependent events in lymphocytes? Science 203:365–367
Greene WC, Parker CM, Parker CW (1976) Calcium and lymphocyte activation. Cell Immunol 25:74–89
Greenspan K, Aronoff MS, Bogdanski DF (1970) Effects of lithium carbonate on turnover and metabolism of norepinephrine in rat brain-correlation to gross behavioral effects. Pharmacology 3:129–136
Haas M, Schooler J, Tosteson DC (1975) Coupling of lithium to sodium transport in human red cells. Nature 258:425–427
Habenicht AJR, Glomset JA, King WC, Nist C, Mitchell CD, Ross R (1981) Early changes in phosphatidylinositol and arachidonic acid metabolism in quiescent Swiss 3T3 cells stimulated to divide by platelet-derived growth factor. J Biol Chem 256:12329–12335
Hallcher LM, Sherman WR (1980) The effects of lithium ion and other agents on the activity of myo-inositol-1-phosphatase from bovine brain. J Biol Chem 255:10896–10901
Hargreaves LNMcF, Hayes PC (1978) The influence of lithium and calcium ions on the aggregation of human blood platelets. Thromb Res 13:79–83
Hart DA (1979) Potentiation of phytohemagglutinin stimulation of lymphoid cells by lithium. Exp Cell Res 119:47–53
Hasegawa-Sasaki H, Sasaki T (1983) Phytohemagglutinin induces rapid degradation of phosphatidylinositol 4,5-bisphosphate and transient accumulation of phosphatidic acid and diacylglycerol in a human T lymphoblastoid cell line, CCRF-CEM. Biochim Biophys Acta 754:305–314
Heisler S, Fast D, Tenenhouse A (1972) Role of Ca^{2+} and cyclic AMP in protein secretion from rat exocrine pancreas. Biochim Biophys Acta 279:561–572

Heisler S, Reisine TD, Hook VYH, Axelrod J (1982) Somatostatin inhibits multireceptor stimulation of cyclic AMP formation and corticotropin secretion in mouse pituitary tumor cells. Proc Natl Acad Sci USA 79:6502–6506
Hellmessen W, Christian AL, Fasold H, Schulz I (1985) Coupled Na^+-H^+ exchange in isolated acinar cells from rat exocrine pancreas. Am J Physiol 249:G125–G136
Hille B (1970) Ionic channels in nerve membranes. Prog Biophys Mol Biol 21:1–32
Hirata M, Suematsu E, Hashimoto T, Hamachi T, Koga T (1984) Release of Ca^{2+} from a non-mitochondrial store site in peritoneal macrophages treated with saponin by inositol 1,4,5-trisphosphate. Biochem J 223:229–236
Hoffmann R, Ristow HJ, Pachowsky H, Frank W (1974) Phospholipid metabolism in embryonic rat fibroblasts following stimulation by a combination of the serum proteins S1 and S2. Eur J Biochem 49:317–324
Hokin LE (1966) Effects of calcium omission on acetylcholine-stimulated amylase secretion and phospholipid synthesis in pigeon pancreas slices. Biochem Biophys Acta 115:219–221
Hokin LE, Hokin MR (1955) Effects of acetylcholine on the turnover of phosphoryl units in individual phospholipids of pancreas slices and brain cortex slices. Biochim Biophys Acta 18:102–110
Hokin LE, Hokin MR (1956) The actions of pancreozymin in pancreas slices and the role of phospholipids in enzyme secretion. J Physiol (Lond) 132:442–453
Hokin LE, Hokin MR (1958 a) Phosphoinositides and protein secretion in pancreas slices. J Biol Chem 233:805–810
Hokin LE, Hokin MR (1958 b) Acetylcholine and the exchange of inositol and phosphate in brain phosphoinositide. J Biol Chem 233:818–821
Hokin MR, Hokin LE (1953) Enzyme secretion and the incorporation of ^{32}P into phospholipids of pancreas slices. J Biol Chem 203:967–977
Hokin MR, Hokin LE (1959) The synthesis of phosphatidic acid from diglyceride and adenosine triphosphate in extracts of brain microsomes. J Biol Chem 234:1381–1386
Hokin MR, Hokin LE (1964) Interconversions of phosphatidylinositol and phosphatidic acid involved in the response to acetylcholine in the salt gland. In: Dawson RMC, Rhodes DN (eds) Metabolism and physiological significance of lipids. Wiley, New York, pp 423–434
Hokin MR, Hokin LE (1967) The formation and continuous turnover of a fraction of phosphatidic acid on stimulation of NaCl secretion by acetylcholine in the salt gland. J Gen Physiol 50:793–811
Honchar MP, Olney JW, Sherman WR (1983) Systemic cholinergic agents induce seizures and brain damage in lithium-treated rats. Science 220:323–325
Hori C, Oka T (1979) Induction by lithium ion of multiplication of mouse mammary epithelium in culture. Proc Natl Acad Sci USA 76:2823–2827
Inhorn RC, Bansal VS, Majerus PW (1987) Pathways for inositol 1,3,4-trisphosphate and 1,4-bisphosphate metabolism. Proc Natl Acad Sci USA 84:2170–2174
Irvine RF, Lander DJ, Letcher AJ, Downes CP (1984 b) Inositol trisphosphates in carbachol-stimulated rat parotid glands. Biochem J 223:237–245
Irvine RF, Änggård EE, Letcher AJ, Downes CP (1985) Metabolism of inositol 1,4,5-trisphosphate and inositol 1,3,4-trisphosphate in rat parotid glands. Biochem J 229:505–511
Irvine RF, Letcher AJ, Heslop JP, Berridge MJ (1986) The inositol tris/tetrakis phosphate pathway-demonstration of inositol (1,4,5) trisphosphate-3-kinase activity in animal tissues. Nature 320:631–634
Irvine RF, Letcher AJ, Lander DJ, Berridge MJ (1986) Specificity of inositol phosphate-stimulated Ca^{2+} mobilization from Swiss-mouse 3T3 cells. Biochem J 240:301–304
Irvine RF, Moor RM (1986) Micro-injection of inositol 1,3,4,5-tetrabisphosphate activates gea urchin eggs by a mechanism dependent on external Ca^{2+}. Biochem J 240:917–920
Joseph SK, Williams RJ (1985) Subcellular localization and some properties of the enzymes hydrolysing inositol polyphosphates in rat liver. FEBS Lett 180:150–154
Kanno T (1972) Calcium-dependent amylase release and electrophysiological measurements in cells of the pancreas. J Physiol (Lond) 226:353–371
Katz B, Miledi R (1967) The timing of calcium action during neuromuscular transmission. J Physiol (Lond) 189:535–544

Katz RI, Kopin IJ (1969) Release of norepinephrine-^{3}H and serotonin-^{3}H evoked from brain slices by electrical-field stimulation-calcium dependency and the effects of lithium, ouabain and tetrodotoxin. Biochem Pharmacol 18:1935–1939
Katz RI, Chase TN, Kopin IJ (1968) Evoked release of norepinephrine and serotonin from brain slices: inhibition by lithium. Science 162:466–467
Kendall DA, Nahorski SR (1985) 5-Hydroxytryptamine-stimulated inositol phospholipid hydrolysis in rat cerebral cortex slices: pharmacological characterization and effects of antidepressants. J Pharmacol Exp Ther 233:473–479
Keynes RD, Swan RC (1959a) The effect of external sodium concentration on the sodium fluxes in frog skeletal muscle. J Physiol (Lond) 147:591–625
Keynes RD, Swan RC (1959b) The permeability of frog muscle fibers to lithium ions. J Physiol (Lond) 147:626–638
Lykouras E, Varsou E, Garelis E, Stefanis CN, Maliaras D (1978) Plasma cyclic AMP in manic-depressive illness. Acta Psychiatr Scand 57:447–453
Lykouras E, Garelis E, Varsou E, Stefanis CN (1979) Physical activity and plasma cyclic adenosine monophosphate levels in manic-depressive patients and healthy adults. Am J Psychiatry 136:540–542
Macara IG, Marinetti GV, Balduzzi PC (1984) Transforming protein of avian sarcoma virus UR2 is associated with phosphatidylinositol kinase activity: possible role in tumorigenesis. Proc Natl Acad Sci USA 81:2728–2732
Mauger JP, Claret M (1986) Mobilization of intracellular calcium by glucagon and cyclic AMP analogues in isolated rat hepatocytes. FEBS Lett 195:106–110
Mendels J, Ramsey TA, Dyson WL, Frazer A (1979) Lithium as an antidepressant. In: Cooper TB, Gershon S, Kline NS, Schou M (eds) Lithium: controversies and unresolved issues. Excerpta Medica, Amsterdam, pp 35–47
Metcalfe JC, Pozzan T, Smith GA, Hesketh TR (1980) A calcium hypothesis for the control of cell growth. Biochem Soc Symp 45:1–26
Michell B (1986) Inositol phosphates. Profusion and confusion. Nature 319:176–177
Michell RH (1975) Inositol lipids and cell surface receptor function. Biochim Biophys Acta 415:81–147
Michell RH (1982) Inositol lipid metabolism in dividing and differentiating cells. Cell Calcium 3:429–440
Michell RH, Kirk CJ, Jones LM, Downes CP, Creba JA (1981) The stimulation of inositol lipid metabolism that accompanies calcium mobilization in stimulated cells: defined characteristics and unanswered questions. Philos Trans R Soc Lond Ser B 296:123–137
Moolenaar WH, Tsien RY, van der Saag PT, de Laat SW (1983) Na^{+}/H^{+} exchange and cytoplasmic pH in the action of growth factors in human fibroblasts. Nature 304:645–648
Murphy DL, Donelly C, Moskowitz J (1973) Inhibition by lithium of prostaglandin E_1 and norepinephrine effects on cyclic adenosine monophosphate production in human platelets. Clin Pharmacol Ther 14:810–814
Nagel W (1977) Influence of lithium upon the intracellular potential of frog skin epithelium. J Membr Biol 37:347–359
Nielsen-Kudsk F, Pedersen AK (1978) Myocardial effects of lithium in vitro. Acta Pharmacol Toxicol 42:311–316
O'Rourke FA, Halenda SP, Zavoico GB, Feinstein MB (1985) Inositol 1,4,5-trisphosphate releases Ca^{2+} from a Ca^{2+}-transporting membrane vesicle fraction derived from human platelets. J Biol Chem 260:956–962
Overton E (1902) Beiträge zur allgemeinen Muskel- und Nervenphysiologie. II. Mitt. Über die Unentbehrlichkeit von Natrium- (oder Lithium) Ionen für den Contractionsact des Muskels. Pflügers Arch 92:346
Pandey GN, Dysken MW, Garver DL, Davis JM (1979) Beta-adrenergic receptor function in affective illness. Am J Psychiatry 136:675–678
Paulus H, Kennedy EP (1960) The enzymatic synthesis of inositol monophosphatide. J Biol Chem 235:1303–1311
Petersen OH, Iwatsuki N (1978) The role of calcium in pancreatic acinar cell stimulus-secretion coupling: an electrophysiological approach. Ann NY Acad Sci 307:599–617
Petersen OH, Ueda N (1976) Pancreatic acinar cells: the role of calcium in stimulus-secretion coupling. J Physiol (Lond) 254:583–606

Pollard AD, Brindley DN (1984) Effects of vasopressin and corticosterone on fatty acid metabolism and on the activities of glycerol phosphate acyltransferase and phosphatidate phosphohydrolase in rat hepatocytes. Biochem J 217:461–469
Prentki M, Biden TJ, Janjic D, Irvine RF, Berridge MJ, Wollheim CB (1984) Rapid mobilization of Ca^{2+} from rat insulinoma microsomes by inositol-1,4,5-trisphosphate. Nature 309:562–564
Ptashne K, Stockdale FE, Conlon S (1980) Initiation of DNA synthesis in mammary epithelium and mammary tumors by lithium ions. J Cell Physiol 103:41–46
Rasmussen H, Barrett PQ (1984) Calcium messenger system: an integrated view. Physiol Rev 64:938–984
Richelson E (1977) Lithium ion entry through the sodium channel of cultured mouse neuroblastoma cells: a biochemical study. Science 196:1001–1002
Rink TJ (1977) The influence of sodium on calcium movements and catecholamine release in thin slices of bovine adrenal medulla. J Physiol (Lond) 266:297–325
Ristow HJ, Frank W, Fröhlich M (1973) Stimulation of embryonic rat cells by calf serum. V. Metabolism of inositol- and choline phospholipids. Stimulierung von Kulturen embryonaler Rattenzellen durch Kälberserum. V. Verhalten der Inosit- und Cholinphospholipide. Z Naturforsch 28c:188–194
Robberecht P, Christophe J (1971) Secretion of hydrolases by perfused fragments of rat pancreas: effect of calcium. Am J Physiol 220:911–917
Rosenblatt JE, Pert CB, Tallman JF, Pert A, Bunney WE (1979) Effect of imipramine and lithium on alpha-receptor and beta-receptor binding in rat brain. Brain Res 160:186–191
Rosoff PM, Stein LF, Cantley LC (1984) Phorbol esters induce differentiation in a pre-B-lymphocyte cell line by enhancing Na^+/H^+ exchange. J Biol Chem 259:7056–7060
Rubin RP (1984) Stimulation of inositol trisphosphate accumulation and amylase secretion by caerulein in pancreatic acini. J Pharmacol Exp Ther 231:623–627
Ryback SM, Stockdale FE (1981) Growth effects of lithium chloride in BALB/c 3T3 fibroblasts and Madin-Darby canine kidney epithelial cells. Exp Cell Res 136:263–270
Santiago-Calvo E, Mule S, Redman CM, Hokin MR, Hokin LE (1964) The chromatographic separation of polyphosphoinositides and studies in their turnover in various tissues. Biochim Biophys Acta 84:550–562
Sawyer ST, Cohen S (1981) Enhancement of calcium uptake and phosphatidylinositol turnover by epidermal growth factor in A-431 cells. Biochemistry 20:6280–6286
Schneyer CA (1974) Autonomic regulation of secretory activity and growth responses of rat parotid gland. In: Thorn NA, Petersen OH (eds) Secretory mechanisms of exocrine glands. Munksgaard, Copenhagen, pp 42–55
Schreurs VVAM, Swarts HGP, De Pont JJHHM, Bonting SL (1976) Role of calcium in exocrine pancreatic secretion. II. Comparison of the effects of carbachol and the ionophore A-23187 on enzyme secretion and calcium movements in rabbit pancreas. Biochim Biophys Acta 419:320–330
Schulz I (1980) Messenger role of calcium in function of pancreatic acinar cells. Am J Physiol 239:G335–G347
Sekar MC, Roufogalis BD (1984a) Differential effects of phenylmethanesulfonyl fluoride (PMSF) on carbachol and potassium stimulated phosphoinositide turnover and contraction in longitudinal smooth muscle of guinea pig ileum. Cell Calcium 5:191–203
Sekar MC, Roufogalis BD (1984b) Muscarinic-receptor stimulation enhances polyphosphoinositide breakdown in guinea-pig ileum smooth muscle. Biochem J 223:527–531
Sherman WR, Leavitt AL, Honchar MP, Hallcher L, Phillips BE (1981) Evidence that lithium alters phosphoinositide metabolism: chronic administration elevates primarily d-myo-inositol-1-phosphate in cerebral cortex of the rat. J Neurochem 36:1947–1951
Sherman WR, Honchar MP, Munsell LY (1985a) Detection of receptor-linked phosphoinositide metabolism in brain of lithium-treated rats. In: Bleasdale JE, Eichberg J, Hauser G (eds) Cyclitols and phosphoinositides, vol 49. Humana, New York
Sherman WR, Munsell LY, Gish BG, Honchar MP (1985b) Effects of systemically administered lithium on phosphoinositide metabolism in rat brain, kidney, and testis. J Neurochem 44:798–807
Sillers PJ, Forer A (1985) Ca^{++} in fertilization and mitosis: the phosphatidylinositol cycle in sea urchin gametes and zygotes is involved in control of fertilization and mitosis. Cell Biol Int Rep 9:275–282

Stern DN, Fieve RR, Neff NH, Costa E (1969) The effect of lithium chloride administration on brain and heart norepinephrine turnover rates. Psychopharmacologia 14:315–322
Stolze H, Schulz I (1980) Effect of atropine, ouabain, antimycin A and A23187 on "trigger Ca^{2+} pool" in exocrine pancreas. Am J Physiol 238:G318–G348
Storey DJ, Shears SB, Kirk CJ, Michell RH (1984) Stepwise enzymatic dephosphorylation of inositol 1,4,5-trisphosphate to inositol in liver. Nature 312:374–376
Streb H, Irvine RF, Berridge MJ, Schulz I (1983) Release of Ca^{2+} from a nonmitochondrial intracellular store in pancreatic acinar cells by inositol-1,4,5-trisphosphate. Nature 306:67–69
Streb H, Bayerdörffer E, Haase W, Irvine RF, Schulz I (1984) Effect of inositol-1,4,5-trisphosphate on isolated subcellular fractions of rat pancreas. J Membr Biol 81:241–253
Suematsu E, Hirata M, Hashimoto T, Kuriyama H (1984) Inositol 1,4,5-trisphosphate releases Ca^{2+} from intracellular store sites in skinned single cells of porcine coronary artery. Biochem Biophys Res Commun 120:481–485
Sugimoto Y, Whitman M, Cantley LC, Erikson RL (1984) Evidence that the Rous sarcoma virus transforming gene product phosphorylates phosphatidylinositol and diacylglycerol. Proc Natl Acad Sci USA 81:2117–2121
Thomas AP, Alexander J, Williamson JR (1984) Relationship between inositol polyphosphate production and the increase of cytosolic free Ca^{2+} induced by vasopressin in isolated hepatocytes. J Biol Chem 259:5574–5584
Toback FG (1980) Induction of growth in kidney epithelial cells in culture by Na^{+}. Proc Natl Acad Sci USA 77:6654–6656
Tomooka Y, Imagawa W, Nandi S, Bern HA (1983) Growth effect of lithium on mouse mammary epithelial cells in serum-free collagen gel culture. J Cell Physiol 117:290–296
Treiser S, Kellar KJ (1979) Lithium effects on adrenergic receptor supersensitivity in rat brain. Eur J Pharmacol 58:85–86
Ullrich KJ, Rumrich G, Kloess S (1976) Active Ca^{2+} reabsorption in the proximal tubule of the rat kidney. Dependence on sodium- and buffer transport. Pflügers Arch 364:223–228
Volpe P, Salviati G, Di Virgilio F, Pozzan T (1985) Inositol 1,4,5-trisphosphate induces calcium release from sarcoplasmic reticulum of skeletal muscle. Nature 316:347–349
Wang YC, Pandey GN, Mendels J, Frazer A (1974) Effect of lithium on prostaglandin E_1-stimulated adenylate cyclase activity of human platelets. Biochem Pharmacol 23:845–855
Waterfield MD, Scrace GT, Whittle N, Stroobant P, Johnsson A, Wasteson A, Westermark B, Heldin CH, Huang JS, Deuel TF (1983) Platelet-derived growth factor is structurally related to the putative transforming protein $p28^{sis}$ of simian sarcoma virus. Nature 304:35–39
Williams RJP (1973) The chemistry and biochemistry of lithium. In: Gershon S, Shopsin B (eds) Lithium: its role in psychiatric research and treatment. Plenum, New York, pp 15–31
Wilson DB, Bross TE, Sherman WR, Berger RA, Majerus PW (1985a) ^{18}O-labeling shows cyclic inositol phosphates are produced by cleavage of polyphosphoinositides with purified sheep seminal vesicle phospholipase C enzymes. Proc Natl Acad Sci USA 82:4013–4017
Wilson DB, Connolly TM, Bross TE, Majerus PW, Sherman WR, Tyler AN, Rubin LJ, Brown JE (1985b) Isolation and characterization of the inositol cyclic phosphate products of polyphosphoinositide cleavage by phospholipase C: physiological effects in permeabilized platelets and limulus photoreceptor cells. J Biol Chem 260:13496–13501
Wood K (1985) The neurochemistry of mania. The effect of lithium on catecholamines, indoleamines and calcium mobilization. J Affective Disord 8:215–223
Yoshikami S, Hagins WA (1970) Ionic basis of dark current and photocurrent of retinal rods. Biophys J 10:60a
Zatz M, Reisine TD (1985) Lithium induces corticotropin secretion and desensitization in cultured anterior pituitary cells. Proc Natl Acad Sci USA 82:1286–1290

CHAPTER 13

Phorbol Esters and Protein Kinase C

H. NAGAMOTO, U. KIKKAWA, and Y. NISHIZUKA

A. Introduction

It is generally accepted that carcinogenesis can be divided into two stages, tumor initiation and promotion. Since these stages can be caused by compounds specific to each, there seem to be two independent processes that are responsible for initiation and promotion. Initial studies of the promotion process utilized croton oil. Today phorbol esters isolated from croton oil as its active components, as well as synthetic analogs with a range of tumor-promoting activity are used for studies of the mechanism of tumor promotion.

Tumor-promoting phorbol esters produce a wide variety of biologic and biochemical changes such as induction of cell proliferation and differentiation, and activation of cellular functions. It has been accepted for a long time that the pleiotropic actions of tumor-promoting phorbol esters are mediated through interaction with their specific cell surface receptors. Reports from this laboratory have provided evidence that phorbol esters bind to and activate protein kinase C directly, both in vivo and in vitro. Subsequent studies in many laboratories confirmed this finding, and it is now clear that many, if not all, of the pleiotropic actions of tumor promoters are mediated through the action of this enzyme. The purpose of this chapter is to describe the experimental procedures for the study of protein kinase C and to summarize the evidence that protein kinase C is a prime target for the action of phorbol esters, and perhaps some other tumor promoters. Many review articles covering the biologic activities of phorbol esters (DIAMOND et al. 1978, 1980; BLUMBERG 1980; BLUMBERG et al. 1984) and several other aspects of protein kinase C (NISHIZUKA 1983, 1984a, b, 1986) are available.

B. Purification and Assay of Protein Kinase C

I. Purification of Protein Kinase C from Rat Brain

To date, severael procedures have been reported to purify protein kinase C from bovine heart (WISE et al. 1982a), pig spleen (SCHATZMAN et al. 1983), rat brain (WOLF et al. 1984), bovine brain (PARKER et al. 1984), rat renal cortex (UCHIDA and FILBURN 1984), and rabbit brain (INAGAKI et al. 1985), but the final preparations obtained by these procedures contain some impurities and several days are needed to complete processing. The method described here is semiautomatic and reproducible, employing three steps of high pressure liquid chromatography (HPLC) (KIKKAWA et al. 1986).

Quickly removed rat brains are homogenized in a polytetrafluoroethylene–glass homogenizer with 6 volumes of extraction buffer: 0.25 *M* sucrose, 10 m*M* ethylene glycol bis(*β*-aminoethylether)-*N*,*N*,*N'*,*N'*-tetraacetic acid (EGTA), 2 m*M* ethylenediamine tetraacetic acid (EDTA), 1 m*M* phenylmethylsulfonylfluoride, 20 m*M* TRIS-HCl, pH 7.5. The supernatant (200 ml, 990 mg protein) obtained by centrifugation at 100000 *g* for 60 min is applied to a DE-52 column (3 × 17 cm) equilibrated with buffer A (0.5 m*M* EGTA, 0.5 m*M* EDTA, 10 m*M* 2-mercaptoethanol, 20 m*M* TRIS-HCl, pH 7.5). After washing the column with 1200 ml buffer A containing 20 m*M* NaCl at a flow rate of 3 ml/min, protein kinase C is eluted batchwise from the column with 340 ml buffer A containing 90 m*M* NaCl. The DE-52 fraction (340 ml, 117 mg protein) is applied directly to a threonine-conjugated Sepharose 4B column (1 × 15 cm) connected to an HPLC (Pharmacia FPLC system), and equilibrated with buffer A. Then, protein kinase C is eluted from the column by an NaCl concentration gradient (0–1.0 *M*) in buffer A, as shown in Fig. 1. The activity of protein kinase C shows two peaks. The minor peak is eluted at 0.3 *M* NaCl (fractions 13–19), and the major peak at 0.6–0.8 *M* NaCl (fractions 37–47). The major peak (55 ml, 1.7 mg protein) is pooled and applied to a TSK gel Phenyl-5PW column (0.75 × 7.5 cm) equilibrated with buffer A containing 1 *M* NaCl. Then, protein kinase C is eluted from the column at a flow rate of 0.5 ml/min. The Pharmacia FPLC system is programed to decrease the concentration of NaCl linearly from 1.0 to 0 *M* in 20 ml buffer A, followed by 20 ml buffer A, as shown in Fig. 2; 1-ml fractions are collected. The enzymatic activity appears as a single peak at about 0.1 *M* NaCl. The TSK gel Phenyl-5PW fraction (fractions 15–20, 0.29 mg protein) is concentrated to 2 ml by an Amicon ultrafiltration cell equipped with a YM-10 filter membrane, and

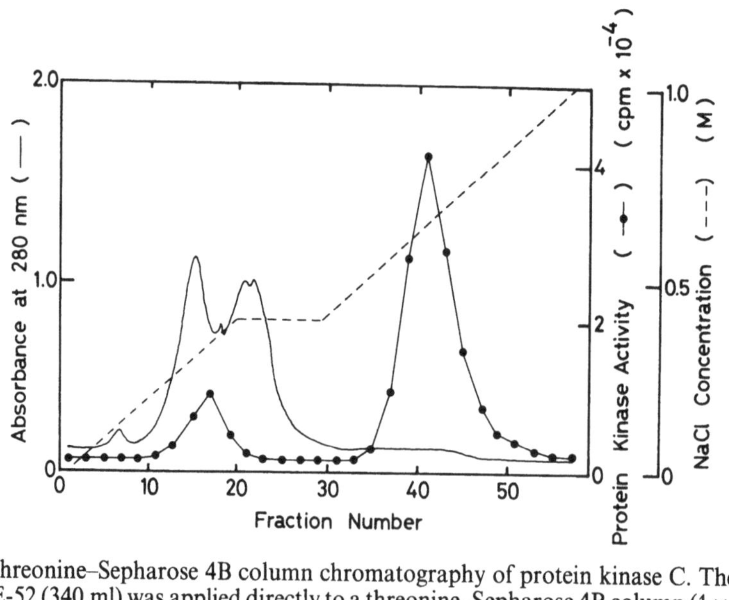

Fig. 1. Threonine–Sepharose 4B column chromatography of protein kinase C. The eluant from DE-52 (340 ml) was applied directly to a threonine–Sepharose 4B column (1 × 15 cm). Protein kinase C was eluted from the column by an NaCl concentration gradient (0–1.0 *M*) in buffer A; 5-ml fractions were collected. The enzymatic activity was assayed under the standard conditions described in Sect. B.II. (Adapted from KIKKAWA et al. 1986)

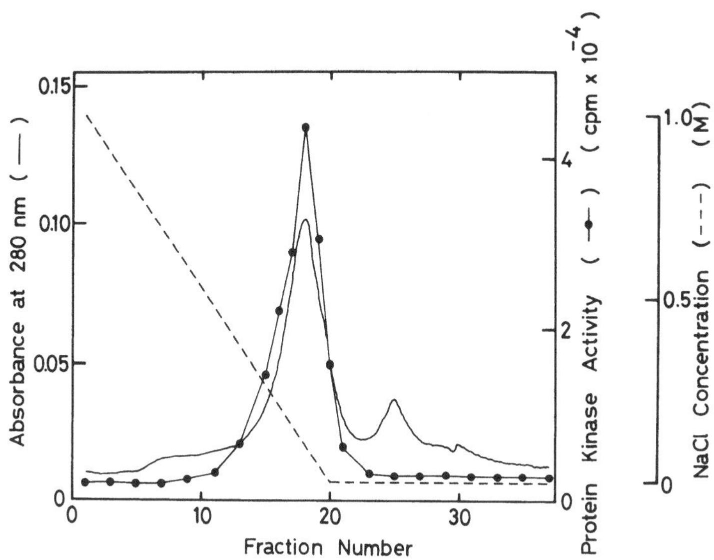

Fig. 2. TSK gel Phenyl-5PW column chromatography of protein kinase C. Fractions 37–47 in Fig. 1 (55 ml) were collected and applied to a TSK gel Phenyl-5PW column (0.75 × 7.5 cm) equilibrated with buffer A containing 1 *M* NaCl. Protein kinase C was eluted from the column by decreasing the concentration of NaCl linearly from 1.0 to 0 *M* in 20 ml buffer A, followed by 20 ml buffer A at a flow rate of 0.5 ml/min; 1-ml fractions were collected. The enzymatic activity was assayed under the standard conditions described in Sect. B.II. (Adapted from KIKKAWA et al. 1986)

applied to a TSK gel G3000SW column (2.15 × 60 cm) equilibrated with buffer A containing 0.1 *M* NaCl. Elution is performed with the same salt buffer at a flow rate of 4.5 ml/min; 4.5-ml fractions are collected. Protein kinase C appears as a sharp symmetric peak, which coincides with the absorption band at 280 nm. This final preparation of the enzyme is almost pure, as judged by silver staining after polyacrylamide gel electrophoresis in the presence of sodium dodecylsulfate.

The results of this procedure are summarized in Table 1. The enzyme is purified approximately 740-fold with an overall yield of at least 15%. The enzyme obtained from brain tissue by this method sometimes exhibits a duplex band on gel

Table 1. Summary of purification of protein kinase C from rat brain. (Adapted from KIKKAWA et al. 1986)

Fraction	Total protein (mg)	Total activity (units $\times 10^3$)	Specific activity (units/mg $\times 10^3$)	Yield (%)	Purification (-fold)
Crude extract	990	1100	1.11	100	1
DE-52	117	908	7.76	83	7
Threonine – Sepharose 4B	1.70	483	284	44	256
TSK gel Phenyl-5PW	0.29	226	779	21	702
TSK gel G3000SW	0.20	164	820	15	739

electrophoresis. The precise nature of this duplex band is not known (KIKKAWA et al. 1986). The purified enzyme is very labile, but can be stored at −80 °C for at least several months without loss of enzyme activity in the presence of glycerol and Triton X-100 in a final concentration of 10% (w/v) and 0.05% (w/v), respectively.

II. Enzyme Assay of Protein Kinase C

Protein kinase C is assayed by measuring the incorporation of ^{32}P into H1 histone from [γ-^{32}P] ATP (KIKKAWA et al. 1982). The reaction mixture (0.25 ml) contains 20 m*M* TRIS-HCl, pH 7.5, 200 μg/ml H1 histone, 10 μ*M* [γ-^{32}P] ATP (50–100 cpm/pmol), 5 m*M* magnesium acetate, 0.1 m*M* $CaCl_2$, 8 μg/ml phosphatidylserine, 0.8 μg/ml dioleoylglycerol, and an appropriate amount of enzyme. Phosphatidylserine and dioleoylglycerol are first mixed in a small volume of chloroform. After the chloroform is removed by N_2 gas, the residue is suspended in 20 m*M* TRIS-HCl, pH 7.5, and then sonicated for 5 min at 0 °C with a Kontes sonifier, K881440. This suspension is employed for the assay. The incubation is carried out for 3 min at 30 °C. The reaction is terminated by the addition of 3 ml 25% (w/v) trichloroacetic acid. The acid-precipitable materials are collected on a nitrocellulose filter (pore size 0.45 μm). The filter is washed with 25% trichloroacetic acid, and the radioactivity trapped on the filter is quantitated by Cerenkov counting using a Packard Tri-Carb liquid scintillation counter, Model 4640. In some cases, e.g., in studies of the relationship between surface pressure and enzyme activity, and to check the molar ratio of phosphatidylserine and diacylglycerol or phorbol ester, it is useful to assay the enzyme with a monomolecular film rather than vesicles of phosphatidylserine.

III. Binding Assay of Protein Kinase C

Although several reports have described the binding assay of [^3H] phorbol-12,13-dibutyrate ([^3H] PDBu) to protein kinase C (KIKKAWA et al. 1983; SHARKEY et al. 1984; TANAKA et al. 1986), here the simplest and most rapid assay is described.

The standard mixture (0.2 ml) in a plastic tube contains 20 m*M* TRIS-maleate, pH 6.8, 100 m*M* KCl, 0.15 m*M* $CaCl_2$, 0.05 m*M* EGTA, 100 μg/ml phosphatidylserine, 30 n*M* [^3H] PDBu (13.4 Ci/mmol), 0.5% dimethylsulfoxide, 4 mg/ml bovine serum albumin, and 0.05–0.1 μg protein kinase C. Phosphatidylserine is suspended in 20 m*M* TRIS-maleate, pH 6.8 by sonication and added to the reaction mixture. After 20 min at 30 °C, 4 ml ice-cold 0.5% dimethylsulfoxide is added. The mixture is poured onto a glass-fiber filter (glass microfiber filter Whatman GF/B), previously soaked in fresh 0.3% polyethyleneimine solution for 1 h (BRUNS et al. 1983), under high vacuum. The filter is washed three times, each time with 3 ml ice-cold 0.5% dimethylsulfoxide, then dried, and the bound radioactivity is determined in a toluene-based liquid scintillation counter.

C. Protein Kinase C and Phorbol Esters

I. Properties

Protein kinase C was first found in 1977 as a proteolytically activated protein kinase capable of phosphorylating histone (INOUE et al. 1977), and later shown to be activated reversibly by association of membrane phospholipids at physiologic concentrations of Ca^{2+} in the presence of diacylglycerol that is transiently produced in membranes from inositol phospholipids in response to extracellular signals (TAKAI et al. 1979a, b; KISHIMOTO et al. 1980). This enzyme is ubiquitously distributed in tissues and organs, with platelets and brain having the highest activity (KUO et al. 1980; MINAKUCHI et al. 1981). In brain tissues, a large quantity of the enzyme is associated with synaptic membranes, whereas in most other tissues the enzyme is present mainly in the soluble fraction as an inactive form. The enzyme was initially thought to be a single entity, but recent analysis of its complementary DNA clones indicates that protein kinase C is a complex family of closely related structures (COUSSENS et al. 1986; KNOPF et al. 1986; OHNO et al. 1987; ONO et al. 1986; PARKER et al. 1986). Comparison of the predicted amino acid sequences reveals that at least four subspecies of the enzyme may exist in mammalian tissues, particularly in brain. The enzymes are encoded by three distinct genes located on different chromosomes, and also derived from alternative splicing of one of the three genes. Preliminary analysis suggests that there is tissue specificity for the expression of the enzyme. Nevertheless, the enzymes so far isolated from various tissues appear to be very similar in their physical, kinetic, and catalytic properties. The molecular weight of all subspecies of the enzyme is roughly 80000. The four species of the enzyme are each composed of a single polypeptide chain with 671, 672, 673, and 697 amino acid residues. The Stokes radius is 42 Å. The sedimentation coefficient is 5.1 S. The frictional ratio of the enzyme is calculated to be 1.6, indicating an asymmetric nature of the molecule. The isoelectric point of the enzyme is pH 5.6. The optimum pH range for activity is 7.5–8.0 with TRIS-acetate as test buffer. Mg^{2+} is essential for the catalytic activity with the optimum range about 5–10 mM. The K_m value for ATP is about 6×10^{-6} M. The enzyme utilizes ATP-γ-S as a phosphate donor with a very slow reaction rate (WISE et al. 1982b), but not GTP (TAKAI et al. 1984). However, it is possible that protein kinase C is a mixture of more than two enzymes with very similar properties, as briefly mentioned already.

II. Biochemical and Physiologic Activation

Protein kinase C is a Ca^{2+}-activated, phospholipid-dependent enzyme that is markedly activated by diacylglycerol. This diacylglycerol is produced in membranes by the receptor-mediated hydrolysis of inositol phospholipids. Kinetic analysis indicates that a small amount of diacylglycerol dramatically increases the affinity of this enzyme for Ca^{2+}, and thereby renders it fully active without net increase in the Ca^{2+} concentration (KAIBUCHI et al. 1981; KISHIMOTO et al. 1980). Thus, the activation of this protein kinase is biochemically dependent on, but physiologically independent of Ca^{2+}. Among phospholipids tested, phosphatidylserine is essential for the enzyme activation.

Diacylglycerols containing the 1,2-*sn* configuration, with various fatty acids of different chain lengths are capable of activating protein kinase C, and those having unsaturated fatty acids are the most active (MORI et al. 1982). However, other stereoisomers, 2,3-*sn*-diacylglycerol and 1,3-diacylglycerol, neither activate nor inhibit the enzyme, suggesting that a highly specific lipid–protein interaction is needed for this enzyme activation (RANDO and YOUNG 1984).

It has been repeatedly shown that a group of hormones, some neurotransmitters, and many other biologically active substances provoke the breakdown of inositol phospholipids in the plasma membrane (MICHELL 1975; MICHELL et al. 1981). Although phosphatidylinositol (PI) was initially regarded as a primary target (HOKIN and HOKIN 1953), phosphatidylinositol-4,5-bisphosphate (PIP_2) rather than PI and phosphatidylinositol-4-phosphate (PIP) has been studied with great interest since PIP_2 is degraded more rapidly in stimulated cells to produce 1,2-diacylglycerol and inositol-1,4,5-trisphosphate (IP_3) (ABDEL-LATIF et al. 1977; AGRANOFF et al. 1983; FISHER et al. 1984). This water-soluble product serves as an inracellular mediator for the release of Ca^{2+} from internal stores (BERRIDGE and IRVINE 1984). A series of studies in this laboratory has provided some evidence that 1,2-diacylglycerol produced in this way initiates the activation of protein kinase C as described, and that information from a variety of extracellular signals is transduced across the membrane to result in protein phosphorylation (NISHIZUKA 1984a, b).

III. Permeable Diacylglycerol and Phorbol Esters

In the studies to explore a link between inositol phospholipid turnover and protein kinase C activation in stimulus–response coupling, a synthetic diacylglycerol, 1-oleoyl-2-acetylglycerol, has been used since it is readily intercalated and dispersed into the membrane phospholipid bilayer, and activates protein kinase C directly (KAIBUCHI et al. 1983). More recently, 1,2-dioctanoylglycerol and 1,2-didecanoylglycerol have been shown to be effective permeable diacylglycerols (LAPETINA et al. 1985). Synthetic diacylglycerols having the 2,3-*sn* configuration are not active for intact cell systems (NOMURA et al. 1986).

Tumor-promoting phorbol esters, such as 12-*O*-tetradecanoylphorbol-13-acetate (TPA), have a structure very similar to diacylglycerol and activate protein kinase C directly, both in vitro and in vivo (CASTAGNA et al. 1982; YAMANISHI et al. 1983). Kinetic analysis indicates that TPA is able to substitute for diacylglycerol at extremely low concentrations and that, like diacylglycerol, TPA dramatically increases the affinity of this enzyme for Ca^{2+}, resulting in its full activation at physiologic concentrations of this divalent cation (YAMANISHI et al. 1983). Several lines of evidence provided by many laboratories seem to indicate that protein kinase C is a prime target of tumor promoters (CASTAGNA et al. 1982; YAMANISHI et al. 1983; NIEDEL et al. 1983; SANDO and YOUNG 1983; LEACH et al. 1983; ASHENDEL et al. 1983; KIKKAWA et al. 1983). The binding studies indicate that [^{3}H] PDBu, which is a potent tumor promoter, but less hydrophobic than TPA, may bind to purified protein kinase C only when both Ca^{2+} and phospholipid are present (KIKKAWA et al. 1983). The apparent dissociation constant K_d of the tumor promoter, 8 nM, is exactly identical with the activation constant K_a for the

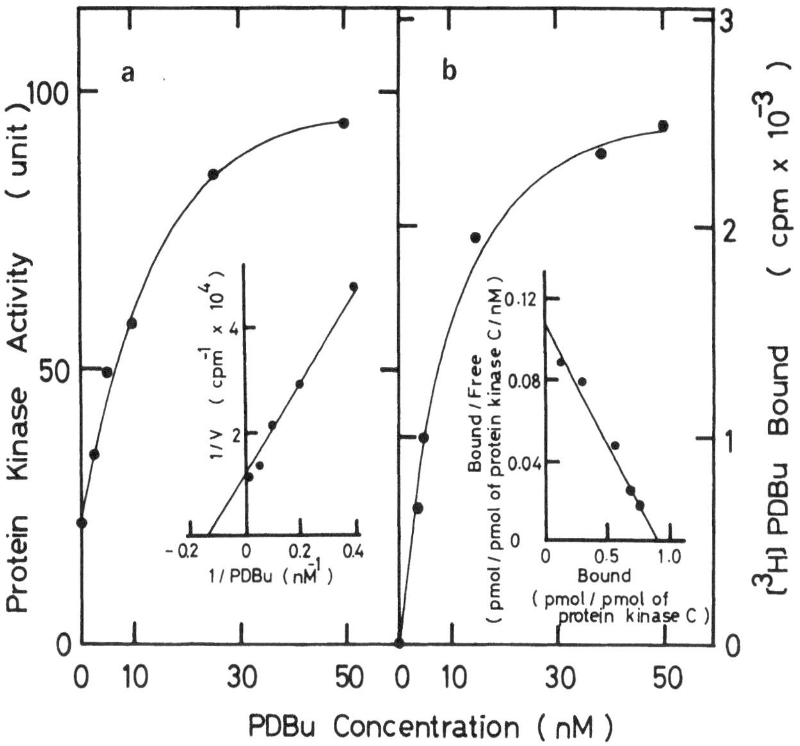

Fig. 3 a, b. Effects of PDBu concentration on activation of protein kinase C and on its binding to protein kinase C, **a** protein kinase activity. Protein kinase C was assayed at various concentrations of PDBu, as indicated, in the presence of 0.1 m*M* $CaCl_2$ and 8 µg/ml phosphatidylserine. *Inset* Lineweaver–Burk double plots, **b** [^{3}H] PDBu binding. Binding assay was carried out at various concentrations of [^{3}H] PDBu as indicated. *Inset* Scatchard plots. Other detailed conditions are described by Kikkawa et al. (1983). (Adapted from Nishizuka 1984a)

enzyme, 8 n*M*, as shown in Fig. 3. These values, obtained with homogeneous protein kinase C and chromatographically pure phosphatidylserine, are remarkably similar to the K_d values described for the specific tumor promoter-binding site in brain particulate fractions, 5.6 n*M* (Dunphy et al. 1981), and on intact cell membranes of rat embryo fibroblasts, 8 n*M* (Horowitz et al. 1981), and mouse epidermal cells, 10 n*M* (Solanki and Slaga 1981). It has been shown for intact cell systems that TPA is a potent competitor for [^{3}H] PDBu binding, whereas phorbol and 4α-phorbol-12,13-didecanoate, which lack tumor-promoting activity, do not interfere with the binding of [^{3}H] PDBu to its receptor. Consistent with these observations, Table 2 shows that TPA, nonradioactive PDBu, and phorbol-12,13-didecanoate, which are all able to promote tumor development in vivo, can activate protein kinase C in vitro, and also inhibit the binding of [^{3}H] PDBu in a competitive manner. Inversely, phorbol derivatives having no tumor-promoting activity are all unable to activate the enzyme and cannot compete with [^{3}H] PDBu for the binding. Under similar conditions, diacylglycerols, such as dioleoylglycerol, also prevent binding of [^{3}H] PDBu to produce the quaternary complex,

Table 2. Effects of various tumor promoters on activation of protein kinase C and inhibition of [^{3}H] PDBu binding to protein kinase C. (Adapted from KIKKAWA et al. 1983)

Tumor promotor	Tumor-promoting activity	Activation or protein kinase C[a] (%)	Inhibition of [^{3}H] PDBu binding[b] (%)
TPA	+++	100	100
PDBu	++	88	100
Phorbol-12,13-didecanoate	++	81	100
Phorbol-12-tetradecanoate	–	0	18
Phorbol-13-acetate	–	0	0
4α-Phorbol-12,13-didecanoate	–	0	0
Phorbol	–	0	0
Mezerein	+	87[c]	ND

[a] Activity of protein kinase C was measured in the presence of 10 μ*M* $CaCl_2$, 20 μg/ml phospholipid, and 10 ng/ml phorbol derivatives.
[b] Inhibition of [^{3}H] PDBu binding to protein kinase C was assayed in the presence of a 100-fold excess of each nonradioactive phorbol derivative. Detailed experimental conditions are described by KIKKAWA et al. (1983).
[c] Mezerein was added at 100 ng/ml.
ND, not determined,

whereas monooleoylglycerol, trioleoylglycerol, and free oleic acid are inactive in this capacity.

Mezerein (MIYAKE et al. 1984; COUTURIER et al. 1984), and teleocidin and *Aplysia* toxin (tumor promoters structurally unrelated to phorbol esters) (FUJIKI et al. 1984; ARCOLEO and WEINSTEIN 1985) also activate protein kinase C, suggesting that a diacylglycerol-like structure is not always essential, and that many tumor promoters so far identified induce a membrane perturbation analogous to that caused by diacylglycerol. It is plausible, therefore, that protein kinase C is a most important receptor for tumor-promoting phorbol esters and some other tumor promoters, such as mezerein and teleocidin, and that the pleiotropic actions of some, if not all, tumor promoters may be mediated through the activation of this protein kinase. Thus, the traditional concept of tumor promoters has been replaced by an explicit biochemical explanation which will make for a better understanding of the role of protein kinase C (NISHIZUKA 1984a).

However, in experiments with phorbol esters, the resulting cellular responses sometimes appear to cast doubt on their suitability for studies on cell biology. Diacylglycerol, the physiologic activator of protein kinase C, is present only transiently in membranes, while TPA is hardly metabolizable and persists for longer periods of time. The possibility arises, therefore, that some limited phase of the cellular response might be extended, so distorting the normal sequence of events. Protein kinase C presumably has a dual action, both positive and negative, depending on the function of its target protein to be phosphorylated. In the early phase of cellular responses, this enzyme appears to act synergistically with Ca^{2+} as part of positive forward, but sometimes negative feedback control, such as down-regulation of some receptors, comes into play immediately (NISHIZUKA

1986). Therefore, it is essential to understand the time sequence of events occurring in different phases of a cellular responses.

In addition, the concentration of phorbol ester employed should be given special consideration when attempting to evaluate its exact contribution to physiologic responses. As emphasized repeatedly elsewhere (NISHIZUKA 1984a, b), at higher concentrations, tumor promoters per se can induce significant biologic effects. There is no proof at present for protein kinase C being the sole target of tumor promoters, and a possibility exists that these compounds act as membrane perturbers or fusigens, particularly at higher concentrations. To interpret any experimental results for phorbol esters correctly, the parallel demonstration of other parameters such as phosphorylation of some endogenous proteins is obviously desirable, as described in our earlier experiments for platelets (YAMANISHI et al. 1983), and lymphocytes (KAIBUCHI et al. 1985). At higher concentrations, permeable diacylglycerols appear to show many nonspecific actions similar to phorbol esters, but it is not surprising to realize that slightly different effects on living cells are obtained from stimulation of either diacylglycerol or phorbol esters (KRUETTER et al. 1985).

D. Conclusion

This chapter summarizes experimental techniques and some of our current knowledge of the relationship between phorbol esters and protein kinase C. The evidence available to date seems to indicate that protein kinase C is a prime target of phorbol esters, and that most of their biologic activities may be mediated through the activation of protein kinase C. However, it seems premature to discuss the precise relationship between the action of phorbol esters and protein kinase C, particularly in long-term cellular responses, such as growth promotion. Several possible functions of this enzyme are currently examined, such as synergistic roles with Ca^{2+}, Ca^{2+}-sensitivity modulation, and down-regulation, and desensitization of receptor functions. Obviously, Ca^{2+} and protein kinase C each appear to play diverse roles in controlling cellular processes, and further exploration of the role of this unique protein kinase may provide clues to better understanding of the biochemical basis of tumor promotion.

Acknowledgments. The authors' research has been supported by grants from the Ministry of Education, Science and Culture, Japan. We are grateful to Mrs. S. Nishiyama for her skillful secretarial assistance.

References

Abdel-Latif AA, Akhtar RA, Hawthorne JN (1977) Acetylcholine increases the breakdown of triphosphoinositide of rabbit iris muscle prelabelled with [^{32}P] phosphate. Biochem J 162:61–73

Agranoff BW, Murthy P, Seguin EB (1983) Thrombin-induced phosphodiesteratic cleavage of phosphatidylinositol bisphosphate in human platelets. J Biol Chem 258:2076–2078

Arcoleo JP, Weinstein IB (1985) Activation of protein kinase C by tumor promoting phorbol esters, teleocidin and aplysiatoxin in the absence of added calcium. Carcinogenesis 6:213–217

Ashendel CL, Staller JM, Boutwell RK (1983) Identification of a calcium- and phospholipid-dependent phorbol ester binding activity in the soluble fraction of mouse tissues. Biochem Biophys Res Commun 111:340–345
Berridge MJ, Irvine RF (1984) Inositol trisphosphate, a novel second messenger in cellular signal transduction. Nature 312:315–321
Blumberg PM (1980) In vitro studies on the mode of action of the phorbol esters, potent tumor promoters. CRC Crit Rev Toxicol 8:153–237
Blumberg PM, Jaken S, Konig B, Sharkey NA, Leach KL, Jeng AY, Yeh E (1984) Mechanism of action of the phorbol ester tumor promoters: specific receptors for lipophilic ligands. Biochem Pharmacol 33:933–940
Bruns RF, Lawson-Wendling K, Pugsley TA (1983) A rapid filtration assay for soluble receptors using polyethylenimine-treated filters. Anal Biochem 132:74–81
Castagna M, Takai Y, Kaibuchi K, Sano K, Kikkawa U, Nishizuka Y (1982) Direct activation of calcium-activated, phospholipid-dependent protein kinase by tumor-promoting phorbol esters. J Biol Chem 257:7847–7851
Coussens L, Parker PJ, Rhee L, Yang-Feng TL, Chen E, Waterfield MD, Francke U, Ullrich A (1986) Multiple, distinct forms of bovine and human protein kinase C suggest diversity in cellular signaling pathways. Science 233:859–866
Couturier A, Bazgar S, Castargna M (1984) Further characterization of tumor-promoter-mediated activation of protein kinase C. Biochem Biophys Res Commun 121:448–455
Diamond L, O'Brien TG, Rovera G (1978) Tumor promoters: effects on proliferation and differentiation of cells in culture. Life Sci 23:1979–1988
Diamond L, O'Brien TG, Baird WM (1980) Tumor promoters and the mechanism of tumor promotion. Adv Cancer Res 32:1–74
Dunphy WG, Kochenburger RJ, Castagna M, Blumberg PM (1981) Kinetics and subcellular localization of specific [^{3}H]phorbol-12,13-dibutyrate binding by mouse brain. Cancer Res 41:2640–2647
Fisher SK, Rooijen LAA Van, Agranoff BW (1984) Renewed interest in the polyphosphoinositides. Trends Biochem Sci 9:53–56
Fujiki H, Tanaka Y, Miyake R, Kikkawa U, Nishizuka Y, Sugimura T (1984) Activation of calcium-activated, phospholipid-dependent protein kinase (protein kinase C) by new classes of tumor promoters: teleocidin, and debromoaplysiatoxin. Biochem Biophys Res Commun 120:339–343
Hokin MR, Hokin LE (1953) Enzyme secretion and the incorporation of ^{32}P into phospholipids of pancreatic slices. J Biol Chem 203:967–977
Horowitz AD, Greenebaum E, Weinstein IB (1981) Identification of receptors for phorbol ester tumor promoters in intact mammalian cells and of an inhibitor of receptor binding in biologic fluids. Proc Natl Acad Sci USA 78:2315–2319
Inagaki M, Watanabe M, Hidaka H (1985) N-(2-aminoethyl)-5-isoquinolinesulfonamide, a newly synthesized protein kinase inhibitor, functions as a ligand in affinity chromatography: purification of Ca^{2+}-activated phospholipid-dependent and other protein kinases. J Biol Chem 260:2922–2925
Inoue M, Kishimoto A, Takai Y, Nishizuka Y (1977) Studies on a cyclic nucleotide-independent protein kinase and its proenzyme in mammalian tissues. II. Proenzyme and its activation by calcium-dependent protease from rat brain. J Biol Chem 252:7610–7616
Kaibuchi K, Takai Y, Nishizuka Y (1981) Cooperative roles of various membrane phospholipids in the activation of calcium-activated, phospholipid-dependent protein kinase. J Biol Chem 256:7146–7149
Kaibuchi K, Takai Y, Sawamura M, Hoshijima M, Fujikura T, Nishizuka Y (1983) Synergistic functions of protein phosphorylation and calcium mobilization in platelet activation. J Biol Chem 258:6701–6704
Kaibuchi K, Takai Y, Nishizuka Y (1985) Protein kinase C and calcium ion in mitogenic response of macrophage-depleted human peripheral lymphocytes. J Biol Chem 260:1366–1369
Kikkawa U, Takai Y, Minakuchi R, Inohara S, Nishizuka Y (1982) Calcium-activated, phospholipid-dependent protein kinase from rat brain. Subcellular distribution, purification, and properties. J Biol Chem 257:13341–13348

Kikkawa U, Takai Y, Tanaka Y, Miyake R, Nishizuka Y (1983) Protein kinase C as a possible receptor protein of tumor-promoting phorbol esters. J Biol Chem 258:11442–11445

Kikkawa U, Go M, Koumoto J, Nishizuka Y (1986) Rapid purification of protein kinase C by high performance liquid chromatography. Biochem Biophys Res Commun 135:636–643

Kishimoto A, Takai Y, Mori T, Kikkawa U, Nishizuka Y (1980) Activation of calcium and phospholipid-dependent protein kinase by diacylglycerol, its possible relation to phosphatidylinositol turnover. J Biol Chem 255:2273–2276

Knopf JL, Lee M-H, Sultzman LA, Kriz RW, Loomis CR, Hewick RM, Bell RM (1986) Cloning and expression of multiple protein kinase C cDNAs. Cell 46:491–502

Kreutter D, Caldwell AB, Morin MJ (1985) Dissociation of protein kinase C activation from phorbol ester-induced maturation of HL-60 leukemia cells. J Biol Chem 260:5979–5984

Kuo JF, Andersson RGG, Wise BC, Mackerlova L, Salomonsson I, Brackett NL, Katoh N, Shoji M, Wrenn RW (1980) Calcium-dependent protein kinase: widespread occurrence in various tissues and phyla of the animal kingdom and comparison of effects of phospholipid, calmodulin, and trifluoperazine. Proc Natl Acad Sci USA 77:7039–7043

Lapetina EG, Reep B, Ganong BR, Bell RM (1985) Exogenous *sn*-1,2-diacylglycerols containing saturated fatty acids function as bioregulators of protein kinase C in human platelets. J Biol Chem 260:1358–1361

Leach KL, James ML, Blumberg PM (1983) Characterization of a specific phorbol ester aporeceptor in mouse brain cytosol. Proc Natl Acad Sci USA 80:4208–4212

Michell RH (1975) Inositol phospholipids and cell surface receptor function. Biochim Biophys Acta 415:81–147

Michell RH, Kirk CJ, Jones LM, Downes DP, Creba JA (1981) The stimulation of inositol lipid metabolism that accompanies calcium mobilization in stimulated cells: defined characteristics and unanswered questions. Philos Trans R Soc Lond [Biol] 296:123–137

Minakuchi R, Takai Y, Yu B, Nishizuka Y (1981) Widespread occurrence of calcium-activated, phospholipid-dependent protein kinase in mammalian tissues. J Biochem 89:1651–1654

Miyake R, Tanaka Y, Tsuda T, Kaibuchi K, Kikkawa U, Nishizuka U (1984) Activation of protein kinase C by non-phorbol tumor promoter, mezerein. Biochem Biophys Res Commun 121:649–656

Mori T, Takai Y, Yu B, Takahashi J, Nishizuka Y, Fujikura T (1982) Specificity of the fatty acyl moieties of diacylglycerol for the activation of calcium-activated, phospholipid-dependent protein kinase. J Biochem 91:427–431

Niedel JE, Kuhn LJ, Vandenbark GR (1983) Phorbol diester receptor copurifies with protein kinase C. Proc Natl Acad Sci USA 80:36–40

Nishizuka Y (1983) Phospholipid degradation and signal translation for protein phosphorylation. Trends Biochem Sci 8:13–16

Nishizuka Y (1984a) The role of protein kinase C in cell surface signal transduction and tumour promotion. Nature 308:693–698

Nishizuka Y (1984b) Turnover of inositol phospholipids and signal transduction. Science 225:1365–1370

Nishizuka Y (1986) Perspectives on the role of protein kinase C in stimulus-response coupling. J Natl Cancer Inst 76:363–370

Nomura H, Ase K, Sekiguchi K, Kikkawa U, Nishizuka Y (1986) Stereospecificity of diacylglycerol for stimulus-response coupling in platelets. Biochem Biophys Res Commun 140:1143–1151

Ohno S, Kawasaki H, Imajoh S, Suzuki K, Inagaki M, Yokokura H, Sakoh T, Hidaka H (1987) Tissue-specific expression of three distinct types of rabbit protein kinase C. Nature 325:161–166

Ono Y, Kurokawa T, Fujii T, Kawahara K, Igarashi K, Kikkawa U, Ogita K, Nishizuka Y (1986) Two types of complementary DNAs of rat brain protein kinase C. Heterogeneity determined by alternative splicing. FEBS Lett 206:347–352

Parker PJ, Stabel S, Waterfield MD (1984) Purification to homogeneity of protein kinase C from bovine brain – identity with the phorbol ester receptor. EMBO J 3:953–959
Parker PJ, Coussens L, Totty N, Rhee L, Young S, Chen E, Stabel S, Waterfield MD, Ullrich A (1986) The complete primary structure of protein kinase C – the major phorbol ester receptor. Science 233:853–859
Rando RR, Young N (1984) The stereospecific activation of protein kinase C. Biochem Biophys Res Commun 122:818–823
Sando JJ, Young MC (1983) Identification of high-affinity phorbol ester receptor in cytosol of EL4 thymoma cells: requirement for calcium, magnesium, and phospholipids. Proc Natl Acad Sci USA 80:2642–2646
Schatzman RC, Raynor RL, Fritz RB, Kuo JF (1983) Purification to homogeneity, characterization and monoclonal antibodies of phospholipid-sensitive Ca^{2+}-dependent protein kinase from spleen. Biochem J 209:435–443
Sharkey NA, Leach KL, Blumberg PM (1984) Competitive inhibition by diacylglycerol of specific phorbol ester binding. Proc Natl Acad Sci USA 81:607–610
Solanki V, Slaga TJ (1981) Specific binding of phorbol ester tumor promoters to intact primary epidermal cells from Sencar mice. Proc Natl Acad Sci USA 78:2549–2553
Takai Y, Kishimoto A, Iwasa Y, Kawahara Y, Mori T, Nishizuka Y (1979a) Calcium-dependent activation of a multifunctional protein kinase by membrane phospholipids. J Biol Chem 254:3692–3695
Takai Y, Kishimoto A, Kikkawa U, Mori T, Nishizuka Y (1979b) Unsaturated diacylglycerol as a possible messenger for the activation of calcium-activated, phospholipid-dependent protein kinase system. Biochem Biophys Res Commun 91:1218–1224
Takai Y, Kikkawa U, Kaibuchi K, Nishizuka Y (1984) Membrane phospholipid metabolism and signal transduction for protein phosphorylation. Adv Cyclic Nucleotide Protein Phosphorylation Res 18:119–158
Tanaka Y, Miyake R, Kikkawa U, Nishizuka Y (1986) Rapid assay of binding of tumor-promoting phorbol esters to protein kinase C. J Biochem 99:257–261
Uchida T, Filburn CR (1984) Affinity chromatography of protein kinase C-phorbol ester receptor on polyacrylamide-immobilized phosphatidylserine. J Biol Chem 259:12311–12314
Wise BC, Raynor RL, Kuo JF (1982a) Phospholipid-sensitive Ca^{2+}-dependent protein kinase from heart. I. Purification and general properties. J Biol Chem 257:8481–8488
Wise BC, Glass DB, Chou CHJ, Raynor RL, Katoh N, Schatzman RC, Turner RS, Kibler RF, Kuo JF (1982b) Phospholipid-sensitive Ca^{2+}-dependent protein kinase from heart. II. Substrate specificity and inhibition by various agents. J Biol Chem 257:8489–8495
Wolf M, Sahyoun N, Levine III H, Cuatrecasas P (1984) Protein kinase C: rapid enzyme purification and substrate-dependence of the diacylglycerol effect. Biochem Biophys Res Commun 122:1268–1275
Yamanishi J, Takai Y, Kaibuchi K, Sano K, Castagna M, Nishizuka Y (1983) Synergistic functions of phorbol ester and calcium in serotonin release from human platelets. Biochem Biophys Res Commun 112:778–786

Calcium and Physiological Function

CHAPTER 14

Drugs Acting on Calcium Channels

T. NARAHASHI

A. Introduction

Drugs acting on calcium channels have been studied extensively. Calcium antagonists or calcium channel blockers have been the subject of both basic and clinical investigations for some time, and some of them have been developed into useful therapeutic drugs for the treatment of hypertension and coronary disorders (FLECKENSTEIN 1977; TRIGGLE 1981). However, most of these studies were performed with classical pharmacologic techniques involving measurement of muscle contractions and tension. It was not until the early 1980s that the drugs acting on calcium channels became a subject of modern electrophysiologic investigations using voltage-clamp techniques (HAGIWARA and BYERLY 1981, 1983; TSIEN 1983; PAPPONE and CAHALAN 1986; REUTER 1983; SKERRITT and MACDONALD 1984; YAKSH and NOUEIHED 1985; WERZ and MACDONALD 1983a, 1985; MACDONALD and WERZ 1986).

The study of the physiology of calcium channels has had a longer history, however. The pioneering work by Hagiwara and his associates on the calcium channels of barnacle muscle fibers has set a milestone in the history of the modern electrophysiology of calcium channels (HAGIWARA et al. 1964; HAGIWARA and NAKA 1964; HAGIWARA and BYERLY 1981). Since then snail neurons have also been used for the study of calcium channels. These neurons possess calcium channels as well as sodium channels to generate action potentials (GEDULDIG and JUNGE 1968, 1970; KERKUT and GARDNER 1967; KOKETSU et al. 1959). Cardiac muscles have also been a subject of modern electrophysiologic investigations with voltage-clamp techniques. Earlier studies, however, encountered formidable difficulties partly because of the complex geometry of the cells (BEELER and MCGUIGAN 1978; REUTER 1979).

In this chapter, some of our recent developments in the study of calcium channel pharmacology will be presented. The study deals with the mechanisms whereby certain ions and drugs block the calcium channels of neuronal membranes. In order to understand the pharmacology of calcium channels, it is imperative to know the basic physiologic properties of calcium channels. Therefore, the physiology of calcium channels is also described briefly.

B. Methods

I. Materials

The experiments were performed with cultured neuroblastoma cells (N1E-115 line) or neuroblastoma × glioma hybrid cells (NG108-15 line). The N1E-115 cells were grown in large Corning tissue culture flasks and were fed every other day during logarithmic growth with Dulbecco's modified Eagle's medium containing 10% fetal of newborn calf serum. Cells were maintained under a humidified atmosphere containing 10% CO_2. At confluence, the cells were first dissociated by mechanical agitation, and placed onto glass coverslips in 60-mm tissue culture dishes which contained 2% dimethylsulfoxide (DMSO). Cells were fed every few days and could be maintained in media containing 2% DMSO for up to 3 weeks. DMSO has been found to cause neuroblastoma cells to extend neurites and develop a highly excitable membrane (KIMHI et al. 1976). Cells were utilized for experiments following growth in DMSO-containing media for 3 days to 2 weeks. The neuroblastoma × glioma hybrid cells were cultured for 5–10 days in the presence of 1 m*M* dibutyryladenosine-3′,5′-cyclic monophosphate to facilitate development of calcium channels and chemical sensitivity (KATO and NARAHASHI 1982).

II. Electrical Recording

A patch electrode voltage-clamp technique was used to record membrane currents from the whole cell (HAMILL et al. 1981; NARAHASHI et al. 1987). A patch electrode with resistance of 0.8–2.0 MΩ was placed onto the cell surface, and the membrane under the pipette tip was ruptured by applying gentle suction to the pipette. The potential inside the cell was kept at ground level. An inverted command voltage, i.e., holding potential plus command pulse, was applied to the external solution via a bath electrode which was composed of 3 *M* KCl agar/Ag–AgCl. Currents were recorded by a current-to-voltage converter which was directly connected to the suction pipette via an Ag–AgCl pellet. Part of the output voltage of the current recording was added to the command voltage to compensate for the series resistance. Pulses were generated with the use of a digital-to-analog converter driven by an 8-bit microcomputer. Unless otherwise stated, linear leakage components determined at membrane potentials ranging between −120 and −60 mV have been substracted from the current records.

III. Solutions

In order to record calcium channel currents, the current components passing through other channels must be eliminated. To suppress the outward potassium current, 25 m*M* tetraethylammonium (TEA) was added to the external media and K^+ was replaced with Cs^+ in both external and internal solutions. Sodium channels were blocked by 0.5 μ*M* tetrodotoxin (TTX) in the external media.

For most of the calcium channel experiments, a high Ba^{2+} (50 m*M*) solution was used as the standard external solution for several reasons: (a) Ba^{2+} is usually more permeant than Ca^{2+} through calcium channels (HAGIWARA and BYERLY

1981); (b) calcium-induced inactivation of calcium channels is much less obvious in Ba^{2+} media than in Ca^{2+} media (ECKERT and CHAD 1984); and (c) external Ba^{2+} suppresses potassium currents through the delayed rectifier (ARMSTRONG et al. 1982) and through the anomalous rectifier (HAGIWARA et al. 1978).

The standard external solution contained: $BaCl_2$ 50 m*M*; NaCl 30 m*M*; CsCl 5 m*M*; glucose 25 m*M*; TEA-Cl 25 m*M*; and TTX 0.5 μ*M*. The pH was adjusted to 7.3–7.4 with 5 m*M* HEPES-Cl. The internal solution contained: Cs-glutamate 130 m*M*; $MgCl_2$ 2.5 m*M*; and glucose 5 m*M*. The pH was adjusted to 7.0–7.1 with 10 m*M* PIPES-Na.

C. Two Types of Calcium Channels

I. Initial Study

Our initial study of calcium channels was performed with N1E-115 neuroblastoma cells using the whole-cell patch-clamp technique combined with intracellular dialysis (QUANDT and NARAHASHI 1980, 1984). Kinetics of both sodium and calcium channels were analyzed in detail in terms of the activation (*m*) and inactivation (*h*) parameters. At that time we noticed interesting behavior of calcium channel currents as a function of time and membrane potential. In the light of information gained later, the calcium channel currents without much inactivation, as observed with large depolarizing steps, are indicative of type II calcium channel currents. However, our analyses at that time were limited to the inactivating or type I calcium channels.

Calcium channel currents were measured after external and internal K^+ was replaced with Cs^+ to eliminate currents through potassium channels and after adding TTX to the external solution to block the sodium channel. Since calcium channel currents are considerably smaller than potassium channel currents, it is important to ascertain that there is no contamination of the recorded calcium channel current with the outward potassium channel current. If there is no such contamination, it would be expected that the time course of calcium current, measured in response to a step depolarization, should be the same as that of the envelope of tail currents, measured at a potential which would tend to reduce any contribution of residual current through the potassium channel. Thus, the tail currents associated with a step repolarization to −80 mV at various times during a step depolarization were measured. The result illustrated in Fig. 1 clearly shows that the time courses of calcium channel current, as measured by the two methods, are identical. This demonstrates that the calcium channel current thus measured is not contaminated by potassium channel current.

The kinetics of the activation and inactivation of calcium channels are illustrated in Fig. 2. The time course of inactivation is described by a single exponential function with a time constant of 15 ms at 0 mV (Fig. 2a). The time course of activation is fitted by a fourth-order process (Fig. 2b). The time constants for activation (*m*) and inactivation (*h*) are plotted as a function of membrane potential in Fig. 2c.

It is well known that the inactivation of calcium channels of certain preparations is dependent on the intracellular calcium concentration (ECKERT and

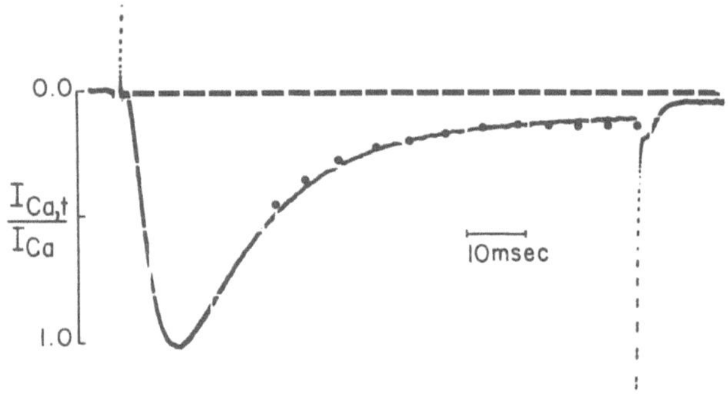

Fig. 1. Comparison of the time course of calcium channel current measured by tail current with that obtained by a step depolarization. A neuroblastoma cell was voltage clamped and perfused internally with Cs^+ internal solution and externally with high Ca^{2+} (20 m*M*) solution. Calcium current was measured in response to a step depolarization to 0 mV. The current as a function of time ($I_{Ca,t}$) for an 80-ms depolarization is plotted after normalizing to the peak current (I_{Ca}). The *circles* indicate the amplitude of initial calcium tail currents in response to the repolarization to −80 mV after various times from the beginning of the depolarization. The tail current marked by the *arrow* was scaled to match $I_{Ca,t}$ which occurred at the corresponding time during the maintained depolarization. All other tail current amplitudes were scaled by this same factor. Note that the envelope of tail currents matches that of the calcium current. Temperature, 30 °C. (QUANDT and NARAHASHI 1984)

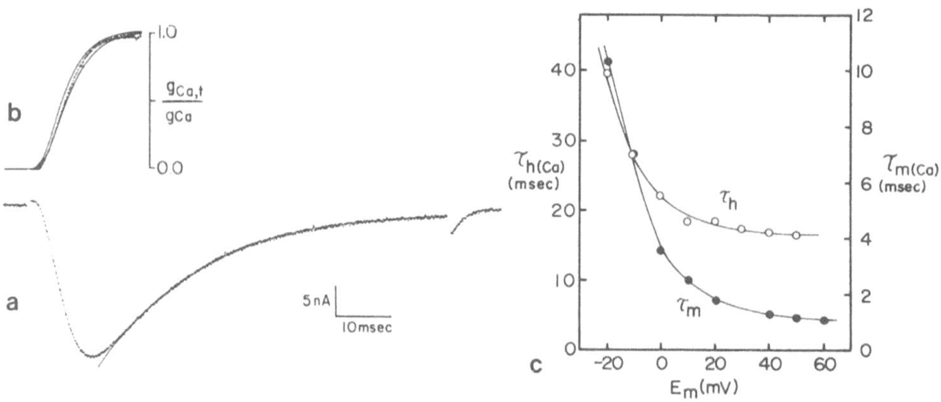

Fig. 2 a–c. Kinetics of type I calcium channel activation and inactivation measured with 80-ms step depolarizations in a neuroblastoma cell. Internal Cs^+ and external high Ca^{2+} solutions were used. Net currents were obtained by substracting capacitive and leakage currents in the presence of 1 m*M* Cd^{2+}. **a** calcium current recorded at 0 mV with a superimposed best estimate of the time course of calcium channel inactivation (*full line*), as determined by a multiexponential, least-squares computeranalysis, **b** the rising phase of calcium conductance ($g_{Ca,t}$) at 0 mV, normalized to its maximum value (g_{Ca}), is shown following the removal of inactivation mathematically. The *full lines* plot two fourth-power functions, with time constants of either 3.5 or 4 ms, **c** the time constants of calcium current activation, $\tau_{m(Ca)}$ (*full circles*), and inactivation, $\tau_{h(Ca)}$ (*open circles*), are given as a function of membrane potential. Curves are drawn by eye. Temperature, 28 °C. (QUANDT and NARAHASHI 1984)

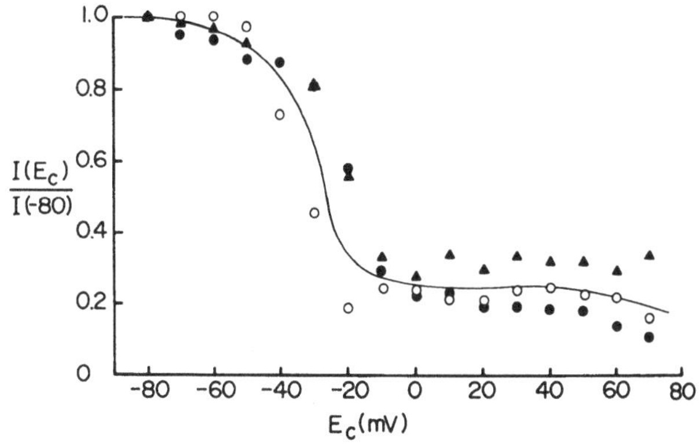

Fig. 3. Voltage-dependent calcium channel inactivation in neuroblastoma cells as measured by a two-pulse protocol. Data from two experiments with one cell (*open circles, full triangles*) and one experiment with another cell (*full circles*) are given. Calcium currents were recorded in high external Ca^{2+} and internal Cs^{+} solutions. An inactivating prepulse 200 ms in duration was followed by an 80-ms test pulse either to 0 mV (*open circles, full circles*) or to +10 mV (*full triangles*), during which peak calcium current ($I(E_c)$) was measured. The interpulse interval was 60 ms (*open circles, full circles*) or 100 ms (*full triangles*), and each pair of pulses was repeated following an interval of 6 s. $I(E_c)$ is normalized to the peak current with the prepulse (E_c) equal to the holding potential. Note that inactivation is not reduced as prepulse potentials approach the zero current potential (typically +60 or +75 mV). The *full line* plots the mean value at each potential. Temperature, 29.5 °C. (QUANDT and NARAHASHI 1984)

EWALD 1982; TILLOTSON 1979; HAGIWARA and BYERLY 1981; ASHCROFT and STANFIELD 1980; BREHM and ECKERT 1978; DUNLAP and FISCHBACH 1981). In neuroblastoma cells, however, three sets of observations support the notion that the observed inactivation is not due to intracellular calcium, but to a voltage-gated mechanism. First, no quantitative differences were found for the calcium currents recorded in the presence and absence of EGTA in the internal media. Second, inactivation did not subside at the membrane potentials at which calcium influx decreased. As is shown in Fig. 3, the inactivation remained almost constant at membrane potentials more positive than −20 mV. Third, inactivation was still observed when Sr^{2+} or Ba^{2+} was used as a carrier instead of Ca^{2+}. Thus, the inactivation of calcium channels in neuroblastoma cells is not due to calcium influx, but to a voltage-gated mechanism.

II. Separation of Two Types of Calcium Channels

During the past few years, it has become increasingly evident that there are at least two types of calcium channels in various neurons. In earlier studies, LLINÁS and YAROM (1981) postulated the existence of two types of calcium channels in mammalian inferior olivary neurons, and FISHMAN and SPECTOR (1981) found a second component of the inward calcium current in neuroblastoma cells. More recently, two or three types of calcium channels have been reported in various neurons, including dorsal root ganglion cells (CARBONE and LUX 1984; NOWYCKY

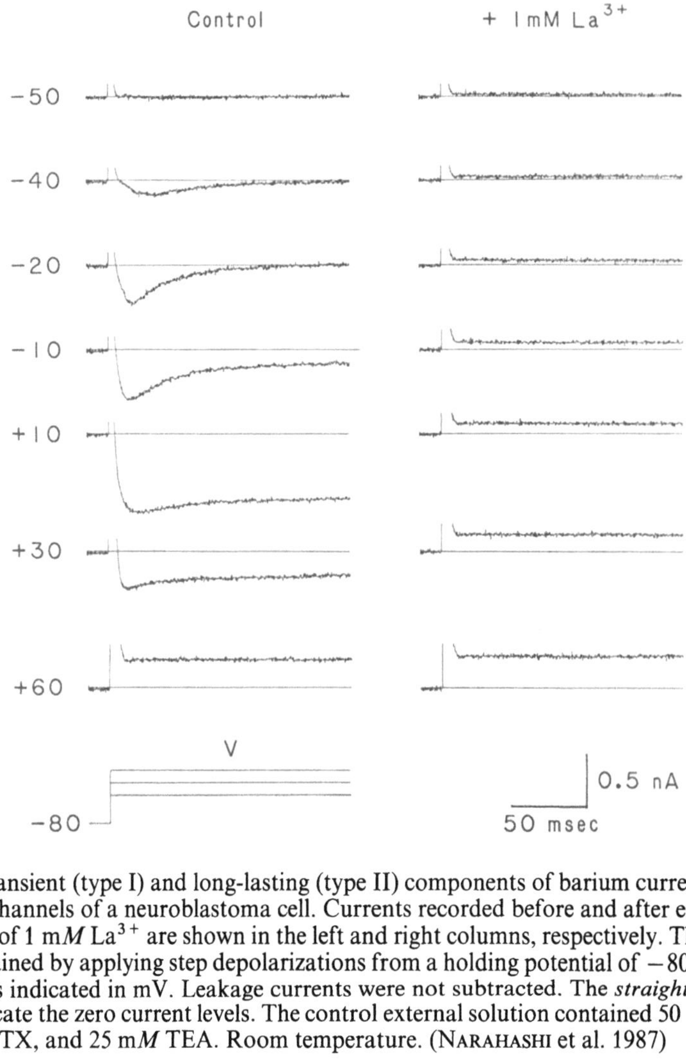

Fig. 4. Transient (type I) and long-lasting (type II) components of barium current through calcium channels of a neuroblastoma cell. Currents recorded before and after external application of 1 m*M* La^{3+} are shown in the left and right columns, respectively. The currents were obtained by applying step depolarizations from a holding potential of −80 mV to the potentials indicated in mV. Leakage currents were not subtracted. The *straight horizontal lines* indicate the zero current levels. The control external solution contained 50 m*M* Ba^{2+}, 0.5 μ*M* TTX, and 25 m*M* TEA. Room temperature. (NARAHASHI et al. 1987)

et al. 1985; FEDULOVA et al. 1985; ARMSTRONG and MATTESON 1985). Two types of calcium channels have also been reported in the egg of the annelid, *Neanthes* (FOX and KASNE 1984) and in the atrial muscle of the heart (BEAN 1985).

Step depolarizations to various membrane potential levels generated a transient and a long-lasting inward barium current (Fig. 4). The transient component of the current appeared at potentials more positive than −50 mV, whereas the long-lasting component appeared only with larger depolarizations, to potentials more positive than −20 mV. Both components were blocked by 1 m*M* La^{3+} (Fig. 4). Therefore, these currents flowed through calcium channels.

Since the transient (type I) calcium channel current inactivates quickly, and also since the long-lasting (type II) calcium channel current is generated at poten-

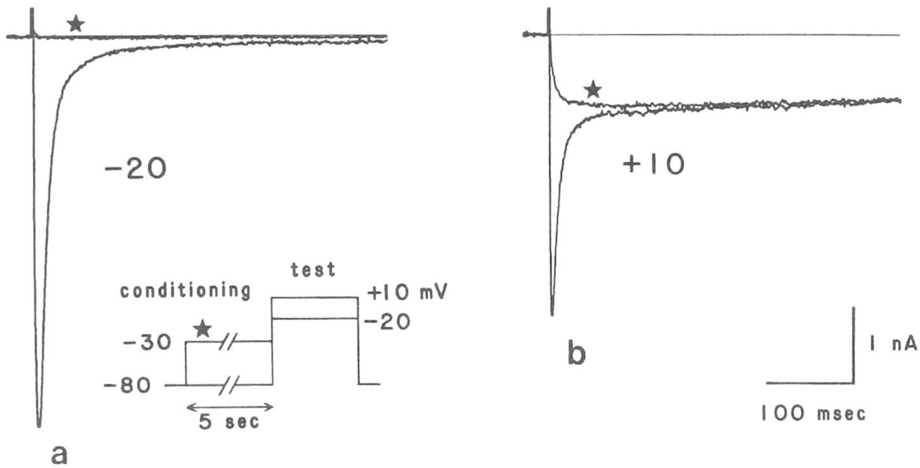

Fig. 5 a, b. Separation of two components of calcium channel currents in a neuroblastoma cell. The transient (type I) component was inactivated by a conditioning depolarization. **a** currents associated with step test depolarizations to −20 mV from a holding potential of −80 mV and from a 5-s conditioning pulse to −30 mV (*star*) are superimposed, **b** same as **a** except that step test depolarizations to +10 mV were used. Leakage currents have been subtracted. Temperature 33.1 °C. (NARAHASHI et al. 1987)

tials more positive than those for activation of the type I current, separation of the two components can be achieved easily by manipulating the holding potential and depolarized level. An example of such a separation is illustrated in Fig. 5. When the membrane was depolarized to −20 mV from a holding potential of −80 mV, a transient inward current was produced and followed by a very small residual current (Fig. 5a). However, when the membrane was held at −30 mV, no measurable inward current was generated during depolarization to −20 mV, because the type I channels were completely inactivated. Membrane depolarization to +10 mV from a holding potential of −80 mV caused both transient and long-lasting inward currents to be generated as both types of calcium channels were activated at that depolarized level (Fig. 5b). When conditioned at −30 mV, however, only the long-lasting component of calcium channel current was produced upon depolarization to +10 mV, because the type I channels were inactivated.

III. Kinetics of Two Types of Calcium Channels

Kinetics of activation and inactivation of the two types of calcium channels have been studied in detail (YOSHII et al. 1985a). A plot of the peak amplitude of transient current against the membrane potential yielded a current–voltage (*I–V*) relationship which was not linear at positive potentials ranging from 0 to +60 mV. Therefore, the constant field theory (GOLDMAN 1943; HODGKIN and KATZ 1949) was used to predict the maximum current amplitude of the type I channel at its full activation (OHMORI and YOSHII 1977). The ratio of the observed current am-

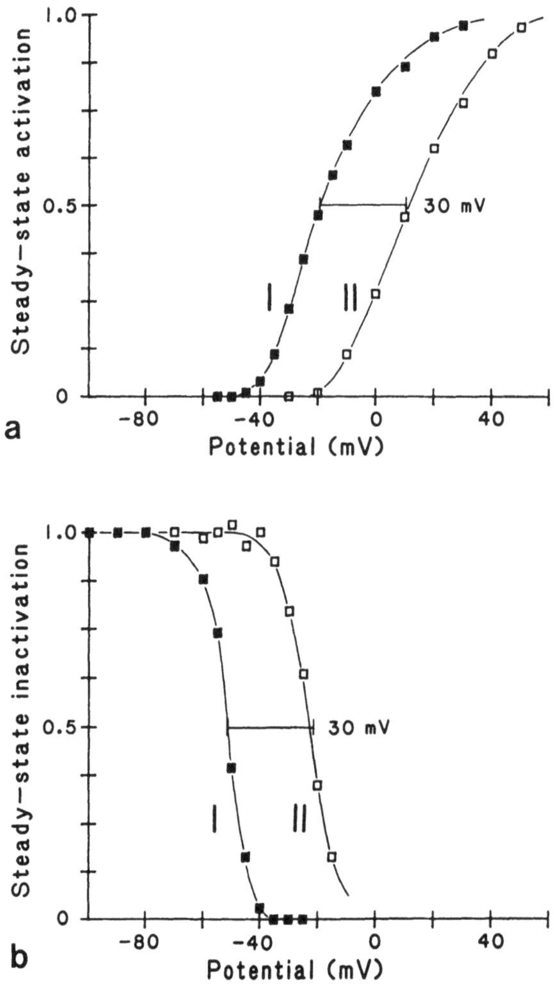

Fig. 6 a, b. Steady state activation (**a**) and steady state inactivation (**b**) of type I (*full squares*) and type II (*open squares*) calcium channels in neuroblastoma cells. **a** for type I channels, the ratio of the current amplitude to the maximum value predicted by the constant field equation is plotted against the membrane potential. For type II channels, the relative amplitude of the instantaneous (tail) current associated with step repolarization is plotted against the membrane potential; *full curves* are drawn by eye, **b** the relative amplitude of the two types of currents is plotted as a function of the membrane potential of the 10-s conditioning pulse. Test pulse: −20 mV for type I, +10 mV for type II; *full curves* are drawn by eye. Temperature, 33.1 °C for type I and 34 °C for type II

plitude to the calculated curve yielded a steady state activation level for a given potential. The activation–membrane potential curve thus calculated is shown in Fig. 6a (*full squares*).

The method of curve fitting as described for the type I calcium channels could not be used for the type II channels because of possible contamination by outward currents at large positive potentials. Therefore, the tail currents associated with repolarizations to a fixed potential (−30 mV) from various levels of conditioning depolarized potentials were measured. The initial amplitude of the tail current thus measured was normalized to the maximum value and is plotted as a function of the conditioning potential in Fig. 6a (*open squares*). The activation curves for type I and type II calcium channels are separated by 30 mV.

The steady state inactivation curve was obtained by plotting the amplitude of current associated with a test depolarizing pulse following a conditioning pulse to various potentials. The type II calcium channel lacked fast inactivation, and inactivated very slowly. Therefore, a conditioning pulse lasting for 10 s was used. For type I channels, the test pulse was selected at −20 mV to observe a maximal current without contamination of the type II current. For type II channels, the test pulse to +10 mV was used to obtain a maximal current. The steady state inactivation curves thus obtained for type I and type II channels are plotted in Fig. 6b as *full* and *open squares*, respectively. Like the activation kinetics, the inactivation curves for the two types are spearated by 30 mV.

IV. Ionic Selectivity

Calcium channels are generally permeable not only to Ca^{2+}, but also to Sr^{2+} and Ba^{2+}, and both types of calcium channels of neuroblastoma cells were no exception. In order to estimate relative permeability to these three divalent cations, the current amplitudes of both types of calcium channels were compared in the presence of 50 m*M* Ca^{2+}, Sr^{2+}, or Ba^{2+} (Yoshii et al. 1985a). The permeability ratios thus obtained were $P_{Ba}:P_{Sr}:P_{Ca} = 1.0:1.0:0.7$ for type I channels and 1.0:0.7:0.3 for type II. Therefore, the ionic selectivity is somewhat different between the two types of channels.

V. Sensitivity to Cyclic AMP

Certain types of calcium channels are known to be modulated by the intracellular level of cyclic AMP. The calcium channels that generate fast current in cardiac cells are stimulated by a β-adrenergic agonist through an increase in the intracellular AMP (Reuter 1983; Tsien 1983; Bean 1985). Cyclic AMP is also required to maintain the activity of type II-like calcium channels in rat sensory neurons (Fedulova et al. 1985).

In our experiments with neuroblastoma cells, dibutyryl cyclic AMP, applied externally at a concentration of 1 m*M*, increased the amplitude of type II calcium channel current without changing the type I current (Narahashi et al. 1987). The results suggest that the activity of type II calcium channels, but not type I calcium channels, depends on the intracellular level of cyclic AMP.

D. Pharmacology of Calcium Channels

I. Polyvalent Cations

It is well established that divalent and trivalent cations such as Cd^{2+}, Ni^{2+}, Co^{2+}, and La^{3+} block calcium channels in various tissues (Hagiwara and Byerly 1981). We have also found that these polyvalent cations block the cal-

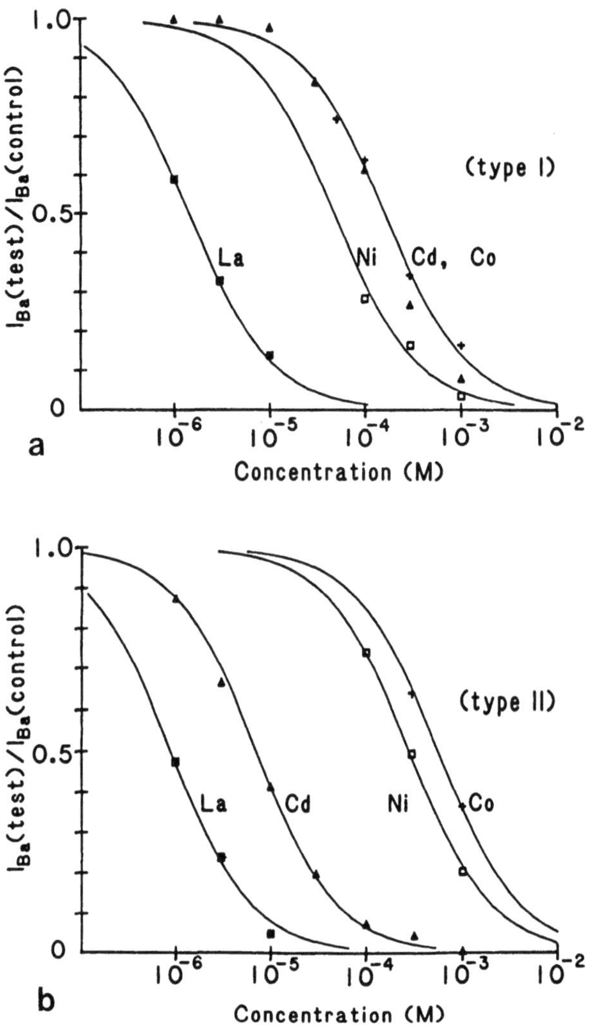

Fig. 7 a, b. Dose–response relationships for the action of polyvalent cations in blocking the two components of barium current through calcium channels in neuroblastoma cells. Relative amplitudes of the transient inward current (type I) (**a**) and long-lasting inward currents (type II) (**b**) recorded at −20 mV and +10 mV, respectively, are plotted against the concentration of the blocking cations. *Full curves* are drawn based on the one-to-one binding stoichiometry. Note a marked difference in blocking potencies of Cd^{2+} between the two types of channels. Room temperature. (NARAHASHI et al. 1987)

cium channels in neuroblastoma cells (NARAHASHI et al. 1987). Whereas both types of channels were blocked by the polyvalent cations, a large difference in blocking potency between the two types of channels was found for Cd^{2+}. Dose–response curves for each channel type are shown in Fig. 7. The sequence of blocking potencies for type I calcium channels is (K_d in μM): La^{3+} (1.5) $\gg$ Ni^{2+} (47) > Cd^{2+} (160) = Co^{2+} (160). The sequence for type II channels is: La^{3+} (0.9) > Cd^{2+} (7.0) $\gg$ Ni^{2+} (280) > Co^{2+} (560). Thus, Cd^{2+} is 23 times more potent on type II channels than type I, whereas Ni^{2+} is 6 times less potent on type II than type I.

The differential blocking action of Cd^{2+} was also found in two kinds of calcium channels in sensory neurons (NOWYCKY et al. 1985). The high potency of Cd^{2+} as compared with Co^{2+} was reported in rat brain synaptosomes (NACHSHEN 1984). The calcium channels in squid presynaptic terminals undergo very slow inactivation (LLINÁS et al. 1981). The last two observations, together with the results of the present study, lend support to the notion that the type II calcium channels are present in the nerve terminals.

II. Opioid Peptides

Calcium channels play an important role in the action of opioids on synaptic transmission. Opioids act on both presynaptic and postsynaptic elements (YAKSH and NOUEIHED 1985; NORTH 1979; WOUTERS and VAN DEN BERCKEN 1980). The presynaptic action is represented by an inhibition of transmitter release, and the postsynaptic action includes membrane hyperpolarization.

At least two major actions of opioids at the cellular and membrane level have been disclosed. One is an inhibition of voltage-gated calcium channels, leading to a shortening of the calcium-dependent action potential, which in turn would cause a decrease in transmitter release from nerve terminals (WERZ and MACDONALD 1982a, b, 1983a, b, 1984, 1985; MACDONALD and WERZ 1986; MUDGE et al. 1979). The inhibition of transmitter release is due to a decrease in quantal content without change in quantal size (KATAYAMA and NISHI 1984; MACDONALD and NELSON 1978). The other is an activation of potassium channels which in turn leads to membrane hyperpolarization (KATAYAMA and NISHI 1984; YOSHIMURA and NORTH 1983; NORTH and WILLIAMS 1985; NORTH et al. 1979), an increase in afterhyperpolarization (TOKIMASA et al. 1981), or a shortening of calcium-dependent action potentials (NORTH and WILLIAMS 1983; WILLIAMS and NORTH 1983).

Patch-clamp experiments with neuroblastoma × glioma hybrid cells (NG108-15) have clearly demonstrated the opioid-induced block of calcium channels (TSUNOO et al. 1986). One of the most important findings is a selective block of type II calcium channels without any effect on type I, nor on sodium and delayed rectifying potassium channels. In order to demonstrate that opioids block a specific receptor, naloxone was tested for its antagonistic effect. Figure 8 shows an example of such an experiment. The control current was evoked by a depolarizing step from a holding potential of -80 mV to $+10$ mV. Thus, the current record includes both type I and type II calcium channel currents. [D-Ala^2, D-Leu^5]-enkephalin at a concentration of 25 nM suppressed the current. After washing out of enkephalin, 1 μM naloxone was applied without causing any effect. Then 25 nM enkephalin was added to the naloxone, again without causing any change in current. In separate experiments in which type I and type II calcium channel currents were recorded separately, the apparent dissociation constant for enkephalin block of type II calcium channels was estimated to be 8.8 nM. Enkephalin had no effect on type I calcium channels, sodium channels, and delayed rectifying potassium channels. In contrast to Leu-enkephalin which is a δ opiate receptor agonist, morphine, a μ opiate receptor agonist, suppressed calcium channel currents

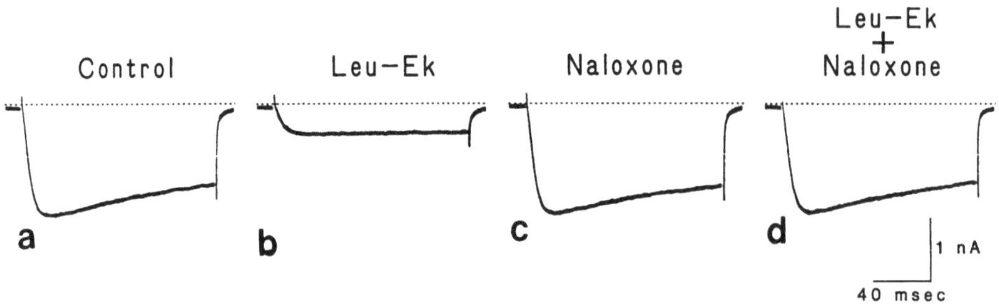

Fig. 8 a–d. Naloxone antagonism of Leu-enkephalin block of calcium channels in a neuroblastoma × glioma hybrid cell. Calcium channel currents were evoked by a depolarization from a holding potential of −80 mV to +10 mV. **a** control, **b** 3 min after exposure to 25 n*M* Leu-enkephalin showing block; then, enkephalin was washed out with a saline solution containing 1 μ*M* naloxone, **c** 5 min after naloxone treatment showing no effect, **d** 3 min after reapplication of 25 n*M* enkephalin in the presence of naloxone showing antagonism. *Broken lines* represent zero current level. Room temperature. (TSUNOO et al. 1986)

only slightly, 1 μ*M* causing only an 18% inhibition of type II calcium channel current and a 21% inhibition of type I current. Thus, Leu-enkephalin suppresses type II calcium channel currents via a δ opiate receptor.

Enkephalin block of type II calcium channels proceeded with a unique time course. An example of such an experiment is shown in Fig. 9. The type I calcium channels were completely inactivated by applying a 10-s conditioning depolarizing pulse to −30 mV. Leu-enkephalin at a concentration of 25 n*M* suppressed the type II current, as can be seen in the records at −10 and 0 mV. At 10, 20, and 30 mV, however, the current in the presence of enkephalin recovered with a slow time course during a depolarizing step. At 40 and 50 mV, while the current was suppressed somewhat by enkephalin during the initial phase of the depolarizing step, it was restored almost completely by the end of the depolarization.

The records of type II calcium channel currents shown in Fig. 9 indicate three important features of enkephalin block. First, when the current amplitude in enkephalin, normalized to the control at each membrane potential, was plotted as a function of the duration of pulse, the current at the beginning of the pulse was suppressed to 27% of the control, regardless of the membrane potential (Fig. 10). This block represents a "resting block" or "closed channel block." Second, during the depolarizing step, the current amplitude recovered toward a steady state, the level of which depended on the membrane potential. Third, the time course of the recovery, which was composed of an initial delay (t^*) and an exponential increase in amplitude, also depended on the membrane potential. The results clearly indicate that the enkephalin block of type II calcium channels is voltage dependent, being intensified with hyperpolarization, and that unblocking upon depolarization occurs slowly with an initial delay followed by an exponential time course in a manner dependent upon the membrane potential.

Analysis of the dose–response relationship for enkephalin block of type II calcium channels indicated that enkephalin interacted with a receptor on a one-to-one stoichiometric basis with an apparent dissociation constant of 8.8 n*M* at 0 mV. Even at 0.5 n*M*, enkephalin caused a noticeable inhibition of the current.

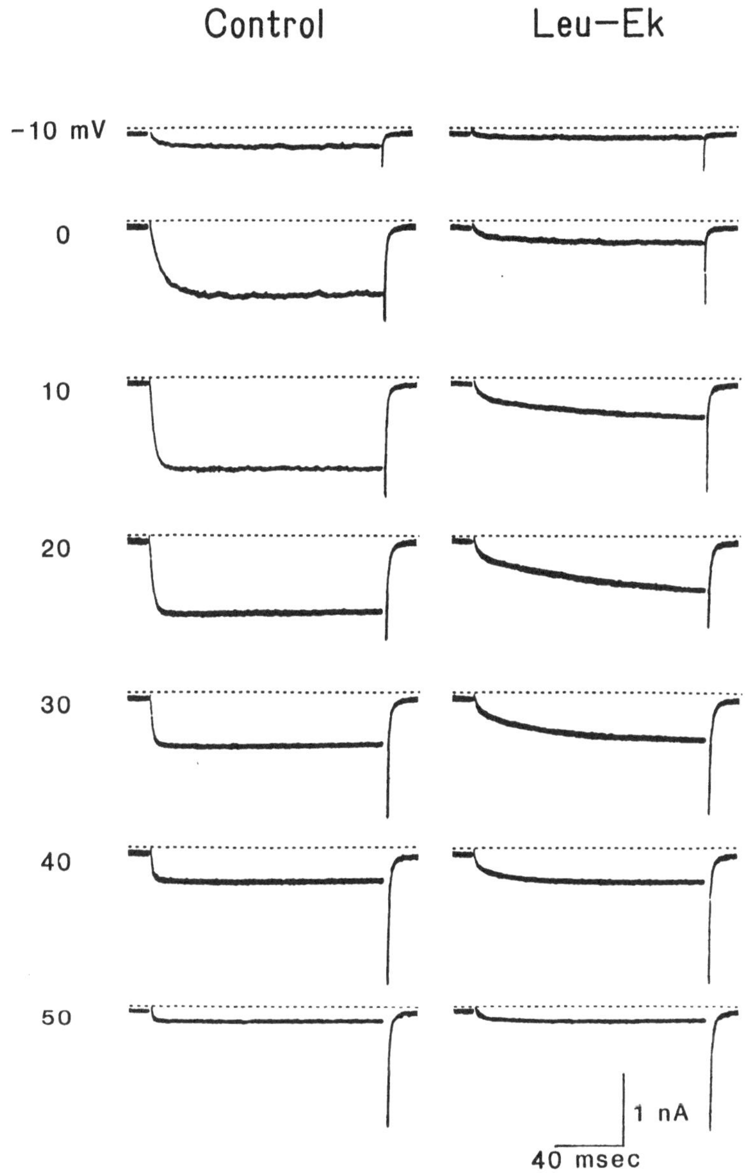

Fig. 9. Effect of Leu-enkephalin on type II calcium channel currents in a neuroblastoma × glioma hybrid cell. Currents were evoked by step depolarizing pulses to the levels indicated on the left following a 10-s conditioning pulse to −30 mV to inactivate type I currents. Left column, control; right column, 3–5 min after bath application of 25 n*M* Leu-enkephalin. *Broken lines* represent zero current level. Note suppression of current and time-dependent recovery in enkephalin at certain membrane potentials. Room temperature. (TSUNOO et al. 1986)

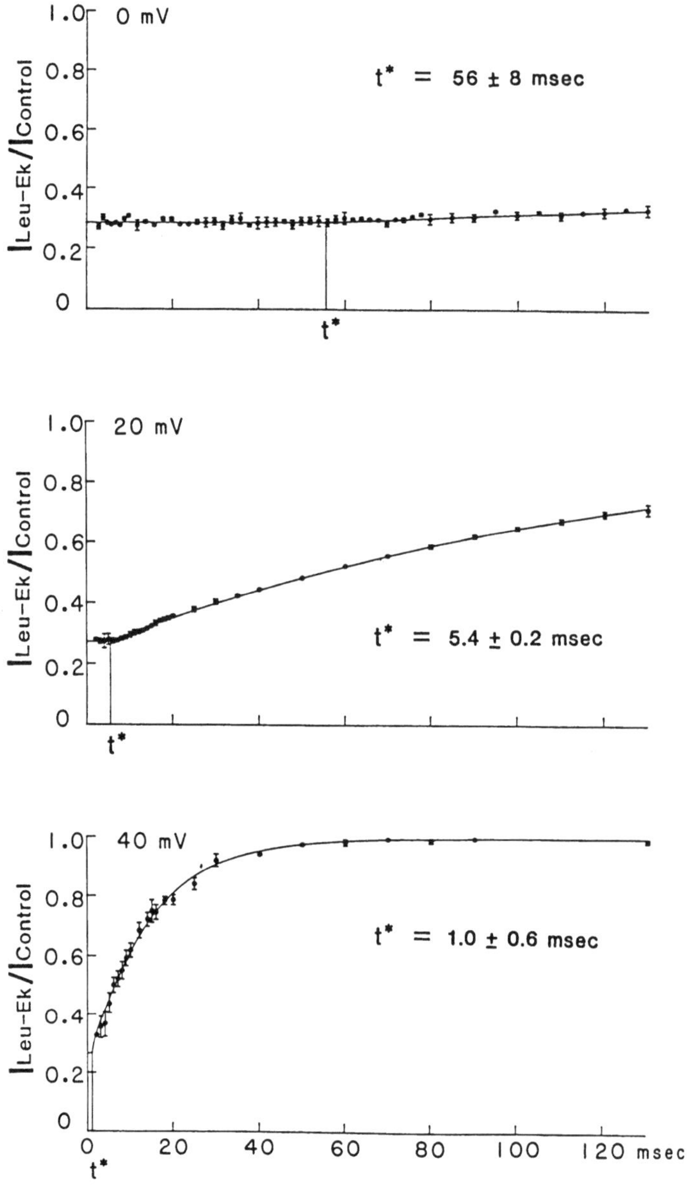

Fig. 10. Time courses of type II calcium channel currents during step depolarizations to the three levels indicated from a holding potential of −30 mV in a neuroblastoma × glioma hybrid cell under the influence of 25 n*M* Leu-enkephalin. Data were taken from current records shown in Fig. 9. The current amplitudes in the presence of enkephalin ($I_{Leu\text{-}Ek}$) relative to those in control ($I_{Control}$) are plotted as a function of time; t^* indicates the delay of onset of recovery. Room temperature. (TSUNOO et al. 1986)

It should be noted that, in view of the voltage-dependent nature of enkephalin block, the apparent dissociation constant would be lower if measurements were made at more negative potentials.

The observed inhibition of type II calcium channels by Leu-enkephalin would have a significant influence on the transmitter release from nerve terminals. It is becoming increasingly evident that type II calcium channels are involved in excitation–secretion coupling (LLINÁS et al. 1981; LLINÁS and YAROM 1981; TSIEN 1983). Therefore, the enkephalin block of type II channels could explain the mechanism underlying the presynaptic inhibition of transmitter release which has been observed before (KATAYAMA and NISHI 1984; MACDONALD and NELSON 1978; YAKSH and NOUEIHED 1985; NORTH 1979; WOUTERS and VAN DEN BERCKEN 1980). The voltage-dependent nature of enkephalin block of type II calcium channels also has an important implication for excitation–secretion coupling. The block is intensified at potentials near the threshold for generation of calcium action potentials. This means that enkephalin can effectively prevent calcium action potentials from being produced.

It is interesting to note that somatostatin has also been found to block the type II calcium channels in neuroblastoma × glioma hybrid cells (TSUNOO et al. 1986). The selectivity, voltage dependence, and potency of the block were similar to those of the enkephalin block already described. The secretory cells are also known to contain type II calcium channels (ARMSTRONG and MATTESON 1985; FENWICK et al. 1982; HAGIWARA and OHMORI 1983; TSIEN 1983). Therefore, the inhibition of various hormones caused by somatostatin is also explicable in terms of the inhibition of type II calcium channels.

III. Phenytoin

One of the suspected anticonvulsant modes of action of phenytoin (diphenylhydantoin, DPH) is its effect on calcium-dependent processes. Phenytoin is known to suppress calcium accumulation in synaptosomes and brain slices (FERRENDELLI and DANIELS-MCQUEEN 1982; PINCUS and LEE 1973; SOHN and FERRENDELLI 1973). It also shortens the duration of the Ca-dependent action potential of the mouse spinal cord neuron (MCLEAN and MACDONALD 1983) and reduces calcium currents in the cardiac fiber (SCHEUER and KASS 1983). We have found that phenytoin blocks type I calcium channels without much effect on type II calcium channels in N1E-115 neuroblastoma cells (TWOMBLY and NARAHASHI 1985). Type I currents were suppressed to 62% of the control after application of 100 μM phenytoin. There was no change in the kinetics of activation and inactivation. The block was voltage dependent and intensified when the holding potential was changed from -80 to -60 mV. Type II calcium channel currents were insensitive to phenytoin, and neither the amplitude nor the kinetics of the current were affected.

The selective suppression of type I calcium channel current by phenytoin is important in the light of the suspected role of the channel in controlling rhythmic activity of cells (JAHNSEN and LLINÁS 1984; ARMSTRONG and MATTESON 1985). Thus, through its effects on the type I calcium channels, phenytoin could limit the rhythmic, paroxysmal depolarizing shifts associated with epileptogenesis, preventing the temporal and spatial propagation of seizure activity.

IV. Pyrethroids

Pyrethroids are synthetic insecticides and have been demonstrated to modify the kinetics of nerve membrane sodium channels. The mean open time of the single sodium channel is greatly prolonged, causing a prolonged sodium current in the nerve membrane. The latter change in turn elevates the depolarizing afterpotential which then reaches the threshold for repetitive discharges. These effects can account for severe symptoms of poisoning in animals – hyperactivity, hypersensitivity, convulsions, and tremors (NARAHASHI 1984, 1985).

Pyrethroids may be divided into two large groups based on the chemical structure and the symptoms of poisoning (NARAHASHI 1985). Type I pyrethroids do not contain a cyano group, and include the natural pyrethrins and many of the conventional synthetic pyrethroids. The symptoms of poisoning are characterized by ataxia, hyperexcitability, convulsions, and eventual paralysis and death. Type II pyrethroids contain a cyano group at the α position, and include newer compounds such as deltamethrin, fenvalerate, and cyphenothrin. The symptoms of poisoning are characterized by hypersensitivity, choreoathetosis, tremors, and eventual paralysis and death. Whereas these two types of pyrethroids modify the kinetics of sodium channels to prolong the mean open time, some differences in action at the channel level can account for the different symptoms of poisoning at the animal level.

We have found that some of the pyrethroids have drastic effects on calcium channels of neuroblastoma cells (YOSHII et al. 1985b). Tetramethrin, a type I pyrethroid, at a concentration of 50 μM caused a progressive block of type I and type II calcium channels. At a steady state, the type I calcium channel current was suppressed to 25% of the control, while the type II current was suppressed only to 70% of the control. The tetramethrin block of both types of calcium channels was time dependent, being enhanced during a 400-ms depolarizing pulse. The time-dependent component of block was easily reversible after washing with drug-free media, while the time-independent component or resting (closed channel) block persisted for a long time. In contrast to the type I pyrethroid tetramethrin, deltamethrin and fenvalerate, type II pyrethroids containing an α-cyano group, at a concentration of 10 μM had no effect on either type of calcium channel currents.

The two components of type I calcium channel block caused by tetramethrin, a time-dependent reversible block and a time-independent irreversible block, suggest two separate sites of action of tetramethrin on the type I calcium channel. It remains to be seen how the calcium channel block by tetramethrin is related to the symptoms of poisoning in animals. In the present study, the calcium channel-blocking action was limited to tetramethrin and not observed with deltamethrin and fenvalerate. It also remains to be studied whether the selective calcium channel block can be extended to other pyrethroids.

E. Summary and Conclusions

Neuroblastoma cells (N1E-115) and neuroblastoma × glioma hybrid cells (NG108-15) possess two kinds of calcium channels. Type I calcium channels inactivate quickly during a step depolarizing pulse, whereas type II calcium channels do not inactivate quickly. The latter inactivates with a very slow time course. The voltage dependence of the activation and inactivation kinetics of the two types of calcium channels are separated by 30 mV, type II requiring larger depolarizations to activate. Thus, the two types of currents can be separated by manipulating the membrane potential. The permeability ratios $P_{Ba}:P_{Sr}:P_{Ca}$ are different in the two types of channels, being 1.0:1.0:0.7 for type I and 1.0:0.7:0.3 for type II. Cyclic AMP stimulates the activity of type II channels, but not type I.

The polyvalent cations Cd^{2+}, Ni^{2+}, Co^{2+}, and La^{3+} block both types of channels. Cd^{2+} shows a much higher potency for type II than type I with a 23-fold difference. The potency ratios (K_d in μ*M*) are: La^{3+} (1.5) $\gg$ Ni^{2+} (47) > Cd^{2+} (160) = Co^{2+} (160) for type I and La^{3+} (0.9) > Cd^{2+} (7.0) $\gg$ Ni^{2+} (280) > Co^{2+} (560) for type II. Leu-enkephalin suppresses type II calcium channels with an apparent dissociation constant of 8.8 n*M* at 0 mV without effect on type I channels. The block is voltage dependent, and intensified with hyperpolarization. This action of enkephalin can account for the presynaptic block which is known to occur.

Somatostatin also blocks type II calcium channels in a manner similar to that of enkephalin. Phenytoin at a concentration of 100 μ*M* suppresses the type I calcium channels without affecting the type II channels. This action accounts for phenytoin suppression of paroxysmal depolarizing shifts associated with epileptogenesis. Tetramethrin, a type I pyrethroid, at a concentration of 50 μ*M* suppresses type I calcium channel currents more effectively than type II currents. However, deltamethrin and fenvalerate, type II pyrethroids containing an α-cyano group, do not affect either type of calcium channels at a concentration of 10 μ*M*. The toxicologic significance of tetramethrin block of calcium channels remains to be seen.

Type I and type II calcium channels appear to be separate entities. Type I channels are likely to be associated with rhythmic activity of neurons, and type II channels associated with transmitter release from nerve terminals. Both types of calcium channels exhibit differential sensitivities to various ions and chemicals, and pharmacologic actions of certain drugs such as opioids and phenytoin are explicable on the basis of a selective blocking action on one of the channel types.

Acknowledgments. The studies quoted in this chapter were supported by grants from the National Institutes of Health, NS14143 and NS14144. I thank Sandra Collins and Janet Henderson for secretarial assistance.

References

Armstrong CM, Matteson DR (1985) Two distinct populations of calcium channels in a clonal line of pituitary cells. Science 227:65–67

Armstrong CM, Swenson RP, Taylor SR (1982) Block of squid axon K channels by internally and externally applied barium ions. J Gen Physiol 80:663–682

Ashcroft FM, Stanfield PR (1980) Calcium dependence of the inactivation of calcium currents in the skeletal muscle fibers of an insect. Science 213:224–226
Bean BP (1985) Two kinds of calcium channels in canine atrial cells. Differences in kinetics, selectivity, and pharmacology. J Gen Physiol 86:1–30
Beeler GW, McGuigan JAS (1978) Voltage clamping of multicellular myocardial preparations: capabilities and limitations of existing methods. Prog Biophys Mol Biol 34:219–254
Brehm P, Eckert R (1978) Calcium entry leads to inactivation of calcium channel in *Paramecium*. Science 202:1203–1206
Carbone E, Lux HD (1984) A low voltage-activated, fully inactivating Ca channel in vertebrate sensory neurones. Nature 310:501–502
Dunlap K, Fischbach GD (1981) Neurotransmitters decrease the calcium conductance activated by depolarization of embryonic chick sensory neurones. J Physiol (Lond) 317:519–535
Eckert R, Chad JE (1984) Inactivation of Ca channels. Prog Biophys Mol Biol 44:215–267
Eckert R, Ewald D (1982) Residual calcium ions depress activation of calcium-dependent current. Science 216:730–733
Fedulova SA, Kostyuk PG, Vaselovsky NS (1985) Two types of calcium channels in the somatic membrane of new-born rat dorsal root ganglion neurones. J Physiol (Lond) 359:431–446
Fenwick EM, Marty A, Neher E (1982) Sodium and calcium channels in bovine chromaffin cells. J Physiol (Lond) 331:599–635
Ferrendelli JA, Daniels-McQueen S (1982) Comparative actions of phenytoin and other anticonvulsant drugs on potassium- and veratridine-stimulated calcium uptake in synaptosomes. J Pharmacol Exp Ther 220:29–34
Fishman MC, Spector I (1981) Potassium current suppression by quinidine reveals additional calcium currents in neuroblastoma cells. Proc Natl Acad Sci USA 78:5245–5249
Fleckenstein A (1977) Specific pharmacology of calcium in myocardium, cardiac pacemakers and vascular smooth muscle. Annu Rev Pharmacol Toxicol 17:149–166
Fox AP, Krasne S (1984) Two calcium currents in *Neanthes arenaceodentatus* egg cell membranes. J Physiol (Lond) 356:491–505
Geduldig D, Junge D (1968) Sodium and calcium components of action potentials in the *Aplysia* giant neurone. J Physiol (Lond) 199:347–365
Geduldig D, Junge D (1970) Voltage clamp of the *Aplysia* giant neurone: early sodium and calcium currents. J Physiol (Lond) 211:217–244
Goldman DE (1943) Potential, impedance and rectification in membranes. J Gen Physiol 27:37–60
Hagiwara S, Byerly L (1981) Calcium channel. Annu Rev Neurosci 4:69–125
Hagiwara S, Byerly L (1983) The calcium channel. Trends Neurosci 6:189–193
Hagiwara S, Naka K (1964) The initiation of spike potential in barnacle muscle fibers under low intracellular Ca^{++}. J Gen Physiol 48:141–162
Hagiwara S, Ohmori H (1983) Studies of single calcium channel currents in rat clonal pituitary cells. J Physiol (Lond) 336:649–661
Hagiwara S, Chichibu S, Nada K (1964) The effects of various ions on resting and spike potentials of barnacle muscle fibers. J Gen Physiol 48:163–179
Hagiwara S, Miyazaki S, Moody W, Patlak J (1978) Blocking effects of barium and hydrogen ions on the potassium current during anomalous rectification in the starfish egg. J Physiol (Lond) 279:167–185
Hamill OP, Marty A, Neher E, Sakmann B, Sigworth FJ (1981) Improved patch-clamp techniques for high-resolution current recording from cells and cell-free membrane patches. Pflugers Arch 391:85–100
Hodgkin AL, Katz B (1949) The effect of sodium ions on the electrical activity of the giant axon of the squid. J Physiol (Lond) 108:37–77
Jahnsen H, Llinás R (1984) Ionic basis for the electroresponsiveness and oscillatory properties of guinea-pig thalamic neurones in vitro. J Physiol (Lond) 349:227–247

Katayama Y, Nishi S (1984) Sites and mechanisms of actions of enkephalin in the feline parasympathetic ganglion. J Physiol (Lond) 351:111–121
Kato E, Narahashi T (1982) Characteristics of the electrical response to dopamine in neuroblastoma cells. J Physiol (Lond) 333:213–226
Kerkut GA, Gardner DR (1967) The role of calcium ions in the action potentials of *Helix aspersa* neurones. Comp Biochem Physiol 20:147–162
Kimhi Y, Palfrey C, Spector I, Barak Y, Littauer UZ (1976) Maturation of neuroblastoma cells in the presence of dimethylsulfoxide. Proc Natl Acad Sci USA 73:462–466
Koketsu K, Cerf JA, Nishi S (1959) Further observations on the activity of frog spinal ganglion cells in sodium-free solutions. J Neurophysiol 22:693–703
Llinás R, Yarom Y (1981) Properties and distribution of ionic conductances generating electroresponsiveness of mammalian inferior olivary neurones in vitro. J Physiol (Lond) 315:569–584
Llinás R, Steinberg IZ, Walton K (1981) Presynaptic calcium currents in squid giant synapse. Biophys J 33:289–322
Macdonald RL, Nelson PG (1978) Specific opiate-induced depression of transmitter release from dorsal root ganglion cells in culture. Science 199:1449–1451
Macdonald RL, Werz MA (1986) Dynorphin A decreases voltage-dependent calcium conductance of mouse dorsal root ganglion neurones. J Physiol (Lond) 377:237–249
McLean MJ, Macdonald RL (1983) Multiple actions of phenytoin on mouse spinal cord neurons in cell culture. J Pharmacol Exp Ther 227:779–789
Mudge AW, Leeman SE, Fischbach GD (1979) Enkephalin inhibits release of substance P from sensory neurons in culture and decreases action potential duration. Proc Natl Acad Sci USA 76:526–530
Nachshen DA (1984) Selectivity of the Ca binding site in synaptosome Ca channels. J Gen Physiol 83:941–967
Narahashi T (1984) Pharmacology of nerve membrane sodium channels. In: Baker PF (ed) The squid axon. Academic, New York, pp 483–516 (Current topics in membranes and transport, vol 22)
Narahashi T (1985) Nerve membrane ionic channels as the primary target of pyrethroids. Neurotoxicology 6:3–22
Narahashi T, Tsunoo A, Yoshii M (1987) Characterization of two types of calcium channels in mouse neuroblastoma cells. J Physiol (Lond) 383:231–249
North RA (1979) Opiates, opioid peptides and single neurones. Life Sci 24:1527–1546
North RA, Williams JT (1983) Opiate activation of potassium conductance inhibits calcium action potentials in rat locus coeruleus neurones. Br J Pharmacol 80:225–228
North RA, Williams JT (1985) On the potassium conductance increased by opioids in rat locus coeruleus neurones. J Physiol (Lond) 364:265–280
North RA, Katayama Y, Williams JT (1979) On the mechanism and site of action of enkephalin on single myenteric neurons. Brain Res 165:67–77
Nowycky MC, Fox AP, Tsien RW (1985) Three types of neuronal calcium channel with different calcium agonist sensitivity. Nature 316:440–443
Ohmori H, Yoshii M (1977) Surface potential reflected in both gating and permeation mechanisms of sodium and calcium channels of the tunicate egg cell membrane. J Physiol (Lond) 267:429–463
Pappone PA, Cahalan MD (1986) Ion permeation in cell membranes. In: Andreoli TE, Hoffman JF, Franestil DD, Schultz SG (eds) Physiology of membrane disorders. Plenum, New York, pp 249–272
Pincus JH, Lee SH (1973) Diphenylhydantoin and calcium. Relation to norepinephrine release from brain slices. Arch Neurol 29:239–244
Quandt FN, Narahashi T (1980) Internal perfusion of neuroblastoma cells and the effects of diphenylhydantoin on voltage-dependent currents. Soc Neurosci Abstr 6:97
Quandt FN, Narahashi T (1984) Isolation and kinetic analysis of inward currents in neuroblastoma cells. Neuroscience 13:249–262
Reuter H (1979) Properties of two inward membrane currents in the heart. Annu Rev Physiol 41:413–424

Reuter H (1983) Calcium channel modulation by neurotransmitters, enzymes and drugs. Nature 301:569–574
Scheuer T, Kass RS (1983) Phenytoin reduces calcium current in the cardiac Purkinje fiber. Circ Res 53:16–23
Skerritt JH, Macdonald RL (1984) Multiple actions of convulsant barbiturates on mouse neurons in cell culture. J Pharmacol Exp Ther 230:82–88
Sohn RS, Ferrendelli JA (1973) Inhibition of Ca^{2+} transport into rat brain synaptosomes by diphenylhydantoin (DPH). J Pharmacol Exp Ther 185:272–275
Tillotson D (1979) Inactivation of Ca conductance dependent on entry of Ca ions in molluscan neurons. Proc Natl Acad Sci USA 76:1497–1500
Tokimasa T, Morita K, North A (1981) Opiates and clonidine prolong calcium-dependent after-hyperpolarizations. Nature 294:162–163
Triggle DJ (1981) Calcium antagonists: basic chemical and pharmacological aspects. In: Weiss GB (ed) New perspectives on calcium antagonists. Waverly, Baltimore, pp 1–18
Tsien RW (1983) Calcium channels in excitable cell membranes. Annu Rev Physiol 45:341–358
Tsunoo A, Yoshii M, Narahashi T (1986) Block of calcium channels by enkephalin and somatostatin in neuroblastoma-glioma hybrid NG108-15 cells. Proc Natl Acad Sci USA 83:9832–9836
Twombly D, Narahashi T (1985) Phenytoin suppresses calcium currents in neuroblastoma cells. Soc Neurosci Abstr 11:518
Werz MA, Macdonald RL (1982a) Opioid peptides decrease calcium-dependent action potential duration of mouse dorsal root ganglion neurons in cell culture. Brain Res 239:315–321
Werz MA, Macdonald RL (1982b) Heterogeneous sensitivity of cultured dorsal root ganglion neurones to opioid peptides selective for μ- and δ-opiate receptors. Nature 299:730–733
Werz MA, Macdonald RL (1983a) Opioid peptides with differential affinity for mu- and delta-receptors decrease sensory neuron calcium-dependent action potentials. J Pharmacol Exp Ther 227:394–402
Werz MA, Macdonald RL (1983b) Opioid peptides selective for mu- and delta-opiate receptors reduce calcium-dependent action potential duration by increasing potassium conductance. Neurosci Lett 42:173–178
Werz MA, Macdonald RL (1984) Dynorphin reduces voltage-dependent conductance of mouse dorsal root ganglion neurons. Neuropeptide 5:253–256
Werz MA, Macdonald RL (1985) Dynorphin and neoendorphin peptides decrease dorsal root ganglion neuron calcium-dependent action potential duration. J Pharmacol Exp Ther 234:49–56
Williams JT, North RA (1983) Opiate activation of G_K inhibits calcium spikes in rat locus coeruleus. Soc Neurosci Abstr 9:1130
Wouters W, van den Bercken J (1980) Effects of met-enkephalin on slow synaptic inhibition in frog sympathetic ganglion. Neuropharmacology 19:237–243
Yaksh TL, Noueihed R (1985) The physiology and pharmacology of spinal opiates. Annu Rev Pharmacol Toxicol 25:433–462
Yoshii M, Tsunoo A, Narahashi T (1985a) Different properties in two types of calcium channels in neuroblastoma cells. Biophys J 47:433a
Yoshii M, Tsunoo A, Narahashi T (1985b) Effects of pyrethroids and veratridine on two types of Ca channels in neuroblastoma cells. Soc Neurosci Abstr 11:518
Yoshimura M, North RA (1983) Hyperpolarization of substantia gelatinosa neurons in vitro by enkephalin and noradrenaline. Soc Neurosci Abstr 9:1129

CHAPTER 15

Calcium and Synaptic Function *

M. P. BLAUSTEIN

A. Introduction

Calcium ions play many important roles in the function of nerve cells. Calcium's involvement in neurotransmitter release is one of its key roles, and was one of the first to be recognized. Early observations demonstrated that extracellular Ca was required for neurotransmitter release (HARVEY and MACINTOSH 1940). Subsequent studies established that transmitter release was evoked by Ca entry and a rise in $[Ca^{2+}]_i$, the intracellular free Ca concentration. This sequence of events accounted for the external Ca dependence (KATZ 1969; BAKER 1972; LLINAS et al. 1981; DRAPEAU and BLAUSTEIN 1983a, b). Moreover, the modulation of synaptic transmission, and related fundamental aspects of transmitter release associated with memory and learning, are all profoundly influenced by Ca.

Calcium ions also have a critical effect on the electrical excitability of neurons. Again, early observations concerned the effects of extracellular Ca concentrations, $[Ca^{2+}]_o$: elevation of $[Ca^{2+}]_o$ increases the firing threshold (i.e., reduces excitability), while reduction of $[Ca^{2+}]_o$ increases excitability. These "membrane-stabilizing" effects of Ca could be explained, at least in part, by calcium's ability to screen surface charges and thereby alter the electrical potential profile between the bulk solution and the surface of the plasma membrane (FRANKENHAEUSER and HODGKIN 1957; FRANKENHAEUSER 1957). Subsequently, intracellular Ca was also found to modulate excitability: for example, by activating certain potassium conductances directly (cf. MEECH 1978), or by controlling Ca-activated protein kinases that, in turn, modulate ionic conductances (cf. RANE and DUNLAP 1986). These actions of Ca on neuronal excitability may directly or indirectly modify neurotransmitter release.

The central role of intracellular Ca in these neuronal functions is readily apparent. Therefore, to gain an understanding of these mechanisms we need to focus first on the mechanisms by which $[Ca^{2+}]_i$ is controlled in nerve cells.

The extracellular free Ca concentration, $[Ca^{2+}]_o$, is of the order of 1 m*M* in most vertebrate organisms (BLAUSTEIN 1974), and is approximately 10000 times greater than $[Ca^{2+}]_i$ (BAKER 1972). As a consequence of this very large concentration gradient and the fact that the cytoplasm of resting neurons is normally about 50–70 mV negative with respect to the extracellular fluid, Ca will tend to leak into the cells, and must be actively extruded. In this chapter we will consider the mechanisms of Ca entry into nerve cells, the disposition of intracellular Ca, and the

* Dedicated to the memory of my student, friend, and colleague, Daniel A. Nachshen, who so carefully elucidated the properties of calcium channels in nerve terminals.

mechanisms involved in Ca extrusion and regulation of $[Ca^{2+}]_i$. We will then examine aspects of synaptic function that are controlled by intracellular Ca.

B. Regulation of Intracellular Calcium in Nerve Cells

I. The Intracellular Free Calcium Concentration in Nerve Cells

The total Ca concentration in the blood plasma of a variety of mammals is of the order of 2 m*M*. About half of this Ca is bound to plasma proteins, and about half is free: that is, $[Ca^{2+}]_o \simeq 1$ m*M* (BLAUSTEIN 1974). In contrast, $[Ca^{2+}]_i$ in "resting" mammalian nerve cells (and most other vertebrate and invertebrate nerve cells) is about 100–300 n*M* (BLINKS et al. 1982; CONNOR 1986; CONNOR et al. 1986). These $[Ca^{2+}]_i$ levels are comparable to those observed in a variety of other types of excitable and nonexcitable cells (e.g., BLINKS et al. 1982; CONWAY 1983). Indeed, the ability of neurons and other types of cells to employ Ca^{2+} as a second messenger (see Sect. C) depends, in part, upon this low $[Ca^{2+}]_i$. The physiologic significance of a low resting $[Ca^{2+}]_i$ is that small changes in the absolute $[Ca^{2+}]_i$ ("signal" Ca, or $\Delta[Ca^{2+}]_i$) can, nevertheless, represent a relatively large signal: background ratio, $(\Delta[Ca^{2+}]_i/[Ca^{2+}]_{i\,(resting)})$, of the order of 5–10 or more (BLAUSTEIN 1985).

The free Ca^{2+} is only a very small fraction of the total intracellular Ca. As discussed in Sect. B.III, most of the intracellular Ca is bound to cytoplasmic proteins or is stored in intracellular organelles. Release of Ca from intracellular storage sites may play a critical role in some Ca-dependent cellular activities – although such Ca-evoked release has not yet been demonstrated in neurons.

II. Calcium Entry into Nerve Cells

Neurons, like many other types of cells, contain a variety of gated, ion-selective channels in their plasma membranes, including some channels that are divalent cation-selective. The latter are, effectively, Ca-selective channels, since they are not very permeable to Mg, the only other common divalent cation in most vertebrate fluids. The best-studied divalent cation-selective channels are the voltage-operated Ca channels (VOCs) that open when the cells are depolarized (e.g., KATZ and MILEDI 1967, 1969; BAKER 1972; LLINAS et al. 1981; NACHSHEN and BLAUSTEIN 1980; NOWYCKY et al. 1985; TSIEN et al. 1987). The distribution of these channels may not be uniform; for example, they may be found at relatively high density in the plasma membrane of axons and/or cell bodies of some neurons (e.g., HAGIWARA and BYERLY 1981; KOSTYUK 1981; NOWYCKY et al. 1985; BOLSOVER and SPECTOR 1986; ROSS et al. 1986). While some neurons have action potentials that are dependent primarily on an inward Na current, other neurons have action potentials that depend primarily upon an inward Ca current, or upon both Na and Ca currents (cf. HILLE 1984). Voltage-operated, Ca-selective channels are often found in abundance in the plasma membrane at presynaptic nerve terminals (NACHSHEN and BLAUSTEIN 1980); at least in some neurons, they tend to be clustered in this region (KATZ and MILEDI 1969; ROSS et al. 1986). This has important implications for (chemical) synaptic transmission.

In addition to the VOCs, there are agonist-operated channels that are permeable to Ca ions as well as other cations. The best-studied of these is the channel associated with the (postsynaptic) nicotinic acetylcholine receptor. This channel, which is relatively selective for monovalent cations (Na and K are nearly equally permeable: TAKEUCHI and TAKEUCHI 1960), also conducts Ca (e.g., TAKEUCHI and TAKEUCHI 1963; MILEDI et al. 1980). The postsynaptic channels activated by glutamate also conduct Ca (EUSEBI et al. 1985). These channels may be models for other agonist receptor-operated channels as well. The potential importance of Ca entry through the postsynaptic channels is widely recognized. We will focus primarily on the presynaptic voltage-gated Ca channels, however, because of their role in transmitter release.

Several types of Ca channels have been described, with different activation and inactivation kinetics, different divalent cation selectivities, and different sensitivities to pharmacologic agents (BOSSU et al. 1985; NOWYCKY et al. 1985; TSIEN et al. 1987). The Ca channels found at mammalian presynaptic nerve terminals appear to be predominantly of the N-type (according to the nomenclature of Nowycky and colleagues). Upon depolarization, they open rapidly, and then inactivate with a half-time of about 0.5–1 s (NACHSHEN and BLAUSTEIN 1980; NACHSHEN 1985). Neuronal Ca channels also appear to be blocked selectively by one of the ω-conotoxins from the venom of the cone snail, *Conus geographus* (REYNOLDS et al. 1986).

The presynaptic Ca channels are permeable to Sr, Ba, and Mn, as well as to Ca, but not to Mg (NACHSHEN and BLAUSTEIN 1982; DRAPEAU and NACHSHEN 1984). Also, in the presence of physiologic concentrations of divalent cations in the extracellular fluid, they are relatively impermeable to Na and K. The flux of ions through the channels is inversely related to the ion charge density, so that the maximum flux rates for the permeable divalent cations in the channels follows the sequence: Ba > Sr > Ca > Mn (NACHSHEN and BLAUSTEIN 1982; NACHSHEN 1984; DRAPEAU and NACHSHEN 1984). Impermeable cations such as Cd and Ni bind strongly to the cation-binding sites in the channels and block the channels, thereby preventing the passage of permeable ions (NACHSHEN 1984). A detailed analysis of Ca channel permeability mechanisms is given by TSIEN et al. (1987).

These neuronal Ca channels are relatively insensitive to verapamil and to the dihydropyridine Ca channel blockers (e.g., NACHSHEN and BLAUSTEIN 1979; SUZKIEW et al. 1986; TSIEN et al. 1987). In contrast, most types of Ca-selective VOCs found in mammalian cardiac and smooth muscle cells are sensitive to submicromolar concentrations of these agents (FLECKENSTEIN 1985).

Ca channels in some neurons (PELLMAR 1981), like those in some other types of cells (REUTER 1983; KAMEYAMA et al. 1986), can be regulated by cyclic AMP-dependent phosphorylation and by dephosphorylation. For example, phosphorylation of the channels may enhance their ability to open when the cells are depolarized (FLOCKERZI et al. 1986). Alternatively, phosphorylation by other protein kinases may prevent Ca channels from opening, as illustrated by the inhibitory actions of a diacylglycerol analog and a phorbol ester (both of which presumably activate a Ca-dependent protein kinase, protein kinase C) on chick sensory neurons (RANE and DUNLAP 1986). Another example is the γ-aminobutyric acid- or norepinephrine-induced, cyclic GMP-mediated inhibition of Ca channels in chick dorsal root ganglion neurons (HOLZ et al. 1986). In some snail neurons, on the other hand, a cyclic GMP-dependent phosphorylation appears to enhance Ca channel activity (PAUPARDIN-TRITSCH et al. 1986). These observations raise the possibility that cyclic AMP-dependent phosphorylations and cyclic GMP-dependent or protein kinase C-dependent phosphorylations may have opposite effects in the same cells, and perhaps even on the same channel molecules (cf. HARTZELL and FISCHMEISTER 1986). Also, some neuronal Ca channels may be modulated by the activation of receptors associated with GTP-binding proteins (SCOTT and DOLPHIN 1986); however, the possible relationship to cyclic AMP, which is regulated by GTP-binding proteins (cf. GILMAN 1987) has not yet been investigated. The specific physiologic roles of these numerous types of modulation of Ca channels in the mammalian nervous system have not yet been explored.

When an action potential invades a nerve terminal, the depolarization transiently (for a few milliseconds) opens the Ca channels. The large Ca electrochemi-

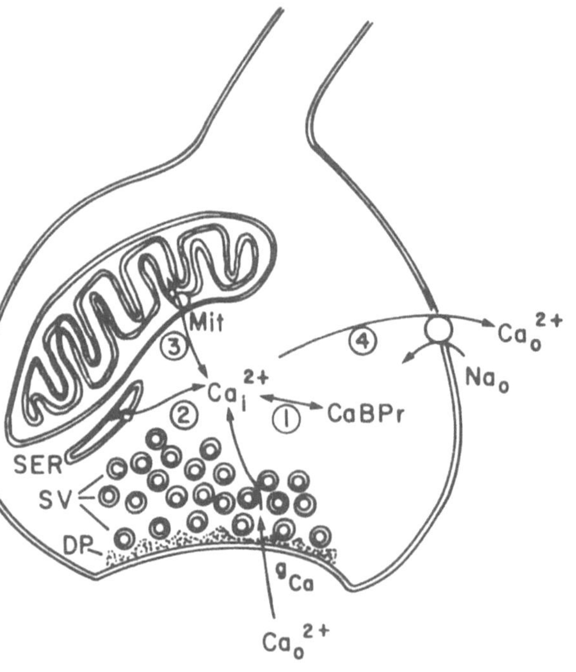

Fig. 1. Pathways for Ca movement and sites of Ca buffering in a presynaptic nerve terminal. Depolarization triggers an increase in the Ca conductance g_{Ca} by opening Ca-selective VOCs in the "active zone" of the plasmalemma (in the regions of synaptic contact where the synaptic vesicles fuse with the plasmalemma). Ca enters the terminal, raising $[Ca^{2+}]_i$, and then rapidly diffuses to the high affinity Ca-binding sites on cytosolic proteins (CaBPr; *reaction 1*). Intraterminal Ca sequestration systems in the smooth endoplasmic reticulum (SER; *reaction 2*), and (to a much lesser extent) in the mitochondria (Mit; *reaction 3*), take up Ca, further lowering $[Ca^{2+}]_i$. Ultimately, the Ca that entered during activity must be extruded from the terminal. The Na/Ca exchange mechanism in the plasmalemma (*reaction 4*) appears to mediate most of this net extrusion (see text), although an ATP-driven Ca pump (not shown) may participate to a small extent. SV synaptic vesicles; DP dense projections in the region of the active zone. (BLAUSTEIN et al. 1980)

cal gradient then promotes net influx of Ca through the channels, and the consequent rise in $[Ca^{2+}]_i$ triggers neurotransmitter release. Upon repolarization, the Ca channels close rapidly, and transmitter release then declines with a half-time of 1 ms or less. This rapid decline in transmitter release is explained by a fall in $[Ca^{2+}]_i$ toward the resting level, with a half-time of <1 ms, as a result of buffering and sequestration mechanisms that remove free Ca from the cytosol (Fig. 1). The transmitter release process, itself, is not inactivated this rapidly; indeed, transmitter release can even be facilitated if the period of Ca entry is prolonged (KATZ and MILEDI 1968).

In the long run, Ca must be extruded from the neurons in order to maintain a steady Ca balance; however, as described in Sects. B.III–V, the Ca extrusion mechanisms can not remove Ca from the cytosol fast enough to account for the very rapid decline in transmitter release rate that is observed. Other mechanisms must therefore very rapidly lower $[Ca^{2+}]_i$ during repolarization.

III. Intracellular Calcium Buffering in Nerve Cells

Several types of mechanisms play a role in intracellular Ca^{2+} buffering (see Fig. 1): sequestration of Ca in smooth endoplasmic reticulum and mitochondria, and binding to cytoplasmic proteins and membrane structures.

1. Mitochondria

Under normal conditions, when $[Ca^{2+}]_i$ is within the dynamic physiologic range (about 0.1–1.0 μ*M*), mitochondria do not sequester much Ca (HANSFORD 1985). Although mitochondria can accumulate Ca, the affinity of the mitochondrial Ca transport system for Ca and the rate of Ca uptake are quite low in the presence of millimolar concentrations of Mg (VINOGRADOV and SCARPA 1973; SCARPA 1976; HUTSON et al. 1976); these are the Mg concentrations normally present in the cytosol (e.g., DE WEER 1976; ALVAREZ-LEEFMANS et al. 1984; HEINONEN and AKERMAN 1986). Thus, mitochondria will not normally take up much Ca from the cytosol unless the other buffering systems are saturated (Figs. 2 and 3). But even with a low affinity for Ca, they may take up small amounts of Ca when $[Ca^{2+}]_i$ rises during cell activation; indeed, this could play a role in coupling mitochondrial enzyme activity to cellular activity (DENTON and MCCORMACK 1985; HANSFORD 1985; MCCORMACK and DENTON 1986). However, when $[Ca^{2+}]_i$ exceeds about 5 μ*M* (i.e., under pathologic conditions of Ca overload; cf. SCHLAEPFER 1977), mitochondria will begin to sequester substantial amounts of Ca (BLAUSTEIN and RASGADO-FLORES 1981; RASGADO-FLORES and BLAUSTEIN 1987 a; and see Fig. 2); this will dissipate the energy that would normally be used for oxidative phosphorylation.

2. Smooth Endoplasmic Reticulum

A second organelle that can sequester Ca is the smooth endoplasmic reticulum. The Ca sequestering properties of this organelle have been most extensively studied in skeletal muscle (sarcoplasmic reticulum: HASSELBACH 1977; INESI 1985).

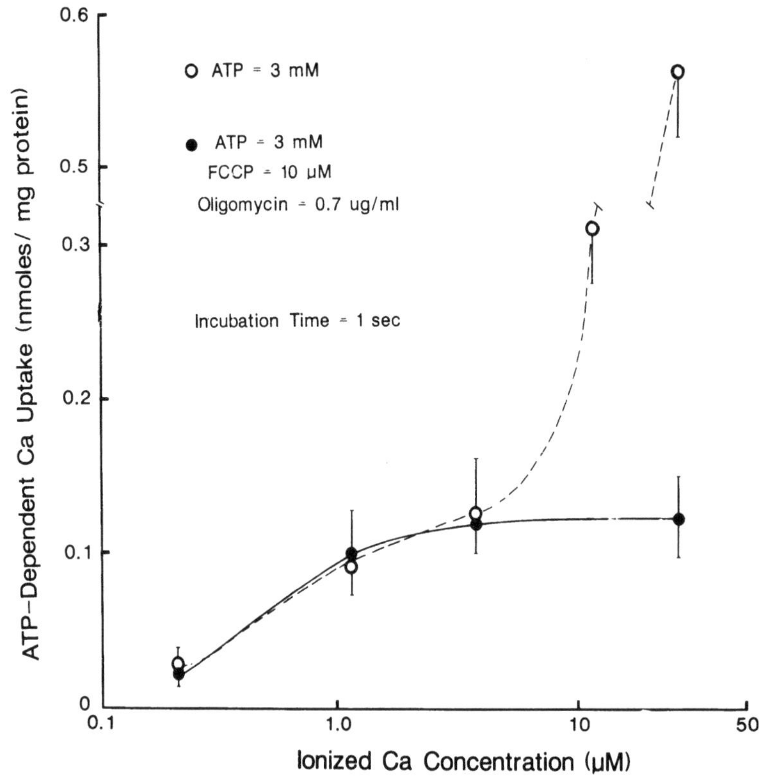

Fig. 2. Effect of mitochondrial blockers on the ATP-dependent Ca uptake into saponin-treated rat brain synaptosomes incubated at different free Ca^{2+} concentrations (as indicated on the abscissa). The saponin renders the plasma membrane leaky, but does not damage the cholesterol-poor mitochondria or smooth endoplasmic reticulum membranes. The leaky synaptosomes were incubated for 1 s with media containing ^{45}Ca and ATP, either without (*open circles*) or with (*full circles*) 10 μ*M* FCCP (a mitochondrial uncoupler) and 0.7 μg/ml oligomycin. The ATP-dependent Ca uptake was obtained by subtracting the Ca uptake in the absence of ATP from the uptake measured in the presence of ATP. Note that, below about 5 μ*M* Ca^{2+}, the ATP-dependent uptake is unaffected by the mitochondrial inhibitors, whereas above this Ca^{2+} concentration the unpoisoned mitochondria take up a large amount of Ca. (Rasgado-Flores and Blaustein 1987a)

Nevertheless, most other types of cells, including neurons, also contain an ATP-driven Ca sequestration system in the endoplasmic reticulum. Indeed, the Ca uptake properties of this system in neurons appear to be very similar, if not identical, to those of muscle sarcoplasmic reticulum (Blaustein et al. 1978, 1980; Rasgado-Flores and Blaustein 1987a).

Activation of muscle cells may trigger the release of Ca from the well-developed sarcoplasmic reticulum; this Ca may contribute to the activation of contraction, even in the absence of extracellular Ca. Although the mechanism of Ca release from the sarcoplasmic reticulum is not completely understood, in some muscles the release appears to be triggered by inositol trisphosphate (cf. Berridge and Irvine 1984; Hashimoto et al. 1986; Somlyo et al. 1985). At nerve terminals,

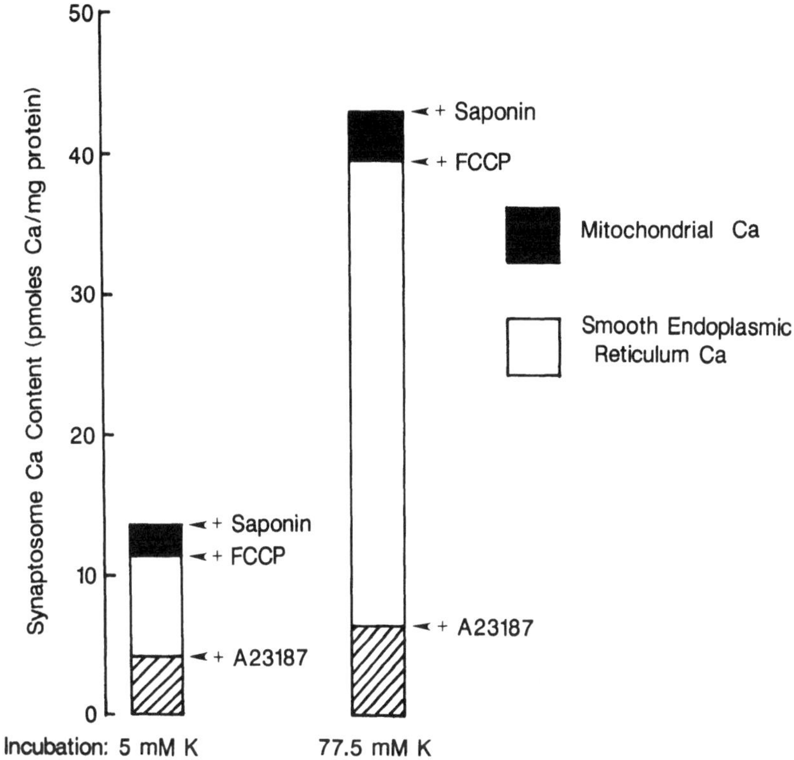

Fig. 3. Distribution of intraterminal ^{45}Ca before and after a small load of Ca. Rat forebrain synaptosomes were incubated at 30 °C for 30 s in standard physiologic salt solution which included 5 m*M* K, 145 m*M* Na, and 20 μ*M* Ca labeled with ^{45}Ca (*left*), or for 5 s in a similar solution, but with 77.5 m*M* K and only 72.5 m*M* Na (*right*). The extracellular ^{45}Ca was then removed by dilution with EGTA-containing (quench) medium and centrifugation. The pelleted ^{45}Ca-loaded synaptosomes were then suspended in a 150 m*M* K, Na- and Ca-free physiologic solution containing 10 μg/ml saponin; in some instances (as indicated) the solution also contained 10 μ*M* FCCP (a mitochondrial uncoupler) or 10 μ*M* FCCP plus 10 μ*M* A23187 (a Ca ionophore). Following a 10-min incubation at 30 °C, these suspensions were filtered and the ^{45}Ca retained in the synaptosomes (on the filters) was determined. The FCCP-sensitive fraction of the ^{45}Ca content (*black bar segments*) corresponds to the Ca in the mitochondria; the A23187-sensitive fraction (*open bar segments*) corresponds to the Ca in the smooth endoplasmic reticulum. The *hatched bar segments* represent residual ^{45}Ca that was not released by these agents. In the "resting" synaptosomes (incubated in 5 m*M* K solution), about 76% of the releasable ^{45}Ca was contained in the endoplasmic reticulum, and only 24% was in the mitochondria. After depolarizing the synaptosomes for 5 s to load them with ^{45}Ca, about 96% of the load was found in the endoplasmic reticulum, and only 4% of the load was found in the mitochondria. The total load before saponin treatment was 90 pmol Ca per milligram protein, which corresponds to the Ca entry during a train of about 15–20 action potentials. With larger Ca loads, the mitochondria store a much larger fraction of the Ca load. (Blaustein et al. 1978; R. V. Parsey 1987, unpublished work)

however, depolarization does not normally trigger transmitter release in the absence of extracellular Ca (e.g., HARVEY and MACINTOSH 1940; KATZ 1969); this implies that little, if any Ca is released from the endoplasmic reticulum under these circumstances. A rise in $[Ca^{2+}]_i$ may also trigger the release of Ca from muscle sarcoplasmic reticulum (Ca-induced Ca release; cf. ENDO 1977; FABIATO 1985). The external Ca dependence of neurotransmitter release could also be explained if a similar Ca-dependent Ca release occurs in nerve terminal endoplasmic reticulum.

During a 1–2 ms action potential, Ca may be expected to enter mammalian brain nerve terminals at a rate of about 5 pmol per milligram protein per millisecond (NACHSHEN 1985). The maximum rate at which the endoplasmic reticulum in the terminals can sequester Ca is about 0.1–0.2 pmol mg^{-1} ms^{-1} (RASGADO-FLORES and BLAUSTEIN 1987a). This is more than an order of magnitude too slow to account for the rate of Ca removal from the cytosol required to terminate transmitter release immediately following an action potential. During a train of action potentials, however, the endoplasmic reticulum will slowly begin to accumulate Ca because Ca extrusion across the plasma membrane lags behind Ca entry (see Sect. B.V). This may be important for processes such as post-tetanic potentiation (PTP): as the endoplasmic reticulum begins to saturate with Ca, it will help to buffer the resting $[Ca^{2+}]_i$ to higher levels. Subsequent action potentials may then possibly trigger the release of substantial amounts of Ca from these stores; this would, of course, significantly enhance the amount of transmitter release evoked by these action potentials (i.e., PTP). These unresolved issues need to be explored.

Inositol trisphosphate, which is formed from phosphatidylinositol during the activation of many tpyes of cells, releases Ca from the endoplasmic reticulum stores in a variety of cells (STREB et al. 1983; BERRIDGE and IRVINE 1984; HASHIMOTO et al. 1986) including neuroblastoma cells (CHUEH and GILL 1986). Recently, GTP has also been found to release Ca from the endoplasmic reticulum of these cells by a mechanism that involves GTP hydrolysis (GILL et al. 1986b); the GTP and inositol trisphosphate act through different pathways (CHUEH and GILL 1986; GILL et al. 1986b). The physiologic role of these phenomena in neurons (including nerve terminals) remains to be elucidated.

3. Cytosolic Buffers

Ca sequestration (in the endoplasmic reticulum and mitochondria) and Ca extrusion (see Sect. B.IV) cannot remove Ca from the transmitter release sites sufficiently rapidly to account for the termination of transmitter release at the end of an action potential. Therefore, we must look elsewhere for the mechanisms responsible. The most likely candidates are neuronal cytoplasmic proteins that have a high affinity for Ca. Several such proteins have been identified, including calmodulin (WOOD et al. 1980; DELORENZO 1982), parvalbumin (HEIZMANN 1984), and vitamin D-dependent Ca-binding proteins (POCHET et al. 1985).

Calmodulin may be regarded as a "Ca-dependent second messenger"; when Ca is bound to calmodulin, the complex can, in turn, bind to, and activate other "receptor" molecules. Parvalbumin may have a different function. It is present in relatively high concentration in fast skeletal muscle cells, and in these cells it appears to serve as a temporary repository for Ca: immediately after the activation of contraction, it removes Ca from troponin C and binds the Ca until it is resequestered in the sarcoplasmic reticulum (PECHERE et al. 1977; SOMLYO et al. 1981; HEIZMANN 1984). Some neurons also contain considerable amounts of parvalbumin (HEIZMANN 1984). This protein may function in neurons in a way comparable to its role in skeletal muscle: it may remove Ca from the transmitter release sites, and then release the bound Ca as it is sequestered (in the endoplasmic reticulum) or extruded across the plasma membrane. This view is supported by the observation that parvalbumin is present in GABAergic neurons that fire at a high frequency, but not in those that fire at a low frequency (CELIO 1986). The parvalbumin may be needed in the high frequency neurons to buffer intracellular Ca in order to keep the Ca from activating Ca-dependent K conductances (see Sect. C.III) that may reduce excitability and prolong the refractory period. The capacity of parvalbumin to bind Ca subtantially exceeds the amount that enters during a single action potential. With a rapid train of action potentials, however, the parvalbumin will become more and more saturated with Ca with each succeeding action potential. Thus, this form of Ca buffering may play a very important role in the potentiation of transmitter release that occurs when an action potential follows rapidly (within a few milliseconds) after a preceding one (facilitation), or after a train of action potentials (PTP).

IV. Calcium Transport Across the Neuronal Plasma Membrane

As noted already, there is a very large electrochemical gradient which favors the entry of Ca across the nerve cell plasma membrane, and there is a net gain of Ca during neuronal activity. Thus, to remain in long-term Ca balance, nerve cells must possess energy-dependent Ca extrusion mechanisms. Like many other types of cells (cf. SHEU and BLAUSTEIN 1986), neurons have two parallel, independent mechanisms in their plasma membranes for extruding Ca (cf. GILL et al. 1981, 1986a): an ATP-driven Ca pump (analogous to the ATP-driven Na pump), and an Na/Ca exchange transport systems (Fig. 4). The latter is a "secondary active" transport system that uses energy available from the Na electrochemical gradient, generated by the ATP-driven Na pump, to export Ca. This transport system is especially interesting, because it can also promote Ca entry when the Na concentration gradient is reduced or the membrane is depolarized (see later in this section). These two transport systems operate in parallel, and we must determine the role that each of them plays in neuronal cell function.

Two general classes of ATP-driven Ca pumps have been recognized: those located in the smooth endoplasmic reticulum, which participate in intracellular Ca sequestration (see Sect. B.III.2), and those located in the plasma membrane, which help to extrude Ca from cells (cf. CARAFOLI 1984; SHEU and BLAUSTEIN 1986). The latter transport systems are modulated by calmodulin, have stoichiometries of 1 Ca ion transported per ATP molecule hydrolyzed, and are mediated by Ca-dependent ATPases with molecular weights of about 140000 (cf. HINCKE and DEMAILLE 1984). In contrast, the endoplasmic reticulum Ca transport systems are not modulated by calmodulin (but see WULFROTH and PELTZELT 1985);

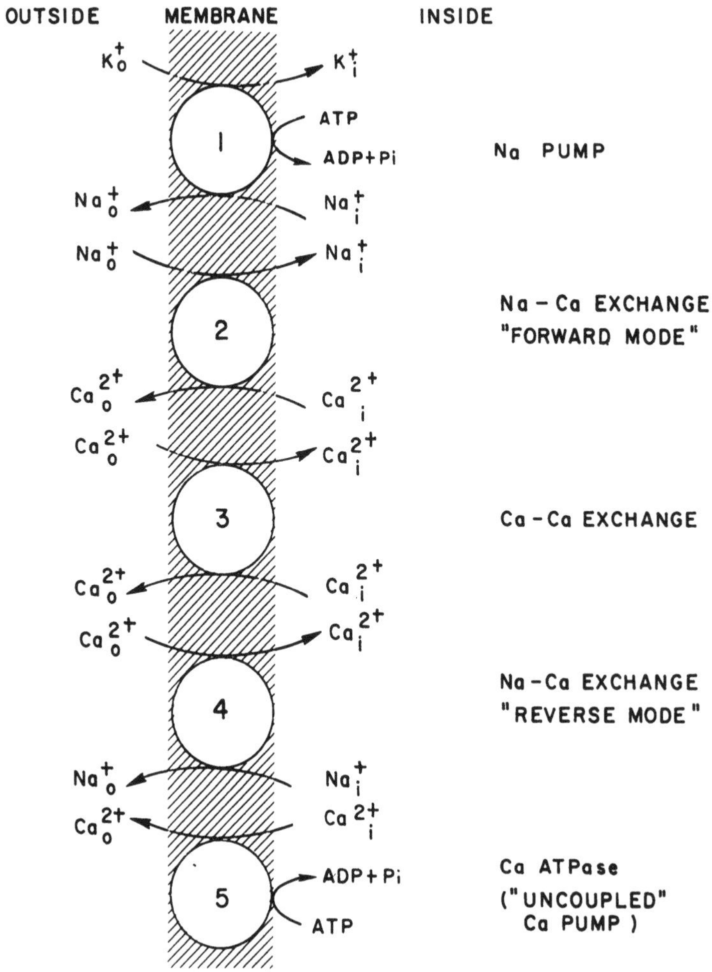

Fig. 4. The Na/Ca exchange (2 and 4) and ATP-driven Ca pump (5) modes of Ca transport across the plasma membrane. Both transport systems can mediate Ca extrusion (2 and 5, respectively). In addition, the Na/Ca exchange system can operate in the "reverse mode" (also a normal mode of operation) – bringing Ca into the cell in exchange for exiting Na; the direction of net Ca movement depends upon the prevailing membrane potential and the Na and Ca equilibrium potentials (see text). In the "reverse mode" of operation, the Na/Ca exchanger moves Ca in parallel with the Ca movement through Ca-selective channels (not shown here; see Fig. 1). The Na/Ca exchanger can also mediate Ca/Ca (tracer) exchange (3). The ATP-driven Na pump (1) is shown at the top of the diagram. (BLAUSTEIN 1984)

they have stoichiometries of 2 Ca ions transported per ATP molecule hydrolyzed, and they are mediated by Ca-dependent ATPases with molecular weights of about 105 000.

In a number of types of cells, including neurons, the plasma membrane ATP-driven Ca pump has a high affinity for Ca (i.e., a K_D of the order of 0.2–0.3 μM), but a low transport capacity (Fig. 5; GILL et al. 1981; CARAFOLI 1984; SANCHEZ-ARMASS and BLAUSTEIN 1987). On the other hand, the bidirectional Na/Ca exchange system may have a somewhat lower affinity for Ca (K_D of about 0.5–1 μ*M*; BLAUSTEIN 1977), but much larger capacity (or maximum velocity of transport) (SANCHEZ-ARMASS and BLAUSTEIN 1987; and see Fig. 4). Also, the Na/Ca exchange system is regulated by (nontransported) intracellular Ca: Ca concentrations within the dynamic physiologic range (~0.1–1.0 μ*M*) are required to promote Ca entry (reverse mode exchange) as well as Ca exit (forward mode exchange) mediated by this transport system (DIPOLO and BEAUGE 1986; RASGADO-FLORES et al. 1986).

The importance of the Na-dependent transport system for extruding Ca from mammalian neurons is illustrated by the data in Fig. 5. When nerve terminals are given a small load of Ca, equivalent to the net Ca entry during about 100–125 action potentials, Ca exit is almost entirely external Na-dependent: the rate of tracer ^{45}Ca efflux is slowed by a factor of about 10 when most of the external Na is replaced by another monovalent cation (in this case, *N*-methylglucamine or choline). In this experiment, Ca was omitted from the external solution during the efflux period to minimize tracer ($^{40}Ca/^{45}Ca$) exchange. A number of monovalent cations were tested (including Na, Li, choline, *N*-methylglucamine, and tetramethylammonium). Under these conditions, only Na appeared to be able to activate Ca efflux. Moreover, Na flux data from neuroblastoma cells (WAKABAYASHI and GOSHIMA 1981) imply that Na is exchanged for the Ca.

Studies with the Ca-sensitive fluorochromes, quin2 and fura-2 (NACHSHEN et al. 1986), provide further evidence that the external Na-dependent Ca efflux plays an important role in $[Ca^{2+}]_i$ regulation in presynaptic nerve terminals. After a period of Ca loading in low Na media, net efflux of Ca and restoration of $[Ca^{2+}]_i$ to its initial low value occur rapidly (within 1 min) when external Na is replaced. Neither the Ca uptake from low Na media, nor the Na-dependent Ca extrusion, is affected by the mitochondrial uncoupler, carbonylcyanide-*p*-trifluoromethoxyphenylhydrazone (FCCP); the implication is that the mitochondria do not normally contain large stores of Ca. However, FCCP, which should markedly reduce intracellular ATP levels, does cause $[Ca^{2+}]_i$ to rise slightly; this may indicate that the ATP-driven Ca pump contributes to the control of resting $[Ca^{2+}]_i$ (see later in this section).

The direction in which the exchanger moves (net) Ca across the plasma membrane is determined by the difference, ΔV, between the membrane potential, V_M, and the reversal potential for the Na/Ca exchanger, $E_{Na/Ca}$ (SHEU and BLAUSTEIN 1986):

$$\Delta V = V_M - E_{Na/Ca} \tag{1}$$

When ΔV is negative, the exchanger will mediate net outward movement of Ca; the converse will be true when ΔV is positive. Thus, the direction of Ca transport is critically dependent upon the stoichiometry of the exchange,

$$(n-2)E_{Na/Ca} = nE_{Na} - 2E_{Ca} \tag{2}$$

where E_{Na} and E_{Ca} are, respectively, the equilibrium potentials for Na and Ca, and n is the number of Na ions exchanged for each Ca ion. In neurons, as well as in other types of cells, n has a value of about 3. For example, there is evidence that

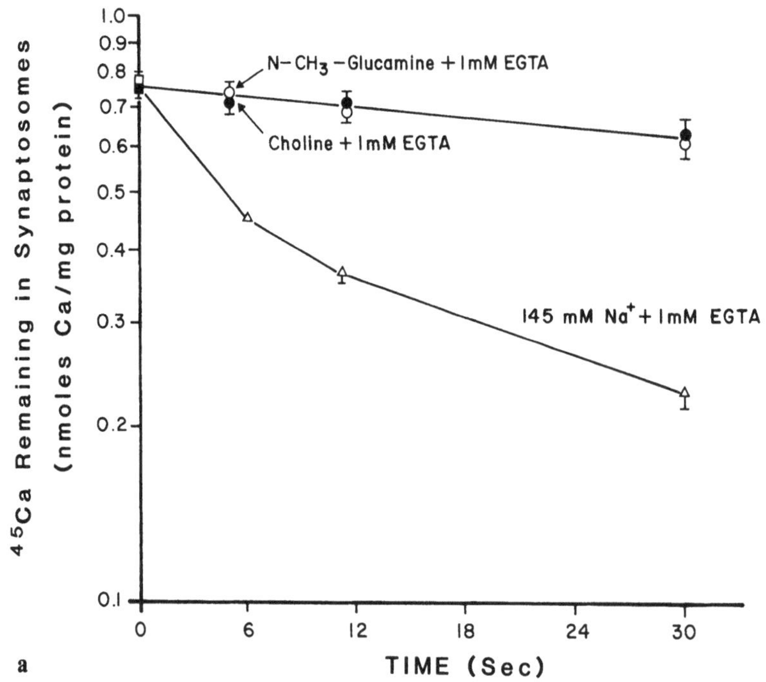

a

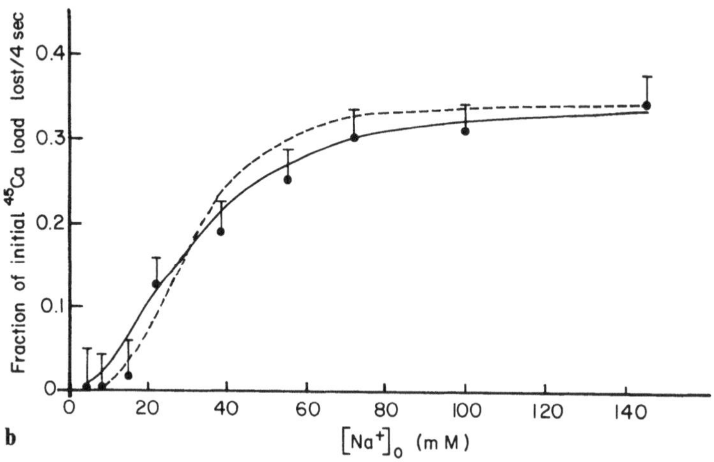

b

Fig. 5 a, b. The effect of external Na on the efflux of ^{45}Ca from tracer-loaded rat brain synaptosomes in the absence of external Na. **a** Time course of the Ca efflux into Ca-free media (with 1 m*M* EGTA) containing 145 m*M* (*triangles*), or only 4.5 m*M* Na and either 140 m*M* choline (*full circles*) or 140 m*M* *N*-methylglucamine (*open circles*), **b** relationship between the external Na concentration (abscissa) and the rate of Ca efflux into Ca-free media. When external Na was reduced below 145 m*M*, it was replaced mole-for-mole by *N*-methylglucamine. The curves are drawn to fit the Hill equation with a Hill coefficient of 2 (*full line*) or 3 (*broken line*). The very flat foot at low $[Na^+]_o$ fits the curve with the Hill coefficient of 3 best; this is consistent with a stoichiometry of 3 Na : 1 Ca. (SANCHEZ-ARMASS and BLAUSTEIN 1987)

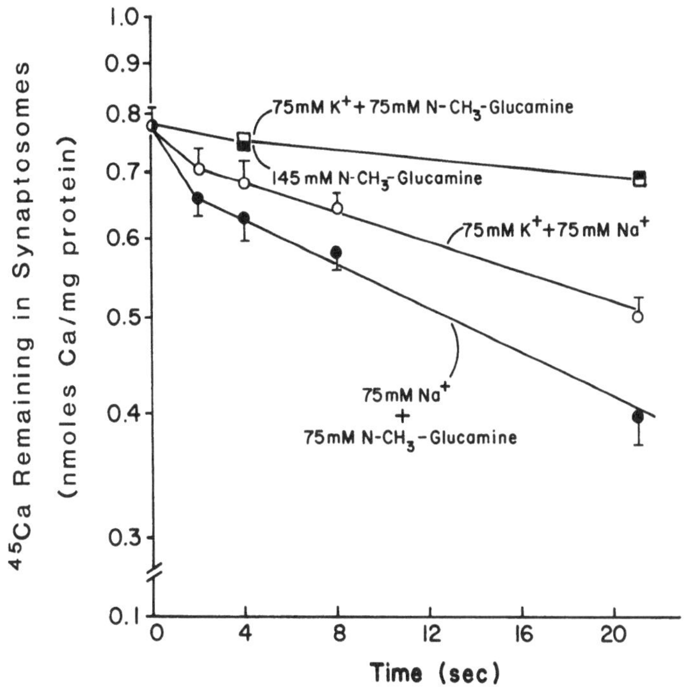

Fig. 6. Effect of depolarization on the efflux of ^{45}Ca from tracer-loaded rat brain synaptosomes. The Ca efflux was initiated by diluting the incubation media with a depolarizing (75 m*M* K) solution, without (*open squares*) or with (*open circles*) Na, or with a low (5 m*M*) K solution without (*full squares*) or with (*full circles*) Na. In all efflux solutions, the sum of the Na + K + *N*-methylglucamine concentrations was 150 m*M*. All the efflux solutions were Ca-free. (SANCHEZ-ARMASS and BLAUSTEIN 1987)

external Na-dependent Ca efflux is voltage sensitive: it is inhibited by depolarization (Fig. 6; BLAUSTEIN et al. 1987; ALLEN and BAKER 1986) and is stimulated by hyperpolarization (MULLINS and BRINLEY 1975; ALLEN and BAKER 1986). The stoichiometry has been measured directly by tracer flux methods in squid axons (BLAUSTEIN and RUSSELL 1975) and barnacle muscle cells (RASGADO-FLORES and BLAUSTEIN 1987b), and more indirectly in several other types of cells (cf. SHEU and BLAUSTEIN 1986), and has been found to be very close to 3 Na : 1 Ca in all cases. Based on this exchange stoichiometry and the measured values for E_{Na}, E_{Ca}, and V_M, ΔV should be slightly negative in resting neurons, so that Ca efflux will be favored under these circumstances (Fig. 7).

In addition to control by these thermodynamic factors, the Na/Ca exchange is also influenced by kinetic factors based on the fractional occupancy of the carriers by transported ions as well as by activating (nontransported) ions (DIPOLO and BEAUGE 1986; KIMURA et al. 1986; RASGADO-FLORES et al. 1986; RASGADO-FLORES and BLAUSTEIN 1987b). For example, studies in squid axons and barnacle muscle cells demonstrate that Ca entry mediated by the Na/Ca exchanger is dependent on (nontransported) intracellular Ca concentrations in the dynamic physiologic range (about 0.1–1.0 μ*M*) (RASGADO-FLORES et al. 1986). Thus, when

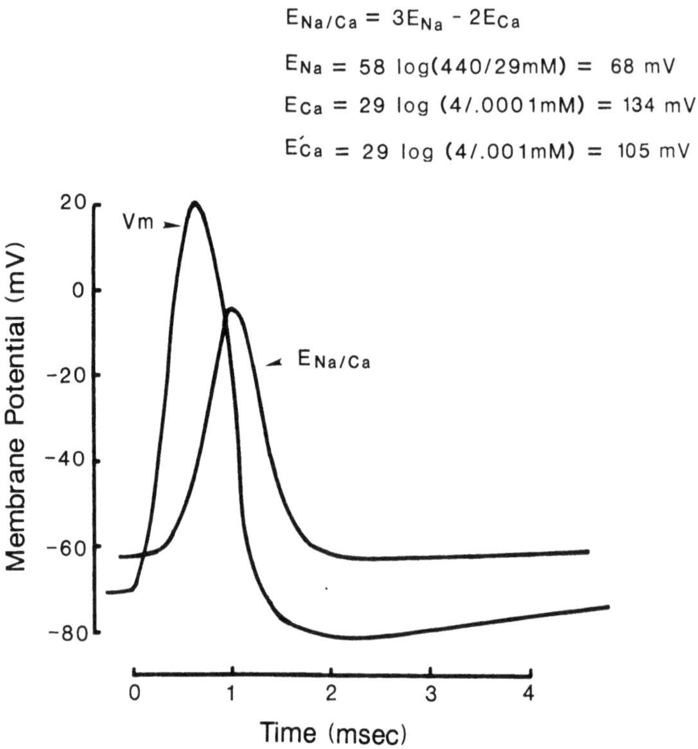

Fig. 7. Graph showing the membrane potential (V_M) and calculated Na/Ca exchanger reversal potential ($E_{Na/Ca}$) for the squid giant synapse during an action potential. $E_{Na/Ca}$ was calculated on the assumption that the concentrations of the free ions at rest are: $[Na^+]_o$ = 440 m*M*, $[Na^+]_i$ = 29 m*M*, $[Ca^{2+}]_o$ = 4 m*M*, and $[Ca^{2+}]_i$ = 0.1 μ*M*. During the action potential, $[Ca^{2+}]_i$ is assumed to increase to 1.0 μ*M* (i.e., E_{Ca} decreases from 134 mV to 105 mV). The exchanger mediates (net) Ca entry during the action potential (i.e., when V_M is more positive than $E_{Na/Ca}$). It mediates (net) Ca exit at the end of the action potential and during the afterhyperpolarization (i.e., when V_M is more negative than $E_{Na/Ca}$). See text for further details. (BLAUSTEIN 1987)

$[Ca^{2+}]_i$ is low, as in the case of resting neurons (~0.1–0.3 μ*M*), the turnover of the exchanger will be very low because the internal Ca sites that participate in Ca extrusion (K_{Ca} ~0.7 μ*M*; BLAUSTEIN 1977), as well as those that activate exchanger-mediated Ca entry (K_{Ca} ~0.6 μ*M*; RASGADO-FLORES et al. 1986) will be largely unoccupied. This means that, under resting conditions, even though the exchanger is poised to move Ca out of the cells (see Fig. 6), much of the net Ca extrusion may be mediated by the ATP-driven Ca pump. Nevertheless, the Na/Ca exchange system will bias the distribution of Ca in the neurons. $[Ca^{2+}]_i$ and the amount of Ca on the buffers and in the endoplasmic reticulum will still be influenced by the Na electrochemical gradient across the plasma membrane.

When the neurons are activated, a different situation prevails. When the membrane is depolarized, ΔV will become positive, and this driving force will now favor Ca entry via the Na/Ca exchanger. Initially, however, the internal Ca activation sites will still be largely unoccupied. But, once the Ca channels open and

Ca enters the cells, the exchanger will be activated and Ca will now also *enter* by (reverse mode) Na/Ca exchange; the relative roles of the Ca channels and the Na/Ca exchange system in mediating Ca entry are not yet clear. Now, with $[Ca^{2+}]_i$ still elevated, when the neurons repolarize, ΔV will become large and negative, and the exchanger will promote Ca extrusion until $[Ca^{2+}]_i$ declines and internal Ca-binding sites on the exchanger are no longer saturated. This anticipated relationship between V_M and $E_{Na/Ca}$ at the presynaptic nerve terminal of the squid giant synapse is shown diagrammatically in Fig. 7. The available data indicte that the Na/Ca exchange system is turned on during depolarization, as a result of positive feedback (by increasing $[Ca^{2+}]_i$), and turned off as a result of negative feedback (by decreasing $[Ca^{2+}]_i$), when the cells repolarize.

V. The "Life Cycle" of Calcium at the Nerve Terminal

We can now obtain a comprehensive picture of how Ca is handled by the terminals during an action potential (see Fig. 1). During the rising phase of the action potential, Ca will enter the terminals via Ca channels and, with a slight delay, via the Na/Ca exchanger. The increase in $[Ca^{2+}]_i$ will trigger transmitter release. Then, as the membrane repolarizes, the Ca channels will close, and some Ca will exit through the Na/Ca exchanger. However, much of the Ca that had just entered will diffuse away from the plasma membrane and will be buffered (perhaps by parvalbumin); these processes will markedly lower $[Ca^{2+}]_i$ in the region of the transmitter release sites with a half-time of about 1 ms. Then, during the next 10 ms or so, the Ca bound to the buffer proteins will be removed and sequestered in the endoplasmic reticulum (Fig. 3). Finally, over the subsequent 0.1–1 s, the Ca extrusion mechanisms (Na/Ca exchange and the ATP-driven Ca pump) will transport, back to the extracellular fluid, the remainder of the Ca that had entered and been sequestered.

During a rapid train of action potentials, $[Na^+]_i$ will tend to rise with time, owing to a slight lag in the ability of the Na pump to extrude the Na that enters during the action potentials (Woodbury 1963). As a result of this reduction in the Na concentration gradient, Na/Ca exchange-mediated Ca entry will be enhanced, and Ca exit will be reduced, thereby causing the interspike $[Ca^{2+}]_i$ to rise slightly. In addition, a slight lag in the sequestration of Ca by the endoplasmic reticulum will cause the cytosolic buffers to begin to saturate. These factors will, in turn, augment neurotransmitter release – as manifested by facilitation and PTP (Charlton and Atwood 1977; Charlton et al. 1980; Atwood et al. 1983; Misler and Hurlbut 1983; Meiri et al. 1986).

C. The Role of Intracellular Calcium in Synaptic Transmission

In the preceding sections of this chapter we examined the various mechanisms involved in the movements of Ca across the plasma membrane and in the regulation of $[Ca^{2+}]_i$. We will now turn our attention to some of the critical physiologic actions of this intracellular Ca (messenger Ca) in neurons.

I. Calcium and Neurotransmitter Release

A rise in $[Ca^{2+}]_i$ is the immediate trigger for secretion in most types of secretory cells in which the secreted substances are stored in vesicles; this is usually the result of Ca entry from the extracellular fluid (cf. DOUGLAS 1974; RUBIN 1982; BURGOYNE 1984). This is also the case for neurotransmitter release: release is critically dependent upon extracellular Ca, and is normally initiated by the opening of Ca channels to permit a large increase in Ca entry and rise in $[Ca^{2+}]_i$ (Fig. 8; and cf. KATZ 1969; BAKER 1972; LLINAS et al. 1981; ZUCKER and LANDO 1986). The rise in $[Ca^{2+}]_i$ promotes fusion of synaptic vesicles with the plasma membrane, thereby inducing the exocytosis of vesicular stores of one or more transmitters, and other molecules, into the synaptic cleft (cf. KATZ 1969; BLAUSTEIN 1978; DRAPEAU and BLAUSTEIN 1983b). Nevertheless, the precise mechanism by which this Ca actually evokes release is uncertain. Data from various neuronal preparations indicate that the cooperative action of several Ca ions is required to activate the release of one packet (quantum) of transmitter, i.e., the amount contained in one synaptic vesicle (DODGE and RAHAMIMOFF 1967; AUGUSTINE et al. 1985; ZUCKER and FOGELSON 1986; but see NACHSHEN and DRAPEAU 1982).

The role of Ca does not appear to be limited to the neutralization of negative surface charges on the inner surface of the plasma membrane and outer surface of the vesicle membrane (cf. PARSEGIAN 1977). While such an effect would allow these two membranes to come together (as observed with high concentrations of Mg or La), such apposition or fusion is not necessarily associated with exocytosis (cf. HEUSER et al. 1971; HEUSER 1977).

Nerve terminals contain relatively high concentrations of contractile proteins and calmodulin, and several authors have suggested that these molecules may play a role in evoked transmitter release (e.g., BERL et al. 1973; BAKER and KNIGHT 1981; BURKE and DELORENZO 1982; DELORENZO 1982, 1983). However, the possible mechanisms by which these molecules may participate in the transmitter release process are not known.

Secretory mechanisms in adrenal chromaffin cells, which are homologous to neurons, are probably very similar to the mechanisms involved in neurotransmitter release. Thus, data on secretion from the adrenal chromaffin cells (e.g., KNIGHT and BAKER 1982, 1985; BURGOYNE 1984) may be directly applicable to neurons. Available data from permeabilized chromaffin cells suggest that the intracellular Ca concentration that promotes half-maximal secretion is about 1 μM; (intracellular) Mg inhibits Ca-dependent secretion with an apparent IC_{50} of about 1 mM (BAKER and KNIGHT 1981; KNIGHT and BAKER 1982). A phosphorylation step appears to be involved in secretion because removal of ATP cause a progressive and reversible inhibition of secretion, and nonhydrolyzable analogs of ATP can not substitute for the ATP (BAKER and KNIGHT 1981). Recently, GTP-binding proteins have also been implicated in Ca-dependent secretion from chromaffin cells (KNIGHT and BAKER 1985).

Attempts to elucidate the role of calmodulin in secretion, by employing calmodulin inhibitors (cf. BAKER and KNIGHT 1981), have been inconclusive – in part because most of the agents employed, such as the phenothiazines, are not very selective. However, recent studies on insulin secretion have led to the suggestion

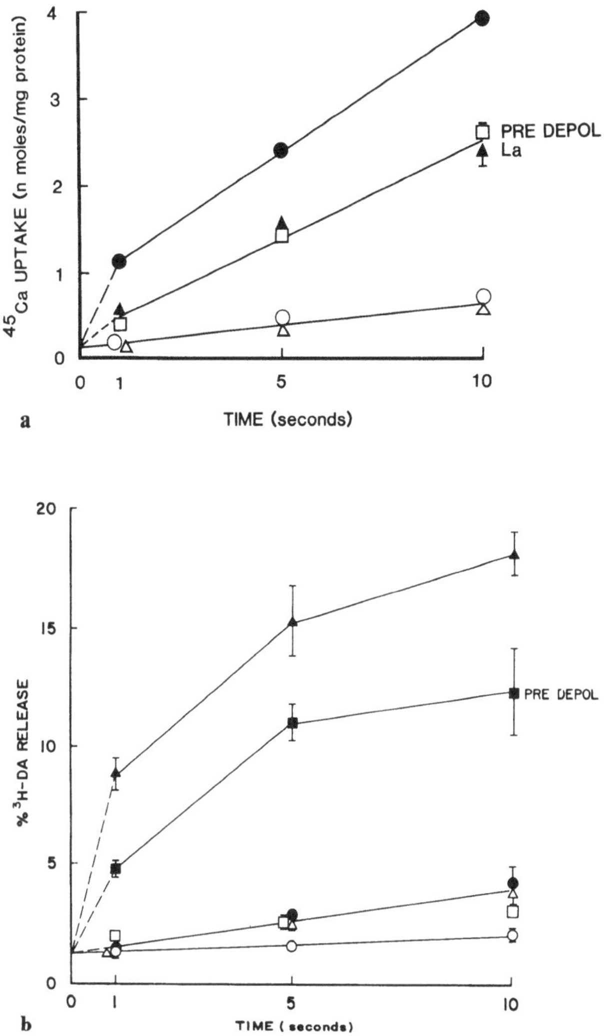

Fig. 8 a, b. Time course of ^{45}Ca uptake **a** and [^{3}H] dopamine (^{3}H-DA) release **b** in synaptosomes from rat brain corpus striatum. **a** Ca uptake was measured in nondepolarizing (145 m*M* Na + 5 m*M* K) media (*open circles, open triangles*) or depolarizing (75 m*M* Na + 75 m*M* K) media (*full circles, full triangles, open squares*). In some instances, the solutions contained 0.1 m*M* La (*open triangles, full triangles*) to block Ca uptake through Ca channels. Some synaptosomes were predepolarized in Ca-free, 75 m*M* K media immediately before the ^{45}Ca uptake was initiated, to inactivate most of the Ca channels (*open squares*), **b** dopamine release (as a percentage of the [^{3}H] dopamine previously taken up by the synaptosomes) under conditions similar to those used for the Ca uptake experiment (in **a**). Dopamine release was measured in 145 m*M* Na + 5 m*M* K media (*open circles, full circles*) or in depolarizing 75 m*M* Na + 75 m*M* K media (*open triangles, full triangles*) that were either Ca-free (*open circles, open triangles*) or contained 2 m*M* Ca (*full circles, full triangles*). Some synaptosomes were predepolarized in Ca-free, 75 m*M* Na + 75 m*M* K media and then incubated (during the [^{3}H] dopamine efflux period) in Ca-free 75 m*M* K media (*open squares*) or in 75 m*M* K media containing 2 m*M* Ca (*full squares*). Note the parallelism, under comparable conditions, between the rates of Ca uptake (**a**) and dopamine release (**b**). (Drapeau and Blaustein 1983 a)

that cyclic AMP and calmodulin may induce secretion by acting at different steps in the secretory process, and that the fusion of secretory granules with the plasma membrane may be a Ca-calmodulin-directed function (Steinberg et al. 1984). Whether or not this is applicable to neurotransmitter release as well, remains to be determined.

All of these observations provide only indirect information about the mechanism of secretion and neurotransmitter release. The precise molecules and mechanisms involved remain an enigma. However, possible clues to the mechanism may come from recent attempts to "reconstitute" the Ca-dependent exocytosis process with isolated secretory vesicles and plasma membrane from sea urchin eggs (Crabb and Jackson 1985). In this study, release of vesicular contents was evoked by Ca-dependent fusion of the secretory vesicles with patches, a "lawn," of plasma membrane. It may be possible to apply similar methods to the study of neurotransmitter release.

II. Dissection of the Steps in Transmitter Release with Toxins and Drugs

A number of selective toxins and drugs have been employed in efforts to dissect the sequence of events that begins with the invasion of the nerve terminal by an action potential and ends with the release of transmitter. The first step is the depolarization-induced opening of Ca channels, followed by net Ca entry. A number of agents, including some tetraalkylamines, aminopyridines, phencyclidines (see Bartschat and Blaustein 1986), and dendrotoxin, a polypeptide toxin component of green mamba venom (Dolly et al. 1984), all selectively block K channels. They thereby prolong the action potential and, thus, enhance Ca entry and evoked transmitter release. Their central effects include the induction of bizarre behavior patterns and seizures, which may be attributed to the excessive release of multiple neurotransmitters (see Bartschat and Blaustein 1986; Dolly et al. 1984).

β-Bungarotoxin, another presynaptically acting polypeptide toxin from a snake venom, whose B-chain has structural homologies to dendrotoxin, increases and then blocks transmitter release. The initial potentiation of release may be due to block of K channels by this toxin (Petersen et al. 1986). β-Bungarotoxin also has a Ca-dependent, Sr-inhibitable phospholipase A_2 activity (Abe and Miledi 1978) that is probably responsible for the block of transmitter release without depletion of vesicular stores of transmitter (Oberg and Kelly 1976).

A polypeptide toxin from black widow spider venom, α-latrotoxin (Nicholls et al. 1982), and crude brown widow spider venom (Pumplin and Reese 1977) transiently increase spontaneous transmitter release; the subsequent inhibition of release is associated with depletion of synaptic vesicles. There is evidence that α-latrotoxin, itself, forms cation-selective channels in lipid bilayers (Robello et al. 1984). However, in a rat pheochromocytoma cell line (PC12), α-latrotoxin opens a non-inactivating cation-selective channel that permits entry of both Na and Ca (Wanke et al. 1986). Thus, in addition to allowing Ca entry directly, the toxin depolarizes terminals and presumably opens VOCs to enable massive Ca entry and transmitter release (Nicholls et al. 1982).

Botulinus and tetanus toxins are particularly interesting: they block transmitter release, but do not prevent Ca entry, nor do they promote synaptic vesicle depletion. Evidence indicates that the action of these toxins involves three steps: (a) binding of the toxin to the presynaptic plasma membrane; (b) transport of the toxin into the cytoplasm; and (c) block of the fusion of the synaptic vesicles with the plasma membrane (SIMPSON 1980; SCHMITT et al. 1981). The toxin translocation step requires the activation of Ca-dependent transmitter release; this step presumably the toxin is then internallized along with recycling synaptic vesicle membrane. The mechanism (or mechanisms) by which the internalized botulinus and tetanus toxins prevent the fusion of synaptic vesicle membrane with plasma membrane is unkown. It must occur at a step after the entry of Ca because Ca entry is not blocked by these toxins (BIGALKE et al. 1981). Furthermore, in the presence of Ca, botulinus toxin inhibits spider venom-induced fusion of synaptic vesicles with the plasma membrane (PUMPLIN and REESE 1977). Elucidation of this mechanism may provide important insight into the way in which Ca normally promotes synaptic vesicle–plasma membrane fusion. The fact that botulinus toxin also inhibits Ca-activated catecholamine exocytosis in permeabilized adrenal medullary cells (KNIGHT et al. 1985) appears to confirm the similarity between the secretory mechanisms in these adrenal cells and nerve terminals.

III. Calcium and the Control of Excitability in Neurons

In Sect. A, we noted that extracellular Ca influences membrane excitability. Under normal physiologic conditions, however, extracellular divalent cation concentrations are usually well regulated within rather narrow limits. In contrast, as we have seen, $[Ca^{2+}]_i$ may vary substantially during cell activity, and this may markedly influence cell excitability.

MEECH (1972) was the first to demonstrate clearly that intracellular Ca can activate a K conductance. Numerous investigators have expanded on these early observations and have shown that a rise in $[Ca^{2+}]_i$ increases membrane conductance and causes hyperpolarization in many types of vertebrate and invertebrate neurons (e.g., MEECH 1978; GORMAN and HERMANN 1982; HIGASHI et al. 1984). Presynaptic nerve terminals in mammalian brain (BARTSCHAT and BLAUSTEIN 1985) and in mammalian peripheral nerve (MALLART 1984) appear to be richly endowed with Ca-activated K channels. Indeed, data from the brain (LANCASTER and ADAMS 1986; BENISHIN et al. 1986) and sympathetic ganglia (PENNEFATHER et al. 1985) indicate that there may be several types of Ca-activated K channels.

Single-channel studies have demonstrated that neuronal membranes contain Ca-activated K channels that are opened by $[Ca^{2+}]_i$ in the dynamic physiologic range (about 0.1–1 μM). In some cases, these channels appear to be blocked by phosphorylation that is promoted by the catalytic subunit of cyclic AMP-dependent protein kinase (BARTSCHAT et al. 1986). In other instances, Ca-activated K conductance is enhanced by cyclic AMP-dependent protein phosphorylation which increases the probability of channel opening in the presence of Ca (DE PEYER et al. 1982; EWALD et al. 1985).

The precise physiologic roles of these forms of channel modulation are not completely understood. Nevertheless, it is apparent that, at least in some types of cells, membrane excitability can be controlled by "second messengers" such as cyclic nucleotides. In turn, the cyclic AMP may be controlled by receptor-regulated GTP-binding proteins, or "G-proteins" (Gilman 1987). Excitability can then be altered either by modulation of the Ca channels (see Sect. B.II) or by modulation of the Ca-activated K channels (and, perhaps, voltage-operated K channels: e.g., Breitwieser and Szabo 1985; Pfaffinger et al. 1985; Sakakibara et al. 1986). These covalent modifications of the ion channels (e.g., phosphorylation: cf. Levitan 1985; Nairn et al. 1985) would be expected to produce relatively long-lasting changes, and are thus a form of "memory." Indeed, this repertoire of ways to modify neuronal membrane excitability provides multiple opportunities for modulating and fine-tuning synaptic transmission.

Intracellular Ca may also modulate other types of ion channels. One such example is its inhibition of the GABA-activated chloride conductance in sensory neurons from the bullfrog dorsal root ganglion. Calcium, when entering the cells through Ca-selective channels, appears to decrease the affinity of the membrane receptor for GABA (Inoue et al. 1986).

The relative contributions of the various types of conductances may differ greatly from cell type to cell type, and this will have a profound effect on the cell function. For example, in some cells, the Ca-activated K conductance may be quite long-lasting, especially after a burst of activity, and this may greatly suppress subsequent firing. Interesting examples are seen in the beating and bursting pacemaker neurons in the abdominal ganglion of the marine mollusk *Aplysia californica* (Gorman and Hermann 1982). The bursting pacemaker (cell R-15), whose cell cycle normally consists of a rapid series of action potentials followed by a long quiescent period, has a larger Ca conductance, and a larger Ca-activated K conductance than does the beating pacemaker (cell L-11). The interburst intervals in the bursting pacemaker can be accounted for by the large, long-lasting hyperpolarization, due to the increase in Ca-activated K conductance, that occurs after a burst of activity (with a relatively large influx of Ca). The implication is that $[Ca^{2+}]_i$ remains elevated for a substantial period of time following the burst. The low frequency, steady repetitive discharge that is normally seen in L-11 can be mimicked in R-15 simply by withdrawing extracellular Ca; this is consistent with the view that the large Ca influx and large Ca-activated K conductance in R-15 is responsible for the bursting activity.

In mammalian cells, it may be possible to determine the role of Ca in regulating activity by intracellular injection of EGTA or of Ca. For example, the inhibition of firing rate followed a period of activation in spontaneously firing locus coeruleus neurons is associated with an afterhyperpolarization that is likely due to a Ca-dependent K conductance. The afterhyperpolarization is abolished, and the spontaneous firing rate and reactivity to sensory stimulation are increased following injection of EGTA into these neurons (Aghajanian et al. 1983).

At many types of synapses, a period of high frequency firing (tetanus) is often followed by a period of enhanced transmitter release in response to subsequent action potentials (i.e., PTP). This is usually attributed to the residual Ca in the cytoplasmas after the burst of activity (see Sect. B.V). PTP is then usually followed by a period of post-tetanic depression. Although this depression has sometimes been attributed to an exhaustion of transmitter stores or inactivation of the release process, a likely alternative is that a large Ca-activated K conductance may cause an afterhyperpolarization and reduce membrane excitability (KRETZ et al. 1982). The latter study, in fact, supports the "residual Ca" hypothesis, and suggests that Ca may be extreded only relatively slowly after a burst of neuronal activity (cf. BOLSOVER and SPECTOR 1986; CONNOR et al. 1986). Alternatively, however, we must also consider the possibility that a relatively brief elevation of $[Ca^{2+}]_i$ might trigger metabolic changes which could then activate K channels for a prolonged period, perhaps as a result of phosphorylation (cf. HIGASHI et al. 1984; MILLER and KENNEDY 1986).

IV. Calcium and Memory

Facilitation, PTP, and afterhyperpolarization are all examples of neuronal activity modulation that are a direct consequence of Ca retention in neurons (cf. BOLSOVER and SPECTOR 1986; CONNOR et al. 1986). We may consider these altered responses, that occur as a result of prior activation of the neurons, manifestations of very brief "memory" – forms of "memory" that last many seconds or a few minutes. Somewhat longer term retention of information from prior neuronal activation (perhaps lasting several minutes or hours), commonly referred to as "short-term memory," may result from covalent modification of ion channels; for example, channel phosphorylation, as mentioned already. In several instances, this long-term potentiation of transmitter release in the mammalian central nervous system has been attributed to increased Ca-dependent release of neurotransmitter (FEASEY et al. 1986). It is possible that this may be the result of alterations in the Ca channels or Ca-activated K channels. In invertebrate systems, some conductance changes produced by protein kinase C activation have been found to mimic those induced by training (FARLEY and AUERBACH 1986). And, in the mammalian central nervous system, phorbol esters, which activate protein kinase C, block a Ca-dependent K conductance (BARABAN et al. 1985) and potentiate synaptic transmission in much the same way as they affect long-term potentiation (MALENKA et al. 1986).

There is also some evidence that long-term potentiation may involve postsynaptic mechanisms. LYNCH and BAUDRY (1984) have suggested that a rise in $[Ca^{2+}]_i$ in the postsynaptic cells may activate a protease (calpain) that degrades a spectrin-like protein (fodrin) and thereby increases, irreversibly, the number of glutamate receptors in the plasma membrane.

Another mechanism by which transient elevation of $[Ca^{2+}]_i$ may lead to a relatively long-term increase in membrane channel phosphorylation involves the Ca-calmodulin-activated autophosphorylation of brain type II protein kinase (MILLER and KENNEDY 1986). With very brief elevation of $[Ca^{2+}]_i$, this protein kinase activity is greatly enhanced by interaction with Ca and calmodulin. With more prolonged elevation of $[Ca^{2+}]_i$, however, the Ca-calmodulin complex promotes autophosphorylation of the kinase and, in this state, the protein kinase activity becomes independent of Ca (and calmodulin). This Ca-independent state may be terminated by dephosphorylation or it may continue for the life of the molecule. Thus, this protein kinase system behaves like a molecular switch that can go from a transiently activatable state to a prolonged or permanently active state.

The forms of memory mentioned do not involve protein synthesis, and are not disrupted by blockers of protein synthesis. In contrast, blockers of protein synthesis do interfere with the development of long-term memory that normally lasts days, weeks, or even years (cf. FLEXNER et al. 1963; DAVIS and SQUIRE 1984). While this implies that activation of genes and induction of messenger RNA and protein synthesis is likely involved in this type of memory, it does not rule out a role for Ca. Indeed, Ca may be needed, either because the induction of short-term memory may be a critical prerequisite for the induction of long-term memory and/or because Ca may be required to activate the genetic mechanisms involved in cell growth and differentiation that are associated with long-term memory. The evidence that short-term and long-term memory may involve the same synaptic connections (cf. GOELET et al. 1986) is consistent with the view that different Ca-dependent mechanisms (see earlier in this section), in the same neurons, may be responsible.

Recent observations in the pheochromocytoma PC12 cell line indicate that the proto-oncogene, c-*fos*, which is known to be involved in cell differentiation and growth (cf. ROSENGURT 1986), is rapidly activated by stimulation of acetylcholine receptors, and by opening of Ca channels, or by growth factors that elevate $[Ca^{2+}]_i$ (GREENBERG et al. 1986; MORGAN and CURRAN 1986). This regulatory gene is normally activated by the same cytosolic messengers that are involved in short-term memory: Ca, cyclic AMP, and diacylglycerol (ROSENGURT 1986). GOELET and KANDEL (1986; and see GOELET et al. 1986) have suggested that activation of some proto-oncogenes may be an early step in the sequence of events that leads to the establishment of long-term memory. While the precise sequence is not clear, there is evidence that long-term memory in both vertebrates (GREENOUGH 1984) and invertebrates (BAILEY and CHEN 1983; LNENICKA et al. 1986) is associated with the growth of synaptic contacts.

D. Summary and Conclusions

In this chapter we have reviewed the subject of Ca transport and regulation of $[Ca^{2+}]_i$ in neurons, and have examined some of the important "second messenger" roles of intracellular Ca in relation to neurotransmitter release. It is clear that intracellular Ca plays a key role in the control of neuronal excitability, and this, in turn, has a profound influence on the release of neurotransmitter substances. The cellular correlates of learning and memory are also critically dependent upon the messenger role of intracellular Ca.

Acknowledgments. The author is grateful to Mr. R. V. Parsey for permission to include data from an unpublished experiment, and to Drs. C. G. Benishin, K. A. Colby, B. K. Krueger, H. Rasgado-Flores and R. G. Sorensen for helpful comments on a preliminary version of this chapter. Some of the research described in this article was supported by a grant from the NIH (NS-16106). This chapter is excerpted and revised from a previously published article (BLAUSTEIN 1987): the figures and portions of the text are reproduced here with the permission of Springer-Verlag.

References

Abe T, Miledi R (1978) Inhibition of β-bungarotoxin action by bee venom phospholipase A_2. Proc R Soc Lond B 200:225–230

Aghajanian GK, Vandermaelen CP, Andrade R (1983) Intracellular studies on the role of calcium in regulating the activity and reactivity of locus coeruleus neurons in vivo. Brain Res 273:237–243

Allen TJA, Baker PF (1986) Influence of membrane potential on calcium efflux from giant axons of *Loligo*. J Physiol (Lond) 378:77–96

Alvarez-Leefmans FJ, Gamino SM, Rink TJ (1984) Intracellular free magnesium in neurones of *Helix aspersa* measured with ion-selective micro-electrodes. J Physiol (Lond) 354:303–317

Atwood HL, Charlton MP, Thompson CS (1983) Neuromuscular transmission in crustaceans is enhanced by a sodium ionophore, monensin, and by prolonged stimulation. J Physiol (Lond) 335:179–195

Augustine GJ, Charlton MP, Smith SJ (1985) Calcium entry and transmitter release at voltage-clamped nerve terminals of squid. J Physiol (Lond) 367:163–181

Bailey CH, Chen M (1983) Morphological basis of long-term habituation and sensitization in *Aplysia*. Science 220:91–93

Baker PF (1972) Transport and metabolism by calcium ions in nerve. Prog Biophys Mol Biol 24:177–223

Baker PF, Knight DE (1981) Calcium control of exocytosis and endocytosis in bovine adrenal medullary cells. Phil Trans R Soc Lond B 296:83–103

Baraban JM, Snyder SH, Alger BE (1985) Protein kinase C regulates ionic conductance in hippocampal pyramidal neurons: electrophysiological effects of phorbol esters. Proc Natl Acad Sci USA 82:2538–2542

Bartschat DK, Blaustein MP (1985) Calcium-activated potassium channels in isolated presynaptic nerve terminals from rat brain. J Physiol (Lond) 361:441–457

Bartschat DK, Blaustein MP (1986) Phencyclidine in low doses selectively blocks a presynaptic voltage-regulated potassium channel in rat brain. Proc Natl Acad Sci USA 83:189–192

Bartschat DK, French RJ, Nairn AC, Greengard P, Krueger BK (1986) Cyclic AMP-dependent protein kinase modulation of single calcium-activated potassium channels from rat brain in planar bilayers. Soc Neurosci Abstr 12:1198

Benishin CG, Krueger BK, Blaustein MP (1986) Low micromolar concentrations of phenothiazines and haloperidol selectively block Ca-activated K channels in rat brain synaptosomes. Soc Neurosci Abstr 12:1199

Berl S, Puszkin S, Nicklas WJ (1973) Actomyosin-like protein in brain. Science 179:441–446

Berridge MJ, Irvine RF (1984) Inositol triphosphate, a novel second messenger in cellular signal transduction. Nature 312:315–321

Bigalke H, Ahnert-Hilger G, Habermann E (1981) Tetanus toxin and botulinum A toxin inhibit acetylcholine release from but not calcium uptake into brain tissue. Naunyn-Schmiedebergs Arch Pharmacol 316:143–148

Blaustein MP (1974) The interrelationship between sodium and calcium fluxes across cell membranes. Rev Physiol Biochem Pharmacol 70:32–82

Blaustein MP (1977) Effects of internal and external cations and of ATP on sodium-calcium exchange in squid axons. Biophys J 20:79–111

Blaustein MP (1978) The role of calcium in catecholamine release from adrenergic nerve terminals. In: Paton DM (ed) The release of catecholamines from adrenergic neurons. Pergamon, Oxford, pp 39–58

Blaustein MP (1984) The energetics and kinetics of sodium-calcium exchange in barnacle muscles, squid axons and mammalian heart: the role of ATP. In: Blaustein MP, Lieberman M (eds) Electrogenic transport: fundamental principles and physiological implications. Raven, New York, pp 129–147

Blaustein MP (1985) Intracellular calcium as a second messenger. What's so special about calcium? In: Rubin RP, Weiss GB, Putney JW Jr (eds) Calcium in biological systems. Plenum, New York, pp 23–33

Blaustein MP (1987) Neuronal cell calcium. In: Nordin BEC (ed) Calcium in human biology. Springer, Berlin Heidelberg New York (Human nutrition reviews, to be published)
Blaustein MP, Rasgado-Flores H (1981) The control of cytoplasmic free calcium in presynaptic nerve terminals. In: Bronner F, Peterlik M (eds) Calcium and phosphate transport across biomembranes. Academic, New York, pp 53–58
Blaustein MP, Russell JM (1975) Sodium-calcium exchange and calcium-calcium exchange in internally dialyzed squid giant axons. J Membrane Biol 22:285–312
Blaustein MP, Russell JM, De Weer P (1974) Calcium efflux from internally dialyzed squid axons: the influence of external and internal cations. J Supermolec Structure 2:558–581
Blaustein MP, Ratzlaff RW, Schweitzer ES (1978) Calcium buffering in presynaptic nerve terminals. II. Kinetic properties of the non-mitochondrial Ca sequestration mechanism. J Gen Physiol 72:43–66
Blaustein MP, McCraw CF, Somlyo AV, Schweitzer ES (1980) How is the cytoplasmic calcium concentration controlled in nerve terminals? J Physiol (Paris) 76:459–470
Blinks JR, Wier WG, Hess P, Prendergast FG (1982) Measurements of Ca^{2+} concentrations in living cells. Prog Biophys Mol Biol 40:1–114
Bolsover SR, Spector I (1986) Measurements of calcium transients in the soma, neurite, and growth cone of single cultured neurons. J Neurosci 6:1934–1940
Bossu JL, Feltz A, Thomann JM (1985) Depolarization elicits two distinct calcium currents in vertebrate sensory neurones. Pflugers Arch 403:360–368
Breitwieser GE, Szabo G (1985) Uncoupling of cardiac muscarinic receptors from ion channels by a guanine nucleotide analogue. Nature 317:538–540
Burgoyne RD (1984) Mechanisms of secretion from adrenal chromaffin cells. Biochim Biophys Acta 779:201–216
Burke BE, DeLorenzo RJ (1982) Ca^{2+}- and calmodulin-dependent phosphorylation of endogenous synaptic vesicle tubulin by a vesicle-bound calmodulin kinase system. J Neurochem 38:1205–1218
Carafoli E (1984) Plasma membrane Ca^{2+} transport, and Ca^{2+} handling by intracellular stores: an integrated picture with emphasis on regulation. In: Donowitz M, Sharp GWG (eds) Mechanisms of intestinal electrolyte transport and regulation by calcium. Liss, New York, pp 121–134
Celio MR (1986) Parvalbumin in most γ-aminobutyric acid-containing neurons of the rat cerebral cortex. Science 231:995–997
Charlton MP, Atwood HL (1977) Modulation of transmitter release by intracellular sodium in squid giant synapse. Brain Res 134:367–371
Charlton MP, Thompson CS, Atwood HL, Farnell B (1980) Synaptic transmission and intracellular sodium loading of nerve terminals. Neurosci Letters 16:193–196
Chueh S-H, Gill DL (1986) Inositol 1,4,5-triphosphate and guanine nucleotides activate calcium release from endoplasmic reticulum via distinct mechanisms. J Biol Chem 261:13883–13886
Connor J (1986) Digital imaging of free calcium changes and of spatial gradients in growing processes in single, mammalian central nervous system cells. Proc Natl Acad Sci USA 83:6179–6183
Connor JA, Kretz R, Shapiro E (1986) Calcium levels measured in a presynaptic neurone of *Aplysia* under conditions that modulate transmitter release. J Physiol (Lond) 375:625–642
Conway AK (1983) Intracellular calcium. Its universal role as regulator. Wiley, Chichester
Crabb JH, Jackson RC (1985) In vitro reconstitution of exocytosis from plasma membrane and isolated secretory vesicles. J Cell Biol 101:2263–2273
Davis HP, Squire LR (1984) Protein synthesis and memory: a review. Psychological Bull 96:518–559
DeLorenzo RJ (1982) Calmodulin in neurotransmitter release and synaptic function. Fed Proc 41:2265–2272
DeLorenzo RJ (1983) Calcium-calmodulin systems in psychopharmacology and synaptic modulation. Psychopharmacol Bull 19:393–397

Denton RM, McCormack JG (1985) Physiological role of Ca^{2+} transport by mitochondria. Nature 315:635
De Peyer JE, Cachelin AB, Levitan IB, Reuter H (1982) Ca^{2+}-activated K^+ conductance in internally perfused snail neurons is enhanced by protein phosphorylation. Proc Natl Acad Sci USA 79:4207–4211
De Weer P (1976) Axoplasmic free magnesium levels and magnesium extrusion from squid giant axons. J Gen Physiol 68:159–178
DiPolo R, Beauge L (1986) Reverse Na/Ca exchange requires internal Ca and/or ATP in squid axons. Biochim Biophys Acta 854:298–306
Dodge F, Rahamimoff R (1967) Co-operative action of calcium ions in transmitter release at the neuromuscular junction. J Physiol (Lond) 193:419–432
Dolly JO, Halliwell JV, Black JD, Williams RS, Pelchen-Matthews A, Breeze AL, Meheraban F, Othman IB, Black AR (1984) Botulinum neurotoxin and dendrotoxin as probes for studies on transmitter release. J Physiol (Paris) 79:280–303
Douglas WW (1974) Involvement of calcium in exocytosis and the exocytosis vesiculation sequence. Biochem Soc Symp 39:1–28
Drapeau P, Blaustein MP (1983a) Initial release of ^3H-dopamine from rat striatal synaptosomes: correlation with Ca entry. J Neurosci 3:703–713
Drapeau P, Blaustein MP (1983b) Calcium and neurotransmitter release: what we know and don't know. In: Kalsner S (ed) Trends in autonomic pharmacology, vol 2. Urban and Schwarzenberg, Baltimore, pp 117–130
Drapeau P, Nachshen DA (1984) Manganese fluxes and manganese-dependent neurotransmitter release in presynaptic nerve endings from rat brain. J Physiol (Lond) 348:493–510
Endo M (1977) Calcium release from the sarcoplasmic reticulum. Physiol Rev 57:71–108
Eusebi F, Miledi R, Parker I, Stinnakre J (1985) Post-synaptic calcium influx at the giant synapse of the squid during activation by glutamate. J Physiol (Lond) 369:183–197
Ewald DA, Williams A, Levitan IB (1985) Modulation of single Ca^{2+}-dependent K^+-channel activity by protein phosphorylation. Nature 315:503–506
Fabiato A (1985) Time and calcium dependence of activation and inactivation of calcium-induced release of calcium from the sarcoplasmic reticulum of skinned cardiac Purkinje cell. J Gen Physiol 85:247–289
Farley J, Auerbach S (1986) Protein kinase C activation induces conductance changes in *Hermissenda* photoreceptors like those seen in associative learning. Nature 319:220–223
Feasey KJ, Lynch MA, Bliss TVB (1986) Long-term potentiation is associated with an increase in calcium-dependent, potassium-stimulated release of [^{14}C]glutamate from hippocampal slices: an ex vivo study in the rat. Brain Res 364:39–44
Fleckenstein A (1985) Calcium antagonism in heart and vascular smooth muscle. Medicinal Res Rev 5:395–425
Flexner JB, Flexner LB, Stellar E (1963) Memory in mice as affected by intracerebral puromycin. Science 141:57–59
Flockerzi V, Oeken H-J, Hofmann F, Pelzer D, Cavalie A, Trautwein W (1986) Purified dihydropyridine-binding site from skeletal muscle t-tubules is a functional sodium channel. Nature 323:66–68
Frankenhaeuser B (1957) The effect of calcium on the myelinated nerve fibre. J Physiol (Lond) 137:245–260
Frankenhaeuser B, Hodgkin AL (1957) The action of calcium on the electrical properties of squid axons. J Physiol (Lond) 137:218–244
Gill DL, Grollman EF, Kohn LD (1981) Calcium transport mechanisms in membrane vesicles from guinea pig brain synaptosomes. J Biol Chem 256:184–192
Gill DL, Chueh S-H, Noel MW, Ueda T (1986a) Orientation of synaptic plasma membrane vesicles containing calcium pump and sodium-calcium exchange activities. Biochim Biophys Acta 856:165–173
Gill DL, Ueda T, Chueh S-H (1986b) Ca^{2+} release from endoplasmic reticulum is mediated by a guanine nucleotide regulatory mechanism. Nature 320:461–464
Gilman AG (1987) G proteins: transducers of receptor-regulated signals. Ann Rev Biochem 56:615–649

Goelet P, Kandel E (1986) Tracking the flow of learned information from membrane receptors to genome. Trends Neurosci 9:492–499
Goelet P, Castellucci V, Schacher S, Kandel ER (1986) The long and the short of long-term memory – a molecular framework. Nature 322:419–422
Gorman ALF, Hermann A (1982) Quantitative differences in the currents of bursting and beating molluscan pace-maker neurones. J Physiol (Lond) 333:681–699
Greenberg ME, Ziff EB, Greene LA (1986) Stimulation of neuronal acetylcholine receptors induces rapid gene transcription. Science 234:80–83
Greenough WT (1984) Possible structural substrates of plastic neuronal phenomena. In: Lynch G, McGaugh JL, Weinberger NM (eds) Neurobiology of learning and memory. Guilford, New York, pp 470–478
Hagiwara S, Byerly L (1981) Calcium channel. Ann Rev Neurosci 4:69–125
Hansford RG (1985) Relation between mitochondrial calcium transport and control of energy metabolism. Rev Physiol Biochem Pharmacol 102:1–72
Hartzell HC, Fischmeister R (1986) Opposite effects of cyclic GMP and cyclic AMP on Ca^{2+} current in single heart cells. Nature 323:273–275
Harvey AM, MacIntosh FC (1940) Calcium and synaptic transmission in a sympathetic ganglion. J Physiol (Lond) 97:408–416
Hashimoto T, Hirata M, Itoh T, Kanmura Y, Kuriyama H (1986) Inositol 1,4,5-triphosphate activates pharmacomechanical coupling in smooth muscle of the rabbit mesenteric artery. J Physiol (Lond) 370:605–618
Hasselbach W (1977) The sarcoplasmic reticulum calcium pump – a most efficient ion translocating system. Biophys Struct Mechanism 3:43–54
Heinonen E, Akerman KEO (1986) Measurement of cytoplasmic, free magnesium concentration with entrapped eriochrome blue in nerve endings isolated from the guinea pig brain. Neurosci Lett 72:105–110
Heizmann CW (1984) Parvalbumin, an intracellular calcium-binding protein; distribution, properties and possible roles in mammalian cells. Experientia 40:910–921
Heuser JE (1977) Synaptic vesicle exocytosis revealed in quick-frozen frog neuromuscular junctions treated with 4-aminopyridine and given a single electric shock. In: Cowan WM, Ferendelli JA (eds) Approaches to the cell biology of neurons. Society for Neuroscience symposia, vol 2. Society for Neuroscience, Bethesda, pp 215–239
Heuser J, Katz B, Miledi R (1971) Structural and functional changes of frog neuromuscular junctions in high calcium solutions. Proc R Soc Lond B 178:407–415
Higashi H, Morita K, North RA (1984) Calcium-dependent after potentials in visceral afferent neurones of the rabbit. J Physiol (Lond) 355:479–492
Hille B (1984) Ionic channels of excitable cells. Sinauer, Boston
Hincke MT, Demaille JG (1984) Calmodulin regulation of the ATP-dependent calcium uptake by inverted vesicles prepared from rabbit synaptosomal plasma membranes. Biochim Biophys Acta 771:188–194
Holz GG, Rane SG, Dunlap K (1986) GTP-binding proteins mediate transmitter inhibition of voltage-dependent calcium channels. Nature 319:670–672
Hutson SM, Pfeifer DR, Lardy HA (1976) Effect of cations and anions on the steady state kinetics of energy-dependent Ca^{2+} transport in rat liver mitochondria. J Biol Chem 251:5251–5258
Inesi G (1985) Mechanism of calcium transport. Ann Rev Physiol 47:573–601
Inoue M, Oomura Y, Yakushiji T, Akaike N (1986) Intracellular calcium ions decrease the affinity of the GABA receptor. Nature 324:156–158
Kameyama M, Hescheler J, Hofmann F, Trautwein W (1986) Modulation of Ca current during the phosphorylation cycle in the guinea pig heart. Pflugers Arch 407:123–128
Katz B (1969) The release of neural transmitter substances. Thomas, Springfield
Katz B, Miledi R (1967) A study of synaptic transmission in the absence of nerve impulses. J Physiol (Lond) 192:407–436
Katz B, Miledi R (1968) The role of calcium in neuromuscular facilitation. J Physiol (Lond) 195:481–492
Katz B, Miledi R (1969) Tetrodotoxin-resistant electrical activity in presynaptic terminals. J Physiol (Lond) 203:459–487

Kimura J, Noma A, Irisawa H (1986) Na-Ca exchange current in mammalian heart cells. Nature 319:596–598
Knight DE, Baker PF (1982) Calcium-dependence of catecholamine release from bovine adrenal medullary cells after exposure to intense electric fields. J Membrane Biol 68:107–140
Knight DE, Baker PF (1985) Guanine nucleotides and Ca-dependent exocytosis. Studies on two adrenal cell preparations. FEBS Lett 189:345–349
Knight DE, Tonge DA, Baker PF (1985) Inhibition of exocytosis in bovine adrenal medullary cells by botulinum toxin type D. Nature 317:719–721
Kostyuk PG (1981) Calcium channels in the neuronal membrane. Biochim Biophys Acta 650:128–150
Kretz R, Shapira E, Kandel ER (1982) Post-tetanic potentiation at an identified synapse in *Aplysia* is correlated with a Ca^{2+}-activated K^+ current in the presynaptic neuron: evidence for Ca^{2+} accumulation. Proc Natl Acad Sci USA 79:5430–5434
Lancaster B, Adams PR (1986) Calcium-dependent current generating the afterhyperpolarization of hippocampal neurons. J Neurophysiol 55:1268–1282
Levitan IB (1985) Phosphorylation of ion channels. J Membrane Biol 87:177–190
Llinas R, Steinberg IZ, Walton K (1981) Relationship between presynaptic calcium current and postsynaptic potential in squid giant synapse. Biophys J 33:323–352
Lnenicka GA, Atwood HL, Marin L (1986) Morphologic transformation of synaptic terminals of a phasic motoneuron by long-term tonic stimulation. J Neurosci 6:2252–2258
Lynch G, Baudry M (1984) The biochemistry of memory: a new and specific hypothesis. Science 224:1057–1063
Malenka RC, Madison DV, Nicoll RA (1986) Potentiation of synaptic transmission in the hippocampus by phorbol esters. Nature 321:175–177
Mallart A (1984) Calcium-activated potassium current in presynaptic terminals. Biomed Res 5:287–290
McCormack JG, Denton RM (1986) Ca^{2+} as a second messenger within mitochondria. Trends Biochem Sci (TIBS) 11:258–262
Meech RW (1972) Intracellular calcium injection causes increased potassium conductance in *Aplysia* nerve cells. Comp Biochem Physiol A 42A:493–499
Meech RW (1978) Calcium-dependent potassium activation in nervous tissues. Ann Rev Biophys Bioeng 7:1–18
Meiri H, Zellingher J, Rahamimoff R (1986) A possible involvement of the Na-Ca exchanger in regulation of transmitter release at the frog neuromuscular junction. In: Rahamimoff R, Katz B (eds) Calcium, neuronal function and transmitter release. Nijhoff, Amsterdam, pp 239–254
Miledi R, Parker I, Schalow G (1980) Transmitter induced calcium entry across postsynaptic membrane at frog end-plates measured using arsenazo III. J Physiol (Lond) 300:197–212
Miller SG, Kennedy MB (1986) Regulation of brain type II Ca^{2+}/calmodulin-dependent protein kinase by autophosphorylation: a Ca^{2+}-triggered molecular switch. Cell 44:861–870
Misler S, Hurlbut WP (1983) Post-tetanic potentiation of acetylcholine release at the frog neuromuscular junction develops after stimulation in Ca^{2+}-free solutions. Proc Natl Acad Sci USA 80:315–319
Morgan JI, Curran T (1986) Role of ion flux in the control of c-*fos* expression. Nature 322:552–555
Mullins LJ, Brinley FJ Jr (1975) Sensitivity of calcium efflux from squid axons to changes in membrane potential. J Gen Physiol 65:135–152
Nachshen DA (1984) Selectivity of the Ca binding site in synaptosome Ca channels. J Gen Physiol 83:941–967
Nachshen DA (1985) The early time course of potassium-stimulated calcium uptake in presynaptic nerve terminals from rat brains. J Physiol (Lond) 361:251–268
Nachshen DA, Blaustein MP (1979) The effects of some organic "calcium antagonists" on calcium influx in presynaptic nerve terminals. Molec Pharmacol 16:579–586

Nachshen DA, Blaustein MP (1980) Some properties of potassium-stimulated calcium influx in presynaptic nerve endings. J Gen Physiol 76:709–728
Nachshen DA, Blaustein MP (1982) The influx of calcium, strontium and barium in presynaptic nerve endings. J Gen Physiol 79:1065–1087
Nachshen DA, Drapeau P (1982) A buffering model for calcium-dependent neurotransmitter release. Biophys J 38:205–208
Nachshen DA, Sanchez-Armass S, Weinstein AM (1986) The regulation of cytosolic calcium in rat brain synaptosomes by sodium-dependent calcium efflux. J Physiol (Lond) 381:17–28
Nairn AC, Hemmings HC Jr, Greengard P (1985) Protein kinases in the brain. Annu Rev Biochem 54:931–976
Nicholls DG, Rugolo M, Scott IG, Meldolesi J (1982) α-Latrotoxin of black widow spider venom depolarizes the plasma membrane, induces massive calcium influx, and stimulates transmitter release in guinea pig brain synaptosomes. Proc Natl Acad Sci USA 79:7924–7928
Nowycky MC, Fox AP, Tsien RW (1985) Three types of neuronal calcium channel with different calcium agonist sensitivity. Nature 316:440–443
Oberg SG, Kelly RB (1976) The mechanism of β-bungarotoxin action. I. Modification of transmitter release at the neuromuscular junction. J Neurobiol 7:129–141
Parsegian VA (1977) Considerations in determining the mode of influence of calcium on vesicle membrane interaction. In: Cowan WM, Ferendelli JA (eds) Approaches to the cell biology of neurons. Society for Neuroscience symposia, vol 2. Society for Neuroscience, Bethesda, pp 161–171
Paupardin-Tritsch D, Hammond C, Gerschenfeld HM, Nairn AC, Greengard P (1986) cGMP-dependent protein kinase enhances Ca^{2+} current and potentiates the serotonin-induced Ca^{2+} current increase in snail neurones. Nature 323:812–814
Pechere J-F, Derancourt J, Haiech J (1977) The participation of parvalbumins in the activation-relaxation cycle of vertebrate fast skeletal muscle. FEBS Lett 75:111–114
Pellmar TC (1981) Ionic mechanism of a voltage dependent current elicited by cyclic AMP. Cell Molec Neurobiol 1:87–97
Pennefather P, Lancaster B, Adams PR, Nicoll RA (1985) Two distinct Ca-dependent K currents in bullfrog sympathetic ganglion cells. Proc Natl Acad Sci USA 82:3040–3044
Petersen M, Penner R, Pierau Fr-K, Dreyer F (1986) β-Bungarotoxin inhibits a non-inactivating potassium current in guinea pig dorsal root ganglion neurones. Neurosci Lett 68:141–145
Pfaffinger PJ, Martin JM, Hunter DD, Nathanson NM, Hille B (1985) GTP-binding proteins couple cardiac muscarinic receptors to a K channel. Nature 317:536–538
Pochet R, Parmentier M, Lawson DEM, Pasteels JL (1985) Rat brain synthesizes two "vitamin D-dependent" calcium-binding proteins. Brain Res 345:251–256
Pumplin DW, Reese TS (1977) Action of brown widow spider venom and botulinum toxin on the frog neuromuscular junction examined with the freeze fracture technique. J Physiol (Lond) 273:443–457
Rane SG, Dunlap K (1986) Kinase C activator 1,2-oleylacetylglycerol attenuates voltage-dependent calcium current in sensory neurons. Proc Natl Acad Sci USA 83:184–188
Rasgado-Flores H, Blaustein MP (1987a) ATP-dependent regulation of cytoplasmic free Ca^{2+} in nerve terminals. Am J Physiol 252 (Cell Physiol 21) C588–C594
Rasgado-Flores H, Blaustein MP (1987b) Na/Ca exchange in barnacle muscle cells has a stoichiometry of 3 Na^{+} : 1 Ca^{2+}. Am J Physiol 252 (Cell Physiol 21) C499–C504
Rasgado-Flores H, Santiago EM, Blaustein MP (1986) Calcium influx and sodium efflux mediated by the Na/Ca exchanger in giant barnacle muscle cells are promoted by intracellular Ca^{2+}. Biophys J 49:546a
Reuter H (1983) Calcium channel modulation by neurotransmitters, enzymes and drugs. Nature 301:569–574
Reynolds IJ, Wagner JA, Snyder SH, Thayer SA, Olivera BM, Miller RJ (1986) Brain voltage-sensitive calcium channel subtypes differentiated by ω-conotoxin fraction GVIA. Proc Natl Acad Sci USA 83:8804–8807

Robello M, Rolandi R, Alema S, Grasso A (1984) *trans*-Bilayer orientation and voltage-dependence of α-latrotoxin-induced channels. Proc R Soc Lond B 220:477–487
Rosengurt E (1986) Early signals in the mitogenic response. Science 234:161–166
Ross WN, Stockbridge LL, Stockbridge NL (1986) Regional properties of calcium entry in barnacle neurons determined with arsenazo III and a photodiode array. J Neurosci 6:1148–1159
Rubin RR (1982) Calcium and cellular secretion. Plenum, New York
Sakakibara M, Alkon DL, DeLorenzo R, Goldenring JR, Neary JT, Heldman E (1986) Modulation of calcium-mediated inactivation of ionic currents by a Ca^{2+}/calmodulin-dependent protein kinase II. Biophys J 50:319–327
Sanchez-Armass S, Blaustein MP (1987) Role of sodium/calcium exchange in the regulation of intracellular Ca^{2+} in nerve terminals. Am J Physiol 252 (Cell Physiol 21) C595–C603
Scarpa A (1976) Kinetic and thermodynamic aspects of mitochondrial calcium transport. In: Packer L, Gomez-Puyou A (eds) Mitochondria. Bioenergetics, biogenesis and membrane structure. Academic, New York, pp 31–45
Schlaepfer WW (1977) Structural alterations of peripheral nerve induced by the calcium ionophore A23187. Brain Res 136:1–9
Schmitt A, Dreyer F, John C (1981) At least three sequential steps are involved in the tetanus toxin-induced block of neuromuscular transmission. Naunyn-Schmiedeberg's Arch Pharmacol 317:326–330
Scott RH, Dolphin AC (1986) Regulation of calcium currents by a GTP analogue: potentiation of (−)-baclophen-mediated inhibition. Neurosci Lett 69:59–64
Sheu S-S, Blaustein MP (1986) Sodium/calcium exchange and the regulation of cell calcium and contractility in cardiac muscle, with a note about vascular smooth muscle. In: Fozzard HA, Haber E, Jennings RB, Katz AM, Morgan HE (eds) The heart and cardiovascular system. Raven, New York, pp 509–535
Simpson LL (1980) Kinetic studies on the interaction between botulinum toxin type A and the cholinergic neuromuscular junction. J Pharmacol Exp Ther 212:16–21
Somlyo AV, Gonzales-Serratos H, Shuman H, McClellan G, Somlyo AP (1981) Calcium release and ionic changes in the sarcoplasmic reticulum of tetanized muscle: an electron probe study. J Cell Biol 90:577–594
Somlyo AV, Bond M, Somlyo AP, Scarpa A (1985) Inositol triphosphate-induced calcium release and contraction in vascular smooth muscle. Proc Natl Acad Sci USA 82:5231–5235
Steinberg JP, Leitner JW, Draznin B, Sussman KE (1984) Calmodulin and cyclic AMP. Possible different sites of action of these two regulatory agents in exocytotic hormone release. Diabetes 33:339–344
Streb H, Irvine RF, Berridge MJ, Schulz I (1983) Release of Ca^{2+} from a nonmitochondrial intracellular store in pancreatic acinar cells by inositol-1,4,5-triphosphate. Nature 306:67–69
Suskiew JB, O'Leary ME, Murawsky, Wang T (1986) Presynaptic calcium channels in rat cortical synaptosomes: fast-kinetics of phasic calcium influx, channel inactivation and relationship to nitrendipine receptors. J Neurosci 6:1349–1357
Takeuchi A, Takeuchi N (1960) On the permeability of end-plate membrane during the action of the transmitter. J Physiol (Lond) 154:52–67
Takeuchi A, Takeuchi N (1963) Effects of calcium on the conductance change of the end-plate during the action of the transmitter. J Physiol (Lond) 167:141–155
Tsien RW, Hess P, McClescky EW, Rosenberg RL (1987) Calcium channels: mechanisms of selectivity, permeation and block. Ann Rev Biophys Biophys Chem 16:265–290
Vinogradov A, Scarpa A (1973) The initial velocities of calcium uptake by rat liver mitochondria. J Biol Chem 248:5527–5531
Wakabayashi S, Goshima K (1981) Kinetic studies on sodium-dependent calcium uptake by myocardial cells and neuroblastoma cells in culture. Biochim Biophys Acta 642:158–172
Wanke A, Ferroni A, Gattanini P, Meldolesi J (1986) α-Latrotoxin of the black widow spider venom opens a small, non-closing cation channel. Biochem Biophys Res Commun 134:320–325

Wood JG, Wallace RW, Cheung WY (1980) Immunochemical studies of the localization of calmodulin and CaM-BP_{80} in brain. In: Cheung WY (ed) Calcium and cell function. Academic, New York, pp 291–303
Woodbury W (1963) Interrelationships between ion transport mechanism and excitatory events. Fed Proc 22:31–35
Wulfroth P, Peltzelt C (1985) The so-called anticalmodulins, fluphenazine, calmidazolium, and compound 48/80 inhibit the Ca^{2+}-transport system of the endoplasmic reticulum. Cell Calcium 6:295–310
Zucker RS, Fogelson AL (1986) Relationship between transmitter release and presynaptic calcium influx when calcium enters through discrete channels. Proc Natl Acad Sci USA 83:3032–3036
Zucker RS, Lando L (1986) Mechanism of transmitter release: voltage hypothesis and calcium hypothesis. Science 231:574–579

CHAPTER 16

Some New Questions Concerning the Role of Ca^{2+} in Exocytosis

S. COCKCROFT and B. D. GOMPERTS

A. Calcium and Cell Activation

The importance of calcium in the activation of cellular processes has been comprehensively documented (e.g. CAMPBELL 1983), but it is only in the case of a very few processes that the precise location of the calcium-sensitive effector has been identified. Two examples stand out. In skeletal muscle it is known beyond doubt that calcium interacts with troponin C, a satellite protein of the actomyosin, and through the consequent activation of myosin ATPase renders the contractile system calcium dependent. In the liver and muscle cascades controlling glycogenolysis, calcium is known to interact with calmodulin, a subunit of phosphorylase kinase (COHEN et al. 1978). Calmodulin and troponin C are closely related, and widely distributed among eukaryotic cells, and can therefore be considered to be reasonable candidates for the sites of calcium action in cellular processes, though there are many others, to be sure (Table 1). Ca^{2+} activation encompasses the whole of animal cell biology. Thus, in addition to the two processes already mentioned, it is known that calcium is involved in initiating the change from G_0 to G_1 and through to S phase in lymphocytes (NISBET-BROWN et al. 1985) (but possibly not in fibroblasts; MCNEIL et al. 1985; ROZENGURT 1985), in the early stages of egg activation (in addition to the cortical granule secretory reactions), in the control of the solute permeability of gap junctions (LOEWENSTEIN 1979) and the K^+ (LEW and FERRIERA 1978) and Na^+ (BLAUSTEIN 1974) permeability of plasma membranes. Ca^{2+} is able to support both exocytotic and nonexocytotic (fluid) secretory processes of a very wide range of cell types, and is also involved in chemotaxis, and phagocytosis. A few examples of such Ca^{2+}-induced cell activation phenomena are surveyed in Table 2.

For many of these cellular processes, a "Ca^{2+}-only signal" (e.g. Ca^{2+} delivered by the Ca^{2+} ionophore A23187) can replace the intermediate steps set in train by activation of receptors by specific ligands. Whether such Ca^{2+}-only signals are actually operative in physiological situations, when the stimulus is initiated at the level of ligand–receptor interactions, is quite another question. To demonstrate a role for Ca^{2+} in a particular cellular activation process, it must either be shown that the level of Ca^{2+} in the cytosol is elevated in response to the physiological stimulus (experiments using intracellular Ca^{2+} indicators) or that the affinity of the effector system for Ca^{2+} becomes enhanced to such a degree that expression of cellular activation occurs at resting Ca^{2+} levels. The criteria for defining a Ca^{2+}-independent process remain somewhat arbitrary since, in the last resort, Ca^{2+} is always present. In view of this, it would be reasonable to ex-

Table 1. Calcium-binding proteins

Name	Molecular weight (k)	Biochemical sources	K_d (μM)	No. of Ca^{2+}-binding sites	Comments	References
Calmodulin	20	Ubiquitous	2.4	4	Inhibited by phenothiazines	MEANS and DEDMAN (1980); WEISS et al (1982); DEDMAN et al. (1977)
F-actin-severing proteins						
Gelsolin	91	Macrophages, neutrophils	1.1	2	Regulates gel – sol transformation	YIN and STOSSEL(1979); YIN et al. (1980)
Villin	95	Intestinal epithelia, microvilli	2.5	1	Regulates gel – sol transformation	GLENNEY et al. (1980)
Fimbrin	95	Platelets				WANG and BRYAN(1981)
Dimeric extended rod-like actin cross-linkers						
α-Actinin	105	Platelets, macrophages				ROSENBERG et al. (1981); BENNETT et al. (1984)
Membrane-associated calcium-binding proteins calelectrins						
p68	68	Lymphocytes	1.2	1		OWENS et al. (1984)
p70	70	Adrenal medulla		>7	No cross-reaction with with synexin	GEISOW et al. (1984); SUDHOF (1985)
p36.5	36.5	Liver				
p32.5	32.5	Brain				
Synexin	47	Adrenal medulla and liver	200		Inhibited by phenothiazines	CREUTZ et al. (1978); POLLARD et al. (1983); SCOTT et al. (1985)
					Facilitates Ca^{2+}-dependent fusion of acidic liposomes	HONG et al. (1981)

Table 2. Ca^{2+}-induced cell activation phenomena

Tissue function	Tissue activation by Ca Ionophore	Activation of permeabilised cells by Ca buffers	Measurements of ligand-stimulated elevation of cytosol Ca
Exocytotic secretion			
Synaptosomes: acetylcholine	Michaelson and Sokolovsky (1978)		Ashley et al. (1984)
Synaptosomes: dopamine	Holz (1975)		
CNS: noradrenaline and GABA	Vargas et al. (1976) Cotman et al. (1976)		
Mast cells: histamine	Foreman et al. (1973)	Bennett et al. (1981) Gomperts et al. (1983)	White et al. (1984)
Basophilic leukaemia cells: mediators (SRS, ECF-A, PAF, histamine)	Lewis et al. (1975)		Beaven et al. (1984)
Neutrophils: lysosomal enzymes	Smith and Ignarro (1975)	Barrowman et al. (1986)	Pozzan et al. (1983)
Macrophages: lysosomal enzymes	Schneider et al. (1978)		Hirata et al. (1984)
Islets of Langerhans: insulin and glucagon	Ashby and Speke (1975)	Yaseen et al. (1982)	Rorsman et al. (1984)
Pituitary: growth hormone	Bicknell and Schofield (1976)		Schlegel and Wollheim (1984)
		Ronning and Martin (1985)	Ronning and Martin (1985)
Pancreas: amylase	Eimerl et al. (1974)	Knight and Koh (1984)	Ochs et al. (1985)
Platelets: ATP	Feinman and Detwiler (1974)	Knight et al. (1984)	Hallam et al. (1984a, b)
Adrenal medulla: catecholamines	Cochrane et al. (1977)	Knight and Baker (1982)	Burgoyne (1984)
Sea urchin eggs: cortical reaction		P.F. Baker et al. (1980)	
Fluid secretion			
Fly salivary gland	Prince et al. (1973)		
Rabbit ileal mucosa	Bolton and Field (1977)		
Rabbit colon	Frizzell (1977)		
Rabbit lacrimal gland	Pholramool and Tangkrisavinont (1976)		
Frog cornea	Candia et al. (1977)		
Rat parotid	Selinger et al. (1974)		Takemura (1985)

Table 2 (continued)

Tissue function	Tissue activation by Ca Ionophore	Activation of permeabilised cells by Ca buffers	Measurements of ligand-stimulated elevation of cytosol Ca
Contractile activity			
Smooth muscle	TRIGGLE et al. (1975)		CAPPONI et al. (1985)
Cardiac muscle	HOLLAND et al. (1975)	MOPE et al. (1980)	POWELL et al. (1984)
Skeletal muscle	MOBLEY (1977)	STEPHENSON (1981)	
Synthesis of inflammatory mediators			
Slow reacting substance of anaphylaxis	JAKSCHIK et al. (1977)		
Prostaglandins in platelets and renal medullar	KNAPP et al. (1977)		
Prostaglandin E in macrophages	GEMSA et al. (1979)		
PGE_2 in renal mesangial cells	PFEILSCHIFTER and BAUER (1986)		PFEILSCHAFTER and BAUER (1986)
Motility			
Leucocyte locomotion	WILKINSON (1975)		
Bacterial locomotion	ORDAL (1977)		
Control of flagella wave motion	HOLWILL and MCGREGOR (1975)		
Proliferative responses			
DNA synthesis in lymphocytes	MAINO et al. (1974)		HESKETH et al. (1983)
DNA synthesis in bone marrow stem cells	GALLIEN-LARTIGUE (1976)		
Parthenogenesis of *Xenopuc laevis* oocyte	BELLE et al. (1977)		
Activation of sea urchin eggs	STEINHARDT and EPEL (1974)	P.F. BAKER et al. (1980)	
Metabolic activation			
Glycogenolysis	KEPPENS et al. (1977)		THOMAS et al. (1984)
Changes in renal gluconeogenesis	MENNES et al. (1978)		
Oxidative metabolism in neutrophils	SCHELL-FREDERICK (1974)		
Glucose oxidation in thyroid	AREM et al. (1984)		RANI et al. (1985)

pect that cellular processes which can proceed at concentrations of Ca^{2+} at and below 10^{-9} *M* are likely to survive the total depletion of Ca^{2+} and thus to be Ca^{2+} independent.

The systems in which Ca^{2+} is most strongly implicated are those in which the activating ligands cause cytosol Ca^{2+} to increase. A case for a Ca^{2+}-only mechanism of cell activation can be sustained only if it can be shown that introduction of Ca^{2+} into the cytosol at physiologically relevant levels can initiate expressions of cellular activity in a manner similar to that triggered by normal ligand–receptor interactions. In many, if not most instances, the activation of the Ca^{2+}-mobilising receptors achieves both an elevation of cytosol Ca^{2+} and an increased affinity to Ca^{2+} of the responsive processes which ensue. While the general conditions which modulate the Ca^{2+} affinity of cellular processes are well known (see Sect. C.V for examples) the molecular basis of this has only been properly explored in the case of cardiac muscle (Mope et al. 1980; McClellan and Winegrad 1980; Robertson et al. 1982; Horowitz and Winegrad 1983). Here, β-adrenergic (cyclic AMP) stimulated phosphorylation of troponin I results in a reduction in the affinity of troponin C for Ca^{2+} and conversely, cholinergic ligands causing dephosphorylation of troponin I increase the affinity of the troponin system, and hence the contractile system, for Ca^{2+}.

B. Exocytosis as an Example of a Cellular Activation Process

We have neither the space nor the competence to consider the whole field of cell activation. Instead we shall concentrate our attention on the exocytotic process, the mechanism by which cells export materials from storage vesicles contained within the cytoplasm. Examples of secretory tissues include the pancreas – having both endocrine (islet cells) and exocrine (acinar) functions, the parotid and the adrenal medulla. In the case of the exocrine tissue secretory cells, the cells are polarised so that the stimulus is sensed by receptors present on the basolateral surface and the secreted products delivered from the apical surface. Somewhat apart from these stands the liver, certainly the main secretory organ of the body when considered in terms of quantity of material secreted. It differs from the other secretory tissues mentioned since, as far as is known, secretion from this organ is constitutive, and not subject to all-or-none on–off triggering by external ligands.

Much of our understanding of the secretory process has come from investigations on cells which until comparatively recently were hardly recognised as such: these include the oocytes, blood platelets, the neutrophils and the mast cells. For all of these tissues and individual cell types, it is clearly understood that exocytosis involves the selective fusion of the membranes surrounding the secretory granules with the plasma membrane, leading to expulsion of their contents. In this chapter we shall consider the mechanism of the secretory process, paying special regard to the mechanism of calcium mobilisation and the possible roles of calcium in the exocytotic process. We shall also consider alternative modes of secretory activation. We shall mainly concentrate on the neutrophils, mast cells and adrenal chromaffin cells as representative examples.

I. Adrenal Chromaffin Cells

The main experimental system for the biochemical investigation of chromaffin cell function is the isolated bovine adrenal medullary cell. The single function of adrenal chromaffin cells is to secrete catecholamines, opioids and other peptides in response to stress. Within a cell there is only one class of secretory granule though there are two sets of catecholamine-secreting cells, the one containing adrenaline, the other noradrenaline. The stimulus–secretion sequences for each class of cell is identical and measurement of either product, or total catecholamine is considered to be fully representative of cell activation (SCHNEIDER et al. 1977).

In vivo, neurostimulation of adrenal chromaffin cell secretion is by acetylcholine delivered by the splanchnic nerve. In bovine cells, acetylcholine activates mainly be stimulating nicotinic receptors (SCHNEIDER et al. 1977) which cause membrane depolarisation and consequent opening of Ca^{2+} channels (FENWICK et al. 1982b). The response to muscarinic stimulation is a slight elevation of cytosol Ca^{2+} (KAO and SCHNEIDER 1985; CHEEK and BURGOYNE 1985; MISBAHUDDIN et al. 1985) insufficient to induce secretion, and possibly inhibitory to the nicotinic response (SWILEM et al. 1983). Muscarinic receptors are generally understood to mobilise Ca^{2+} from intracellular sources following the activation of polyphosphoinositide phosphodiesterase (PPI-pde) and the generation of inositol-1,4,5-trisphosphate (IP_3) although such experiments have yet to be reported on this particular cell system.

In other species (cat, dog, guinea pig, chicken) muscarinic receptors predominate, but regardless of the receptors involved, cholinergic stimulation always leads to a rise in cytosol Ca^{2+}. Adrenal medullary cells of primates can also be stimulated to secrete in response to corticotropin-releasing factor (CRF, a key hormone in the integrated response to stress). In contrast to activation by the cholinergic agonists, the CRF peptide receptors are linked to the activation of adenylate cyclase (UDELSMAN et al. 1986).

II. Neutrophils

The neutrophils represent the main phagocytic arm of the defence against invading organisms, and they are multifunctional in their response to cell activation. The initial function is one of locomotion towards a site of infection in response to soluble stimuli released by bacteria (SCHIFFMANN et al. 1975; BENNETT et al. 1980c) or anaphylatoxins generated by interaction of cell wall material with the alternative pathway of complement activation. The synthetic formyltripeptide fMetLeuPhe, which turns out to be the main neutrophil chemotactic product generated in cultures of *Escherichia coli* (MARASCO et al. 1984), is commonly used in laboratory experiments. Neutrophils can also be stimulated by anaphylatoxins present in serum which has been activated by treatment with yeast cell wall material.

On arrival at a site of infection, the function of the cell is phagocytosis, and here the stimulus is the aggregated IgG or complement-derived C3b, coating invading organisms (SILVERSTEIN et al. 1977). Phagosomes thus formed fuse with lysosomes, effectively by a directed exocytotic process in which the secreted materials (initially the contents of the specific granules and subsequently of lysosomal and other granules; BAINTON 1973; A. W. SEGAL et al. 1980) are retained in phagolysosomes. The killing phase which ensues involves the generation of oxygen radicals owing to the activation of the NAD(P)H oxidase system. There are situations such as within the inflamed synovium of the arthritic joint in which neutrophils can be confronted by a phagocytic stimulus which is too large to engulf (BECKER et al. 1974). In this case lysosomal materials are released to the extracellular space and cause uncontrolled tissue damage.

In the laboratory, in order to study early events leading to neutrophil activation, many of these complexities can be obviated by treating the cells with cytochalasin B (GOLDSTEIN et al. 1973; BECKER et al. 1974; BENNETT et al. 1980a). This effectively paralyses the motile machinery and ensures that the exocytotic process is now directed externally so that the neutrophil can be treated as a secretory cell releasing easily quantifiable products to the exterior. Although the cytochalasin B-treated neutrophil can be said to behave as a straightforward secretory system, there remain complexities due to the presence of at least two sets of granules. The lysosomes (also known as azurophilic or primary granules) contain the acid hydrolases. It is common to measure the release of β-glucuronidase for which there are available fluorogenic and chromogenic substrates. The smaller specific (secondary) granules contain vitamin B_{12}-binding protein, lactoferrin and the bulk of the cell lysozyme.

III. The Mast Cells

The physiological stimulus for activation of these cells is the cross-linking of the cell surface receptors for IgE (D. M. SEGAL et al. 1977), and secretion of histamine (and other mediators of the immediate type allergic response) occurs when Ca^{2+} (in the range 0.1–1 m*M*) is present in the external medium (FOREMAN and MONGAR 1972). Since ligand-induced secretion can be prevented by applying low concentrations of La^{3+} (FOREMAN and MONGAR 1973) and since cromoglycate and other antiallergic drugs (FOREMAN et al. 1975; FEWTRELL and GOMPERTS 1977) and agents which have the effect of elevating cyclic AMP are inhibitory to IgE mediated, but not A23187-mediated secretion, it has long been held that the cross-linking of IgE receptors has the effect of opening plasma membrane ion channels which permit the entry of Ca^{2+} down its concentration gradient into the cytosol. On stimulation with physiological ligands the concentration of cytosol Ca^{2+} increases (WHITE et al. 1984) owing to early events involving inositide metabolism (COCKCROFT and GOMPERTS 1979b; ISHIZUKA and NOZAWA 1983) and, as with the neutrophils, these processes can be inhibited by pretreatment with pertussis toxin (NAKAMURA and UI 1985). In all respects this would appear to be a cell having Ca^{2+}-mobilising receptors, and in which an elevation of Ca^{2+} would be expected to provide a stimulus to exocytosis.

The plasma membrane protein to which cromoglycate binds has been isolated from a related (rat basophilic leukaemia, RBL) cell line (MAZUREK et al. 1982) and this, when reconstituted in phospholipid bilayer membranes and cross-linked by specific anti-cromoglycate-binding protein antibodies causes an increase in membrane conductance, possibly through the opening of Ca^{2+} channels (MAZ-

UREK et al. 1984). Such conductance can be inhibited by application of cromoglycate. It has been held that the cromoglycate-binding protein forms such Ca^{2+} channels as a consequence of IgE receptor cross-linking in the mast cell plasma membrane, but it is doubtful whether these are instrumental to the pathway of cellular activation (LINDAU and FERNANDEZ 1986b).

C. Direct Manipulation of Cytosol Ca^{2+}

I. Calcium Ionophores

While the importance of Ca^{2+} in cell and tissue activation has been long recognised, it was only with the advent of the Ca^{2+} ionophores A23187 (REED and LARDY 1972) and ionomycin (LIU and HERMANN 1978), lipid-soluble Ca^{2+} carriers, that it became possible to elevate Ca^{2+} in the cytosol of microscopic cells in a manner not involving receptor activation. With regard to secretory systems there appeared the possibility of testing the proposition that Ca^{2+} can provide the coupling in the stimulus–secretion coupling process (DOUGLAS and RUBIN 1961) and it is true to say that few secretory systems have been found which fail to respond (see Table 2). However, while secretion from most cell types will occur under circumstances in which cytosol calcium is artificially elevated, the problem still remains of determining whether physiological (receptor) stimulation elevates Ca^{2+} to levels which alone would be sufficient to act as a trigger. Certainly the neutrophils, mast cells and the adrenal chromaffin cells all respond to ionophore stimulation (FOREMAN et al. 1973; BENNETT et al. 1980b; COCKCROFT et al. 1981; COCHRANE et al. 1977; GARCIA et al. 1975; CARVALHO et al. 1982; WRIGHT et al. 1977), but it has become apparent that these responses, in particular those of the neutrophil, are far from straightforward and allow of no simple conclusions.

Unlike fMetLeuPhe-induced neutrophil activation which results from the mobilisation of intracellular Ca^{2+} stores (NACCACHE et al. 1977), the A23187-induced process has a requirement for extracellular Ca^{2+} unless steps are taken to exclude extracellular Mg^{2+} (DI VIRGILIO and GOMPERTS 1983). Ionomycin-induced secretion is dependent on extracellular Ca^{2+} under all circumstances. A delay of 30–40 s (COCKCROFT et al. 1981) in the response to ionomycin points to the activation of a succession of processes, especially as the rise in cytosol Ca^{2+} is effectively immediate (POZZAN et al. 1983).

Ionophore-mediated activation of neutrophils leads to the release of arachidonic acid metabolites, e.g. platelet-activating factor (PAF), leukotriene B_4 (LTB_4), 5-hydroxyeicosatetraenoic acid (5-HETE), thromboxanes and prostaglandins (BOKOCH and REED 1980; SMITH et al. 1982). Of these PAF and LTB_4 are independent agonists in their own right and act in a manner involving cytosol Ca^{2+} (P. D. LEW et al. 1984a; SERHAN et al. 1982; SWENDSEN et al. 1983; GOETZL 1980) while 5-HETE potentiates their actions (O'FLAHERTY 1985). When applied together these lipid agonists desensitise the neutrophil responses to ionophore stimulation, but not that due to fMetLeuPhe or C5a (O'FLAHERTY 1985; O'FLAHERTY et al. 1984). This suggests that activation by Ca^{2+} ionophores is not simply due to elevation of cytosol Ca^{2+}, but results from the involvement of receptors activated by a complex array of cell-derived arachidonic acid metabolites.

In view of the complexities revealed in the activation pathway of ionophore-stimulated neutrophils, it can not be assumed that the calcium ionophores deliver a simple Ca^{2+}-only stimulus to a secretory process. Conclusions regarding the role of Ca^{2+} in the activation of these and other secretory cells had to await the development of new techniques. In part, these have now been provided by the development of fluorescent indicators capable of monitoring changes in cytosol Ca^{2+} in response to stimulation by receptor-directed ligands (TSIEN 1980; GRYNKIEWICZ et al. 1985) and by techniques which allow the control of cytosol Ca^{2+} at precise and predetermined levels in permeabilised, but in other respects fully viable cells. These techniques have been applied to many secretory cell types (see Table 2) including the three selected for discussion.

Measurements using the cytosol Ca^{2+} indicator quin2 have shown that in the resting situation, neutrophils and adrenal chromaffin cells maintain Ca^{2+} on average at about 10^{-7} *M* (KNIGHT and KESTEVEN 1983; BURGOYNE 1984; POZZAN et al. 1983) in agreement with all other systems so far tested. Following stimulation with appropriate ligands applied at concentrations which cause maximal secretion, the level of Ca^{2+} typically rises towards 800 n*M*. An effective stimulus due to Ca^{2+} ionophores requires the elevation of cytosol Ca^{2+} beyond 1 μ*M*, saturating the quin2 signal. In the absence of extracellular Ca^{2+}, treatment with the Ca^{2+} ionophores or agonists causes the level of cytosol Ca^{2+} to increase to a reduced extent, but only the agonists are able to promote secretion (POZZAN et al. 1983; BURGOYNE 1984). This points to the possibility of other processes, set in train by receptor activation which may cooperate with the lower level Ca^{2+} signal.

II. Manipulation of Cytosol Ca^{2+} in Permeabilised Cells

Whilst one could demonstrate the possibility of Ca^{2+}-induced secretory and other cellular functions using Ca^{2+} ionophores, there are only very few instances in which this approach has been refined to an extent which allows the concentration of Ca^{2+} delivered to the cytosol to be precisely clamped (P. D. LEW et al. 1984b, 1986). The difficulties can be obviated in experiments with permeabilised cells, in which calcium buffers (typically Ca–EGTA) can be introduced directly into the cytosol to fix Ca^{2+} at predetermined concentrations. There are a number of techniques which allow of such cell permeabilisation in a manner permitting survival in the short term, provided precautions are taken to replace required metabolites such as MgATP which would otherwise leak from the permeabilised cells (P. F. BAKER et al. 1985; GOMPERTS 1985; GOMPERTS and FERNANDEZ 1985).

III. Methods of Plasma Membrane Permeabilisation

Sendai virus, which binds to the ubiquitous cell surface glycoproteins (BACHI et al. 1978) and glycolipids (HAYWOOD 1974) has been used as a permeabilising agent for a wide variety of cells (PASTERNAK et al. 1985) and we have used it to control the composition of the cytosol of both mast cells (GOMPERTS et al. 1983) and the neutrophils (GOMPERTS et al. 1986; BARROWMAN et al. 1986). In an alternative method cells are exposed to brief high voltage discharges to cause membrane permeabilisation and by suitable adjustment of the discharge potential the intracellular membranes enclosing secretory granules and other organelles remain intact (KNIGHT and BAKER 1982). In addition to adrenal chromaffin cells, this technique for membrane permeabilisation has been applied to problems of Ca^{2+} buffering in platelets (KNIGHT and SCRUTTON 1980; KNIGHT et al. 1984), islet cells (PACE et al. 1980; YASEEN et al. 1982), pancreatic acinar cells (KNIGHT and KOH 1984) and pituitary (GH_3) cells (RONNING and MARTIN 1985).

In the case of the mast cell (COCKCROFT and GOMPERTS 1979 a), and a limited number of other cell types, mainly transformed cell lines (HEPPEL et al. 1985), it is possible to permeabilise the plasma membrane by applying extracellular ATP under conditions in which it is fully dissociated in the form ATP^{4-} (BENNETT et al. 1981). ATP^{4-}-permeabilised cells can be treated in exactly the same way as cells treated with Sendai virus or high voltage discharge, but additionally, since the permeability lesions undergo spontaneous repair on removal of the tetrabasic ligand (i.e. on conversion of ATP^{4-} to its Mg^{2+} salt) this method of cell permeabilisation can be used to trap exogenous solutes in the cytosol of otherwise fully intact cells (GOMPERTS 1983).

IV. Ca^{2+}-Induced Secretion from Permeabilised Neutrophils

Cytochalasin B-treated neutrophils permeabilised by treatment with Sendai virus secrete both lysozyme and β-glucuronidase as the concentration of Ca^{2+} is brought towards the micromolar range (BARROWMAN et al. 1987). There is a requirement for provision of MgATP at concentrations in excess of 0.5 m*M* and the secretion occurs without loss of the cytoplasmic protein lactate dehydrogenase. The precise concentrations of Ca^{2+} which induce secretion of the two granules represented by lysozyme and β-glucuronidase are different. Whilst lysozyme secretion is complete with Ca^{2+} at 1 μM (midpoint 0.3 μM), secretion of β-glucuronidase is only maximal with Ca^{2+} at concentrations in excess of 5 μM (midpoint 0.7 μM). This result contrasts with those derived from permeabilised platelets in which two secretory granules (secreting amines and lysosomal enzymes) are mobilised in response to the same concentrations of Ca^{2+} yet respond to different concentrations of the stimulating ligand thrombin (KNIGHT et al. 1982).

The secretion of lysozyme and other specific granule markers occurs at concentrations of Ca^{2+} compatible with the levels achieved by receptor-directed agonists as indicated by measurements using the indicator quin2 (Fig. 1; POZZAN et al. 1983) and so it is possible that the ligand-induced stimulus to secretion of the specific granules could arise simply from the elevation of cytosol Ca^{2+}. Secretion of specific granule markers also occurs on treatment of neutrophils with subthreshold concentrations of Ca^{2+} ionophores together with activators of protein kinase C, indicating the possibility of synergy by the two pathways of activation (DI VIRGILIO et al. 1984). This also occurs in the generation of superoxide (DALE and PENFIELD 1984; PENFIELD and DALE 1984). Such synergy has commonly been observed in the activation of cellular processes (NISHIZUKA 1984; KAJIKAWA et al. 1983; DELBEKE et al. 1984; SINGH 1985; RUSKOAHO et al. 1985; TRUNEH et al. 1985).

The situation with secretion of β-glucuronidase is quite different. Maximal elevation of cytosol Ca^{2+} occurs at concentrations of ligands well below those which stimulate secretion (Korchak et al. 1984; P. D. Lew et al. 1984a) and for this reason it is considered that the complete stimulus delivered by the receptors requires both the mobilisation of Ca^{2+} and release of diacylglycerol to cause activation of protein kinase C. However, it is only under closely defined experimental conditions that such a synergy between ionophore and phorbol ester occurs (Naccache et al. 1985). These involve omission of extracellular Mg^{2+}, a step which was already pointed out, is by itself permissive of A23187-induced secretion due to mobilisation of Ca^{2+} from intracellular stores (Di Virgilio and Gomperts 1983). Rather, it is our experience that the major action of phorbol ester is to suppress β-glucuronidase secretion due to the Ca^{2+}-only stimulus (Barrowman et al. 1986). It would appear that the Ca^{2+} delivered by agonist stimulation is insufficient to cause lysosomal enzyme secretion, and since protein kinase C signals do not synergise, the possibility has to be considered that it is dependent on yet another signal, as yet unidentified.

V. Ca^{2+}-Induced Secretion from Permeabilised Adrenal Chromaffin Cells

On stimulation of bovine adrenal chromaffin cells with a cholinergic ligand the level of cytosol Ca^{2+} increases from a resting level of about 100 n*M* to about 600 n*M* (Knight and Kesteven 1983; Burgoyne and Cheek 1985; Kao and Schneider 1985). A very small part of this Ca^{2+} rise is due to mobilisation of intracellular Ca^{2+} stores owing to activation of muscarinic receptors. The extracellular environment contributes the major source of Ca^{2+} and this enters the cell through voltage-dependent calcium channels activated by nicotinic depolarisation. The concentration of Ca^{2+} required in a Ca^{2+}-only stimulus (using permeabilised cells) is higher than that achieved by ligand stimulation: maximal secretion is only attained when the level of Ca^{2+} approaches 10 μ*M* (midpoint 1.2 μ*M*) (Fig. 1; Knight and Baker 1982). Another way to open the Ca^{2+} channels is by depolarising the cells by increasing the concentration of extracellular K^+. This elevates cytosol Ca^{2+} to higher levels than those achieved by nicotinic stimulation, but like the Ca^{2+} ionophores, it provides a poor stimulus to secretion (Burgoyne and Cheek 1985).

As with the neutrophils, the possibility of involvement of protein kinase C, to lower the requirement for cytosol Ca^{2+} in the stimulation of these cells by cholinergic ligands has been considered. In this case, the signal for cholinergic generation of the endogenous activator, diacylglycerol, would have to be delivered through the muscarinic-receptors, and exogenous activators of protein kinase C such as phorbol ester certainly appear to shift the midpoint Ca^{2+} requirement for secretion down to about 0.5 μ*M* in permeabilised cells (Knight and Baker 1983; Brocklehurst and Pollard 1985). However, the nicotinic ligands appear to deliver a complete stimulus to secretion and experiments with intact cells provide no evidence of enhancement by simultaneous muscarinic stimulation (Cheek and Burgoyne 1985) or by phorbol ester (Burgoyne 1984; Burgoyne and Norman 1984). There are also reports that muscarinic stimulation even leads to inhibition

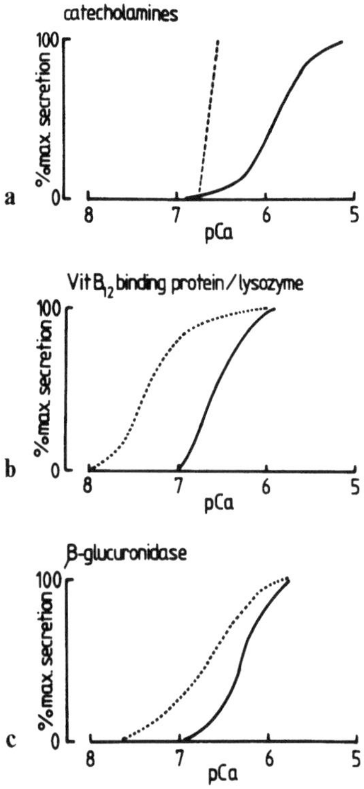

Fig. 1 a–c. Calcium dependence of secretion. The graphs compare the concentrations of cytosol Ca^{2+} required to induce secretion from **a** adrenal chromaffin cells, and **b, c** neutrophils due to a Ca^{2+}-only stimulus (*full lines*) and due to ligand stimulation (*broken lines*). To generate Ca^{2+}-only stimulation, permeabilised cells were loaded with Ca–EGTA buffers. The experimental approach to measuring the Ca^{2+} dependence of ligand stimulation of the two cell types differed and the results are therefore not strictly comparable. Adrenal chromaffin cells were loaded with quin2 and the change in Ca^{2+} consequent on stimulation with carbamoylcholine was measured. The estimates of cytosol Ca^{2+} (commensurate with zero and maximal secretion) thus represent averages of values which may vary widely among individual cells. Neutrophils were loaded with high concentrations of quin2 such that the dye could also be used as a buffer to regulate cytosol Ca^{2+} due to the Ca^{2+} ionophore ionomycin. Over the range of Ca^{2+}, the ionophore had only a small effect on secretion. With Ca^{2+} clamped in this manner, the concentration of Ca^{2+} commensurate with secretion of **b** lysozyme and **c** β-glucuronidase due to the ligand fMetLeuPhe was measured. Data taken from **a** KNIGHT and BAKER (1982) and BURGOYNE (1984); **b** P. D. LEW et al. (1986) and BARROWMAN et al. (1987)

of the nicotinic responses (SWILEM et al. 1983; DEROME et al. 1981; LEMAIRE et al. 1981) and here we can envisage that protein phosphorylation by protein kinase C might switch off the signal to exocytosis. There are precedents for this such as the ablation by phorbol ester of G_i (inhibitory) signals to adenylate cyclase (JAKOBS et al. 1985) and inhibition of β-adrenergic receptor function (SIBLEY et al. 1984, 1985).

The very evident difference between the Ca^{2+}-clamped cells and the response of intact cells to ligand stimulation calls for a modulator of Ca^{2+} affinity. This

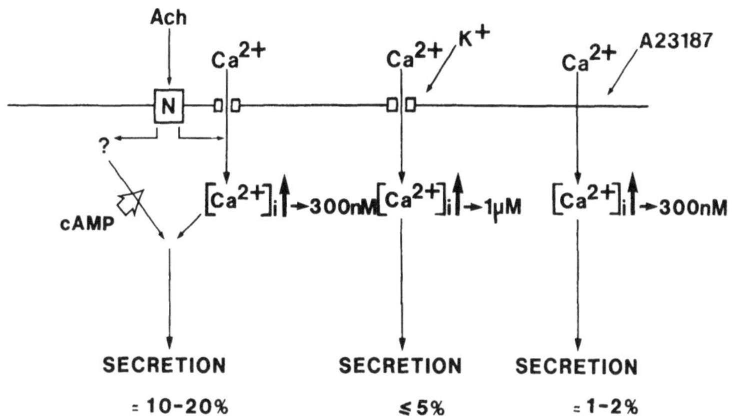

Fig. 2. Schematic illustration of the relationship between cytosol Ca^{2+} and catecholamine secretion from adrenal chromaffin cells. Maximal stimulation of catecholamine secretion by nicotinic ligands is achieved when cytosol Ca^{2+} is increased to around 300 n*M*. A concentration of ionophore (A23187) sufficient to elevate cytosol Ca^{2+} to the same level causes a very low level of secretion, while K^+ depolarisation, which can increase cytosol Ca^{2+} beyond 1 µ*M* also provides a comparatively poor stimulus to secretion. Agents which have the effect of elevating cyclic AMP selectively inhibit secretion due to nicotinic stimulation, and do so without affecting the entry or the buffering of Ca^{2+}. (Courtesy of R. D. BURGOYNE)

could act through protein kinase C. Such a modulator is unlikely to be Ca^{2+} induced since stimulation by K^+ depolarisation and Ca^{2+} ionophores (in contrast to cholinergic ligands) induce a low level of secretion in the face of a much greater elevation of cytosol Ca^{2+} (BURGOYNE and CHEEK 1985; CHEEK and BURGOYNE 1985; SCHNEIDER et al. 1977). It is well known that nicotinic stimulation does not cause turnover of the inositol phospholipids (MICHELL et al. 1977; FISHER et al. 1981) and so there is no obvious mechanism for the consequent generation of diacylglycerol, the only known physiological activator of protein kinase C. Further evidence for the idea of a modulator of nicotinic receptor-directed exocytosis comes from the finding that elevation of cyclic AMP (by treating adrenal cells with forskolin) inhibits selectively the secretion due to nicotinic agonists while having no effect on the elevation of cytosol Ca^{2+} (E. M. BAKER et al. 1985). It has no effect on the secretion due to K^+ depolarisation or Ca^{2+} ionophores. The relationships between cytosol Ca^{2+} and the stimulus due to nicotinic agonists, K^+ depolarisation and ionophore A23187 are summarised in Fig. 2.

Though highly conserved in structure and function, there do exist differences in nicotinic receptors between species, and between different locations within a single species. An important characteristic of the nicotinic receptor is the α-bungarotoxin-binding site and while this is present on the adrenal receptor, α-bungarotoxin is without effect on secretion (WILSON and KIRSCHNER 1977; TRIFARO and LEE 1977) and channel activity (FENWICK et al. 1982a). In this respect the adrenal nicotinic receptor is similar to those present in ganglia, but not those of the neuromuscular junction. Such differences should just alert us to the possibility of other variations in receptor function which might be invisible to the electrophysiologist, yet provide for the generation of diffusible modulators which could control the affinity for Ca^{2+} in the exocytotic process.

D. Exocytotic Secretion Without Elevation of Cytosol Ca^{2+}

I. Phorbol Ester

While ligand stimulation of adrenal chromaffin cells, neutrophils and mast cells causes an elevation in the level of cytosol calcium, there are other stimuli which do not. An example of these is the phorbol ester, phorbol myristate acetate (PMA). In neutrophils, not only is there no elevation in cytosol Ca^{2+} on treatment with PMA (SHA'AFI et al. 1983), but cell activation (respiratory burst and secretion of vitamin B_{12}-binding protein and other components of specific granules) still occurs when the cells are loaded with high concentrations of quin2 sufficient to repress and maintain cytosol Ca^{2+} at levels around 10^{-9} *M* (DI VIRGILIO et al. 1984) pointing to the possibility of cell activation occurring as a consequence of protein phosphorylations by both Ca^{2+} and protein kinase C pathways. If experiments with phorbol ester are held to be physiologically irrelevant then the reality of Ca^{2+}-independent cell activation comes from the observation that stimulation of phagocytosis by antibody-coated latex beads can occur in a manner not involving the elevation of cytosol Ca^{2+} (CAMPBELL and HALLETT 1983; P. D. LEW et al. 1985). Similarly, platelets stimulated with phorbol undergo exocytosis by a mechanism involving an increase in the affinity of the exocytotic process to resting levels of Ca^{2+} (RINK et al. 1983; RINK and SANCHEZ 1984; KNIGHT et al. 1984).

II. Guanine Nucleotides

Another possible series of modulators of Ca^{2+} dependence are the guanine nucleotides. In a number of systems it has now been shown that GTP (and its so-called nonhydrolysable analogues) can induce or modulate cell activation phenomena in a manner probably unrelated to its well-recognised roles in the regulation of adenylate cyclase (RODBELL 1980) and the treadmilling of tubulin (TIMASHEFF and GRISHAM 1980; HAMEL et al. 1984). These include the control of ion channels (PFAFFINGER et al. 1985; BREITWEISER and SZABO 1985), an involvement in excitation–contraction coupling in skeletal muscle (VOLPE et al. 1986; DI VIRGILIO et al. 1986) and the control or modulation of a number of secretory processes.

The first investigation which clearly pinpointed a role for GTP in an exocytotic process concerned the mast cell (GOMPERTS 1983). In these cells it is possible to introduce, and then trap exogenous solutes in the cytosol by treating them successively with low concentrations of ATP in its tetrabasic acid form (to permeabilise) and then with Mg^{2+} (to cure the ATP^{4-}-induced permeability lesions). With GTP analogues (i.e. GTP-γ-S, GppNHp and $GppCH_2p$) trapped in the cytosol these cells undergo a normal exocytotic reaction, secreting histamine, when confronted with extracellular Ca^{2+} at concentrations in the millimolar range. This experiment pointed to a role for a guanine nucleotide-binding protein (G-protein) at some stage (or stages) in the stimulus–secretion coupling process, and the finding that guanine nucleotide-loaded mast cells become refractory to such stimulation following treatment with pertussis toxin (known to inhibit the functions of G-proteins by covalent ADP ribosylation) focused attention on the signal transduction sequence of reactions (NAKAMURA and UI 1983).

In neutrophils, as in other systems in which Ca^{2+} is mobilised from intracellular stores, the ligand–receptor interaction is followed by the activation of PPI-pde which hydrolyses PIP_2, so generating IP_3 and diacylglycerol (BERRIDGE 1984; DOWNES and MICHELL 1985). Coupling between the receptor and the catalytic unit of PPI-pde is provided by a G-protein (G_p), thus establishing one target of guanine nucleotide action in these secretory cells (COCKCROFT and GOMPERTS 1985). From studies with intact neutrophils it has been found that pertussis toxin inhibits all ligand-induced functional responses (MOLSKI et al. 1984; SPANGRUDE et al. 1985) and early events, including the rise in cytosol Ca^{2+} and the metabolism of the inositol lipids. It inhibits the breakdown of the polyphosphoinositides and the consequent generation of inositol phosphates and diacylglycerol. It is thus more than likely that pertussis toxin, by the ADP ribosylation of G_p inhibits the signal transduction pathway in these cells. Subsequent events are unaffected and pertussis toxin is without effect on the activation due to calcium ionophores (BECKER et al. 1985, 1986), or to Ca^{2+} introduced into the cytosol of permeabilised cells (BARROWMAN et al. 1986a).

E. A Role for G-Protein in Exocytosis?

It now appears that the G-proteins are involved at two stages in the stimulus secretion sequence. The experiments on which this conclusion is based were prompted by observations made with permeabilised platelets (HASLAM and DAVIDSON 1984). Here it was shown that GTP has the effect of enhancing the affinity to Ca^{2+} of the exocytotic process such that Ca^{2+} at concentrations as low as 10^{-8} *M* becomes capable of delivering an effective stimulus. A similar change, but of smaller magnitude is seen in permeabilised chicken adrenal cells (KNIGHT and BAKER 1985). While such shifts in Ca^{2+} affinity can probably be explained on the basis of interactions of GTP with G_p and the resulting generation of diacylglycerol, there are other effects induced by GTP and its analogues in permeabilised secretory cells which certainly can not.

When GTP and its analogues are incorporated alongside Ca^{2+} buffers into the cytosol of permeabilised bovine adrenal cells and neutrophils, a second role for G-proteins becomes apparent, though the effects generated in the two cells are exerted in opposite directions. It is likely that these effects of GTP are exerted at a site removed from the signal transduction system, but close to the catalytic site controlling exocytosis. While GTP-γ-S (guanosine-5′-*O*-(3-thiotriphosphate), a stable analogue of GTP, suppresses catecholamine secretion from Ca^{2+}-clamped bovine adrenal chromaffin cells (KNIGHT and BAKER 1985), in the neutrophils the secretion of lysosomal enzymes due to all concentrations of Ca^{2+} in the range 10^{-7} -10^{-5} *M* is enhanced (BARROWMAN et al. 1986). Unlike the platelets, in neither of these systems is there any shift in the sensitivity to Ca^{2+}, and in the neutrophils

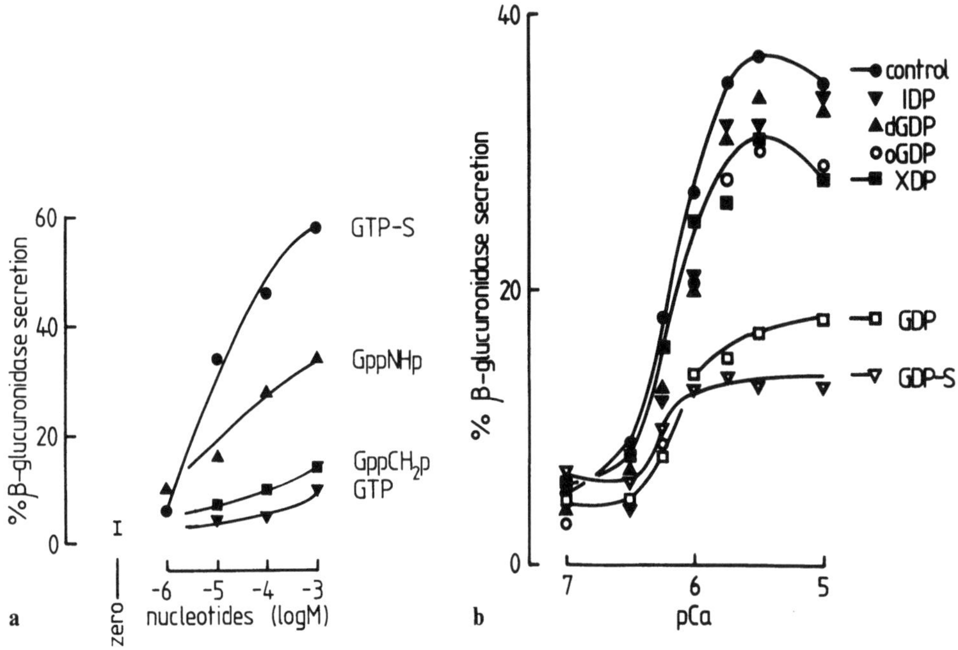

Fig. 3. a GTP and GTP analogues can induce Ca^{2+}-independent secretion from permeabilised neutrophils. Neutrophils were permeabilised with Sendai virus in the presence of EGTA (3 m*M*) to deplete cytosol Ca^{2+}; 2.5 min later GTP or GTP analogues were added at the concentrations shown. These nucleotides are capable of supporting secretion from lysosomal granules, even in the face of 30 m*M* EGTA or BAPTA. **b** GDP and GDP-β-S inhibit lysosomal enzyme secretion elicited by Ca^{2+} buffers. Secretion of β-glucuronidase from permeabilised neutrophils can be inhibited by GDP and GDP-β-S at concentrations down to 1 μ*M*. The experiment illustrated shows that other structurally related nucleotides have little inhibitory effect in the Ca^{2+}-clamped cells, even when applied at 100 μ*M*. IDP is inosine diphosphate; dGDP is 2′-deoxy-GDP; oGDP is periodate-oxidised GDP; XDP is xanthosine diphosphate. (Barrowman et al. 1986)

GTP (analogues) support secretion even after pretreatment of the permeabilised cells with high concentrations of the Ca^{2+}-specific chelators EGTA and BAPTA (1,2-bis(2-aminophenoxy)ethane-*N,N,N′,N′*-tetraacetic acid) (Fig. 3 a).

There are good reasons for believing that this effect of GTP and its analogues is not due to interactions with G_p. In the first place, GTP is unable to cause Ca^{2+}-independent secretion of lysozyme, a marker of specific granules. Next, the sensitivity of PPI-pde to Ca^{2+} is quite distinct from that of the secretory mechanism, and there are conditions under which GTP-γ-S induces secretion while causing no generation of diacylglycerol. Pertussis toxin is without effect on exocytosis of lysosomal enzymes due to GTP analogues (Fig. 4), and further, phorbol ester rather than synergising with Ca^{2+}, is inhibitory in this system. Finally it was shown that GDP and its synthetic analogue GDP-β-S could inhibit the secretion due to buffered Ca^{2+} (see Fig. 3 b). This last effect is truly specific since closely related nucleoside diphosphates such as IDP and dGDP are without effect on Ca^{2+}-induced secretion.

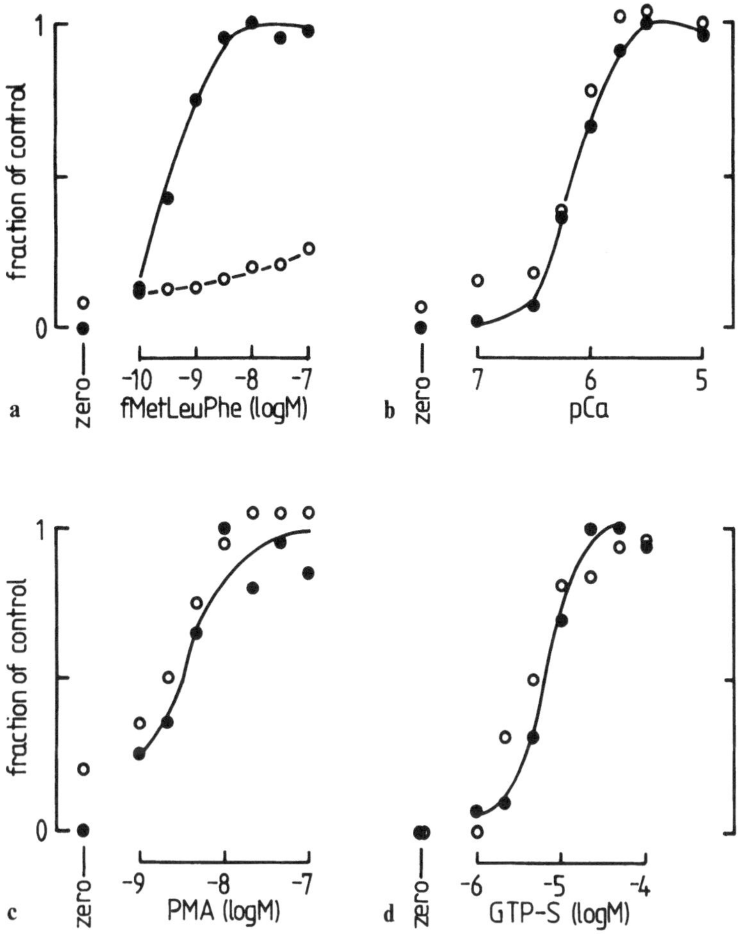

Fig. 4 a–d. Effect of pertussis toxin on lysosomal enzyme secretion from cytochalasin B-treated rabbit neutrophils. Pertussis toxin prevents secretion due to **a** ligand (fMetLeuPhe) stimulation, but is without effect on secretion from permeabilised cells due to **b** Ca^{2+} buffers, **c** phorbol ester and **d** GTP-γ-S. There is strong evidence that signals emanating from receptors are mediated by a G-protein (G_p) which is susceptible to ADP ribosylation by pertussis toxin. The finding that GTP-γ-S-induced secretion is refractory to pertussis toxin indicates that the G-protein (G_e) mediating exocytosis is distinct from G_p. (BARROWMAN et al. 1986)

On the basis of these observations we have suggested that there exists a second G-protein, which we have termed G_e, situated distal to the site of action of Ca^{2+} in the exocytosis of acid hydrolases. The finding of inhibition of Ca^{2+}-induced secretion by GDP certainly points to the possibility of G_e having physiological relevance, but in the knowledge that the concentration of GTP in cells does not fluctuate greatly as a consequence of stimulation, it is reasonable to ask whether G_e plays any role as a regulator of ligand-induced secretion? If, as we have tentatively suggested, G_e is located on the cytosolic surface of the acid hydrolase-secreting granules (GOMPERTS et al. 1986), then one might consider the possibility of its being controlled by another diffusible messenger, as yet unidentified. Alternatively, if it is located on the plasma membrane as a recognition site for lysosomal enzyme secretion it could be under direct control of the receptor: this however would appear to be an improbable location in a cell the physiological function of which is phagocytosis, not exocytosis. One can envisage that G_e is normally occupied by GTP and that this is the condition under which Ca^{2+} is able to trigger the exocytotic mechanism.

F. Questions Concerning the Role of Ca^{2+} in Exocytosis

There are clearly difficulties involved in reconciling the dependence on Ca^{2+} for secretion of catecholamines and lysosomal enzymes from adrenal cells and neutrophils, with the levels of Ca^{2+} actually achieved in these selfsame cells by ligand stimulation. We have seen that while it is possible to shift the dependence on Ca^{2+} of secretion from permeabilised adrenal cells towards somewhat lower concentrations, the means whereby this can be achieved has no effect when applied to fully intact cells subject to stimulation by ligands. Similarly in neutrophils, while phorbol ester has been reported to enhance β-glucuronidase secretion when the cells are treated with ionophores in rather unusual conditions, its effect on Ca^{2+}-clamped cells is inhibitory. Guanine nucleotides, which shift the affinity of the exocytotic reaction in permeabilised platelets and the adrenal cells of some species to low Ca^{2+}, are inhibitory to secretion from bovine adrenal cells and support Ca^{2+}-independent secretion of acid hydrolases from neutrophils. Regardless of its different effects in different systems, GTP would appear to be a poor candidate to act as a diffusible "second messenger" since its concentration in the cytosol is likely to be controlled, not by events at the receptor, but by transphosphorylation within the general pool of purine and pyrimidine nucleotides. In order to control triggered processes such as exocytosis, one would expect the affinity of the G-protein for its guanine nucleotide ligand to be modulated, and for such control to be exerted at sites removed from the plasma membrane, yet another second messenger would seem to be required.

In this light we should reconsider the question of the concentrations of Ca^{2+} actually achieved on stimulation, and ask also whether Ca^{2+} necessarily provides the physiological stimulus in a system in which a Ca^{2+}-only stimulus is capable of causing exocytosis. Recent experiments with mast cells have provided some unexpected answers to these dilemmas, and some new problems to go with them. It has been shown that secretion stimulated by cross-linking of IgE receptors occurs without the opening of Ca^{2+} channels (LINDAU and FERNANDEZ 1986b), and that changes in the concentration in cytosol Ca^{2+} are at best only indirectly related to the exocytotic event (NEHER and ALMERS 1986).

I. Single Cells

1. Membrane Capacitance Changes in Exocytosis

The progress of exocytosis in individual secretory cells has been monitored by measuring changes in membrane capacitance. The large "patch pipette" with which the electrical measurements are made is sealed onto a cell, and the membrane occluding the end of the pipette (approximately 1 μm diameter) is ruptured so that the cytosol is now controlled by the composition of the pipette filling solution with which it is in direct communication. In this way calcium buffers, nucleotides and even proteins can be introduced into the cytosol (Fig. 5a). Originally this technique was applied to the exocytotic process of adrenal chromaffin cells (NEHER and MARTY 1982). When the pipette contained Ca^{2+} at around 1 μ*M* (and MgATP) the capacitance increased, indicating enlargement of the area of the cell membrane. Similarly, so long as the internal solutions were only lightly buffered with EGTA, the capacitance increased, indicating membrane expansion, following application of depolarising potentials linked to inward Ca^{2+} currents. The increase in membrane capacitance occurred as a series of discrete step-like changes in the range $0.4–0.8 \times 10^{-15}$ F, a magnitude which correlates with predicted accretions in membrane area due to the fusion of individual chromaffin granules.

The mast cells, by virtue of their much larger secretory granules (average surface area 1.5 μm^2) are particularly well suited to this technique, and also have the advantage that the exocytotic degranulation can easily, and simultaneously be visualised under the light microscope. However, when Ca^{2+} buffers (to regulate Ca^{2+} at micromolar concentrations) were introduced into the cytosol through the recording patch pipettes, these cells signally failed to respond. Nor did the pipette-attached mast cells respond to receptor-directed ligands which were perfectly effective in stimulating an exocytotic response from neighbouring (but unattached) cells in the field of view. Only when GTP (and GTP analogues) were introduced together with EGTA into the cytosol was a capacitance increase elicited (typically corresponding to an approximately fourfold increase in membrane area), and as with the adrenal cells this was found to occur in a series of steps (centred on 15×10^{-15} F), again correlating well with the predictions based on individual exocytotic fusions between secretory granules and the plasma membrane (FERNANDEZ et al. 1984). The mechanism of GTP-induced exocytosis in these cells probably occurs though an involvement of G_e, similar to the GTP-induced secretion of lysosomal enzymes from permeabilised neutrophils already mentioned.

The failure to respond to Ca^{2+} and Ca^{2+}-mobilising ligands was rationalised according to the possibility of leakage of Ca^{2+}-binding proteins into the effectively infinite volume of the recording pipette (approximately 25 μl dialysing the intracellular aqueous space of approximately 0.6 pl) (BENNETT et al. 1984). A new development now allows agonist stimulation of pipette-attached mast cells and has provided strong evidence that the IgE receptor system operates by a mechanism not involving the entry of Ca^{2+} through ion channels (LINDAU and FERNANDEZ 1986a). Here, the patch pipette is sealed to the membrane in the normal way, but instead of rupturing the occluding area of membrane, it is permeabilised selectively to molecules of low molecular weight by the presence of ATP^{4-} in the pipette filling solution (Fig. 5b). In this way, no high molecular weight proteins can leak from the cytosol, yet electrical communication allowing the measurement of current and capacitance is still maintained. These cells can support exocytotic degranulation in response to cross-linking of IgE receptors by specific antigens, yet display no increase in membrane conductance over and above that

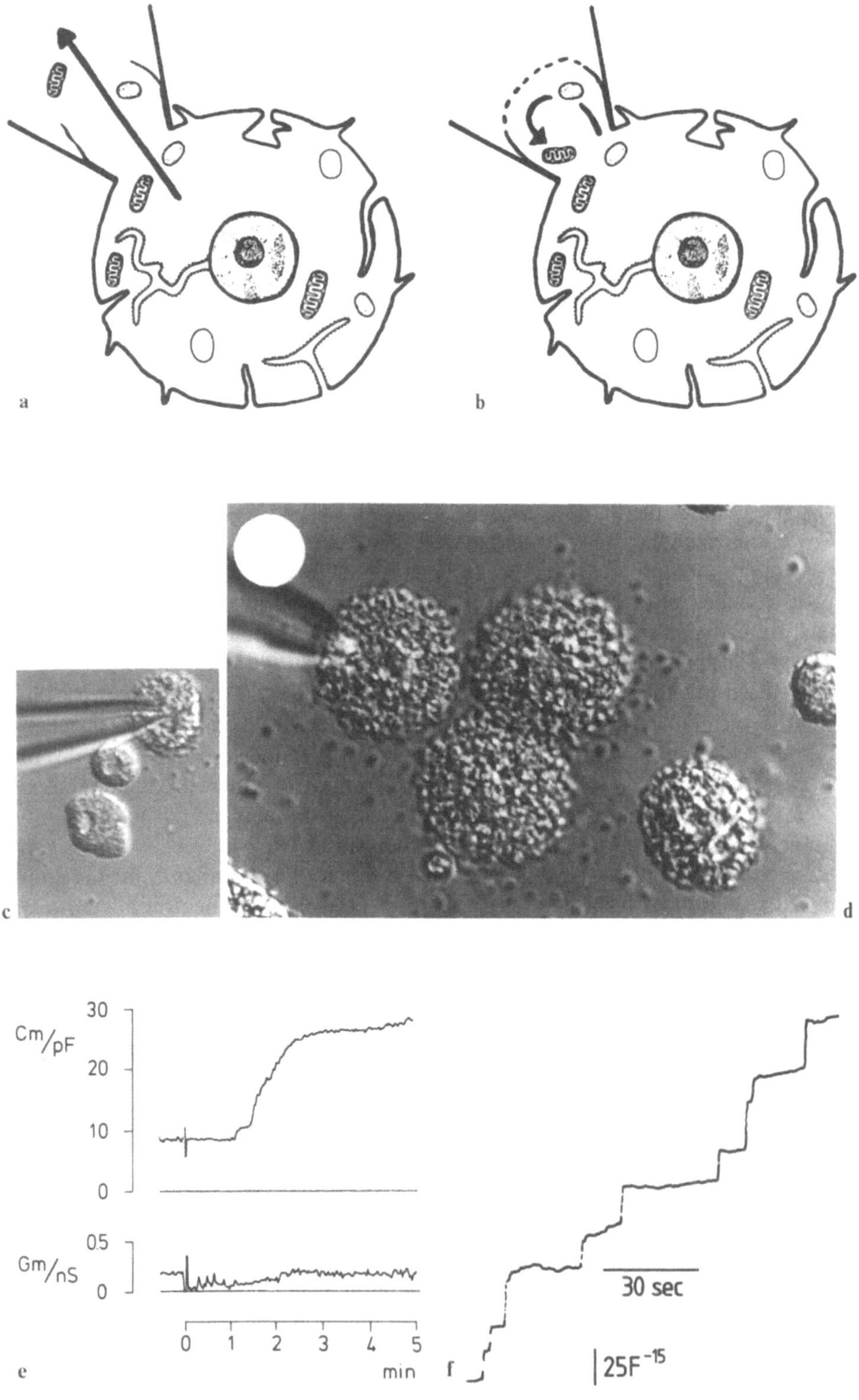
a
b
c
d
Cm/pF
30
20
10
0
Gm/nS
0.5
0
0 1 2 3 4 5
min
e
f
30 sec
25F^{-15}

which would be expected from the considerable increase in membrane area. This would seem to rule out the possibility, so long assumed, of Ca^{2+} mobilisation occurring by a mechanism involving influx through plasma membrane Ca^{2+} channels.

2. Fast Ca^{2+} Transients Are not Sufficient to Trigger Exocytosis

Old prejudices concerning the role of Ca^{2+} in the IgE-mediated activation of mast cells have also been challenged recently by experiments indicating that while elevation of cytosol Ca^{2+} tends to enhance the secretory process, it is by no means sufficient to cause secretion, and may not even be mandatory (NEHER and ALMERS 1986). These observations have been based on analysis of cytosol Ca^{2+} in single mast cells loaded with the fluorescent dye fura-2 (GRYNKIEWICZ et al. 1985) together with visual assessment of the progress of degranulation. Unlike the monotonic pattern of cytosol Ca^{2+} elevation due to averaging among large numbers of cells inevitable in cuvette experiments (WHITE et al. 1984), treatment of single cells allows the resolution of rapid events. Cytosol Ca^{2+} following stimulation with IgE-receptor-directed ligands is often, but not invariably seen to fluctuate as random transients of a few seconds duration, and may reach 5 μM, sometimes falling back to a plateau, sometimes returning to the basal level. The high concentrations, if maintained, would be more than sufficient to cause secretion. Instead, there seems to be little correlation between the level of cytosol Ca^{2+} and the initiation of exocytosis in any particular cell (Fig. 6). Following stimulation with IgE-directed agonists or with compound 48/80, individual cells were seen that degranulated during the falling phase of a Ca^{2+} transient and on occasions even at

Fig. 5 a–f. Monitoring single mast cell degranulation with patch pipettes. **a, b** schematic illustrations of "whole-cell patch-clamp" and "slow whole-cell patch-clamp" techniques. In the whole-cell patch-clamp technique (**a**), a pipette with a polished tip is fixed to the cell surface by gentle suction to form a tight electrical seal having resistance in excess of 10^9 Ω. By applying a strong suction pulse, the area of membrane occluding the tip of the pipette is ruptured, so forming a continuity between the pipette and the interior of the cell. This method can be used to control the composition of the cytosol and to make measurements of membrane capacitance and conductance, but one should be aware that soluble macromolecules as well as electrolytes and metabolites will be diluted into the effectively infinite volume of the recording pipette. In the slow whole-cell approach (**b**), the pipette is loaded with an agent (in the case of treating mast cells, ATP^{4-}) which can generate permeability lesions of nanometre dimensions. In this way the cytosol can be dialysed against the contents of the pipette filling solution while retaining all endogenous macromolecules. **c** Mast cells attached in the whole cell mode are refractory to receptor-directed ligands and introduction of Ca^{2+} buffers, yet undergo exocytotic degranulation when injected with GTP analogues in the presence of EGTA. Only the injected cell degranulates. (Courtesy of Dr. Manfred Lindau.) **d** Mast cells attached to patch pipettes in the slow whole-cell mode can be stimulated to undergo exocytotic degranulation by normal receptor-directed ligands. All the cells in the field are responsive to the stimulus. **e** The membrane capacitance of antigen-stimulated mast cells (attached in the slow whole-cell mode) increases, indicating fusion of secretory granules, yet there is no evidence of increased conductance indicative of Ca^{2+} channels. **f** Part of an expanded capacitance trace from a GTP-γ-S-injected cell showing the step changes due to individual granule–plasma membrane fusion events. (FERNANDEZ et al. 1984; LINDAU and FERNANDEZ 1986b)

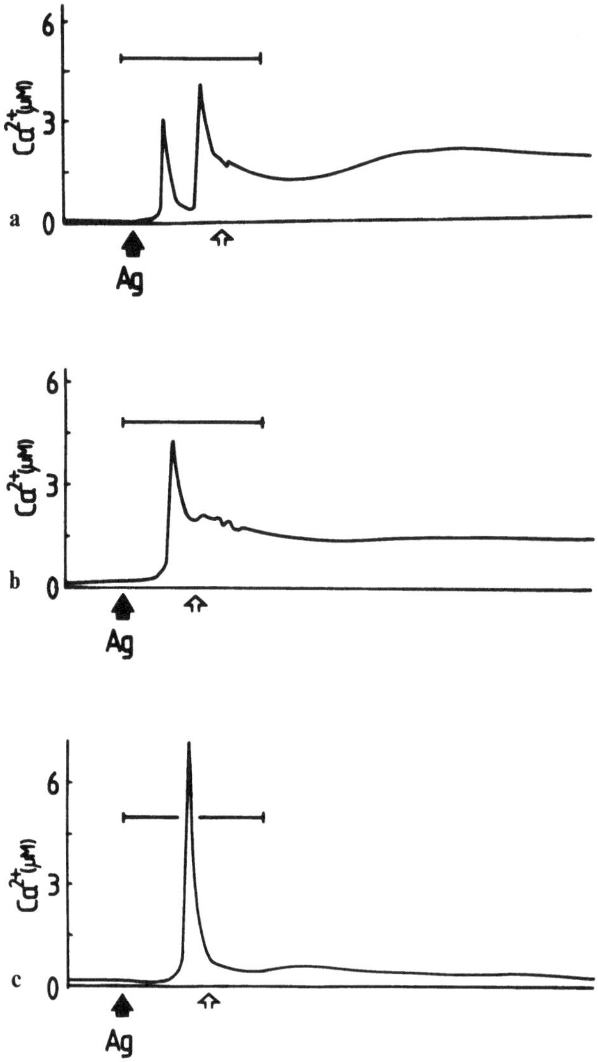

Fig. 6 a–c

Fig. 6 a–f. Examples of Ca^{2+} transients in stimulated mast cells. Mast cells were loaded with the Ca^{2+}-sensing dye fura-2 and the fluorescence of individual cells was monitored using dual wavelength excitation. Exocytosis was assessed as degranulation by observation of the cells under Nomarski optics. In general, secretion is more vigorous when the level of Ca^{2+} is elevated and maintained above baseline. However, there seems to be little correlation between the timing or the extent of the Ca^{2+} transients and the onset of exocytosis. *Full arrows* indicate times of additions; *open arrows* indicate onset of degranulation. Traces (**a–c**) show the effects of adding an antigen to mast cells treated with phosphatidylserine (PS) which enhances the degree of secretion due to ligands acting through IgE receptors. The onset of degranulation typically occurs 30–40 s after adding the stimulus. Traces **d–f** show that PS can initiate Ca^{2+} transients without causing degranulation. The almost instantaneous degranulation due to stimulation with compound 48/80 (**f**) occurred without any detectable elevation of cytosol Ca^{2+}. The *horizontal bars* indicate 1 min. (NEHER and ALMERS 1986)

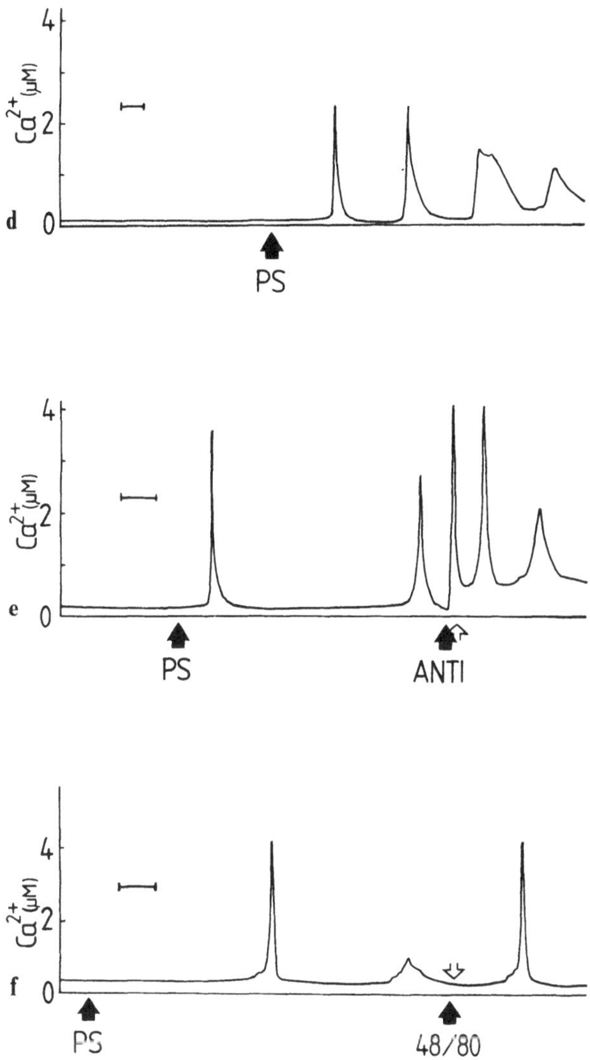

Fig. 6 d–f

times preceding any ligand-induced transients. Degranulation seemed to be more vigorous when an initial transient was followed by a period of elevated Ca^{2+}. The over-riding impression is that while increased Ca^{2+} enhances the degree of antigen- or compound 48/80-induced secretion, it is by no means a sufficient stimulus.

G. Towards Reconstitution of Exocytosis in Cell-Free Systems

At the time that Ca^{2+} was understood to deliver a complete stimulus to the secretory process, it was natural that attempts were made to understand the mechanism in terms of the Ca^{2+}-binding proteins of the cytosol and the contractile elements of the cytoskeleton providing the motile force for the exocytotic membrane fusions. The aim has been to reconstruct an in vitro exocytosis system and most experiments in this area have been carried out using isolated chromaffin granules. There are a number of reports indicating that catecholamines can be released from adrenal chromaffin granules in a manner dependent on the provision of Ca^{2+}, ATP and isolated plasma membranes. The problem however is that of distinguishing between lysis of the granules and a fusion process which involves release of granule contents at the "extracellular" surface of isolated plasma membranes.

Certainly there are examples of Ca^{2+}-induced fusion in which the possibility of lysis does not arise. Thus, liver Golgi-derived vesicles containing proalbumin and converting enzyme fuse together in the presence of micromolar Ca^{2+} (GRATZL and DAHL 1978) to generate the mature form of albumin (JUDAH and QUINN 1978; QUINN and JUDAH 1978). There is no requirement for ATP. Since such vesicles are prepared in the presence of chelating agents it is likely that they are devoid of Ca^{2+}-dependent binding proteins and that the Ca^{2+}-sensitive site for membrane fusion is intrinsic to the membranes themselves. This would appear to be a clear example of a Ca^{2+}-induced membrane fusion process. The secretion of albumin however is likely to be a constitutive function of liver cells which occurs regardless of the state of receptor activation. Here, Ca^{2+} is a requirement, not a trigger for the exocytotic process. In the examples we have been discussing, in which the secretory process is under the control of external ligands, Ca^{2+} and other diffusible second messengers activate exocytosis in a manner requiring ATP (or other nucleoside triphosphates). These comprise the minimum requirements for an in vitro experiment having relevance to the stimulated exocytotic process.

To date, the attempts to reconstitute an in vitro exocytosis system have all been based on the assumption that Ca^{2+} provides a complete signal. These have generally involved the mixing of isolated secretory granules, plasma membrane vesicles, calcium, ATP, etc. Many of these experiments have unfortunately suffered from technical and conceptual defects. There have been uncertainties concerning the precise levels of Ca^{2+} since inadequate Ca^{2+} buffering systems have been applied (LAZARUS and DAVIS 1979; KONINGS and DE POTTER 1981). More importantly, the criterion by which the in vitro exocytosis has commonly been tested has been based on the release of secretory materials into the bulk phase and it has been impossible to distinguish this from lysis (POLLARD et al. 1977, 1979). Consideration of the topography of the interaction between the secretory granule and the cytosol surface of the plasma membrane in an exocytotic reconstruction should surely require that the secreted product be not released into the environment, but retained within inside-out plasma membrane vesicles.

It seems to us that new approaches are needed, requiring developments of technique and an awareness that exocytosis is not simply a matter of elevating Ca^{2+}. Even the limited view obtainable from the consideration of three cell types indicates that Ca^{2+} is unlikely to be a unique activator of secretion in most physiologically relevant situations. Yet whatever the activators, the mechanism of phospholipid reorganisation which results in membrane fusion and exocytosis is likely to be the same. Membrane phospholipids are maintained in a stable bilayer configuration by the presence of water which is present at 55 *M*, and the affinity of membrane surfaces for their hydration layers is of a very high order.

This can simply be understood when it is realised that cells may aggregate, but do not fuse when subjected to the extremes of force in the ultracentrifuge. When it is appreciated that the pressure required to fuse two phospholipid vesicles is of the order of 1000 atmospheres (Parsegian and Rand 1983; Rand and Parsegian 1984) one can see that the problem of membrane fusion in exocytosis must involve catalysis. This is the point at which to seek to understand the roles of Ca^{2+} and the other effectors of exocytosis, diacylglycerol and GTP, and the requirement for ATP. Any experiment or scheme devised to understand that role of Ca^{2+} in stimulus–secretion coupling without accounting for its modulation by other effectors and the necessity for protein phosphorylation is simply evading the point.

References

Arem R, Chayoth R, Shenkenberg T, Field JB (1984) Role of calcium in acetylcholine-induced desensitization in dog thyroid slices. Arch Biochem Biophys 230:168–177

Ashby JP, Speke RN (1975) Insulin and glucagon secretion from isolated islets of Langerhans: the effects of calcium ionophores. Biochem J 150:89–96

Ashley RH, Brammer MJ, Marchbanks R (1984) Measurement of intrasynaptosomal free calcium by using the fluorescent indicator quin-2. Biochem J 219:149–158

Bachi T, Eichenberger G, Hauri HP (1978) Sendai virus hemolysis: influence of lectins and analysis by immune fluorescence. Virology 85:518–530

Bainton DF (1973) Sequential degranulation of the two types of polymorphonuclear leukocyte granules during phagocytosis of microorganisms. J Cell Biol 58:249–264

Baker EM, Cheek TR, Burgoyne RD (1985) Cyclic AMP inhibits secretion from bovine adrenal chromaffin cells evoked by carbamylcholine but not by high K^+. Biochim Biophys Acta 846:388–393

Baker PF, Knight DE, Whitaker MJ (1980) The relation between ionized calcium and cortical granule exocytosis in eggs of the sea urchin Echinus esculentus. Proc R Soc Lond [Biol] 207:149–161

Baker PF, Knight DE, Umbach JA (1985) Calcium clamp of the intracellular environment. Cell Calcium 6:5–14

Barrowman MM, Cockcroft S, Gomperts BD (1986) Two roles for guanine nucleotides in stimulus secretion sequence of neutrophils. Nature 319:504–507

Barrowman MM, Cockcroft S, Gomperts BD (1987) Differential control of azurophilic and specific granule exocytosis in Sendai virus-permeabilized rabbit neutrophils. J Physiol (Lond) 383:115–124

Beaven MA, Rogers J, Moore JP, Hesketh TR, Smith GA, Metcalfe JC (1984) The mechanism of the calcium signal and correlation with histamine release in 2H3 cells. J Biol Chem 259:7129–7136

Becker E, Showell H, Henson P, Hsu L (1974) The ability of chemotactic factors to induce lysosomal enzyme release: the characteristics of the release, the importance of surfaces and the relation of enzyme release to chemotactic responsiveness. J Immunol 112:2047–2054

Becker EL, Kermode JC, Naccache PH, Yassin R, Marsh ML, Munoz JJ, Sha'afi RJ (1985) The inhibition of neutrophil granule secretion and chemotaxis by pertussis toxin. J Cell Biol 100:1641–1646

Becker EL, Kermode JC, Naccache PH, Yassin R, Munoz JJ, Marsh ML, Huang CK, Sha'afi RI (1986) Pertussis toxin as a probe of neutrophil activation. Fed Proc 45:2151–2155

Belle RJ, Ozon R, Stinnakre J (1977) Free calcium in full grown Xenopus laevis oocyte following treatment with the ionophore A23187 or progesterone. Mol Cell Endocrinol 8:65–72

Bennett JP, Cockcroft S, Gomperts BD (1980a) Use of cytochalasin B to distinguish between early and late events in neutrophil activation. Biochim Biophys Acta 601:584–591

Bennett JP, Cockcroft S, Gomperts BD (1980b) Ionomycin stimulates mast cell histamine secretion by forming a lipid soluble calcium complex. Nature 282:851–853

Bennett JP, Hirth KP, Fuchs E, Sarvas M, Warren GB (1980c) The bacterial factors which stimulate neutrophils may be derived from procaryote signal peptides. FEBS Lett 116:57–61

Bennett JP, Cockcroft S, Gomperts BD (1981) Rat mast cells permeabilised with ATP secrete histamine in response to calcium ions buffered in the micromolar range. J Physiol (Lond) 317:335–345

Bennett JP, Zaner KS, Stossel TP (1984) Isolation and some properties of macrophage a-actinin: evidence that is not an actin gelling-protein. Biochemistry 23:5081–5086

Berridge MJ (1984) Inositol trisphosphate and diacylglycerol as second messengers. Biochem J 220:345–360

Bicknell RJ, Schofield JG (1976) Mechanism of action of somatostatin: Inhibition of ionophore induced release of growth hormone from dispersed bovine pituitary cells. FEBS Lett 68:23–26

Blaustein MP (1974) The interrelationship between sodium and calcium fluxes across cell membranes. Rev Physiol Biochem Pharmacol 70:33–82

Bokoch GM, Reed PW (1980) Stimulation of arachidonic acid metabolism in the polymorphonuclear leukocyte by an N-formylated peptide. J Biol Chem 255:10223–10226

Bolton JE, Field M (1977) Calcium ionophore stimulated ion secretion in rabbit ileal mucosa: relation to actions of 3′,5′-AMP and carbamylcholine. J Membr Biol 35:159–173

Breitweiser GE, Szabo G (1985) Uncoupling of cardiac muscarinic and β-adrenergic receptors from ion channels by a guanine nucleotide analogue. Nature 317:538–540

Brocklehurst KW, Pollard BP (1985) Enhancement of Ca^{2+}-induced catecholamine release by the phorbol ester TPA in digitonin-permeabilized cultured bovine adrenal chromaffin cells. FEBS Lett 183:107–110

Burgoyne RD (1984) The relationship between secretion and intracellular free calcium in bovine adrenal chromaffin cells. Biosci Rep 4:605–611

Burgoyne RD, Cheek TR (1985) Is the transient nature of the secretory response of chromaffin cells due to inactivation of calcium channels? FEBS Lett 182:115–118

Burgoyne RD, Norman KM (1984) Effect of calmidazolium and phorbol ester on catecholamine secretion from adrenal chromaffin cells. Biochim Biophys Acta 805:37–43

Campbell AK (1983) Intracellular calcium: its universal role as regulator. John Wiley, Chichester, 556 pp

Campbell AK, Hallett MB (1983) Measurement of intracellular free calcium ions and oxygen radicals in polymorphonuclear leukocyte-erythrocyte "ghost" hybrids. J Physiol (Lond) 338:537–550

Candia DA, Montoreano R, Podos SM (1977) Effect of the ionophore A23187 on chloride transport across isolated frog cornea. Am J Physiol 233:F94–F101

Capponi AM, Lew PD, Vallotton MB (1985) Cytosolic free calcium levels in monolayers of cultured rat aortic smooth muscle cells: effects of angiotensin II and vasopressin. J Biol Chem 260:7836–7842

Carvalho MH, Prat JC, Garcia AG, Kirpekar SM (1982) Ionomycin stimulates secretion of catecholamines from cat adrenal gland and spleen. Am J Physiol 242:E137–E145

Cheek TR, Burgoyne RD (1985) Effect of activation of muscarinic receptors on intracellular free calcium and secretion in bovine adrenal chromaffin cells. Biochim Biophys Acta 846:167–173

Cochrane DE, Douglas WW, Mouri T, Nakazato Y (1977) Calcium and stimulus-secretion coupling in the adrenal medulla: contrasting stimulating effects of the ionophores X537A and A23187 on catecholamine output. J Physiol (Lond) 252:363–378

Cockcroft S, Gomperts BD (1979a) ATP induces nucleotide permeability in rat mast cells. Nature 279:541–542

Cockcroft S, Gomperts BD (1979b) Evidence for a role of phosphatidylinositol turnover in stimulus-secretion coupling: studies with rat peritoneal mast cells. Biochem J 178:681–687

Cockcroft S, Gomperts BD (1985) Guanine nucleotides activate polyphosphoinositide phosphodiesterase, the effector for Ca^{2+} mobilisation and protein kinase C activation. Nature 314:534–536
Cockcroft S, Bennett JP, Gomperts BD (1981) The dependence on Ca^{2+} of phosphatidylinositol breakdown and enzyme secretion in rabbit neutrophils stimulated by formylmethionylleucylphenylalanine or ionomycin. Biochem J 200:501–508
Cohen P, Burchell A, Foulkes JG, Cohen PTW, Vanaman TC, Nairn AC (1978) Identification of the Ca^{2+}-dependent modulator protein as the fourth subunit of rabbit skeletal muscle phosphorylase kinase. FEBS Lett 92:287–293
Cotman CW, Haycock JW, White WF (1976) Stimulus-secretion coupling in brain: analysis of noradrenaline and gamma-amino-butyric acid release. J Physiol (Lond) 254:475–506
Creutz CE, Pazoles CJ, Pollard HB (1978) Identification and purification of an adrenal medullary protein (synexin) that causes calcium-dependent aggregation of isolated chromaffin granules. J Biol Chem 253:2858–2866
Dale M, Penfield A (1984) Synergism between phorbol ester and A23187 in superoxide production by neutrophils. FEBS Lett 175:170–172
Dedman JR, Potter JD, Jackson RL, Johnson JD, Means AR (1977) Physicochemical properties of rat testis Ca^{2+}-dependent regulator protein of cyclic nucleotide phosphodiesterase: relationship of Ca^{2+}-binding, conformational changes, and phosphodiesterase activity. J Biol Chem 252:8415–8422
Delbeke D, Kojima I, Dannies PS, Rasmussen H (1984) Synergistic stimulation of prolactin release by phorbol ester, A23187 and forskolin. Biochem Biophys Res Commun 123:735–741
Derome G, Tseng R, Mercier P, Lemaire I, Lemaire S (1981) Possible muscarinic regulation of catecholamine secretion mediated by cyclic GMP in isolated bovine adrenal chromaffin cells. Biochem Pharmacol 30:855–860
Di Virgilio F, Gomperts BD (1983) Cytosol Mg^{2+} modulates Ca^{2+} ionophore induced secretion from rabbit neutrophils. FEBS Lett 163:315–318
Di Virgilio F, Lew DP, Pozzan T (1984) Protein kinase C activation of physiological processes in human neutrophils at vanishingly small cytosolic Ca^{2+} levels. Nature 310:691–693
Di Virgilio F, Salviati G, Pozzan T, Volpe P (1986) Is a guanine nucleotide binding protein involved in excitation-contraction coupling in skeletal muscle? EMBO J 5:259–262
Douglas WW, Rubin RP (1961) The role of calcium in the secretory response of the adrenal medulla to acetylcholine. J Physiol (Lond) 159:40–57
Downes CP, Michell RH (1985) Inositol phospholipid breakdown as a receptor-controlled generator of second messengers. In: Cohen P, Houslay MD (eds) Molecular mechanisms of transmembrane-signalling. Elsevier, Amsterdam, pp 3–56
Eimerl S, Savion N, Heichal O, Selinger Z (1974) Induction of enzyme secretion in rat pancreatic slices using the ionophore A23187 and Ca^{2+}. J Biol Chem 249:3991–3993
Feinman RD, Detwiler PC (1974) Platelet secretion induced by divalent cation ionophores. Nature 249:172–173
Fenwick EM, Marty A, Neher E (1982a) A patch-clamp study of bovine chromaffin cells and of their sensitivity to acetylcholine. J Physiol (Lond) 331:577–597
Fenwick EM, Marty A, Neher E (1982b) Sodium and calcium channels in bovine chromaffin cells. J Physiol 331:599–635
Fernandez JM, Neher E, Gomperts BD (1984) Capacitance measurements reveal stepwise fusion events in degranulating mast cells. Nature 312:453–455
Fewtrell CMS, Gomperts BD (1977) Quercetin: a novel inhibitor of Ca^{2+} influx and exocytosis in rat peritoneal mast cells. Biochim Biophys Acta 469:52–60
Fisher SK, Holz RW, Agranoff BW (1981) Muscarinic receptors in chromaffin cell cultures mediate enhanced phospholipid labelling but not catecholamine secretion. J Neurochem 37:491–497
Foreman JC, Mongar JL (1972) The role of the alkaline earth ions in anaphylactic histamine secretion. J Physiol (Lond) 224:753–769

Foreman JC, Mongar JL (1973) The action of lanthanum and manganese on anaphylactic histamine secretion. Br J Pharmacol 48:527–537
Foreman JC, Mongar JL, Gomperts BD (1973) Calcium ionophores and movement of calcium ions following the physiological stimulus to a secretory process. Nature 245:249–251
Foreman JC, Mongar JL, Gomperts BD, Garland LG (1975) A possible role for cyclic AMP in the regulation of histamine secretion and the action of cromoglycate. Biochem Pharmacol 74:538–540
Frizzell RA (1977) Active chloride secretion by rabbit colon: calcium dependent stimulation by ionophore A23187. J Membr Biol 35:175–187
Gallien-Lartigue O (1976) Calcium and ionophore A23187 as initiators of DNA replication in the pluripotent stem cell. Cell Tissue Kinet 9:533–540
Garcia AG, Kirpekar SM, Prat JC (1975) A calcium ionophore stimulating the secretion of catecholamines from the cat adrenal. J Physiol (Lond) 244:253–262
Geisow M, Childs J, Dash B, Harris A, Panayotou G, Sudhof T, Walker JH (1984) Cellular distribution of three mammalian Ca^{2+}-binding proteins related to Torpedo calelectrin. EMBO J 3:2969–2974
Gemsa D, Seitz M, Kramer W, Grimm W, Till G, Resch K (1979) Ionophore A23187 raises cyclic AMP levels in macrophages by stimulating prostaglandin E formation. Exp Cell Res 118:55–62
Glenney JR, Bretscher A, Weber K (1980) Calcium control of the intestinal microvillus cytoskeleton: its implications for the regulation of microfilament organizations. Proc Natl Acad Sci USA 77:6458–6462
Goetzl EJ (1980) Novel effects of 1-O-hexadecyl-2-acyl-SN-glycero-3-phosphorylcholine mediators of human leukocyte function: delineation of the specific roles of the acyl substituents. Biochem Biophys Res Commun 94:881–888
Goldstein I, Hoffstein S, Gallin J, Weissmann G (1973) Mechanisms of lysosomal enzyme release from human leukocytes: microtubule assembly and membrane fusion induced by a component of complement. Proc Nat Acad Sci 70:2916–2920
Gomperts BD (1983) Involvement of guanine nucleotide-binding protein in the gating of Ca^{2+} by receptors. Nature 306:64–66
Gomperts BD (1985) Manipulation of the cytosolic composition of mast cells: a study of early events in stimulus-secretion coupling. In: Dean RT, Stahl P (eds) Developments in cell biology, vol 1. Butterworths, London, pp 18–37
Gomperts BD, Fernandez JM (1985) Techniques for membrane permeabilisation. Trends Biochem Sci 10:414–417
Gomperts BD, Baldwin JM, Micklem KJ (1983) Rat mast cells permeabilised with Sendai virus secrete histamine in response to Ca^{2+} buffered in the micromolar range. Biochem J 210:737–745
Gomperts BD, Barrowman MM, Cockcroft S (1986) Dual role for guanine nucleotides in stimulus-secretion coupling: an investigation of mast cells and neutrophils. Fed Proc 45:2156–2161
Gratzl M, Dahl G (1978) Fusion of secretory vesicles isolated from rat liver. J Membr Biol 40:343–364
Grynkiewicz G, Poenie M, Tsien RY (1985) A new generation of Ca^{2+} indicators with greatly improved fluorescence properties. J Biol Chem 260:3440–3450
Hallam TJ, Sanchez A, Rink TJ (1984a) Stimulus-response coupling in human platelets: changes evoked by platelet-activating factor in cytoplasmic free calcium monitored with the fluorescent calcium indicator quin2. Biochem J 218:819–827
Hallam T, Thompson N, Scrutton M, Rink T (1984b) The role of cytoplasmic free calcium in the responses of quin2-loaded human platelets to vasopressin. Biochem J 221:897–901
Hamel E, Lustbader J, Lin CM (1984) Deoxyguanosine nucleotide analogues: potent stimulators of microtubule nucleation with reduced affinity for the exchangeable nucleotide site of tubulin. Biochemistry 23:5314–5325
Haslam RJ, Davidson MML (1984) Guanine nucleotides decrease the free $[Ca^{2+}]$ required for secretion of serotonin from permeabilized blood platelets: evidence of a role for a GTP-binding-protein in platelet activation. FEBS Lett 174:90–95

Haywood AM (1974) Characteristics of Sendai virus receptors in a model membrane. J Mol Biol 83:427–436
Heppel LA, Weisman GA, Friedberg I (1985) Permeabilization of transformed cells in culture by external ATP. J Membr Biol 86:189–196
Hesketh TR, Bavetta S, Smith GA, Metcalfe JC (1983) Duration of the calcium signal in the mitogenic stimulation of thymocytes. Biochem J 214:575–579
Hirata M, Hashimoto T, Hamachi T, Koga T (1984) Changes of intracellular free Ca^{2+} in macrophages following N-formyl chemotactic peptide stimulation. Direct measurement by the loading of quin2. J Biochem (Tokyo) 96:9–16
Holland DR, Steinberg MI, Armstrong WM (1975) A23187: a calcium ionophore that directly increases cardiac contractility. Proc Soc Exp Biol Med 148:1141–1145
Holwill MJ, McGregor JL (1975) Control of flagella wave movement in Crithidia oncopelti. Nature 255:157–158
Holz RW (1975) The release of dopamine from synaptosomes from rat brain striatum by the ionophores X537A and A23187. Biochim Biophys Acta 375:138–152
Hong K, Duzgunes N, Papahadjopoulos D (1981) Role of synexin in membrane fusion: enhancement of calcium dependent fusion of phospholipid vesicles. J Biol Chem 256:3641–3644
Horowitz R, Winegrad S (1983) Cholinergic regulation of calcium sensitivity in cardiac muscle. J Mol Cell Cardiol 15:277–280
Ishizuka Y, Nozawa Y (1983) Concerted stimulation of PI-turnover, Ca^{2+}-influx and histamine release in antigen-activated rat mast cells. Biochem Biophys Res Commun 117:710–717
Jakobs KH, Bauer S, Watanabe Y (1985) Modulation of adenylate cyclase of human platelets by phorbol ester: impairment of the hormone-sensitive inhibitory pathway. Eur J Biochem 151:425–430
Jakschik BA, Kulczycki A, MacDonald HH, Parker CW (1977) Release of slow reacting substance (SRS) from rat basophilic leukemia cells. J Immunol 119:618–622
Judah JD, Quinn PS (1978) Calcium ion-dependent vesicle fusion in the conversion of proalbumin to albumin. Nature 271:384–385
Kajikawa K, Kaibuchi K, Matsubara T, Kikkawa U, Takai Y, Nishizuka Y (1983) A possible role of protein kinase C in signal-induced lysosomal enzyme release. Biochem Biophys Res Commun 116:743–750
Kao LS, Schneider AS (1985) Muscarinic receptors on bovine chromaffin cells mediate a rise in cytosolic calcium that is independent of extracellular calcium. J Biol Chem 260:2019–2022
Keppens S, van Den Heede JR, de Wulf H (1977) On the role of calcium as second messenger in liver for the hormonally induced activation of glycogen phosphorylase. Biochim Biophys Acta 496:448–457
Knapp HR, Oelz O, Roberts BJ, Sweetman BJ, Oates JA, Reed PW (1977) Ionophores stimulate prostaglandin and thromboxane biosynthesis. Proc Natl Acad Sci USA 74:4251–4255
Knight DE, Baker PF (1982) Calcium-dependence of catecholamine release from bovine adrenal medullary cells after exposure to intense electric fields. J Membr Biol 68:107–140
Knight DE, Baker PF (1983) The phorbol ester TPA increases the affinity of exocytosis for calcium in "leaky" adrenal medullary cells. FEBS Lett 160:98–100
Knight DE, Baker PF (1985) Guanine nucleotides and Ca-dependent exocytosis. FEBS Lett 189:345–349
Knight DE, Kesteven NT (1983) Evoked transient intracellular free Ca^{2+} changes and secretion in isolated bovine adrenal medullary cells. Proc R Soc Lond [Biol] 218:177–199
Knight DE, Koh E (1984) Ca^{2+} and cyclic nucleotide dependence of amylase release from isolated rat pancreatic acinar cells rendered permeable by intense electric fields. Cell Calcium 5:401–418
Knight DE, Scrutton MC (1980) Direct evidence for a role for Ca^{2+} in amine storage granule secretion by human platelets. Thromb Res 20:437–446

Knight DE, Hallam TJ, Scrutton MC (1982) Agonist selectivity and second messenger concentration in Ca^{2+}-mediated secretion. Nature 296:256–257
Knight DE, Niggli V, Scrutton MC (1984) Thrombin and activators of protein kinase C modulate secretory responses of permeabilised human platelets induced by Ca^{2+}. Eur J Biochem 143:437–446
Konings F, Potter W De (1981) Calcium-dependent in vitro interaction between bovine adrenal medullary cell membranes and chromaffin granules as a model for exocytosis. FEBS Lett 126:103–106
Korchak HM, Wilkenfeld C, Rich AM, Radin AR, Vienne K, Rutherford LE (1984) Stimulus response coupling in the human neutrophils: differential requirements for receptor occupancy in neutrophil responses to a chemoattractant. J Biol Chem 259:7439–7445
Lazarus NR, Davis B (1979) Some events at the islet cell plasma membrane that may be associated with exocytotic insulin release. In: Hopkins CR, Duncan CJ (eds) Secretory mechanisms. Cambridge University Press, Cambridge, pp 299–321
Lemaire S, Derome G, Tseng R, Mercier P, Lemaire I (1981) Distinct regulations by calcium of cyclic GMP levels and catecholamine secretion in isolated bovine adrenal chromaffin cells. Metabolism 30:462–468
Lew PD, Dayer JM, Wollheim CB, Pozzan T (1984a) Effect of leukotriene B_4, prostaglandin E_2 and arachidonic acid on cytosolic-free calcium in human neutrophils. FEBS Lett 166:44–48
Lew PD, Wollheim CB, Waldwogel FA, Pozzan T (1984b) Modulation of cytosolic-free calcium transients by changes in intracellular calcium-buffering capacity: correlation with exocytosis and O_2-production in human neutrophils. J Cell Biol 99:1212–1220
Lew PD, Andersson T, Hed J, Di Virgilio F, Pozzan T, Stendahl O (1985) Ca^{2+}-dependent and Ca^{2+}-independent phagocytosis in human neutrophils. Nature 315:509–511
Lew PD, Monod A, Waldwogel FA, Dewald B, Baggiolini M, Pozzan T (1986) Quantitative analysis of the cytosolic free calcium dependency of exocytosis from three subcellular compartments in intact human neutrophils. J Cell Biol 102:2197–2204
Lew VL, Ferreira HG (1978) Calcium transport and the properties of a calcium-activated potassium channel in red cell membranes. Curr Top Membranes Transp 10:217–277
Lewis RA, Goetzl E, Wasserman SI, Valone FH, Rubin RH, Austen KF (1975) The release of four mediators of immediate hypersensitivity from human leukemic basophils. J Immunol 114:87–92
Lindau M, Fernandez JM (1986a) A patch clamp study of histamine secreting cells. J Gen Physiol 88:349–368
Lindau M, Fernandez JM (1986b) IgE mediated degranulation of mast cells does not require opening of ion channels. Nature 319:150–153
Liu CM, Hermann TE (1978) Characterization of ionomycin as a calcium ionophore. J Biol Chem 253:5892–5894
Loewenstein WR (1979) Junctional intercellular communication and the control of growth. Biochim Biophys Acta 560:1–65
Maino CV, Green NM, Crumpton MJ (1974) The role of divalent cations in initiating transformation of lymphocytes. Nature 251:34–37
Marasco WA, Phan SH, Krutzsch H, Showell HJ, Feltner DE, Nairn R, Becker EL, Ward PA (1984) Purification and identification of formyl-methionyl-leucyl-phenylalanine as the major peptide neutrophil chemotactic factor produced by Escherichia coli. J Biol Chem 259:5430–5439
Mazurek N, Bashkin P, Pecht I (1982) Isolation of a basophilic membrane protein binding the anti-allergic drug cromolyn. EMBO J 1:585–590
Mazurek N, Schindler H, Schurholz T, Pecht I (1984) The cromolyn binding protein constitutes the Ca^{2+} channel of basophils opening upon immunological stimulus. Proc Natl Acad Sci USA 81:6841–6845
McClellan GB, Winegrad S (1980) Cyclic nucleotide regulation of the contractile proteins in mammalian cardiac muscle. J Gen Physiol 75:283–295
McNeil P, McKenna MP, Taylor DL (1985) A transient rise in cytosolic calcium follows stimulation of quiescent cells with growth factors and is inhibitable with phorbol myristate acetate. J Cell Biol 100:1325–1332

Means AR, Dedman JR (1980) Calmodulin – an intracellular calcium receptor. Nature 285:73–77
Mennes VC, Yates J, Klahr S (1978) Effects of ionophore A23187 and external Ca^{2+} concentrations on renal gluconeogenesis. Proc Soc Exp Biol Med 157:168–174
Michaelson DM, Sokolovsky M (1978) Induced acetylcholine release from active purely cholinergic torpedo synaptosomes. J Neurochem 30:217–230
Michell RH, Jones LM, Jafferji SS (1977) The relationship between agonist-stimulated phosphatidylinositol metabolism and the mechanisms of receptor systems. In: Case RM, Goebell H (eds) Stimulus-secretion coupling in the gastrointestinal tract. MTP Press, Lancaster, pp 89–103
Misbahuddin M, Isosaki M, Houchi H, Oka M (1985) Muscarinic receptor-mediated increase in cytoplasmic free Ca^{2+} in isolated bovine adrenal medullary cells: effects of TMB-8 and phorbol ester TPA. FEBS Lett 190:25–28
Mobley BA (1977) Calcium ionophores and tension production in skinned frog muscle fibers. Eur J Pharmacol 45:101–104
Molski TFP, Naccache PH, Marsh ML, Kermode J, Becker EL, Sha'afi RI (1984) Pertussis toxin inhibits the rise in intracellular concentration of free calcium that is induced by chemotactic factors in rabbit neutrophils. Biochem Biophys Res Commun 124:644–650
Mope L, McClellan GB, Winegrad S (1980) Calcium sensitivity of the contractile system and phosphorylation of troponin in hyperpermeable cardiac cells. J Gen Physiol 75:271–282
Naccache PH, Showell HJ, Becker EL, Sha'afi RI (1977) Transport of sodium, potassium, and calcium across rabbit polymorphonuclear leukocyte membranes: effect of chemotactic factor. J Cell Biol 73:428–444
Naccache PH, Molski TFP, Borgeat P, White JR, Sha'afi RI (1985) Phorbol esters inhibit the fMet-Leu-Phe- and leukotriene B_4-stimulated calcium mobilization and enzyme secretion in rabbit neutrophils. J Biol Chem 260:2125–2131
Nakamura T, Ui M (1983) Islet activating factor, pertussis toxin inhibits Ca^{2+}-induced and guanine nucleotide dependent releases of histamine and arachidonic acid from rat mast cells. FEBS Lett 173:414–418
Nakamura T, Ui M (1985) Simultaneous inhibitions of inositol phospholipid breakdown, arachidonic acid release, and histamine secretion in mast cells by islet-activating protein, pertussis toxin: a possible involvement of the toxin-specific substrate in the Ca^{2+}-mobilizing receptor-mediated biosignaling system. J Biol Chem 260:3584–3593
Neher E, Almers W (1986) Fast calcium transients in rat peritoneal mast cells are not sufficient to trigger exocytosis. EMBO J 5:51–53
Neher E, Marty A (1982) Discrete changes of cell membrane capacitance observed under conditions of enhanced secretion in bovine adrenal chromaffin cells. Proc Natl Acad Sci USA 79:6712–6716
Nisbet-Brown E, Cheung RK, Lee JWW, Gelfand EW (1985) Antigen-dependent increase in cytosolic free calcium in specific human T-lymphocyte clones. Nature 316:545–547
Nishizuka Y (1984) The role of protein kinase C in cell surface signal transduction and tumour promotion. Nature 308:693–698
O'Flaherty JT (1985) Neutrophil degranulation: evidence pertaining to its mediation by the combined effects of leukotriene B_4, platelet-activating factor, and 5-HETE. J Cell Physiol 122:229–239
O'Flaherty JT, Wykle RL, Thomas MJ, McCall CE (1984) Neutrophil degranulation responses to combinations of arachidonate metabolites and platelet-activating factor. Res Commun Chem Pathol Pharmacol 43:3–22
Ochs DL, Korenbrot JI, Williams JA (1985) Relation between free cytosolic calcium and amylase release by pancreatic acini. Am J Physiol 249:G389–G398
Ordal GW (1977) Calcium ion regulates chemotactic behaviour in bacteria. Nature 270:66–67
Owens RJ, Gallagher J, Crumpton MJ (1984) Cellular distribution of p68, a new calcium-binding protein from lymphocytes. EMBO J 3:945–952
Pace CS, Tarvin JT, Neighbors AS, Pirkle JA, Greider MH (1980) Use of a high voltage technique to determine the molecular requirements for exocytosis in islet cells. Diabetes 29:911–918

Parsegian VA, Rand RP (1983) Membrane interaction and deformation. Ann NY Acad Sci 416:1–12
Pasternak CA, Alder GM, Bashford CL, Buckley CD, Micklem KJ, Patel K (1985) Cell damage by viruses, toxins and complement: common features of pore-formation and its inhibition by Ca^{2+}. Biochem Soc Trans 50:247–264
Penfield A, Dale M (1984) Synergism between A23187 and 1-oleoyl-2-acetyl-glycerol in superoxide production by human neutrophils. Biochem Biophys Res Commun 125:332–336
Pfaffinger PJ, Martin JM, Hunter DD, Nathanson NM, Hille B (1985) GTP-binding proteins couple cardiac muscarinic receptors to a K channel. Nature 317:536–538
Pfeilschifter J, Bauer C (1986) Pertussis toxin abolishes angiotension II induced phosphoinositide hydrolysis and prostaglandin synthesis in rat renal mesangial cells. Biochem J 236:289–294
Pholramool C, Tangkrisanavinont V (1976) A calcium ionophore induced secretion in rabbit lacrimal gland in vivo. Life Sci 19:381–388
Pollard HB, Pazoles CJ, Creutz CE, Ramu A, Strott CA, Ray P, Brown EM, Aurbach GD, Tack-Goldman KM, Shulman NR (1977) A role for anion transport in the regulation of release from chromaffin granules and exocytosis from cells. J Supramol Struct 7:277–285
Pollard HB, Pazoles CJ, Creutz CE, Zinder O (1979) The chromaffin granule and possible mechanisms of exocytosis. Int Rev Cytol 58:159–197
Pollard HB, Scott JH, Creutz CE (1983) Inhibition of synexin activity and exocytosis from chromaffin cells by phenothiazine drugs. Biochem Biophys Res Commun 113:908–915
Powell T, Tatham PER, Twist VW (1984) Cytoplasmic free calcium measured by quin2 fluorescence in isolated ventricular myocytes at rest and during potassium-depolarisation. Biochem Biophys Res Commun 122:1012–1020
Pozzan T, Lew DP, Wollheim CB, Tsien RY (1983) Is cytosolic ionized calcium regulating neutrophil activation? Science 221:1413–1415
Prince WT, Rasmussen H, Berridge MJ (1973) The role of Ca^{2+} in fly salivary gland secretion analysed with the ionophore A23187. Biochim Biophys Acta 329:98–107
Quinn PS, Judah JD (1978) Calcium-dependent Golgi-vesicle fusion and cathepsin B in the conversion of proalbumin into albumin in rat liver. Biochem J 172:301–309
Rand RP, Parsegian VA (1984) Physical force considerations in model and biological membranes. Can J Biochem Cell Biol 62:752–759
Rani CSS, Boyd AE, Field JB (1985) Effects of acetylcholine, TSH and other stimulators on intracellular calcium concentration in dog thyroid cells. Biochem Biophys Res Commun 131:1041–1047
Reed PW, Lardy HA (1972) A23187: a divalent cation ionophore. J Biol Chem 247:6970–6977
Rink T, Sanchez A (1984) Effects of prostaglandin I_2 and forskolin on the secretion from platelets evoked at basal concentrations of cytoplasmic free calcium by thrombin, collagen, phorbol ester and exogenous diacylglycerol. Biochem J 222:833–936
Rink T, Sanchez A, Hallam T (1983) Diacylglycerol and phorbol ester stimulate secretion without raising cytoplasmic free calcium in human platelets. Nature 305:317–319
Robertson SP, Johnson JD, Holroyde MJ, Kranias EG, Potter JD, Solaro R (1982) The effect of troponin-i phosphorylation on the Ca^{2+}-binding properties of the Ca^{2+}-regulatory site of bovine cardiac troponin. J Biol Chem 257:260–263
Rodbell M (1980) The role of hormone receptors and GTP-regulatory proteins in membrane transduction. Nature 284:17–22
Ronning SA, Martin TFJ (1985) Prolactin secretion in permeable GH_3 pituitary cells is stimulated by Ca^{2+} and protein kinase C activators. Biochem Biophys Res Commun 130:524–532
Rorsman P, Abrahamsson H, Gylfe E, Hellman B (1984) Dual effects of glucose on the cytosolic free Ca^{2+} activity of mouse pancreatic beta-cells. FEBS Lett 170:196–200
Rosenberg S, Stracher A, Burridge K (1981) Isolation and characterization of a calcium-sensitive alpha-actinin-like protein from human platelets. J Biol Chem 256:12986–12981

Rozengurt E (1985) The mitogenic response of cultured 3T3 cells: integration of early signals and synergistic effects in a unified framework. In: Cohen P, Houslay MD (eds) Molecular mechanisms of transmembrane signalling. Elsevier, Amsterdam, pp 429–452

Ruskoaho H, Toth M, Lang RE (1985) Atrial natriuretic peptide secretion: synergistic effects of phorbol ester and A23187. Biochem Biophys Res Commun 133:581–588

Schell-Frederick E (1974) Stimulation of oxidative metabolism of polymorphonuclear leucocytes by the Ca^{2+} ionophore A23187. FEBS Lett 48:37–40

Schiffmann E, Showell HJ, Corcoran BA, Ward PA, Smith E, Becker EL (1975) The isolation and partial characterization of neutrophil chemotactic factors from *Escherichia coli*. J Immunol 114:1831–1837

Schlegel W, Wollheim CB (1984) Thyrotropin-releasing hormone increases cytosolic free Ca^{2+} in clonal pituitary cells (GH_3 cells): direct evidence for the mobilization of cellular calcium. J Cell Biol 99:83–87

Schneider AS, Herz R, Rosenheck K (1977) Stimulus-secretion coupling in chromaffin cells isolated from bovine adrenal medulla. Proc Natl Acad Sci USA 11:5036–5040

Schneider C, Gennaro R, de Nicola G, Romeo D (1978) Secretion of granule enzymes from alveolar macrophages: regulation by intracellular Ca^{2+} buffering capacity. Exp Cell Res 112:249–256

Scott JH, Creutz CE, Pollard HB, Ornberg R (1985) Synexin binds in a calcium-dependent fashion to oriented chromaffin cell plasma membranes. FEBS Lett 180:17–23

Segal AW, Dorling J, Coade S (1980) Kinetics of fusion of the cytoplasmic granules with phagocytic vacuoles in human polymorphonuclear leukocytes. J Cell Biol 85:42–59

Segal DM, Taurog JD, Metzger H (1977) Dimeric immunoglobulin E serves as a unit signal for mast cell degranulation. Proc Natl Acad Sci USA 74:2993–2997

Selinger Z, Eimerl S, Schramm M (1974) A calcium ionophore stimulating the action of epinephrine on the a-adrenergic receptor. Proc Natl Acad Sci USA 71:128–131

Serhan CN, Radin A, Smolen JE, Korchak H, Samuelsson B, Weissmann G (1982) Leukotriene B_4 is a complete secretogogue in human neutrophils: a kinetic analysis. Biochem Biophys Res Commun 107:1006–1012

Sha'afi RI, White JR, Molski TFP, Shefczyk J, Volpi M, Naccache PH, Feinstein MB (1983) Phorbol 12-myristate 13-acetate activates rabbit neutrophils without an apparent rise in the level of intracellular free calcium. Biochem Biophys Res Commun 114:638–645

Sibley DR, Nambi P, Peters JR, Lefkowitz RJ (1984) Phorbol diesters promote β-adrenergic receptor phosphorylation and adenylate cyclase desensitization in duck erythrocytes. Biochem Biophys Res Commun 121:973–979

Sibley DR, Strasser RH, Caron MG, Lefkowitz RJ (1985) Homologous desensitization of adenylate cyclase is associated with phosphorylation of the β-adrenergic receptor. J Biol Chem 260:3883–3886

Silverstein SC, Steinman RM, Cohn ZA (1977) Endocytosis. Ann Rev Biochem 46:665–722

Singh J (1985) Phorbol ester (TPA) potentiates noradrenaline and acetylcholine-evoked amylase secretion in the rat pancreas. FEBS Lett 180:191–195

Smith RJ, Ignarro LJ (1975) Bioregulation of lysosomal enzyme secretion from human neutrophils: roles of guanosine 3′:5′-monophosphate and calcium in stimulus-secretion coupling. Proc Natl Acad Sci USA 72:108–112

Smith RJ, Sun FF, Bowman BJ, Iden SS, Smith HW, McGuire JC (1982) Effect of 6,9-deepoxy-6,9-(phenylimino)-delta^{6,8}-prostaglandin I_1 (U-60,257), an inhibitor of leukotriene synthesis, on human neutrophil function. Biochem Biophys Res Commun 109:943–949

Spangrude GJ, Sacchi F, Hill HR, Epps DE van, Daynes RA (1985) Inhibition of lymphocyte and neutrophil chemotaxis by pertussis toxin. J Immunol 135:4135–4143

Steinhardt RA, Epel D (1974) Activation of sea urchin eggs by a calcium ionophore. Proc Natl Acad Sci USA 71:1915–1919

Stephenson EW (1981) Activation of fast skeletal muscle: contributions of studies of skinned fibers. Am J Physiol 240:C1–C19

Sudhof TC, Walker JH, Fritsche U (1985) Characterization of calelectrin, a Ca^{2+}-binding protein isolated from the electric organ of Torpedo marmorata. J Neurochem 44:1302–1307

Swendsen CL, Ellis JM, Chilton FH, O'Flaherty JT, Wykle RL (1983) 1-*O*-Alkyl-2-acyl-*SN*-glycero-3-phosphocholine: a novel source of arachidonic acid in neutrophils stimulated by the calcium ionophore A23187. Biochem Biophys Res Commun 113:72–79

Swilem AMF, Hawthorne JN, Azila N (1983) Catecholamine secretion by perfused bovine adrenal medulla in response to nicotinic activation is inhibited by muscarinic receptors. Biochem Pharmacol 32:3873–3874

Takemura H (1985) Changes in free cytosolic calcium concentration in isolated rat parotid cells by cholinergic and β-adrenergic agonists. Biochem Biophys Res Commun 131:1048–1055

Thomas A, Alexander J, Williamson J (1984) Relationship between inositol polyphosphate production and the increase of cytosolic free Ca^{2+} induced by vasopressin in isolated hepatocytes. J Biol Chem 259:5574–5584

Timasheff SN, Grisham LM (1980) In vitro assembly of cytoplasmic microtubules. Annu Rev Biochem 49:565–591

Trifaro JM, Lee RWH (1977) Morphological characteristics and stimulus-secretion coupling in bovine adrenal chromaffin cell cultures. Neuroscience 5:1533–1546

Triggle CR, Grant WF, Triggle DJ (1975) Intestinal smooth muscle contraction and the effects of cadmium and A23187. J Pharmacol Exp Ther 194:182–190

Truneh A, Albert F, Golstein P, Schmitt-Verhulst AM (1985) Early steps of lymphocyte activation bypassed by synergy between calcium ionophores and phorbol ester. Nature 313:318–320

Tsien TY (1980) New calcium indicators and buffers with selectivity against magnesium and protons: design, synthesis and properties of prototype structures. Biochemistry 19:2396–2404

Udelsman R, Harwood JP, Millan MA, Chrousos GP, Goldstein DS, Zimlichman R, Catt KJ, Aguilera G (1986) Functional corticotropin releasing factor receptors in the primate peripheral sympathetic nervous system. Nature 319:147–150

Vargas O, Miranda R, Orrego F (1976) Effects of Na^{+}-deficient media and of a calcium ionophore (A23187) on the release of ^{3}H-noradrenaline, ^{14}C-alpha-aminobutyrate and ^{3}H-gamma-aminobutyrate from superfused slices of rat neocortex. Neuroscience 1:137–145

Volpe P, Virgilio F Di, Pozzan T, Salviati G (1986) Role of inositol-1,4,5-trisphosphate in excitation-contraction coupling in skeletal muscle. FEBS Lett 197:1–4

Wang LL, Bryan J (1981) Isolation of calcium-dependent platelet proteins that interact with actin. Cell 25:637–649

Weiss B, Prozialeck WC, Wallace TL (1982) Interaction of drugs with calmodulin: biochemical, pharmacological and clinical implications. Biochem Pharmacol 31:2217–2226

White JR, Ishizaka T, Ishizaka K, Sha'afi RI (1984) Direct demonstration of increased intracellular concentration of free calcium measured by quin-2 in stimulated rat mast cells. Proc Natl Acad Sci USA 81:3978–3982

Wilkinson PC (1975) Leucocyte locomotion and chemotaxis: influence of divalent cations and cation ionophores. Exp Cell Res 93:420–426

Wilson SP, Kirschner N (1977) The acetylcholine receptor of the adrenal medulla. J Neurochem 28:687–695

Wright DG, Bralove DA, Gallin JI (1977) The differential mobilisation of human neutrophil granules. Effects of phorbol myristate and ionophore A23187. Am J Pathol 87:273–283

Yaseen MA, Pedley KC, Howell SL (1982) Regulation of insulin secretion from islets of Langerhans rendered permeable by electric discharge. Biochem J 206:81–87

Yin HL, Stossel TP (1979) Control of cytoplasmic actin gel-sol transformation by gelsolin, a calcium-dependent regulatory protein. Nature 281:583–586

Yin HL, Zaner KS, Stossel TP (1980) Ca^{2+} control of actin gelation: interaction of gelsolin with actin filaments and regulation of actin gelation. J Biol Chem 255:9494–9500

CHAPTER 17

Exo-Endocytosis: Mechanisms of Drug and Toxin Action

J. MELDOLESI, T. POZZAN and B. CECCARELLI

A. Introduction

All eukaryotic cells (except red blood cells) share the ability to segregate intracellularly (within membrane-bounded organelles) their specific hydrophilic secretion products (e.g., proteins, peptides, classical neurotransmitters, such as amines, acetylcholine, and amino acids) and to discharge these products by a process known as exocytosis. The term exocytosis was originally proposed to describe the release in bulk of the contents of a membrane-bounded release organelle (named, depending on its size, a granule or a vesicle). The sequence of events leading to the release process includes the fusion of the organelle membrane with the plasmalemma, followed by the fission of the fused membranes. Such a fission, which proceeds by the sequential elimination of membrane layers, creates continuity between the intraorganelle compartment and the extracellular space. With time, the meaning of the term exocytosis has expanded to include not only the release, but also the preceding fusion–fission steps. Such an extended meaning is used throughout this chapter (PALADE 1975; CECCARELLI and HURLBUT 1980a; MELDOLESI and CECCARELLI 1981).

As a mechanism of secretion, exocytosis is convenient for at least three reasons: (a) large quantities of secretory products can be prestored within the cell and released when needed without large energy expenditure; (b) release can take place at specific, strategically located portions of the plasmalemma (polarized release), so that high concentrations of secretory products can be achieved in the regions of the extracellular space immediately adjacent to the release sites; and (c) release can be quickly activated, graded, and inactivated without delay. However, exocytosis implies the incorporation of the organelle membrane into the plasma membrane. It follows that secretory cells are faced with two major problems: (a) maintaining the population of the storage organelles; and (b) conserving the size and identity of the plasma membrane. These problems are solved together by the coupling of exocytosis to a process of membrane retrieval by endocytosis. In other words, following its incorporation into the plasma membrane, the organelle membrane is retrieved to the cytoplasm to be reused for further storage and release. The term exo-endocytosis encompasses the entire process: fusion–fission, release, and membrane retrieval. Inasmuch as this type of endocytosis is the compensatory limb of the membrane addition to the plasma membrane occurring during secretion (MELDOLESI and CECCARELLI 1981; HERZOG 1981; FARQUHAR 1982), it has to be differentiated from other types of endocytosis that subserve different functions. The latter are usually referred to as fluid phase and receptor-mediated (or adsorptive) endocytosis.

The present chapter is focused primarily on synapses and neurosecretory cells. Various reasons have dictated our decision to concentrate the discussion on these cellular systems. At synapses and in neurosecretory cells, exo-endocytosis has evolved to the highest degree of efficacy and complexity, and therefore represents an important paradigm with respect to other types of secretory cells. In addition, a vast battery of agents is available for studying synapses and neurosecretory cells. Classical pharmacology of synaptic transmission is by far more comprehensive and elaborate than that of any other system. Moreover, a number of natural toxins, precisely selected by evolution, are specifically addressed to synapses. In the hands of investigators these toxins have turned from defense and aggression devices into valuable experimental tools that are used to unravel specific aspects of the process of neurotransmitter release and of its regulation.

B. Exocytosis

Exocytosis, first described in pancreatic acinar cells, has been subsequently demonstrated in a variety of glandular and other secretory cells. At cholinergic synapses, the classical studies by Katz and associates led to the proposal of the synaptic vesicle as the structural correlate of the acetylcholine quantum whose discharge induces the unitary postsynaptic event, the miniature end-plate potential (KATZ 1966). Subsequent studies have lent considerable support to this concept, but alternative views have also been proposed (TAUC 1982). The recent demonstration (by rapid freezing electron microscopy) of a temporal coincidence between the appearance of vesicle fusion and the evoked end-plate potential in electrically stimulated frog neuromuscular junctions (Fig. 1; TORRI-TARELLI et al. 1985) rules out an important argument raised in the past against the classical interpretation of Katz. Nonexocytotic, molecular release of acetylcholine has also been shown to occur at the neuromuscular junction, and to predominate under resting conditions (KATZ and MILEDI 1977; GORIO et al. 1978). At catecholaminergic cells and synapses, overwhelming evidence (both morphological and biochemical) identifies exocytosis as the process responsible for quantal release (DOUGLAS 1974; SMITH and WINKLER 1972; THURESON-KLEIN 1983). Nonexocytotic background release also exists in these systems, and can become important during a variety of pharmacologic treatments.

Besides glandular, endocrine, neuronal, and other secretory cells that are capable of regulated, intermittent discharge of their secretion products (see Sect. B.I), other systems are known that appear to release continuously. These systems include both "professional" secretors (e.g., plasma cells) and other cells whose secretory activity is more limited and concerns primarily the proteins and proteoglycans of the extracellular matrix. The recent work of Kelly and associates has yielded interesting developments in this field. It is now clear that the nonregulated (or constitutive) secretory pathway also exists in cells capable of regulated release. In other words, the regulated pathway had developed in these cells in addition to the constitutive pathway. Release from both pathways is believed to occur by exocytosis. The constitutive pathway, however, does not include a large storage compartment. The typical, easily recognized granules and vesicles of se-

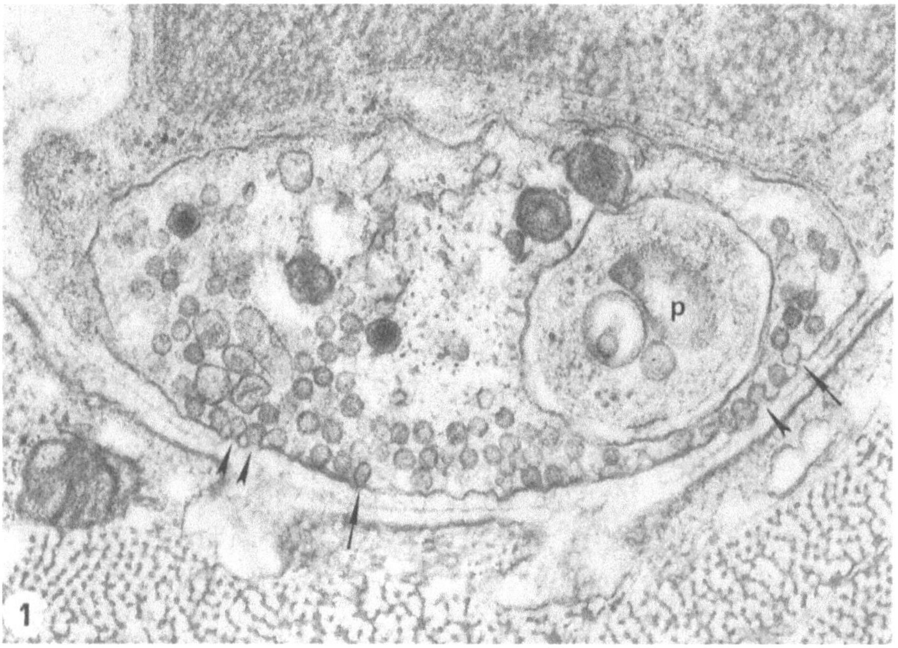

Fig. 1. Electron micrograph of a neuromuscular junction from cutaneus pectoris nerve–muscle preparation of frog quick-frozen 2.5 ms after a single stimulus in 1 m*M* 4-aminopyridine. In this cross section at the level of an active zone, different degrees of association between synaptic vesicles and the prejunctional membrane are evident. *Arrowheads* indicate clear openings whereas *arrows* indicate images suggesting intermediate steps between fusion and fission. The total transmission time to the edge of the muscle where the specimens were collected was about 3 ms. Thus, vesicle openings were caught by this physical fixation precisely at the time the quanta of acetylcholine were released. The symbol *p* indicates the Schwann cell process. (TORRI-TARELLI et al. 1985) × 70000

cretory cells and synapses belong therefore to the regulated pathway. The information that is responsible for the targeting of proteins to either pathway is apparently written in their sequence, and this is beginning to be understood. Only a few proteins are known that do not discriminate between the two pathways, and are therefore transported by both (KELLY 1985).

Many neurons are able to release independently two types of secretion products, a classical neurotransmitter and one or more peptides, that are stored within organelles of different morphology. In these synapses the degree of complexity of the secretion process appears even greater than we have described. Two regulated pathways appear to operate in parallel (LUNDBERG and HOKFELT 1983). One, analogous to the pathway in endocrine cells, is responsible for peptide neurotransmitters; the second, for classical neurotransmitters, is peculiar inasmuch as it can be filled locally in the nerve terminal and is characterized by fast exo-endocytotic cycling (see Sects. B.II and C.I). Interestingly, the release organelles of the two regulated pathways of neurons might possess specific membrane-associated components. All the vesicles containing classical neurotransmitters expose at

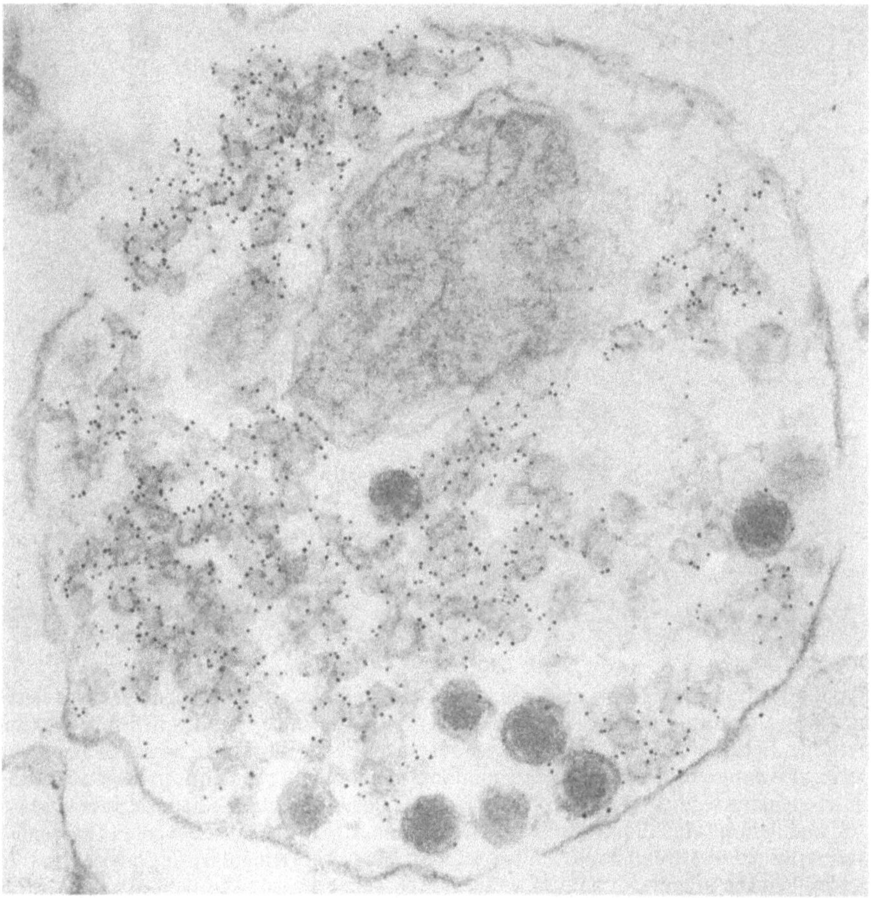

Fig. 2. Synapsin I immunoreactivity in a rat brain synaptosome. The specific localization of synapsin I around clear synaptic vesicles is marked by ferritin particles. Notice that dense-cored vesicles, mitochondria as well as the plasma membrane, are unlabeled. (NAVONE et al. 1984) ×115000

their cytosolic surface a peripheral phosphoprotein, named synapsin I (Fig. 2; DE CAMILLI et al. 1983; NAVONE et al. 1984) that might be involved at some initial step in the regulation of the release process (LLINÁS et al. 1985, see Sect. B.II).

I. Second Messenger Control of Regulated Exocytosis

Since the discovery of exocytosis, the problem of its control has attracted a great deal of uninterrupted interest. Up to a few years ago, however, progress in the field was slow and fragmentary. In particular, different secretory systems were widely believed to be under the control of distinct regulatory mechanisms. The neuromuscular junction and the parotid acinar cells could be mentioned as typical examples. In the first system, the key role played by the cytosolic Ca^{2+} concentration $[Ca^{2+}]_i$ was initially inferred from the strict dependency on extracellular Ca^{2+} of the acetylcholine quantal release evoked by electrical stimulation, and this was later confirmed by a large body of additional experimental evidence (KATZ 1966; SILINSKY 1985). In contrast, secretion of salivary enzymes from the parotid acinar cells was shown to be under the control of cAMP (SCHRAMM and SELINGER 1975). Even in these two systems, however, evidence had been obtained suggestive of a greater complexity of the exocytotic control, with possible involvement of multiple second messengers. Thus, at the neuromuscular junction several treatments were found to cause exocytosis, even when applied in Ca^{2+}-free media (HURLBUT and CECCARELLI 1979; GINSBORG and JENKINSON 1976; REICHARDT and KELLY 1983); at the parotid acinar cells slight release was obtained, even after application of agents that induce the rise not of cAMP, but of $[Ca^{2+}]_i$ (SCHRAMM and SELINGER 1975).

During the last 5 years, the concept of multiple mechanisms regulating the exocytotic discharge has gained considerable momentum. This progress has been the result of a number of important contributions. For the first time it has become possible to evaluate separately the role of various second messengers. In addition, comparison of the results obtained in different secretory systems has yielded interesting clues to the problem. Among the major advances in the field we wish to mention the recent understanding of the role played by a receptor-triggered metabolic reaction, the hydrolysis of polyphosphoinositides. Such a reaction is now known to be responsible for both the redistribution of Ca^{2+} from a "microsomal" store to the cytosol, and for the activation of an important phosphorylating enzyme, protein kinase C (Chaps. 12 and 13; BERRIDGE and IRVINE 1984). Other advances include the introduction in secretion studies of the techniques of cell permeabilization, which offer the unique opportunity of direct access to the cytosol from the extracellular medium (BAKER and KNIGHT 1981; KNIGHT and BAKER 1982) and the development of ingenious techniques for measuring $[Ca^{2+}]_i$ in cells of any size by means of trapped fluorescence indicators (quin2 and, more recently, fura-2). These same indicators can be used to clamp $[Ca^{2+}]_i$ at defined levels, for example in the range 10^{-8}–10^{-6} M, from 0.1 to 10 times the resting $[Ca^{2+}]_i$ level in most cell types (TSIEN et al. 1982, 1984; GRYNKYEVICZ et al. 1985); the development of a new electrophysiologic technique (patch-clamp) by which a wealth of information can be obtained concerning, for example, the conductance of single channels; the currents carried by individual ions; the changes of plasma membrane capacitance underlying the addition and removal of vesicle membranes (SAKMANN and NEHER 1984; NEHER and MARTY 1982; FERNANDEZ et al. 1984).

The information on the regulation of exocytosis that has been gathered in a variety of cell types following these technical and intellectual developments confirms that in most secretory systems changes of $[Ca^{2+}]_i$ are signals sufficient to trigger exocytotic responses. $[Ca^{2+}]_i$ transients have been accurately measured in permeabilized as well as intact cells, and found to vary in the range 0.3–1 μM (see inter alia KNIGHT and BAKER 1982, 1983; KNIGHT and SCRUTTON 1985; POZZAN et al. 1984). An unexpected finding was the transiency of the release responses induced by $[Ca^{2+}]_i$ rise in both permeabilized and intact cells. In the latter cells the causative $[Ca^{2+}]_i$ transients were found to be much more prolonged. These results point to the existence of one or more inactivation mechanisms that render exocytosis insensitive to the intracellular second messenger. In intact cells, however, the distribution of Ca^{2+} in the cytosol is expected to be uneven after stimulation. In particular, localized Ca^{2+} gradients in the cytosolic regions immediately adjacent to the plasmalemma might play a key role in the triggering of exocytotic responses. This problem, which in the past has been extensively considered at the theoretical level (for a comprehensive review see SILINSKI 1985), is now beginning to be investigated experimentally (WILLIAMS et al. 1985; KEITH et al. 1985).

The greater responses induced in parotid acinar cells by isoproterenol and other agents that induce rises of cAMP in comparison with those working through $[Ca^{2+}]_i$ has already been mentioned (SCHRAMM and SELINGER 1975). In pancreatic acinar cells, the efficacy of the two second messengers is reversed (SCHULZ and STOLZE 1980). It should be noted, however, that the effect of cAMP on secretion is not always stimulatory, but can be inhibitory, for example in mast cells. In contrast, as far as we know, no inhibition of exocytosis has ever been observed after the application of activators of protein kinase C, such as diacylglycerol and phorbol esters. By themselves, these agents are without effect on $[Ca^{2+}]_i$ levels. In a variety of secretory systems (e.g., platelets: RINK et al. 1983; pancreatic acinar cells: GUNTHER 1981; PC12 cells: POZZAN et al. 1984), protein kinase C activators trigger exocytotic responses, even when administered alone. In granulocytes, platelets, and, to a smaller extent, in PC12 cells, this stimulation was maintained, even when $[Ca^{2+}]_i$ was clamped to very low levels, i.e., $<10^{-8}$ *M* (RINK et al. 1982; DI VIRGILIO et al. 1984; POZZAN et al. 1984). In contrast, in other systems (e.g., bovine chromaffin cells KNIGHT and BAKER 1983; frog neuromuscular junctions HAIMANN et al. 1987), the effect of protein kinase C activators alone was negligible; but large, synergistic effects were observed when these agents were administered together with agents that increase $[Ca^{2+}]_i$, for example, Ca^{2+} ionophores. These synergistic responses were not only greater, but also more persistent than those elicited by $[Ca^{2+}]_i$-increasing agents alone (POZZAN et al. 1984; KOLESNIK and GESHENGORN 1985). In other words, the activators of protein kinase C appear to prevent the inactivation of the stimulatory effect of $[Ca^{2+}]_i$ on exocytosis.

Activation of the receptors coupled to polyphosphoinositide hydrolysis results in both the rise of $[Ca^{2+}]_i$ and the generation of diacylglycerol, the physiologic activator of protein kinase C. Study of the release responses elicited by the activation of these receptors has yielded interesting results. For example, in pancreatic acini the stimulation of the muscarinic receptor results in a rapid, but transient ($\sim$5 min) rise of $[Ca^{2+}]_i$, and in a much more prolonged ($>$1 h) stimulation of exocytosis. The latter response is entirely dependent on the activation of the receptor, inasmuch as it can be quickly shut off by the application of the muscarinic antagonist, atropine (PANDOL et al. 1985). Thus, the two second messengers generated by receptor activation appear to play coordinated roles. Raised $[Ca^{2+}]_i$ triggers the response, and stimulated protein kinase C keeps it going for long periods of time.

The results that we have summarized emphasize the complexity of the interaction between two intracellular signals, Ca^{2+} and protein kinase C. As discussed by BAKER (1984), various models can be envisaged to account for the observed results. The two signals could work separately, by triggering parallel stimulatory processes; alternatively, one of them could play a facilitatory role with respect to the other (see also Sect. B.II). Evidence suggestive of an even greater complexity of the intracellular control processes has begun to accumulate. Some of the results now reported (see for example VARA and ROZENGURT 1985; COOKE and HALLET 1985; HALLAM et al. 1985) appear to be difficult to explain by the mechanisms discussed so far (as well as by cGMP, a second messenger system that is now attracting interest; see HOUSLAY 1985). At the moment, we are facing a deluge of new information. Discovery of new messengers, or recognition of new functions to be assigned to already known signaling pathways might therefore be expected in the near future.

II. Membrane Fusion–Fission in Exocytosis

The time between the application of the stimulus and the appearance of the ensuing exocytosis varies considerably in different cell systems, and in relation to the second messenger involved. At the neuromuscular junction and other synapses, it has been calculated that the delay from the opening of the voltage-gated, presynaptic Ca^{2+} channels to the postsynaptic recording of the evoked end-plate potential is only a fraction of a millisecond (LLINÁS et al. 1981), during which the rate of quantal transmitter release can increase over 1000-fold (TORRI-TARELLI et al. 1985). Such a time is probably too short to include an enzyme-catalyzed reaction (REICHARDT and KELLY 1983). The nature of the nonenzymatic Ca^{2+} effect is still undefined. Screening of negative fixed charges at strategic sites of the cytosolic surface of vesicle and plasma membranes could play an important role by breaking the electrostatic energy barrier for membrane approach and, possibly, by facilitating membrane fusion (BLIOCH et al. 1968). To overcome the even more significant hydration energy barrier (BASS and MOORE 1966; SILINSKI 1985), Ca^{2+} could cause conformational changes of one or more membrane proteins. By analogy with the better known membrane fusion–fission mediated by the influenza virus hemagglutinin, the conformational change responsible for exocytosis has been proposed to involve the exposure of a hydrophobic domain at the surface of one or more fusigenic proteins (REICHARDT and KELLY 1983). A few candidate fusigenic proteins have been envisaged, but none identified with any certainty. A calmodulin-like protein was shown to be an integral component of brain synaptic vesicles (HOOPER and KELLY 1984). Indeed, anti-calmodulin drugs and antibodies were reported to block exocytosis (BAKER and KNIGHT 1981; KENIGSBERG et al. 1982; KENIGSBERG and TRIFARÒ 1985). However, in view of the poor specificity of these drugs (which inhibit a variety of other enzymes as well, including protein kinase C) and of the central role played by calmodulin in a variety of functions other than exocytosis, these observations still appear inconclusive. Another Ca^{2+}-binding protein, synexin, studied in chromaffin cells, might be involved more in the attachment than in the fusion of membranes (POLLARD et al. 1981). Synapsin I, a phosphoprotein phosphorylated by both Ca^{2+}-calmodulin and cAMP-dependent protein kinases, appears to be involved in the regulation of neurotransmitter release at steps preceding exocytosis, such as the movement of the vesicles containing classical neurotransmitters from the deep cytoplasm of the nerve ter-

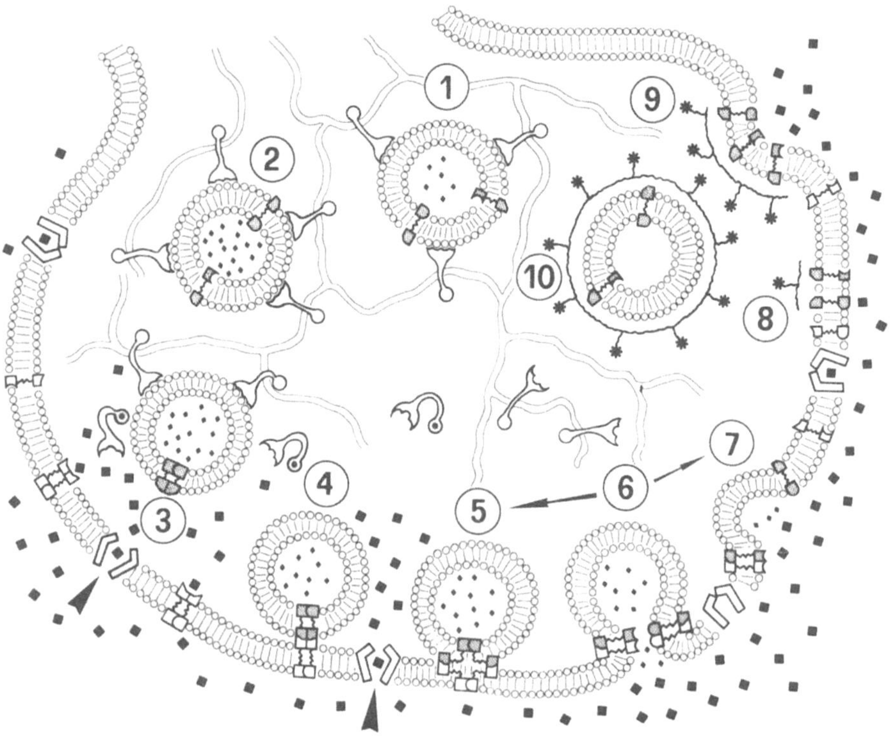

Fig. 3. Schematic representation of the Ca^{2+}-triggered exocytosis and subsequent endocytosis at a synapse. No attempt has been made to draw the various structures and components to scale. Two synaptic vesicles removed from the presynaptic membrane (*1, 2*) are represented in successive stages of loading with the transmitter (*full rhombi*). These vesicles are shown to interact with the cytoskeleton by means of the peripheral protein, synapsin I (drawn as a *goblet*). Opening of the voltage-gated Ca^{2+} channels (*arrowheads*) results in the influx of Ca^{2+} ions (*solid squares*) that cause phosphorylation, detachment, and deformation of synapsin I and exposure of the putative fusigenic protein complexes in both the vesicle and presynaptic membranes (*3*). Interaction of these two complexes leads to close apposition of the two membranes (*4*), followed by rearrangement of the interacting proteins and lipids (*5*), and by fission (i.e., physical opening), which allows transmitter molecules to escape into the synaptic cleft (*6*). Reversal of this last step (*long arrow*) leads to closure of the opening and quick recapture of the vesicle (from 6 to 3). Alternatively (*short arrow*), the vesicle membrane could collapse and flatten down (*7*), with dispersal of its components in the presynaptic membrane. Subsequently, endocytosis occurs by the involvement of a coating mechanism (*8, 9*). The resulting coated vesicles (*10*) would ultimately shed their coat to regenerate uncoated, functional synaptic vesicles (from *10* to *1*)

minal to regions near the specific sites of fusion on the presynaptic membrane (NAVONE et al. 1984; LLINÁS et al. 1985). Finally, recent indirect evidence in permeabilized chromaffin cells and granulocytes suggests the involvement in exocytotic fusion–fission of a GTP-binding protein that might be localized in either one of the interacting membranes (KNIGHT and BAKER 1985; BARROWMAN et al. 1986). A possible interpretation of the events occurring at synapses during Ca^{2+}-triggered release of neurotransmitter is schematically drawn in Fig. 3.

The rates of exocytosis induced by agents that are believed to work exclusively through the activation of protein kinases (cAMP-dependent and protein kinase C) are much slower than those induced by Ca^{2+}. A plausible hypothesis is that phosphorylation by these enzymes of specific proteins increases the probability of the fusion process, possibly by increasing affinity for Ca^{2+} (BAKER 1984; KNIGHT and BAKER 1983), or, alternatively, by decreasing the number of Ca^{2+} ions that need to bind in order to activate the process (HAIMANN et al. 1987). These models are attractive because they can explain readily the stimulation of exocytosis without concomitant increase of $[Ca^{2+}]_i$, and the synergism of the responses induced by the application of a protein kinase C activator together with a $[Ca^{2+}]_i$-raising agent, discussed in Sect. B.I. Some existing experimental results, however, appear difficult to reconcile with this unifying hypothesis. Among these are the stimulation of exocytosis by either protein kinase C activators or GTP analogs at vanishingly low $[Ca^{2+}]_i$, well below the K_d for Ca^{2+} of any known Ca^{2+}-regulated proteins (DI VIRGILIO et al. 1984); the greater efficacy of cAMP with respect to increases in $[Ca^{2+}]_i$ in the parotid acinar cells (SCHRAMM and SELINGER 1975); and the complete dissociation between the stimulatory effect of protein kinase C activators and the inhibition brought about by Ca^{2+} in parathyroid cells (SHOBAK et al. 1984; BROWN et al. 1984). As far as the substrates of protein kinases go, increased phosphorylations have indeed been observed following stimulation in a variety of cell systems (among recent reports see NIGGLI et al. 1984; WRENN 1984; HUTTON et al. 1984; QUISSEL et al. 1985). In no case, however, has evidence of a direct involvement of these phosphorylations in the modulation of exocytosis been convincingly demonstrated. Thus, almost 30 years after the discovery of exocytosis, its molecular mechanism of control remains largely mysterious.

III. Drugs

In the preceding sections we have discussed some important features of the cell biology and physiology of exocytosis that provide a good basis for the understanding of pharmacologic interference in the process. Thus, it should be quite clear now that a large variety of drugs or agents able to change directly or indirectly the levels or the general homeostasis of the second messengers are unavoidably expected to activate, inhibit, or regulate the final step of the secretory process. This long list includes agonists and antagonists of ionophoric receptors (such as the acetylcholine nicotinic receptors of chromaffin and other neuroendocrine cells) and of the receptors coupled to the hydrolysis of polyphosphoinositides, such as the muscarinic M_1, adrenergic α_1, serotonergic S_2, histaminergic H_2, and many peptidergic receptors (e.g., the receptors for angiotensin, vasopressin, CCK, gastrin, etc.). Besides causing redistribution of Ca^{2+} from the stores and activation of protein kinase C, as discussed in Sect. B.I, the activation of these receptors appears coupled to the opening of voltage-independent channels of the plasmalemma specific for Ca^{2+} and/or K^+. At synapses and excitable cells, many other drugs affect exocytosis by acting at the level of the voltage-gated channels. Examples are veratridine and congeners that delay the inactivation of the Na^+ channel and thus prolong the action potential and Ca^{2+} currents. 4-Aminopyridine and congeners induce the same effect by inhibiting the K^+ channel. Blockers (verapamil, the various dihydropyridines) and activators (Bay K 8644) of the voltage-gated Ca^{2+} channel are effective on exocytosis in various neuroendocrine cell systems. It should be noted, however, that Ca^{2+} channels have recently been shown to be heterogeneous (NOWYCKY et al. 1985). Evidence indicates that at least part of the channels in the presynaptic membrane are not of the L type (the targets of dihydropyridines), but possibly of the N type, for which no pharmacologic information exists at the present time. Ca^{2+} homeostasis can also be perturbed intracellularly. Several drugs (e.g., TMB-8, theophylline) have been reported to block Ca^{2+} release from the "microsomal" store. These drugs, however, are not specific for this effect.

Ca^{2+} ionophores (e.g., A23187 and ionomycin), which induce the electroneutral exchange of Ca^{2+} and H^+ across membranes, have been employed extensively to manipulate the intracellular Ca^{2+} homeostasis. These drugs are not easy to use because of both their high cytotoxic potential and their ability to get inserted not only in the plasmalemma, but also in the intracellular membranes, and therefore to affect Ca^{2+} fluxes both at the cell surface and at the cytoplasmic stores. Less specific ionophores, such as X-537A which transports mono- and divalent cations, and the ionophores specific for monovalent cations are even more problematic because they induce the collapse of voltage and proton gradients across membranes, the nonexocytotic release of neurotransmitters such as catecholamines, and many other nonspecific effects.

The agents that stimulate exocytosis via the activation of protein kinases can induce in the cells a variety of other effects, some of which interact with their secretagogue effect. Channel modulation by cAMP-dependent phosphorylations has been reported in a variety of cell systems (see KOSTYUK 1984; REUTER 1983). Knowledge about protein kinase C activators (phorbol esters and diacylglycerol), on the other hand, is more recent, but already considerable. These drugs have been found to cause the desensitization of various types of receptors, including many receptors coupled to polyphosphoinositide hydrolysis. In addition, they inhibit the voltage-gated Ca^{2+} influx in various cell types (PC12 pheochromocytoma, RINF5m insulinoma DI VIRGILIO et al. 1986; dorsal root ganglion neurons RANE and DUNLAP 1986). Protein kinases appear therefore to regulate exocytosis at multiple levels: directly, by stimulating the process; indirectly, by affecting causative and control events at the plasma membrane.

A final series of drugs and treatments are known to stimulate exocytosis at synapses, even in the absence of extracellular Ca^{2+}. The underlying process (or processes) responsible for these effects have not been identified with certainty. Perturbation of the Na^+ homeostasis (induced for example by ouabain BAKER and CRAWFORD 1975; HAIMANN et al. 1985; or Li^+ CRAWFORD 1975) are believed to ultimately affect $[Ca^{2+}]_i$, but direct proof is still lacking. The same is true for divalent and trivalent cations, Ba^{2+} and La^{3+}, that could act by themselves or by displacing Ca^{2+} at strategic extracellular or intracellular sites (SILINSKI 1985; SEGAL et al. 1985). Hypertonicity, and treatment with ethanol and a variety of other chemicals are also effective (GINSBORG and JENKINSON 1976).

IV. Toxins

Toxins that, in one system or another (most frequently in neurons), are able to affect exocytosis are overwhelming in number, and their mechanisms of action are exceedingly diverse. Only a few of them will be mentioned here, and a detailed discussion will be restricted to those that have profound effects on the exocytotic process. For additional information the reader is referred to various books and exhaustive reviews (CECCARELLI and CLEMENTI 1979; HOWARD and GUNDERSEN 1980; HUCHO and OVCHINNICHOV 1983; HARRIS 1986).

1. Toxins Targeted to Channels and Receptors

A whole wealth of natural toxins targeted to channels and receptors can indirectly affect exocytosis by mechanisms analogous to those of some drugs discussed in the preceding section. Some of these toxins (e.g., bungarotoxin and batrachotoxin for the nicotinic receptor; strychnine for the glycine receptor; tetrodotoxin and saxitoxin, sea anemone and North African scorpion toxins, for the Na^+ channel) are well-known pharmacologic tools that have been employed for many years. Recent developments in this field include the characterization of new Na^+ channel toxins (obtained from South American scorpions) that differ from their old-world counterparts in their high potency and different mechanism of action (block of early Na^+ conductance with no effect on the channel inactivation kinetics JAIMOVICH et al. 1982); and the recognition of toxins addressed to the K^+ channel. Dendrotoxin and other toxins from the venom of the green and black mamba snakes facilitate the release of acetylcholine at the neuromuscular junction and induce convulsions when injected intraventricularly. The effects of dendrotoxin appear due to the attenuation of the A current at the voltage-gated K^+ channel (DOLLY et al. 1984; WELLER et al. 1985). Interestingly, some connection appears to exist between dendrotoxin and the phospholipase presynaptic toxins, in particular β-bungarotoxin. The inhibition of transmitter release brought about by the latter toxin at cholinergic and GABAergic synapses could thus be mediated by the activation of a presynaptic K^+ channel.

Toxins addressed to the voltage-gated Ca^{2+} channel would obviously be of great importance in the field of exocytosis (MILLER 1984). Maitotoxin (from a dinoflagellate) has been suggested to act by activating the Ca^{2+} channel (presumably, of the L type) (TAKAHASHI et al. 1983). However, some doubts still exist as to this interpretation. In contrast, the targeting to the Ca^{2+} channel of the inhibitory toxin from the venom of the giant snail, *Conus geographus*, has been convincingly documented (KERR and YOSHIKAMI 1984).

2. Clostridium Toxins: Inhibitors of Exocytosis

Work carried out during the last decade has led to the unexpected conclusion that neurologic syndromes as diverse as tetanus and botulinus paralysis, induced by poisoning with the potent *Clostridium* toxins, are probably due to a similar or identical action, the blockade of exocytosis, but occurring at different types of synapses (SELLIN 1985; SIMPSON 1986). Botulinus and tetanus toxins constitute two very similar families of peptide toxins. Each member of these families is composed of two disulfide-bonded chains, $M_r = 100$ and 50 kdalton, generated by the cleavage of a precursor. The effects of the toxins of both families appear after a delay of at least 30 min. The time is now known to be needed for the binding and transmembrane translocation of the toxins. Various gangliosides (G_{D1b}, G_{T1b}, G_{D1a}, and possibly others) share the ability to bind *Clostridium* toxins, and these gangliosides were believed to be the toxin receptors. Recent evidence, however, demonstrates that the number of botulinus toxin receptors in the presynaptic membrane is much ($>10^3$-fold) lower than the number of ganglioside molecules. In addition, different botulinus toxins that have affinity for the same gangliosides bind to different receptors (DOLLY et al. 1984; BLACK and DOLLY 1986a). As discussed by MONTECUCCO (1986), the role of gangliosides could consist in concentrating the toxin molecules at the external surface of the target membranes, thereby increasing their apparent affinity for the specific protein receptors.

Recent evidence from the group led by Dolly (DOLLY et al. 1984; BLACK and DOLLY 1986a, b) indicates that at the specific target, the neuromuscular junction, botulinus toxin binds to the receptor, and the complex thus formed is then redistributed to an intraterminal, endosomal compartment. Thus, at the neuromuscular junction, the uptake of the toxin appears to proceed by receptor-mediated endocytosis. The low pH of the endosomal compartment could promote the conformational rearrangement of the heavier toxin chain, with generation of a pore in the endosomal membrane, through which the lighter chain is translocated (BLACK and DOLLY 1986a, b; HOCH et al. 1985; SIMPSON 1986). Interestingly, many synapses of the central nervous system bind botulinus toxin with high affinity, but lack the ability to translocate the lighter subunit efficiently. These synapses are affected only by high concentrations of the toxin (BLACK and DOLLY 1986b). Even more striking is the recent demonstration by KNIGHT et al. (1985) that bovine chromaffin cells become blocked when cultured for days in the presence of high concentrations of D-type botulinus toxin, with no concomitant inhibition of Na^+ and Ca^{2+} fluxes. Such inhibition was also seen with $[Ca^{2+}]_i$ up to 10 μM in permeabilized chromaffin cells. It can thus be concluded that many (possibly all) synapses and neurosecretory cells represent potential targets of tetanus and botulinus toxins, and that both the expression of the specific receptors, and the presence of effective transmembrane transport processes, represent the determinants of the toxin target specificity.

The mechanism (or mechanisms) whereby *Clostridium* toxins, once translocated to the cytosol, inhibit exocytosis remain undefined. Several hypotheses have been put forth. By analogy with the other toxins (e.g., diphtheria, cholera, pertussis toxins) that share similar binding–internalization processes, tetanus and botulinus toxin lighter chains have been proposed to possess enzyme activity (ADP-ribosylase), or to be able to affect the activity of enzymes in nerve terminals. Alternatively, the toxins could bind to and inactivate strategic site (or sites) for exocytosis (SIMPSON 1986; SELLIN 1985). In this latter respect, however, it should be noted that the thorough radioautographic studies of BLACK and DOLLY (1986a, b) and DOLLY et al. (1984) have failed to reveal concentration of botulinus toxin at discrete intraterminal sites of mouse neuromuscular junctions. In any event, the block of exocytosis by *Clostridium* toxins appears to occur at a step (or steps) distal to the action potential-induced opening of Ca^{2+} channels that occurs normally in toxin-poisoned terminals. Increasing the $[Ca^{2+}]_i$ signal (for example, by the use of 4-aminopyridine) results in a partial relief of the block (SELLIN 1985; SIMPSON 1986). In addition, the stimulation of exocytosis brought about by α-latrotoxin is almost unaffected by *Clostridium* toxins. This problem is further discussed in the subsequent section.

3. α-Latrotoxin and Congeners: Stimulators of Exocytosis

α-Latrotoxin is a high M_r (130 kdalton) protein contained in the venom of the black widow spider. Work carried out more than 15 years ago revealed that at the frog neuromuscular junction the action of this toxin consists in a massive stimulation of quantal acetylcholine release, i.e., of synaptic vesicle exocytosis, leading within 20–60 min to complete vesicle depletion in the nerve terminal (HURLBUT and CECCARELLI 1979; MELDOLESI et al. 1986). The latter effect is due to blockade of endocytosis, also induced by the toxin (see Sect. C.II). Subsequent studies revealed that sensitivity to α-latrotoxin is not a property of the motor endplate only, but of all synaptic terminals of vertebrates so far investigated. Among neurosecretory cells, those of the PC12 line are sensitive, whereas rat chromaffin cells are unaffected by the toxin in the intact animal, and become responsive when cultured in the presence of nerve growth factor (MELDOLESI et al. 1986).

The mechanism of α-latrotoxin action has been unraveled in part. The initial step is the binding of the toxin to a specific, high affinity receptor (a high M_r, integral membrane protein), that can now be considered as a surface marker of nerve terminals (VALTORTA et al. 1984; SCHEER and MELDOLESI 1985). The receptor is also expressed by the few other targets of the toxin. In contrast to *Clostridium* toxins, α-latrotoxin is not internalized, but remains bound to its receptor exposed to the external surface of the presynaptic membrane (VALTORTA et al. 1985; MELDOLESI et al. 1986). In PC12 cells, the binding interaction activates a small (~15 pS), non-closing cation channel (WANKE et al. 1986), and this entails depolarization, increased Ca^{2+} influx, and persistent elevation of the $[Ca^{2+}]_i$ (GRASSO et al. 1980; MELDOLESI et al. 1984). These phenomena have also been recently reproduced in a reconstituted system composed of liposomes bearing the receptor exposed to α-latrotoxin. The nature of the channel (whether it is composed of the receptor, or of a hydrophobic domain of the toxin molecule inserted across the membrane after binding to the receptor) remains to be clarified (MELDOLESI et al. 1986).

The activation of the cation channel represents only part of the α-latrotoxin action. In fact, at variance with the ionophoric drugs and other depolarizing agents, the toxin is known to stimulate exocytosis, even when applied in extracellular medium where the concentration of Ca^{2+} is maintained in the nanomolar range by the addition of EGTA, provided that another divalent cation, such as Mg^{2+}, is present in millimolar concentrations. This phenomenon is particularly striking at the frog neuromuscular junction (HURLBUT and CECCARELLI 1979). In this system, recent accurate analyses have confirmed that the toxin-induced responses are massive both with and without Ca^{2+}, but distinct differences exist in the pattern and time course of the release. Specifically, with Ca^{2+} the pattern of the release is bursting, and the response is more prolonged (FESCE et al. 1986).

Also at the synapses of the central nervous system and PC12 cell the effect of α-latrotoxin in a Ca^{2+}-free medium is pronounced. The possibility that the Ca^{2+}-free effect of α-latrotoxin is only apparent, due to redistribution of Ca^{2+} from the stores to the cytosol, has not been substantiated by direct experiments with the fluorescent Ca^{2+} indicator, quin2 (MELDOLESI et al. 1984). Thus, it may be concluded that exocytosis is stimulated by α-latrotoxin through the activation of at least two parallel mechanisms: one dependent, the other independent of $[Ca^{2+}]_i$. The activation of this second mechanism (which so far remains undefined) might account for the unique ability of α-latrotoxin to surmount the blocking effects of *Clostridium* toxins, as mentioned in the preceding section (see HURLBUT and CECCARELLI 1979).

In addition to α-latrotoxin, the venom of the black widow spider contains at least two more toxic principles, up to now only partially characterized, that induce in their targets (crustacean and insect synapses) effects very similar to those brought about by α-latrotoxin at vertebrate synapses (HURLBUT and CECCARELLI 1979; FRITZ and MAURO 1982). Another congener, leptinotoxin h (purified from the hemolymph of an insect) resembles α-latrotoxin in its action, and has partially overlapping target specificity (MADEDDU et al. 1985). Together with the *Clostridium* blockers, these stimulator toxins promise to be valuable tools to solve some of the many mysteries that still remain in the field of exocytosis and its control.

C. Endocytosis

As already mentioned in Sect. A, this chapter is focused exclusively on the form of endocytosis that follows exocytosis in secretory cells. The existence of this process was initially inferred from the results of two types of experiment: tracer experiments, that showed increased labeling of cytoplasmic vesicles in cells and synapses stimulated to secrete in the presence of an extracellular marker; and turnover experiments, demonstrating the longer half-life (and thus, the reutilization) of granule/vesicle membrane proteins with respect to secretory proteins (CECCARELLI and HURLBUT 1980a; HERZOG 1981; MELDOLESI and CECCARELLI 1981; FARQUHAR 1982). More recent studies have provided evidence that the membrane recycled from the cell surface might be molecularly identical to the granule/vesicle membrane incorporated by exocytosis, and therefore different from the bulk of the plasma membrane. Results along these lines were first obtained by freeze-fracture (parotid acinar cells, DE CAMILLI et al. 1976) and then by immunocytochemistry (chromaffin cells), using antibodies raised against specific components of the granule membrane (PATZAK et al. 1984; PATZAK and WINKLER 1986).

I. Membrane Sorting in Endocytosis

The specific composition of the recycled membrane with respect to the plasma membrane raises the problem of the sorting mechanism (or mechanisms). Essentially two models can be envisaged. First, the components of the two membranes, granule/vesicle and plasma membranes, could have little tendency to rapid intermixing, and would therefore remain segregated, at least for some time, after exocytosis. Indeed, the freeze-fracture and immunocytochemical studies of DE CAMILLI et al. (1976) and PATZAK and WINKLER (1986) have documented that in stimulated cells, components of the granule/vesicle membrane remain segregated in patches at the cell surface. Rapid endocytosis restricted to these patches could easily preserve the molecular identity of the two participating membranes. Another possibility is that a certain degree of intermixing does occur, but this is corrected specifically though a proces of "molecular filtration" by the retrieval organelles, the coated pits. Such a process would consist in the exclusion from the pits of the components specific for the plasma membrane, with concomitant accumulation of those specific for the granule/vesicle membrane (BRETSCHER et al. 1980).

Studies on the frog neuromuscular junction suggest that these two models of endocytosis are not necessarily alternative, but can coexist, at least in that system. The experimental evidence shows that, as long as secretion evoked by electrical stimulation is maintained at rates up to 400 quanta per second, the incorporation of vesicles in the presynaptic membrane and the ensuing recycling are kept in balance, i.e., no decrease in vesicle number or enlargement of the presynaptic surface area takes place at the nerve terminal. In these terminals, the number of coated pits or vesicles is only slightly increased (CECCARELLI et al. 1973; CECCARELLI and HURLBUT 1975, 1980a). These results indicate that, under the conditions of the experiment, the residence time of the vesicle membrane at the terminal surface is very short, possibly because each vesicle undergoes a quick exo-endocytotic cycle, without even flattening down in the presynaptic membrane. As discussed in detail elsewhere (CECCARELLI and HURLBUT 1980a), endocytosis without previous vesicle flattening would be energetically favorable because it does not require work for membrane invagination (for alternative interpretations of secretion at the frog neuromuscular junction see HEUSER and REESE 1973; MILLER and HEUSER 1984). Only when the secretion rate is high, or treatments are used that impair recycling (see Sect. C.II), the presynaptic surface area increases, and some coated vesicles accumulate (MELDOLESI and CECCARELLI 1981). Therefore, at the neuromuscular junction, the sorting mechanism might be based primarily on the rapidity of the exo-endocytotic cycle. The membrane-filtering ability of coated pits (BRETSCHER et al. 1980) would become important only when recycling is slowed down (see Sect. C.II). The coexistence at synapses of the two mechanisms of vesicle membrane recycling is illustrated schematically in Fig. 3.

Also in bovine chromaffin cells results have been obtained suggesting the coexistence of fast and slow mechanisms of endocytosis. In this system, membrane capacitance measurements by NEHER and MARTY (1982) demonstrated that endocytosis can follow exocytosis within seconds. However, if the cells are heavily stimulated, the residence time of the specific granule components at the cell surface (studied by immunocytochemistry) appears much longer ($t_{1/2} \sim 10$ min, PATZAK et al. 1984). Under the latter conditions, membrane recycling seems to occur by means of coated vesicles. The coated vesicle-mediated type of recycling appears to predominate in exocrine glandular cells and mast cells (MELDOLESI and CECCARELLI 1981; HERZOG 1981; THILO 1985). However, in the latter system the possibility of quick membrane recapture has also been suggested (FERNANDEZ et al. 1984). In conclusion, it should be admitted that our knowledge of the molecular mechanisms of endocytosis is still very limited. The possibility that endocytosis changes its rate and even mechanism, depending on the nature and intensity of the stimuli applied to cause exocytosis, calls for great caution in the interpretation of the available information. This is particularly true because some of the results so far obtained come from experiments performed under conditions of exhaustive stimulation.

II. Regulation of Endocytosis

As summarized in the preceding section, recycling of granule/vesicle membrane from the cell surface can occur at different rates, at least in some secretory systems. This, however, does not necessarily mean that the type of endocytosis at hand is independently regulated by second messenger mechanisms. At the neuromuscular junction, and possibly at other synapses and neurosecretory cells, the coupling between exo- and endocytosis appears tight at low rates of secretion, and becomes loose at high rates. This is tantamount to saying that the V_{max} of exocytosis exceeds that of (rapid) endocytosis. The enlargement of the surface area that follows the uncoupling of the two processes might be the trigger of the slow, coated vesicle-mediated recycling in these systems (CECCARELLI and HURLBUT 1980a; MELDOLESI and CECCARELLI 1981).

Up to now, the pharmacology of endocytosis has attracted only limited attention. At the neuromuscular junction, a variety of treatments have been shown ultimately to induce depletion of synaptic vesicles. In some cases the measurement of the total quanta release demonstrated that depletion was achieved only after the original vesicle pool had turned over several times. Under these conditions, vesicle depletion could result from the uncoupling of exo- and endocytosis, as already mentioned. With two treatments, however, ouabain and high concentrations of the black widow spider venom toxin, α-latrotoxin (see Sect. B.IV.3), depletion of vesicles is rapid, the nerve terminals ultimately become swollen, and the number of quanta secreted is equal to the number present in the terminal at rest, i.e., endocytosis is blocked (HAIMANN et al. 1985; FESCE et al. 1986). The increase of intracellular Na^+ that are expected to occur in both these treatments might be responsible for the blockade. Indeed, when α-latrotoxin was applied in a medium where Na^+ had been replaced by glucosamine$^+$, swelling and vesicle depletion failed to appear (GORIO et al. 1978).

Studies on frog neuromuscular junctions exposed to low concentrations of α-latrotoxin (CECCARELLI and HURLBUT 1980b) and parotid acinar cells stimulated with isoproterenol (KOIKE and MELDOLESI 1981) revealed that endocytosis is impaired when Ca^{2+} is withdrawn from the incubation medium. With other treatments, however (e.g., La^{3+}), only slight differences were found at neuromuscular junctions incubated with and without Ca^{2+} in the medium (SEGAL et al. 1985; FESCHE et al. 1986). A role for Ca^{2+} in the regulation of endocytosis is therefore possible, but its contribution remains to be defined.

Acknowledgments. We are indebted to Dr. R. Fesce for helpful discussions and suggestions in preparing Fig. 3. We gratefully acknowledge the assistance of S. Avogadro who typed the different versions of the manuscript.

References

Baker PF (1984) Multiple controls for secretion? Nature 310:619–620
Baker PF, Crawford AC (1975) A note on the mechanism by which inhibitors of the sodium pump accelerate spontaneous release of transmitter from motor nerve terminals. J Physiol (Lond) 247:209–226
Baker PF, Knight DE (1981) Calcium control of exocytosis and endocytosis in bovine adrenal medullary cells. Philos Trans R Soc Lond [Biol] 296:83–104
Barrowman MM, Cockcroft S, Gomperts BD (1986) Two roles for guanine nucleotides in the stimulus-secretion sequence of neutrophiles. Nature 319:504–507
Bass L, Moore WJ (1966) Electrokinetic mechanism of miniature postsynaptic potentials. Proc Natl Acad Sci USA 55:1214–1217
Berridge MJ, Irvine RF (1984) Inositol tris-phosphate, a novel second messenger in cellular signal transduction. Nature 312:315–320
Black JD, Dolly JO (1986a) Interaction of ^{125}I-labelled botulinum neurotoxins with nerve terminals. I. Ultrastructural autoradiographic localization and quantitation of distinct membrane acceptors for type A and B on motor nerves. J Cell Biol 103:521–534
Black JD, Dolly JO (1986b) Interaction of ^{125}I-labelled botulinum neurotoxin with nerve terminals. II Autoradiographic evidence for its uptake into motor nerves by acceptor-mediated endocytosis. J Cell Biol 103:535–544
Blioch ZL, Glagoleva IM, Liberman EA, Nenashev VA (1968) A study of the mechanism of quantal transmitter release at a chemical synapse. J Physiol (Lond) 199:11–35
Bretscher MS, Thompson JN, Pierce BMF (1980) Coated pits act as molecular filters. Proc Natl Acad Sci USA 77:4156–4159
Brown EM, Redgrave J, Thatcher J (1984) Effect of phorbol ester TPA on PTH secretion. Evidence for a role for protein kinase C in the control of PTH release. FEBS Lett 175:72–75
Ceccarelli B, Clementi F (1979) Neurotoxins, tools in neurobiology. Raven, New York
Ceccarelli B, Hurlbut WP (1975) The effects of prolonged repetitive stimulation in hemicholinium on the frog neuromuscular junction. J Physiol (Lond) 247:163–188
Ceccarelli B, Hurlbut WP (1980a) Vesicle hypothesis of the release of quanta of acetylcholine. Physiol Rev 60:396–441
Ceccarelli B, Hurlbut WP (1980b) Ca^{2+}-dependent recycling of synaptic vesicles at the frog neuromuscular junction. J Cell Biol 87:297–303
Ceccarelli B, Hurlbut WP, Mauro A (1973) Turnover of transmitter and synaptic vesicles at the frog neuromuscular junction. J Cell Biol 57:499–524
Cooke E, Hallett MB (1985) The role of C-kinase in the physiological activation of the neutrophil oxidase. Biochem J 232:323–327
Crawford HC (1975) Lithium ions and the release of transmitter at the frog neuromuscular junction. J Physiol (Lond) 246:109–142
De Camilli P, Peluchetti D, Meldolesi J (1976) Dynamic changes of the lumenal plasmalemma in stimulated parotid acinar cells. J Cell Biol 70:59–74
De Camilli P, Harris SM Jr, Huttner WB, Greengard P (1983) Synapsin I (protein I), a nerve terminal-specific phosphoprotein. II. Its specific association with synaptic vesicles demonstrated by immunocytochemistry in agarose-embedded synaptosomes. J Cell Biol 96:1355–1373
Di Virgilio F, Lew DP, Pozzan T (1984) Protein kinase C activation of physiological processes in human neutrophils at vanishingly small cytosolic Ca^{2+} levels. Nature 310:691–693
Di Virgilio F, Pozzan T, Wollheim CB, Vicentini LM, Meldolesi J (1986) Tumor promoter phorbol myristate acetate inhibits Ca^{2+} influx through voltage-gated Ca^{2+} channels in two secretory cell lines, PC 12 and RIN m5F. J Biol Chem 261:32–36
Dolly JO, Haliwell JV, Black JD, Williams RS, Pelchen-Matthews A, Breeze AL, Mehraban F, Othman IB, Black AR (1984) Botulinum neurotoxin and dendrotoxin as probes for studies on transmitter release. J Physiol (Paris) 79:280–303

Douglas WW (1974) Exocytosis and the exocytosis-vesiculation sequence: with special reference to neurohypophysis, chromaffin and mast cells, calcium and calcium ionophores. In: Thorn NA, Petersen OH (ed) Secretory mechanisms in exocrine glands. Munksgaard, Copenhagen, pp 116–136
Farquhar MG (1982) Membrane recycling in secretory cells: pathways to the Golgi complex. CIBA Found Symp 92:157–183
Fernandez JM, Neher E, Gomperts BD (1984) Capacitance measurements reveal stepwise fusion events in degranulating mast cells. Nature 312:453–455
Fesce R, Segal JR, Ceccarelli B, Hurlbut WP (1986) Effects of black widow spider venom and Ca^{2+} on quantal secretion at the frog neuromuscular junction. J Gen Physiol 88:737–752
Fritz LC, Mauro A (1982) The ionic dependence of black widow spider venom action at the stretch receptor neuron and neuromuscular junction of crustaceans. J Neurobiol 13:385–401
Ginsborg BL, Jenkinson DH (1976) Transmission of impulses from nerve to muscle. In: Zaimis E (ed) Neuromuscular junction. Springer, Berlin Heidelberg New York, pp 229–364 (Handbook of experimental pharmacology, vol 42)
Gorio A, Hurlbut WP, Ceccarelli B (1978) Acetylcholine compartments in mouse diaphragm: a comparison of the effects of black widow spider venom, electrical stimulation and high concentrations of potassium. J Cell Biol 78:716–733
Gorio A, Rubin LL, Mauro A (1978) Double mode of action of black widow spider venom on frog neuromuscular junction. J Neurocytol 7:193–205
Grasso A, Alema S, Rufini S, Senni MI (1980) Black widow spider toxin-induced calcium fluxes and transmitter release in a neurosecretory cell line. Nature 283:774–776
Grynkyevicz G, Poenie M, Tsien RY (1985) A new generation of Ca^{2+} indicators with greatly improved fluorescence properties. J Biol Chem 260:3440–3450
Gunther GR (1981) Effect of 12-O-tetradecanoyl-phorbol-13-acetate on Ca^{2+} efflux and protein discharge in pancreatic acini. J Biol Chem 256:12040–12045
Haimann C, Torri-Tarelli F, Fesce R, Ceccarelli B (1985) Measurement of quantal secretion induced by ouabain and its correlation with depletion of synaptic vesicles. J Cell Biol 101:1953–1965
Haimann C, Meldolesi J, Ceccarelli B (1987) The phorbol ester enhances the evoked quantal release of acetylcholine at frog neuromuscular junction. Pflügers Arch 408:27–31
Hallam TJ, Daniel JL, Kendrick-Jones R, Rink TJ (1985) Relationship between cytoplasmic free calcium and myosin light chain phosphorylation in intact platelets. Biochem J 232:373–377
Harris TB (1986) Proceedings of the 8th world congress of animal, plant and microbial toxins. Oxford University Press, Oxford
Herzog V (1981) Endocytosis in secretory cells. Philos Trans R Soc Lond [Biol] 269:67–72
Heuser JE, Reese TS (1973) Evidence for recycling of synaptic vesicle membrane during transmitter release at the frog neuromuscular junction. J Cell Biol 57:315–344
Hoch T, Tomero-Mira M, Ehrich BE, Finkelstein A, Das Gupta BR, Simpson LL (1985) Channels formed by botulinum, tetanus and diphtheria toxins in planar lipid bilayers. Relevance of translocation of proteins across membranes. Proc Natl Acad Sci USA 82:1682–1696
Hooper JE, Kelly RB (1984) Calmodulin is tightly associated with synaptic vesicles independent of calcium. J Biol Chem 259:148–153
Houslay MD (1985) Renaissance of cyclic GMP. Trends Biochem Sci 10:465–466
Howard BD, Gundersen CB (1980) Effects and mechanisms of polypeptide neurotoxins that act presynaptically. Ann Rev Pharmacol Toxicol 20:307–336
Hucho F, Ovchinnichov YA (1983) Toxins as tools in neurochemistry. de Gruyter, Berlin New York
Hurlbut WP, Ceccarelli B (1979) Use of black widow spider venom to study the release of neurotransmitters. In: Ceccarelli B, Clementi F (eds) Neurotoxins, tools in neurobiology. Raven, New York, pp 87–115

Hutton JC, Peshavaria M, Brocklehurst KW (1984) Phorbol ester stimulation of insulin release and secretory granule protein phosphorylation in a transplantable rat insulinoma. Biochem J 224:483–490

Jaimovich E, Ildefonse M, Barhanin J, Rougier O, Lazdunski M (1982) Centruroides toxin, a selective blocker of surface Na^+ channels in skeletal muscle: voltage clamp analysis and biochemical characterization of the receptor. Proc Natl Acad Sci USA 79:3896–3900

Katz B (1966) Nerve, muscle, and synapse. McGraw-Hill, New York

Katz B, Miledi R (1977) Transmitter leakage from motor nerve endings. Proc R Soc Lond [Biol] 196:59–72

Keith CH, Ratan R, Maxfield FR, Bajer A, Shelanski ML (1985) Local cytoplasmic calcium gradients in living mitotic cells. Nature 316:848–850

Kelly RB (1985) Pathway of protein secretion in eukaryotes. Science 230:25–31

Kenigsberg RL, Trifarò JM (1985) Microinjection of calmodulin antibodies into cultured chromaffin cells blocks catecholamine release in response to stimulation. Neuroscience 14:335–347

Kenigsberg RL, Coté A, Trifarò JM (1982) Trifluoperazine, a calmodulin inhibitor, blocks secretion in cultured chromaffin cells at a step distal from calcium entry. Neuroscience 7:2277–2286

Kerr LM, Yoshikami D (1984) A venom peptide with a novel presynaptic blocking action. Nature 308:282–284

Knight DE, Baker PF (1982) Calcium dependence of catecholamine release from bovine adrenal medullary cells after exposure to intense electric field. J Membr Biol 68:107–140

Knight DE, Baker PF (1983) The phorbol ester TPA increases the affinity of exocytosis for calcium in leaky adrenal medullary cells. FEBS Lett 160:98–100

Knight DE, Baker PF (1985) Guanine nucleotides and Ca^{2+}-dependent exocytosis. Studies on two adrenal cell preparations. FEBS Lett 189:345–349

Knight DE, Scrutton MC (1985) The relationship between intracellular second messengers and platelet secretion. Biochem Soc Trans 12:969–971

Knight DE, Tonge DA, Baker PF (1985) Inhibition of exocytosis in bovine adrenal medullary cells by botulinum toxin type D. Nature 317:719–721

Koike H, Meldolesi J (1981) Post-stimulation retrieval of lumenal surface membrane in parotid acinar cells is Ca^{2+}-dependent. Exp Cell Res 134:377–388

Kolesnick RN, Geshenghorn MG (1985) Direct evidence that burst but not sustained secretion of prolactin stimulated by TRH is dependent on elevation of cytoplasmic Ca. J Biol Chem 260:5217–5220

Kostynk PG (1984) Metabolic control of ionic channels in the neuronal membrane. Neuroscience 13:983–990

Llinás R, Steinberg Z, Walton K (1981) Relationship between calcium current and postsynaptic potential in squid giant synapse. Biophys J 33:323–352

Llinás R, McGuiness TLM, Leonard CS, Sugimori M, Greengard P (1985) Intraterminal injection of synapsin I or calcium/calmodulin-dependent protein kinase II alters neurotransmitter release at the squid giant synapse. Proc Natl Acad Sci USA 82:3035–3039

Lundberg JM, Hökfelt T (1983) Coexistence of peptides and classical neurotransmitters. Trends Neurosci 6:325–333

Madeddu L, Saito I, Hsiao TH, Meldolesi J (1985) Leptinotoxin-h action in synaptosomes and neurosecretory cells. Stimulation of neurotransmitter release. J Neurochem 45:1719–1730

Meldolesi J, Ceccarelli B (1981) Exocytosis and membrane recycling. Philos Trans R Soc Lond [Biol] 296:55–65

Meldolesi J, Huttner WB, Tsien RY, Pozzan T (1984) Free cytoplasmic Ca^{2+} and neurotransmitter release: studies on PC12 cells and synaptosomes exposed to α-latrotoxin. Proc Natl Acad Sci USA 81:620–624

Meldolesi J, Scheer H, Madeddu L, Wanke E (1986) On the mechanism of action of α-latrotoxin, the presynaptic stimulatory toxin of the black widow spider venom. Trends Pharm Sci 7:151–155
Miller RJ (1984) Toxin probes for voltage-sensitive calcium channels. Trends Neurosci 7:309
Miller TM, Heuser JE (1984) Endocytosis of synaptic vesicle membrane at the frog neuromuscular junction. J Cell Biol 98:685–698
Montecucco C (1986) How do tetanus and botulinum toxin bind to neuronal membranes? Trends Biochem Sci 11:314–316
Navone F, Greengard P, Camilli P De (1984) Synapsin I in nerve terminals: selective association with small synaptic vesicle. Science 226:1209–1211
Neher E, Marty A (1982) Discrete changes of cell membrane capacitance observed under conditions of enhanced secretion in bovine adrenal chromaffin cells. Proc Natl Acad Sci USA 79:6712–6716
Niggli V, Knight DE, Baker PF, Vigny A, Henry JP (1984) Thyrosine hydroxylase in leaky adrenal medullary cells. Evidence for in situ phosphorylation by separate Ca^{2+} and cyclic AMP-dependent enzymes. J Neurochem 43:646–658
Nowycky MC, Fox AP, Tsien RW (1985) Three types of neuronal calcium channel with different calcium agonist sensitivity. Nature 316:440–443
Palade GE (1975) Intracellular aspects of the process of protein secretion. Science 189:347–358
Pandol SJ, Schoeffield MS, Sachs G, Muallem S (1985) Role of free Ca^{2+} in secretagogue stimulated amylase release from dispersed acini from guinea pig pancreas. J Biol Chem 260:10081–10086
Patzak A, Winkler H (1986) Exocytotic exposure and recycling of membrane antigens of chromaffin granules: ultrastructural evaluation after immunolabelling. J Cell Biol 102:510–515
Patzak A, Bock G, Fischer-Colbrie R, Schanchenstein K, Schmidt W, Lingg G, Winkler H (1984) Exocytotic exposure and retrieval of membrane antigens of chromaffin granules: quantitative evaluation of immunofluorescence on the surface of chromaffin cells. J Cell Biol 98:1817–1824
Pollard HB, Pozoles CJ, Creutz CE (1981) Mechanism of calcium action and release of vesicle-bound hormones during exocytosis. Rec Prog Horm Res 37:299–332
Pozzan T, Gatti G, Dozio N, Vicentini LM, Meldolesi J (1984) Ca^{2+}-dependent and -independent release of neurotransmitters from PC12 cells. A role for protein kinase C activation? J Cell Biol 99:628–638
Quissel DO, Deisher LM, Barzen KA (1985) The rate determining step in cAMP-mediated exocytosis in the rat parotid and submandibular glands appears to involve analogous 26 kDa integral membrane phosphoprotein. Proc Natl Acad Sci USA 82:3237–3241
Rane SG, Dunlap K (1986) Kinase C activator 1,2-oleoyl acetylglycerol attenuates voltage-dependent calcium current in sensory neurons. Proc Natl Acad Sci USA 83:184–188
Reichardt LF, Kelly RB (1983) A molecular description of nerve terminal function. Annu Rev Biochem 52:871–926
Reuter H (1983) Calcium channel modulation by neurotransmitters, enzymes and drugs. Nature 301:569–574
Rink TJ, Smith SW, Tsien RY (1982) Cytoplasmic free Ca^{2+} in human platelets: Ca^{2+} thresholds and Ca-independent activation for shape changes and secretion. FEBS Lett 148:21–26
Rink TJ, Sanchez A, Hallam TJ (1983) Diacylglycerol and phorbol ester stimulate secretion without raising cytoplasmic free calcium in human platelets. Nature 305:317–319
Sakmann B, Neher E (1984) Patch clamp techniques for studying ionic channels in excitable membranes. Ann Rev Physiol 46:455–472
Scheer H, Meldolesi J (1985) Purification of the putative α-latrotoxin receptor from bovine synaptosomal membranes in an active binding form. EMBO J 4:323–327
Schramm M, Selinger Z (1975) The functions of cyclic AMP and calcium as alternative second messengers in parotid gland and pancreas. J Cycl Nucl Res 1:181–192
Schulz I, Stolze H (1980) The exocrine pancreas: the role of secretatogues, cyclic nucleotides and calcium in enzyme secretion. Ann Rev Physiol 42:127–156

Segal JR, Ceccarelli B, Fesce R, Hurlbut WP (1985) Miniature endplate potential frequency and amplitude determined by an extension of Campbell's theorem. Biophys J 47:183–202

Sellin LC (1985) The pharmacological mechanism of botulism. Trends Pharmacol Sci 6:80–83

Shoback DM, Thatcher J, Leombruno R, Brown EM (1984) Relationship between parathyroid hormone secretion and cytosolic calcium concentration in dispersed parathyroid cells. Proc Natl Acad Sci USA 81:3113–3117

Silinski EM (1985) The biophysical pharmacology of calcium-dependent acetylcholine secretion. Pharmacol Rev 37:81–132

Simpson LL (1986) Molecular pharmacology of botulinum toxin and tetanus toxin. Ann Rev Pharmacol Toxicol 26:427–454

Smith AD, Winkler H (1972) Fundamental mechanisms of release of catecholamines. In: Blasko H, Muscholl E (eds) Springer, Berlin Heidelberg New York, pp 538–617 (Handbook of experimental pharmacology, vol 33)

Takahashi M, Tatsumi M, Ohizumi Y, Yasumoto T (1983) Ca^{2+} channel activating function of maitotoxin, the most potent marine toxin known, in clonal rat pheochromocytoma cells. J Biol Chem 258:10944–10947

Tauc L (1982) Non-vesicular release of neurotransmitter. Physiol Rev 62:857–893

Thilo L (1985) Selective internalization of granule membrane after secretion in mast cells. Proc Natl Acad Sci USA 82:1716–1720

Thureson-Klein A (1983) Exocytosis from large and small dense-cored vesicles in noradrenergic nerve terminals. Neuroscience 10:245–259

Torri-Tarelli F, Grohovaz F, Fesce R, Ceccarelli B (1985) Temporal coincidence between synaptic vesicle fusion and quantal secretion of acetylcholine. J Cell Biol 101:1386–1399

Tsien RY, Pozzan T, Rink TJ (1982) Calcium homeostasis in intact lymphocytes: cytoplasmic free calcium monitored by a new, intracellularly trapped fluorescent indicator. J Cell Biol 94:325–334

Tsien RY, Pozzan T, Rink TJ (1984) Measuring and manipulating cytosolic Ca^{2+} with trapped indicators. Trends Biochem Sci 9:263–266

Valtorta F, Madeddu L, Meldolesi J, Ceccarelli B (1984) Specific localization of the α-latrotoxin receptor in the nerve terminal plasma membrane. J Cell Biol 99:124–132

Vara F, Rozengurt E (1984) Stimulation of Na^+/H^+ antiport activity of epidermal growth factor and insulin occurs without activation of protein kinase C. Biochim Biophys Res Commun 130:646–653

Wanke E, Ferroni A, Gattanini F, Meldolesi J (1986) α-latrotoxin of the black widow spider venom opens a small, non closing cation channel. Biochim Biophys Res Commun 134:320–325

Weller U, Bernhardt U, Siemen D, Dreyer F, Vogel W, Haberman E (1985) Electrophysiological and neurobiochemical evidence for the blockade of a potassium channel by dendrotoxin. Naunyn-Schmiedebergs Arch Pharmacol 330:77–83

Williams DA, Fogarty KE, Tsien RY, Fay FS (1985) Calcium gradients in single smooth muscle cells revealed by the digital imaging microscope using Fura-2. Nature 318:558–561

Wrenn RW (1984) Phosphorylation of a pancreatic zymogen granule membrane protein by endogenous calcium-phospholipid dependent protein kinase. Biochim Biophys Acta 775:1–6

CHAPTER 18

Pharmacology of Calcium Metabolism in Smooth Muscle

T. GODFRAIND

A. Introduction

Calcium may be considered as the ultimate intracellular messenger of the signals received at the level of smooth muscle pericellular membrane to produce a change in the active tone. The calcium responsible for the increase in free cytoplasmic levels is potentially available from the intracellular space and also from several compartments having various affinities for calcium. Several cellular mechanisms maintain the calcium gradient across the plasmalemmal membrane. They are associated with the plasma membrane itself and with intracellular organelles. Differences between muscles are to some extent due to differences in the importance and the properties of the different mechanisms operating at rest and during activation. Pharmacologic agents influence calcium movements by acting on those mechanisms responsible for calcium homeostasis at the cellular level (Table 1). Understanding of the mode of action of those agents requires a proper analysis of the mechanisms of this calcium homeostasis.

B. Calcium Entry and Calcium Regulation at Rest: Action of Pharmacologic Agents

The existence of a slow calcium turnover at rest has been demonstrated in various tissues by radiochemical techniques. Calcium entry at rest ranges between 10 and 100×10^{-15} mol cm^{-2} s^{-1} (GODFRAIND-DE BECKER and GODFRAIND 1980). This slow entry must be compensated for by an equal efflux in order to avoid cellular calcium overload. Two separate mechanisms have been shown to mediate the efflux of calcium from cells, an Na,Ca exchange system that was first described in squid axon (BAKER et al. 1969) and in heart (REUTER 1974), and a specific Ca-ATPase, first described in red blood cells (SCHATZMANN and VINCENZI 1969). These mechanisms have been shown to be present in most cells, but their relative importance varies among tissues.

An appropriate method to characterize the calcium pumps is to isolate the plasma membrane. The usual tissue fractionation techniques do not allow a proper separation of plasma membrane from other cellular fractions. Therefore, such studies have been mostly performed with erythrocyte membranes. In an alternative approach, we have investigated calcium transport activities in relatively crude microsomal fractions, which contained a fair proportion of the plasma membrane present in the homogenate and have established the subcellular location of those activities by analytic density gradient centrifugation. The selective

Table 1. Calcium modulators.[a] Agents affecting Ca^{2+} movements (calcium antagonists and calcium agonists)

A. Inhibitors: calcium antagonists
1. Agents acting at the plasma membrane
1.1 Calcium entry blockers
Group I: Selective calcium entry blockers
Subgroup I.A: Agents selective for slow calcium channels in myocardium (slow channel blockers)
Phenylalkylamines: verapamil, gallopamil (D600); under investigation: anipamil, desmethoxyverapamil (D888), emopamil, falipamil (AQ-A-39), ronipamil
Dihydropyridines: nifedipine, nicardipine, niludipine, nimodipine, nisoldipine, nitrendipine, ryosidine; under investigation: amlodipine, azodipine, dazodipine (PY 108-068), felodipine, flordipine, FR 34235, iodipine, isrodipine, mesudipine, ni(l)vadipine, oxodipine, PN 200-110, riodipine
Benzothiazepines: diltiazem; under investigation: fostedil (KB-944)
Subgroup I.B: Agents with no perceived actions on slow calcium channels in myocardium
Diphenylpiperazines: cinnarizine and flunarizine
Group II: Nonselective calcium entry blockers
Subgroup II.A: Agents acting at similar concentrations on calcium channels and fast sodium channels
Bencyclane, bepridil, caroverine, etafenone, fendiline, lidoflazine, perhexiline, prenylamine, SKF 525A, terodiline, tiapamil
Subgroup II.B: Agents interacting with calcium channels while having another primary site of action
They include, among others: agents acting on sodium channels (local anesthetics, phenytoin); on catecholamine receptors (benextramine, nicergoline, phenoxybenzamine, phenothiazines, pimozide, propranolol, WB-4101, yohimbine derivatives); on benzodiazepine rezeptors (diazepam, flurazepam); on opiate receptors (loperamide, fluperamide); on clyclic nucleotide phosphodiesterases (amrinone, cromoglycate, papaverine); barbiturates; cyproheptadine; indomethacin; reserpine

1.1 Sodium calcium exchange inhibitors
Amiloride and derivatives

2. Agents acting within the cell
2.1 Acting on sarcoplasmic reticulum
Dantrolene, TMB-8
2.2 Acting on mitochondria
Ruthenium red
2.3 Calmodulin antagonists
Phenothiazines: trifluoperazine, chlorpromazine
Naphthalene derivatives: W7
Local anesthetics: dibucaine
Dopamine antagonists: pimozide, haloperidol
Calmidazolium (R24571)

B. Facilitators
1. Agents acting at the plasma membrane
1.1 Calcium agonists
Dihydropyridines: Bay K 8644, CGP 28392, YC 170

2. Agents acting on sarcoplasmic reticulum
Inositol-1,4,5-trisphosphate
Caffeine

3. Ionophores
A23187, ionomycin

[a] For references, see Godfraind et al. (1986).

effects of digitonin on plasmalemmal elements proved to be particularly useful in our studies. Digitonin forms a highly insoluble, equimolar complex with cholesterol and this reaction has long been used for cholesterol determination. When added at a low concentration to a subcellular fraction from a tissue, digitonin binds preferentially to the plasmalemmal elements containing cholesterol and markedly increases their equilibrium density in a sucrose gradient, whereas other types of membranes, such as endoplasmic reticulum (ER) or mitochondrial membranes, which are poor in cholesterol, are not affected (Fig. 1). This digitonin shift therefore allows an activity associated with the plasma membrane to be identified. This has allowed us to separate the oxalate-sensitive CaATP pump of the ER from the plasmalemmal pump (MOREL et al. 1981; WIBO et al. 1981). The Ca-ATPase of smooth muscle plasma membrane shares the main properties of the red cell and heart sarcolemmal enzymes: it has a high affinity for calcium that is increased in the presence of calmodulin. It is more sensitive to vanadate inhibition than the ATPase from the sarcoplasmic reticulum and its calcium-binding capacity is destabilized by a high digitonin concentration.

Such studies have also allowed us to identify an Na,Ca exchanger in the plasma membrane of the smooth muscle (MOREL and GODFRAIND 1982, 1984). We have observed that the rate of calcium release from plasmalemmal vesicles previously loaded with calcium was faster in the presence of sodium than lithium or potassium. The apparent equilibrium constant for sodium was 28 m*M* with a Hill coefficient of 1.9. In both heart and smooth muscle, the Na,Ca carrier shows a lower affinity than the ATP-dependent calcium pump (Fig. 2). However, cardiac and smooth muscles are quite different as far as the calcium transport rate and the capacity of the Na or ATP-dependent Ca^{2+} transport mechanisms are concerned. In the heart, as reported by CARONI et al. (1980) and CARONI and CARAFOLI (1983) in dog membranes and confirmed by ourselves in guinea pig membranes, Na,Ca exchange is characterized by a very high calcium transport velocity and by a calcium accumulation capacity that is about twice that of the ATP-dependent calcium pump. Therefore, the Na,Ca exchange mechanism appears to be most important for calcium regulation in cardiac cells. By contrast, the activity of the smooth muscle plasmalemmal Na,Ca exchange would be weak. The maximum amount of calcium accumulation is one-fifth of the maximum ATP-dependent calcium uptake. Weak activity of Na,Ca exchange in smooth muscle is in agreement with some studies in intact tissue, which indicate that an inwardly oriented sodium gradient is not required for the maintenance of a low calcium concentration in smooth muscle (CASTEELS and VAN BREEMEN 1975). Interestingly, the physiologic importance of Na,Ca exchange may vary according to the type of smooth muscle, as illustrated by HIRATA et al. (1981), who showed that this calcium extrusion system was absent from single cells prepared from the pig coronary artery, whereas it was operative in cells from guinea pig teniae coli.

In heart sarcolemmal preparations, the Na,Ca system operates electrogenically, exchanging three or more Na^+ ions for one Ca^{2+} (CARONI et al. 1980; PHILIPSON and NISHIMOTO 1980; PITTS 1979; REEVES and SUTKO 1979). It has therefore been suggested that in heart, Na,Ca exchange could contribute not only to calcium extrusion during relaxation, but also to calcium entry during depolariza-

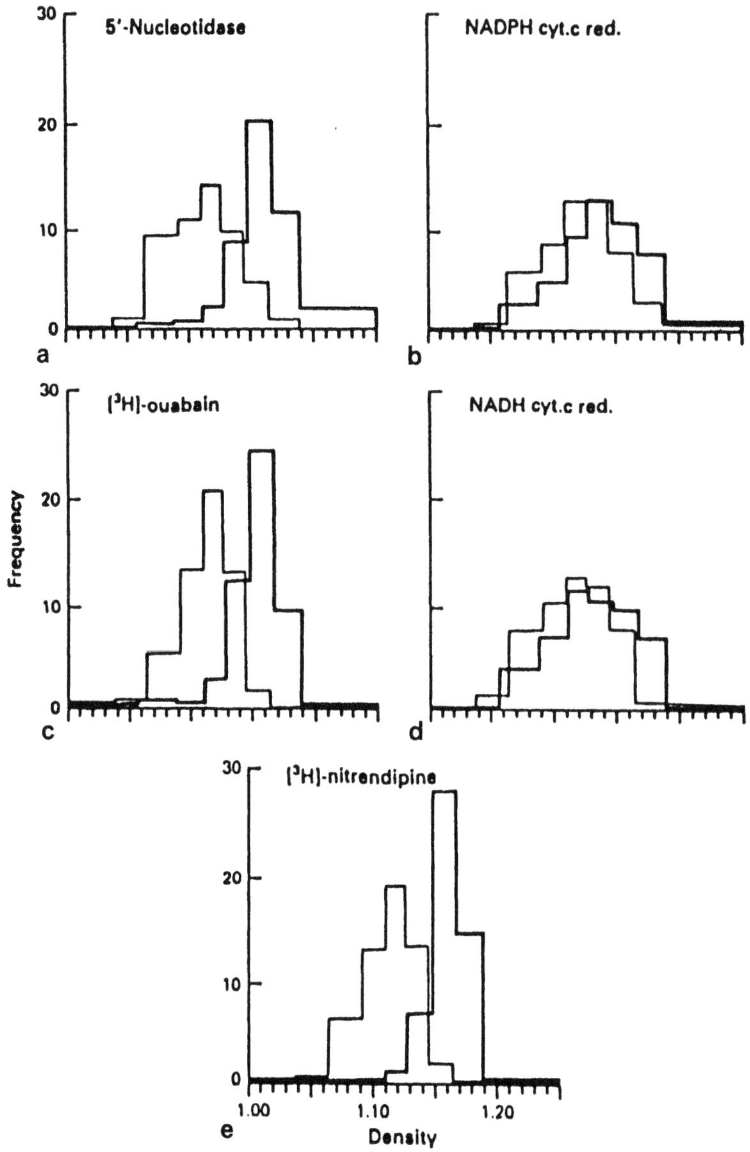

Fig. 1 a–e. Density distribution histograms obtained from an untreated microsomal fraction (*thin line*) and from a microsomal fraction that had been treated with digitonin (*thick line*). Specific binding or enzyme activities in the total microsomal fraction, related to 1 g tissue wet weight, were: 5′-nucleotidase 2.08 units; NADPH : cytochrome *c* reductase (*cyt. c red.*) 0.083 units; [^{3}H] ouabain 17 pmol (radioligand concentration 30 n*M*); NADH : cytochrome *c* reductase 4.5 units; [^{3}H] nitrendipine 1.38 pmol (radioligand concentration 0.25 n*M*); protein 4.08 mg. Specific binding or enzyme activities recovered in gradient subfractions, expressed as percentages of the total microsomal fraction, were: 5′-nucleotidase 100% (untreated sample) and 95% (digitonin-treated sample) : NADPH : cytochrome *c* reductase 87% and 105%; [^{3}H] ouabain 85% and 64%; NADH : cytochrome *c* reductase 81% and 92%; [^{3}H] nitrendipine 84% and 67%. The frequency is the fractional amount of specific binding, or of enzyme activity, recovered in a given subfraction, divided by the density increment across this subfraction. The abscissae of **a–d** have the same density scale as **e**. (GODFRAIND and WIBO 1985)

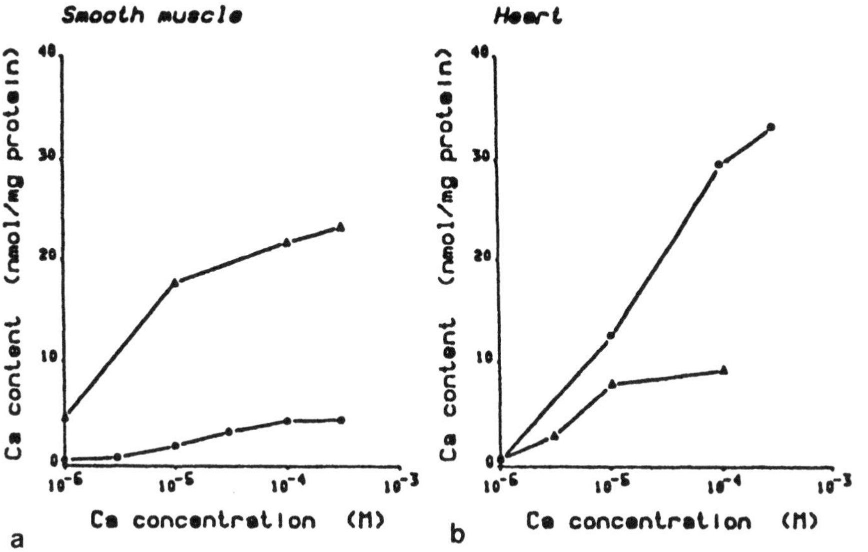

Fig. 2a, b. Comparison between ATP-dependent (*triangles*) and Na,Ca exchange-dependent (*circles*) transport of calcium in plasma membrane of smooth muscle (**a**) and cardiac muscle (**b**). (MOREL and GODFRAIND 1982)

tion. Our results suggest that in smooth muscle also, more than one Na^+ is required to compete with one Ca^{2+} (MOREL and GODFRAIND 1982, 1984). However, owing to the weak efficacy of the system, it is unlikely to contribute to calcium entry during smooth muscle depolarization, a conclusion in agreement with HIRATA et al. (1981).

Theoretically, two groups of agents may affect calcium movements in a resting cell: those acting on Na,Ca exchange and those acting on the CaATP pumps. The existence of an interaction between calcium and sodium indicates that the sodium pump activity may interfere with calcium metabolism. This question has mainly been discussed for the heart where it has been shown that an increase in intracellular sodium interferes with Na,Ca exchange and has important consequences for the contractile state of the heart (REUTER 1982). In smooth muscle, the situation is more controversial because the activity of the CaATP pumps is such that it could buffer any change in intracellular calcium activity occurring through a mechanism that has undoubtedly a lower transport capacity. Therefore, the action of cardiac glycosides in smooth muscles is unimportant in the presence of antagonists for neurotransmitters released as a consequence of blockade of the Na pump of nerve terminals. Similarly, the pharmacologic action of amiloride and dichlorobenzamil (its derivative with a more selective effect on the Na,Ca exchanger) is also unimportant in smooth muscle, by contrast to an impressive action observed in cardiac tissue (WINQUIST 1984).

Because the ATP-dependent calcium pumps are activated by the calcium–calmodulin complex, calmodulin antagonists could modify calcium turnover in a resting smooth muscle. Nevertheless, experimental data are lacking because these agents have been studied more in acellular preparations and in contracting muscles than in resting muscles (see HIDAKA and HARTSHORNE 1985).

C. Calcium Movements During Excitation: Their Sensitivity to Pharmacologic Agents

Direct measurements of ^{45}Ca movements are not feasible in smooth muscle because they are not indicative of changes that could occur within the cell. Indeed, a large amount of calcium is bound to extracellular sites. Several attempts have been made to identify the biologically active calcium fraction and the most successful are based on the use of lanthanum. Lanthanum replaces Ca^{2+} on superficial binding sites and does not penetrate the cell. Because lanthanum blocks transmembrane fluxes of calcium, it has been proposed that the calcium content of a muscle washed in a lanthanum solution could provide an estimate of cellular calcium. The ^{45}Ca fluxes across the smooth muscle cell membrane may be estimated in the rat aorta by measuring the ^{45}Ca turnover in the La^{3+}-resistant Ca fraction. This fraction corresponds to the amount of Ca that is not displaced when the tissue is soaked in 50 mM La^{3+} solution (GODFRAIND 1986b).

Norepinephrine increases the rate of uptake of ^{45}Ca into this calcium fraction. There is no net gain in tissue calcium, and calcium efflux increases in a similar manner (Fig. 3). The increase of the rate of ^{45}Ca uptake is dose dependent. In the presence of the α-antagonist phentolamine, the dose–response curves for the effect of norepinephrine on ^{45}Ca uptake are displaced to the right in a manner suggesting a competitive antagonism, the pA_2 for phentolamine being 7.8. Because of the resemblance of this pA_2 value to that obtained for the antagonistic effects of phentolamine on the contractile responses to norepinephrine, it appears that the activation of α-adrenoceptors is responsible for both the increased rate of Ca entry and the contraction. These observations suggest that α-adrenoceptor activation does open Ca pathways in the membrane; the accepted view is that these pathways could consist of calcium channels closely associated with the receptors, which have been termed ROCs (receptor-operated channels). High K solutions also allow an increased influx of ^{45}Ca that presents some differences from the agonist-dependent influx, namely its sensitivity to calcium antagonists. Those channels controlled by the membrane potential have been termed POCs (potential-operated channels). The slow turnover of intracellular calcium at rest is likely to occur through Ca leak channels (BOLTON 1979; GODFRAIND and MILLER 1983; MEISHERI et al. 1981).

Another major means by which to analyze the properties of the calcium channels is the study of calcium entry blockers. The term calcium entry blocker has been introduced in order to specify among the group of calcium antagonists those whose pharmacologic actions result from the blockade of calcium entry through membrane channels opened during cell excitation (GODFRAIND 1986a). They inhibit calcium entry evoked by vasoconstriction stimuli, but barely affect Ca efflux (Fig. 3). The concept of calcium antagonism emerged from earlier studies with the antianginal drug lidoflazine and with cinnarizine, which was introduced as an antihistaminic (GODFRAIND et al. 1968; GODFRAIND and POLSTER 1968). On the

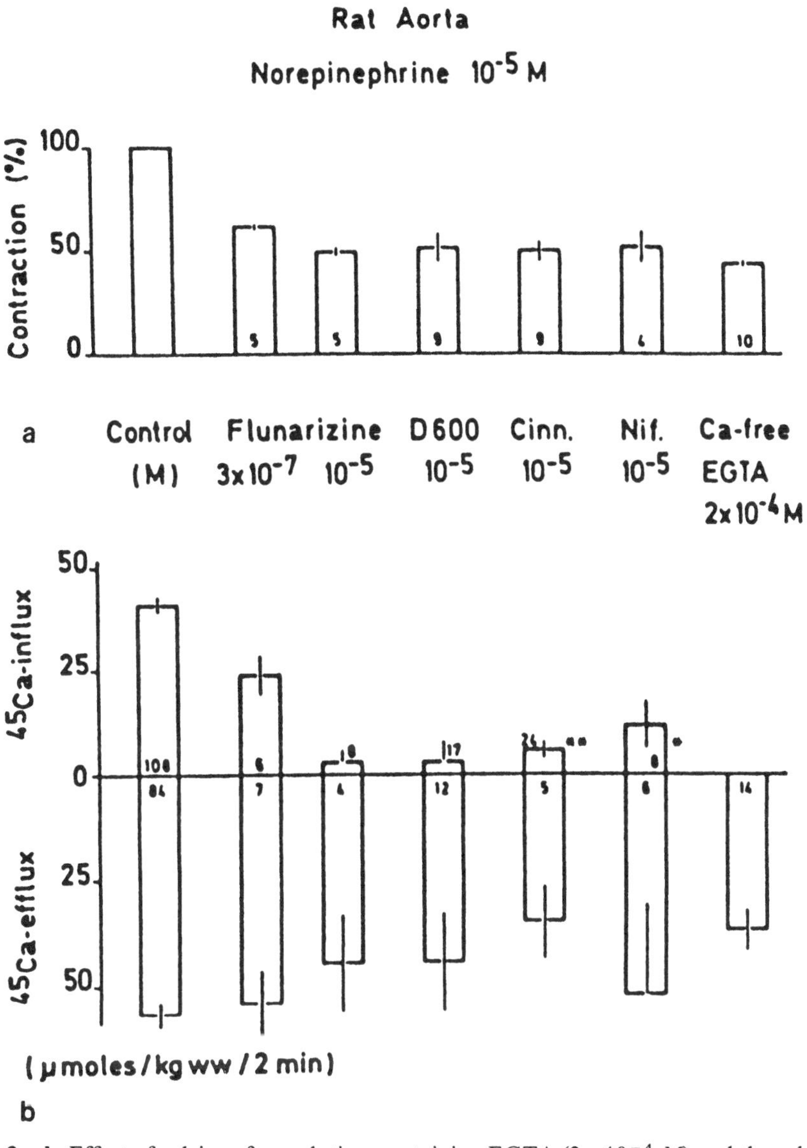

Fig. 3 a, b. Effect of calcium-free solution containing EGTA (2×10^{-4} *M*) and the calcium entry-blocking drugs, flunarizine (3×10^{-7} and 10^{-5} *M*), D600 (10^{-5} *M*), cinnarizine (10^{-5} *M*), and nifedipine (10^{-5} *M*), on contractions (**a**) and $^{45}Ca^{2+}$ influx and efflux (**b**) produced by norepinephrine (10^{-5} *M*) in rat aorta. Tissues were incubated with Ca^{2+} entry blockers for 90 min before contractions were induced with norepinephrine. Also shown is the contraction induced by norepinephrine after the aorta was washed in Ca^{2+}-free solution for 5 min. At this time, the tissue does not contract in response to depolarization (100 m*M* K^+). All contractions are expressed as percentages of the maximal contraction induced by norepinephrine in the absence of blockers. Numbers in colomns indicate the number of determinations. *Vertical bars* represent SEM. (GODFRAIND and MILLER 1983)

other hand, studies on the action of verapamil (see FLECKENSTEIN 1973) on the heart have greatly improved our views on those drugs that have a broad spectrum of clinical use (Table 1; GODFRAIND 1985).

The interaction of calcium entry blockers with calcium channels may be examined in binding studies and in functional studies. Functional studies have revealed striking differences in the blockade of POCs and ROCs by calcium entry blockers. In contractile and in ^{45}Ca flux experiments, it was observed that the sensitivity of POCs is time dependent and that their blockade can be complete (Fig. 4). On the other hand, ROC blockade is not increased by the duration of the stimulus and is apparently not complete, since ^{45}Ca entry is still observed at maximum effective concentration (Fig. 3; GODFRAIND 1983). Because most of the drugs so far used are highly lipophilic, binding studies are generally performed on preparations enriched with the plasma membrane, rather than on intact tissue. When the binding of [^3H] nitrendipine is examined in microsomal fractions isolated from smooth muscle, only one class of binding site can be detected. This is at variance with cardiac tissue where two binding sites have been characterized. Subfractionation of microsomes by isopyknic density gradient centrifugation after digitonin treatment, as discussed in Sect. B, has established that the binding sites are located in the plasma membrane, as this was expected for drugs acting on calcium channels (see Fig. 1; GODFRAIND and WIBO 1985).

The apparent affinity of calcium entry blockers for calcium channels may be estimated from experiments in which their potency to displace a radioactive ligand is studied. With intestinal smooth muscle, extensive studies have been carried out to compare affinities of dihydropyridine derivatives for their binding site and potencies in pharmacologic experiments (BOLGER et al. 1982, 1983; GODFRAIND and WIBO 1985). Essentially 1:1 correlations have been obtained between inhibition of [^3H] nitrendipine binding and inhibition of K^+ depolarization-induced mechanical response. The response evoked by activation of the muscarinic receptor was less sensitive to dihydropyridine inhibitors, so that IC_{50} values for inhibition of the tonic component of contraction were tenfold greater than the values for inhibition of binding. The *o*-NCS (isothiocyanate) analog of nifedipine irreversibly blocked both the mechanical response to muscarinic agonists and [^3H] nitrendipine binding (VENTER et al. 1983).

Similar, but less extensive, studies have been performed on arterial and myometrial tissues. With pig coronary arteries, DE POVER et al. (1982) showed that K_D of [^3H] nitrendipine corresponded to its EC_{50} for relaxation of K^+-depolarized arteries. With rat aorta, the K_i values of several nifedipine analogs, including nimodipine stereoisomers, were very similar to their IC_{50} values for inhibition of agonist-evoked contraction and ^{45}Ca entry (Table 2; GODFRAIND et al. 1985b; WILLIAMS 1980). However, with mesenteric arteries from rat or dog, K_D of [^3H] nitrendipine was 20 times lower than IC_{50} of nitrendipine for inhibition of K^+-evoked contraction (TRIGGLE et al. 1982). In rat myometrium, the inhibitory potency of dihydropyridines on K^+-stimulated ^{45}Ca uptake and contraction agreed with their binding affinity (BATRA 1985; GROVER and OAKES 1985).

The stereoselectivity of D600 in [^3H] nitrendipine binding experiments (BOLGER et al. 1983) on ileal smooth muscle was comparable to its selectivity in mechanical experiments (JIM et al. 1981). *d-cis*-Diltiazem, the pharmacologically

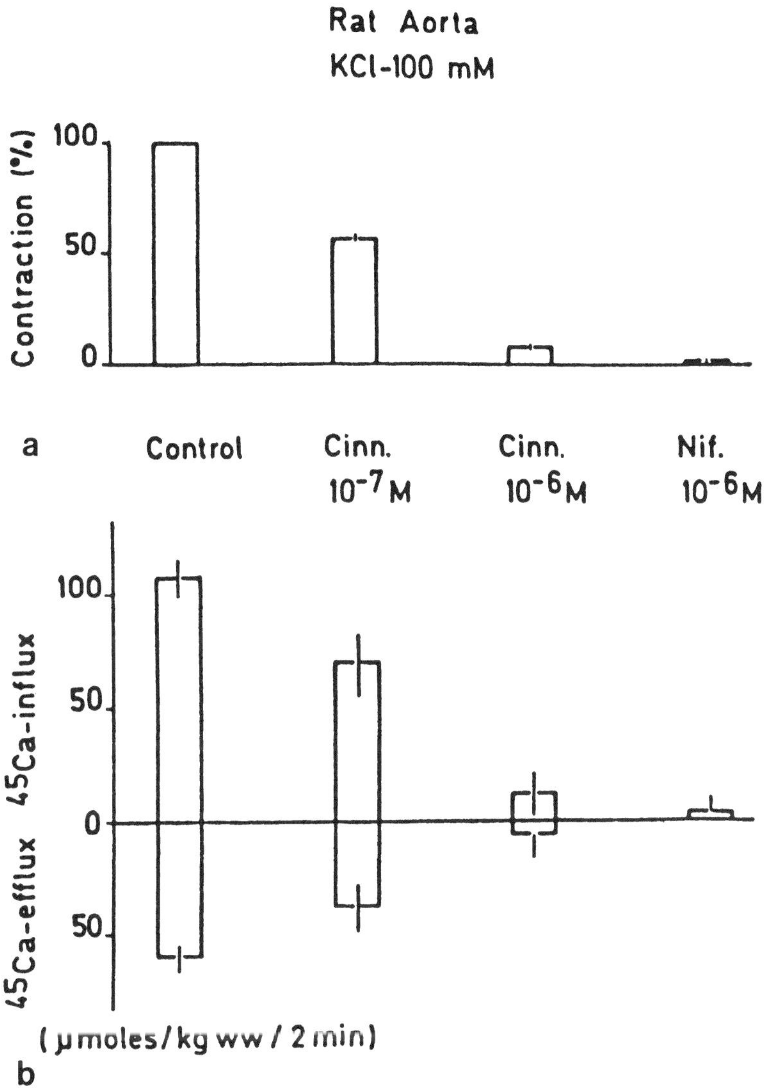

Fig. 4. a Inhibition of maximal (100 m*M* K^+) depolarization-induced contractions of rat aorta produced by cinnarizine (10^{-7} and 10^{-6} *M*) and by nifedipine (10^{-6} *M*). **b** Inhibition of depolarization- (100 m*M* K^+)-stimulated cellular $^{45}Ca^{2+}$ influx and efflux (units: µ*M* Ca^{2+} per kilogram wet weight in 2 min) by cinnarizine (10^{-7} and 10^{-6} *M*) and by nifedipine (10^{-6} *M*). In all cases, effects of the calcium entry-blocking agents were assessed after a 90-min preincubation period. *Vertical bars* represent SEM of at least six determinations. (GODFRAIND and MILLER 1983)

active isomer, enhanced [^{3}H] nitrendipine binding to pig coronary arteries, whereas *l-cis*-diltiazem was inactive (DE POVER et al. 1982). The stimulatory effects of *d-cis*-diltiazem on binding occurred in the same concentration range as its relaxant effect on depolarized coronary arteries.

In the absence of extracellular calcium or when the calcium channels are completely blocked by lanthanum, a contraction can still be evoked by various ago-

Table 2. Concentration of dihydropyridines producing 50% of the maximum reduction of the contractile response (IC_{50}) or Ca influx (I_{50}) in rat aorta, of maximum [^{3}H] nitrendipine binding (K_i) in rat aorta light microsomal fraction

Ca antagonist		IC_{50}(n*M*)	I_{50}(n*M*)	K_i(n*M*)
Nifedipine	Norepinephrine	19	17	
	Clonidine	5.3	7.2	
	KCl 2 min	4.2		
	KCl 35 min	1.3	1.6	
	Binding			4.5
(−)-Nimodipine	$PGF_{2\alpha}$	0.73	0.9	
	KCl 2 min	0.8	1.1	
	KCl 35 min	0.2		
	Binding			1.2
Nisoldipine	Norepinephrine	0.82	0.62	
	KCl 2 min	0.41	0.74	
	KCl 35 min	0.03		
	Binding			0.93

nists. It has been proposed for a long time that this is due to the release of calcium from intracellular stores (GODFRAIND and KABA 1969). Recent experiments show that this release is blocked both in intact and skinned muscles by high concentrations of diltiazem (SAIDA and VAN BREEMEN 1983). It is likely associated with ER.

D. Heterogeneity of Excitation–Contraction Coupling Mechanisms

When the action of calcium entry blockers is studied in a wide variety of muscles, great variations are observed in the degree of inhibition of the contraction. For instance, in earlier studies, we have reported that the contraction evoked by norepinephrine in rabbit aorta is insensitive to cinnarizine, which is a potent inhibitor of the response of rabbit mesenteric artery (BROEKAERT and GODFRAIND 1979; GODFRAIND et al. 1968). More recent studies (CAUVIN et al. 1982; GODFRAIND 1983) with flunarizine, nifedipine, and diltiazem have confirmed these observations of differences between aorta and mesenteric arteries stimulated by norepinephrine or K depolarization (Fig. 5). Differences are also found between cardiac and smooth muscles. IC_{50} values (concentrations producing 50% inhibition of contraction) for nifedipine are 1 n*M* in rat aorta and 2700 in cat papillary muscle; corresponding values for flunarizine are respectively 220 and >21 000 n*M* (VAN NUETEN 1982). A similar observation has been done by comparing human heart and coronary artery sensitivity to nifedipine (GODFRAIND et al. 1984).

These examples are among those which clearly establish the existence of tissue selectivity for pharmacologic control of excitation–contraction coupling (GODFRAIND 1985). We may summarize the various factors involved in these differences by considering the influence of the stimulus, the nature of the inhibitor, and

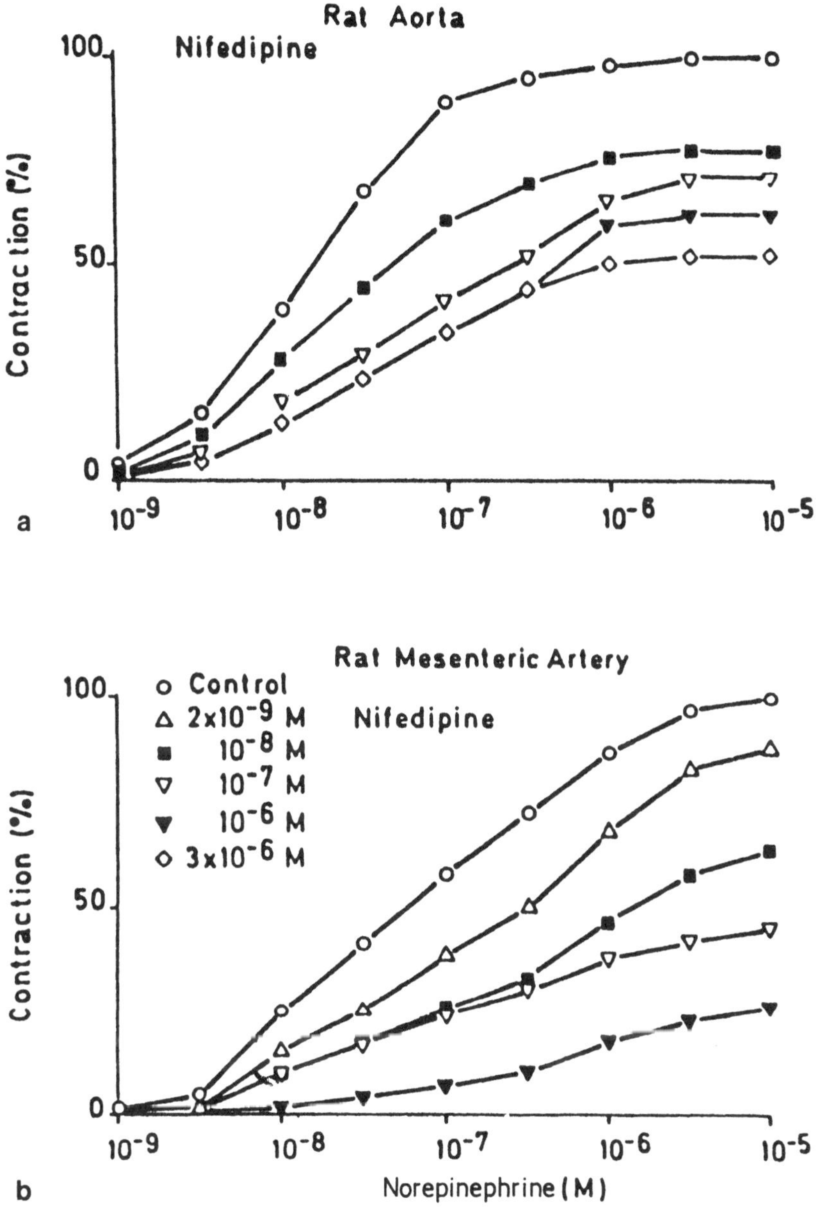

Fig. 5 a, b. Cumulative concentration–response curves in rat aorta (**a**) and superior mesenteric arteries (**b**) to norepinephrine in the absence (*open circles*) and presence of nifedipine at the concentrations indicated. Tissues were incubated with nifedipine for 90 min before contractions were induced with norepinephrine. Each curve is the mean of at least four determinations. (GODFRAIND 1983)

the characteristics of the tissue. This will result in a characterization of the various factors responsible for the increase in the concentration of intracellular activator calcium. This can be produced by the activation of three main pathways: calcium channels, release from intracellular stores, and entry through the Na,Ca exchanger. The first pathway has already been discussed. Calcium release from intracellular stores operates in smooth muscle and in cardiac muscle. In cardiac muscle, this release operates as an amplification mechanism known as calcium-induced calcium release, such that a small amount of calcium entering through the slow Na,Ca channel triggers a release of a much larger quantitiy from the sarcoplasmic reticulum (FABIATO and FABIATO 1979). In smooth muscle, the mechanism by which several agonists are able to release calcium from the ER is not yet established, although there are indications that activation of phosphatidylinositol-4,5-bisphosphate breakdown, leading to inositol-1,4,5-trisphosphate (IP_3) may occur during receptor activation by agonists. IP_3 releases calcium from intracellular store sites (SOMLYO et al. 1985; SUEMATSU et al. 1984). This mechanism is insensitive to calcium entry blockers. It is obvious from experiments with calcium entry blockers that the magnitude of releasable intracellular calcium varies between vessels. This can be identified by measuring the proportion of the agonist-evoked contraction that is resistant to calcium entry blockade. For instance, 100% is resistant in rabbit aorta, 50% in rat aorta, 10% in rat mesenteric arteries, and 0% in human coronary arteries (GODFRAIND et al. 1984).

The Na,Ca antiporter that contributes to calcium extrusion from cardiac cells operates electrogenically and could also contribute to calcium entry during depolarization (REUTER 1982). We have already shown that the contribution of this system is weak in smooth muscle. However, it could be that the density of Na,Ca antiporters could be variable in the plasma membrane of smooth muscle cells. In some smooth muscles, Na substitution by Li has been reported to lead to an increase in muscle tension; these observations have been interpreted on the basis of the Na,Ca exchange model (BRADING 1981). Aortas isolated from normal Wistar rats do not contract in low Na solutions, which also fail to increase their ^{45}Ca influx significantly. These data are in agreement with the observations done with microsomal fractions, suggesting that Ca extrusion in aorta is not predominantly dependent on the Na gradient. Interestingly, the degree of participation of the Na,Ca exchange in Ca homeostasis may vary according to the origin of the tissue, as indicated by the observation that a high Li solution causes contraction and enhances ^{45}Ca entry in aortas isolated from SHR rats (T. GODFRAIND and N. MOREL, unpublished work).

E. A Pharmacologic Example: Contraction of Vascular Smooth Muscle, Role of Endothelium, and Action of Dihydropyridines and Diphenylpiperazines

The simplest experimental procedure to assess the relation between calcium entry and the development of contraction is to preincubate a smooth muscle in Ca-free physiologic solution, to depolarize it with Ca-free KCl-rich solution, and gradually to increase the calcium concentration in the bathing solution. This will result in an increase in tension, dependent on Ca^{2+} concentration (GODFRAIND and KABA 1969). When an agonist is used to stimulate the tissue, the calcium dependency of the response may also be examined. It must be pointed out that the response of a vessel to vasoconstrictors may depend on the interaction of the agonist with its receptors, but that it may be modulated by factors extrinsic to the muscle cell. Some of those factors may depend on the endothelium. The essential role of endothelium as a mediator of relaxant responses induced in isolated vascular tissues by acetylcholine and other vasodilators is now well established (FURCHGOTT 1984). We have examined the role of the endothelial tissue as a modulator of contractile effects of agonists on vascular smooth muscle. We first examined the influence of endothelium of the contractile response evoked by the adrenoceptor agonists norepinephrine and clonidine in rat isolated aorta (EGLÈME et al. 1984). Norepinephrine concentration–effect curves obtained in the absence of endothelium were shifted to the left as compared with the concentration–effect curves obtained in the presence of endothelium. A similar shift was observed with other agonists, such as serotonin, PGF_2, and phenylephrine. On the other hand, removal of endothelium evoked a dramatic effect on the contractile responsiveness of rat aorta to clonidine. The maximum response increased by a factor of 9, and the EC_{50} value decreased. A similar observation has been made with other α_2-agonists such as UK 14-304-18 and oxymetazoline. It therefore appears that endothelium reduces the action of vasoconstrictors (GODFRAIND et al. 1985a, b).

I. Inhibition of Contraction

The action of dihydropyridines and diphenylpiperazines has been examined in depolarized aorta and in aorta stimulated by adrenoceptor agonists in the presence and the absence of endothelium. In depolarized arteries, different experimental protocols have been followed. Calcium dose–effect curves have been established in the presence of various concentrations of nifedipine, nimodipine (the racemic mixture and the two isomers), cinnarizine and flunarizine, the results were similar except that the potency was different between these compounds. They produced a slight shift to the right and a progressive depression of the maximum response, mainly noticeable at high doses. Such an effect is graphically similar to the interaction of some agonists with antagonists and justifies the use of the term "calcium antagonist" to characterize those drugs (Fig. 6).

Another protocol consisted in the analysis of the relaxing action of diphenylpiperazines and dihydropyridines on the contraction evoked by a K-depolarizing solution containing 100 mM KCl and 1.25 mM $CaCl_2$. In such experiments, it was observed that relaxation is dose dependent (GODFRAIND and MILLER 1983) and that the contraction can be completely abolished. In this respect, the pioneer experiments performed with cinnarizine and lidoflazine (GODFRAIND and POLSTER 1968) have been confirmed, since the relaxation is reduced by increasing the concentration of calcium in the perfusion fluid. In this series of experiments, differences between drugs were not only related to concentration, but also to the rate at which relaxation was achieved, as shown by a comparison between nifedipine and flunarizine (Fig. 7).

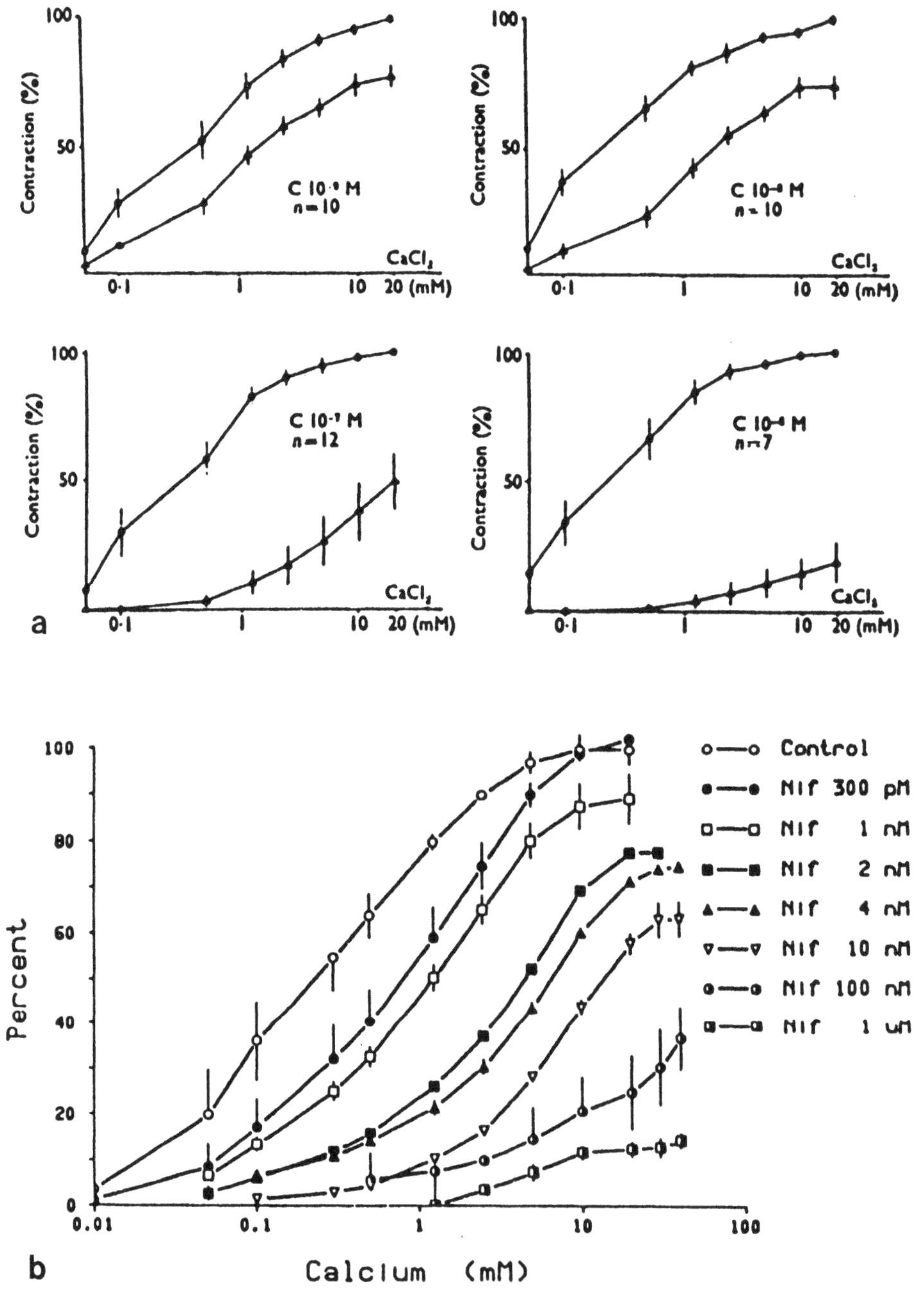

Fig. 6 a, b. Illustration of the action of Ca^{2+} antagonists on Ca^{2+}-evoked contraction in depolarized vascular smooth muscle, showing typical families of dose–effect curves that can be observed in two different tissue preparations. The difference is not due to the drug used, but to the tissue. **a** cumulative dose–effect curves evoked by Ca^{2+} in K^+-depolarized mesenteric arteries before (*open circles*) and after (*open triangles*) the addition of cinnarizine at the concentrations indicated, **b** cumulative dose–effect curves evoked by Ca^{2+} in K^+-depolarized rat aorta before and after the addition of nifedipine at the concentrations indicates. (GODFRAIND 1985)

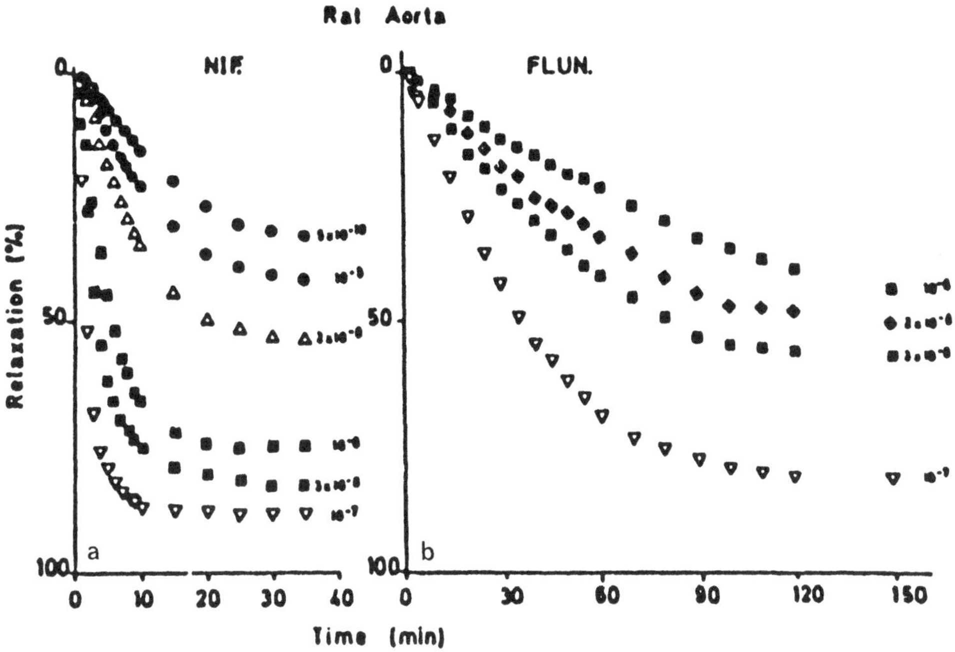

Fig. 7 a, b. Time-dependent relaxation of depolarization (100 mM K^+)-stimulated maximal contractions of rat aorta produced by nifedipine (**a**) and flunarizine (**b**). Nifedipine and flunarizine, at the final concentrations indicated, were added to the bath at time zero and relaxation was followed for the next 150 min. Points at each concentration are the means of at least three determinations. (GODFRAIND and MILLER 1983)

The third protocol is illustrated in Fig. 8 for three dihydropyridines. Contraction of rat aorta was evoked by K-depolarizing solution containing 100 mM KCl and 1.25 mM $CaCl_2$, then the preparations were incubated in various concentrations of nifedipine, nimodipine, and nisoldipine, as shown in Fig. 8. The time course of the contraction was different in controls and in treated preparations. For example, the contraction in the presence of 300 mM nisoldipine shows an initial phasic response followed by a secondary plateau; such a contractile pattern was initially described by GODFRAIND and DIEU (1981) for flunarizine. Since the inhibitor effect increased with the duration of the K depolarization, we have proposed that it could be attributed to a (use)–time–voltage-dependent process. Indeed, this increased inhibitory effect can not be attributed to an insufficient preincubation time with the drug (it was 90 min). After washout of the K solution by physiologic solution containing nisoldipine, a second cycle of K depolarization-evoked contraction shows the same pattern. The increase in IC_{50} resulting from this use dependence follows the order nisoldipine > nimodipine > nifedipine (see also Table 2). This increase in the inhibitory potency of calcium entry blockers with the duration of the depolarization has been interpreted assuming that, during prolonged depolarization, calcium channels are inactivated and, in this state, show a higher affinity for the blockers (GODFRAIND 1986 a).

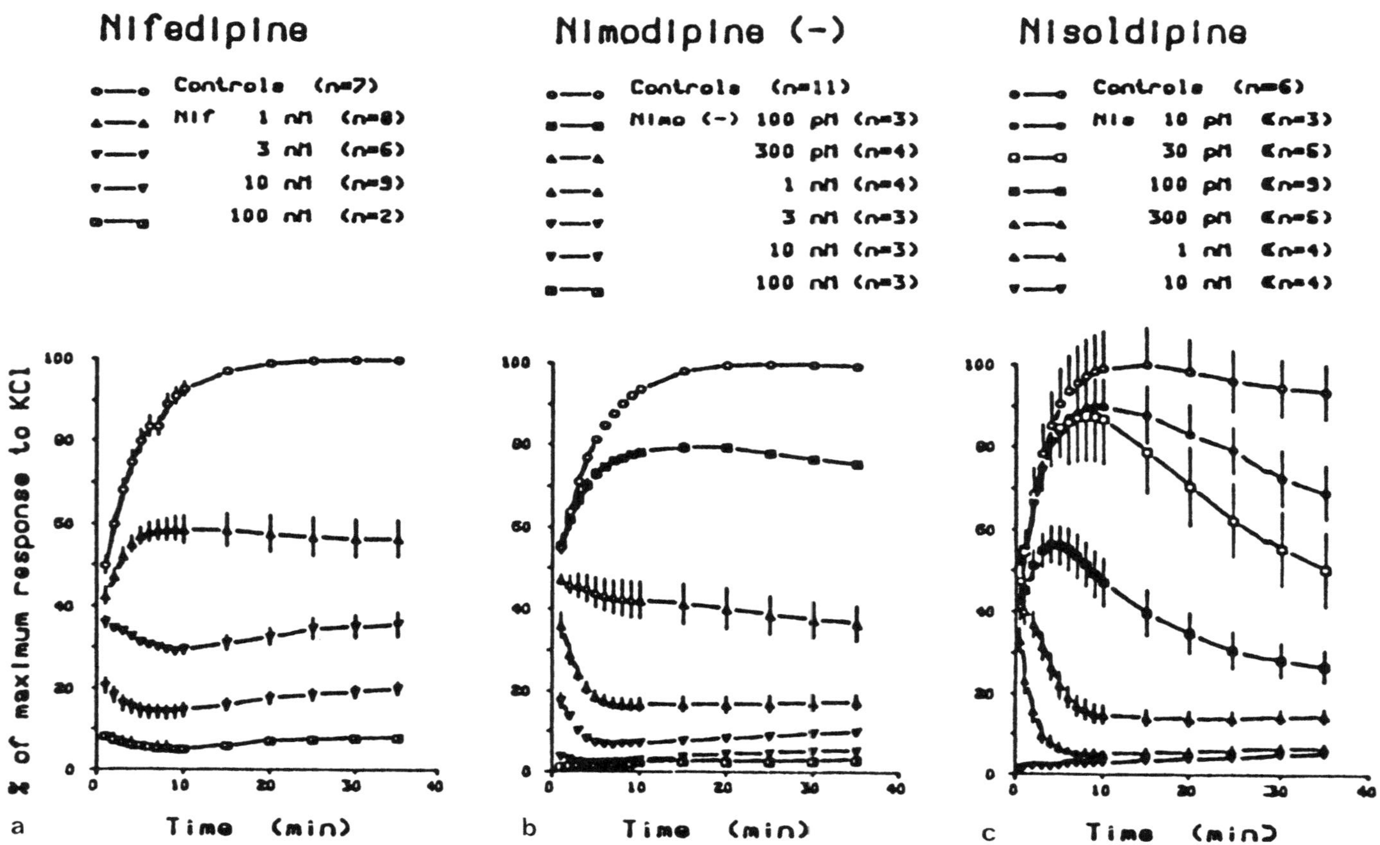

Fig. 8 a–c. Time course of the contraction of rat aorta evoked by K depolarization before (control) and after 90-min preincubation with various concentrations of nifedipine (**a**), (−)-nimodipine (**b**), or nisoldipine (**c**). Note use-dependence mainly observable with nisoldipine. (GODFRAIND et al. 1985b)

II. Role of Endothelium

The action of dihydropyridines has been studied in rat aorta with and without endothelium, stimulated by α_1- and α_2-agonists. It has been previously shown by GODFRAIND et al. (1985a) that the maximal inhibition of the response to α_2-agonists by calcium entry blockers was much greater than was the maximal inhibition of norepinephrine and phenylephrine response. As Fig. 9 illustrates, in the absence of endothelium, the portion of the contraction resistant to a maximum effective concentration of nifedipine is lower with clonidine than with norepinephrine. In addition, the norepinephrine response resistant to nifedipine is influenced by the presence of endothelium. With submaximum concentrations of norepinephrine, close to those likely obtained after endogenous release, the enhanced response recorded in the absence of endothelium is highly sensitive to nifedipine, but the response resistant to calcium entry blockers is also increased (Fig. 10).

The calcium agonist Bay K 8644 evokes a vasoconstriction in vivo; as shown by SCHRAMM et al. (1983), it can evoke the contraction of a vascular smooth muscle only when the K concentration of the medium is increased from 6 to 12 m*M* in order to depolarize the preparation slightly. In aortas without endothe-

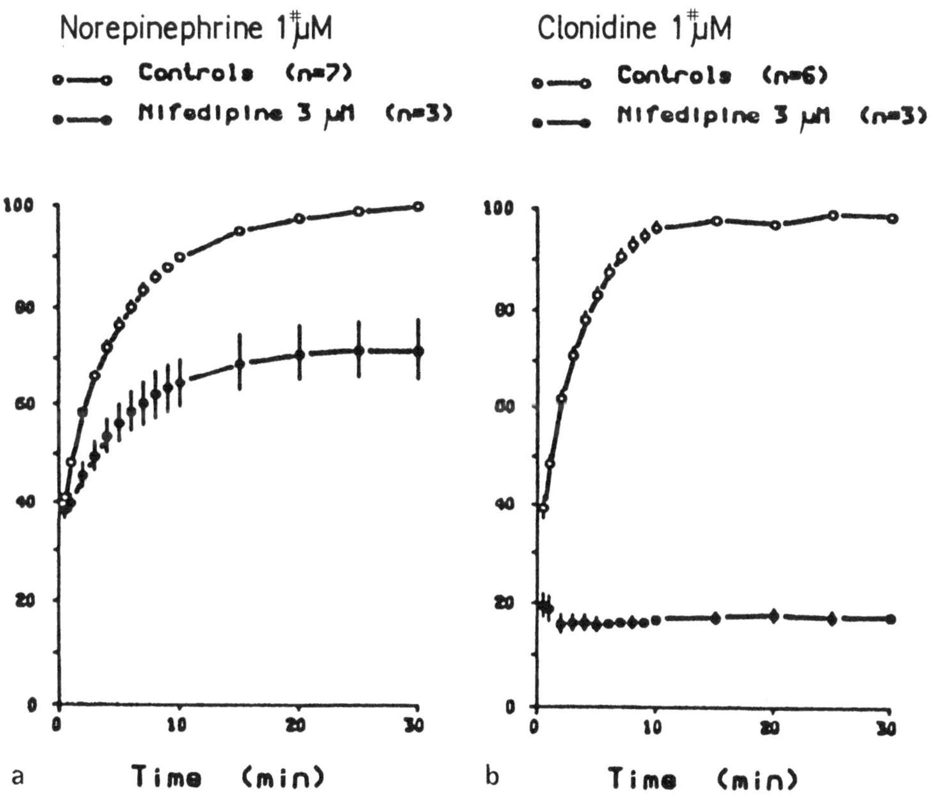

Fig. 9 a, b. Time course of the contraction evoked by (**a**) norepinephrine and (**b**) clonidine in rat aorta without endothelium, before and after treatment with a maximum effective concentration of nifedipine. (GODFRAIND et al. 1985b)

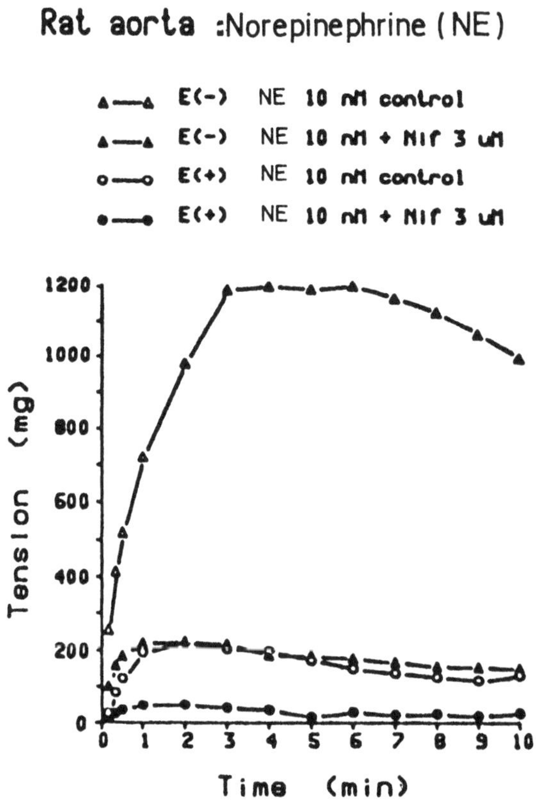

Fig. 10. Response of rat aorta to 10 n*M* norepinephrine, with (E+) and without (E−) endothelium; inhibition by 3 µ*M* nifedipine. (GODFRAIND et al. 1985b)

lium, Bay K 8644 evokes a contraction in the presence of 6 m*M* KCl and therefore does not require a KCl concentration above the physiologic level. This confirms the existence of an endothelial modulation of the contractile response of the adjacent smooth muscle (GODFRAIND et al. 1985a, b).

It has been reported that stimulation of receptors located on endothelial cells release a factor (EDRF) that activates the guanylate cyclase activity of adjacent smooth muscle cells, leading to an increase of intracellular cGMP content. Methylene blue, known as an inhibitor of guanylate cyclase, inhibits the relaxing effect of acetylcholine and of other agents, producing an endothelium-dependent relaxation of contracted smooth muscle. On the other hand, bromo-cGMP has an opposite effect to methylene blue (ALOSACHIE and GODFRAIND 1986). Experimental data show that cGMP attenuates the importance of norepinephrine-dependent calcium influx, indicating that this nucleotide could modulate the opening of ROCs (GODFRAIND 1986b). MEISHERI and VAN BREEMEN (1982) have shown that cAMP could inhibit the opening of POCs. Although these studies are in their early stages, they allow the conclusion that cyclic nucleotides control opening of calcium channels in the membrane. Therefore, calcium movements in smooth muscle may be influenced by β-adrenoceptor agonists and by inhibitors of phos-

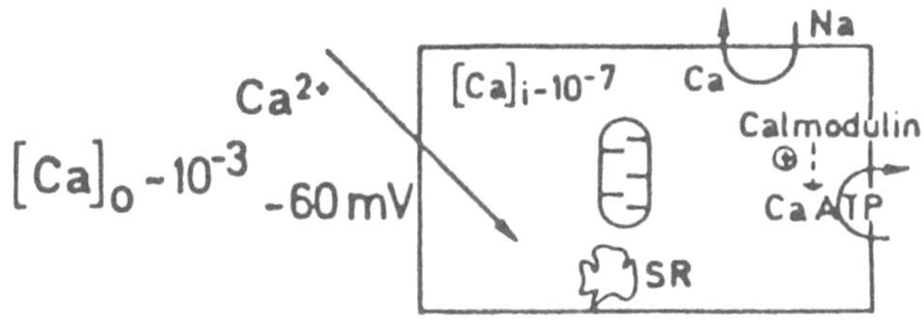

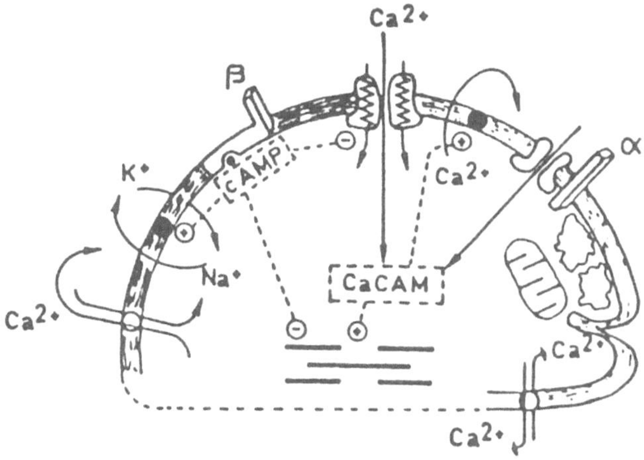

Fig. 11 a, b. Schematic representation of cellular processes controlling cellular Ca^{2+} metabolism in smooth muscle at rest (**a**) and during stimulation (**b**). **a** although the Ca^{2+} electrochemical gradient is oriented inward, the low intracellular Ca^{2+} concentration is maintained by the low permeability of the cell membrane to Ca^{2+} and by the activity of Ca^{2+} buffering systems described in the text. The various mechanisms illustrated here are insensitive to specific Ca^{2+} entry blockers, **b** illustration of the various processes activated by smooth muscle stimulation. From left to right: (1) Na^+,Ca^{2+} exchange mechanism; (2) Na^+,K^+ pump, (3) β-adrenoceptor coupled to adenylate cyclase responsible for the cyclization of ATP into cAMP – the latter has a stimulatory action on the Na^+,K^+ pump, a negative action on the contractile machinery, and has been proposed to impair the opening of the potential-operated channels; (4) potential-operated Ca^{2+} channels. Their opening is dependent on the level of the membrane potential. Once they are open, Ca^{2+} entry occurs as a function of the Ca^{2+} electrochemical gradient. Inside the cell, Ca^{2+} forms a complex with calmodulin (CaCAM). This complex activates the contractile machinery and the Ca^{2+} extrusion pumps; (5) ATP-dependent Ca^{2+} pump stimulated by calmodulin; (6) α-adrenoceptor associated with receptor-operated channel and with intracellular Ca^{2+} stores, most likely the endoplasmic reticulum. Once the channel is open, Ca^{2+} entry is also passive (see 4). The association formed by mitochondria and endoplasmic reticulum is responsible for intracellular Ca^{2+} buffering. (8) Ca^{2+},Ca^{2+} exchange mechanism occurring through a leak channel insensitive to Ca^{2+} entry blockers. Explanations are given in the text. (GODFRAIND 1985)

phodiesterase. On the other hand, LINCOLN (1983) has suggested that increased levels of cGMP might enhance calcium uptake into sarcoplasmic reticulum, a mechanism that could contribute to its relaxant effect on norepinephrine-evoked contraction.

F. Concluding Remarks

The various factors controlling cellular metabolism in smooth muscle are summarized in Fig. 11. Several pharmacologic agents affect calcium movements in smooth muscle cell by an action that can be located on different subcellular sites (see Table 1). Inhibitors of calcium movements have an inhibitory effect on the contraction evoked by agonists and by depolarization. Facilitators of calcium movements may produce a contraction in the absence of another agent when the calcium supply is high enough to activate the contractile machinery.

Among inhibitors, calcium entry blockers are of considerable interest for the treatment of several cardiovascular and neurologic diseases, including angina pectoris, hypertension, migraine, etc. In this group, dihydropyridines show a great potential value since small modifications of their structures change their interaction with calcium channels, giving calcium channel opening or blockade (GODFRAIND et al. 1986).

References

Alosachie I, Godfraind T (1986) Role of cyclic GMP in the modulation by endothelium of the adrenolytic action of prazosin in rat isolated aorta. Br J Pharmacol 80:525–532

Baker PF, Blaustein MP, Hodgkin AL, Steinhard RA (1969) The influence of Ca on Na efflux in squid axons. J Physiol (Lond) 200:431–458

Batra S (1985) Characterization of [^{3}H]-nitrendipine binding to uterine smooth muscle plasma membrane and its relevance to inhibition of calcium entry. Br J Pharmacol 85:767–774

Bolger GT, Gengo PJ, Luchowski EM, Siegel H, Triggle DJ, Janis RA (1982) High affinity binding of a calcium channel antagonist to smooth and cardiac muscle. Biochem Biophys Res Commun 104:1604–1609

Bolger GT, Gengo P, Klockowski R, Luchowski E, Siegel H, Janis RA, Triggle AM, Triggle DJ (1983) Characterization of binding of the Ca^{++} channel antagonist, [^{3}H]nitrendipine, to guinea-pig ileal smooth muscle. J Pharmacol Exp Ther 225:291–309

Bolton TB (1979) Mechanisms of action of transmitters and other substances on smooth muscle. Physiol Rev 59:606–718

Brading AF (1981) Ionic distribution and mechanism of transmembrane ion movements in smooth muscle. In: Bülbring EE, Brading AF, Jones AW, Tomita T (eds) Smooth muscle. Arnold, London, pp 65–92

Broekaert A, Godfraind T (1979) A comparison of the inhibitory effect of cinnarizine and papaverine on the noradrenaline and calcium-evoked contraction of isolated rabbit aorta and mesenteric arteries. Eur J Pharmacol 53:281–288

Caroni P, Carafoli E (1983) The regulation of the Na^+-Ca^{2+} exchanger of heart sarcolemma. Eur J Biochem 132:451–460

Caroni P, Reinlib L, Carafoli E (1980) Charge movements during the Na^+-Ca^{2+} exchange in heart sarcolemmal vesicles. Proc Natl Acad Sci (USA) 77:6354–6358

Casteels R, van Breemen C (1975) Active and passive Ca^{2+} fluxes across cell membranes of the guinea-pig taenia coli. Pflügers Arch 359:197–207

Cauvin C, Saida K, van Breemen C (1982) Effects of Ca antagonists on Ca fluxes in resistance vessels. J Cardiovasc Pharmacol 4:S287–290
De Pover A, Matlib MA, Lee SW, Dube GP, Grupp IL, Grupp G, Schwartz A (1982) Specific binding [^{3}H] nitrendipine to membranes from coronary arteries and heart in relation to pharmacological effects. Paradoxical stimulation by diltiazem. Biochem Biophys Res Commun 108:110–117
Egleme C, Godfraind T, Miller RC (1984) Enhanced responsiveness of rat isolated aorta to clonidine after removal of the endothelial cell. Br J Pharmacol 81:16–18
Fabiato A, Fabiato F (1979) Calcium and cardiac excitation contraction coupling. Ann Rev Physiol 41:473–484
Fleckenstein A (1973) Calcium antagonism in heart and smooth muscle. Wiley, New York
Furchgott RF (1984) The role of endothelium in the responses of vascular smooth muscle to drugs. Ann Rev Pharmacol Toxicol 24:175–197
Godfraind T (1976) Calcium exchange in vascular smooth muscle, action of noradrenaline and lanthanum. J Physiol (Lond) 260:21–35
Godfraind T (1983) Actions of nifedipine on calcium fluxes and contraction in isolated rat arteries. J Pharmacol Exp Ther 224:443–450
Godfraind T (1985) Cellular and subcellular approaches to the mechanism of action of calcium antagonists. In: Rubin RP, Weiss GB, Putney JW (eds) Calcium in biological systems. Plenum, New York, pp 411–421
Godfraind T (1986a) Calcium entry blockade and excitation contraction coupling in the cardiovascular systems (with an attempt of pharmacological classification). Acta Pharmacol Toxicol 58 (Suppl 2):5–30
Godfraind T (1986b) EDRF and cyclic GMP control gating of receptor-operated calcium channels in vascular smooth muscle. Eur J Pharmacol 126:341–343
Godfraind T, Dieu D (1981) The inhibition by fluanrizine on the norepinephrine evoked contraction and calcium influx in rat aorta and mesenteric arteries. J Pharmacol Exp Ther 217:510–515
Godfraind T, Kaba A (1969) Blockade or reversal of contraction induced by calcium and adrenaline in depolarized arterial smooth muscle. Br J Pharmacol 36:549–560
Godfraind T, Miller RC (1983) Specificity of action of Ca^+ entry blockers. A comparison of their actions in rat arteries and in human coronary arteries. Circ Res 52 (Suppl 1):81–91
Godfraind T, Polster P (1968) Etude comparative de médicaments inhibant la réponse contractile de vaisseaux isolés d'origine humaine et animale. Thérapie 23:1209–1220
Godfraind T, Wibo M (1985) Subcellular localization of [^{3}H]-nitrendipine binding sites in guinea-pig ileal smooth muscle. Br J Pharmacol 85:335–340
Godfraind T, Kaba A, Polster P (1968) Differences in sensitivity of arterial smooth muscle to inhibition of their contractile response to depolarization by potassium. Arch Int Pharmacodyn Ther 172:235–239
Goodfraind T, Finct M, Socrates Lima J, Miller R (1984) Contractile activity of human coronary arteries and human myocardium in vitro and their sensitivity to calcium entry blockade by nifedipine. J Pharmacol Exp Ther 230:514–518
Godfraind T, Egleme C, Alosachie I (1985a) Role of endothelium in the contractile response of rat aorta to alpha-adrenergic agonists. Clin Sci 68:65s–71s
Godfraind T, Egleme C, Wibo M (1985b) Effects of dihydropyridines on human and animal isolated vessels. In: Fleckenstein A, Breemen C van, Gross R, Hoffmeister F (eds) Cardiovascular effects of dihydropyridine-type calcium antagonists and agonists. Springer, Berlin Heidelberg New York Tokyo, pp 309–325 (Bayer symposium, vol 9)
Godfraind T, Miller RC, Wibo M (1986) Calcium antagonism and calcium entry blockade. Pharmacol Rev 38:321–416
Godfraind-De Becker A, Godfraind T (1980) Calcium transport system: a comparative study in different cells. Int Rev Cytol 67:141–170
Grover AK, Oakes PJ (1985) Calcium channel antagonist binding and pharmacology in rat uterine smooth muscle. Life Sci 37:2187–2192

Hidaka H, Hartshorne DJ (1985) Calmodulin antagonists and cellular physiology. Academic, Orlando, p 543
Hirata M, Itoh T, Kuriyama H (1981) Effects of external cations on calcium efflux from single cells of guinea-pig taenia coli and porcine coronary artery. J Physiol (Lond) 310:321–336
Jim K, Harris A, Rosenberger LB, Triggle DJ (1981) Stereoselective and non-stereoselective effects of D 600 (methoxyverapamil) in smooth muscle preparations. Eur J Pharmacol 76:67–72
Lincoln TM (1983) Effects of nitroprusside and 8-bromo-cyclic GMP on the contractile activity of the rat aorta. J Pharmacol Exp Ther 224:100–107
Meisheri KD, van Breemen C (1982) Effects of beta-adrenergic stimulation on calcium movements in rabbit aortic smooth muscle: relation with cyclic AMP. J Physiol (Lond) 331:429–441
Meisheri KD, Hwang O, van Breemen C (1981) Evidence for two separate Ca^{2+} pathways in smooth muscle-plasmalemma. J Membr Biol 59:19–25
Morel N, Godfraind T (1982) Na-Ca exchange in heart and smooth muscle microsomes. Arch Int Pharmacodyn Ther 258:319–321
Morel N, Godfraind T (1984) Sodium/Calcium exchange in smooth-muscle microsomal fractions. Biochem J 218:421–427
Morel N, Wibo M, Godfraind T (1981) A calmodulin-stimulated Ca^{2+} pump in rat aorta plasma membranes. Biochim Biophys Acta 644:82–88
Philipson KD, Nishimoto AY (1980) Na^{+}-Ca^{2+} exchange is affected by membrane potential in cardiac sarcolemmal vesicles. J Biol Chem 255:6880–6882
Pitts BJ (1979) Stoichiometry of sodium-calcium exchange in cardiac sarcolemmal vesicles. J Biol Chem 254:6232–6235
Reeves JP, Sutko JL (1979) Sodium-calcium ion exchange in cardiac membrane vesicles. Proc Natl Acad Sci (USA) 76:590–594
Reuter H (1974) Exchange of calcium ions in the mammalian myocardium. Mechanisms and physiological significance. Circ Res 34:599–609
Reuter H (1982) Na-Ca countertransport in cardiac muscle. In: Martonosi A (ed) Membrane and transport, vol 1. Plenum, New York, pp 623–631
Saida K, van Breemen C (1983) Mechanism of Ca^{2+} antagonist-induced vasodilation. Intracellular actions. Circ Res 52:137–142
Schatzmann HJ, Vincenzi FF (1969) Calcium movements across the membrane of human red cells. J Physiol (Lond) 201:369
Schramm M, Thomas G, Towart R, Franckowiak G (1983) Novel dihydropyridines with positive inotropic action through activation of Ca^{2+} channels. Nature 303:535–537
Somlyo AV, Bond M, Somloy AP, Scarpa A (1985) Inositol trisphosphate-induced calcium release and contraction in vascular smooth muscle. Proc Natl Acad Sci USA 82:5231
Suematsu E, Hirata M, Hashimoto T, Kuriyama H (1984) Inositol 1,4,5-trisphosphate releases Ca^{2+} from intracellular store sites in skinned single cells of porcine coronary artery. Biochem Biophys Res Commun 120:481
Triggle CR, Agrawal DK, Bolger GT, Daniel EE, Kwan CY, Luchowski EM, Triggle DJ (1982) Calcium channel antagonist binding to isolated vascular smooth muscle membranes. Can J Physiol Pharmacol 60:1738–1741
Van Nueten JM (1982) Selectivity of calcium entry blockers. In: Godfraind T, Albertini A, Paoletti R (eds) Calcium modulators. Elsevier, Amsterdam, pp 199–208
Venter JC, Fraser CM, Schaber JS, Jung CY, Bolger S, Triggle DJ (1983) Molecular properties of the slow inward calcium channel. Molecular weight determinations by radiation inactivation and covalent affinity labeling. J Biol Chem 258:9344–9348
Wibo M, Morel N, Godfraind T (1981) Differentiation of Ca^{2+} pumps linked to plasma membrane and endoplasmic reticulum in the microsomal fraction from intestinal smooth muscle. Biochim Biophys Acta 649:651–660
Williams JA (1980) Regulation of pancreatic acinar cell function by intracellular calcium. Am J Physiol 238:G269–G279
Winquist RJ (1984) Modulators of intracellular calcium. Drug Dev Res 4:241–256

CHAPTER 19

Drugs Affecting Cardiac Calcium Metabolism

P. HONERJÄGER

A. Introduction

The electrical, mechanical, and metabolic activities of the heart are either directly controlled or profoundly affected by transmembrane movements and the intracellular concentration of calcium ions. This important role of the calcium ion is reflected by the fact that most of the drugs that affect these functions do so via more or less direct effects on calcium metabolism of the cardiac cells. The present chapter reviews this class of cardioactive drugs which comprises both clinically used agents, such as the positive inotropic drugs, antiarrhythmic, and some antianginal drugs, and experimentally used substances which can help in clarifying physiologic and biochemical aspects of cardiac calcium metabolism. While it is impossible to present a detailed pharmacologic description of all drugs that affect cardiac calcium metabolism within the space of a handbook chapter, it is attempted to present: (a) a rather comprehensive list of the various subcellular sites and mechanisms of action of prototypical compounds; and (b) a discussion of how this primary action relates to the drug effect on cellular and global heart function.

B. The Calcium Signal

One of the most challenging problems posed by cardiac calcium metabolism has been the measurement, in the presence of 2 m*M* extracellular calcium, of the rapid rise and subsequent decline of the free cytosolic calcium concentration from submicromolar to micromolar levels and back. This triggers, by interaction with troponin C, the mechanical interaction of actin and myosin. This central event of the cardiac cycle, termed the calcium transient or calcium signal, serves to couple sarcolemmal excitation to contraction and is a major determinant of the force of myocardial contraction and of the myocardial ATP consumption resulting from myosin ATPase activity and Ca-ATPase activity of the sarcoreticular pump. Only after the successful detection of the calcium signal by use of the calcium-sensitive photoprotein aequorin (BLINKS et al. 1982) has it been possible to determine unequivocally whether inotropic drug effects can be related to parallel changes of the calcium transient or whether a change in sensitivity to intracellular calcium of the contractile proteins has to be inferred.

As shown by Fig. 1, the positive inotropic effect of a cardiotonic steroid, strophanthidin, can be related to a corresponding change of the calcium transient in cat papillary muscle, and the same is true for the negative inotropic effect of a cal-

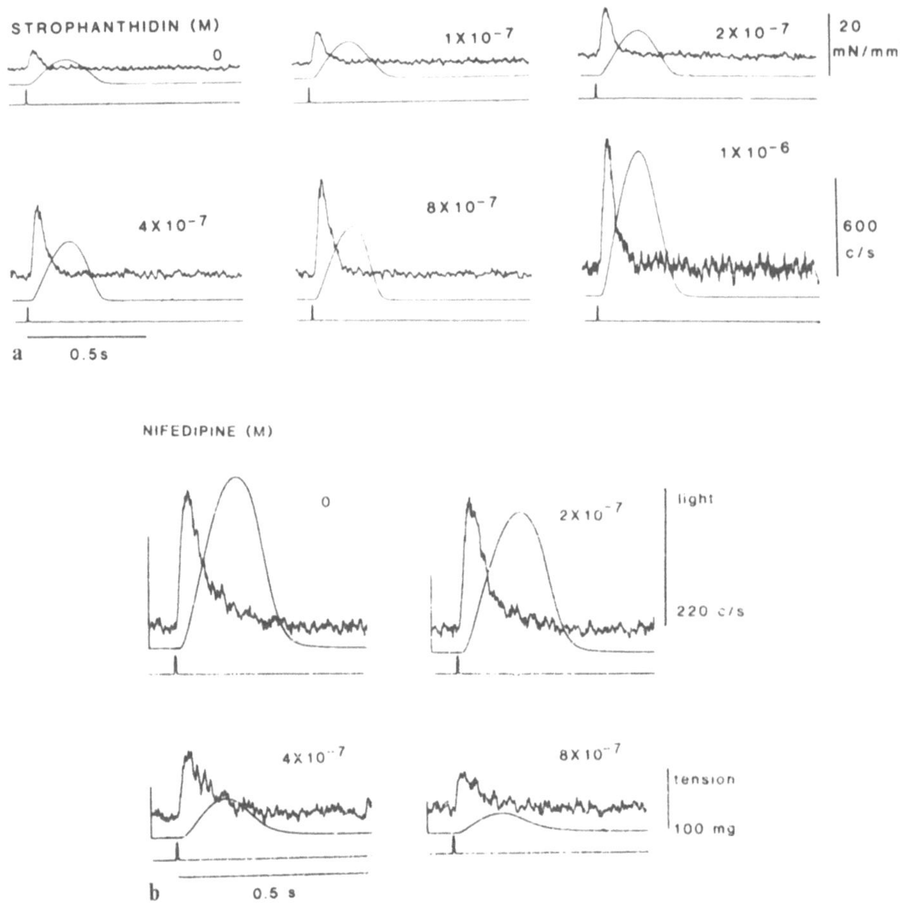

Fig. 1 a, b. Positive and negative inotropic drug effects resulting from changes of the intracellular calcium signal (calcium transient) as detected by light emission from intracellularly injected aequorin. Influence of cumulatively increasing concentrations of strophanthidin (**a**) or nifedipine (**b**) on isometric force of contraction and aequorin signal (noisy traces) of cat papillary muscle stimulated at 4-s intervals. Stimulus marked on lowest trace; *c/s* denotes photon counts per second. **a** MORGAN and BLINKS (1982); **b** MORGAN et al. (1983) by permission of the authors, the Canadian Journal of Physiology and Pharmacology, and the American Heart Association, Inc.

cium channel inhibitor, nifedipine. The importance of such direct information is stressed by Table 1 which lists those cardioactive drugs whose effect on force of contraction and calcium transient has been simultaneously measured by the aequorin technique. These findings support the idea that most of the well-known inotropic drugs exert their effect primarily through an effect on intracellular calcium. A notable exception is provided by the metabolic inhibitor cyanide (or hypoxia) which decreases the force of contraction without changing the size of the aequorin light signal (ALLEN and ORCHARD 1983), probably because the associ-

Table 1. Survey of positive and negative inotropic drugs whose inotropic effect can be related to a corresponding alteration of the intracellular calcium transient in isolated ventricular myocardium, as measured with intracellularly injected aequorin

Cardioactive drug	Species	Reference
A. Positive inotropic drugs which increase the calcium transient		
Epinephrine	Cat	ALLEN and KURIHARA (1980)
Norepinephrine	Cat	MORGAN and BLINKS (1982)
Isoprenaline	Cat	BLINKS et al. (1980)
Dopamine	Rabbit	ENDOH and BLINKS (1984a)
Glucagon	Cat	BLINKS et al. (1980)
Cholera toxin	Cat	BLINKS et al. (1980)
Forskolin	Ferret	GWATHMEY and MORGAN (1985)
Dibutyryl-cAMP	Ferret	GWATHMEY and MORGAN (1985)
Theophylline	Cat	MORGAN and BLINKS (1982)
Papaverine	Cat	BLINKS et al. (1980)
Isobutylmethylxanthine	Ferret	GWATHMEY and MORGAN (1985)
Amrinone	Cat	MORGAN et al. (1986)
Milrinone	Ferret	GWATHMEY and MORGAN (1985)
Fenoximone	Dog	ENDOH and BLINKS (1984b)
Piroximone	Ferret	GWATHMEY and MORGAN (1985)
OPC 8212	Dog	ENDOH et al. (1986)
Sulmazole	Dog	BLINKS and ENDOH (1984)
Phenylephrine[a] (+bupranolol)	Rabbit	BLINKS and ENDOH (1986)
Strophanthidin	Cat	MORGAN and BLINKS (1982)
Bay K 8644	Dog	ENDOH et al. (1986)
B. Negative inotropic drugs which decrease the calcium transient		
Nifedipine	Cat	
Verapamil	Cat	
Gallopamil	Cat	MORGAN et al. (1983)
Diltiazem	Cat	
Perhexiline	Cat	
Carbachol[b]	Dog	ENDOH et al. (1986)

[a] Minimal increase of the calcium transient.
[b] In the presence of forskolin.

ated intracellular acidification reduces the calcium sensitivity of the contractile proteins. Caffeine prolongs the time to peak force and increases the peak force of contraction of cat ventricular muscle, but decreases the amplitude of the calcium transient (ALLEN and KURIHARA 1980). The latter event correlates with a caffeine-induced delay in the onset of contraction, but the augmented late peak of contraction is probably caused by a drug-induced increase of the calcium sensitivity of the contractile proteins (EISNER and VALDEOLMILLOS 1985). These examples show that it may be misleading to infer changes in intracellular calcium solely on the basis of observed changes in force of contraction. Direct measurement of calcium flux across the various cellular membranes by electrophysiologic, tracer, and other indicator techniques and measurement of the intracellular calcium distribution would ideally be required to fully characterize drug effects on cardiac calcium metabolism.

C. Classification and Selection of Drugs

As in other cells, the control of transmembrane flow and cytosolic concentration of calcium ions by cardiac cells rests on the high affinity complexation of calcium by specific proteins. A drug may influence cellular calcium homeostasis either by direct chemical interaction with the calcium ion (as chelating agent or ionophore) or by direct or indirect modification of one or more of the cellular proteins that regulate calcium metabolism, subsequently referred to as calcium-binding proteins. Most cardioactive drugs belong to this second category.

There are two different functional classes of calcium-binding proteins (CARAFOLI 1985): (a) calcium-*transporting* proteins, which are integral components of cell or organelle membranes; and (b) calcium-*modulated* proteins which are either dissolved in the cell or, like troponin C, part of a nonmembranous intracellular structure. Consequently, drugs affecting cardiac calcium-binding proteins can be subdivided depending on whether they act on calcium-transporting or calcium-modulated proteins. Furthermore, the drug may bind directly to the calcium-binding protein, or the primary drug–receptor interaction may initiate one or more intermediate reactions which finally results in modification of a calcium-binding protein.

Calcium-transporting proteins have been documented in the sarcolemma, the sarcoplasmic reticulum, and mitochondria of heart cells (Fig. 2). The sarcolemma contains voltage-gated calcium channels, a sodium/calcium exchanger, and a calcium-pumping ATPase. The sarcoplasmic reticulum membrane contains a calcium-releasing mechanism (channel?) and a calcium-pumping ATPase. Calcium transport across the inner mitochondrial membrane (the outer membrane is freely permeable for calcium ions and larger solutes) occurs via an electrophoretic uniport and a sodium/calcium exchanger. The major cardiac calcium-modulated proteins are troponin C and calmodulin. On this basis, cardioactive drugs are classified in Table 2 according to their target calcium-binding protein and further subdivided according to the nature of the interaction with this protein. Mechanisms by which drugs may modify the function of calcium-transporting proteins are summarized in Fig. 2. Drugs that selectively modify the sarcolemmal calcium pump or the calcium-modulated proteins calmodulin or troponin C, when applied to intact cardiac tissue, are yet unknown. Although calmodulin antagonists have been developed and studied (HIDAKA and HARTSHORNE 1985), and although, in the heart, calcium-activated calmodulin is thought to stimulate the phosphodiesterase I isozyme and the sarcoreticular calcium pump (WALSH et al. 1980), no specific cardiac drug effects relating to calmodulin antagonism are known. Drugs that increase the affinity of troponin C for calcium would constitute a new class of cardiotonic agents devoid of the danger of inducing the arrhythmogenic and detrimental metabolic effects of calcium overload (HERZIG 1984; BLINKS and ENDOH 1986; RÜEGG 1986). While such an effect has been proposed, based mainly on experiments with skinned cardiac fibers, for a number of drugs (sulmazole, APP 201-533, pimobendan, DPI 201-106), one of these has been found to increase the calcium transient contrary to the proposal (sulmazole BLINKS and ENDOH 1984) and the others were found to act by classical inotropic mechanisms (see Table 2). The α-adrenoceptor-mediated positive inotropic effect of phenyleph-

Table 2. Classification of drugs affecting cardiac calcium metabolism, according to their target calcium-transporting protein and mode of interaction

Cardioactive drug	Target calcium-transporting protein	Effect	Mode of interaction	Reference
A. Sarcolemma				
"Inhibitory" dihydropyridines	Calcium channel (for different sensitivities of subtypes, see text)	Inhibition	Direct ("dihydropyridine receptor")	JANIS and TRIGGLE (1983)
Nifedipine				
Nimodipine				
Nitrendipine				
(R)-Bay K 8644				FRANCKOWIAK et al. (1985)
(R)-202-791				KONGSAMUT et al. (1985), WILLIAMS et al. (1985)
"Activating" dihydropyridines		Activation		
(S)-Bay K 8644				FRANCKOWIAK et al. (1985)
(S)-202-791				KONGSAMUT et al. (1985), WILLIAMS et al. (1985)
CGP 28392				KOKOBUN and REUTER (1984)
Papaverine derivatives		Inhibition	Direct	JANIS and TRIGGLE (1983)
Verapamil				
Gallopamil				
AQA 39				
Benzothiazepine: Diltiazem				
Toxins		Activation	Direct (?)	
Maitotoxin				KOBAYASHI et al. (1985)
Atrotoxin				HAMILTON et al. (1985)
Angiotensin II			Receptor-mediated (messenger?)	ROGERS et al. (1986)
α-Adrenoceptor agonist				BRÜCKNER and SCHOLZ (1984)
Phenylephrine				
Potassium channel inhibitors			By prolongation of action potential plateau	
4-Aminopyridine				WOLLMER et al. (1981)
Clofilium				STEINBERG and MICHELSON (1985)
Sotalol				CARMELIET (1985)

Table 2. (continued)

Cardioactive drug	Target calcium-transporting protein	Effect	Mode of interaction	Reference
β-Adrenoceptor agonists		Activation (by cAMP-catalyzed phosphorylation)	Receptor-mediated stimulation of adenylate cyclase	COLUCCI et al. (1986)
Parenteral				
Norepinephrine				
Epinephrine				
Isoprenaline				
Dopamine				
Dobutamine				
Oral				
Prenalterol				
Pirbuterol				
Xamoterol				
TA-064				
Histamine H_2-receptor agonists				
Histamine				MCNEILL (1984)
Dimaprit				BAUMANN et al. (1981)
Impromidine				
Peptide receptor agonists				
Glucagon				FARAH (1983)
Secretin				CHRISTOPHE et al. (1984)
Vasoactive intestinal peptide				
Cholera toxin			Adenylate cyclase stimulation via stimulatory coupling protein	ENDOH et al. (1984)
Forskolin			Direct stimulation of catalytic subunit of adenylate cyclase	BRISTOW et al. (1984)
Cyclic AMP			Substitution for endogenous cAMP	KAMEYAMA et al. (1985)
Cyclic AMP analogs				KORTH and ENGELS (1987)

Phosphodiesterase inhibitors		Accumulation of cAMP	WEISHAAR et al. (1985)
Nonselective			
Theophylline			
Isobutylmethylxanthine			
Sulmazole (also inhibits Na, K-ATPase)			
Selective for isozyme III			
Amrinone			
Milrinone			
Cilostamide			
Fenoximone			
Piroximone			
CI 914			
CI 930			
Isozyme activity unknown			
Ro 7-2956			OSBORNE et al. (1971)
Buquineran			ALABASTER et al. (1977)
OPC 8212			TAIRA et al. (1984)
Pimobendan (UD-CG 115)			HONERJÄGER et al. (1984)
UD-CG 212			BERGER et al. (1985)
Ro 13-6438			HOLCK et al. (1984)
APP 201-533			SCHOLTYSIK et al. (1985a)
Isomazole (EMD 41064, LY 175326)			HAYES et al. (1985); HONERJÄGER et al. (1985)
BM 14.478			STREIN et al. (1986)
Muscarinic cholinoceptor agonists	Inhibition of cAMP-induced activation	Receptor-mediated inhibition of adenylate cyclase or cAMP-induced events	SULAKHE et al. (1986)
Acetylcholine			
Carbachol			
Adenosine R-receptor agonists			HOSEY et al. (1984)
Adenosine			BÖHM et al. (1985a)
PIA			
NECA			

Table 2. (continued)

Cardioactive drug	Target calcium-transporting protein	Effect	Mode of interaction	Reference
Dichlorobenzamil	Sodium/calcium exchange	Inhibition	Direct	SIEGL et al. (1984)
Cardiotonic steroids, e.g., ouabain		Increased Ca influx	Intracellular Na accumulation due to inhibition of Na pump	LEE (1985)
Erythrophleum alkaloids, e.g., cassaine				AKERA et al. (1981)
Prednisolonebisguanylhydrazone				
BIIA (isoquinoline)				
Imidazopyridines				HONERJÄGER et al. (1980, 1985)
AR-L 57				
Sulmazole (also inhibits phosphodiesterase)				
EMD 46512				
Toxins			Intracellular Na accumulation due to prolonged activation of Na channels	HONERJÄGER (1982)
1. Alkaloids				
Ceveratrum alkaloids, e.g., veratridine				
Batrachotoxin				
Aconitine				HONERJÄGER and MEISSNER (1983)
2. Diterpenoids				
Grayanotoxins				
3. Polypeptides				
Sea anemone toxins				ISENBERG and RAVENS (1984)
Scorpion toxins				
Goniopora toxin				
4. Pyrethrins and pyrethroids				BERLIN et al. (1984)
5. Cyclic polyethers				RODGERS et al. (1984)
Brevetoxin-B				
Brevetoxin-A				
Synthetic cardiotonic vasodilator DPI 201-106				BUGGISCH et al. (1985)
Monensin			Ionophore-induced intracellular Na accumulation	SUTKO et al. (1977)

Muscarinic cholinoceptor agonist			
Carbachol (1–300 μmol/l; ventricular myocardium)		Receptor-mediated intracellular Na accumulation	KORTH and KÜHLKAMP (1985)
Toxins	Decreased Ca influx	Decrease in intracellular Na due to block of Na channels	FOZZARD and WASSERSTROM (1985)
Tetrodotoxin			
Saxitoxin			
Antiarrhythmic drugs			SHEU and LEDERER (1985)
Class 1a			HONERJÄGER et al. (1986)
Quinidine			
Procainamide			
Disopyramide			
Class 1b			
Phenytoin			
Lidocaine			
Mexiletine			
Tocainide			
Class 1c			
Ajmaline			
Aprindine			
Propafenone			
Lorcainide			
Flecainide			
Encainide			
Indecainide			
Amiloride analogs		Decrease in intracellular Na due to inhibition of Na influx via Na/H exchange	LAZDUNSKI et al. (1985)
Amiloride			
Ethylisopropylamiloride			

Table 2. (continued)

Cardioactive drug	Target calcium-transporting protein	Effect	Mode of interaction	Reference
B. Sarcoplasmic reticulum				
Ryanodine	Putative calcium release channel	Opened by low, blocked by high concentration	Direct (?)	HUNTER et al. (1983) SUTKO et al. (1985)
Caffeine		Opened	Direct (?)	CHAPMAN (1979)
Procaine Tetracaine Ruthenium red Trifluoperazine Chlorpromazine		Blocked	Direct (?)	VOLPE et al. (1983) CHAMBERLAIN et al. (1984)
All drugs that elevate cAMP (see above)	Calcium pump	Activation (by cAMP-catalyzed phosphorylation)	Various mechanisms that elevate cAMP (see above)	TADA and KATZ (1982)
Pumiliotoxin B High concentrations of: Quinidine Propranolol Verapamil Tetracaine		Inhibition	Direct	DALY et al. (1985) BALZER (1972)
C. Inner mitochondrial membrane				
Ruthenium red	Calcium uniport	Block	Direct (?)	MOORE (1971)
Diltiazem Prenylamine Fendiline Nifedipine Verapamil	Sodium/calcium exchange	Inhibition	Direct (?)	VÁGHY et al. (1982)

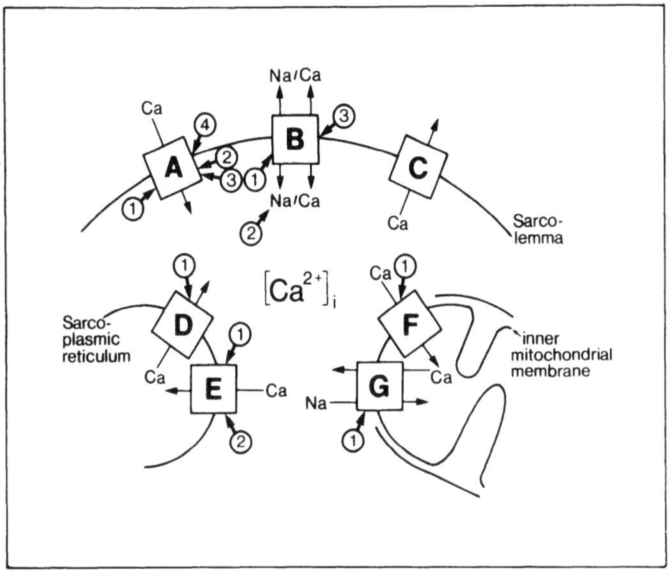

Fig. 2. Simplified scheme of seven calcium-transporting proteins (*A–G*) of the cardiac cell and some changes of their function (*circled numbers*) induced by cardioactive drugs. *A* voltage-gated calcium channel: *1* activation by cAMP-catalyzed phosphorylation; *2* modulation by dihydropyridines; *3* inhibition by nondihydropyridines; *4* indirect modulation by drug effects on membrane potential. *B* sodium/calcium exchange: *1* inhibition by amiloride analogs; *2* indirect modulation by drug effects on intracellular sodium; *3* indirect modulation by drug effects on membrane potential. *C* calcium pump (Ca,Mg-ATPase). *D* putative calcium release channel: *1* activation (?) by ryanodine, caffeine; inhibition by local anesthetics. *E* calcium pump (Ca,Mg-ATPase): *1* activation by cAMP-catalyzed phosphorylation; *2* inhibition (various drugs in high concentration). *F* calcium uniport: *1* inhibition by ruthenium red. *G* sodium/calcium exchange: *1* inhibition by diltiazem

rine, on the other hand, is associated with only a minimal increase of the calcium transient, strongly suggesting that the major part of the inotropic response involves enhanced calcium sensitivity of the contractile proteins (BLINKS and ENDOH 1986).

D. Drugs Affecting Sarcolemmal Calcium-Transporting Proteins

I. Calcium Channels

1. Direct Inhibitors and Activators: "Calcium Antagonists" and "Calcium Agonists"

Two classes of synthetic compounds block cardiac calcium channels: 1,4-dihydropyridines (e.g., nifedipine) and a heterogeneous group of aromatic amines. Some members of the latter class are related to papaverine (e.g., verapamil), one is of the benzothiazepine type (diltiazem). Certain members of the dihydropyridine class, but none of the aromatic amines, induce the opposite effect: they activate calcium channels by promoting their open state. An activation of cardiac calcium channels is also produced by two toxins of higher molecular weight, maitotoxin and the peptide atrotoxin, but the detailed mechanism of action of these compounds is as yet unknown. Recent work has revealed that sarcolemmal calcium channels do not represent a homogeneous population of ion channels. The cardiac plasmalemma (like that of other excitable cells) contains at least three different types of calcium channels, characterized by different conductances and gating parameters, and only one of these, the predominant classical "slow" channel, was found to be typically affected by pharmacologic channel modifiers. The "fast" calcium channel subtype is insensitive to dihydropyridines and isoprenaline (BEAN 1985; NILIUS et al. 1985), the "very slow" one is not affected by isoprenaline (LEE et al. 1984).

Although calcium channel inhibitors do not differ with regard to their final effect, namely block of ion flow through the channel (LEE and TSIEN 1983), binding studies suggest the presence of different, though interlinked, receptors for dihydropyridines, verapamil, and diltiazem, respectively (JANIS and TRIGGLE 1984; GLOSSMANN et al. 1985; SPEDDING 1985). The putative receptor for dihydropyridines is of special significance, because its occupation may result in either closure or prolonged opening of the calcium channel, depending only on the steric conformation of thc dihydropyridinc. Thus, the (S)-enantiomers of Bay K 8644 and of the dihydropyridine 202-791 are calcium channel activators, while the corresponding (R)-enantiomers are inhibitors competing for the same binding site (HOF et al. 1985; FRANCKOWIAK et al. 1985; KONGSAMUT et al. 1985; WILLIAMS et al. 1985).

At the level of the single calcium channel, the two dihydropyridine types act by stabilizing different modes of gating behavior (HESS et al. 1984). Mode 0, characterized by no channel openings during a test depolarization, is favored by nitrendipine and nimodipine, while the long-lasting openings with very brief closure defining mode 2 are favored by Bay K 8644. Mode 1, expressed as current records with brief openings, occurs less often in the presence of the dihydropyridines, but with unaltered kinetics. In contrast, drugs that act via cAMP alter the rate constants within mode 1 (Sect. D.I.2). The binding to calcium channels of dihydropyridine-, verapamil-, and diltiazem-type inhibitors, as assessed in electrophysiologic experiments, depends critically on whether the channel is open, inactivated, or resting (UEHARA and HUME 1985) in line with the modulated receptor theory that was developed to explain the interaction of local anesthetics with sodium channels (HONDEGHEM and KATZUNG 1984). Calcium channel inhibitors, like local anesthetics, bind more strongly to the inactivated conformation of the channel than to rested channels. In the rhythmically active heart cell, the effect of these substances thus represents an equilibrium between drug association to inactivated calcium channels during the action potential and drug dissociation from resting channels between action potentials. The selective effect of certain dihydropyridines on vascular smooth muscle relative to cardiac muscle may relate partly to the different fraction of time the calcium channels spend in a given state in these tissues (BEAN 1984). The generally accepted view that these drugs act directly on the calcium channel or a closely associated protein has received support from the finding that calcium channels derived from cardiac sarcolemmal vesicles and incorporated into planar lipid bilayers are sensitive to channel modifiers of the dihydropyridine and aralkylamine type (EHRLICH et al. 1986).

The negative chronotropic and negative dromotropic cardiac effects of calcium channel inhibitors are explained by the dependence on calcium channels of both the diastolic depolarization in the sinoatrial pacemaker cells and conduction through atrioventricular nodal cells (IRISAWA and NOMA 1986). Owing to the central role of calcium channels in excitation–contraction coupling, calcium channel inhibitors produce a negative inotropic effect (FLECKENSTEIN 1977). While the inotropic action of nonspecific inhibitors such as prenylamine, which blocks both sodium channels and calcium channels, may involve more than one mechanism, certain dihydropyridines and verapamil-like calcium channel inhibitors appear to depress myocardial contractility solely through their effect on sarcolemmal calcium channels, as indicated by the following evidence:

1. The negative inotropic effect of verapamil, gallopamil, nifedipine, diltiazem, and perhexiline is associated with decreased intracellular calcium release as shown by the reduction of the light signal emitted by intracellular aequorin (cf. Fig. 1, Table 1).

2. The variation with membrane potential and excitation frequency of calcium channel block induced by verapamil and related agents is paralleled by a similar or identical variation of the negative inotropic effect (TRAUTWEIN et al. 1981; PELZER et al. 1982).

3. The observation that both the effect on calcium channels and the inotropic effect are inverted by a change of only the steric conformation of certain dihydropyridines (as discussed already) argues strongly for a cause–effect relationship between calcium channel modification and inotropic effect induced by the dihydropyridines.

In addition, other positive inotropic mechanisms have been ruled out for Bay K 8644. Its inotropic effect on guinea pig papillary muscle occurs without a change in cAMP content (BÖHM et al. 1985 b) which eliminates the possibility that the drug acts as a phosphodiesterase inhibitor. Other dihydropyridines were found to act as cardiac phosphodiesterase inhibitors (NORMAN et al. 1983). Bay K 8644 does not affect the calcium sensitivity or calcium release of skinned cardiac fiber preparations (THOMAS et al. 1985).

The direct inhibitors of cardiac calcium channels have found many therapeutic and experimental applications. In addition to the vasodilating effect (useful in the treatment of hypertension, vasospastic disorders, and coronary heart disease), the direct negative chronotropic and negative inotropic effects of some agents (e.g., verapamil, diltiazem) contribute to their antianginal action by reducing myocardial oxygen consumption. Inhibition by these drugs of the atrioventricular node, where conduction occurs via calcium channel-mediated action potentials, may result in interruption of supraventricular tachyarrhythmias or slowing of ventricular rate in atrial fibrillation. Calcium channel inhibitors protect the myocardium against ischemic injury (FLECKENSTEIN 1977). The calcium channel-activating dihydropyridines, as well as maitotoxin and atrotoxin, form a distinct class of cardiotonic drugs. In the case of Bay K 8644 (SCHRAMM et al. 1983; PREUSS et al. 1985; WADA et al. 1985), therapeutic use in heart failure patients would probably be severely limited by the extracardiac vasoconstrictor effect as well as by restriction of coronary blood flow due to a direct vasoconstrictor effect, slowed ventricular relaxation, and the shortened diastole resulting from the positive chronotropic effect. Experimentally, direct modifiers of cardiac calcium channels have been used to isolate and characterize the ion channel with biochemical and electrophysiologic techniques and to assess the physiologic and pathophysiologic consequences of transmembrane calcium influx through calcium channels.

2. Drugs Acting via Cyclic AMP

Cardiac calcium metabolism is drastically affected by an increased level of the second messenger cAMP within the cytoplasm. Among the various proteins which are phosphorylated owing to activation of cAMP-dependent protein kinases, the sarcolemmal calcium channels and the sarcoplasmic reticulum calcium pump are the major recognized targets responsible for the change in subcellular calcium transport. The different drugs which induce an increase in cardiac cAMP are discussed in the present section, because the ensuing modification of calcium channels relates most directly to the positive chronotropic, dromotropic, and inotropic effects. As mentioned in Sect. C.I.1, only "slow" calcium channels are sensitive to cAMP-induced modification.

Cardiac cAMP content rises as a consequence of enhanced production or diminished breakdown. Thus, stimulators of the membrane-bound adenylate cyclase and inhibitors of phosphodiesterase elevate the cAMP level. Activation of the sympathetic nervous system raises cardiac cAMP both by release of the neurotransmitter norepinephrine and liberation of epinephrine from adrenal medullary cells, because these catecholamines bind to cardiac β-adrenoceptors, resulting in activation of adenylate cyclase. In addition to the various natural and synthetic β-adrenoceptor agonists, histamine and histamine H_2-receptor agonists, glucagon, secretin, and vasoactive intestinal peptide cause receptor-mediated activation of cardiac adenylate cyclase. Agonists at β-adrenoceptors or H_2-receptors are competitively antagonized by β-adrenoceptor blockers and H_2-receptor blockers, respectively. Cardiac sarcolemmal adenylate cyclase is coupled not only to stimulatory, but also to inhibitory receptors, namely muscarinic cholinoceptors (SULAKHE et al. 1986) and adenosine receptors (HOSEY et al. 1984). Stimulation of the latter receptors inhibits the activation of adenylate cyclase induced by β-adrenoceptor agonists or any other of the agonists already mentioned. Other subunits of the cardiac receptor-linked adenylate cyclase system are also subject to pharmacologic modification, such as the catalytic subunit itself (which is stimulated by the plant diterpene forskolin) or the stimulatory coupling protein (which is modified by cholera toxin, resulting in permanent activation of adenylate cyclase). Cyclic nucleotide phosphodiesterase inhibitors comprise a large and chemically diverse group of drugs (AMER and KREIGHBAUM 1975), only some of which, listed in Table 2, exert cAMP-related cardiac effects. In the mammalian heart, three isoenzymes of phosphodiesterase have been separated by anion exchange chromatographic techniques (WEISHAAR et al. 1985). Only one of these, phosphodiesterase III, shows high affinity as well as specificity for cAMP. The phosphodiesterases I and II hydrolyze both cAMP and cGMP. In addition, the activity of phosphodiesterase I differs from the other two in its dependence on calmodulin. Both nonspecific phosphodiesterase inhibitors such as the methylxanthines (which inhibit all three fractions) and specific inhibitors of phosphodiesterase III such as amrinone (see Table 2) elevate cardiac cAMP and induce the characteristic physiologic responses. There is little information on the cardiac effects of inhibitors with selectivity for fraction I or II phosphodiesterase.

The available evidence is consistent with the concept that an increase in the cellular cAMP concentration resulting from stimulation of β-adrenoceptors, or any other of the drugs mentioned so far, leads to phosphorylation of a calcium channel-related protein which, in turn, promotes the open state of calcium channels during the plateau of the cardiac action potential (KAMEYAMA et al. 1985). A dynamic equilibrium between protein phosphorylation and dephosphorylation, mediated by an as yet unknown phosphatase, has to be assumed in order to account for the rapid reversibility of cAMP-mediated effects. The phosphorylation appears to increase both the number of functioning calcium channels and the probability of opening of a given calcium channel (CACHELIN et al. 1986). The phosphorylation hypothesis implies that effects similar to those produced by cAMP-elevating drugs would be produced by chemical alteration of enzymes that link cAMP to the calcium channel, namely direct activation of the protein kinase or inhibition of the channel-associated phosphatase. Specific drugs of this type have apparently not yet been found.

The dependence on calcium channel activity of both the diastolic depolarization in the sinoatrial pacemaker cells and conduction through atrioventricular nodal cells (IRISAWA and NOMA 1986) explains the characteristic positive chronotropic and positive dromotropic cardiac actions of cAMP-elevating drugs. The positive inotropic effect is causally related to the enhanced calcium inward current (REUTER 1974). In addition to its effect on sarcolemmal calcium channels, cAMP-dependent protein kinase catalyzes phosphorylation of phospholamban, a membrane protein of the sarcoplasmic reticulum, with resultant increases in the rate and amount of calcium accumulation into this organelle (TADA and KATZ 1982). While the relative importance of these two actions for the inotropic effect is not known, the accelerated intracellular calcium sequestration explains the decrease in relaxation time induced by this class of agents. The effect is physiologically important in that it preserves diastolic ventricular filling and coronary flow at the high heart rate induced by sympathetic stimulation. Such a beneficial effect on relaxation is not observed with any other class of positive inotropic drugs. The dependence on cAMP of both heart rate and force explains that most of the cAMP-elevating drugs indeed exert positive inotropic and positive chronotropic effects at similar doses. There are, however, some as yet unexplained exceptions to this rule, notably among phosphodiesterase inhibitors. Buquineran lacks a positive chronotropic effect (ALABASTER et al. 1977), and fenoximone causes a significantly smaller increase in right atrial rate than inotropically equieffective concentrations of isoprenaline (ROEBEL et al. 1982). In contrast, the related phosphodiesterase inhibitor Ro 7-2956 has a predominant positive chronotropic effect (OSBORNE et al. 1971). The ratio between inotropic and chronotropic activity of a drug depends on differential direct effects on the sinus node and ventricles, respectively, as well as on various extracardiac, e.g., vascular, effects related to the reflex control of heart rate. The inotropic selectivity of dobutamine, a synthetic β-adrenoceptor agonist, may result from additional agonist effects on vascular α-adrenoceptors, counterbalancing its vasodilator effect, and myocardial α-adrenoceptors, adding to its positive inotropic, but not chronotropic, activity (RUFFOLO et al. 1981).

Two β-adrenoceptor agonists, dopamine and dobutamine, and the phosphodiesterase inhibitor amrinone are used in short-term intravenous treatment of acute or severe chronic heart failure. Several orally effective β-adrenoceptor agonists and phosphodiesterase inhibitors, listed in Table 2, have been developed for application in heart failure patients. The reader is referred to recent reviews on this subject (FARAH et al. 1984; MANCINI et al. 1985; COLUCCI et al. 1986). Although these drugs have desirable properties, including myocardial relaxant and vasodilating effects, their therapeutic dose range is limited by two effects inherent to cAMP elevation, tachycardia and tachyarrhythmias, which reflect the enhanced activity of pacemaker cells and/or calcium overload.

3. Other Drugs

Angiotensin II binds to specific receptors in ventricular myocardium without stimulating adenylate cyclase (ROGERS et al. 1986) and produces a direct positive inotropic effect that is insensitive to blockade of α- or β-adrenoceptors (DEMPSEY et al. 1971). The absence of a shortening effect on relaxation time (KOCH-WESER 1964) is further evidence against involvement of cAMP. The octapeptide increases calcium inward current (KASS and BLAIR 1981). The messenger or mechanism coupling angiotensin II receptors to calcium channels is unknown. Atrial myocardium and the sinoatrial node are only slightly affected (KOCH-WESER 1964). Similarly, stimulation of α-adrenoceptors in ventricular myocardium, e.g., by phenylephrine, induces a positive inotropic effect, an increase in calcium inward current (BRÜCKNER and SCHOLZ 1984), a slight increase of the intracellular calcium transient, no change in cAMP content, and a prolongation of relaxation time. The mechanism that links α-adrenoceptors to calcium channel function or, more importantly, to the increased calcium sensitivity of the contractile proteins (BLINKS and ENDOH 1986) is not known. Additional aspects of α-adrenoceptor activation are mentioned in Sect. E.

Serotonin (5-hydroxytryptamine, 5-HT) produces a direct positive inotropic effect on ventricular myocardium that is associated with a prolongation of the action potential and relaxation (KAUMANN 1985). It is not known whether calcium channels are modified directly or only indirectly by the prolonged repolarization. The receptors mediating this response to relatively high serotonin concentrations have been termed 5-HT_4 by KAUMANN (1985).

Some experimental agents (4-aminopyridine, clofilium) and the clinically used β-adrenoceptor blocker sotalol prolong the repolarization phase of the action potential by blocking potassium channels. As discussed elsewhere in more detail (HONERJÄGER 1986), these agents produce a moderate positive inotropic effect that may result from a prolonged activation of the voltage-dependent calcium channels and/or from a prolongation of calcium influx via sodium/calcium exchange during the action potential. These drugs have to be distinguished from the dihydropyridines and sodium channel modifiers which prolong repolarization by directly activating calcium channels and sodium channels, respectively. Despite similar effects on action potential configuration, the latter two classes exert more powerful inotropic effects (STEINBERG and MICHELSON 1985).

II. Sodium/Calcium Exchange

1. Direct Inhibitors

Cardiac sarcolemmal sodium/calcium exchange, which mediates the exchange of one calcium ion for three or more sodium ions, is held responsible both for the maintenance of the transsarcolemmal calcium concentration gradient and for part of the calcium influx during the plateau phase of the action potential (MULLINS 1981; EISNER and LEDERER 1985). The pyrazine diuretic amiloride and its more potent derivative dichlorobenzamil inhibit sodium/calcium exchange in cardiac sarcolemma, probably by competing with sodium ions at the sodium-binding sites of the exchanger. When applied to isolated papillary muscle, dichlorobenzamil produces a negative inotropic effect and inhibits the positive inotropic effect of two agents that elevate intracellular sodium – veratridine, and ouabain (SIEGL et al. 1984). These actions are consistent with an asymmetric effect on sodium/calcium exchange of dichlorobenzamil involving the predominant inhibition of the sodium efflux/calcium influx mode that is thought to operate during the action potential (KIMURA et al. 1986). At higher concentrations, dichlorobenzamil reverses its negative inotropic effect and induces contracture. Inhibition of sodium gradient-dependent calcium efflux could account for these effects (SIEGL et al. 1984).

2. Drugs Affecting Intracellular Sodium

Most of the drugs affecting myocardial sodium/calcium exchange act indirectly, by increasing or decreasing the intracellular sodium concentration (activity). Intracellular sodium is altered by drugs acting on any of three recognized sarcolemmal sodium-transporting mechanisms: the sodium pump, sodium channels, and sodium/hydrogen exchange, and by sodium ionophores.

Cardiotonic steroids (cardiac glycosides) constitute the most important class of pharmacologic sodium pump inhibitors. Their selectivity of action is unsurpassed by any other drug. It has been firmly established that inotropically effective concentrations increase intracellular sodium (LEE 1985), increase unidirectional calcium influx (KAZAZOGLOU et al. 1983), and increase the intracellular calcium signal that initiates contraction (see Fig. 1; MORGAN and BLINKS 1982). Cardiotonic steroids exemplify a pharmacologic modification of cardiac calcium metabolism that involves the intracellular messenger sodium rather than a direct action on calcium-binding proteins. Other positive inotropic sodium pump inhibitors include erythrophleum alkaloids and various synthetic nitrogenous compounds (see Table 2).

Drugs that increase or decrease the influx of sodium ions through sodium channels change the level of the intracellular sodium concentration by establishing a new equilibrium between passive sodium influx and active efflux via the sodium pump. In ventricular myocardial cells, sodium influx through sodium channels is normally restricted to the first new milliseconds of the action potential. A chemically heterogeneous group of plant and animal toxins, listed in Table 2, specifically prolongs the open state of sodium channels, thereby augmenting the sodium influx associated with each action potential. The positive inotropic effect of veratridine, which has been most extensively studied, involves the following steps: (a) modification of the sodium channels (no effect if these are blocked by tetrodotoxin); (b) intracellular sodium accumulation (BRILL and WASSERSTROM 1986) associated with frequent modified action potentials (no effect at low stimulation rates); and (c) increase in unidirectional calcium influx in exchange for intracellular sodium (FOSSET et al. 1977). Recently, a synthetic lipophilic indole derivative, DPI 201-106, was similarly found to produce a positive inotropic effect by drastically slowing the inactivation of sodium channels (BUGGISCH et al. 1985; KOHLHARDT et al. 1986). The cardiovascular effects of this drug, which are considered useful for the treatment of heart failure (SCHOLTYSIK et al. 1985b), suggest an additional blocking effect on vascular and cardiac calcium channels (HOF and HOF 1985; BUGGISCH et al. 1985). This additional property, which requires higher concentrations than the effect on sodium channels, might inhibit the arrhythmias and calcium overload which usually hinder therapeutic use of the cardiotonic property of this class of drugs.

Sodium channel blockers from the largest class, class 1, of antiarrhythmic drugs and are thought to act by prolonging the refractory period (VAUGHAN WILLIAMS 1984). In addition, their cardiac action probably involves an effect on sodium/calcium exchange mediated by a decrease in intracellular sodium activity that is expected and proven for lidocaine in Purkinje fibers (FOZZARD and WASSERSTROM 1985; SHEU and LEDERER 1985). A sodium-mediated decrease in calcium influx might explain, in part, the negative inotropic effect that is typically produced by these agents (HONERJÄGER et al. 1986), and it might also be involved in some antiarrhythmic effects, because a decrease in intracellular calcium prevents arrhythmogenic oscillatory afterdepolarizations induced by calcium overload (SHEU and LEDERER 1985). Additional direct block of calcium channels is a common property of local anesthetics (CHAPMAN and LÉOTY 1981) and class 1 antiarrhythmic agents (NAWRATH 1981), and this contributes to their effect on calcium metabolism, particularly at higher doses. Two experimental antiarrhythmic drugs, bepridil (ANNO et al. 1984) and BRL 31660 (BRÜCKNER et al. 1985), have similar blocking potencies for myocardial sodium and calcium channels. The selective sodium channel blocker tetrodotoxin (NARAHASHI 1984) is an invaluable experimental tool, even though cardiac sodium channels, with the exception of the chick heart (MARCUS and FOZZARD 1981), are less sensitive than neural sodium channels.

Recent work has established the presence of a continuously active sodium/hydrogen exchange mechanism in the contracting cardiac cell, mediating proton efflux in exchange for sodium influx to compensate for the intracellular generation of metabolic acid (MAHNENSMITH and ARONSON 1985). The best inhibitors identified to date are amiloride derivatives. There is little information on how cardiac function is altered by selective inhibition of this exchanger. A negative inotropic effect would be expected to result from the ensuing intracellular acidosis and decrease in intracellular sodium. Amiloride itself produces a positive inotropic effect that can neither be related to inhibition of sodium/hydrogen exchange nor to inhibition of sodium/calcium exchange, but implies additional actions (KENNEDY et al. 1986). Experiments on cultured cardiac cells with the most active inhibitor of sodium/hydrogen exchange, ethylisopropylamiloride, indicate a sizeable sodium influx via this route. The inhibitor diminishes the ability of ouabain to increase intracellular sodium and calcium (FRÉLIN et al. 1984). The same group has proposed that sodium entry via this route, followed by sodium/calcium exchange, is an important factor in calcium overload following reperfusion of the ischemic heart (LAZDUNSKI et al. 1985).

In addition to inhibiting cAMP-related effects and in contrast to its negative inotropic effect on atrial muscle, carbachol exerts a direct positive inotropic effect on ventricular myocardium that is associated with an increased intracellular sodium activity (KORTH and KÜHLKAMP 1985). The inotropic effect, which is mediated by muscarinic cholinoceptors, thus probably involves sodium/calcium exchange, but the cause of the change in intracellular sodium is as yet unknown.

E. Drugs Affecting Sarcoreticular Calcium-Transporting Proteins

In mammalian myocardium, the sarcoplasmic reticulum is considered to be the major site of release and of sequestration of the calcium ions that transiently activate the contractile proteins (WINEGRAD 1982; CHAPMAN 1983; FABIATO 1983). Drugs may affect sarcoreticular calcium transport by modifying the putative calcium release channel or the calcium pump (Ca,Mg-dependent ATPase) that mediates calcium sequestration. In contrast to the rich pharmacology of sarcolemmal calcium-binding proteins, few agents act selectively on intracellular organelles when applied to the intact myocardial cell.

The plant alkaloid ryanodine is the most potent and apparently highly specific drug for the sarcoplasmic reticulum (SUTKO et al. 1985). At nanomolar concentrations, ryanodine eliminates a number of mechanical phenomena that are generally attributed to calcium handling by the sarcoplasmic reticulum: the strong rested state contraction of atrial muscle, an early component of contraction in ventricular muscle, postextrasystolic potentiation, and oscillatory aftercontractions. These effects are consistent with a ryanodine-induced calcium "leak" in the sarcoreticular membrane, perhaps resulting from permanent activation of the putative calcium release channel. Ryanodine-induced depletion of a calcium pool has been observed with tracer techniques in the perfused heart (HUNTER et al. 1983). In contrast, inotropically effective low concentrations of ryanodine did not affect calcium transport by isolated sarcoplasmic reticulum, but inhibited calcium release at high concentrations (BESCH 1985). Low cardioactive concentrations of ryanodine failed to alter the contractions of skinned cardiac muscle attributed to calcium-induced calcium release from the sarcoplasmic reticulum (FABIATO 1985). These discrepancies are as yet unresolved; further analysis of the ryanodine effect will yield important information on cardiac excitation–contraction coupling. Sarcoreticular calcium uptake and release are modified by other drugs, including methylxanthines and local anesthetics (see Chap. 11), but the cardiac effect of these agents is confounded by additional effects on sarcolemmal calcium transport.

Drugs that elevate the cAMP concentration in cardiac cells abbreviate systole by enhancing the rate of relaxation. This effect is attributed to cAMP-catalyzed phosphorylation of the sarcoreticular protein phospholamban which activates the calcium pump (TADA and KATZ 1982) as has been discussed in Sect. D.I.2. In addition, an enhanced rate of calcium dissociation from troponin C, resulting from phosphorylation of tropinin I, has been proposed to contribute to the relaxant effect (ENGLAND 1986).

The endoplasmic reticulum of noncardiac, e.g., smooth muscle cells, releases calcium in response to elevation of the second messenger inositol trisphosphate which is generated by agonist-stimulated breakdown of plasmalemmal phosphoinositides (BERRIDGE and IRVINE 1984). Muscarinic cholinergic or α-adrenergic stimulation of myocardial cells results in phosphoinositide hydrolysis (BROWN et al. 1985). Inositol trisphosphate, if applied directly to skinned cardiac muscle fibers, appears to increase the amount of calcium released cyclically by the sarcoplasmic reticulum (NOSEK et al. 1986).

F. Drugs Affecting Mitochondrial Calcium-Transporting Proteins

In contrast to the previously held view that the role of mitochondria in cardiac calcium metabolism is to serve as a calcium-buffering store, more recent evidence suggests that the intramitochondrial calcium concentration is similar to that in the cytoplasm, and may play a messenger role in adjusting ATP supply to demand by acting on calcium-sensitive intramitochondrial enzymes (DENTON and MCCORMACK 1985; HANSFORD 1985). Within this hypothesis, the main role of the cardiac calcium-transporting proteins in the inner mitochondrial membrane, namely the electrophoretic uniport and the sodium/calcium antiport, providing calcium uptake and egress, respectively, is to relay changes in cytoplasmic calcium into the mitochondrial matrix. Drugs which stimulate ATP-requiring processes such as muscle contraction by an increased intracellular calcium signal could in this way also increase intramitochondrial oxidative metabolism and hence promote the replenishment of ATP.

Mitochondrial calcium uptake is inhibited by the glycoprotein stain ruthenium red (MOORE 1971), mitochondrial sodium/calcium exchange by diltiazem and other inhibitors of sarcolemmal calcium channels (VÁGHY et al. 1982). In the intact cardiac cell, the sarcolemmal effect of diltiazem on calcium influx predominates, making it difficult to assess the consequences of mitochondrial sodium/calcium exchange inhibition. Drugs that change intracellular sodium (Sect. D.II.2) might affect mitochondrial sodium/calcium exchange in addition to sarcolemmal sodium/calcium exchange, thereby altering mitochondrial metabolism. Two cardiac effects of ruthenium red have been ascribed to its inhibitory influence on mitochondrial calcium uptake. The substance blocks the activation of the mitochondrial pyruvate dehydrogenase complex induced by epinephrine in rat hearts without diminishing the positive inotropic affect (McCormack and England 1983), which is consistent with the hypothesis already outlined. Administration of ruthenium red diminishes the losses of ATP and myocardial function following ischemia and subsequent reperfusion (Ferrari et al. 1982), supporting the view that massive accumulation of calcium in the mitochondria is a critical factor in this pathophysiologic situation. Since ruthenium red modifies the function of plasmalemmal sodium and calcium channels in neuronal cells (Stimers and Byerly 1982) and inhibits calcium-induced calcium release from isolated cardiac sarcoplasmic reticulum (Chamberlain et al. 1984), further experiments may be necessary to firmly establish that the cardiac effects described are exclusively due to blockage of the mitochondrial calcium uniport.

G. Conclusions

This survey of drugs affecting cardiac calcium metabolism or cardioactive drugs in general has shown clearly that most of the known agents and all of the described drugs in clinical use exert functionally important effects by acting on only two or three of the seven known calcium-transporting proteins in the cardiac cell, namely on sarcolemmal calcium channels, sodium/calcium exchange, and, if the second messenger cAMP is involved, also on the sarcoreticular calcium pump. The extent to which the molecular mechanism of action has been revealed differs widely among drugs. The reactions, for example, that link occupation of a β-adrenoceptor by norepinephrine to increased probability of opening of single sarcolemmal calcium channels are known in considerable detail, while none of the events that couple myocardial angiotensin II or α-adrenoceptor occupancy to calcium channel activation are known. Knowledge of the functional pharmacology of modulators of sodium/calcium exchange or sodium/hydrogen exchange is extremely limited. In the absence of a selective inhibitor of the sarcolemmal calcium pump (ATPase), neither the physiologic nor the possible pharmacologic role of this enzyme can be assessed. Similar considerations apply to the calcium transport systems in sarcoreticular and mitochondrial membranes.

References

Akera T, Fox AL, Greeff K (1981) Substances possessing inotropic properties similar to cardiac glycosides. In: Greeff K (ed) Cardiac glycosides. Springer, Berlin Heidelberg New York, pp 459–486 (Handbook of experimental pharmacology, vol 56)

Alabaster CT, Blackburn KJ, Joice JR, Massingham R, Scholfield PC (1977) UK-14,275, a novel orally-active cardiac stimulant. Br J Pharmacol 60:284P–285P

Allen DG, Kurihara S (1980) Calcium transients in mammalian ventricular muscle. Eur Heart J 1 (Suppl A):5–15

Allen DG, Orchard CH (1983) Intracellular calcium concentration during hypoxia and metabolic inhibition in mammalian ventricular muscle. J Physiol (Lond) 339:107–122

Amer MS, Kreighbaum WE (1975) Cyclic nucleotide phosphodiesterases: properties, activators, inhibitors, structure-activity relationships, and possible role in drug development. J Pharm Sci 64:1–37

Anno T, Furuta T, Itho M, Kodama I, Toyama J, Yamada K (1984) Effects of bepridil on the electrophysiological properties of guinea-pig ventricular muscles. Br J Pharmacol 81:589–597
Balzer H (1972) The effect of quinidine and drugs with quinidine-like action (propranolol, verapamil and tetracaine) on the calcium transport system in isolated sarcoplasmic reticulum vesicles of rabbit skeletal muscle. Naunyn-Schmiedebergs Arch Pharmacol 274:256–272
Baumann G, Felix SB, Schrader J, Heidecke CD, Rieß G, Erhardt WD, Ludwig L, Loher U, Sebening F, Blömer H (1981) Cardiac contractile and metabolic effects mediated via the myocardial H_2-receptor adenylate cyclase system. Res Exp Med (Berl) 179:81–98
Bean BP (1984) Nitrendipine block of cardiac calcium channels: high-affinity binding to the inactivated state. Proc Natl Acad Sci USA 81:6388–6392
Bean BP (1985) Two kinds of calcium channels in canine atrial cells. J Gen Physiol 86:1–30
Berger C, Meyer W, Scholz H, Starbatty J (1985) Effects of the benzimidazole derivatives pimobendan and 2-(4-hydroxyphenyl)-5-(5-methyl-3-oxo-4,5-dihydro-2H-6-pyridazinyl)-benzimidazole HCl on phosphodiesterase activity and force of contraction in guinea-pig hearts. Arzneimittelforsch 35:1668–1673
Berlin JR, Akera T, Brody TM, Matsumura F (1984) The inotropic effects of a synthetic pyrethroid decamethrin on isolated guinea pig atrial muscle. Eur J Pharmacol 98:313–322
Berridge MJ, Irvine RF (1984) Inositol trisphosphate, a novel second messenger in cellular signal transduction. Nature 312:315–321
Besch HR (1985) Effects of ryanodine on cardiac subcellular membrane fractions. Fed Proc 44:2960–2963
Blinks JR, Endoh M (1984) Sulmazol (AR-L 115 BS) alters the relation between $[Ca^{2+}]$ and tension in living canine ventricular muscle. J Physiol (Lond) 253:63P
Blinks JR, Endoh M (1986) Modification of myofibrillar responsiveness to Ca^{++} as an inotropic mechanism. Circulation 73 (Suppl III):III-85–III-98
Blinks JR, Lee NKM, Morgan JP (1980) Ca^{++} transients in mammalian heart muscle: effects of inotropic agents on aequorin signals. Fed Proc 39:854
Blinks JR, Wier WG, Hess P, Prendergast FG (1982) Measurement of Ca^{++} concentrations in living cells. Prog Biophys Mol Biol 40:1–114
Böhm M, Brückner R, Meyer W, Nose M, Schmitz W, Scholz H, Starbatty J (1985a) Evidence for adenosine receptor-mediated isoprenaline-antagonistic effects of the adenosine analogs PIA and NECA on force of contraction in guinea-pig atrial and ventricular cardiac preparations. Naunyn-Schmiedebergs Arch Pharmacol 331:131–139
Böhm M, Burmann H, Meyer W, Nose M, Schmitz W, Scholz H (1985b) Positive inotropic effect of Bay K 8644: cAMP-independence and lack of inhibitory effect of adenosine. Naunyn-Schmiedebergs Arch Pharmacol 329:447–450
Brill DM, Wasserstrom JA (1986) Intracellular sodium and the positive inotropic effect of veratridine and cardiac glycoside in sheep Purkinje fibers. Circ Res 58:109–119
Bristow MR, Ginsburg R, Strosberg A, Montgomery W, Minobe W (1984) Pharmacology and inotropic potential of forskolin in the human heart. J Clin Invest 74:212–223
Brown JH, Buxton IL, Brunton LL (1985) α_1-Adrenergic and muscarinic cholinergic stimulation of phosphoinositide hydrolysis in adult rat cardiomyocytes. Circ Res 57:532–537
Brückner R, Scholz H (1984) Effects of α-adrenoceptor stimulation with phenylephrine in the presence of propranolol on force of contraction, slow inward current and cyclic AMP content in the bovine heart. Br J Pharmacol 82:223–232
Brückner R, Schmitz W, Scholz H (1985) Effects on transmembrane action potential, slow inward current and force of contraction in ventricular cardiac muscle of BRL 31660, a new antiarrhythmic drug with class I and class IV activity. Naunyn-Schmiedebergs Arch Pharmacol 329:86–93
Buggisch D, Isenberg G, Ravens U, Scholtysik G (1985) The role of sodium channels in the effects of the cardiotonic compound DPI 201-106 on contractility and membrane potentials in isolated mammalian heart preparations. Eur J Pharmacol 118:303–311

Cachelin AB, Kokubun S, de Peyer JE (1986) Basic properties of sarcolemmal ion channels. In: Rupp H (ed) The regulation of heart function. Thieme, New York, pp 106–120
Carafoli E (1985) The homeostasis of calcium in heart cells. J Mol Cell Cardiol 17:203–212
Carmeliet E (1985) Electrophysiologic and voltage clamp analysis of the effects of sotalol on isolated cardiac muscle and Purkinje fibers. J Pharmacol Exp Ther 232:817–825
Chamberlain BK, Volpe P, Fleischer S (1984) Inhibition of calcium-induced calcium release from purified cardiac sarcoplasmic reticulum vesicles. J Biol Chem 259:7547–7553
Chapman RA (1979) Excitation-contraction coupling in cardiac muscle. Prog Biophys Mol Biol 35:1–52
Chapman RA (1983) Control of cardiac contractility at the cellular level. Am J Physiol 245:H535–H552
Chapman RA, Leoty C (1981) The effects of tetracaine on the membrane currents and contraction of frog atrial muscle. J Physiol (Lond) 317:475–486
Christophe J, Waelbroeck M, Chatelain P, Robberecht P (1984) Heart receptors for VIP, PHI and secretin are able to activate adenylate cyclase and to mediate inotropic and chronotropic effects. Species variations and physiopathology. Peptides 5:341–353
Colucci WS, Wright RF, Braunwald E (1986) New positive inotropic agents in the treatment of congestive heart failure. N Engl J Med 314:290–299, 349–358
Daly JW, McNeal ET, Overman LE, Ellison DH (1985) A new class of cardiotonic agents: structure-activity correlations for natural and synthetic analogues of the alkaloid pumiliotoxin B. J Med Chem 28:482–486
Dempsey PJ, McCallum ZT, Kent KM, Cooper T (1971) Direct myocardial effects of angiotensin II. Am J Physiol 220:477–481
Denton RM, McCormack JG (1985) Ca^{2+} transport by mammalian mitochondria and its role in hormone action. Am J Physiol 249:E543–E554
Ehrlich BE, Schen CR, Garcia ML, Kaczorowski GJ (1986) Incorporation of calcium channels from cardiac sarcolemmal membrane vesicles into planar lipid bilayers. Proc Natl Acad Sci USA 83:193–197
Eisner DA, Lederer WJ (1985) Na-Ca exchange: stoichiometry and electrogenicity. Am J Physiol 248:C189–C202
Eisner DA, Valdeolmillos M (1985) The mechanism of the increase of tonic tension produced by caffeine in sheep cardiac Purkinje fibres. J Physiol (Lond) 364:313–326
Endoh M, Blinks JR (1984a) Regulation of the intracellular Ca^{++} transient and Ca^{++}-sensitivity of myofibrils via α- and β-adrenoceptors in the rabbit papillary muscle. Cell Calcium 5:301
Endoh M, Blinks JR (1984b) Comparison of effects of five new inotropic drugs on calcium transients and contractions of canine ventricular muscle. IUPHAR 9th international congress of pharmacology. Macmillan, London, Abstract 88
Endoh M, Maruyama M, Yanagisawa T (1984) Cholera toxin-induced changes in force of contraction and cyclic AMP levels in canine ventricular myocardium: inhibition by carbachol. Life Sci 35:2397–2406
Endoh M, Yanagisawa T, Taira N, Blinks JR (1986) Effects of new inotropic agents on cyclic nucleotide metabolism and calcium transients in canine ventricular muscle. Circulation 73 (Suppl III):III-117–III-133
England PJ (1986) The phosphorylation of cardiac contractile proteins. In: Rupp H (ed) The regulation of heart function. Thieme, New York, pp 223–233
Fabiato A (1983) Calcium-induced calcium release of calcium from the cardiac sarcoplasmic reticulum. Am J Physiol 245:C1–C14
Fabiato A (1985) Effects of ryanodine in skinned cardiac cells. Fed Proc 44:2970–2976
Farah AE (1983) Glucagon and the circulation. Pharmacol Rev 35:181–217
Farah AE, Alousi AA, Schwarz RP (1984) Positive inotropic agents. Ann Rev Pharmacol Toxicol 24:275–328
Ferrari R, di Lisa F, Raddino R, Visioli O (1982) The effects of ruthenium red on mitochondrial function during postischaemic reperfusion. J Mol Cell Cardiol 14:737–740

Fleckenstein A (1977) Specific pharmacology of calcium in myocardium, cardiac pacemakers, and vascular smooth muscle. Ann Rev Pharmacol Toxicol 17:149–166

Fosset M, de Barry J, Lenoir MC, Lazdunski M (1977) Analysis of molecular aspects of Na^+ and Ca^{2+} uptakes by embryonic cardiac cells in culture. J Biol Chem 252:6112–6117

Fozzard HA, Wasserstrom JA (1985) Voltage dependence on intracellular sodium and control of contraction. In: Zipes DP, Jalife J (eds) Cardiac electrophysiology and arrhythmias. Grune and Stratton, Orlando, pp 51–57

Franckowiak G, Bechem M, Schramm M, Thomas G (1985) The optical isomers of the 1,4-dihydropyridine Bay K 8644 show opposite effects on Ca channels. Eur J Pharmacol 114:223–226

Frélin C, Vigne P, Lazdunski M (1984) The role of the Na^+/H^+ exchange system in cardiac cells in relation to the control of the internal Na^+ concentration. J Biol Chem 259:8880–8885

Glossmann H, Ferry DR, Goll A, Striessnig J, Zernig G (1985) Calcium channels and calcium channel drugs: recent biochemical and biophysical findings. Arzneimittelforsch 35:1917–1935

Gwathmey JK, Morgan JP (1985) The effects of milrinone and piroximone on intracellular calcium handling in working myocardium from the ferret. Br J Pharmacol 85:97–108

Hamilton SL, Yatani A, Hawkes MJ, Redding K, Brown AM (1985) Atrotoxin: a specific agonist for calcium currents in heart. Science 229:182–184

Hansford RG (1985) Relation between mitochondrial calcium transport and control of energy metabolism. Rev Physiol Biochem Pharmacol 102:1–72

Hayes JS, Pollock GD, Wilson H, Bowling N, Robertson DW (1985) Pharmacology of LY175326: a potent cardiotonic agent with vasodilator activities. J Pharmacol Exp Ther 233:318–326

Herzig JW (1984) Contractile proteins: possible targets for drug action. Trends Pharmacol Sci 5:296–300

Hess P, Lansman JB, Tsien RW (1984) Different modes of Ca channel gating behaviour favoured by dihydropyridine Ca agonists and antagonists. Nature 311:538–544

Hidaka H, Hartshorne DJ (eds) (1985) Calmodulin antagonists and cellular physiology. Academic, Orlando

Hof RP, Hof A (1985) Mechanism of the vasodilator effects of the cardiotonic agent DPI 201-106. J Cardiovasc Pharmacol 7:1188–1192

Hof RP, Rüegg UT, Hof A, Vogel A (1985) Stereoselectivity at the calcium channel: opposite action of the enantiomers of a 1,4-dihydropyridine. J Cardiovasc Pharmacol 7:689–693

Holck M, Thorens S, Muggli R, Eigenmann R (1984) Studies on the mechanism of positive inotropic activity of Ro 13-6438, a structurally novel cardiotonic agent with vasodilating properties. J Cardiovasc Pharmacol 6:520–530

Hondeghem LM, Katzung BG (1984) Antiarrhythmic agents: the modulated receptor mechanism of action of sodium and calcium channel-blocking drugs. Ann Rev Pharmacol Toxicol 24:387–423

Honerjäger P (1982) Cardioactive substances that prolong the open state of sodium channels. Rev Physiol Biochem Pharmacol 92:1–74

Honerjäger P (1986) Regulation of myocardial force of contraction by sarcolemmal ion channels, the sodium pump, and sodium-calcium exchange. In: Rupp H (ed) The regulation of heart function. Thieme, New York, pp 159–177

Honerjäger P, Meissner A (1983) The positive inotropic effect of aconitine. Naunyn-Schmiedebergs Arch Pharmacol 322:49–58

Honerjäger P, Reiter M, Baker PF (1980) Inhibition of the sodium pump in squid axons by the cardioactive drug AR-L 57. Mol Pharmacol 17:350–355

Honerjäger P, Heiss A, Schäfer-Korting M, Schönsteiner G, Reiter M (1984) UD-CG 115 – a cardiotonic pyridazinone which elevates cyclic AMP and prolongs the action potential in guinea-pig papillary muscle. Naunyn-Schmiedebergs Arch Pharmacol 325:259–269

Honerjäger P, Klockow M, Schönsteiner G (1985) Imidazopyridines: a heterogeneous class of cardiotonic drugs. Naunyn-Schmiedebergs Arch Pharmacol 329:R50

Honerjäger P, Loibl E, Steidl I, Schönsteiner G, Ulm K (1986) Negative inotropic effects of tetrodotoxin and seven class 1 antiarrhythmic drugs in relation to sodium channel blockade. Naunyn-Schmiedebergs Arch Pharmacol 332:184–195

Hosey MM, McMahon KK, Green RD (1984) Inhibitory adenosine receptors in the heart: characterization by ligand binding studies and effects on beta-adrenergic receptor stimulated adenylate cyclase and membrane protein phosphorylation. J Mol Cell Cardiol 16:931–942

Hunter DR, Haworth RA, Berkoff HA (1983) Modulation of cellular calcium stores in the perfused rat heart by isoproterenol and ryanodine. Circ Res 53:703–712

Irisawa H, Noma A (1986) Origin and regulation of cardiac rhythmicity. In: Rupp H (ed) The regulation of heart function. Thieme, New York, pp 95–105

Isenberg G, Ravens U (1984) The effects of the *Anemonia sulcta* toxin (ATX II) on membrane currents of isolated mammalian myocytes. J Physiol (Lond) 357:127–149

Janis RA, Triggle DJ (1983) New developments in Ca^{2+} channel antagonists. J Med Chem 26:775–785

Janis RA, Triggle DJ (1984) 1,4-Dihydropyridine Ca^{2+} channel antagonists and activators: a comparison of binding characteristics with pharmacology. Drug Dev Res 4:257–274

Kameyama M, Hofmann F, Trautwein W (1985) On the mechanism of β-adrenergic regulation of the Ca channel in the guinea-pig heart. Pflügers Arch 405:285–293

Kass RS, Blair ML (1981) Effects of angiotensin II on membrane current in cardiac Purkinje fibers. J Mol Cell Cardiol 13:797–809

Kaumann AJ (1985) Two classes of myocardial 5-hydroxytryptamine receptors that are neither 5-HT_1 nor 5-HT_2. J Cardiovasc Pharmacol 7 (Suppl 7):S76–78

Kazazoglou T, Renaud J-F, Rossi B, Lazdunski M (1983) Two classes of ouabain receptors in chick ventricular cardiac cells and their relation to (Na^+,K^+)-ATPase inhibition, intracellular Na^+ accumulation, Ca^{2+} influx, and cardiotonic effect. J Biol Chem 258:12163–12170

Kennedy RH, Berlin JR, Ng YC, Akera T, Brody TM (1986) Amiloride: effects on myocardial force of contraction, sodium pump and Na^+/Ca^{2+} exchange. J Mol Cell Cardiol 18:177–188

Kimura J, Noma A, Irisawa H (1986) Na-Ca exchange current in mammalian heart cells. Nature 319:596–597

Kobayashi M, Ohizumi Y, Yasumoto T (1985) The mechanism of action of maitotoxin in relation to Ca^{2+} movements in guina-pig and rat cardiac muscles. Br J Pharmacol 86:385–391

Koch-Weser J (1964) Myocardial actions of angiotensin. Circ Res 14:337–344

Kohlhardt M, Fröbe U, Herzig JW (1986) Modification of single cardiac Na^+ channels by DPI 201-106. J Membr Biol 89:163–172

Kokubun S, Reuter H (1984) Dihydropyridine derivatives prolong the open state of Ca channels in cultured cardiac cells. Proc Natl Acad Sci USA 81:4824–4827

Kongsamut S, Kamp TJ, Miller RJ, Sanguinetti MC (1985) Calcium channel agonist and antagonist effects of the stereoisomers of the dihydropyridine 202-791. Biochem Biophys Res Commun 130:141–148

Korth M, Engels J (1987) Inotropic and electrophysiological effects of 8-substituted cyclic AMP analogues on guinea-pig papillary muscle. Naunyn-Schmiedebergs Arch Pharmacol 335:77–85

Korth M, Kühlkamp V (1985) Muscarinic receptor-mediated increase of intracellular Na^+-ion activity and force of contraction. Pflügers Ach 403:266–272

Lazdunski M, Frélin C, Vigne P (1985) The sodium/hydrogen exchange system in cardiac cells: its biochemical and pharmacological properties and its role in regulating internal concentrations of sodium and internal pH. J Mol Cell Cardiol 17:1029–1042

Lee CO (1985) 200 years of digitalis: the emerging central role of the sodium ion in the control of cardiac force. Am J Physiol 249:C367–C378

Lee KS, Tsien RW (1983) Mechanism of calcium channel blockade by verapamil, D600, diltiazem and nitrendipine in single dialysed heart cells. Nature 302:790–794

Lee KS, Noble D, Lee E, Spindler AJ (1984) A new calcium current underlying the plateau of the cardiac action potential. Proc R Soc Lond [Biol] 223:35–48

Mahnensmith RL, Aronson PS (1985) The plasma membrane sodium-hydrogen exchanger and its role in physiological and pathophysiological processes. Circ Res 57:773–788
Mancini DM, Keren G, Aogaichi K, LeJemtel TH, Sonnenblick EH (1985) Inotropic drugs for the treatment of heart failure. J Clin Pharmacol 25:540–554
Marcus NC, Fozzard H (1981) Tetrodotoxin sensitivity in the developing and adult chick heart. J Mol Cell Cardiol 13:335–340
McCormack JG, England PJ (1983) Ruthenium red inhibits the activation of pyruvate dehydrogenase caused by positive inotropic agents in the perfused rat heart. Biochem J 214:581–585
McNeill JH (1984) Histamine and the heart. Can J Physiol Pharmacol 62:720–726
Moore CL (1971) Specific inhibition of mitochondrial calcium transport by ruthenium red. Biochem Biophys Res Commun 42:298–305
Morgan JP, Blinks JR (1982) Intracellular Ca^{2+} transients in the cat papillary muscle. Can J Physiol Pharmacol 60:524–528
Morgan JP, Wier WG, Hess P, Blinks JR (1983) Influence of Ca^{++}-channel blocking agents on calcium transients and tension development in isolated mammalian heart muscle. Circ Res 52 (Suppl I):47–52
Morgan JP, Gwathmey JK, DeFeo TT, Morgan KM (1986) The effects of amrinone and related drugs on intracellular calcium in isolated mammalian cardiac and vascular smooth muscle. Circulation 73 (Suppl III):III-65–III-77
Mullins LJ (1981) Ion transport in heart. Raven, New York
Narahashi T (1984) Drug-ionic channel interactions: single-channel measurements. Ann Neurol 16 (Suppl):S39–S51
Nawrath H (1981) Action potential, membrane currents and force of contraction in mammalian heart muscle fibers treated with quinidine. J Pharmacol Exp Ther 216:176–182
Nilius B, Hess P, Lansman JB, Tsien RW (1985) A novel type of cardiac calcium channel in ventricular cells. Nature 316:443–446
Norman JA, Ansell J, Phillipps MA (1983) Dihydropyridine Ca^{2+} entry blockers selectively inhibit peak I cAMP phosphodiesterase. Eur J Pharmacol 93:107–112
Nosek TM, Williams MF, Zeigler ST, Godt RE (1986) Inositol trisphosphate enhances calcium release in skinned cardiac and skeletal muscle. Am J Physiol 250:C807–C811
Osborne MW, Wenger JJ, Moe RA (1971) Hemodynamic effects of 4-(3,4-dimethoxybenzyl)-2-imidazolidinone (Ro 7-2956): a nonadrenergic myocardial stimulant. J Pharmacol Exp Ther 176:174–183
Pelzer D, Trautwein W, McDonald TF (1982) Calcium channel block and recovery from block in mammalian ventricular muscle treated with organic channel inhibitors. Pflügers Arch 394:97–105
Preuss KC, Brooks HL, Gross GJ, Warltier DC (1985) Positive inotropic actions of the calcium channel stimulator, Bay K 8644, in awake, unsedated dogs. Basic Res Cardiol 80:276–332
Reuter H (1974) Localization of *beta* adrenergic receptors, and effects of noradrenaline and cyclic nucleotides on action potentials, ionic currents and tension in mammalian cardiac muscle. J Physiol (Lond) 242:429–451
Rodgers RL, Chou HN, Temma K, Akera T, Shimizu Y (1984) Positive inotropic and toxic effects of brevetoxin-B on rat and guinea pig heart. Toxicol Appl Pharmacol 76:296–305
Roebel LE, Dage RC, Cheng HC, Woodward JK (1982) Characterization of the cardiovascular activities of a new cardiotonic agent, MDL 17043. J Cardiovasc Pharmacol 4:721–729
Rogers TB, Gaa ST, Allen IS (1986) Identification and characterization of functional angiotensin II receptors on cultured heart myocytes. J Pharmacol Exp Ther 236:438–444
Rüegg JC (1986) Effects of new inotropic agents on Ca^{++} sensitivity of contractile proteins. Circulation 73 (Suppl):78–84
Ruffolo RR, Spradlin TA, Pollock GD, Waddell JE, Murphy RT (1981) Alpha and beta adrenergic effects of the stereoisomers of dobutamine. J Pharmacol Exp Ther 219:447–452

Scholtysik G, Honerjäger P, Markstein R, Bormann G (1985a) Positive inotropic and electrophysiological effects of APP 201-533 can be explained by an increase of cardiac cyclic AMP. J Cardiovasc Pharmacol 7:597–603
Scholtysik G, Salzmann R, Berthold R, Herzig JW, Quast U, Markstein R (1985b) DPI 201-106, a novel cardioactive agent. Combination of cAMP-independent positive inotropic, negative chronotropic, action potential prolonging and coronary dilatory properties. Naunyn-Schmiedebergs Arch Pharmacol 329:316–325
Schramm M, Thomas G, Towart R, Franckowiak G (1983) Novel dihydropyridines with positive inotropic action through activation of Ca^{2+} channels. Nature 303:535–537
Sheu SS, Lederer WJ (1985) Lidocaine's negative inotropic and antiarrhythmic actions. Dependence on shortening of action potential duration and reduction of intracellular sodium activity. Circ Res 57:578–590
Siegl PKS, Cragoe EJ, Trumble MJ, Kaczorowski GJ (1984) Inhibition of Na^{+}/Ca^{2+} exchange in membrane vesicle and papillary muscle preparations from guinea pig heart by analogs of amiloride. Proc Natl Acad Sci USA 81:3238–3242
Spedding M (1985) Activators and inactivators of Ca^{++} channels: new perspectives. J Pharmacol (Paris) 16:319–343
Steinberg MI, Michelson EL (1985) Cardiac electrophysiologic effects of specific class III substances. In: Reiser HJ, Horowitz LN (ed) Mechanisms and treatment of cardiac arrhythmias; relevance of basic studies to clinical management. Urban and Schwarzenberg, Baltimore, pp 263–281
Stimers JR, Byerly L (1982) Slowing of sodium current inactivation by ruthenium red in snail neurons. J Gen Physiol 80:485–497
Strein K, Honerjäger P, Jäger H, Freund P (1986) In-vitro effects of the novel cardiotonic BM 14.478 – indications for its mode of action. Naunyn-Schmiedebergs Arch Pharmacol 332:R49
Sulakhe PV, Jagadeesh G, Braun AP (1986) Cardiac autonomic receptors and adenylate cyclase in health and disease. In: Rupp H (ed) The regulation of heart function. Thieme, New York, pp 71–94
Sutko JL, Besch HR, Bailey JC, Zimmerman G, Watanabe AM (1977) Direct effects of the monovalent cation ionophores monensin and nigericin on myocardium. J Pharmacol Exp Ther 203:685–700
Sutko JL, Ito K, Kenyon JL (1985) Ryanodine: a modifier of sarcoplasmic reticulum calcium release in striated muscle. Fed Proc 44:2984–2988
Tada M, Katz AM (1982) Phosphorylation of the sarcoplasmic reticulum and sarcolemma. Ann Rev Physiol 44:401–423
Taira N, Endoh M, Iijima T, Satoh K, Yanagisawa T, Yamashita S, Maruyama M, Kawada M, Morita T, Wada Y (1984) Mode and mechanism of action of OPC-8212, a novel positive inotropic drug, on the dog heart. Arzneimittelforsch 34:347–355
Thomas G, Groß R, Pfitzer G, Ruegg JC (1985) The positive inotropic dihydropyridine Bay K 8644 does not affect calcium sensitivity or calcium release of skinned cardiac fibres. Naunyn-Schmiedebergs Arch Pharmacol 328:378–381
Trautwein W, Pelzer D, McDonald TF, Osterrieder W (1981) AQA 39, a new bradycardic agent which blocks myocardial calcium (Ca) channels in a frequency- and voltage-dependent manner. Naunyn-Schmiedebergs Arch Pharmacol 317:228–232
Uehara A, Hume JR (1985) Interactions of organic calcium channel antagonists with calcium channels in single frog atrial cells. J Gen Physiol 85:621–647
Vághy PL, Johnson JD, Matlib MA, Wang T, Schwartz A (1982) Selective inhibition of Na^{+}-induced Ca^{2+} release from heart mitochondria by diltiazem and certain other Ca^{2+} antagonist drugs. J Biol Chem 257:6000–6002
Vaughan Williams EM (1984) A classification of antiarrhythmic actions reassessed after a decade of new drugs. J Clin Pharmacol 24:129–147
Volpe P, Palade P, Costello B, Mitchell RD, Fleischer S (1983) Spontaneous calcium release from sarcoplasmic reticulum. Effect of local anesthetics. J Biol Chem 258:12434–12442
Wada Y, Satoh K, Taira N (1985) Cardiovascular profile of Bay K 8644, a presumed calcium channel activator, in the dog. Naunyn-Schmiedebergs Arch Pharmacol 328:382–387

Walsh MP, Le Peuch CJ, Vallet B, Cavadore JC, Demaille JG (1980) Cardiac calmodulin and its role in the regulation of metabolism and contraction. J Mol Cell Cardiol 12:1091–1101

Weishaar RE, Cain MH, Bristol JA (1985) A new generation of phosphodiesterase inhibitors: multiple molecular forms of phosphodiesterase and the potential for drug selectivity. J Med Chem 28:537–545

Williams JS, Grupp IL, Grupp G, Vághy PL, Dumont L, Schwartz A, Yatani A, Hamilton S, Brown AM (1985) Profile of the oppositely acting enantiomers of the dihydropyridine 202-791 in cardiac preparations: receptor binding, electrophysiological, and pharmacological studies. Biochem Biophys Res Commun 131:13–21

Winegrad S (1982) Calcium release from cardiac sarcoplasmic reticulum. Ann Rev Physiol 44:451–462

Wollmer P, Wohlfahrt B, Khan AR (1981) Effect of 4-aminopyridine on contractile response and action potential of rabbit papillary muscle. Acta Physiol Scand 113:183–187

CHAPTER 20

Hormonal Control of Extracellular Calcium

I. MACINTYRE, M. ZAIDI, C. MILET, and P. J. R. BEVIS

A. Introduction

Despite wide differences in calcium intake, plasma calcium is maintained with remarkable constancy. Approximately half of total plasma calcium is ionised and the remainder is largely bound to plasma albumin. Extracellular calcium concentration is more than 1000-fold greater than the critical intracellular concentration; a major change in the former would almost certainly be followed by gross disturbances in cellular metabolism, eventually leading to death. To prevent this, a set of interlocking hormonal mechanisms have evolved in vertebrates. These have been reviewed here with particular reference to the major clinical consequences of their disturbance and therapeutic approaches used for their correction.

B. Parathyroid Hormone

Parathyroid hormone (PTH) is exclusively found in amphibians and land vertebrates, including humans. In essence, its action involves conservation of calcium and elimination of phosphorus.

I. Chemistry

The structures of bovine, porcine and human PTH are known (Fig. 1; KEUTMAN et al. 1978) and a functional map of the hormone is shown in Fig. 2 (ROSENBLATT et al. 1980). It is clear that while full activity requires positions 1–34, only the 27 amino terminal residues are essential for biological activity.

II. Biosynthesis

PTH is synthesised by ribosomes as a 115 amino acid polypeptide (ROSENBLATT et al. 1980). As the amino terminus of the chain emerges from the ribosome, the two methionines are removed, while the 23 amino acid hydrophobic leader sequence associates with the membranes of the endoplasmic reticulum in accord with the signal hypothesis. The leader sequence is then cleaved off enzymatically. The remaining 90 amino acid peptide is transported into the Golgi apparatus, where the amino terminal hexapeptide is removed, leaving the mature 84 amino

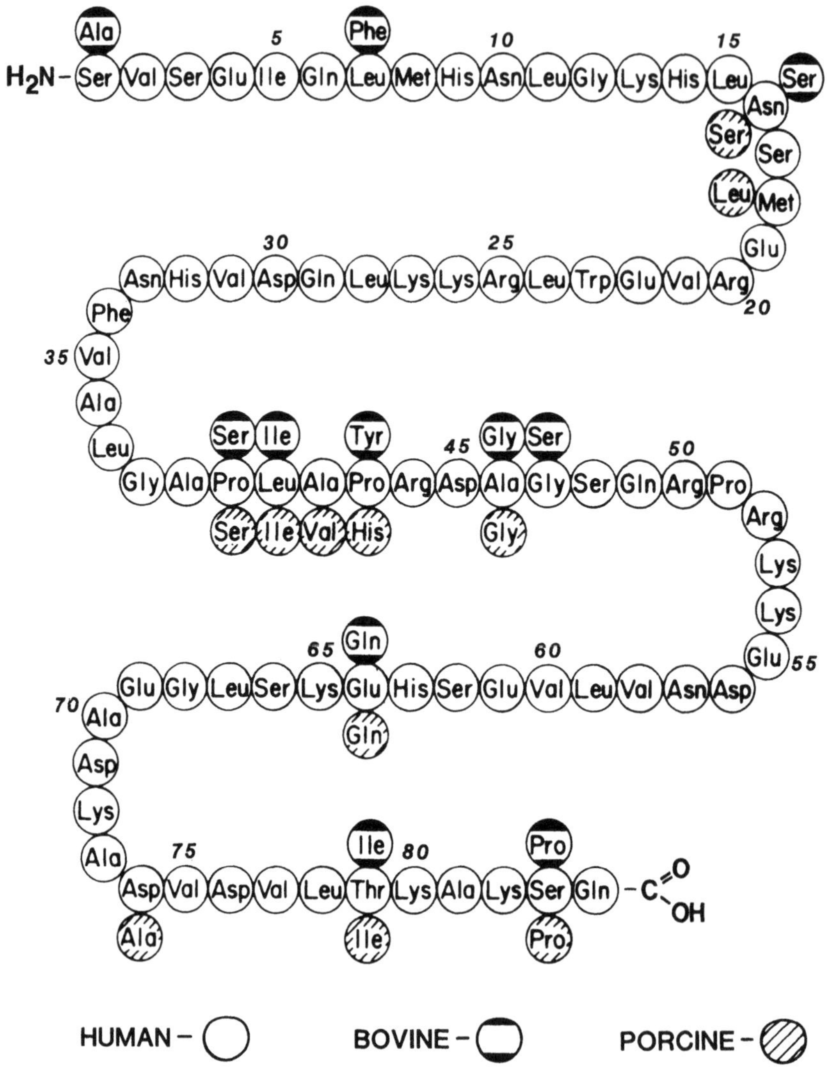

Fig. 1. The complete amino acid sequence of native human parathyroid hormone (PTH). Nonidentical amino acid residues of bovine and porcine PTH (mature forms) are indicated beside the appropriate positions of the human sequence. (MacIntyre et al. 1978)

acid form of the hormone to be incorporated into secretory granules. The latter are then transported to the periphery of the cell and excreted (Habener et al. 1984).

III. Secretion and Metabolism

Little PTH is stored. The main factor controlling its secretion is plasma calcium (Habener et al. 1984). The action is mediated via an exquisitely sensitive adenylate cyclase system: the enzyme is stimulated by hypocalcaemia to cause cyclic

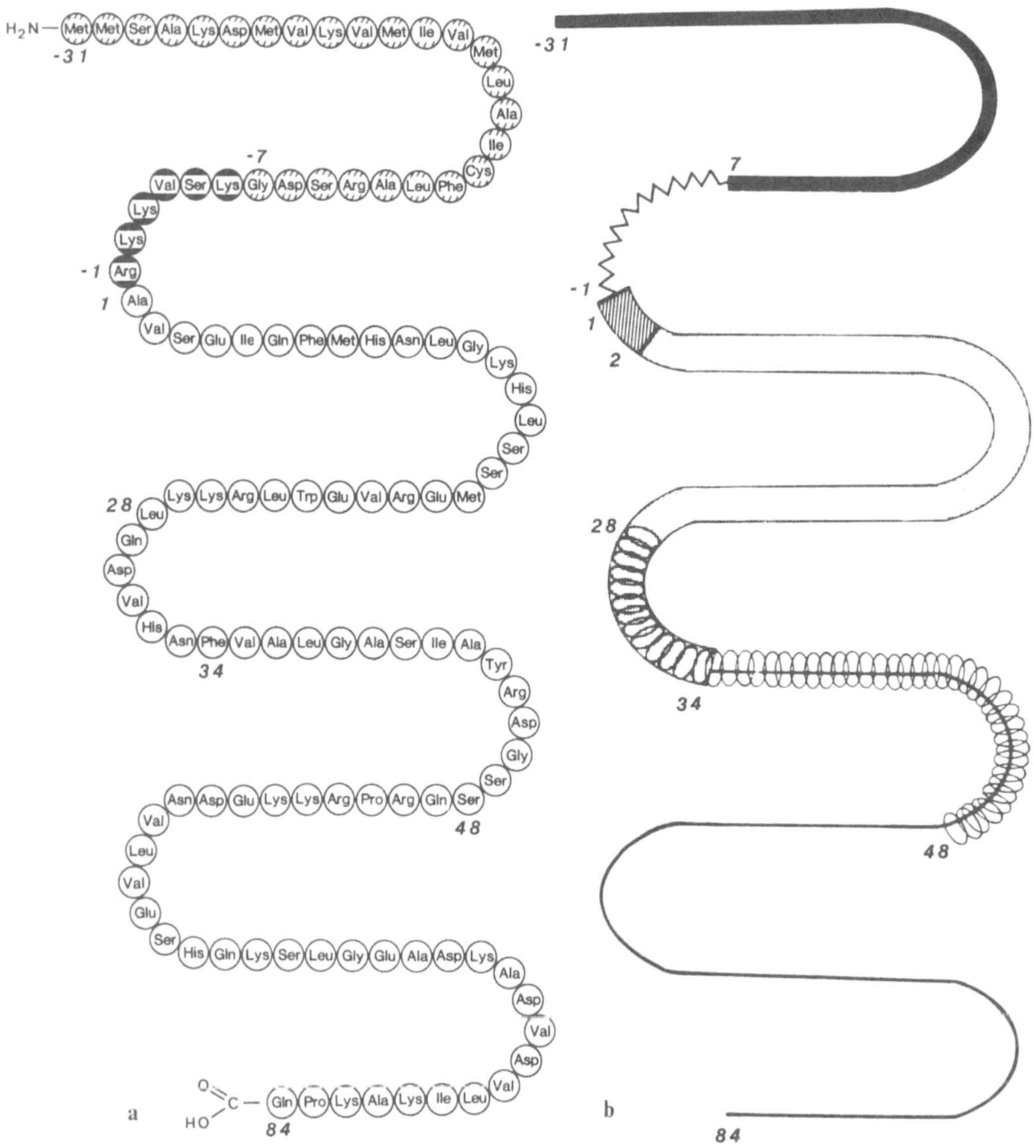

Fig. 2 a, b. Functional map of bovine preproparathyroid hormone, showing the complete amino acid sequence (**a**) and primary sequences required for physiological function or metabolism (**b**). They comprise a leader sequence (positions −31 to −7; *wide full line*), proparathyroid hormone-specific sequence (positions −6 to −1; *saw-tooth line*), activating sequences (positions 1 and 2; *hatched region*), a region determining receptor binding (positions 3 to 34; *strippled region*) and a region recognised by metabolising enzyme (positions 28 to 48; *coiled region*). The active molecule comprises positions 1 to 34 and inactive carboxy terminal sequences are at position 34 to 81 (*narrow full line*). (MacIntyre et al. 1978)

AMP generation, protein phosphorylation and finally peptide secretion. The reaction requires magnesium and severe hypomagnesaemia may grossly impair PTH secretion. Receptors for the active hormonal form of vitamin D, 1,25-dihydroxycholecalciferol (or calcitriol), have been identified in the parathyroid gland cells. These appear to be associated with inhibition of secretion. Catecholamines are also known to modulate PTH secretion: this is, however, of little importance under physiological conditions. Furthermore, elevated calcitonin levels, found in medullary thyroid carcinoma, are often associated with increased plasma PTH levels. Though this may suggest a direct action of calcitonin on PTH release, it is certainly not an important action at normal circulating levels of calcitonin. There is also some evidence to account for the control of PTH secretion by transcriptional or translational modulation, in response to plasma calcium levels.

There are two generally accepted mechanisms for the control of PTH secretion: plasma calcium-dependent hyperplasia or atrophy of the glands in the long term and enhancement or inhibition of intracellular hormone degradation in the short term. There is never a complete inhibition of parathyroid secretion even by marked hypercalcaemia (GOLDSMITH et al. 1973). Once the parathyroid glands have greatly enlarged, secretion of PTH remains abnormally elevated, despite hypercalcaemia.

Mature hormone is cleaved between positions 34 and 37 in the Kupffer cells of the liver, with liberation of amino terminal and carboxy terminal fragments (FISCHER et al. 1974). However, in addition, similar fragments are also secreted by the gland itself. The carboxy terminal fragment (molecular weight 6000, HABENER et al. 1984) has the longest half-life in the circulation and is thus the dominant hormonal form. It is removed mainly by the kidney, and renal failure is, thus, associated with elevated levels of this species. Further, it is due to this extreme heterogenaeity in circulating forms that different immunoassays produce different results and that interpretation of the assay is often fraught with difficulties (CANTERBURY and REISS 1972; ARNAUD et al. 1974). Most assays measure the dominant carboxy terminal fragment and therefore reflect long-term changes in hormone secretion or removal, rather than the more rapid physiological or pathological variations. The principal uses of the assays are to distinguish hyperparathyroidism from other causes of hypercalcaemia and to detect parathyroid hyperplasia in chronic renal failure.

IV. Biological Actions

The hormone acts on three major sites: kidney, bone and intestine. In the kidney, PTH markedly enhances phosphate excretion, increases calcium resorption and accelerates the conversion of 25-hydroxycholecalciferol to its active metabolic form, 1,25-dihydroxycholecalciferol (calcitriol) (AVIOLI and KRANE 1978; GARABEDIAN et al. 1972; HORIUCHI et al. 1976). The combination of these effects on the kidney produces the basic action of PTH: enhancement of plasma calcium. These three actions are brought about by two second messengers, cyclic AMP and intracellular calcium (AURBACH and CHASE 1976; CHASE and AURBACH 1967, 1968). An increase in plasma PTH levels is followed by stimulation of renal adenylate cyclase, reflected by an increase in intracellular cyclic AMP and a rapid spillover of cyclic AMP into the urine (CHASE and AURBACH 1967; KAMINSKY et al. 1970). Evidence for the involvement of intracellular calcium is less clear.

On bone, PTH has two effects. It enhances osteoclastic bone resorption by a primary action on osteoblasts (MCSHEEHY and CHAMBERS 1986a). The latter then secrete a local activator of osteoclastic activity (MCSHEEHY and CHAMBERS 1986b), the nature of which has not been established. This action produces an increased rate of bone destruction. Paradoxically, there is also some evidence that PTH may have an anabolic effect on bone to increase net bone mass (PARSONS 1976; PARSONS and ZANELLI 1980). This action is at the basis of the proposed use of PTH to enhance bone mass in osteoporosis. At present, the importance of a physiological anabolic role for PTH and what would be its role in the treatment of osteoporosis is under further investigation. The cellular mechanisms involved in the action of PTH on bone presumably resemble those in the kidney. Enhancement of intestinal absorption of calcium and phosphorus follows experimental administration of PTH. This effect is similar to that seen in hyperparathyroidism in humans. In both cases, however, the effect is an indirect one, exerted via increased renal production of calcitriol.

V. Pathophysiology

1. Primary Hyperparathyroidism

One in every three thousand adult patients attending a general hospital outpatient department present with primary hyperparathyroidism. The aetiology of this disorder is still completely unknown. The condition is characterised by an increased number of parathyroid gland cells occurring predominantly in one or several glands. These cells still retain nearly normal responsiveness to plasma calcium levels. As there is an irreducible minimum of PTH secretion by each cell, despite elevated plasma calcium, increased cell number provides an excess of total hormone secretion.

The crucial diagnostic finding is a repeatedly elevated plasma calcium. To reach a positive diagnosis, it is wise to obtain at least three plasma calcium estimations, separated by intervals of several weeks, and to make sure that the samples were collected under suitable basal conditions. As half of plasma calcium is bound to albumin, an alteration of plasma proteins by 1 g per 100 ml changes total plasma calcium by about 0.2 mmol/l. It is therefore essential to take into account abnormalities in plasma protein concentration during interpretation of results. Besides hypercalcaemia other chemical abnormalities of the disease include hypophosphataemia, hyperphosphaturia and increased levels of cAMP in the urine.

The plasma levels of calcitriol (1,25-dihydroxycholecalciferol) tend to be elevated, but the circulating levels of immunoassayable PTH are often within the normal range. Immunoassay of PTH levels can be helpful in distinguishing primary hyperparathyroidism from the hypercalcaemia of malignancy, as there is almost never ectopic production of PTH by malignant cells. The hypercalcaemia of malignant disease is due to the direct destruction of the skeleton either by malignant cells or by osteoclasts activated by factors from cancer cells. Many different factors can be involved in osteoclast activation including transforming growth factors, tumour necrosis factor and interleukin-1 (T. J. Chambers 1986, personal communication).

2. Renal Failure

Chronic renal failure is almost always associated with some degree of parathyroid hyperplasia. This is due to the hyperplastic drive directed by a combination of low plasma calcium with low plasma calcitriol. Further, as damaged kidneys are ineffective in removing the dominant circulating carboxy terminal PTH fragment, immunoassayable levels of PTH are often markedly increased.

C. Hormone from the Corpuscle of Stannius

Corpuscles of Stannius (Stannius 1839) are small endocrine bodies restricted to holosteans and teleosteans. They are involved in osmoregulation, essentially in the calcium balance.

I. Chemistry and Biosynthesis

The hormone from the corpusclc of Stannius (CS) has some immunological similarity to PTH. This has been confirmed by radioimmunoassay of CS extracts and immunocytochemistry of the CS, where the molecule cross-reacts with antisera to PTH (Milet et al. 1980; MacIntyre et al. 1981; Lopez et al. 1984a). Moreover, the cell types described in mammalian parathyroids have also been identified in the corpuscles of Stannius of the eel (Lopez et al. 1984b). Thus, the provisional name "parathyrin of the corpuscles of Stannius" (PCS) (Milet et al. 1980) reflects both its glandular source and its immunological resemblance to the PTH family.

Very little is known of the biochemical properties of this molecule. It has been isolated from eel corpuscles of Stannius and partially purified by anion exchange chromatography and reverse-phase HPLC. The biologically active peak has been found to be 500-fold more potent than the crude extract (Milet 1986) and has been used to identify the main product (34 kdalton) by polyacrylamide gel electrophoresis. However, studies involving immunoprecipitation of mRNA-translated products reveal that its synthesis occurs via an mRNA encoding a 45 kdalton precursor (Milet 1986). A 3 kdalton glycopeptide isolated from salmon corpuscles of Stannius has also been purified. This has been found to be active on gill Ca-ATPase (Ma and Copp 1978). However, several authors have suggested that the active molecule has an apparent molecular weight greater than 10 kdalton (Fenwick 1982).

II. Biological Action

The gill is the main target organ for the hormone secreted by these glands. Surgical removal of the corpuscles leads to dramatic hypercalcaemia (Fontaine 1964) due to decreased calcium outflow from gills. Furthermore, saline extracts of the corpuscles have been shown to increase calcium outflow in a perfused gill preparation (MILET et al. 1979a, b). This effect further explains the retention of plasma calcium observed in eels deprived of corpuscles of Stannius (LWOWSKI 1978). Saline extracts of eel corpuscles of Stannius have also been found to cause hypercalcaemia, to stimulate osteoclastic bone resorption (shown by histology of the rat femur) and to cause a dose-dependent release of radioactive calcium from foetal rat bones and mouse calvaria (LAFEBER et al. 1984).

A significance of the presence of a big PTH-like molecule in fish has not been explained. However, a hypothetical relationship can be drawn. The primary transcript of the human PTH gene has been shown to be approximately 800 nucleotides long (HEINRICH et al. 1984). It could theoretically encode a 30 kdalton protein, were processing of this transcript not to occur. Furthermore, mRNA encoding for a 30 kdalton protein has been extracted from tumours associated with hypercalcaemia. The extracted protein has been found to possess PTH-like bioactivity and immunoreactivity (BROADUS et al. 1985). One can therefore assume that the mechanisms for mRNA processing might be deficient in tumour cells and lower vertebrates.

D. Vitamin D: Endocrine System

Vitamin D is improperly named. In normal conditions, its inclusion in the diet is unnecessary for full health; it is formed in the skin and functions as a precursor to an extremely potent secosteroid calcium-regulating hormone (for review see MACINTYRE et al. 1978).

I. Chemistry and Biosynthesis

Vitamin D_3 (cholecalciferol; 9,10-seco-5,7,10(19)-cholestatrien-3β-ol) is a 9-10-secosteroid derived from a saturated hydrocarbon, 5α-cholestane (Fig. 3). Its immediate precursor, 7-dehydrocholesterol (Fig. 4) is present in the skin. Exposure to ultraviolet light leads to the cleavage of the bond between carbons 9 and 10 in ring *B* followed by the rotation of ring *A* through 180° around the bond between carbons 6 and 7. A double bond is thus formed between carbons 10 and 19 and the 3β-hydroxyl group of vitamin D_3 projects below the plane of ring *A* because of the rotation. It is still designated as having the β configuration because this was its position in the parent steroid 5α-cholestane (Fig. 5).

II. Metabolism

Vitamin D_3 by itself is almost inactive, and to produce metabolic effects on calcium must undergo two successive hydroxylations. The first stage is 25-hydroxylation in the liver and other tissues to produce the major circulating form, 25-hydroxycholecalciferol, having only modest biological effects. The second and critical hydroxylation occurs in the distal part of the proximal convoluted tubule of the kidney: a mitochondrial enzyme inserts a hydroxyl group at portion 1, converting 25-dihydroxycholecalciferol to the biologically active 1α,25-dihydroxycholecalciferol (1,25$(OH)_2D_3$, or calcitriol) (Fig. 6). This compound is a thousand

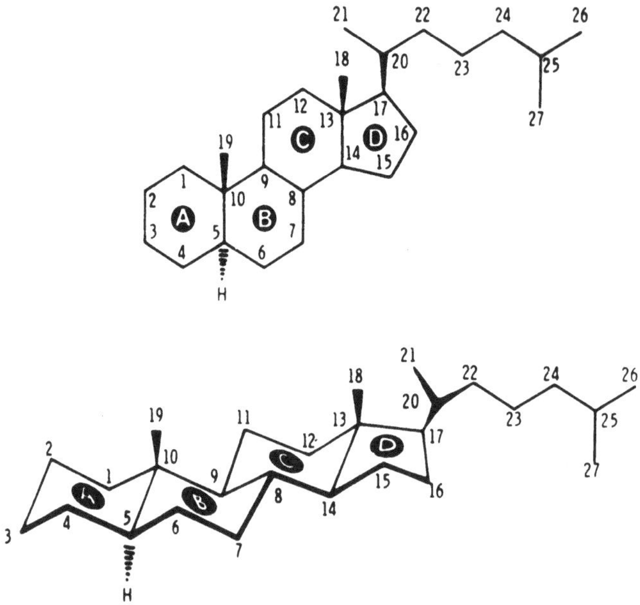

Fig. 3. Conventional and perspective formulae of 5α-cholestane. (MacIntyre et al. 1978)

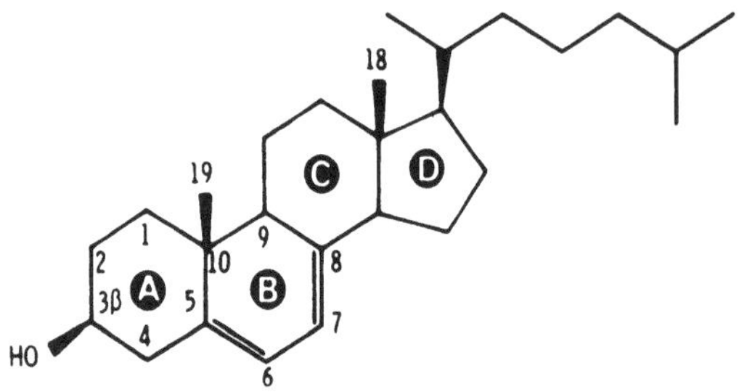

Fig. 4. 7-Dehydrocholesterol. (MacIntyre et al. 1978)

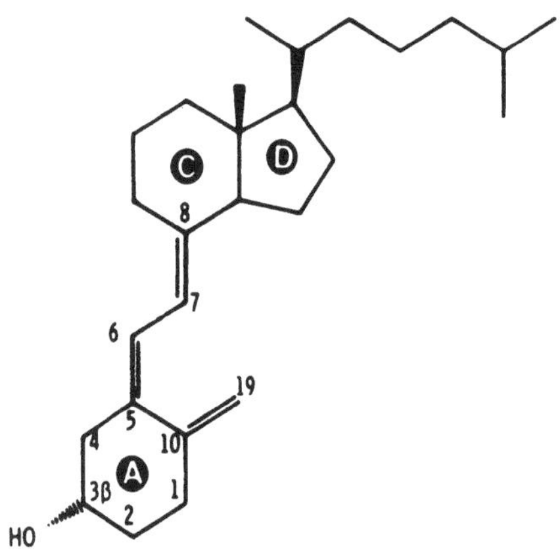

Fig. 5. Cholecalciferol (vitamin D_3) (MacIntyre et al. 1978). (Figs. 3 to 5 reproduced with kind permission from: Annals of the New York Academy of Sciences, vol 307, p 347)

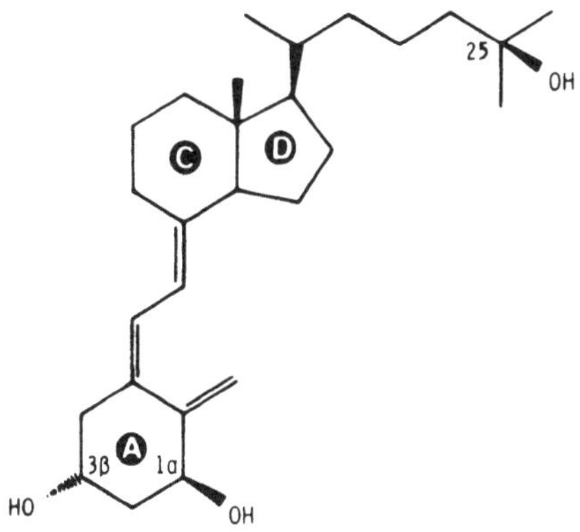

Fig. 6. 1α,25-Dihydroxycholecalciferol (calcitriol) (MACINTYRE et al. 1978). (Reproduced with kind permission from: Annals of the New York Academy of Sciences, vol 307, p 347)

times more potent than its precursors and is responsible for all the biological effects of vitamin D_3 on calcium metabolism. The formation of a second and poorly active renal steroid, 24,25-dihydroxycholecalciferol, is regarded as an alternative metabolic pathway when further production of $1{,}25(OH)_2D_3$ is not required. The factors which control the secretion of the active form of vitamin D_3 ($1{,}25(OH)_2D_3$) or calcitriol may be classified as mineral regulation or hormonal regulation.

1. Mineral Regulation

Hypocalcaemia stimulates and hyperphosphataemia inhibits the production of calcitriol. These actions do not require any hormonal intervention and it is not clear exactly how these effects are produced (LARKINS et al. 1973; HAUSSLET et al. 1976; BOYLE et al. 1971). There is always a reciprocal effect of factors controlling the metabolism of $1{,}25(OH)_2D_3$ on $24{,}25(OH)_2D_3$: when the production of one is stimulated, the other is inhibited.

2. Hormonal Regulation

a) Parathyroid Hormone

PTH stimulates the production of calcitriol in the presence of hypocalcaemia, serving to potentiate the action of calcium deficiency. However, in hyperparathyroidism when secretion of PTH coincides with hypercalcaemia, this stimulating effect may be overcome. All cases of hyperparathyroidism are, thus, not associated with increased calcitriol secretion. Nevertheless, PTH is an important physiological regulator of the hormone.

b) Growth Hormone and Prolactin

Growth hormone and prolactin may be the major physiological hormonal regulators of calcitriol. Both these hormones stimulate calcitriol production. During periods of calcium stress, namely growth, pregnancy and lactation, plasma calcitriol levels tend to be elevated owing to increased production of the 1,25 form.

c) Calcitonin

Calcitonin directly stimulates the production of calcitriol, acting at a site in the proximal tubule, adjacent to, but distinct from, that acted upon by PTH (GALANTE et al. 1972; KAWASHIMA et al. 1981). This action may have no physiological significance. It is, however, possible that the elevated levels of calcitonin occurring during periods of increased calcium requirement may play some role, together with growth hormone and prolactin.

d) Vitamin D Metabolites

Metabolites of vitamin D, mainly calcitriol, exert a negative feedback effect on calcitriol production, simultaneously inducing the 24-hydroxylase enzyme to increase the production of the inactive metabolite. These probably depend on transcriptional changes induced by the steroids. Importantly, they minimise the likelihood of excess, unwanted production of calcitriol and maximise its production when the plasma levels are low. Other tissues may also synthesise very small amounts of calcitriol, specifically in pregnancy when the placenta may also contribute.

III. Biological Actions

Saturable, high affinity binding sites for calcitriol (presumably receptors) are found in cells (SUDA et al. 1983) of the known calcium-transporting organs: gut, kidneys, bone, lactating breast and shell glands of birds. Similar "receptors" are also found in cells of the immune system and in certain nerve cells, and are especially abundant in eel brain. In the calcium-transporting organs, we know that the active hormone causes increased calcium transport. For example, in the gut there is enhancement of calcium and phosphorus absorption. The precise mechanism via which this is accomplished is unknown.

In some cells of the immune system, calcitriol induces differentiation; a similar effect can be observed in human myeloid leukaemia cells (HL60) which differentiate to monocytes or macrophages. Although FRASER (1980) has suggested that all these effects may be a consequence of calcium transport, it may be best, at present, to consider calcitriol as having two sets of actions: specific effects on the calcium-transporting organs and multiple permissive effects on a wide range of body cells (as with cortisol). Another class of saturable high affinity receptors are found within glands producing peptides which regulate vitamin D_3 metabolism (parathyroid cells, C cells, pituitary). Presumably, these are part of feedback loops. For example, occupation of the calcitriol receptor in the parathyroid gland and C cell inhibits parathyroid hyperplasia and stimulates calcitonin secretion, respectively, thus minimising the hypercalcaemia consequent upon the actions of $1,25(OH)_2D_3$ on the skeleton.

IV. Pathophysiology

Rickets in children or osteomalacia in adults is the consequence of inadequate supplies of the active hormonal form of vitamin D, calcitriol. The disease is essentially due to an impaired mineralisation rate of the protein matrix of bone. Abnormally large uncalcified seams are seen when decalcified bone sections are examined. The common form of rickets is due to consumption of a diet too low in vitamin D for the prevailing exposure to sunlight. It normally occurred in poor families living in industrial cities with a polluted atmosphere. Pigmented skin forms vitamin D more slowly than white skin and Asian immigrants have shown a remarkably high incidence of the disease owing to a combination of factors such as pigmented skin, clothes that cover most of the body and an adaptation to a diet appropriate to areas with adequate sunlight, but inadequate for more northern climates. Though it has been hypothesised that the development of highly pigmented skin in Africans and Asians was an evolutionary response to the need to avoid sunburn and it was also suggested that dark skin could abolish the supposed risk of vitamin D poisoning, it is now clear that vitamin D poisoning can never occur under any sunlight exposure because vitamin D itself is not toxic. Enhanced synthesis of cholecalciferol will not be followed by increased calcitriol production owing to the efficient control of calcitriol synthesis by the kidney. The question should be phrased: why are Europeans white? and not, why are Asians and Africans dark? Unpigmented skin was obviously an evolutionary advantage as peoples moved north to less sunlit regions. It is interesting to consider that white societies arose by evolutionary pressure on their Asian ancestors to adapt to lower ultraviolet exposure.

However, the major action of vitamin D in enhancing mineralisation, in the cure of rickets, appears to be indirect. As such, mineralisation of rachitic or osteomalacic bone can be achieved by provision of adequate calcium in the absence of vitamin D supply. Moreover, rickets is healed by enhancing calcium absorption from the gut. This surprising conclusion accords with the known facts, but may still have to be modified by future unmasking of a still undiscovered direct mineralising effect of the vitamin on bone. There is no doubt that vitamin D indirectly enhances bone osteoclastic activity (T. J. Chambers 1985, personal communication); this action is responsible for the hypercalcaemia produced by toxic doses of the vitamin or its metabolites. Familial forms of rickets also occur, some of which are due to renal defects involving phosphorus loss.

Two other interesting forms of rickets exist: types I and II. In type I, the kidney does not produce calcitriol, plasma levels are low and calcitriol is curative. In type II, specific receptors are defective or lacking and despite high plasma levels of calcitriol, gross rickets is present. Some response may be seen with very large doses of calcitriol. In renal osteodystrophy inadequate production of calcitriol by the damaged kidney is a major factor in the bone disease, calcium malabsorption and parathyroid hyperplasia seen in chronic renal disease. Cautious administration of calcitriol may directly inhibit parathyroid hyperplasia, thus helping to prevent further bone disease and the necessity of parathyroid surgery.

E. The Calcitonin Gene Peptides

I. Discovery

1. Calcitonin

The calcitonin gene complex encodes a small family of peptides; calcitonin, katacalcin and calcitonin gene-related peptide (CGRP). The existence of calcitonin was first postulated by Copp and colleagues in 1961 (Copp et al. 1962) who suggested that this principle came from the parathyroid glands. The existence of the peptide was soon confirmed by the Hammersmith group (Kumar et al. 1963), but the Hammersmith experiments made the thyroid equally likely as a possible origin. Simultaneously, Hirsch et al. (1963) had shown that cautery of the rat thyroid or acid extraction yielded a calcium-lowering principle. The matter was resolved by further experiments from Hammersmith (Foster et al. 1964) which clearly showed that calcitonin was a thyroid hormone.

2. Katacalcin

Katacalcin was the name given by MacIntyre and colleagues (Hillyard et al. 1983) to the carboxy terminal flanking peptide from the calcitonin precursor. This peptide is secreted in equimolar amounts with calcitonin. It was first thought to have a plasma calcium-lowering effect (MacIntyre et al. 1982). However, katacalcin has no acute plasma calcium-lowering effect in normal humans (J.C. Stevenson and I. MacIntyre 1984, unpublished work) and although tissue culture experiments do show some action at high concentrations (MacIntyre et al. 1982), the plasma calcium-lowering effects in the rat have proved not to be reproducible (MacIntyre et al. 1984). Thus, the function of katacalcin, if any, is quite unknown.

3. Calcitonin Gene-Related Peptide

The nucleotide sequence of the mRNA of CGRP was predicted by Rosenfeld and colleagues (Rosenfeld et al. 1982, 1983; Amara et al. 1982) from studies with rat medullary carcinoma cells in culture. The first isolation of any CGRP-like molecule from any source was completed by Morris et al. (1984). They not only confirmed the actual existence of the predicted peptide, but also established its existence in humans.

II. Biosynthesis

The present state of knowledge can be summarised as follows. There appear to be two calcitonin genes (Fig. 7, Table 1). The α-gene which was the first to be recognised and about which most is known, is located on the short arm of chromosome 11 (Hoppener et al. 1984; Kittur et al. 1985). The primary nuclear transcript of the α-gene is processed to produce two separate messengers encoding two distinct precursors. One precursor peptide encodes calcitonin and its carboxy terminal flanking peptide, katacalcin, and the other encodes CGRP. Both precursors have common amino terminal sequences. Recently, (Steenbergh et al. 1985;

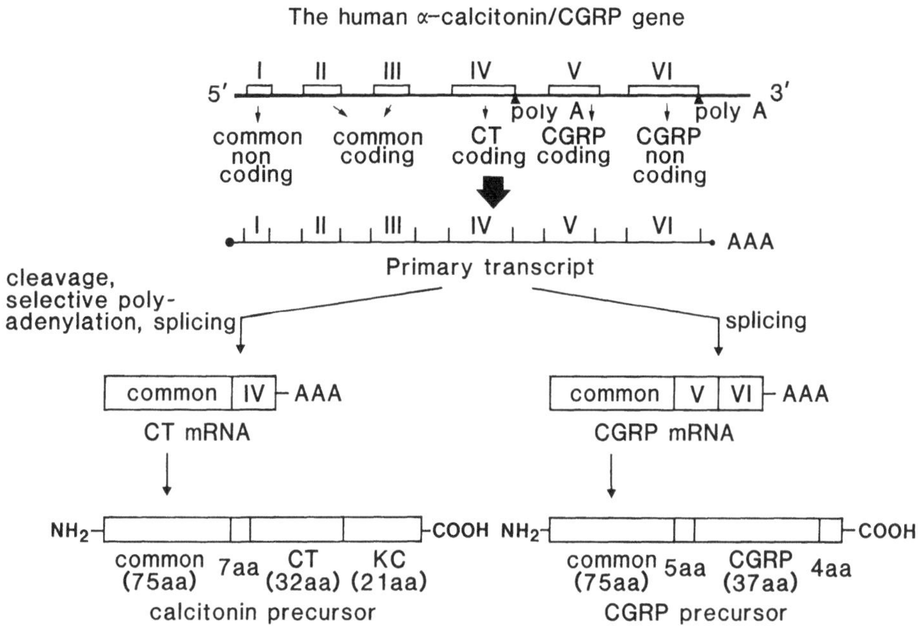

Fig. 7. Organisation of the human α calcitonin/CGRP gene. The same primary transcript is alternatively processed to form mRNA encoding either calcitonin or CGRP. The processing mechanism involves the selective use of the polyadenylation site, cleavage and splicing. It appears that the organisation of the β gene is similar and the coding region has a high degree of homology (≧90%) with the α gene. The 5′ and 3′ non-coding regions exhibit a 30%–40% divergence. (MACINTYRE 1987)

Table 1. Location of the α and β human calcitonin/CGRP genes on the short arm of chromosome 11 (MACINTYRE 1987)

Gene	Location
α	Chromosome 11 (p13–p15) between catalase (16cM) and PTH (8cM)
β	Chromosome 11 (p12–qter)

HOPPENER et al. 1985; AMARA et al. 1985) evidence has been produced that there is a second calcitonin gene, usually referred to as the β-gene. This gene encodes a second CGRP-like peptide which differs in three amino acid residues from α-CGRP. There is also a sequence which hybridises to α-calcitonin probes, but whether this sequence is translated to produce a second calcitonin remains to be determined.

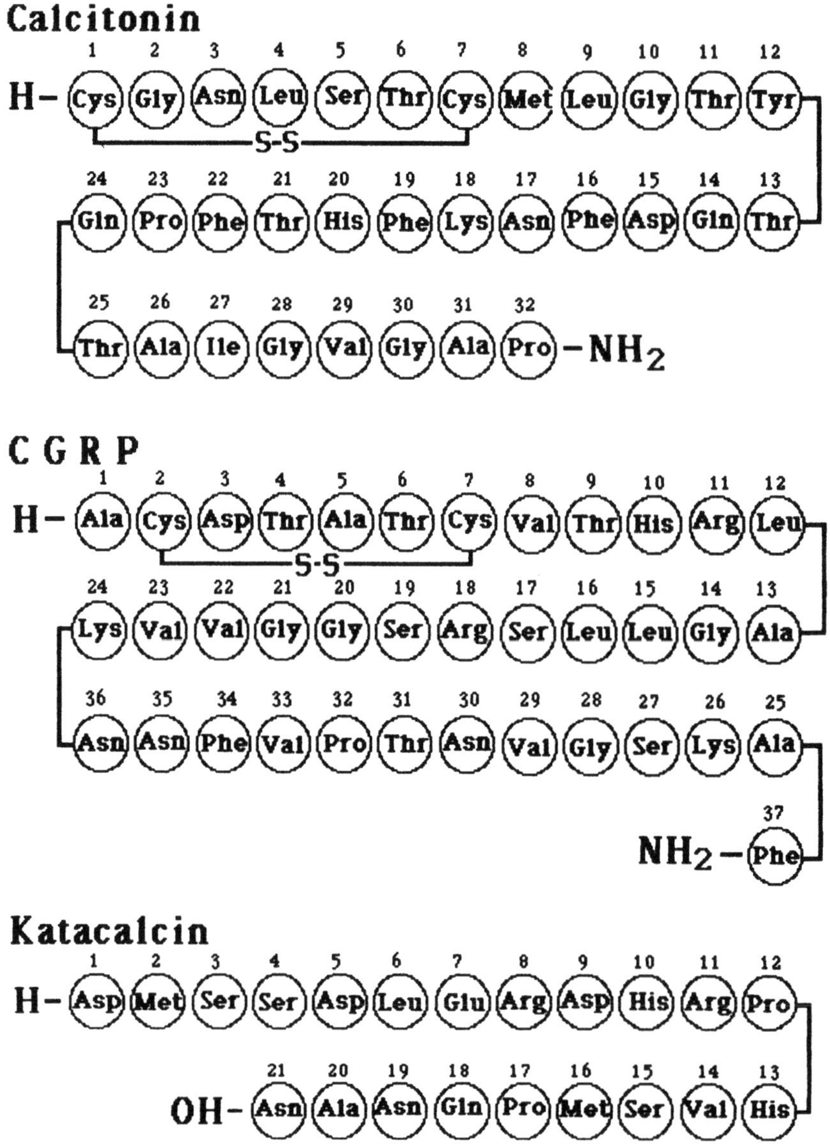

Fig. 8. Sequences of the three major peptides (calcitonin, calcitonin gene-related peptide and katacalcin) from the human α calcitonin/CGRP gene

III. Chemistry

The sequences of the three peptides are shown in Fig. 8. The sequences of calcitonin from eight different species are shown in Table 2. The invariant residues are clustered at the two ends of the molecule, suggesting their importance in biological activity. Figure 9 shows that there are homologies in structure between salmon calcitonin and CGRP. This might suggest that both peptides have arisen

Table 2. Amino acid sequences of the calcitonins[a]

Human 1	Rat	Salmon 2	Salmon 3	Salmon 1	Eel	Chicken[b]	Porcine	Bovine	Ovine	Human 2[b]
H										
|										
1 Cys	—	—	—	—	—	—	—	—	—	Tyr
2 Gly		Ser	Ser	Ser	Ser	Ala	Ser	Ser	Ser	Ser
3 Asn						Ser				
4 Leu	—	—	—	—	—	—	—	—	—	—
5 Ser	—	—	—	—	—	—	—	—	—	—
6 Thr	—	—	—	—	—	—	—	—	—	—
7 Cys	—	—	—	—	—	—	—	—	—	—
8 Met			Val	Val	Val	Val	Val	Val	Val	Leu
9 Leu	—	—	—	—	—	—	—	—	—	Gln
10 Gly							Ser	Ser	Ser	
11 Thr		Lys	Lys	Lys	Lys	Lys	Ala	Ala	Ala	
12 Tyr		Leu	Leu	Leu	Leu	Leu				
13 Thr		Ser	Ser	Ser	Ser	Ser	Trp	Trp	Trp	Leu
14 Gln							Arg	Lys	Lys	
15 Asp				Glu	Glu	Glu	Asn			Tyr
16 Phe	Leu	Leu	Leu	Leu	Leu	Leu	Leu	Leu	Leu	Leu
17 Asn		His	His	His	His	His				Lys
18 Lys							Asn	Asn	Asn	Asn
19 Phe		Leu	Leu	Leu	Leu	Leu		Tyr	Tyr	
20 His		Gln	Gln	Gln	Gln	Gln				
21 Thr							Arg	Arg	Arg	Met
22 Phe				Tyr	Tyr	Tyr			Tyr	
23 Pro						Ser	Ser	Ser		
24 Gln		Arg	Arg	Arg	Arg	Arg	Gly	Gly	Gly	Gly
25 Thr							Met	Met	Met	Ile
26 Ala	Ser	Asn	Asn	Asn	Asp	Asp	Gly	Gly	Gly	Asn
27 Ile		Thr	Thr	Thr	Val	Val	Phe	Phe	Phe	Phe
28 Gly	—	—	—	—	—	—	—	—	—	—
29 Val		Ala	Ala	Ser	Ala	Ala	Pro	Pro	Pro	Pro
30 Gly						Glu	Glu	Glu	Glu	Gln
31 Ala		Val	Val	Thr	Thr	Thr	Thr	Thr	Thr	Ile
32 Pro	—	—	—	—	—	—	—	—	—	—
|										
NH_2										

[a] Amino acid substitutions are shown; *horizontal lines* represent invariant residues at positions 1, 4, 5, 6, 7, 9, 28 and 30. All calcitonins contain at least one acidic amino acid (glutamic or aspartic acid at position 15 and/or 30) and at least two basic residues (arginine, lysine or histidine) and/or an amide at positions 14, 17, 18 and 20. Hydrophobic residues (leucine, phenylalanine and tyrosine) are distributed regularly, occupying positions 4, 9, 12, 16, 19 and 27.

[b] Chicken and human 2 sequences (ALEVIZAKI et al., 1986) are predicted.

by gene duplication. The distribution of calcitonin and CGRP are, however, quite different. Calcitonin is abundant only in the thyroid; CGRP is present not only in the thyroid, but is widespread in the central and peripheral nervous systems (MACINTYRE 1984; TSCHOPP et al. 1984). It is abundant in perivascular nerves, in

```
         1   2   3   4   5   6   7   8   9   10  11  12  13  14  15  16  17  18
s-CT     Cys-Ser-Asn-Leu-Ser-[Thr]-[Cys]-[Val]-Leu-Gly-Lys-[Leu]-Ser-Gln-Glu-[Leu]-His-Lys-
                         *                             *         *                    *
h-CGRP   Ala-Cys-Asp-Thr-Ala-[Thr]-[Cys]-[Val]-Thr-His-Arg-[Leu]-Ala-Gly-Leu-[Leu]-Ser-Arg-

19  20  21  22  23  24  25  26  27  28  29  30  31  32  33  34  35  36  37
                    Leu-Gln-Thr-Tyr-Pro-Arg-[Thr]-[Asn]-Thr-[Gly]-[Ser]-Gly-Thr-Pro-NH2
                            *   *       *                             *
Ser-Gly-Gly-Val-Val-Lys-Asn-Asn-Phe-Val-Pro-[Thr]-[Asn]-Val-[Gly]-[Ser]-Lys-Ala-Phe-NH2
```

Fig. 9. Comparison of the sequences of salmon calcitonin (s-CT) and human α CGRP (h-CGRP). There is approximately 30% homology in structure: a gap of five amino acids (positions 19–23) has been introduced to maximise homology. The *boxed* amino acids are similar in both peptides; those marked by *asterisks* differ only by one base pair. (MacIntyre 1987)

various sites in the brain and in the spinal cord. CGRP appears to be colocalised with substance P in some neural tissues (Wanaka et al. 1987; Lee et al. 1985).

IV. Secretion

Calcitonin and katacalcin are cosecreted in equimolar proportions from C cells (Fig. 10). The major control of calcitonin and katacalcin secretion is probably the plasma calcium concentration; there are also other factors which are less well understood. Oestrogen may be one of these. It is known that deficit of ovarian hormones, after artificial or natural menopause, is followed by a fall in calcitonin secretion (Stevenson et al. 1982; Taggert et al. 1982; Torring et al. 1984; Galan Galan et al. 1985). Chronic oestrogen administration elevates katacalcin and calcitonin (Morimoto et al. 1980; Stevenson et al. 1981). However, it is unclear as to whether this effect represents a direct action on the C cells.

As with parathyroid cells, receptors for $1,25(OH)_2D_3$ have also been identified in C cells (Freake and MacIntyre 1982). It is most likely that vitamin D also has some controlling influence on calcitonin secretion. Other secretagogues for calcitonin are also known. For example, gastrin and cholecystokinin produce a marked increase of calcitonin secretion (Cooper et al. 1971; Care 1970). It has been suggested that secretion of such gastrointestinal hormones following a meal induces calcitonin secretion, thus conserving skeletal calcium. It is uncertain whether this is true for humans. CGRP is mainly secreted from perivascular nerve terminals and in some instances from hyperplastic C cells (Zaidi et al. 1985 a, b, 1986 a).

Like most peptide hormones, plasma immunoreactive calcitonin is heterogeneous and there has been wide disagreement over circulating levels. Two facts are reasonably established. The first is that plasma levels in women are much lower than in men (Fig. 11); and the second is that ovarian failure is followed by a decline in calcitonin levels. However, there is disagreement over the supposed decline of plasma calcitonin with age.

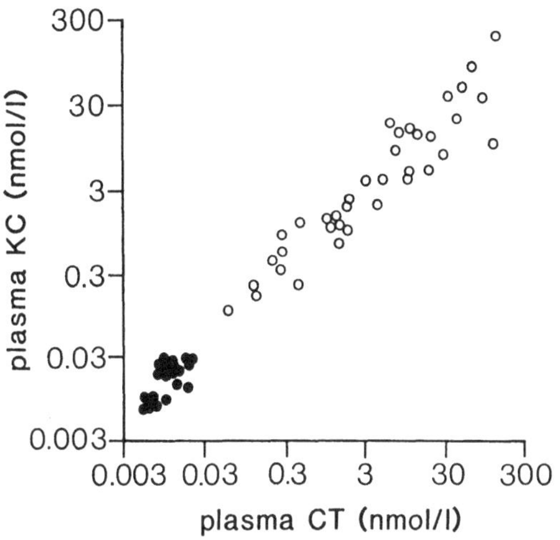

Fig. 10. Correlation between plasma calcitonin (CT) and katacalcin (KC) levels in normal volunteers (*full circles*) and in patients with medullary carcinoma of the thyroid (*open circles*). (MacIntyre 1987)

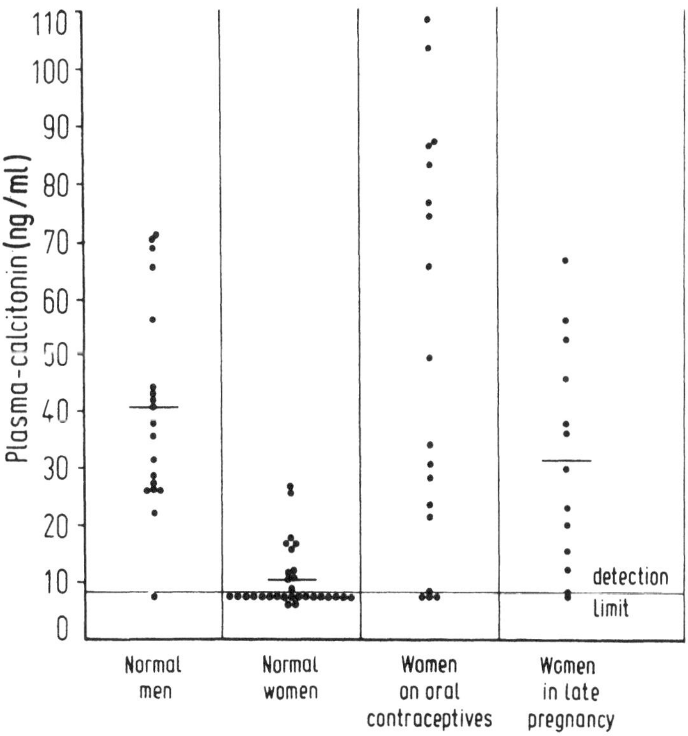

Fig. 11. Plasma calcitonin levels in normal men and three groups of women. *Horizontal bars* represent mean plasma calcitonin levels in each group. Normal women had a significantly ($P \leq 0.001$) lower mean plasma level than the other three groups. (Hillyard et al. 1978)

V. Actions

1. Calcitonin

Calcitonin has a range of effects. Its main action, and certainly the one of major physiological importance, is on the osteoclast. It has only recently been conclusively established by elegant studies from Chambers and colleagues (CHAMBERS et al. 1984, 1985 a, b; CHAMBERS and MOORE 1983; ZAIDI et al. 1987 a) that calcitonin acts directly to produce an inhibition of the bone resorbing activity of the osteoclast (Figs. 12 and 13). Although this action had been deduced for many years (ROBINSON et al. 1967; FRIEDMAN and RAISZ 1965; MILHAUD et al. 1965), direct evidence had been lacking until recently. Though it appears that this effect is mediated via adenylate cyclase activation, intracellular calcium may be another mechanism. Calcitonin has a second effect on the osteoclast which is of particular importance when the drug is used to treat Paget's disease. The hormone produces

a 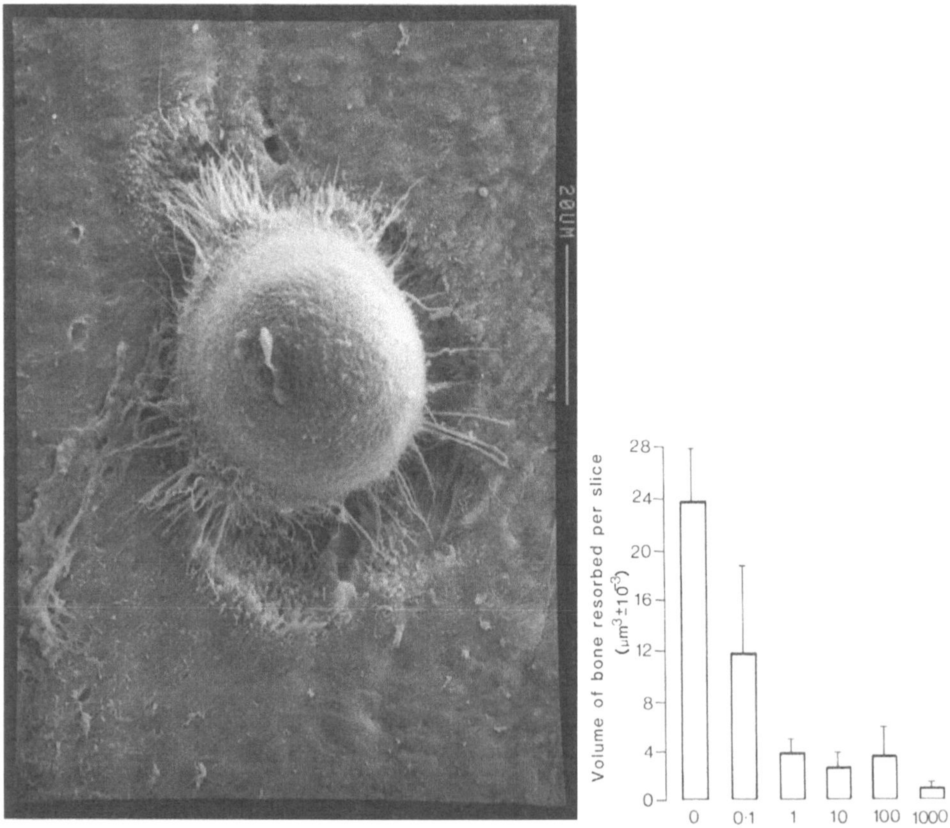 b

Fig. 12. a Scanning electron micrograph of an osteoclast after 8 h incubation on a human cortical bone slice. The cell is associated with an area of resorption to which it is attached by dorsal microvilli. (CHAMBERS 1982) ×1200. **b** Volume of bone resorbed by osteoclasts after incubation for 8 h in varying concentrations of salmon calcitonin. (After CHAMBERS et al. 1985 a)

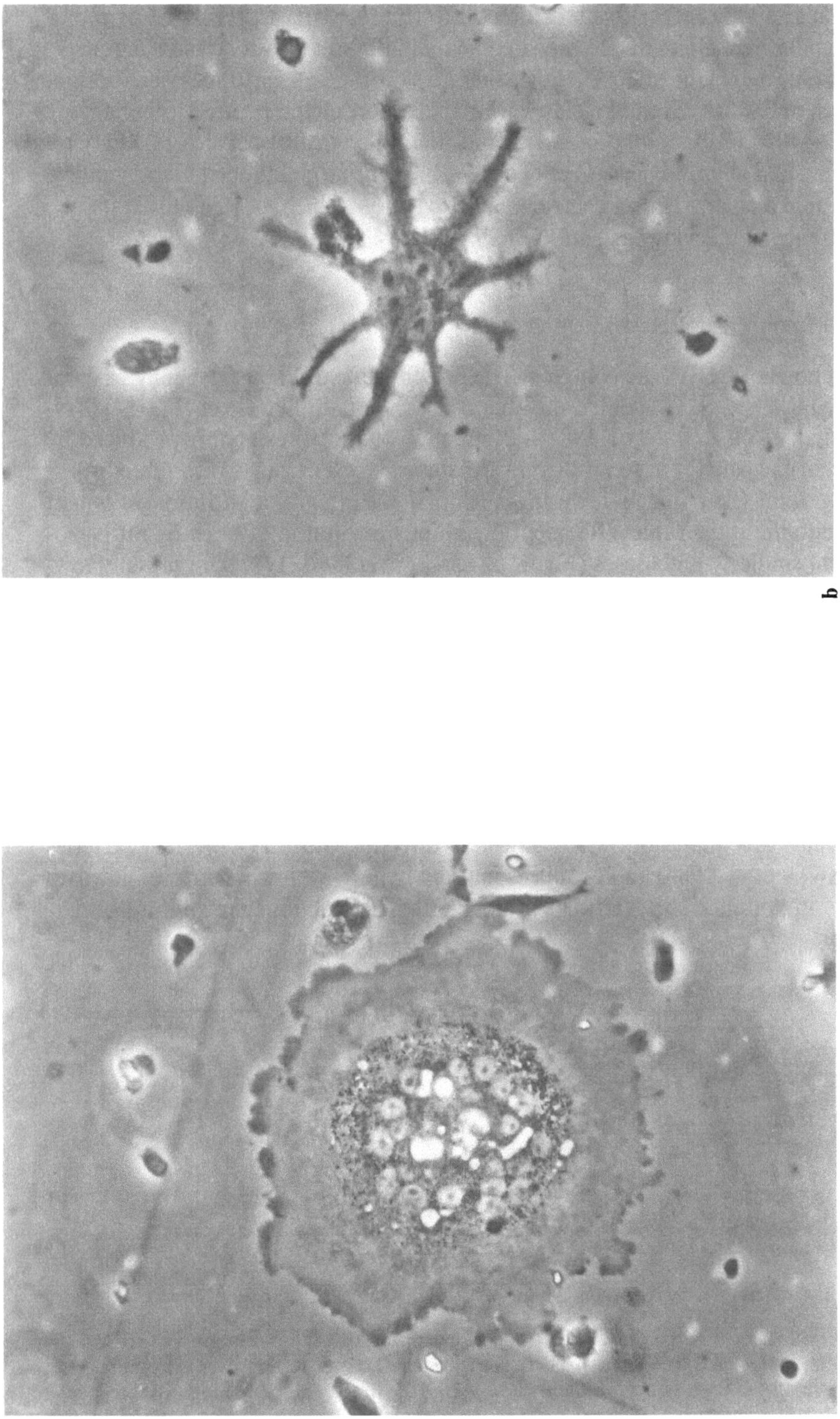

a b

Fig. 13. a Osteoclast after 2 h in control medium, showing a lobulated periphery with pseudopodial ruffling activity. Phase contrast, × 320. **b** Osteoclast after incubation for 2 h with salmon calcitonin. There is an abrupt cessation of all movement followed by gradual fragmentation and pseudopodial retraction to an appearance characteristic of immotility. (Courtesy Professor T.J. Chambers, St. George's Hospital, London)

a marked diminution in osteoclast numbers when administered chronically. It is quite unknown whether this is a direct effect on osteoclast precursors or represents an indirect effect due to repeated acute inhibition of osteoclastic action.

At pharmacological doses, calcitonin acts on the kidney to increase sodium, calcium and water excretion. However, the action of the peptide in enhancing $1{,}25(OH)_2D_3$ production (Galante et al. 1972; Kawashima et al. 1981) may represent a real physiological effect of importance during childhood and pregnancy. Calcitonin also possesses a central analgesic action, which may be of therapeutic importance in relieving the pain of bone diseases.

2. Calcitonin Gene-Related Peptide

The actions of CGRP are complex. The present state of knowledge may be summarised as follows. CGRP has a rather weak calcitonin-like effect (Fig. 14) (Tippins et al. 1984; Zaidi et al. 1987a; Bevis et al. 1986c), presumably produced by action at the calcitonin receptor on the osteoclast. CGRP is two to three orders of magnitude less potent in lowering calcium in the rat, although almost as potent as calcitonin in the rabbit. The peptide inhibits resorption of bone by rat osteoclasts at similarly high doses (Fig. 15; Zaidi et al. 1986d, 1987b, in press). However, CGRP has a second effect on calcium metabolism. In larger doses in the rabbit and also in the chick, CGRP mimics the action of PTH hormone (Tippins et al. 1984). Most probably, it has a receptor at the osteoblast (Crawford et al. 1986), but it seems unlikely that these effects on bone cells could be exerted via circulating CGRP. Nevertheless, it is possible that these effects do reflect a physiological role of CGRP in the local modulation of bone cell function (Zaidi et al. 1986b).

CGRP is a very potent vasodilator (Brain et al. 1985, 1986). It also produces profound vasodilatation when injected directly into the coronary arteries (McEwan et al. 1985) and produces marked hypotension when infused intravenously in volunteers (Struthers et al. 1985). These actions are still not under-

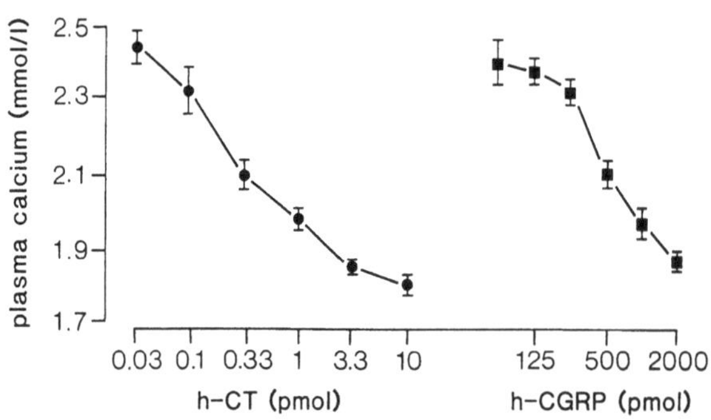

Fig. 14. Effect of a range of doses of human calcitonin (h-CT) and human CGRP(α) (h-CGRP) administered to 50-g rats. There is a significant regression of response on log dose; h-CT is about 1000-fold more potent than h-CGRP, though slopes of log dose–response lines are not identical. (MacIntyre 1987)

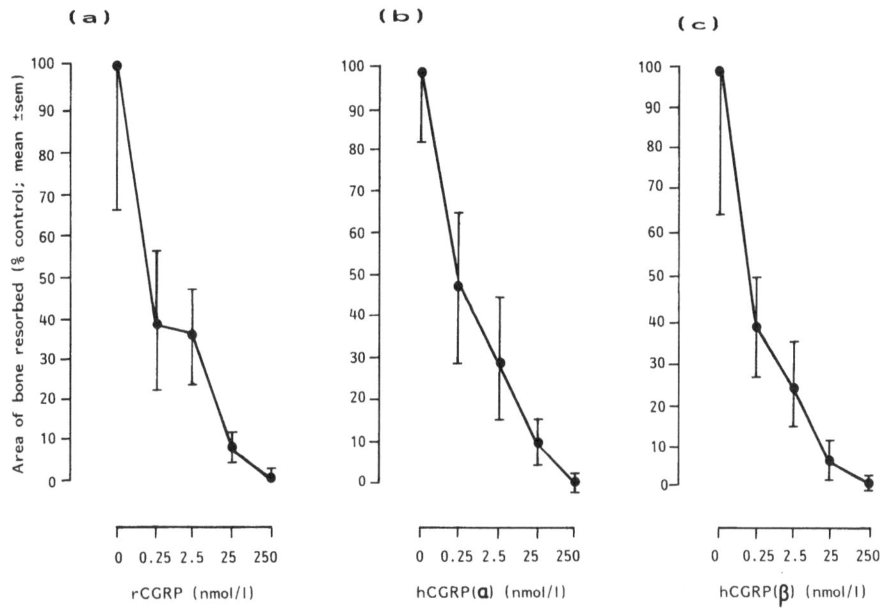

Fig. 15. a Effect of a range of doses of rat calcitonin gene-related peptide (rCGRP), **b** human calcitonin gene-related peptide (α) (hCGRP(α)) and **c** human calcitonin gene-related peptide (β) (hCGRP(β)) on bone resorption by "isolated" osteoclasts

stood in detail. CGRP also appears to be an important sensory neurotransmitter or neuromodulator. It is also likely that the peptide has a range of important actions in the brain. Although receptors have been mapped and identified throughout the central nervous system (TSCHOPP et al. 1985; GOLTZMAN and MITCHELL 1985), their functional significance, if any, remain to be determined.

VI. Physiological Role

1. Calcitonin

There is increasing agreement that the major role of calcitonin is to protect the skeleton during periods of calcium stress, namely growth, pregnancy and lactation. Circulating levels of the peptide are increased during these times (MACINTYRE et al. 1979; STEVENSON et al. 1979). Conversely, there is evidence that failure of ovarian function at natural or artificial menopause is followed by a fall in circulating calcitonin levels (STEVENSON et al. 1982; TAGGERT et al. 1982) also accompanied by evidence of increased osteoclastic activity. It seems reasonable to conclude that this is one factor, perhaps an important one, in postmenopausal osteoporosis. The fall in circulating calcitonin levels is reversed by oestrogen administration (MORIMOTO et al. 1980; STEVENSON et al. 1983), and there is evidence that calcitonin can reduce postmenopausal bone loss (I. MACINTYRE 1983, unpublished work). It is clear that the hormone has a normal physiological role in maintaining skeletal integrity. The impairment of this role is a factor in the pathogenesis of osteoporosis.

2. Calcitonin Gene-Related Peptide

Besides being a neuromodulator, the major peripheral role of CGRP appears to be the control of blood flow. The factors which influence CGRP release from the peripheral nerves are unclear, but it may be relevant to mention the circulating levels (Fig. 16). There is no age or sex difference with CGRP (Girgis et al. 1985; Zaidi et al. 1986a) as is found with plasma calcitonin levels (Hillyard et al. 1978). Studies with colchicine in experimental animals show that administration of this drug dramatically lowers plasma CGRP levels (Fig. 17), suggesting that plasma CGRP is dependent on axonal transport (Zaidi et al. 1985b; Bevis et al. 1986a, b). Partial depletion of plasma CGRP levels in rats also occurs owing to deafferentation produced by neonatal capsaicin lesioning: there is a sharp rise following acute nerve terminal depolarisation (Zaidi et al. 1986b, c; Bevis et al. 1986b). No change is produced in calcitonin levels. The conclusion from these studies is that, unlike calcitonin, which is thyroidal in origin, circulating CGRP comes mainly from the perivascular nerves. Calcitonin is a circulating hormone

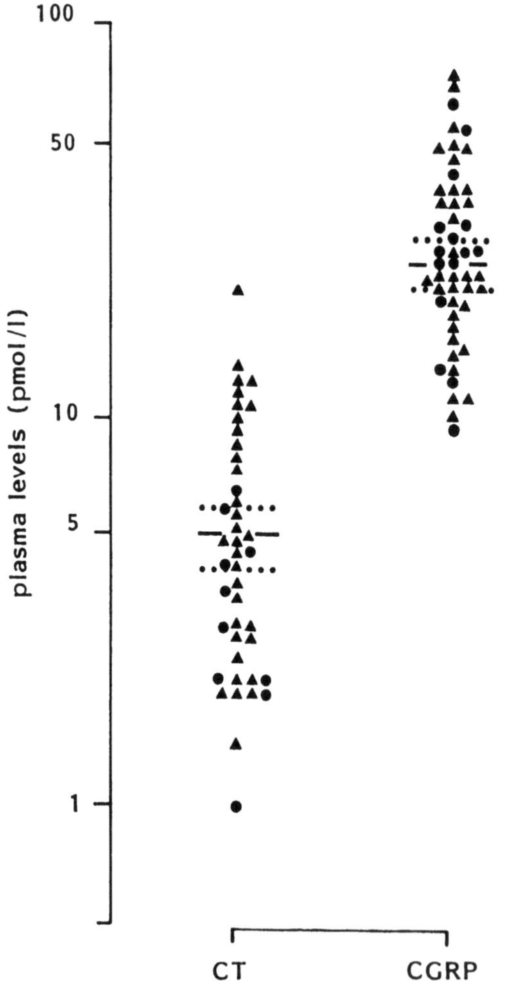

Fig. 16. Plasma calcitonin (CT) and calcitonin gene-related peptide (CGRP) levels in normal males (*triangles*) and females (*circles*). *Horizontal bars* and *dotted lines* represent mean and standard error of mean of the groups, respectively. (Girgis et al. 1985)

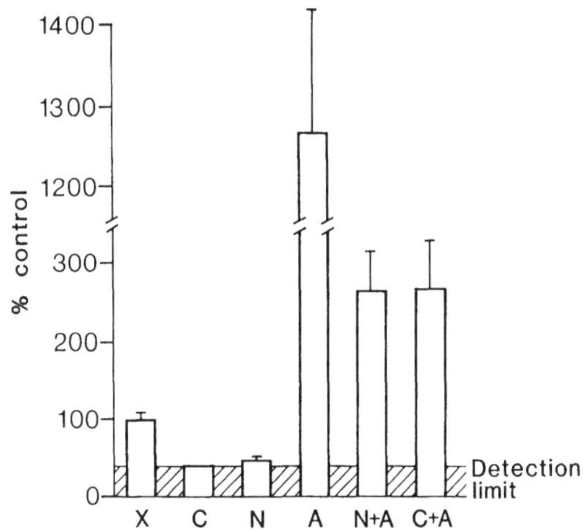

Fig. 17. Effect of capsaicin or colchicine treatment on plasma calcitonin gene-related peptide of 40-day-old rats (expressed as percentages of control values). Each *bar* represents a group of rats treated as follows: *X* no treatment; *C* colchicine (10 mg/kg intraperitoneal, bled after 6 h); *N* capsaicin (to neonatal rats, 50 mg/kg subcutaneous, bled as adults); *A* capsaicin (to adult rats, 10 mg/kg intraperitoneal, bled after 7.5 min); *N+A* capsaicin treatment to neonatally capsaicinised adults (protocol as for *N* and *A*); *C+A* capsaicin to colchicine pretreated adults (protocol as for *C* and *A*). (MacIntyre 1987)

controlling osteoclastic activity; plasma CGRP is the consequence of release from perivascular nerves which modulate vessel tone and regulate blood flow.

VII. Pathophysiology: Medullary Carcinoma of the Thyroid

Plasma calcitonin and katacalcin are nearly always elevated in medullary carcinoma of the thyroid. Measurement of these peptides may be useful in judging complete operative removal of the tumour, in predicting relapse and in excluding the familial form of the disease. Marginally elevated levels must be repeatedly confirmed prior to surgery. CGRP is often, though not always, elevated in this tumour and recent evidence suggests that it may be a useful index for differentiating the familial from sporadic forms.

VIII. Therapeutic Considerations

1. Calcitonin

a) Paget's Disease

In our view calcitonin is an essential part of the treatment of Paget's disease. The hormone produces a remission and partial healing of the disease, although the biochemistry rarely returns to normal completely. Though EHDP (etidronate sodium, ethane-1-hydroxy-1,1-diphosphonate disodium) produces a much more complete and sustained effect on biochemistry it also causes the formation of osteoid seams resembling osteomalacia; to use it may therefore be dangerous in osteolytic forms of the disease (Nagant de Deuxchaisnes et al. 1979). These reservations however, do not apply to some of the newer diphosphonates. In general it is unlikely that they would be introduced because of fears of bone marrow effects. The fact that calcitonin may have to be given by daily injection, almost indefinitely, represents the main drawback of its use. However, mild forms of the disease may require shorter courses or administration, thrice weekly.

b) Osteoporosis

Much smaller doses of calcitonin than those necessary for Paget's disease are necessary for the treatment of osteoporosis (10 units thrice weekly as compared with 100 units for Paget's disease) (I. MacIntyre and J. C. Stevenson 1984, unpublished work). Though there is marked inhibition of postmenopausal bone loss, it is not likely that calcitonin alone can reverse severe established osteoporosis. Calcitonin therapy is however rational in its prevention. It should also be part of a therapeutic programme which attempts to reverse established disease.

c) Hypercalcaemia

Calcitonin is effective in the treatment of hypercalcaemia of malignancy. As the effect wears off in a couple of days, it needs to be combined with agents such as Cl_2MDP (dichloromethylene diphosphonate) or ADP (3-amino-1-hydroxypropylidene-1,1-bisphosphonate) which have a more persistent effect.

2. Calcitonin Gene-Related Peptide

As the most powerful vasodilator known, CGRP may have some therapeutic potential in vascular disease, especially involving the coronary or cerebral vessels. In our view, CGRP may come to have a role in the treatment of acute vascular insufficiency, but is less likely to be a major agent for chronic administration. The necessity of administration by injection and the cost of such a programme would be the main impediments. When the mode of its action and the receptors with which it interacts are better understood, simpler synthetic agents may be devised having the effects of CGRP without its disadvantages.

Acknowledgments. The authors acknowledge the secretarial assistance of Mrs. Dorothy Simmonds and Mrs. Brenda Salvage.

References

Amara SG, Jonas V, Rosenfeld MG, Ong ES, Evans RM (1982) Alternative RNA processing in calcitonin gene expression generates mRNAs encoding different polypeptide products. Nature 298:240–244

Amara SG, Arriza JL, Leff SE, Swanson LW, Evans RM, Rosenfeld MG (1985) Expression in brain of messenger RNA encoding a novel neuropeptide homologous to calcitonin gene-related peptide. Science 229:1094–1097

Arnaud CD, Goldsmith RS, Bordier PJ, Sizemore GW (1974) Influence of immunoheterogenaeity of circulating parathyroid hormone on results of radioimmunoassay of serum in man. Am J Med 56:785–793

Aurbach GD, Chase LD (1976) Cyclic nucleotides and biochemical actions of parathyroid hormone and calcitonin. In: Greep RO, Astwood EB (eds) Handbook of physiology. Endocrinology. American Physiological Society, Washington DC, pp 353–381
Avioli LV, Krane SM (1978) Metabolic bone disease, vol 2. Academic, New York
Bevis PJR, MacIntyre I, Morris HR, Zaidi M (1986a) On the presence of a potent vasodilator calcitonin gene-related peptide in rat plasma. J Physiol (Lond) 376:24
Bevis PJR, MacIntyre I, Zaidi M (1986b) Further evidence for the neural release of plasma calcitonin gene-related peptide. Br J Pharmacol 88:314
Bevis PJR, Zaidi M, Lynch C, Beacham J, Chambers TJ, MacIntyre I (1986c) Human beta-calcitonin gene related peptide inhibits osteoclastic bone resorption and lowers plasma calcium. Bone (in press)
Boyle IT, Gray RW, DeLuca HF (1971) Regulation by calcium of in vivo synthesis of 1,25-dihydroxycholecalciferol and 24,25-dihydroxycholecalciferol. Proc Natl Acad Sci USA 58:2131–2134
Brain SD, Williams TJ, Tippins JR, Morris HR, MacIntyre I (1985) Calcitonin gene-related peptide is a potent vasodilator. Nature 313:54–56
Brain SD, MacIntyre I, Williams TJ (1986) A second form of human calcitonin gene-related peptide which is a potent vasodilator. Eur J Pharmacol 124:349–352
Broadus AE, Goltzman D, Webbs AC, Kronenberg HM (1985) Messenger ribonucleic acid from tumors associated with humoral hypercalcemia of malignancy directs the synthesis of a secretory parathyroid hormone-like peptide. Endocrinology 117:1661–1666
Canterbury JM, Reiss E (1972) Multiple immunoreactive molecular forms of parathyroid hormone in human serum. Proc Soc Exp Biol Med 140:1393–1398
Care AD (1970) The effects of pancreozymin and secretin on calcitonin release. Fed Proc 29:253
Chambers TJ (1982) Pathobiology of the osteoclast. J Clin Pathol 38:214–252
Chambers TJ, Moore A (1983) The sensitivity of isolated osteoclasts to morphological transformation by calcitonin. J Clin Endocrinol Metab 57:819–824
Chambers TJ, Athanasou NA, Fuller K (1984) Effect of parathyroid hormone and calcitonin on the cytoplasmic spreading of isolated osteoclasts. J Endocrinol 102:281–286
Chambers TJ, McSheehy PMJ, Thomson BM, Fuller K (1985a) The effect of calcium-regulating hormones and prostaglandins on bone resorption by osteoclasts disaggregated from neonatal rabbit bones. Endocrinology 60:234–239
Chambers TJ, Fuller K, McSheehy PMJ, Pringle JAS (1985b) The effect of calcium regulating hormones on bone resorption by isolated human osteoclastoma cells. J Pathol 145:297–305
Chase LR, Aurbach GD (1967) Parathyroid function of the renal excretion of 3′,5′-adenylic acid. Proc Natl Acad Sci USA 58:518–525
Chase LR, Aurbach GD (1968) Renal adenyl cyclase: anatomical separation of sites sensitive to parathyroid hormone and vasopressin. Science 159:545–547
Cooper CW, Schwesinger WH, Mahgoub AM, Ontjes DA (1971) Thyrocalcitonin stimulation of secretion by pentagastrin. Science 172:1238–1240
Copp DH, Cameron EC, Cheney BA, Davidson AGF, Henze KG (1962) Evidence for calcitonin – a new hormone from the parathyroid that lowers blood calcium. Endocrinology 70:638–649
Crawford A, Evans DB, Skjodt H, Beresford JN, MacIntyre I, Russell RGG (1986) Effect of human calcitonin gene-related peptide on human bone-derived cells in culture. Bone 7:157–158
Fenwick JC (1982) Some evidence concerning the nature of the hypocalcemic factor in the Stannius corpuscles. In: Oguro C, Pang PKT (eds) Comparative endocrinology of calcium regulation. Japan Scientific Societies Press, Tokyo, pp 167–172
Fischer JA, Binswanger U, Dietrich FM (1974) Human parathyroid hormone: immunological characterization of antibodies against a glandular extract and the synthetic amino-terminal fragments 1–12 and their use in the determination of immunoreactive hormone in human sera. J Clin Invest 54:1382–1394
Fontaine M (1964) Corpuscles de Stannius et régulation ionique (Ca, K, Na) du milieu interieur de l'anguille (*Anguilla anguilla* L.). CR Seances Acad Sci [III] 259:875–878

Foster GV, Baghdiantz A, Kumar MA, Slack E, Soliman HA, MacIntyre I (1964) Thyroid origin of calcitonin. Nature 202:1303–1305
Fraser DR (1980) Regulation of the metabolism of vitamin D. Physiol Rev 60:551–607
Freake HC, MacIntyre I (1982) Specific binding of 1,25-dihydroxycholecalciferol in human medullary thyroid carcinoma. Biochem J 206:181–184
Friedman J, Raisz LG (1965) Thyrocalcitonin inhibitor of bone resorption in tissue culture. Science 150:1465–1467
Galan Galan F, Perez Cano R, Rodriguez R et al. (1985) Deficit de calcitonia en la osteoporosis posmenopausica. Med Clin (Barc) 85:221
Galante L, Colston KW, MacAuley SJ, MacIntyre I (1972) Effect of calcitonin on vitamin D metabolism. Nature 238:271–273
Garabedian M, Holick MF, DeLuca HF, Boyle IT (1972) Control of 25-hydroxycholecalciferol metabolism by parathyroid glands. Proc Natl Acad Sci USA 69:1673–1676
Girgis SI, Macdonald DWR, Stevenson JC, Bevis PJR, Lynch C, Wimalawansa SJ, Self CH, Morris HR, MacIntyre I (1985) Calcitonin gene-related peptide: potent vasodilator and major product of calcitonin gene. Lancet 2:14–16
Goldsmith RS, Furszyfer J, Johnson WJ, Fournier AE, Sizemore GW, Arnaud CD (1973) Etiology of hyperparathyroidism and bone disease during chronic hemodialysis. III. Evaluation of parathyroid suppressibility. J Clin Invest 52:173–180
Goltzman D, Mitchell J (1985) Interaction of calcitonin and calcitonin gene-related peptide at receptor sites in target tissues. Science 227:1343–1345
Habener JF, Rosenblatt M, Potts JT Jr (1984) Parathyroid hormone: biochemical aspects of biosynthesis, secretion, action and metabolism. Physiol Rev 64:985–1053
Hausslet MR, Bayling DJ, Hughes MR, Brumbaugh PF, Wergedal JE, Shen FH, Nielsen RL, Counts SJ, Bursac KM, McCain TA (1976) The assay of 1α,25-dihydroxyvitamin D_3: physiologic and pathologic modulation of circulating hormone levels. Clin Endocrinol (Oxf) 5:151S–165S
Heinrich G, Kronenberg HM, Potts JT Jr, Habener JF (1984) Gene encoding parathyroid hormone. Nucleotide sequence of the rat gene and deduced aminoacid sequence of rat preproparathyroid hormone. J Biol Chem 259:3320–3329
Hillyard CJ, Stevenson JC, MacIntyre I (1978) Relative deficiency of plasma-calcitonin in normal women. Lancet 1:961–962
Hillyard CJ, Myers C, Abeyasekera G, Stevenson JC, Craig RK, MacIntyre I (1983) Katacalcin: a new plasma calcium-lowering hormone. Lancet 1:846–848
Hirsch PF, Gauthier GF, Munson PL (1963) Thyroid hypocalcaemic principle and recurrent laryngeal nerve injury as factors affecting the response to parathyroidectomy in rats. Endocrinology 73:244–252
Hoppener JWM, Steenbergh PH, Bakker E, Pearson PL, Geurts van Kessel AHM, Jansz HS, Lipps CJM (1984) Localization of the polymorphic human calcitonin gene on chromosome 11. Hum Genet 66:309–312
Hoppener JWM, Steenbergh PH, Geurts van Kessel AHM, Baylin SB, Nelkin BD, Jansz HS, Lips CJM (1985) The second human calcitonin/CGRP gene is located on chromosome 11. Hum Genet 70:259–263
Horiuchi N, Suda T, Sasaki S, Takahashi H, Shimazawa E, Ogata E (1976) Absence of regulatory effects of 1,25-dihydroxyvitamin D_3 on 25-hydroxyvitamin D metabolism in rats constantly infused with parathyroid hormone. Biochem Biophys Res Commun 73:869–875
Kaminsky NI, Broadus AE, Hardman JG, Jones DJ Jr, Ball JH, Sutherland EW, Liddle GW (1970) Effects of parathyroid hormone plasma and urinary adenosine 3′,5′-monophosphate in man. J Clin Invest 49:2387–2395
Kawashima H, Torikai S, Kurokawa K (1981) Calcitonin selectively stimulates 25-hydroxyvitamin D_3-1-hydroxylase in proximal straight tubule of rat kidney. Nature 291:327–319
Keutman HT, Sauer RM, Hendy GN, O'Riordan JLH, Potts JT Jr (1978) Complete amino acid sequence of human parathyroid hormone. Biochemistry 17:5723–5729
Kittur SD, Hoppener JWM, Antonaratis SE, Daniels JDJ, Meyers DA, Maestri NE, Jansen M, Kormeluk RG, Neltin BD, Kazazian HH (1985) Linkage map of the short

arm of human chromosome 11: location of the genes for catalase, calcitonin and insulin-like growth factor II. Proc Natl Acad Sci USA 82:5064–5067
Kumar MA, Foster GV, MacIntyre I (1963) Further evidence for calcitonin, a rapid-acting hormone which lowers plasma calcium. Lancet 2:480–482
Lafeber F, Hermann-Erlee MPM, van der Meer JM, Flick G, Verbost P, Wendelaar Bonga SE (1984) Effects of products released by trout corpuscles of Stannius on calcium mobilization and cAMP formation of mouse calvaria in vitro and on plasma calcium in eels. Calcif Tissue Int [Suppl 2] 3:50
Larkins RG, MacAuley SJ, Colston KW, Evans IMA, Galante LS, MacIntyre I (1973) Regulation of vitamin D metabolism without parathyroid hormone. Lancet 2:289–291
Lee Y, Kawai Y, Shiosaka S, Takami K, Kiyama H, Hillyard CJ, Girgis S, MacIntyre I, Emson PC, Tohyama M (1985) Coexistence of calcitonin gene-related peptide and substance P-like peptide in single cells of the trigeminal ganglion of the rat: immunohistochemical analysis. Brain Res 330:194–196
Lopez E, Tisserand-Jochem EM, Eyquem A, Milet C, Hillyard C, Lallier F, Vidal B, MacIntyre I (1984a) Immunocytochemical detection in eel corpuscles of Stannius of a mammalian parathyroid-like hormone. Gen Comp Endocrinol 53:28–36
Lopez E, Tisserand-Jochem EM, Vidal B, Milet C, Lallier F, MacIntyre I (1984b) Les corpuscles de Stannius sont-ils les glandes parathyroides des poissons téléostéens? Arguments ultrastructuraux, cytologiques et immunocytochimiques. CR Seances Acad Sci [III] 298:359–364
Lwowski ES (1978) The corpuscles of Stannius and calcified tissues in the eel, *Anguilla rostrata* L. Comp Biochem Physiol 59A:183–187
Ma SWY, Copp DH (1978) Purification, properties and action of a glycopeptide from the corpuscles of Stannius which affects calcium metabolism in the teleost. In: Gaillard PJ, Boer HH (eds) Comparative endocrinology. Elsevier/North Holland, Amsterdam, pp 283–286
MacIntyre I (1984) The calcitonin gene peptide family and the central nervous system. In: Labrie F, Proulx L (eds) Proceedings of the 7th International Congress of Endocrinology, Quebec, Canada, July 1984. Excerpta Medica, Amsterdam, pp 930–933
MacIntyre I (1987) Calcitonin gene peptides. In: De Groot L (ed) Endocrinology, 2nd edn. Grune and Stratton, Orlando
MacIntyre I, Colston KW, Szelke M, Spanos E (1978) A survey of hormonal factors that control calcium metabolism. In: Scarpa A, Carafoli E (eds) Calcium transport and cell function. Ann NY Acad Sci 307:345–355
MacIntyre I, Arnett TR, Brown DJ, Galan Galan F, Girgis SI, Rogers RM, Spanos E, Stevenson JC, Bone Q (1979) The interrelation of the calcium regulating hormones: some recent findings. In: MacIntyre I, Szelke M (eds) Molecular endocrinology. Elsevier/North Holland, Amsterdam, pp 193–201
MacIntyre I, Milet C, Arnett TR, Coghlan JP, Hillyard CJ, Girgis S, Martelly E, Niall HD, Lopez E (1981) The eel corpuscles of Stannius secrete a molecule resembling mammalian parathyroid hormone (abstr 165). 63rd Annual Meeting of the Endocrine Society USA (1981)
MacIntyre I, Hillyard CJ, Murphy PK, Reynolds JJ, Gaines Das RE, Craig RK (1982) A second plasma calcium lowering peptide from the human calcitonin precursor. Nature 300:460–462
MacIntyre I, Hillyard CJ, Reynolds JJ, Gaines Das RE, Craig RK (1984) A second plasma calcium lowering peptide from the human calcitonin precursor: a re-evaluation. Nature 308:84
MacIntyre I, Alerizaki M, Bevis PJR, Zaidi M (1987) Calcitonin and peptides from the calcitonin gene. Clin Orthop and Rel Res 217:45–54
McEwan J, Chierchia S, Davies G, Stevenson JC, Brown M, Maseri A, MacIntyre I (1985) Coronary vasodilatation by calcitonin gene-related peptide. Br Heart J 54:643–645
McSheehy PMJ, Chambers TJ (1986a) Osteoblastic cells mediate osteoclastic responsiveness to PTH. Endocrinology 118:824–827
McSheehy PMJ, Chambers TJ (1986b) Osteoblast-like cells in the presence of parathyroid hormone release soluble factor that stimulates osteoblastic bone resorption. Endocrinology 119 (in press)

Milet C (1986) PhD Dissertation, University of Paris (in press)

Milet C, Lopez E, Chartier MM, Martelly E, Lallier F, Vidal B (1979a) A new calcium regulating hormone from the corpuscles of Stannius. In: MacIntyre I, Szelke M (eds) Molecular endocrinology. Elsevier/North Holland, Amsterdam, pp 341–348

Milet C, Peignoux-Deville J, Martelly E (1979b) Gill calcium fluxes in the eel, *Anguilla anguilla* L. Effects of Stannius corpuscles and ultimobranchial body. Comp Biochem Physiol 63A:63–70

Milet C, Hillyard CJ, Martelly E, Girgis S, MacIntyre I, Lopez E (1980) Similitudes structurales entre l'hormone hypocalcémiante des corpuscles de Stannius (PCS) de l'anguille (*Anguilla anguilla* L.) et l'hormone parathyroidienne mammalienne. CR Seances Acad Sci [III] 291:977–980

Milhaud G, Perault A-M, Moukhtar MA (1965) Etude du mécanisme de l'action hypocalcémiante de la thyrocalcitonine. CR Seances Acad Sci [III] 261:813–816

Morimoto S, Tsuji M, Okada Y, Onishi T, Kumahara Y (1980) The effect of oestrogens on human calcitonin secretion after calcium infusion in elderly female subjects. Clin Endocrinol (Oxf) 13:135–143

Morris HR, Panico M, Etienne T, Tippins J, Girgis SI, MacIntyre I (1984) Isolation and characterisation of human calcitonin gene-related peptide. Nature 308:746–748

Nagant de Deuxchaisnes C, Rombouts-Lindemans C, Huaux JP, Devogelaer JP, Malghem J, Maldague B (1979) Roentgenologic evaluation of the action of the diphosphonate EHDP and of combined therapy (EHDP and calcitonin) in Paget's disease of bone. In: MacIntyre I, Szelke M (eds) Molecular endocrinology. Elsevier/North Holland, Amsterdam, pp 405–433

Parsons JA (1976) Parathyroid physiology and the skeleton. In: Bourne GH (ed) The biochemistry and physiology of bone. Academic, New York, pp 159–225

Parsons JA, Zanelli JM (1980) Physiological role of the parathyroid glands. In: Kuhlencordt F, Bartelheimer H (eds) Klinische Osteologie. Springer, Berlin Heidelberg New York, pp 135–172 (Handbuch der inneren Medizin, vol 6/1)

Robinson CJ, Martin TJ, Matthews EW, MacIntyre I (1967) Mode of action of thyrocalcitonin. J Endocrinol 39:71–79

Rosenblatt M, Beaudette NV, Fasman GD (1980) Conformational studies of the synthetic precursor-specific region of pre-proparathyroid hormone. Proc Natl Acad Sci USA 77:3983–3987

Rosenfeld MG, Lin CR, Amara SG, Stolarsky L, Roos BA, Ong ES, Evans RM (1982) Calcitonin mRNA polymorphism: peptide switching associated with alternative RNA splicing events. Proc Natl Acad Sci USA 79:1717–1721

Rosenfeld MG, Mermod J-J, Amara SG, Swanson LW, Sawchenko PE, Rivier J, Vale WW, Evans RM (1983) Production of a novel neuropeptide encoded by the calcitonin gene via tissue-specific RNA processing. Nature 304:129–135

Stannius FH (1839) Ueber Nebennieren bei Knochenfischen. Arch Anat Physiol 91–101

Steenbergh PH, Hoppener JWM, Zandberg J, Lips CJM, Jansz HS (1985) A second human calcitonin/CGRP gene. FEBS Lett 183:403–407

Stevenson JC, Hillyard CJ, MacIntyre I, Cooper H, Whitehead MI (1979) Physiological role for calcitonin: protection for the maternal skeleton. Lancet 2:769–770

Stevenson JC, Abeyasekera G, Hillyard CJ, Phang KG, MacIntyre I, Campbell S, Townsend PT, Young O, Whitehead MI (1981) Calcitonin and the calcium-regulating hormones in postmenopausal women: effect of oestrogens. Lancet 1:693–695

Stevenson JC, MacIntyre I, Whitehead MI (1982) Impaired calcitonin secretion after premature menopause. Calcif Tissue Int [Suppl 1] 34:821

Stevenson JC, Abeyasekera G, Hillyard CJ, Phang KG, MacIntyre I, Campbell S, Lane E, Townsend PT, Young O, Whitehead MI (1983) Regulation of calcium-regulating hormones by exogenous sex steroids in early postmenopause. Eur J Clin Invest 13:481–487

Struthers AD, Brown MJ, Macdonald DWR, Beacham JL, Stevenson JC, Morris HR, MacIntyre I (1985) Human CGRP: the most potent endogenous vasodilator in man. Clin Sci 70:389–393

Suda T, Abe E, Miyaura C et al. (1983) Does vitamin D have a specific role in cell growth and differentiation? In: Cohn DV, Fujita T, Potts JT Jr, Talmage RV (eds) Endocrine control of bone and calcium metabolism, vol 8A. Excerpta Medica, Amsterdam, pp 308–315

Taggert HMc, Chesnut CH, Ivey JL, Baylink DJ, Sisom K, Huber MB, Roos BA (1982) Deficient calcitonin response to calcium stimulation in postmenopausal osteoporosis. Lancet 1:475–478

Tippins JR, Morris HR, Panico M, Etienne T, Bevis P, Girgis S, MacIntyre I, Azria M, Attinger M (1984) The myotropic and plasma-calcium modulating effects of calcitonin gene-related peptide (CGRP). Neuropeptides 4:425–434

Torring O, Bucht E, Sjöberg HE (1984) Can a relative calcitonin deficiency contribute to the development of postmenopausal osteoporosis? In: Christiansen C, Arnaud CD, Nordin BEC, Parfitt AM, Peck WA, Riggs BL (eds) Osteoporosis. Glostrup Hospital, Glostrup, pp 393–395

Tschopp FA, Tobler PH, Fischer JA (1984) Calcitonin gene-related peptide in the human thyroid pituitary and brain. Mol Cell Endocrinol 36:53–57

Tschopp FA, Henke H, Petermann JB, Tobler TH, Janzer R, Hokfelt T, Lundberg JM, Cuells C, Fischer JA (1985) Calcitonin gene related peptide and its binding site in the human central nervous system and pituitary. Proc Natl Acad Sci USA 82:248–252

Wanaka A, Matsuyama T, Yoneda S, Kimura K, Kamada T, MacIntyre I, Emson PC, Tohyama M (1987) Origins and distribution of calcitonin gene-related peptide-containing nerves in the wall of the cerebral arteries of the guinea pig with special reference to the coexistence with substance P. Brain Res (in press)

Zaidi M, Bevis PJR, Abeyasekera G, Girgis SI, MacIntyre I (1985a) Studies on circulating calcitonin gene-related peptide in the rat. J Endocrinol 107:43

Zaidi M, Bevis PJR, Girgis SI, Lynch C, Stevenson JC, MacIntyre I (1985b) Circulating CGRP comes from the perivascular nerves. Eur J Pharmacol 117:283–284

Zaidi M, Bevis PJR, Abeyasekera G, Girgis SI, Wimalawansa SJ, Morris HR, MacIntyre I (1986a) The origin of calcitonin gene-related peptide in the rat. J Endocrinol 110:185–190

Zaidi M, Bevis PJR, Diez Guerra J, Lynch C, Wimalawansa SJ, Emson PC, MacIntyre I (1986b) Circulation of neurally derived calcitonin gene related peptide (abstr). Neuroendocrinology 74 (Special Issue)

Zaidi M, Bevis PJR, MacIntyre I (1986c) The dual origin of plasma calcitonin gene related peptide (abstr). 68th Annual Meeting of the Endocrine Society

Zaidi M, Bevis PJR, Lynch C, McSheehy PMJ, Chambers TJ, MacIntyre I (1986d) The calcitonin-like effect of α and β human CGRP is mediated via direct osteoclastic inhibition. J Bone Min Res 1:357

Zaidi M, Fuller K, Bevis PJR, Gaines Das RE, Chambers TJ, MacIntyre I (1987a) Calcitonin gene related peptide inhibits osteoclastic bone resorption: a comparative study. Calcif Tissue Int 40:149–154

Zaidi M, Chambers TJ, Gaines Das RE, Morris HR, MacIntyre I (1987b) A direct effect of human calcitonin gene related peptide on isolated osteoclasts. J Endocrinol 115 (in press)

Note added in proof: Very recently COGHLAN, BUTKUS et al. (personal communication) have cloned a cDNA from messenger RNA extracts from the corpuscle. This appears to reflect the most abundant species present and despite absence of similarity to PTH may represent the active hormone. The predicted molecular weight is 32,000. The immunoreactivity of the CS to PTH antisera may therefore reflect conformational similarity rather than homology of sequence. The peptide predicted by COGHLAN and BUTKUS, however, still requires extensive bioassays of selected synthetic portions of the predicted sequence to establish this molecule as the active hormone.

References added in proof: Alevizaki M, Shiraishi A, Rassool FV, Ferrier GJM, MacIntyre I, Legon S (1986) The calcitonin-like sequence of the -CGRP gene. FEBS Lett 206:47–41

CHAPTER 21

Bisphosphonates: A New Class of Drugs in Diseases of Bone and Calcium Metabolism

H. FLEISCH

A. Introduction

The bisphosphonates are a new class of drugs which have been developed in the past two decades for use in various diseases of bone, tooth, and calcium metabolism. Three are on the market today, while others are under clinical or preclinical investigation. This chapter will cover chemical and experimental aspects as well as clinical applications of these compounds. Literature is essentially limited to one reference, usually the original report. For two recent reviews, weighted somewhat differently and with more extensive references, see FLEISCH (1983) and FRANCIS and MARTODAM (1983).

B. Chemistry and General Characteristics

Bisphosphonates, previously erroneously called diphosphonates, are compounds characterized by two C–P bonds. If the two bonds are located on the same carbon atom, the compounds are called geminal bisphosphonates. They are therefore analogs of pyrophosphate, which contains an oxygen instead of a carbon atom.

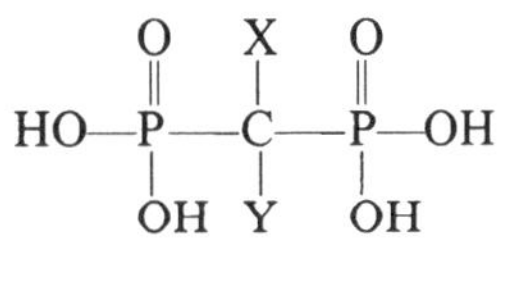

Bisphosphonic acid

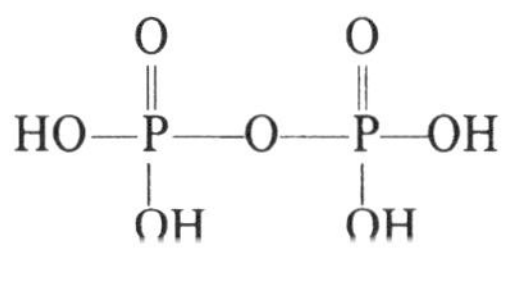

Pyrophosphoric acid

In this chapter, only the geminal bisphosphonates will be discussed, since these have been shown to have the strongest activity in vivo and are the only ones used clinically today, the other types of bisphosphonates being less active or not active at all on bone and calcification. For the sake of simplicity, they will be called just bisphosphonates, although it must be emphasized that this is not entirely correct.

The basic structure P–C–P allows a great number of possible variations, either by changing the two lateral chains on the carbon atom or by esterifying the phosphate groups. A large number of bisphosphonates can be and has been synthesized. Some of them have been tested biologically. It has emerged that each bisphosphonate has its own physicochemical and biologic characteristics, and this is of great interest in the light of the future development of these compounds. This means, however, that it is not possible to extrapolate from the results of one compound to others without great caution, and that it is not correct to talk generally of the effects of bisphosphonates. It is necessary to consider each bisphosphonate on its own and always to restrict statements to specific compounds.

The P–C–P bond of the bisphosphonates is relatively stable to heat and most chemical reagents and completely resistant to enzymatic hydrolysis. The four dissociation constants vary with the two side chains on the carbon atom, pK_1 being mostly between 1 and 2, pK_2 between 2 and 3, pK_3 between 5 and 8, and pK_4 between 8 and 12 (CURRY and NICHOLSON 1972). The bisphosphonates have a strong affinity for metal ions such as calcium, magnesium, and especially iron (CURRY and NICHOLSON 1972). Some uncertainty still exists as to their state when in solution. Indeed, they are only partially ultrafiltrable in aqueous solutions as well as in plasma (WIEDMER et al. 1983). In the presence of calcium and other metals, the hydroxybisphosphonates can make polynuclear complexes not only at alkaline pH (GRABENSTETTER and CILLEY 1971), but also at physiologic pH (LAMSON et al. 1984). Whether this occurs in vivo, and whether this is the cause of the impaired ultrafiltrability, is unknown.

The following bisphosphonates have been investigated in humans, the first three being on the market in certain countries, the former two as an agent against bone disease, the latter for dental use.

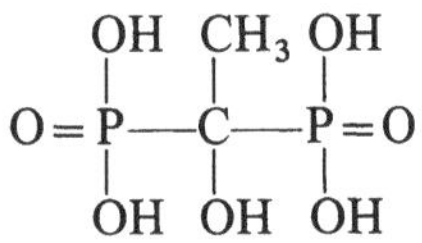

1-Hydroxyethylidene-1,1-bisphosphonic acid (HEBP), previously called ethane-1-hydroxy-1,1-diphosphonic acid (EHDP)

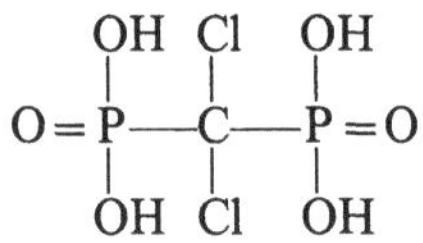

Dichloromethylenebisphosphonic acid (Cl_2MBP), previously called dichloromethylenediphosphonic acid (Cl_2MDP)

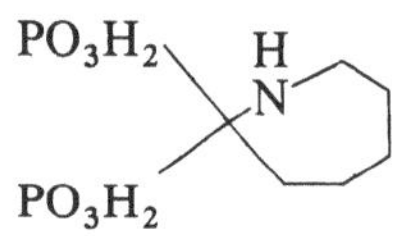

Azacycloheptylidene-2,2-bisphosphonic acid

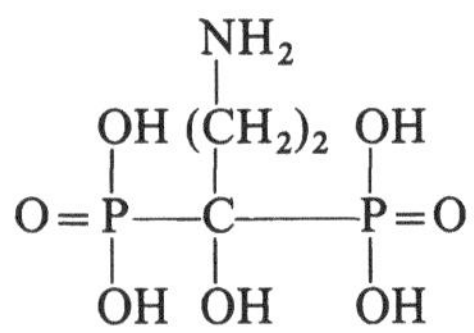

3-Amino-1-hydroxypropylidene-1,1-bisphosphonic acid (AHPrBP), previously called 3-amino-1-hydroxypropane-1,1-diphosphonic acid (APD)

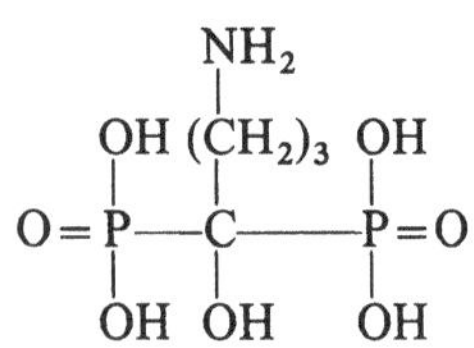

4-Amino-1-hydroxybutylidene-1,1-bisphosphonic acid (AHBuBP)

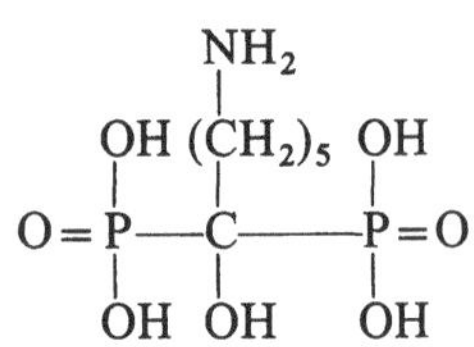

6-Amino-1-hydroxyhexylidene-1,1-bisphosphonic acid (AHHexBP)

C. Synthesis

The bisphosphonates can be synthesized in a variety of ways (for review see CURRY and NICHOLSON 1972; WORMS and SCHMIDT-DUNKER 1976). The commonest method of obtaining 1-hydroxy-1,1-bisphosphonates is the reaction of the corresponding carboxylic acid with a mixture of H_3PO_4 and PCl_3. The products obtained under these anhydrous conditions are condensates, i.e., two or more molecules of the bisphosphonate condensed via the removal of a molecule of water. These condensates can then be converted by heating in water or in 6 *M* HCl. This technique is one of the simplest ways of producing the bisphosphonates HEBP or AHPrBP (BLASER et al. 1971). Another method consists in using a Michaelis–Arbuzov-like reaction (HARVEY and DE SOMBRE 1964), whereby a carboxylic acid chloride is made to react with a trialkylphosphite. The resultant acylphosphonate reacts under slightly basic conditions with a dialkylphosphite to yield a bisphosphonte tetraalkyl ester, which can then be hydrolyzed with HCl to the corresponding free acid. The 1-amino-1,1-bisphosphonates are made by reacting a nitrile or an amide with H_3PO_4 and a phosphorus trihalide and hydrolyzing the product with water. The reaction can also be carried out directly in the presence of water (WORMS and BLUM 1979).

D. Methods of Determination

Various methods are available to determine the structure and the purity of bisphosphonates and are used in the synthesis of these compounds. They include, among others, nuclear magnetic resonance techniques (H-NMR, ^{13}C-NMR, and P-NMR) and gel electrophoresis. The measurement in biologic fluids or in tissues is, however, more difficult and far from satisfactory. The difficulty arises from the small concentrations present and from the large amounts of inorganic phosphate and various organic phosphates found in biologic samples. For this reason, pharmacodynamics have been performed almost exclusively with radioactively labeled compounds and pharmacokinetic data with unlabeled compounds are very scanty in humans.

One technique (BISAZ et al. 1975a), which permits measurement of HEBP down to 1 nmol, is based on a partial purification of HEBP from organic and inorganic phosphate and measurement of inorganic phosphate production under UV exposure. Another (LIGGETT 1973) which is, however, much less sensitive, involves titration with thorium diaminocyclohexanetetraacetate in the presence of xylanol orange after purification by means of precipitation on calcium phosphate or calcium hydroxide. Finally, a technique developed for Cl_2MBP (CHESTER et al. 1981) and which allows the measurement of about 7 nmol is based on flame photometry detection after purification by adsorption onto calcium phosphate and ion exchange chromatography.

E. Monophosphonates

Since the bisphosphonates are chemically two monophosphonates, a few comments on these latter compounds are relevant. Whereas until now no bisphosphonates have been found to occur naturally in animals or humans, monophosphonates, that is, compounds which have one P–C bond, occur throughout the animal world (HILDEBRAND and HENDERSON 1983). The first and probably the most important, was detected only in 1959 and identified as 2-aminoethylphosphonic acid (AEP) (HORIGUCHI and KANDATSU 1959). This monophosphonate is present both as a free molecule and incorporated in lipids as phospholipids, and in other macromolecules. It occurs in plants as well as in many animal species, mostly in membranes.

The biologic role of the monophosphonates is still unknown. One possibility is that they may make the molecule in which they are incorporated more resistant to enzymatic degradation. Possibly they also act as a phosphorus source in lower organisms. The synthesis of the C–P bond seems to be restricted to lower organisms such as bacteria, phytoplankton, protozoa, and invertebrates. The monophosphonates present in vertebrates come most probably from the diet or in cattle from rumen organisms as well (SMITH 1983).

Monophosphonates have been used for a variety of purposes. One biologic application has been their use as analogs of natural phosphates for the elucidation of biochemical mechanisms. This property to act as analogs has also been made use of for chemotherapeutic purposes. Thus, certain phosphonates such as phosphonoformate and phosphonoacetate act as antimetabolites and are effective clinically against herpes and Epstein–Barr viruses by inhibiting DNA polymerase and therefore virus proliferation (ENGEL 1983). Certain streptomyces make a monophosphonate called fosfomycin which has antibiotic properties (HENDLIN et al. 1969) and is used for this purpose in medicine. Other phosphonates act as insecticides, plant growth regulators, and chemical warfare agents on the basis of their inhibitory action on acetylcholinesterase (HILDEBRAND 1983). Industrially, they have been used as adhesives, antioxidants, catalysts, corrosion inhibitors, flame and fire retardants, gelling agents, heat and light stabilizers, discoloration inhibitors, hydraulic fluid and fuel additives, ion exchange resins, lubricants, plasticizers, in photography, etc. (DRAKE and CALAMARI 1983).

F. History

Not only monophosphonates, but also bisphosphonates have been known for some time in industry because of their various industrial applications, as water softeners (DRAKE and CALAMARI 1983) among others. This effect is due to their calcium-chelating property and their inhibition of calcium carbonate precipitation through crystal growth poisoning. However, our knowledge of their biologic characteristics dates from the last 20 years only. This has been derived from earlier studies on inorganic pyrophosphate (FLEISCH and RUSSELL 1970). We had found that biologic fluids such as plasma and urine contain compounds inhibiting calcium phosphate precipitation, and that part of this inhibitory activity is due to inorganic pyrophosphate, a compound which had not been described previously in these media (FLEISCH and NEUMAN 1961; FLEISCH and BISAZ 1962; RUSSELL et al. 1971). Pyrophosphate was shown to impair the crystallization of calcium phosphate from solution (FLEISCH and NEUMAN 1961) as well as the dissolution of these crystals (FLEISCH et al. 1966). When given in vivo, pyrophosphate inhibits ectopic calcification induced by various means in tissues such as arteries, kidneys, and skin (SCHIBLER et al. 1968). These effects are, however, present only when the compound is given parenterally, not when it is given orally. On the other hand, bone resorption is not influenced, possibly because pyrophosphate is hydrolyzed too quickly. On the basis of these results, it was suggested that pyrophosphate might be a physiologic regulator of calcification and perhaps also of decalcification in vivo, its local concentration being determined by the activity of local pyrophosphatases (FLEISCH et al. 1966).

With the exception of the dental field, the therapeutic use of this compound in diseases where ectopic calcification or increased bone resorption occurs was, however, not likely to be successful, because of its failure to act when given orally, and because of its rapid hydrolysis when given parenterally. This prompted a search for analogs which would display similar physicochemical activity, but which would resist enzymatic hydrolysis and would therefore not be broken down metabolically. The bisphosphonates fulfilled these conditions.

G. Mode of Action

I. Physicochemical Effects

As was anticipated in view of the structural similarity between the bisphosphonates and pyrophosphate, the physicochemical effects of most of the bisphosphonates are very similar to those of pyrophosphate. Thus, many of the bisphosphonates inhibit the precipitation of calcium phosphate from clear solution, even at very low concentration (FLEISCH et al. 1970; MEYER and NANCOLLAS 1973), block the transformation of amorphous calcium phosphate into hydroxyapatite (FRANCIS 1969; FRANCIS et al. 1969), and delay the aggregation of apatite crystals into larger clusters (HANSEN et al. 1976). They also slow down the dissolution of these crystals (FLEISCH et al. 1969; RUSSELL et al. 1970). All these effects appear to be related to the marked affinity of these compounds for solid phase calcium phosphate. Thus, they bind to the surface by chemisorption onto calcium (JUNG et al. 1973), especially at screw dislocations and kink sites of growth, and then act as a crystal poisons on both growth and dissolution. The binding is thought to be either bidentate, by means of the phosphate groups, or tridentate in the case of the hydroxybisphosphonates, the hydroxyl group making the third link (BARNETT and STRICKLAND 1979). Bisphosphonates also inhibit the formation (FRASER et al. 1972; MEYER et al. 1977) and the aggregation (ROBERTSON et al. 1973) of calcium oxalate crystals.

II. Effect on Calcification In Vivo

Like pyrophosphate, bisphosphonates also inhibit calcification in vivo very efficiently. Thus, they prevent experimentally induced calcification of the arteries, kidneys, skin, and heart (FLEISCH et al. 1970; CASEY et al. 1972; ROSENBLUM et al. 1977) among others. In contrast to pyrophosphate, which acts only when given parenterally, they are also active when administered orally. Interestingly, in the arteries they decrease not only mineral deposition, but also the accumulation of cholesterol, elastin, and collagen (HOLLANDER et al. 1978; KRAMSCH and CHAN 1978). Bisphosphonates can also inhibit the calcification of bioprosthetic heart valves (LEVY et al. 1985) and platelet deposition onto them (DEVANJEE et al. 1984), as well as the formation of experimental urinary stones (FRASER et al. 1972). Finally, topical administration leads to a decreased formation of dental calculus (BRINER et al. 1971). Certain bisphosphonates, among others HEBP, inhibit not only ectopic calcification, but can in certain cases also inhibit ectopic ossification, when given systemically (PLASMANS et al. 1978), or locally in slow-release form (AHRENGART and LINDGREN 1986).

Bisphosphonates can, if administered in sufficient doses, also impair the mineralization of normal calcified tissues such as bone (JOWSEY et al. 1970; KING et al. 1971; SCHENK et al. 1973), cartilage (SCHENK et al. 1973), and dentin (LARSSON 1974). This can lead to an impairment of fracture healing (LENEHAN et al. 1985). The inhibition is eventually reversed after discontinuation of the drug (FLORA et al. 1980). In growing animals, the radiologic picture resembles that induced by vitamin D deficiency. However, the two conditions vary histologically (BISAZ et al. 1975b), indicating that the mechanism of action leading to the inhibition is different. The doses required to induce the block of mineralization vary according to the bisphosphonate used, the animal species, the length of the treatment, and the route of administration. The effect on bone occurs at a lower dose than that on cartilage.

There is a close relationship between the ability of an individual bisphosphonate to inhibit the formation of calcium phosphate in vitro and its effectiveness on ectopic calcification in vivo (FLEISCH et al. 1970; SHINODA et al. 1983), suggesting that the latter can be explained in terms of a physicochemical mechanism. The geminal bisphosphonates (P–C–P bond) are the most effective, although the activity varies from compound to compound. The vicinal bisphosphonates (P–C–C–P bond) are less effective or not effective and the monophosphonates (C–P bond) are not effective at all.

The mechanism of the inhibition of normal mineralization is at present not completely clear, although it is likely to be due to a physicochemical mechanism. All compounds tested so far which are effective in vivo are also good inhibitors in vitro (TRECHSEL et al. 1977). The opposite is, however, not true. Cl_2MBP displays relatively little inhibitory activity on bone mineralization in spite of its strong inhibition of crystal growth in vitro and of soft tissue calcification in vivo (SCHENK et al. 1973). The explanation for this discrepancy is not yet clear. Differences in availability is a possibility.

III. Inhibition of Bone Resorption

The fact that bisphosphonates inhibit calcium phosphate crystal dissolution in vitro led us to hypothesize that these compounds might also act on bone resorption in vivo. The hypothesis proved correct, but the reason from which it was derived was not. Bisphosphonates proved to be very powerful inhibitors of bone resorption when tested in a variety of conditions, both in vitro and in vivo. In vitro, they block bone resorption induced by various means (REYNOLDS et al. 1972), as well as dissolution of bone particles by macrophages (CHAMBERS 1980; REITSMA et al. 1982). In growing rats, they block the degradation of both bone and cartilage and thus arrest the remodeling of the metaphysis which becomes club-shaped and radiologically more dense than normal (SCHENK et al. 1973; REITSMA et al. 1980). This effect is currently used as a model to study the potency of new compounds (SCHENK et al. 1986). In the mouse, a similar effect has been found, leading to a picture similar to that seen in gray-lethal congenital osteopetrotic mice (REYNOLDS et al. 1973). The inhibition of endogenous bone resorption has also been documented by ^{45}Ca kinetic studies and by hydroxyproline excretion (GASSER et al. 1972).

Bisphosphonates also impair bone resorption induced by various agents. They blunt the effect of parathyroid hormone (FLEISCH et al. 1969; RUSSELL et al. 1970) as well as that of retinoids. They also prevent various types of osteoporosis, such as that induced by heparin (HAEHNEL et al. 1973), corticosteroids (JEE et al. 1981), by immobilization (MUEHLBAUER et al. 1971), and ovariectomy (WINK et al. 1985). Finally, the inhibit tumoral invasion of bone (JUNG et al. 1984; POLLARD and LUCKERT 1985; RADL et al. 1985), as well as various types of tumoral hypercalcemia (JOHNSON et al. 1982; MARTODAM et al. 1983). In the dental field, they prevent periodontal destruction in rice rats (LEONARD et al. 1979).

The degree of activity of individual bisphosphonates varies greatly from compound to compound (SHINODA et al. 1983). If the aliphatic carbon backbone is lengthened, there is an increase in activity up to nine carbon atoms, but a decrease in activity occurs with additional length. Adding a hydroxyl group to the carbon atom at position 1 increases potency. Amino derivatives with an amino group at the end of the side chain are extremely active. Here again, the length of the side chain is very important, the highest activity being found with a length of four carbons (4-amino-1-hydroxybutylidene-1,1-bisphosphonate AHBuBP), which is active at doses as low as 1 µg phosphorus per kilogram per day s.c. in the rat (SCHENK et al. 1986). Lately, a series of cyclic geminal bisphosphonates has been synthesized (BENEDICT et al. 1985a, b), some of them such as 2-(2-pyridinyl)ethylidenebisphosphonate being active at 1 µg phosphorus per kilogram per day s.c. (BEVAN et al. 1985). Of the five bisphosphonates tested clinically, AHBuBP is the most potent in the rat, followed by AHPrBP and AHHexBP, then by Cl_2MBP, HEBP being the least effective (SCHENK et al. 1986). It is therefore evident that the intensity of the effect is exquisitely dependent on the side chain. While the P–C–P structure is a prerequisite for the activity, it is not sufficient by itself. It is also of interest that there is no relationship at all between the intensity of the inhibition of mineralization and of the inhibition of resorption.

It was hoped that the decrease in resorption would be accompanied by a conspicuous increase in calcium balance and in the mineral content of bone. Although this is sometimes the case, especially in growing animals, this increase is in most cases smaller than predicted (GASSER et al. 1972). In addition, it is only transient (REITSMA et al. 1980), since after a certain time, bone formation decreases almost in parallel with the change in resorption, probably because of the well-known coupling between formation and resorption. The main effect is therefore a reduction of bone turnover. The greatest increase in retention has been found with Cl_2MBP, the aminobisphosphonates, and 1-hydroxypentylidene-1,1-bisphosphonate.

Originally, it was thought that by analogy with what occurs in mineralization, the bisphosphonates would act on bone resorption through their inhibitory effect on calcium phosphate crystal dissolution. This proved, however, to be a wrong assumption. Indeed, no correlation has been found between the inhibition of bone resorption in vivo and the inhibition of crystal dissolution in vitro (SHINODA et al. 1983). Furthermore, bisphosphonates alter the morphology of osteoclasts both in vitro and when administered in vivo (SCHENK et al. 1973; MILLER and JEE 1979). Finally, the fact that extremely small amounts are acting in vivo also makes a physicochemical effect unlikely. Thus, it seems that the action in vivo is mediated through other mechanisms, most probably cellular.

IV. Biochemical and Cellular Effects

A great number of different biochemical effects of bisphosphonates have been described. These effects vary greatly and can sometimes even go in opposite directions with different compounds, or with the same compound at different concentrations. At present, no clear picture of a structure–effect relationship has been ascertained. Some of the effects may be relevant to bone resorption. Certain bisphosphonates (especially Cl_2MBP and HEBP) reduce lactic acid production through decreased glycolysis both in intact calvaria (MORGAN et al. 1973) and in isolated bone and cartilage cells (FAST et al. 1978). Conversely, however, other bisphosphonates, such as AHPrBP or long chain bisphosphonates, increase lactic acid production, possibly because of a toxic action (ENDE 1979; SHINODA et al. 1983). Various bisphosphonates, especially Cl_2MBP, inhibit certain lysosomal enzymes in vitro (FELIX et al. 1976) and diminish their activity when added to cultured calvaria (MORGAN et al. 1973; DELAISSÉ et al. 1985) or when given in vivo (DOTY et al. 1972; ENDE 1979). Certain bisphosphonates such as Cl_2MBP and HEBP inhibit prostaglandin synthesis by bone cells or calvaria, both when added in vitro and when given in vivo (FELIX et al. 1981; OHYA et al. 1985). Finally, various bisphosphonates inhibit the multiplication of bone macrophages in vitro, even at very low concentrations (CECCHINI et al. 1987). Furthermore, they show cytotoxic and migration inhibitory effects on peritoneal macrophages (STEVENSON and STEVENSON 1986). Thus, macrophages seem to be specially sensitive to bisphosphonates.

Numerous additional cellular actions have been described for individual bisphosphonates. These include: an increase of fatty acid oxidation (FELIX and FLEISCH 1981) and amino oxidation (ENDE 1979); a stimulation of the citric acid cycle (ENDE 1979); an increase in cellular content of glycogen (FELIX et al. 1980); an increase in the production of alkaline phosphatase (FELIX and FLEISCH 1979); an increase in the biosynthesis of bone and cartilage collagen (GUENTHER et al. 1981 a, b), possibly by impaired intracellular collagenolysis (GALLAGHER et al. 1982); an impairment of dentin and cementum formation (BEERTSEN et al. 1985); an increase in the synthesis of proteoglycans in vitro (GUENTHER et al. 1979), but a decrease when administered in vivo (LARRSON 1976); a reduction in the release of calcium from kidney mitochondria in vitro (GUILLAND et al. 1974) and an increase in calcium of mitochondria in vivo (PLASMANS et al. 1980); contradictory effects on cAMP production (PILCZYK et al. 1972; GEBAUER et al. 1976); a decrease or an increase in cellular multiplication (FAST et al. 1978; EVÊQUOZ et al. 1985); an inhibition of amebal phosphofructokinase, which raised the possibility of their use in amebiasis (EUBANK and REEVES 1982); in the case of certain thiobisphosphonates an inhibition of interleukin-1-induced enzymes release (EDMONDS-ALT et al. 1985); an inhibition of the action of mitogens on mononuclear function and on the lymphoblastic response (DE VRIES et al. 1982); and an inhibition of the influence of antilymphocyte serum on T-lymphocytes (ZERNOV et al. 1979).

These findings suggest that bisphosphonates enter mammalian cells. This has been confirmed by studies in vitro, both for HEBP and Cl_2MBP (FAST et al. 1978). The cellular uptake was mostly in the cytosol and the concentration expressed in terms of cellular water can be severalfold higher than in the medium, indicating that the bisphosphonates are accumulating within specific cellular compartments (FELIX et al. 1984). Cells with phagocytic properties can also take up bisphosphonates, with especial avidity if the compounds are bound to apatite crystals (CHAMBERS 1980), or when they are encapsulated into liposomes, in which case they are also taken up in vivo by the spleen (VAN ROOIJEN et al. 1985).

V. Mode of Action in the Inhibition of Bone Resorption

From this variety of biochemical and cellular effects, some may be candidates to play a role in the inhibition of bone resorption. The decrease in lactic acid production may be relevant, since it is thought that acid production plays an important role in crystal dissolution. The same might be true for the inhibition of lysosomal enzymes, as these are thought to be important in matrix degradation, and for the inhibition of the synthesis of prostaglandins, powerful bone resorbers. Since osteoclasts originate from the monocyte macrophage system, the inhibitory effect on the multiplication of bone macrophages might lead to a decrease in osteoclast recruitment. The initial hypothesis of a physicochemical effect can not be rejected completely either. Finally, some compounds, especially at high doses, may act through a toxic effect on the relevant cells.

Unfortunately, studies performed up to now with various bisphosphonates have not shown a correlation between any of these biologic effects and the inhibition of bone resorption in vivo (SHINODA et al. 1983). Thus, none of these mechanisms is the *only* explanation for the inhibition of bone resorption, although, it can not be ruled out that some or all of them may contribute and that bisphosphonates have various modes of action. Recent results suggest that certain bisphosphonates, such as Cl_2MBP, lead to a decrease of osteoclasts, while others, such as the aminobisphosphonates, induce an increased number of osteoclasts, this in spite of the fact that bone resorption is blocked (STUTZER et al. 1987b).

In view of the large array of cellular effects of bisphosphonates, it might be found surprising that they act almost exclusively on calcified tissues. This selectivity probably stems from the strong affinity of these compounds for calcium phosphate which allows them to be cleared very rapidly from blood and to be incorporated into calcified tissues, especially bone (JUNG et al. 1973; BISAZ et al. 1978). Whenever the latter is resorbed, the compounds will be released into the surrounding solution, either in free form or bound onto the surface of apatite crystals. They can thus reach high local concentration in the vicinity of the resorbing cells and be taken up by these, either as free compounds or with the crystals. This may explain why one administration of bisphosphonates can be active for long periods of time, both in animals (STUTZER et al. 1987a) and in humans. It now appears that the role of the P–C–P moiety of the bisphosphonates is mainly to confer on them their affinity for the mineralized tissues and thus target the effect on these. On the other hand, their cellular effect seems to be conferred by the whole molecule, and is thus dependent on the structure of the side chains. This opens the exciting possibility of the synthesis of new compounds with stronger antiresorbing effects.

VI. Other Effects

Besides the effects on mineralized tissues, some other actions have been described in vivo, mostly, however, after large doses. A few concern the immune system. A decrease in the formation of antibody-secreting cells in response to immunization and an impaired delayed and immediate hypersensitivity has been found (KOMISSARENKO et al. 1977). Cl_2MBP given to newborn mice, leads to atrophy of the thymus (MILHAUD et al. 1983), to the disappearance of certain thymus-dependent macrophages (LABAT et al. 1983), of natural killer cells (LABAT et al. 1984), and to a diminution of the response of T-lymphocytes to mitogens (MILHAUD et al. 1983). It is interesting that the latter changes have also been found in osteopetrotic mice. However, the doses of bisphosphonates which lead to these changes are very high, so that it is not clear if the latter are relevant to the effects of the compounds on bone. In this line it is also of interest that a thiobisphosphonate has been found to inhibit passive cutaneous anaphylaxis (BARBIER et al. 1985) and that Cl_2MBP, HEBP, and a thiobisphosphonate inhibit some of the changes seen in adjuvant-induced polyarthritis (FRANCIS et al. 1972; FLORA 1979; BARBIER et al. 1986). Various bisphosphonates also inhibit the 1,25-dihydroxy vitamin D-induced increase in plasma osteocalcin in rats (STRONSKI et al. 1985).

Other effects of bisphosphonates are probably secondary to their skeletal action. Large doses of HEBP decrease the intestinal absorption of calcium (BONJOUR et al. 1973) because of a decrease in the formation of 1,25-dihydroxy vitamin D_3 (HILL et al. 1973; BAXTER et al. 1974), while low doses induce an increase in the hormone formation (GUILLAND et al. 1975). These changes are likely to be due to an indirect homeostatic mechanism aimed at adapting intestinal calcium absorption to the needs of the organism. The large doses act via their inhibition of mineralization, while with the smaller doses, where no such inhibition of calcification occurs, the triggering mechanism is the decreased bone resorption. A similar indirect mechanism is probably also the cause of the decreased capacity of the kidney to reabsorb phosphate found in the rat under HEBP treatment (BONJOUR et al. 1978).

H. Pharmacokinetics

Bisphosphonates are synthetic compounds which have not yet been found to occur naturally in animals or humans. No enzymes able to cleave the P–C–P bond have yet been described. According to our current knowledge, the administered bisphosphonates are absorbed, stored, and excreted unaltered in animals. Thus, bisphosphonates appear to be nonbiodegradable, both in animals and in solution (STEBER and WIERICH 1986). Most of the pharmacokinetic data on bisphosphonates have been obtained with HEBP or Cl_2MBP. The intestinal absorption lies between 1% and 10% of an oral dose, is generally higher in the young, higher when ingestion increases, and shows a great inter- and intraspecies variation (MICHAEL et al. 1972; RECKER and SAVILLE 1973; GURAL 1975). In humans, absorption is 1%–9% for HEBP (FOGELMAN et al. 1986), and 1%–2% for Cl_2MBP (YAKATAN et al. 1982). Absorption occurs, at least in the rat, mostly in the small intestine (WASSERMAN et al. 1973) and is diminished by the presence of calcium (see FRANCIS and MARTODAM 1983). This is the reason why bisphosphonates have always to be given to humans before meals and never together with milk products. The absorption is inversely related to the size of the compounds, some of the large polyphosphonates being virtually nonabsorbable (ANBAR et al. 1973).

Between 20% and 50% of the absorbed HEBP is localized to bone, the remainder being rapidly excreted in the urine (MICHAEL et al. 1972). The excreted part is somewhat larger for Cl_2MBP (CONRAD and LEE 1981). The renal clearance of HEBP and Cl_2MBP exceeds that of inulin, suggesting the presence of a secretory pathway (TROEHLER et al. 1975). Sometimes, bisphosphonates can deposit in other organs such as the liver (WINGEN and SCHMAEHL 1985). This might be due to the formation of particles after too rapid intravenous injection, which are then phagocytosed by the reticuloendothelial system.

The half-life of circulating bisphosphonates is only of the order of minutes, the rate of entry into bone being very fast, similar to that of calcium and phosphate (BISAZ et al. 1978; CONRAD and LEE 1981; YAKATAN 1982). Therefore, soft tissues will be exposed to these compounds for only short periods. On the other hand, the half-time of the skeletal retention is long and depends on the turnover rate of the skeleton itself.

J. Toxicology

Toxicologic animal studies reported to date deal almost exclusively with HEBP and Cl_2MBP. Acute, subacute, and chronic administration in several animal species have revealed little toxicity. Teratogenic, mitogenic, and carcinogenic tests were negative (NOLEN and BUEHLER 1971; NIXON et al. 1972). The acute toxicity appears to be due to the formation of complexes with calcium, leading to hypocalcemia. When the compound is administered intravenously, it varies with the speed of infusion (FRANCIS and SLOUGH 1984). The first chronic side effect is the appearance of fractures at about 0.15 mg phosphorus per kilogram s.c. for HEBP and 0.7 mg phosphorus per kilogram s.c. for Cl_2MBP (FLORA et al. 1980). With HEBP at 0.6 mg phosphorus per kilogram an inhibition of mineralization occurs.

This is not seen with Cl_2MBP at doses at high as 7 mg phosphorus per kilogram (FLORA et al. 1980). At higher doses, renal lesions appear. Furthermore, inflammatory gastric lesions (even when the compounds are given parenterally) as well as pulmonary lesions have been reported. However, it must be stressed that these results can not necessarily be extrapolated to other bisphosphonates. Indeed, in our experience, toxicity, both in culture and in vivo, varies greatly from one compound to another, so that great caution has to be applied when using new compounds clinically.

K. Drug Interactions

No interactions have been described up to now.

L. Clinical Use

The studies described have led to trials of various bisphosphonates in human diseases. Clinical applications have focused upon three main areas: (a) use as skeletal markers in the form of ^{99m}Tc derivatives for diagnostic purposes in nuclear medicine; (b) therapeutic use in patients with ectopic calcification and ossification; and (c) therapeutic use in patients with increased bone destruction. Only the therapeutic aspects will be discussed here, the application in scintigraphy being beyond the scope of this chapter. Until now, three bisphosphonates, namely HEBP, Cl_2MBP, and AHPrBP have been investigated on a larger scale in humans. First results have been reported with AHBuBP and AHHexBP. Only HEBP and Cl_2MBP are on the market for clinical use. In the dental field, another bisphosphonate, azacycloheptylidene-2,2-bisphosphonate, is on sale in a toothpaste.

I. Ectopic Calcification and Ossification

Based on the preventive effect of bisphosphonates, especially HEBP, on ectopic calcifications and ectopic ossifications in animals, it was hoped that a similar effect would be found in human diseases involving abnormal mineralization. Generally speaking, the results are ambiguous in ectopic calcification, but encouraging in ectopic ossification.

1. Soft Tissue Calcification

HEBP has been given in some cases of scleroderma (METZGER et al. 1974; RABENS and BETHUNE 1975), dermatomyositis (METZGER et al. 1974; STEINER et al. 1974), and calcinosis universalis (CRAM et al. 1971). Efficacy is uncertain since these disorders often show spontaneous remissions.

2. Urolithiasis

The hope that the inhibitory effect on crystal growth and aggregation of both calcium phosphate and calcium oxalate would be useful for the prevention of urinary stones has not been fulfilled, at least with HEBP. Although pilot studies showed a certain effect in chronic stone formers (BAUMANN et al. 1978; BONE et al. 1979), the necessary dose to obtain inhibition of crystal growth in urine is high, about 1600 mg/day orally (OHATA and PAK 1974; BAUMANN et al. 1978), so that it will also induce skeletal effects. Furthermore, the clinical benefit is uncertain, other large-scale studies having failed to show efficacy. Bisphosphonates can not therefore be recommended for use in urolithiasis.

3. Dental Calculus

Many investigations have shown that topical application of HEBP diminishes the development of dental calculus (MUEHLEMANN et al. 1970; STURZENBERGER et al. 1971). Azacycloheptylidene-2,2-bisphosphonate is added to a toothpaste which has been marketed in various countries.

4. Fibrodysplasia Ossificans Progressiva

This disease, also called myositis ossificans progressiva, is the first in which a bisphosphonate, HEBP, has been investigated in humans (BASSETT et al. 1969). Despite a series of further investigations (GEHO and WHITESIDE 1973; REINER et al. 1980), it remains to be established whether this drug is really active in decreasing ectopic bone formation. It appears that some retardation in the evolution of the disease can occur, but that a complete standstill is rarely obtained. Already formed lesions are not influenced. Despite this uncertainty, in view of the outcome of the disease and the lack of alternative treatment, the use of HEBP in a dosage of 20 mg kg^{-1} day^{-1} seems advisable. Since effective doses also inhibit mineralization of normal bone and can, at least in children, lead to rickets (REINER et al. 1980), the drug should not be given for longer than 3 months, but better for shorter periods of a few weeks, and only when a new exacerbation occurs.

5. Other Heterotopic Ossifications

Results seem more encouraging with other types of heterotopic ossifications. HEBP has been found to diminish the appearance of ossifications in patients with spinal cord injury (FINERMAN and STOVER 1981), after cranial trauma (SPIELMAN et al. 1983), and after total hip replacement (SLOOFF et al. 1974; FINERMAN and STOVER 1981). In the latter, although ectopic bone formation reappears, at least partially, after discontinuation of the drug, the mobility of the hip seems nevertheless to be improved in the HEBP-treated patients. Recently, however, these results have been questioned (THOMAS and AMSTUTZ 1985).

Although its efficacy is not absolutely ascertained, it seems nevertheless justifiable to administer HEBP preventively, especially to those patients who are particularly liable to develop ectopic ossifications, for example, patients who require a second operation after total hip replacement because of ossifications after the first intervention. The daily oral dosage lies between 800 and 1600 mg, given for 3 months and starting just before the operation. A longer treatment should not be given because of the inhibition of normal mineralization.

II. Abnormally Increased Bone Resorption

1. Paget's Disease

The main clinical use of bisphosphonates today is in patients with Paget's disease, a condition characterized by an increased skeletal turnover. All three bisphosphonates, HEBP, Cl_2MP, and AHPrBP, are effective in this disease in decreasing turnover, the largest number of investigations having been performed with HEBP. HEBP decreases both serum alkaline phosphatase, an index of bone formation, as well as urinary hydroxyproline excretion, an index of bone destruction (SMITH et al. 1971; ALTMAN et al. 1973; RUSSELL et al. 1974; DE VRIES and BIJVOET 1974). The effect on hydroxyproline usually precedes that on alkaline phosphatase, suggesting that the primary effect of the bisphosphonate is on bone resorption, the effect on bone formation being possibly secondary to the coupling between the two processes. The action of HEBP on bone turnover is also illustrated by ^{45}Ca kinetic studies (GUNCAGA et al. 1974; DE VRIES and BIJVOET 1974) and by morphological studies (RUSSELL et al. 1974; GUNCAGA et al. 1974; DE VRIES and BIJVOET 1974; MEUNIER et al. 1975). It is interesting that the bone formed under treatment is lamellar, in contrast to the woven bone typical of this disease. However, HEBP does not affect the virus-induced measles nucleocapsid-like inclusions in the nuclei of the osteoclasts, nor the measles-type viral antigens in the osteoclasts (BASLE et al. 1984). HEBP decreases the elevated cardiac output (HENLEY et al. 1979) and the subjective symptoms such as bone pain (ALTMAN et al. 1973; MEUNIER et al. 1975). On the other hand, no improvement in the X-ray picture can be obtained (DE VRIES and BIJVOET 1974; CANFIELD et al. 1977). In fact, in certain cases, the drug induces the appearance of radiolucent areas (NAGANT DE DEUXCHAISNES et al. 1979), which probably reflect locations of impaired mineralization.

Indeed, oral doses between 800 and 1600 mg have been shown to induce an inhibition of mineralization of both pagetic and normal bone (RUSSELL et al. 1974; GUNCAGA et al. 1974; DE VRIES and BIJVOET 1974; MEUNIER et al. 1975). This inhibition is reversible when the treatment is stopped. Oral doses of 400 mg do not lead to this generalized inhibition of mineralization, but can occasionally produce focal osteomalacia (BOYCE et al. 1984). Fractures have been described under HEBP therapy. However, fractures are also more frequent in untreated patients, so that the role of treatment is difficult to assess (JOHNSTON et al. 1983). Since the margin between the dose of HEBP which inhibits resorption and that which inhibits mineralization is small, and since intestinal absorption of the compound is not only small, but very variable from patient to patient, the optimal dosage is difficult to define. The generally recommended dose is either 1600 mg orally for maximally 3 months or 800 mg orally for 6 months to 1 year. Other modes of oral treatment such as intermittent dosage (SIRIS et al. 1980a) or an association with calcitonin have not been investigated well enough yet. It has been shown that treatment for 1 month with 1600 mg/day is as effective as treatment for 6 months with the same dose, and more effective that treatment for 6 months with 400 mg/day (PRESTON et al. 1986).

Also, Cl_2MBP has been found to be very active in decreasing bone turnover in Paget's disease, although the clinical experience is much smaller than for

HEBP. Oral daily doses between 800 and 1600 mg decrease both urinary hydroxyproline and serum alkaline phosphatase (MEUNIER et al. 1979; DOUGLAS et al. 1980). In analogy to what was found in animals, these doses of Cl_2MBP induce, contrary to HEBP, no inhibition of bone mineralization (MEUNIER et al. 1979). Treatment for 1 month (CHAPUY et al. 1983) is as effective as treatment for 6 months. This is also true for 5 days intravenous administration of 300 mg/day (YATES et al. 1985), a treatment which produces a maximal effect on urinary hydroxyproline after 4–5 days.

The third bisphosphonate investigated in Paget's disease is AHPrBP. As in animals, this is the most active of the three compounds and is effective at an oral daily dose of 200–1600 mg. Again, both parameters of bone turnover, urinary hydroxyproline and serum alkaline phosphatase, are decreased (FRIJLINK et al. 1979; HEYNEN et al. 1982). All the bisphosphonates maintain their action months and even years after discontinuation of the therapy (RUSSELL et al. 1974; SIRIS et al. 1980a; CHAPUY et al. 1983), in contrast to calcitonin which is active only as long as it is administered.

2. Primary Hyperparathyroidism

While the oral administration of HEBP has been found ineffective in this condition (KAPLAN et al. 1977), Cl_2MBP at oral daily doses between 1200 and 3200 mg decreases the degree of hypercalcemia, the level of serum alkaline phosphatase, and the urinary excretion of hydroxypoline and calcium (DOUGLAS et al. 1980; SHANE et al. 1981).

3. Hypercalcemia of Malignancy and Tumoral Bone Destruction

A variety of tumors can produce bone destruction and hypercalcemia. The mechanism involves either the production of humoral osteolytic cytokines by a tumor situated outside the bone, or the local destruction of the bone by the tumor in situ. Both types respond to bisphosphonates (see GARATTINI 1985). HEBP has been found to be of little use when given orally, but to lead to a rapid decrease in calcemia and often to a normalization within 3–4 days when given intravenously at a daily dose of around 500 mg (JUNG 1982; RYZEN et al. 1985).

Cl_2MBP on the other hand, displays activity on a variety of tumors when given orally at a dose between 1600 and 3200 mg/day (CHAPUY et al. 1980; SIRIS et al. 1980b, 1983). Calcemia can be restored to normal and calciuria is decreased dramatically. Intravenous administration of 120–1200 mg decreases calcemia as soon as 2 days after starting the treatment, normal values being obtained within 5–7 days (JUNG 1982; SIRIS et al. 1980b; JACOBS et al. 1981). Of great clinical interest is the finding that the bisphosphonate also inhibits the development of new metastases (ELOMAA et al. 1983). Also, AHPrBP is effective in various tumors and again it appears to be the most active of the three. Doses of 300–1500 mg orally (VAN BREUKELEN et al. 1979, 1982) and 1.6–24 mg intravenously (SLEEBOOM et al. 1983) are effective. Even a single administration appears to be effective (THIÉBAUD et al. 1986).

4. Osteoporosis

There have been only a few studies so far using bisphosphonates in patients with osteoporosis. HEBP did not significantly improve calcium balance in senile osteoporosis, although bone turnover was cut down by about 50% (HEANEY and SAVILLE 1976). However, the dosage used was too high, at a level which induces inhibition of mineralization, which could have masked a positive effect. No effect on calcium balance was found either during bed rest, a condition which is associated with enhanced skeletal turnover and negative calcium balance (LOCKWOOD et al. 1975). Recently, however, it has been reported that discontinuous oral administration of 400 mg/day for 2 weeks every 15 weeks over a period of 2 years leads to a 7% increase of trabecular bone in the spine of postmenopausal women, while controls lost 5% (SOERENSEN et al. 1987). Cl_2MBP at an oral dose of 1600 mg/day partially prevented the osteoporosis of immobilization in paraplegic patients (MINAIRE et al. 1981). This result suggests that bisphosphonates might be used in the future for the prevention of at least high turnover osteoporosis.

M. Side Effects

As in animals, studies in humans have revealed few important side effects. The first and major complication under HEBP therapy is the inhibition of normal skeletal mineralization. This effect appears at daily oral doses between 800 and 1600 mg (RUSSELL et al. 1974; JOWSEY et al. 1971; GUNCAGA et al. 1974; DE VRIES and BIJVOET 1974). Fractures have occurred in children (REINER et al. 1980) and possibly also in adults, although in the latter the cause–effect relation has not yet been proven with certainty (JOHNSTON et al. 1983). In children, long-term treatment at an oral dosage of 20 mg/kg may induce proximal muscular weakness leading to an abnormal gait, similar to that seen in rickets (REINER et al. 1980). HEBP also causes a conspicuous rise in plasma phosphate, often to high levels, both in healthy persons and in patients. The change is associated with an increase in renal tubular reabsorption of phosphate (RECKER et al. 1973; WALTON et al. 1975). It seems that hyperphosphatemia shows a correlation with the degree of inhibition of bone mineralization, but the reason for this correlation is as yet unclear. Caution must be taken with the intravenous administration of HEBP, since a rapid injection has led to renal failure (BOUNAMEAUX et al. 1983), possibly because of the precipitation of calcium bisphosphonate or the formation of calcium bisphosphonate aggregates in the blood.

No proven side effects have been described as yet for Cl_2MBP. Contrary to HEBP, this compound does not inhibit mineralization of bone and does not induce hyperphosphatemia. In the course of the clinical evaluation of this compound, some of the treated patients have developed acute leukemia. However, further evaluation cast doubts that the disease was induced by the drug so that it is quite possible that this finding was fortuitous.

Finally, AHPrBP does not induce an inhibition of bone mineralization at doses active on bone resorption. This compound does not lead at this dosage to an increase in plasma phosphate. On the other hand, AHPrBP induces during the first few days a transient increase in temperature and a leukopenia (BIJVOET et al. 1980). Furthermore, this substance appears to lead to alterations of the oral and digestive mucosa under certain circumstances when given orally.

N. Future Prospects

The bisphosphonates present a most interesting new development in the field of the treatment of bone diseases. It is probable that this is only the beginning of a new area of therapy. Indeed, since the effects vary greatly from one bisphosphonate to another, it is quite possible that new bisphosphonates may be synthesized which are superior to those known up to now. It will be especially desirable to develop compounds with a greater margin between the dose inhibiting bone resorption and that inhibiting mineralization. For such a development it would be advantageous to know the mechanisms of action and to be in possession of a structure–effect relationship for the compounds. It might also be possible in future to find bisphosphonates which act more on ectopic calcification than on normal mineralization. Recent results suggest that certain bisphosphonates might also be active in rheumatic diseases. Finally, with the exception of ^{99m}Tc in nuclear medicine, the possible use of bisphosphonates as carriers of drugs acting on the skeleton has not been examined at all and may also be an interesting development.

References

Ahrengart L, Lindgren U (1986) Prevention of ectopic bone formation by local application of ethane-1-hydroxy-1,1-diphosphonate (EHDP): an experimental study in rabbits. J Orthop Res 4:18–26

Altman RD, Johnston CC, Khairi MRA, Wellman H, Serafini AN, Sankey RR (1973) Influence of disodium etidronate on clinical and laboratory manifestations of Paget's disease of bone (osteitis deformans). N Engl J Med 289:1379–1384

Anbar M, Newell GW, St John GA (1973) Fate and toxicity of orally administered polyethylene polyphosphonates. Food Cosmetics Tox 11:1001–1010

Barbier A, Brelière JC, Paul RP, Roncucci R (1985) Comparative study of etidronate and SR 41319, a new diphosphonate, on passive cutaneous anaphylaxis and phospholipase A_2 activity. Agents Actions 16:41–42

Barbier A, Brelière JC, Remandet B, Roncucci R (1986) Studies on the chronic phase of adjuvant arthritis: effect of SR 41319, a new diphosphonate. Ann Rheum Dis 45:67–74

Barnett BL, Strickland LC (1979) Structure of disodium dihydrogen 1-hydroxyethylidenediphosphonate tetrahydrate: a bone growth regulator. Acta Crystall B35:1212–1214

Basle MF, Rebel A, Renier JC, Audran M, Filmon R, Malkani K (1984) Bone tissue in Paget's disease treated by ethane-1-hydroxy-1,1 diphosphonate (EHDP). Clin Orthop Rel Res 184:281–288

Bassett CAL, Donath A, Macagno F, Preisig R, Fleisch H, Francis MD (1969) Diphosphonates in the treatment of myositis ossificans. Lancet 2:845

Baumann JM, Bisaz S, Fleisch H, Wacker M (1978) Biochemical and clinical effects of ethane-1-hydroxy-1,1-diphosphonate on calcium nephrolithiasis. Clin Sci Mol Med 54:509–516

Baxter LA, DeLuca HF, Bonjour JP, Fleisch H (1974) Inhibition of vitamin D metabolism by ethane-1-hydroxy-1,1-diphosphonate. Arch Biochem Biophys 164:655–662

Beertsen W, Niehof A, Everts V (1985) Effects of 1-hydroxyethylidene-1,1-bisphosphonate (HEBP) on the formation of dentin and the periodontal attachment apparatus in the mouse. Am J Anat 174:83–103

Benedict JJ, Degenhardt CR, Perkins CM, Johnson KY, Bevan JA, Olson HM (1985a) Cyclic geminal bis(phosphonates) as inhibitors of bone resorption. Calcif Tissue Int 38 [Suppl]:S31

Benedict JJ, Johnson KY, Bevan JA, Perkins CM (1985b) A structure/activity study of nitrogen heterocycle containing bis(phosphonates) as bone resorption inhibiting agents. Calcif Tissue Int 38 [Suppl]:S31

Bevan JA, Johnson KY, Slough C, Benedict J, Fleisch H, Black J (1985) Skeletal effects of 2-(2-pyridinyl)-ethylidene-bis(phosphonate) in acute and subchronic rat studies. Calcif Tissue Int 38 [Suppl]:S31

Bijvoet OLM, Frijlink WB, Jie K, Linden H van der, Meijer CJLM, Mulder H, Paassen HC van, Reitsma PH, Velde J te, Vries E de, Wey JP van der (1980) APD in Paget's disease of bone. Role of the mononuclear phagocyte system? Arthritis Rheum 23:1193–1204

Bisaz S, Felix R, Fleisch H (1975a) Quantitative determination of ethane-1-hydroxy-1,1-diphosphonate in urine and plasma. Clin Chim Acta 65:299–307

Bisaz S, Schenk R, Kunin AS, Russell RGG, Mühlbauer R, Fleisch H (1975b) The comparative effects of vitamin D deficiency and ethane-1-hydroxy-1,1-diphosphonate administration on the histology and glycolysis of chick epiphyseal and articular cartilage. Calcif Tissue Res 19:139–152

Bisaz S, Jung A, Fleisch H (1978) Uptake by bone of pyrophosphate, diphosphonates and their technetium derivatives. Clin Sci Mol Med 54:265–272

Blaser B, Worms KH, Germscheid HG, Wollmann K (1971) Ueber 1-Hydroxyalkan-1,1-diphosphonsäuren. Z Anorg Allg Chem 381:247–259

Bone HG, Zerwekh JE, Britton F, Pak CYC (1979) Treatment of calcium urolithiasis with diphosphonate: efficacy and hazards. J Urol 121:568–571

Bonjour JP, Russell RGG, Morgan DB, Fleisch H (1973) Intestinal calcium absorption, Ca-binding protein, and Ca-ATPase in diphosphonate-treated rats. Am J Physiol 224:1011–1017

Bonjour JP, Tröhler U, Preston C, Fleisch H (1978) Parathyroid hormone and renal handling of Pi: effect of dietary Pi and diphosphonates. Am J Physiol 234:F487–F505

Bounameaux HM, Schifferli J, Montani JP, Jung A, Chatelanat F (1983) Renal failure associated with intravenous diphosphonate. Lancet 1:471

Boyce BF, Fogelman I, Ralston S, Smith L, Johnston E, Boyle IT (1984) Focal osteomalacia due to low-dose diphosphonate therapy in Paget's disease. Lancet 1:821–824

Briner WW, Francis MD, Widder JS (1971) The control of dental calculus in experimental animals. Int Dent J 21:61–73

Canfield R, Rosner W, Skinner J, McWorther J, Resnick L, Feldman F, Kammerman S, Ryan K, Kunigonis M, Bohne W (1977) Diphosphonate therapy of Paget's disease of bone. J Clin Endocrinol Metab 44:96–106

Casey PA, Casey G, Fleisch H, Russell RGG (1972) The effect of polyphloretin phosphate, polyestradiol phosphate, a diphosphonate and a polyphosphate on calcification induced by dihydrotachysterol in skin, aorta and kidney of rats. Experientia 28:137–138

Cecchini M, Felix R, Cooper PH, Fleisch H (1987) Effect of bisphosphonates on proliferation and viability of mouse bone marrow-derived macrophages. J Bone Min Res 2:135:142

Chambers TJ (1980) Diphosphonates inhibit bone resorption by macrophages in vitro. J Pathol 132:255–262

Chapuy MC, Meunier PJ, Alexandre CM, Vignon EP (1980) Effects of disodium dichloromethylene diphosphonate on hypercalcemia produced by bone metastases. J Clin Invest 65:1243–1247

Chapuy MC, Charhon SA, Meunier PJ (1983) Sustained biochemical effects of short treatment of Paget's disease of bone with dichloromethylene diphosphonate. Metab Bone Dis Rel Res 4:325–328

Chester TL, Lewis EC, Benedict JJ, Sunberg RJ, Tettenhorst WC (1981) Determination of (dichloromethylene)diphosphonate in physiological fluids by ion-exchange chromatography with phosphorus-selective detection. J Chromatogr Sci 225:17–25
Conrad KA, Lee SM (1981) Clodronate kinetics and dynamics. Clin Pharmacol Ther 30:114–120
Cram RL, Barmada R, Geho WB, Ray RD (1971) Diphosphonate treatment of calcinosis universalis. N Engl J Med 285:1012–1013
Curry JD, Nicholson DA (1972) Oligophosphonates. In: Griffith J, Grayson M (eds) Topics in phosphorus chemistry, vol 7. Wiley, New York, p 37
Delaissé J-M, Eeckhout Y, Vaes G (1985) Biphosphonates and bone resorption: effects on collagenase and lysosomal enzyme excretion. Life Sci 37:2291–2296
Vries HR De, Bijvoet OLM (1974) Results of prolonged treatment of Paget's disease of bone with disodium ethane-1-hydroxy-1,1-diphosphonate (EHDP). Neth J Med 17:281–298
Vries E De, Weij JP van der, Veen CJP v d, Paassen HC van, Jager MJ, Sleeboom HP, Bijvoet OLM, Cats A (1982) In vitro effect of (3-amino-1-hydroxypropylidene)-1,1-bisphosphonic acid (APD) on the function of mononuclear phagocytes in lymphocyte proliferation. Immunology 47:157–163
Dewanjee MK, Didisheim P, Kaye MP, Solis E, Zollman PE, Francis MD, Torianni M, Trastek VS, Tago M, Edwards WD (1984) Platlet deposition on and calcification of bovine pericardial valve. Eur Heart J 5 [Suppl D]:1–5
Doty SB, Jones R, Finerman GA (1972) Diphosphonate influence on bone cell structure and lysosomal activity. J Bone Joint Surg 54:1128–1129
Douglas DL, Russell RGG, Preston CJ, Prenton MA, Duckworth T, Kanis JA, Preston FE, Woodhead JS (1980) Effect of dichloromethylene diphosphonate in Paget's disease of bone and in hypercalcaemia due to primary hyperparathyroidism or malignant disease. Lancet 1:1043–1047
Drake GL, Calamari TA (1983) Industrial uses of phosphonates. In: Hilderbrand RL (ed) The role of phosphonates in living systems. CRC Press, Boca Raton, p 171
Elomaa I, Blomqvist C, Gröhn P, Porkka L, Kairento AL, Selander K, Lamberg-Allardt C, Holmström T (1983) Long-term controlled trial with diphosphonate in patients with osteolytic bone metastases. Lancet 1:146–149
Emonds-Alt X, Brelière J-C, Roncucci R (1985) Effects of 1-hydroxyethylidene-1,1 bisphosphonate and (chloro-4 phenyl) thiomethylene bisphosphonic acid (SR 41319) on the mononuclear cell factor-mediated release of neutral proteinases by articular chondrocytes and synovial cells. Biochem Pharmacol 34:4043–4049
Ende JJ (1979) Effects of some diphosphonates on the metabolism of bone in vivo and in vitro. Thesis, University of Leiden
Engel R (1983) Phosphonic acids and phosphonates as antimetabolites. In: Hilderbrand RL (ed) The role of phosphonates in living systems. CRC Press, Boca Raton, p 97
Eubank WB, Reeves RE (1982) Analog inhibitors for the pyrophosphate-dependent phosphofructokinase of Entamoeba histolytica and their effect of culture growth. J Parasitol 68:599–602
Evêquoz V, Trechsel U, Fleisch H (1985) Effect of bisphosphonates on production of interleukin 1-like activity by macrophages and its effect on rabbit chondrocytes. Bone 6:439–444
Fast DK, Felix R, Dowse C, Neumann WF, Fleisch H (1978) The effects of diphosphonates on the growth and glycolysis of connective-tissue cells in culture. Biochem J 172:97–107
Felix R, Fleisch H (1979) Increase in alkaline phosphatase activity in calvaria cells cultured with diphosphonates. Biochem J 183:73–81
Felix R, Fleisch H (1981) Increase in fatty acid oxidation in calvaria cells cultured with diphosphonates. Biochem J 196:237–245
Felix R, Russell RGG, Fleisch H (1976) The effect of several diphosphonates on acid phosphohydrolases and other lysosomal enzymes. Biochim Biophys Acta 429:429–438
Felix R, Fast DK, Sallis JD, Fleisch H (1980) Effect of diphosphonate on glycogen content of rabbit ear cartilage cells in culture. Calcif Tissue Int 30:163–166

Felix R, Bettex JD, Fleisch H (1981) Effect of diphosphonates on the synthesis of prostaglandins in cultured calvaria cells. Calcif Tissue Int 33:549–552
Felix R, Guenther HL, Fleisch H (1984) The subcellular distribution of ^{14}C dichloromethylenebisphosphonate and ^{14}C 1-hydroxyethylidene-1,1-bisphosphonate in cultured calvaria cells. Calcif Tissue Int 36:108–113
Finerman GAM, Stover SL (1981) Heterotopic ossification following hip replacement or spinal cord injury. Two clinical studies with EHDP. Metab Bone Dis Rel Res 4:337–342
Fleisch H (1983) Bisphosphonates: mechanisms of action and clinical applications. In: Peck WA (ed) Bone and mineral research, annual 1. Excerpta Medica, Amsterdam, 319–357
Fleisch H, Bisaz S (1962) Isolation from urine of pyrophosphate, a calcification inhibitor. Am J Physiol 203:671–675
Fleisch H, Neuman WF (1961) Mechanisms of calcification: role of collagen, polyphosphates, and phosphatase. Am J Physiol 200:1296–1300
Fleisch H, Russell RGG (1970) Pyrophosphate and polyphosphate. In: Encyclopaedia (Int) of pharmacology and therapeutics, section 51. Pharmacology of the endocrine system and related drugs. Pergamon, Oxford, p 61
Fleisch H, Russell RGG, Straumann F (1966) Effect of pyrophosphate on hydroxyapatite and its implications in calcium homeostasis. Nature 212:901–903
Fleisch H, Russell RGG, Francis MD (1969) Diphosphonates inhibit hydroxyapatite dissolution in vitro and bone resorption in tissue culture and in vivo. Science 165:1262–1264
Fleisch H, Russell RGG, Bisaz S, Mühlbauer RC, Williams DA (1970) The inhibitory effect of phosphonates on the formation of calcium phosphate crystals in vitro and on aortic and kidney calcification in vivo. Eur J Clin Invest 1:12–18
Flora L (1979) Comparative antiinflammatory and bone protective effects of two diphosphonates in adjuvant arthritis. Arthritis Rheum 4:340–346
Flora L, Hassing GS, Parfitt AM, Villanueva AR (1980) Comparative skeletal effects of two diphosphonates in dogs. Metab Bone Dis Rel Res 2:389–407
Fogelman I, Smith L, Mazess R, Wilson MA, Bevan JA (1986) Absorption of oral disphosphonate in normal subjects. Clin Endocrinol 24:57–62
Francis MD (1969) The inhibition of calcium hydroxyapatite crystal growth by polyphosphates. Calcif Tissue Res 3:151–162
Francis MD, Martodam RR (1983) Chemical, biochemical, and medicinal properties of the diphosphonates. In: Hilderbrand RL (ed) The role of phosphonates in living systems. CRC Press, Boca Raton, Florida, p 55
Francis MD, Slough CL (1984) Acute intravenous infusion of disodium dihydrogen (1-hydroxyethylidene)diphosphonate: mechanism of toxicity. J Pharm Sci 73:1097–1100
Francis MD, Russell RGG, Fleisch H (1969) Diphosphonates inhibit formation of calcium phosphate crystals in vitro and pathological calcification in vivo. Science 165:1264–1266
Francis MD, Flora LF, King WF (1972) The effects of disodium ethane-1-hydroxy-1,1-diphosphonate on adjuvant induced arthritis in rats. Calcif Tissue Res 9:109–121
Fraser D, Russell RGG, Pohler O, Robertson WG, Fleisch H (1972) The influence of disodium ethane-1-hydroxy-1,1-diphosphonate (EHDP) on the development of experimentally induced urinary stones in rats. Clin Sci 42:197–207
Frijlink WB, Velde J te, Bijvoet OLM, Heynen G (1979) Treatment of Paget's disease with (3-amino-1-hydroxypropylidene)-1,1-bisphosphonate (A.P.D.). Lancet 1:799
Gallagher JA, Guenther HL, Fleisch H (1982) Rapid intracellular degradation of newly synthesized collagen by bone cells. Effect of dichloromethylenebisphosphonate. Biochim Biophys Acta 719:349–355
Garattini S (1985) Bone resorption, metastasis, and diphosphonates. Monographs of the Mario Negri Institute for pharmacological research. Raven, New York
Gasser AB, Morgan DB, Fleisch HA, Richelle LJ (1972) The influence of two diphosphonates on calcium metabolism in the rat. Clin Sci 43:31–45

Gebauer U, Russell RGG, Touabi M, Fleisch H (1976) Effect of diphosphonates on adenosine 3' : 5' cyclic monophosphate in mouse calvaria after stimulation by parathyroid hormone in vitro. Clin Sci Mol Med 50:473–478
Geho WB, Whiteside JA (1973) Experience with disodium etidronate in diseases of ectopic calcification. In: Frame B, Parfitt AM, Duncan H (eds) Clinical aspects of metabolic bone disease. Excerpta Medica, Amsterdam, p 506
Grabenstetter RJ, Cilley WA (1971) Polynuclear complex formation in solutions of calcium ion and ethane-1-hydroxy-1,1-diphosphonic acid. I. Complexometric and pH titrations. J Phys Chem 75:676–682
Guenther HL, Guenther HE, Fleisch H (1979) Effects of 1-hydroxyethane-1,1-diphosphonate and dichloromethane-diphosphonate on rabbit articular chondrocytes in culture. Biochem J 184:203–214
Guenther HL, Guenther HE, Fleisch H (1981 a) The effects of 1-hydroxyethane-1,1-diphosphonate and dichloromethanediphosphonate on collagen synthesis by rabbit articular chondrocytes and rat bone cells. Biochem J 196:293–301
Guenther HL, Guenther HE, Fleisch H (1981 b) The influence of 1-hydroxyethane-1,1-diphosphonate and dichloromethanediphosphonate on lysine hydroxylation and cross-link formation in rat bone, cartilage and skin collagen. Biochem J 196:303–310
Guilland DF, Sallis JD, FleischH (1974) The effect of two diphosphonates on the handling of calcium by rat kidney mitochondria in vitro. Calcif Tissue Res 15:303–314
Guilland D, Trechsel U, Bonjour JP, Fleisch H (1975) Stimulation of calcium absorption and apparent increased intestinal, 1,25-dihydroxycholecalciferol in rats treated with low doses of ethane-1-hydroxy-1,1-diphosphonate. Clin Sci Mol Med 48:157–160
Guncaga J, Lauffenburger R, Lentner C, Dambacher MA, Haas HG, Fleisch H, Olah AJ (1974) Diphosphonate treatment of Paget's disease of bone. A correlated metabolic, calcium kinetic and morphometric study. Horm Metab Res 6:62–69
Gural RP (1975) Pharmacokinetics and gastrointestinal absorption behavior of etidronate. Dissertation, University of Kentucky
Hähnel H, Mühlbach R, Lindenhayn K, Schaetz P, Schmidt UJ (1973) Zum Einfluss von Diphosphonat auf die experimentelle Heparinosteopathie. Z Altersforsch 27:289–292
Hansen NM Jr, Felix R, Bisaz S, Fleisch H (1976) Aggregation of hydroxyapatite crystals. Biochim Biophys Acta 451:549–559
Harvey RG, Sombre ER De (1964) The Michaelis-Arbuzov and related reactions. In: Grayson M, Griffith EJ (eds) Topics in phosphorus chemistry, vol I. 3. Wiley, New York, p 57
Heaney RP, Saville PD (1976) Etidronate disodium in postmenopausal osteoporosis. Clin Pharmacol Ther 20:593–604
Hendlin D, Stapley EO, Jackson M, Wallick H, Miller AK, Wolf FJ, Miller TW, Chaiet L, Kahan FM, Foltz EL (1969) Phosphonomycin, a new antibiotic produced by strains of streptomyces. Science 166:122–123
Henley JW, Croxson RS, Ibbertson HK (1979) The cardiovascular system in Paget's disease of bone and the response to therapy with calcitonin and diphosphonate. Aust NZ J Med 9:390–397
Heynen G, Delwaide P, Bijvoet OLM, Franchimont P (1982) Clinical and biological effects of low doses of (3 amino-1-hydroxypropylidene)-1,1-bisphosphonate (APD) in Paget's disease of bone. Eur J Clin Invest 11:29–35
Hilderbrand RL (1983) The effects of synthetic phosphonates on living systems. In: Hilderbrand RL (ed) The role of phosphonates in living systems. CRC Press, Boca Raton, p 139
Hilderbrand RL, Henderson TO (1983) Phosphonic acids in nature. In: Hilderbrand RL (ed) The role of phosphonates in living systems. CRC Press, Boca Raton, p 5
Hill LF, Lumb GA, Mawer EB, Stanbury SW (1973) Indirect inhibition of the biosynthesis of 1,25-dihydroxycholecalciferol in rats treated with a diphosphonate. Clin Sci 44:335–347
Hollander W, Prusty S, Nagraj S, Kirkpatrick B, Paddock J, Colombo M (1978) Comparative effects of cetaben (PHB) and dichloromethylene diphosphonate (Cl_2MBP) on the development of atherosclerosis in the cynomolgus monkey. Atherosclerosis 31:307–325

Horiguchi M, Kandatsu M (1959) Isolation of 2-aminoethane phosphonic acid from rumen Protozoa. Nature 184:901–902
Jacobs TP, Siris ES, Bilezikian JP, Baquiran DC, Shane E, Canfield RE (1981) Hypercalcemia of malignancy: treatment with intravenous dichloromethylene diphosphonate. Ann Intern Med 94:312–316
Jee WSS, Black HE, Gotcher JE (1981) Effect of dichloromethane disphosphonate on cortisol-induced bone loss in young adult rabbits. Clin Orthop Rel Res 156:39–51
Johnson KY, Wesseler MA, Olson HM, Martodam RR, Poser JW (1982) The effects of diphosphonates on tumor-induced hypercalcemia and osteolysis in Walker carcinosarcoma 256 (W-256) of rats. In: Donath A, Courvoisier B (eds) Diphosphonates and bone. Editions Médecine et Hygiène, Genève, p 386
Johnston CC Jr, Altman RD, Canfield RE, Finerman GAM, Taulbee JD, Ebert ML (1983) Review of fracture experience during treatment of Paget's disease of bone with etidronate disodium (EHDP). Clin Orthop 172:186–194
Jowsey J, Holley KE, Linman JW (1970) Effect of sodium etidronate in adult cats. J Lab Clin Med 76:126–133
Jowsey J, Riggs BL, Kelly PJ, Hoffman DL, Bordier P (1971) The treatment of osteoporosis with disodium ethane-1-hydroxy-1,1-diphosphonate. J Lab Clin Med 78:574–584
Jung A (1982) Comparison of two parenteral diphosphonates in hypercalcemia of malignancy. Am J Med 72:221–226
Jung A, Bisaz S, Fleisch H (1973) The binding of pyrophosphate and two diphosphonates on hydroxyapatite crystals. Calciv Tissue Res 11:269–280
Jung A, Bornand J, Mermillod B, Edouard C, Meunier PJ (1984) Inhibition by diphosphonates of bone resorption induced by the Walker tumor of the rat. Cancer Res 44:3007–3011
Kaplan RA, Geho WB, Pointdexter C, Haussler M, Dietz GW, Pak CYC (1977) Metabolic effects of diphosphonate in primary hyperparathyroidism. J Clin Pharmacol 17:410–419
King WR, Francis MD, Michael WR (1971) Effect of disodium ethane-1-hydroxy-1,1-diphosphonate on bone formation. Clin Orthop 78:251–270
Komissarenko SV, Zhuravskii NI, Karlova NP, Gulyi MF (1977) Inhibition of hypersensitivity of delayed and immediate types in guinea pigs by methylenediphosphonic acid. Bull Exp Biol Med 84:1322–1323
Kramsch DM, Chan CT (1978) The effect of agents interfering with soft tissue calcification and cell proliferation on calcific fibrous fatty plaques in rabbits. Circ Res 42:562–571
Labat ML, Tzehoval E, Moricard Y, Feldmann M, Milhaud G (1983) Lack of a T-cell dependent subpopulation of macrophages in (dichloromethylene) diphosphonate-treated mice. Biomed Pharmacother 37:270–276
Labat ML, Florentin I, Davigny M, Moricard Y, Milhaud G (1984) Dichloromethylene diphosphonate (Cl_2MDP) reduces natural killer (NK) cell activity in mice. Metab Bone Dis Rel Res 5:281–287
Lamson ML, Fox JL, Huguchi WI (1984) Calcium and 1-hydroxyethylidene-1,1-bisphosphonic acid: polynuclear complex formation in the physiological range of pH. Int J Pharmaceut 21:143–154
Larsson A (1974) The short-term effects of high doses of ethylene-1-hydroxy-1,1-diphosphonates upon early dentin formation. Calcif Tissue Res 16:109–127
Larsson SE (1976) The metabolic heterogeneity of glycosaminoglycans of the different zones of the epiphyseal growth plate and the effect of ethane-1-hydroxy-1,1-diphosphonate (EHDP) upon glycosaminoglycan synthesis in vivo. Calcif Tissue Res 21:67–82
Lenehan TM, Balligand M, Nunamaker DM, Wood FE Jr (1985) Effect of EHDP on fracture healing in dogs. J Orthop Res 3:499–507
Leonard EP, Reese WV, Mandel EJ (1979) Comparison of the effects of ethane-1-hydroxy-1,1-diphosphonate and dichloromethane diphosphonate upon periodontal bone resorption in rice rats. Arch Oral Biol 24:707–708

Levy RJ, Wolfrum J, Schoen FJ, Hawley MA, Lund SA (1985) Inhibition of calcification of bioprosthetic heart valves by local controlled-release diphosphonate. Science 228:190–192

Liggett SJ (1973) Determination of ethane-1-hydroxy-1,1-diphosphonic acid (EHDP) in human feces and urine. Biochem Med 7:68–77

Lockwood DR, Vogel JM, Schneider VS, Hulley SB (1975) Effect of the diphosphonate EHDP on bone mineral metabolism during prolonged bed rest. J Clin Endocrinol Metab 41:533–541

Martodam RR, Thornton KS, Sica DA, Souza SM, Flora L, Mundy GR (1983) The effects of dichloromethylene diphosphonate on hypercalcemia and other parameters of the humoral hypercalcemia of malignancy in the rat Leydig cell tumor. Calcif Tissue Int 35:512–519

Metzger AL, Singer FR, Bluestone R, Pearson CM (1974) Failure of disodium etidronate in calcinosis due to dermatomyositis and scleroderma. N Engl J Med 291:1294–1296

Meunier P, Chapuy MC, Courpron P, Vignon E, Edouard C, Bernard J (1975) Effets cliniques, biologiques et histologiques de l'éthane-1-hydroxy-1,1-diphosphonate (EHDP) dans la maladie de Paget. Rev Rhum Mat Osteoartic 42:699–705

Meunier PJ, Alexandre C, Edouard C, Mathieu L, Chapuy MC, Bressot C, Vignon E, Trechsel U (1979) Effects of disodium dichloromethylenediphosphonate on Paget's disease of bone. Lancet 2:489–492

Meyer JL, Nancollas GH (1973) The influence of multidentate organic phosphonates on the crystal growth of hydroxyapatite. Calcif Tissue Res 13:295–303

Meyer JL, Lee KE, Bergert JH (1977) The inhibition of calcium oxalate crystal growth by multidentate organic phosphonates. Calcif Tissue Res 23:83–86

Michael WR, King WR, Wakim JM (1972) Metabolism of disodium ethane-1-hydroxy-1,1-diphosphonate (disodium etidronate) in the rat, rabbit, dog and monkey. Toxicol Appl Pharmacol 21:503–515

Milhaud G, Labat ML, Moricard Y (1983) (Dichloromethylene) diphosphonate-induced impairment of T-lymphocyte function. Proc Natl Acad Sci USA 80:4469–4473

Miller SC, Jee WSS (1979) The effect of dichloromethylenediphosphonate, a pyrophosphate analog, on bone and bone cell structure in the growing rat. Anat Rec 193:439–462

Minaire P, Bérard E, Meunier PJ, Edouard C, Goedert G, Pilonchéry G (1981) Effects of disodium dichloromethylene diphosphonate on bone loss in paraplegic patients. J Clin Invest 68:1086–1092

Morgan DB, Monod A, Russell RGG, Fleisch H (1973) Influence of dichloromethylene diphosphonate (Cl_2MDP) and calcitonin on bone resorption, lactate production and phosphatase and pyrophosphatase content of mouse calvaria treated with parathyroid hormone in vitro. Calcif Tissue Res 13:287–294

Mühlbauer RC, Russell RGG, Williams DA, Fleisch H (1971) The effects of diphosphonates, polyphosphates, and calcitonin on immobilisation osteoporosis in rats. Eur J Clin Invest 1:336–344

Mühlemann HR, Bowles D, Schatt A, Bernimoulin JP (1970) Effect of diphosphonate on human supragingival calculus. Helv Odont Acta 14:31–33

Nagant de Deuxchaisnes C, Rombouts-Lindemans C, Huaux JP, Devogelaer JP, Malghem J, Madlague B (1979) Roentgenologic evaluation of the action of the diphosphonate EHDP and of combined therapy (EHDP and calcitonin) in Paget's disease of bone. Mol Endocrinol 1:405–433

Nixon GA, Buehler EV, Newmann EA (1972) Preliminary safety assessment of disodium etidronate as an additive to experimental oral hygiene products. Toxicol Appl Pharmacol 22:661–671

Nolen GA, Buehler EV (1971) The effects of disodium etidronate on the reproductive functions and embryogeny of albino rats and New Zealand rabbits. Toxicol Appl Pharmacol 18:548–561

Ohata M, Pak CY (1974) Preliminary study of the treatment of nephrolithiasis (calcium stones) with diphosphonate. Metabolism 23:1167–1173

Ohya K, Yamada S, Felix R, Fleisch H (1985) Effect of bisphosphonates on prostaglandin synthesis by rat bone cells and mouse calvaria in culture. Clin Sci 69:403–411
Pilczyk R, Sutcliffe H, Martin TJ (1972) Effects of pyrophosphate and diphosphonates on parythyroid hormone- and fluoride-stimulated adenylate cyclase activity. FEBS Lett 24:225–228
Plasmans CMT, Kuypers W, Slooff TJJH (1978) The effect of ethane-1-hydroxy-1,1-diphosphonic acid (EHDP) on matrix induced ectopic bone formation. Clin Orthop Rel Res 132:233–243
Plasmans CMT, Jap PHK, Kujipers W, Slooff TJJH (1980) Influence of diphosphonate on the cellular aspect of young bone tissue. Calcif Tissue Int 32:247–256
Pollard M, Luckert PH (1985) Effects of dichloromethylene diphosphonate on the osteolytic and osteoplastic effects of rat prostate adenocarcinoma cells. JNCI 75:949–954
Preston CJ, Yates AJP, Beneton MNC, Russell RGG, Gray RES, Smith R, Kanis JA (1986) Effective short term treatment of Paget's disease with oral etidronate. Br Med J 292:79–80
Rabens SF, Bethune JE (1975) Disodium etidronate therapy for dystrophic cutaneous calcification. Arch Dermatol 111:357–361
Radl J, Croese JW, Zurcher C, Enden-Vieveen MHM Van Den, Brondijk RJ, Kazil M, Haaijman JJ, Reitsma PH, Bijvoet OLM (1985) Influence of treatment with ADP-bisphosphonate on the bone lesions in the mouse 5T2 multiple myeloma. Cancer 55:1030–1040
Recker RR, Saville PD (1973) Intestinal absorption of disodium ethane-1-hydroxy-1,1-diphosphonate (disodium etidronate) using a deconvolution technique. Toxicol Appl Pharmacol 24:580–589
Recker RR, Hassing GS, Lau JR, Saville PD (1973) The hyperphosphatemic effect of disodium ethane-1-hydroxy-1,1-diphosphonates (EHDP™): renal handling of phosphorus and the renal response to parathyroid hormone. J Lab Clin Med 81:258–266
Reiner M, Sautter V, Olah A, Bossi A, Largiadèr U, Fleisch H (1980) Diphosphonate treatment in myositis ossificans progressiva. In: Caniggia A (ed) Etidronate. Instituto Gentili, Pisa, p 237
Reitsma PH, Bijvoet OLM, Verlinden-Ooms H, Wee-Pals LJA van der (1980) Kinetic studies of bone and mineral metabolism during treatment with (3-amino-1-hydroxy-propylidene)-1,1-bisphosphonate (APD) in rats. Calcif Tissue Int 32:145–147
Reitsma PH, Teitelbaum SL, Bijvoet OLM, Kahn AJ (1982) Differential action of the bisphosphonates (3-amino-1-hydroxypropylidene)-1,1-bisphosphonate (APD) and disodium dichloromethylidene bisphosphonate (Cl_2MDP) on rat macrophage-mediated bone resorption in vitro. J Clin Invest 70:927–933
Reynolds JJ, Minkin C, Morgan DB, Spycher D, Fleisch H (1972) The effect of two diphosphonates on the resorption of mouse calvaria in vitro. Calcif Tissue Res 10:302–313
Reynolds JJ, Murphy H, Mühlbauer RC, Morgan DB, Fleisch H (1973) Inhibition by diphosphonates of bone resorption in mice and comparison with grey lethal osteopetrosis. Calcif Tissue Res 12:59–71
Robertson WG, Peacock M, Nordin BEC (1973) Inhibitors of the growth and aggregation of calcium oxalate crystals in vitro. Clin Chim Acta 43:31–37
Rosenblum IY, Black HE, Ferrell JF (1977) The effects of various diphosphonates on a rat model of cardiac calciphylaxis. Calcif Tissue Res 23:151–159
Russell RGG, Mühlbauer RC, Bisaz S, Williams DA, Fleisch H (1970) The influence of pyrophosphate, condensed phosphates, phosphonates and other phosphate compounds on the dissolution of hydroxyapatite in vitro and on bone resorption induced by parathyroid hormone in tissue culture and in thyroparathyroidectomised rats. Calcif Tissue Res 6:183–196
Russell RGG, Bisaz S, Donath A, Morgan DB, Fleisch H (1971) Inorganic pyrophosphate in plasma in normal persons and in patients with hypophosphatasia, osteogenesis imperfecta and other disorders of bone. J Clin Invest 50:961–969
Russell RGG, Smith R, Preston C, Walton RJ, Woods CG (1974) Diphosphonates in Paget's disease. Lancet 1:894–898

Ryzen E, Martodam RR, Troxell M, Benson A, Paterson A, Shepard K, Hicks R (1985) Intravenous etidronate in the management of malignant hypercalcemia. Arch Intern Med 145:449–452

Schenk R, Merz WA, Mühlbauer R, Russell RGG, Fleisch H (1973) Effect of ethane-1-hydroxy-1,1-diphosphonate (EHDP) and dichloromethylene diphosphonate (Cl_2MDP) on the calcification and resorption of cartilage and bone in the tibial epiphysis and metaphysis of rats. Calcif Tissue Res 11:196–214

Schenk R, Eggli P, Felix R, Fleisch H, Rosini S (1986) Quantitative morphometric evaluation of the inhibitory activity of new aminobisphosphonates on bone resorption in the rat. Calcif Tissue Int 38:342–349

Schibler D, Russell RGG, Fleisch H (1968) Inhibition of pyrophosphate and polyphosphate of aortic calcification induced by vitamin D_3 in rats. Clin Sci 35:363–372

Shane E, Baquiran DC, Bilezikian JP (1981) Effect of dichloromethylene diphosphonate on serum and urinary calcium in primary hyperparathyroidism. Ann Intern Med 95:23–27

Shindoa H, Adamek G, Felix R, Fleisch H, Schenk R, Hagan P (1983) Structure-activity relationship of various bisphosphonates. Calcif Tissue Int 35:87–99

Siris ES, Canfield RE, Jacobs TP, Baquiran DC (1980 a) Long-term therapy of Paget's disease of bone with EHDP. Arthritis Rheum 23:1177–1183

Siris ES, Sherman WH, Baquiran DC, Schlatterer JP, Osserman EF, Canfield RE (1980 b) Effects of dichloromethylene diphosphonate on skeletal mobilization of calcium in multiple myeloma. N Engl J Med 302:310–315

Siris ES, Hyman GA, Canfield R (1983) Effects of dichloromethylene diphosphonate in woman with breast carcinoma metastatic to the skeleton. Am J Med 74:401–406

Sleeboom HP, Bijvoet OLM, van Oosterom AT, Gleed JH, O'Riordan JLH (1983) Comparison of intravenous (3-amino-1-hydroxypropylidene)-1,1-bisphosphonate and volume repletion in tumor-induced hypercalcemia. Lancet 2:239–243

Slooff TJJH, Feith R, Bijvoet OLM, Nollen AJG (1974) The use of a disphosphonate in para-articular ossification after total hip replacement. A clinical study. Acta Orthop Belg 40:820–828

Smith JD (1983) Metabolism of phosphonates. In: Hilderbrand RL (ed) The role of phosphonates in living systems. CRC Press, Boca Raton, p 31

Smith R, Russell RGG, Bishop M (1971) Diphosphonates and Paget's disease of bone. Lancet 1:945–947

Spielman G, Gennarelli TA, Rogers CR (1983) Disodium etidronate: its role in preventing heterotopic ossification in severe head injury. Arch Phys Med Rehabil 64:539–542

Steber J, Wierich P (1986) Properties of hydroxyethane disphosphonate affecting its environmental fate: degradability, sludge adsorption, mobility in soils, and bioconcentration. Chemosphere 15:929–945

Steiner RM, Glassman L, Schwartz MW, Vanace P (1974) The radiological findings in dermatomyositis of childhood. Pediatr Radiol 111:385–393

Stevenson PH, Stevenson JR (1986) Cytotoxic and migration inhibitory effects of bisphosphonates on macrophages. Calcif Tissue Int 38:227–233

Stronski St A, Trechsel U, Fleisch H (1985) Plasma osteocalcin: lack of relation with bone resorption and effect of bisphosphonates in rats. Calcif Tissue Int 38 [Suppl]:S39

Sturzenberger OP, Swancar JR, Reiter G (1971) Reduction of dental calculus in humans through the use of a dentifrice containing a crystal-growth inhibitor. J Periodontol 42:416–419

Stutzer A, Fleisch H, Trechsel U (1987 a) Long and short term effects of a single administration of bisphosphonates on retinoid induced bone resorption. J Bone Min Res 2 Suppl 1:416

Stutzer A, Trechsel U, Fleisch H, Schenk R (1987 b) Effect of bisphosphonates on osteoclast number and bone resorption in the rat. J Bone Min Res 2 Suppl 1:266

Thiébaud D, Jaeger P, Jacquet AF, Burckhardt P (1986) A single-day treatment of tumor-induced hypercalcemia by intravenous amino-hydroxypropylidene bisphosphonate. J Bone Min Res 1:555–562

Thomas BJ, Amstutz HC (1985) Results of the administration of diphosphonate for the prevention of heterotopic ossification after total hip arthroplasty. J Bone Joint Surg 67A:400–403

Trechsel U, Schenk R, Bonjour JP, Russell RGG, Fleisch H (1977) Relation between bone mineralization, Ca absorption, and plasma Ca in phosphonate-treated rats. Am J Physiol 232:E298–E305

Troehler U, Bonjour JP, Fleisch H (1975) Renal secretion of diphosphonates in rats. Kidney Int 8:6–13

Van Breukelen FJM, Bijovet OLM, Oosterom AT (1979) Inhibition of osteolytic bone lesions by (3-amino-1-hydroxypropylidene)-1,1-bisphosphonate (A.P.D.). Lancet I:803–805

Van Breukelen FJM, Bijvoet OLM, Frijlink WB, Sleebloom HP, Mulder H, von Oosterom AT (1982) Efficacy of amino-hydroxypropylidene bisphosphonate in hypercalcemia: observations on regulation of serum calcium. Calcif Tissue Int 34:321–327

Van Rooijen N, van Nieuwmegen R, Kamperdijk EWA (1985) Elimination of phagocytic cells in the spleen after intravenous injection of liposome-encapsulated dichloromethylene diphosphonate. Virchows Arch (Zel) 49:375–383

Walton RJ, Russell RGG, Smith R (1975) Changes in the renal and extrarenal handling of phosphate induced by disodium etidronate (EHDP) in man. J Clin Sci Mol Med 49:45–56

Wasserman RH, Bonjour JP, Fleisch H (1973) Ileal absorption of disodium ethane-1-hydroxy-1,1-diphosphonate (EHDP) and disodium dichloromethylene diphosphonate (Cl_2MDP) in the chick. Experientia 29:1110–1111

Wiedmer WH, Zbinden AM, Trechsel U, Fleisch H (1983) Ultrafiltrability and chromatographic properties of pyrophosphate, 1-hydroxyethylidene-1,1-bisphosphonate, and dichloromethylenebisphosphonate in aqueous buffers and in human plasma. Calcif Tissue Int 35:397–400

Wingen F, Schmähl D (1985) Distribution of 3-amino-1-hydroxypropane-1,1-diphosphonic acid in rats and effects on rat osteosarcoma. Arzneimittelforsch 35:1565–1571

Wink CS, Onge MSt, Parker B (1985) The effects of dichloromethylene bisphosphonate on osteoporotic femora of adult castrate male rats. Acta Anat 124:117–121

Worms KH, Blum H (1979) Umsetzungen von 1-Aminoalkal-1,1-diphosphonsäuren mit salpetriger Säure. Z Anorg Allg Chem 457:209–213

Worms KH, Schmidt-Dunker M (1976) Phosphonic acids and derivatives. In: Kosolapoff GM, Maier L (eds) Organic phosphorus compounds, vol 7. Wiley, New York, p 1

Yakatan GJ, Poynor WJ, Talbert RL, Floyd BF, Slough CL, Ampulski RS, Benedict JJ (1982) Clodronate kinetics and bioavailability. Clin Pharmacol Ther 31:402–410

Yates AJP, Percival RC, Gray RES, Urwin GH, Hamdy NAT, Preston CJ, Beneton MNC, Russell RGG, Kanis JA (1985) Intravenous clodronate in the treatment and retreatment of Paget's disease of bone. Lancet 1:1474–1477

Zernov IM, Stefani DV, Vel'tischev YE (1979) Assessment of the protective action of diphosphonate compound against damage to T-lymphocytes by antilymphocytic serum. Bull Exp Biol Med 87:253–254

CHAPTER 22

Calcium and Hypertension

R. D. BUKOSKI and D. A. MCCARRON

A. Introduction

Essential hypertension is a health problem of wide dimensions, with at least 40 million people currently at risk in the United States alone. Epidemiologic studies have demonstrated a significant correlation between blood pressure and cardiovascular disease (PAGE 1983) and there is little doubt that this results in a considerable burden in both personal and economic terms. With these simple facts in mind, it is obvious that an understanding of the physiologic processes underlying this disease and a safe means of preventing or alleviating abnormal elevations in blood pressure need to be developed.

Systemic hypertension is believed to result from an increase in peripheral vascular resistance and as such may be thought of as a disease affecting resistance vessels. Among the hypotheses that have been put forward to explain this malady are an alteration in sodium metabolism of the organism of the smooth muscle cell (MACGREGOR 1985), structural alterations in the vasculature (FOLKOW 1982), autonomic or central nervous system disturbances (BRODY et al. 1983), or a basic defect related to calcium metabolism at the level of the smooth muscle cell (DANIEL and KWAN 1981; MCCARRON 1985). It is the latter hypothesis that will serve as the primary focus of this chapter.

It is understood that contraction and relaxation of the vascular smooth muscle cell are ultimately mediated by a rise or fall in the concentration of free intracellular calcium. The means by which intracellular Ca^{2+} is increased is dependent upon the activating stimulus while several independent systems are thought to play a role in lowering elevated levels of intracellular calcium. Given this delicately balanced system, it has been suggested that a primary defect in essential hypertension may be an elevation of ionized Ca^{2+} within the cell that gives rise to an increase in basal tone of the muscle (MCCARRON 1985). Obvious candidates for the cellular lesion are an increased leak of Ca^{2+} into the cell or a defect in Ca^{2+} extrusion mechanisms.

Further insight into the problem may be gained from the observed effectiveness of calcium channel blockers as antihypertensive agents. The fact that these compounds have been shown to block entry of Ca^{2+} into smooth muscle cells and to lower blood pressure suggests that a basal influx of Ca^{2+} into the cell that is linked to active tone may exist in hypertension. Furthermore, this influx of calcium may be perfectly balanced by extrusion in vessels of normotensive subjects, but unmatched by efflux in the hypertensive state (SPIVAK et al. 1983). The observation that elevated dietary calcium intake can have a blood pressure lowering effect in a population of "calcium-sensitive" humans (MCCARRON and MORRIS 1985) and in experimental hypertension (AYACHI 1979; LAU et al. 1984a; MCCARRON et al. 1985; BUKOSKI and MCCARRON 1986) may also be of significance. Although this finding may at first glance appear contrary to conclusions derived from the results of experiments employing Ca^{2+} channel blockers, examination of the evidence indicates that dietary intake of calcium may well alter reactivity of vascular smooth muscle.

This chapter will critically discuss the experimental evidence for the blood pressure lowering effect of dietary calcium and the postulated mechanisms involved, against a background of the role of altered calcium metabolism in the genesis and maintenance of hypertension. Comparisons will be made between studies of essential hypertension and the widely used laboratory model for this disease, the spontaneously hypertensive rat (SHR). This discussion will hopefully give insight into the mechanism (or mechanisms) by which people with essential hypertension respond to increases in dietary calcium.

B. Role of Ca^{2+} in Vascular Smooth Muscle Contraction

Before embarking on a discussion of calcium and hypertension, a brief discussion of current views of calcium metabolism of the vascular smooth muscle cell is warranted. For in-depth treatment of the subject, the reader is referred to Chaps. 3 and 18 of this volume and to other reviews (VAN BREEMEN et al. 1979; GROVER and DANIEL 1985). As we have noted, it is currently understood that contraction of vascular smooth muscle is mediated by an increase in the concentration of free Ca^{2+} in the cell. As Ca^{2+} rises above 0.1 μM it binds to calmodulin in an allosteric manner to form a complex that is capable of activating the enzyme myosin light chain kinase, which in turn catalyzes phosphorylation of myosin light chains (CHACKO et al. 1985). Myosin ATPase activity is greatly stimulated in its phosphorylated state and interacts with action through a sliding filament mechanism to cause shortening of the cell and formation of a "latch state" (DILLON et al. 1981). In the latch state, cross-bridge attachments are maintained in the presence of continued slow cycling of the myosin ATPase. In addition, a second mechanism, believed to be highly sensitive to Ca^{2+}, is also thought to play a role in the maintenance of tension (AKSOY et al. 1983; CHATTERJEE and MURPHY 1983). Relaxation is believed to result from a lowering of the concentration of intracellular Ca^{2+} to less than 0.1 μM with resultant dephosphorylation of the myosin light chains to basal levels and detachment of cross-bridges.

The means by which intracellular Ca^{2+} is increased to induce contraction is highly dependent on the activating stimulus. It is believed that Ca^{2+} can enter the cell either through receptor-operated channels (ROCs) (BOLTON 1979) or via voltage-operated channels (VOCs) (CAUVIN et al. 1983). In addition, there is evidence consistent with calcium being released from an internal store, probably the sarcoplasmic reticulum, either through a Ca^{2+}-induced Ca^{2+} release mechanism (SAIDA and VAN BREEMEN 1983) or through a mechanism linked to phosphatidyl inositol metabolism (CAMPBELL et al. 1985; SOMLYO et al. 1985; WILLIAMSON 1986). In the latter process, occupation of a specific receptor by an agonist such as norepinephrine can induce formation of inositol trisphosphate (IP_3) which has been demonstrated to cause release of Ca^{2+} from sarcoplasmic reticulum in skinned cultured vascular smooth muscle cells (SUEMATSU et al. 1984; YAMAMOTO and VAN BREEMEN 1985). In addition to IP_3, diacylglycerol is also formed and this metabolite has been shown to stimulate the C kinase enzyme which is capable of activating a variety of systems in vascular smooth muscle (FORDER et al. 1985).

Normal functioning of this system requires that a means of tightly regulating free intracellular Ca^{2+} at rest must exist, as well as a system for removing excess Ca^{2+} after stimulation by a contractile agonist. It is believed that Ca^{2+} is removed from the cell by active extrusion across the cell membrane by the Ca^{2+} pump, the enzymatic manifestation of which is believed to be the Ca^{2+}-activated ATPase present in sarcolemma. Evidence for the existence of this pump in vascular smooth muscle has largely been the observation that ATP can stimulate Ca^{2+} uptake in isolated membrane vesicles (BHALLA et al. 1978; KWAN et al. 1979; WUYTACK and CASTEELS 1980). A passive mechanism, which has not yet met with total acceptance as a regulator of intracellular Ca^{2+} levels in vascular smooth muscle is an Na^+,Ca^{2+} exchange mechanism, similar to that described in cardiac muscle (REUTER et al. 1973; OZAKI and URAKAWA 1979). At least two reports have demonstrated this process in isolated membrane fractions of vascular smooth muscle (DANIEL et al. 1982; MATLIB et al. 1985), but one of them has calculated that the activity of the carrier is probably too low to contribute significantly to Ca^{2+} removal from the vascular smooth muscle cell (GROVER 1985). This conclusion is in agreement with results obtained using intact tissue which indicate that little or no role may be played by this mechanism in vascular smooth muscle (VAN BREEMEN et al. 1979).

A third mechanism for Ca^{2+} removal from the interior of the cell is active uptake by the sarcoplasmic reticulum. Although purification of this system from vascular tissue has presented technical difficulties in the past, results of newer approaches using saponin-skinned preparations that remove the cell membrane as a barrier, but leave the network of sarcoplasmic reticulum intact, have demonstrated active uptake by this system (YAMAMOTO and VAN BREEMEN 1985). In addition to the systems mentioned, it is recognized that mitochondria may play a role in regulating intracellular Ca^{2+}. Unfortunately, with the exception of the report by VALLIERES et al. (1975) few preparations of mitochondria isolated from vascular smooth muscle, which demonstrate characteristics of well-coupled oxidative phosphorylation and substrate-supported Ca^{2+} uptake, have been described. It is thus difficult to judge the extent to which this system may play a role in vascular smooth muscle.

In addition to these various pathways for Ca^{2+} influx and extrusion, it is believed that Ca^{2+} can modulate both its own entry and fluxes of monovalent cations across the vascular smooth muscle cell membrane, therapy exerting a membrane-stabilizing effect (WEBB and BOHR 1978). Evidence which is consistent with such an effect includes observations of an increase in vessel tone in response to Ca^{2+} challenge after removal of extracellular Ca^{2+} (BOHR and WEBB 1985; BUKOSKI and MCCARRON 1986), increased peremeability of smooth muscle to monovalent cations after removal of extracellular Ca^{2+} (JONES 1974), and a relaxing effect of extracellular Ca^{2+} added to contracted vessels (WEBB and BOHR 1978).

It should be noted, however, that this latter effect only occurs at nonphysiologic concentrations of Ca^{2+}. Whether Ca^{2+} acts at a specific Ca^{2+} channel to exert this stabilizing action or binds to specific stabilizing sites which allosterically alter monovalent cation flux is currently not known.

C. Role of Altered Ca^{2+} Metabolism in Hypertension

I. Human Studies

It should be apparent that active wall stress and therefore resistance to blood flow are, at least in part, dependent upon intracellular levels of free Ca^{2+} and the enzymatic mechanisms which regulate its influx and efflux. With this in mind, an obvious hypothesis explaining increased peripheral resistance in essential hypertension is that there exists an elevated concentration of free intracellular calcium in the smooth muscle cell which gives rise to an increase in basal tone of the resistance vessel or increases reactivity of the vessel. Although no direct measurements of intracellular free calcium have been made in vascular smooth muscle cells from subjects with essential hypertension, studies using the fluorescent indicators quin2 and fura-2 in platelets have shown elevated levels of Ca^{2+} (ERNE et al. 1984; BRUSHI et al. 1985; SANG ct al. 1986). In agreement with those observations, RESINK et al. (1986) found that Ca^{2+}-stimulated ATPase activity of platelets from essential hypertensive subjects was increased although fractional calmodulin-stimulated Ca^{2+}-ATPase activity was depressed, a finding also noted by other investigators in red blood cell membranes derived from hypertensive subjects (VINCENZI et al. 1986). These authors have suggested that the increased enzyme activity is an adaptive response to a Ca^{2+} leak into the cell. Similar changes in intracellular Ca^{2+} concentration have also been described in platelets obtained from an experimental model of genetic hypertension, the SHR (BRUSHI et al. 1985; SKUHERSKA et al. 1986). It should be noted, however, that this finding is not universal with respect to cell type since intracellular Ca^{2+} in neutrophils from humans with essential hypertension has not been found to be elevated (LEW et al. 1985). While these data do not establish a link between intracellular free Ca^{2+} in vascular smooth muscle and hypertension, they are suggestive that such a defect may exist.

Obvious candidates for the lesion in essential hypertension are an increase in either passive or stimulated Ca^{2+} flux into the cell, a defect in the Ca^{2+} extrusion or sequestration processes, or an alteration in the membrane-stabilizing properties of extracellular Ca^{2+}. Such defects have been described in red blood cells of essential hypertensives (POSTNOV and ORLOV 1985). It has been reported that there is a decrease in the capacity of the red cell membranes to bind Ca^{2+}, which may contribute to increases in intracellular Ca^{2+} (POSTNOV et al. 1977, 1984).

This concept has been more fully developed in what has been termed the "urinary calcium leak" hypothesis. Several reports have now demonstrated that there are depressed levels of serum ionized calcium in humans with essential hypertension (McCARRON 1982a; RESNICK et al. 1983; FOLSOM et al. 1984) as well as an increase in urinary calcium excretion and circulating levels of parathyroid hormones (McCARRON et al. 1980; STRAZZULLO et al. 1983). It has been proposed that a defect in renal Ca^{2+} transport underlies the hypercalciuria and this is causally related to the observed low levels of serum ionized Ca^{2+} and elevated levels of parathyroid hormone. These findings, in subjects with essential hypertension, have experimental support in similar observations made in the SHR (McCARRON et al. 1981; WRIGHT and RANKIN 1982; LAU et al. 1984 a, b; BINDELS et al. 1984).

This hypothesis, however, remains controversial. LAU et al. (1984 b) have suggested that the urinary calcium leak observed in the SHR is actually secondary to increased intestinal absorption of Ca^{2+} and is unrelated to a renal tubule Ca^{2+} transport defect. Careful examination of the substantiating data, however, shows that intestinal uptake of Ca^{2+} is increased in the mature SHR, but is depressed in the young SHR at a time when blood pressure is increased. This latter finding is consistent with results of SCHEDL et al. (1984, 1986) and McCARRON et al. (1985) who demonstrated depressed intestinal absorption of Ca^{2+} in the young SHR. These data make it seem unlikely that enhanced Ca^{2+} uptake by the small bowel plays a role in the development of hypertension, although the relationship between uptake and urinary calcium excretion in the older animal is not clear.

In addition, there are at least two reports which have demonstrated that serum ionized calcium is positively correlated with blood pressure in essential hypertension (FOGH-ANDERSON et al. 1984; HUNT et al. 1984). It should be noted that in the latter studies, some of the subjects were receiving thiazide diuretic therapy, which is known to elevate serum ionized Ca^{2+}. This may account for the dissimilar results. A reason for these discrepant results may be found in the results of RESNICK et al. (1983) who found that serum ionized Ca^{2+} levels are positively correlated with plasma renin activity. Therefore, population samples that were not normally distributed with respect to plasma renin activity could explain these contrasting results, although a major sampling error would have to occur since high renin patients comprise only a small percentage of the population. It is clear that additional investigative effort will be required before these issues can be resolved.

II. The Spontaneously Hypertensive Rat

While experimental evidence for lesions in calcium metabolism in human vascular tissue is scarce, numerous defects have been described in vascular smooth muscle isolated from an experimental model of essential hypertension, the SHR. One consistent finding has been that ATP-stimulated uptake of Ca^{2+} by either microsomal fractions of aorta (AOKI et al. 1974; WEBB and BHALLA 1976; BHALLA et al. 1978) or cell membrane-enriched fractions of mesenteric arteries (KWAN et al. 1980; KWAN and DANIEL 1982) is depressed when compared with tissue from the normotensive genetically related Wistar Kyoto rat (WKY). In contrast, the one study that has examined Na^{+},Ca^{2+} exchange in cell membranes isolated from mesenteric arteries of the SHR and WKY was unable to find any differences in this parameter (MATLIB et al. 1985). These data suggest that the isolated Na,Ca exchange carrier does not exhibit an abnormality in the SHR. Unfortunately, biochemical studies such as the latter one can not assess behavior of this system in the intact cell. The other system for Ca^{2+} removal, the sarcoplasmic reticulum, has not been studied rigorously using the newer methods of skinned smooth muscle cells, so its involvement in hypertension remains in question.

Ca^{2+} entry into the smooth muscle cell of the SHR has also been shown to be disturbed. Increased permeability of the cell membrane to Ca^{2+} that is linked to increases in basal tone in mesenteric resistance vessels of the SHR have been reported by CAUVIN and VAN BREEMEN (1985). Similarly, NOON et al. (1978) have reported increased membrane permeability to Ca^{2+} using indirect measurements in aorta isolated from the SHR. In addition, an increase in permeability of the cell membrane to Na^+ and K^+ in the aorta of the SHR has been demonstrated and this is enhanced upon removal of extracellular Ca^{2+} (JONES 1974), a finding that is consistent with an alteration in the membrane-stabilizing action of extracellular Ca^{2+} (WEBB and BOHR 1978). This finding is also supported by results of studies in which the response of aortas isolated from SHRs and WKYs to addition of Ca^{2+} was determined after incubation in K^+-free, Ca^{2+}-free solution (BOHR and WEBB 1985; BUKOSKI and MCCARRON 1986). It was observed that the magnitude of the contraction was greater in aortas of the SHR than in the normotensive control, indicating that differences in the effect of Ca^{2+} on membrane fluxes may be present.

III. Ca^{2+} Antagonists and Hypertension

Although experimental evidence obtained using vascular tissue isolated from humans is lacking, valuable insight into the possibility that altered Ca^{2+} metabolism in vascular smooth muscle may underlie vascular changes observed in hypertension may be gleaned from an examination of the role that Ca^{2+} channel blockers have assumed in antihypertensive therapy. These drugs are potent vasodilators and are so named because they inhibit Ca^{2+} entry into cells through what is thought to be a competitive action at Ca^{2+} channels (BRAUNWALD 1982). In antihypertensive therapy, they are currently used in conjunction with other agents, including beta-blockers and diuretics, for reasons of both pharmacokinetics and side effects such as water retention and tachycardia (FRISHMAN and CHARLAP 1984).

Verapamil and nifedipine, two of the three calcium channel blockers that are currently used in the United States, have been shown to relax blood vessels isolated from both humans (MIKKELSON et al. 1978a, b) and experimental animals (TRIGGLE and JANIS 1985). Interestingly, they are more effective on vessels in which active tone has been induced by exposure to a depolarizing stimulus (such as high K^+) than with an agonist that acts through an ROC (norepinephrine) (VAN BREEMAN et al. 1981). Furthermore, studies examining ^{45}Ca flux in isolated arterial preparations have shown that influx induced by both the depolarizing stimulus, KCl, and the ROC-mediated stimulus, norepinephrine, is inhibited by diltiazem and the verapamil analog, D600, in a dose-dependent manner (VAN BREEMEN et al. 1981; CAUVIN et al. 1984a, b). These results provide a basis for understanding the mechanism by which Ca^{2+} antagonists relax vascular smooth muscle.

Given this mechanism of action, it would seem that these vasodilators would act only in vessels with previously existing basal or active tone – which could be induced neurogenically or by a mismatch of basal Ca^{2+} influx and Ca^{2+} extrusion/sequestration processes. That this may be the case is suggested by the observations that people with elevated blood pressure respond to nifedipine and verapamil with a decrease in peripheral resistance while normotensive individuals show little or no response (BUHLER et al. 1982; ROBINSON et al. 1982; AOKI et al. 1985; FRISHMAN et al. 1986). Along these same lines, it has been demonstrated that aortas of the SHR, when contracted with norepinephrine, show a greater sensitivity to the relaxing effect of nifedipine than do arteries isolated from the normotensive WKY (PEDERSEN et al. 1978; NYBORG et al. 1985).

Another important clinical observation has been that not all patients with essential hypertension respond favorably to calcium channel blockers, with values ranging from 40%–70% (FRISHMAN et al. 1986). The profile of the best responder is the older person with low plasma renin activity (ERNE et al. 1983). These observations would favor the hypothesis that a subset of individuals with a specific type of vascular lesion (perhaps related to altered Ca^{2+} metabolism) may respond favorably to these drugs. As will be discussed in Sect. D, these observations, arrived at using Ca^{2+} channel blockers, have some parallels in the recently described findings that dietary calcium supplementation has a blood pressure lowering action in a subpopulation of people with essential hypertension.

D. Role of Dietary Calcium in Hypertension

I. Human Studies

Within the past decade there has been an emerging connection between consumption of calcium and blood pressure in humans. In the initial prospective epidemiologic report, a population of 90 people in the Portland, Oregon area were examined and it was found that an inverse correlation between calcium intake and risk of having hypertension existed (MCCARRON et al. 1982). A subsequent report from this group (MCCARRON et al. 1984) examined data from 10372 individuals who were surveyed and included in a data base for the National Center for Health Statistics, Health and Nutrition Examination Survey I (HANES I). The results of this study indicate there was an inverse relationship between dietary calcium intake and risk of being hypertensive that was independent of such variables as body mass index, age, sex, or race.

In addition to this study, numerous reports have now demonstrated a similar relationship in several populations of humans (BELIZAN and VILLAR 1980; ACKLEY et al. 1983; GARCIA-PALMIERI 1984; SOWERS et al. 1985), including the same HANES I data base (HARLAN et al. 1984, 1985; SEMPOS et al. 1985). These studies have received criticism for several reasons, including the means by which interactions with other confounding variables (i.e., body mass index, consumption of ethanol and nutrients other than Ca^{2+}) were assessed (FEINLEIB et al. 1984). In addition, the report by FEINLEIB et al. (1984) demonstrated that results contrary to those reported by MCCARRON et al. (1984) can be obtained from the HANES I data base if the entry data are weighted differently. The latter observation illustrates the technical difficulties that are inherent in this type of study, and the fact remains that the majority of studies have been consistent in supporting the initial observations that calcium intake and blood pressure are inversely related.

Clinical studies based in part on these epidemiologic results have now demonstrated that supplementation of the diet with calcium results in a reduction of blood pressure in a subpopulation of humans with essential hypertension (BELIZAN et al. 1983; RESNICK et al. 1984, 1985 b; JOHNSON et al. 1985; McCARRON and MORRIS 1985; LUFT et al. 1986). In an early study, JOHNSON et al. (1985) examined the effect of Ca^{2+} supplementation on bone density of women over a 4-year period and, upon retrospective examination of blood pressure data, found that those with a supplemented diet had a lower blood pressure than the unsupplemented group. This study was confounded, however, by the fact that concurrent antihypertensive medications was also taken by these women. Nevertheless, these results are consistent with those of RESNICK et al. (1984, 1985 a, b) who have demonstrated that oral calcium was able to lower blood pressure, both over a short-term period (5 days) and over longer periods, i.e., 1 and 5 months. An interesting finding in the latter study was that a subset of subjects responded with larger drops in blood pressure and this correlates with their having both lower plasma renin levels and lower initial ionized serum calcium values (RESNICK et al. 1985 b). This represents a striking parallel to the responsiveness of patients in their trial to the Ca^{2+} channel blocker, verapamil (RESNICK et al. 1985 a).

Recently, McCARRON and MORRIS (1985) have demonstrated an overall decrease in blood pressure in essential hypertensives with 1 g supplementation of their diets with elemental calcium. This study was carried out in a double-blind, placebo-controlled crossover design and thus rigorously controlled for a placebo effect in each subject. In addition to a small but significant overall effect on blood pressure, a subset of the group (21 of 48 patients) was observed to respond to calcium with at least a 10-mmHg lowering of blood pressure. In this regard, it should be noted that some individuals showed no blood pressure response to calcium while others exhibited a pressor response. These data again suggest that a group of calcium-sensitive humans with essential hypertension exists, which may account for as much as 40% of the total population.

A small but significant overall decline in systolic blood pressure was also observed in a clinical trial carried out by LUFT et al. (1986). A main goal of this group was to examine Na^+ balance in humans receiving calcium supplementation and no evidence for natriuresis or associated volume change was observed. These authors concluded that an effect of dietary calcium on Na^+ balance plays no role in the short-term blood pressure response of essential hypertensives.

II. The Spontaneously Hypertensive Rat

Aside from the study discussed in Sect. D.I, little is currently known about the mechanism of the blood pressure lowering effect of dietary calcium in humans. This is partly due to the fact that the response to calcium has only been recognized for a few years and that tissue for studies is not easily obtained from human subjects. In contrast to the human situation, much more work has been carried out examining the mechanism of action of dietary calcium using the SHR as an experimental model. There have been numerous reports which demonstrate that di-

Table 1. Effect of dietary Ca on blood pressure in the rat

Reference	Strain	Calcium (%)	Entry age (weeks)	Time on diet (weeks)	Results (mm Hg)	Age at which BP falls (weeks)
AYACHI (1979)	SHR	1.2 vs 2.5	4	15	210 vs 190[a]	7
MCCARRON et al. (1981)	SHR	0.5 vs 4	10	20	210 vs 185[a]	34
MCCARRON (1982B)	WKY	0.25 vs 4	8–10	24	123 vs 102[a]	12
LAU et al. (1984a)	SHR	0.22 vs 4.3	9	2	184 vs 166[a]	11
STERN et al. (1984)	SHR	0.4 vs 2.8	6	4	144 vs 150	Not applicable
TENNER et al. (1985)	SHR	0.02, 0.1, 0.5, 2.5	5	20	Significant inverse correlation	Not available
BUKOSKI and MCCARRON (1986)	SHR	1 vs 2	6	15	159 vs 148[a]	14
STERN et al. (1986)	SHR	0.4 vs 2.8	11	4	199 vs 180[a]	Not available

[a] Indicates a statistically significant difference in blood pressure.

Table 2. Mechanisms by which calcium loading may lower blood pressure

Decreased reactivity of vascular smooth muscle
Depressed cardiovascular function secondary to hypophosphatemia
Induced natriuresis and volume contraction
Depressed activity of the central or autonomic nervous system
Composite hypothesis – correction of an underlying defect in cellular Ca^{2+} metabolism

etary calcium can attenuate the rise in blood pressure in the SHR (AYACHI 1979; MCCARRON et al. 1981, 1985; LAU et al. 1984a; BUKOSKI and MCCARRON 1986; STERN et al. 1986). The data listed in Table 1 illustrate this point. Listed are the amount of calcium in the diets; the ages at which the interventions were begun, when the effect on blood pressure occurs, and the magnitude of the responses. The reader should note the age dependency of the blood pressure lowering effect as well as the intralaboratory differences in the diets, ages, and conditions of the experiments, since it is becoming apparent that both the level of calcium used and the age at which the dietary intervention is begun are crucial in terms of the time of onset and perhaps the magnitude of the response. These results may have important implications with respect to maintenance of proper calcium balance during human development.

The current hypotheses that have been put forward to explain the blood pressure lowering effect of dietary calcium are listed in Table 2 and include an effect on vascular smooth muscle, an effect on phosphate metabolism with a subsequent deleterious effect on blood pressure, an effect on Na^+ balance with resulting volume contraction, and an effect on the autonomic nervous system. In addition to these, a composite hypothesis has been formulated that explains the actions of calcium on the ability of this cation to correct for a widespread disturbance of

Ca^{2+} metabolism in the organism which manifests itself as defects in vascular smooth muscle, intestinal Ca^{2+} absorption, renal Ca^{2+} handling, and in Ca^{2+} homeostasis in other cell types. Each of these hypotheses will be discussed in turn.

E. Postulated Mechanisms

I. Effect of Dietary Calcium on Vascular Smooth Muscle

One site that has been the focus of experiments examining the mechanism of the blood pressure lowering effect of dietary calcium is the vascular smooth muscle cell. BUKOSKI and MCCARRON (1986) have examined aortic reactivity in tissue isolated from SHRs fed diets with either 1% or 2% calcium for periods of either 8 or 15 weeks, beginning at 6 weeks of age. No differences were observed at the earlier time, when the blood pressure effect first emerged. However, after 15 weeks on the diets when the pressure effect had been established for approximately 2 months, there was a decrease in apparent sensitivity to KCl, a decrease in response to addition of Ca^{2+} after incubation in Ca^{2+}-free, K^+-free physiologic salt solution, and an attenuation in age-related increases in wall stiffness. These data were interpreted to indicate that a decrease in aortic reactivity and vascular wall stiffness are associated with the blood pressure lowering effect of calcium in the SHR. In a parallel study, these authors examined Ca^{2+} metabolism in aortic rings isolated from the same rats as described already (BUKOSKI et al. 1986). After 8 weeks on these diets, there were no differences in Ca^{2+} influx, whereas after 15 weeks on the diets, Ca^{2+} influx after 1 min exposure to label was significantly elevated in the 2% calcium diet group. Since no changes in extracellular binding were observed, and reactivity was found to be decreased, it was concluded that exposure to increased dietary calcium induces an increase in the ability of the aortic smooth muscle cell to sequester Ca^{2+}, perhaps at a site that stabilizes the cell via an effect on Na^+ influx or a Ca^{2+}-activated K^+ efflux (Fig. 1).

In addition to establishing that a measurable effect of dietary calcium occurs in smooth muscle, another important finding of this study was the observation that it takes approximately 8 weeks dietary supplementation to result in a lowering of blood pressure in the SHR. This is not consistent with an acute effect of the diet intervention, rather it implies that long-term adaptive responses to the elevated Ca^{2+} load may be taking place. These findings also give rise to the possibility that regulators of calcium homeostasis in the vascular smooth muscle cell (Ca^{2+} pumps, binding proteins) may not be fixed, but may actively respond to the extracellular calcium load overtime. Such a mechanism would not be unlike the increase in the number of sodium pump sites which occurs in toad bladder in response to aldosterone and sodium load presented to the animal (EDELMAN 1981; KATZ 1982; MARVER and KOKKO 1983).

The data from this study are in contrast to two preliminary reports that also examined vascular response after feeding SHRs diets with elevated amounts of Ca^{2+}. STERN et al. (1986) examined responses of caudal artery to norepinephrine and transmural nerve stimulation and found no differences between arteries isolated from animals fed 0.4% or 2.8% Ca^{2+}. Similarly, TENNER et al. (1985) were

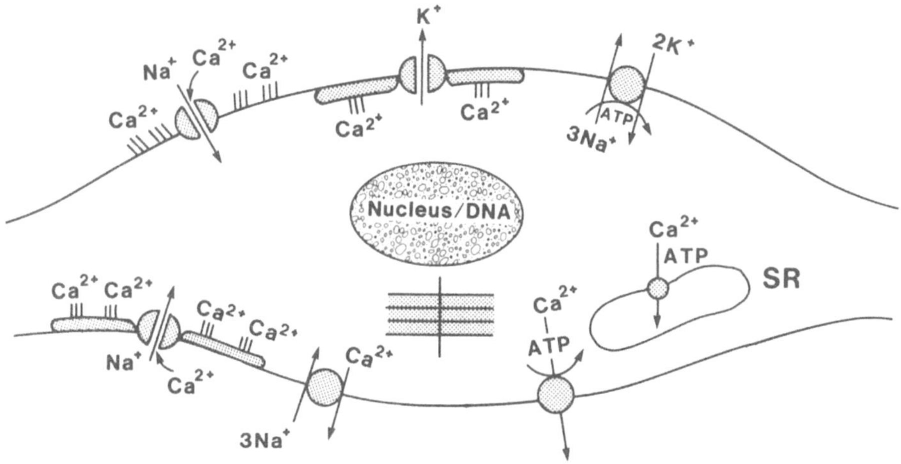

Fig. 1. Model illustrating possible cellular sites at which increased Ca^{2+} binding could depress vascular reactivity. On the left side of the figure are channels located within the cell membrane where Ca^{2+} can modulate Na^+ entry into the cell either by competition at a channel, or by binding to membrane-associated Ca^{2+}-binding sites in the vicinity of the channel. Also depicted are a Ca^{2+}-activated K^+ efflux channel and the sodium pump. Depicted at the bottom are an Na,Ca exchange carrier: the cell membrane-associated Ca^{2+} pump and intracellular sarcoplasmic reticular network (SR). Our data are consistent with an effect of dietary Ca^{2+} mediated by an increase in the ability of the inner cell membrane to sequester Ca^{2+} and stabilize the cell membrane, perhaps through altered Na^+ permeability or increased K^+ efflux

also unable to show changes in reactivity of rat tail artery, although blood pressure was inversely proportional to Ca^{2+} intake. It should be pointed out, however, that the study by BUKOSKI and MCCARRON (1986) was also unable to demonstrate changes in maximal force development in the aorta, although other parameters were altered. It seems prudent that properties other than classic concentration–response profiles to common agonists also be examined before an action of dietary calcium on vascular reactivity is ruled out. In addition, examination of reactivity of resistance-caliber vessels is also warranted.

The means by which the action of dietary calcium on vascular smooth muscle is exerted remains to be answered. RESNICK et al. (1985 a) have suggested that hormones involved in whole animal Ca^{2+} homeostasis (PTH, vitamin D, calcitonin) may have effects on vascular smooth muscle. In any consideration, it should be noted that levels of both PTH and 1,25-dihydroxyvitamin D_3 are decreased with increased dietary intake. Therefore, PTH, which has been demonstrated to be a vasodilator (NICHOLS et al. 1986), would be unlikely to mediate the blood pressure lowering response. On the other hand, 1,25-dihydroxyvitamin D_3 has been suggested to act as a vasoconstrictor Ca^{2+} ionophore. Suppression of 1,25-dihydroxyvitamin D_3, therefore, could mediate the Ca^{2+} effect (RESNICK et al. 1983; BARAN and MILNE 1986). While this hormone has been shown to alter Ca^{2+} metabolism in cultured skeletal muscle cells (BOLAND et al. 1983; BOLAND and BOLAND 1985), and appropriate receptors have been shown to exist in myoblast cells

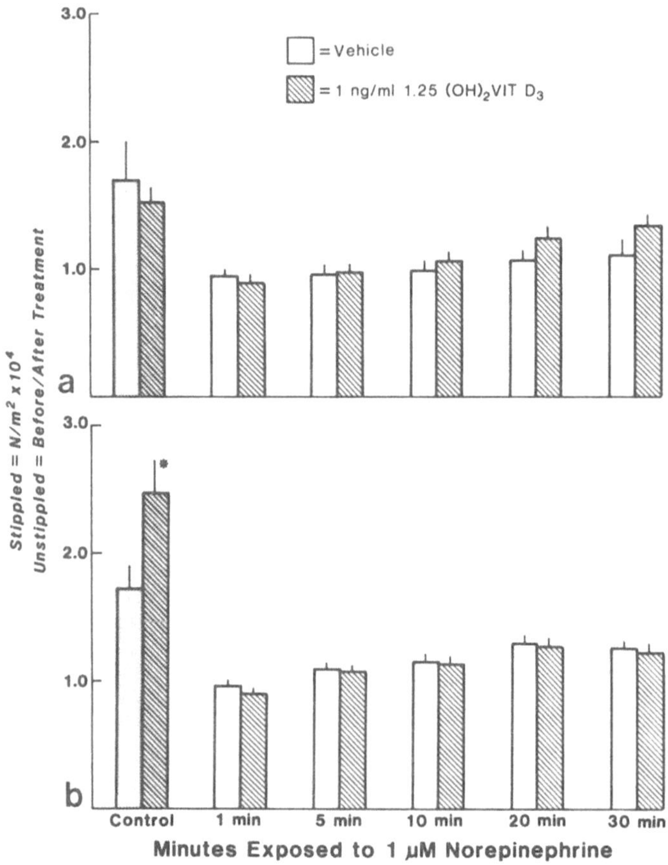

Fig. 2 a, b. Effect of a 3-h treatment of aortic rings with 1 μg/ml 1,25-dihydroxyvitamin D_3 on time course and magnitude of response to 1 μ*M* norepinephrine. **a** Wistar Kyoto rats $n=5$, **b** spontaneously hypertensive rats $n=8$. *Hatched bars* represent vitamin D_3 treatment and *open bars* are vehicle (ethanol). *Shaded areas* are controls, where units are N/m^2 ($\times 10^4$) and *unshaded areas* are the ratio of response before and after exposure to 1,25-dihydroxyvitamin D_3 or ethanol (H. XUE, R. D. BUKOSKI and D. A. McCARRON 1986, unpublished work). *Asterisk* indicates a difference at $P<0.05$

(SIMPSON et al. 1985), our laboratory has examined the acute effect of 1,25-dihydroxyvitamin D_3 on aorta isolated from the SHR and WKY and has observed no apparent effect, either as a contractile agonist or as a permissive ionophore (Fig. 2). In addition to these hormones, a long-term effect of ionized Ca^{2+} acting on vascular smooth muscle has been proposed to alter vascular calcium metabolism, but has not yet been thoroughly examined (McCARRON 1985). Much work remains to be completed in this area and it is complicated by the fact that these hormones are difficult if not impossible to manipulate independently of one another in the intact organism.

II. Dietary Calcium and Phosphate Metabolism

The fact that calcium and phosphate absorption in the gut and their subsequent metabolism within the organism are closely linked has led to the hypothesis that dietary calcium loading exerts its blood pressure lowering action by inducing hypophosphatemia which has a subsequent deleterious effect on cardiovascular performance (LAU et al. 1984a). The report upon which this hypothesis is based compared the effect of diets containing 4.3% or 1.2% calcium on blood pressure and several other relevant physiologic parameters in the SHR. In the study, 22-week-old animals were fed the 4.3% calcium diet over a 10-day period and a fall in serum phosphate from 6.4 to 2.8 mg per 100 ml was observed along with a fall in blood pressure. Another group of animals on the same diet received daily intraperitoneal injections of Na_3PO_4 to prevent PO_4 depletion, and although serum PO_4 values were not normalized, 2.8 vs 3.7 mg per 100 ml, the calcium-induced fall in blood pressure was blunted.

This hypothesis is tenable since other reports (FULLER et al. 1978; SAGELIKES et al. 1985) have shown that similar levels of diet-induced phosphate depletion severely compromise cardiovascular function and reduce mean arterial pressure, cardiac output and pressor responses to agonists such as norepinephrine and angiotensin II. The mechanism by which phosphate depletion acts has been suggested to be interference with the cell's ability to maintain stores of high energy phosphate intermediates, i.e., ATP. Additional work is required before this latter point can be validated. One drawback of the phosphate hypothesis is that the level of calcium used by LAU et al. (1984a) was quite high (4.3%) in comparison with levels that have been found to decrease blood pressure in other rat studies (usually ~2%, see Table 1). It is possible that with high levels of calcium, a pathophysiologic state is induced which does lower blood pressure. However, the relevance of this physiologic state to normal levels of dietary calcium intake remains in question. Additionally, the observation that intraperitoneal N_3PO_4 blunted the fall in blood pressure induced by calcium must be interpreted with caution, since serum PO_4 levels were not normalized and it was not determined whether similar treatments have a pressor effect in normal animals that is unrelated to the PO_4 depletion. These points need to be clarified before a major role for this mechanism can be assumed in the blood pressure lowering effect of calcium.

III. Dietary Calcium, Sodium Metabolism, and Fluid Balance

It has been suggested that dietary calcium can lower blood pressure by inducing a natriuresis through an effect on the renal tubule cell and a subsequent volume contraction (AYACHI 1979). The basis of this hypothesis is logical since it is well recognized that renal handling of Ca^{2+} and Na^+ are closely linked and that increases in filtered Ca^{2+} load can lead to an increased Na^+ excretion. Evidence that this occurs, however, is controversial. LAU et al. (1984a), in studies examining the effect of 4.3% calcium on the SHR, found no change in sodium balance after a 14-week period on the diets. This type of study, of course, can not rule out changes which occur early in the course of the diet intervention. This latter point was examined by LUFT et al. (1986) using humans with essential hypertension who received supplemental calcium for 8 days; no changes in sodium balance were observed. In addition, MCCARRON et al. (1985) observed that the blood pressure lowering action of dietary calcium loading persisted in the presence of moderate Na^+ loading of the SHR (1.0% vs 0.45% in control experiments). These results argue against a natriuretic-induced volume contraction mechanism over either short- or long-term supplementation with calcium. Data by STERN et al. (1986), however, do not agree. These authors showed that 2% dietary calcium was associated with a greater sodium excretion after 4 weeks dietary exposure and suggest this may alter plasma volume and blood pressure. No direct measurements of plasma volume were made in these experiments as a separate means of verification.

A consequence of a natriuretic mechanism, if it were to exist, would be a compensatory increase in plasma renin activity and an increase in angiotensin II and aldosterone levels. These hormones would eventually be expected to induce a vasoconstriction. The finding that blood pressure is decreased in these rats, however, makes this mechanism unlikely, unless dietary calcium is also suppressing release of renin, its action on angiotensinogen, or the subsequent effect of angiotensin II on aldosterone production. Finally, it should be kept in mind that the blood pressure lowering effect of calcium supplementation has been observed after both short-term (days) and long-term (weeks) exposure. It is conceivable that a natriuretic mechanism occurs after a few days while an adaptive response underlies the more chronic response.

IV. Dietary Calcium and the Autonomic Nervous System

It has also been postulated that there is an effect of Ca^{2+} in the diet on autonomic nervous system function (HATTON et al. 1986). An accepted marker of sympathetic activation has classically been levels of circulating plasma catecholamines and these have been shown to be elevated both in humans with essential hypertension (ENGELMAN et al. 1970; DEQUATTRO and CHAN 1972; LOUIS et al. 1973) and in the SHR (NAGAOKA and LOVENBERG 1976). If the blood pressure lowering effect of dietary calcium is mediated through a suppression of the autonomic nervous system, then a depression of the levels of catecholamines might be observed. This has been examined both in humans (LUFT et al. 1986) and in the SHR (STERN et al. 1986) and no effect was observed on either norepinephrine or epinephrine. These results are not consistent with the hypothesis that dietary calcium acts by diminishing levels of circulating catecholamines secondary to diminished sympathetic output.

Evidence that the autonomic nervous system may be involved in the blood pressure lowering action of dietary calcium has been presented by HATTON et al. (1986). These authors examined the effect of dietary calcium on both resting blood pressure in conscious unrestrained rats and on a classically conditioned pressor response to an electrical shock. It was observed that there was an attenuated response of rats maintained on the elevated calcium diets (2% calcium) and an enhanced response of animals on Ca^{2+}-restricted diets (0.1%). The discrepancies between these results and those already cited that examined levels of circulating catecholamines indicate that dietary calcium may be acting centrally or at the peripheral neurovascular junction. More research in this area needs to be conducted before definite conclusions can be formulated.

V. Composite Hypothesis

A unifying hypothesis has been proposed which suggests that there is a basic defect in Ca^{2+} handling that is manifest in humans with essential hypertension who are "calcium responders" and in the SHR and that this defect is corrected by the increased Ca^{2+} load presented to the animal with diet supplementation (McCarron et al. 1984). Evidence for this hypothesis in essential hypertensives is that there is a decrease in serum levels of ionized calcium (McCarron 1982a; Resnick et al. 1983; Folsom et al. 1984), a renal leak of Ca^{2+} (McCarron et al. 1980; Strazzullo et al. 1983), and the previously mentioned increases in free intracellular Ca^{2+} platelets (Erne et al. 1984; Brushi et al. 1985; Sang et al. 1986). Experimental findings in the SHR include a depressed intestinal absorption of calcium (Schedl et al. 1984, 1986; McCarron et al. 1985), increased levels of ionized Ca^{2+} in platelets (Brushi et al. 1985; Skuherska et al. 1986), a renal leak of Ca^{2+} (McCarron et al. 1981; Wright and Rankin 1982; Bindels et al. 1984; Lau et al. 1984a, b), and the aforementioned defects in Ca^{2+} metabolism in vascular smooth muscle (Aoki et al. 1974; Webb and Bhalla 1976; Bhalla et al. 1978; Kwan et al. 1980; Kwan and Daniel 1982).

Calmodulin, a highly conserved Ca^{2+}-binding protein found in all mammalian tissue, is known to play a role in many of the Ca^{2+}-regulating systems. It has been suggested that there is a defect in calmodulin which decreases the affinity of this essential protein for Ca^{2+}. Subsequently, its ability to activate Ca^{2+} extrusion processes (i.e., the Ca-ATPase of sarcolemma or sarcoplasmic reticulum) is hindered and this may account for increases in intracellular levels of free Ca^{2+} and a decrease in intestinal absorption and renal reabsorption of the cation.

To date there is little biochemical evidence for such a lesion in the SHR. Pokudin et al. (1985) has made a comparative study of calmodulin isolated from the brains of SHRs and normotensive controls and has found no differences in either the affinity of this binding protein for Ca^{2+} or in its ability to activate several calmodulin-dependent enzymes. Using a different approach to this problem Nojima et al. (1986) have sequenced a gene coding for a form of calmodulin from the brain of the SHR and have concluded from a four-dimensional model of the peptide sequence that one of the four Ca^{2+}-binding sites has a mutation which would alter the affinity of the protein for Ca^{2+}. It remains to be determined, however, whether this gene is expressed in the SHR and, if so, what percentage of the total calmodulin complement it comprises. Criticism of this hypothesis to date has largely been teleological in nature and argues that a lesion in an activator protein that is involved in so many essential enzymatic processes would not be compatible with life.

Another possibility along these same lines is that there is a defect in a Ca^{2+}-binding protein which is not calmodulin, but which is involved in sequestering calcium and helps to stabilize cell membranes. Production of this protein could be dependent on the amount of calcium that is presented to the animal in its diet and regulated by the hormones that regulate whole animal calcium homeostasis.

It has already been established that there are vitamin D-dependent calcium-binding proteins in both intestinal and renal cells.

Whatever the mechanism of action, there must be a link between the presentation of calcium to the intestinal surface and a subsequent effect on resistance vessels. An obvious candidate is serum levels of ionized calcium which are altered with dietary interventions. This does not appear to be an acute effect however, since ionized Ca^{2+} has been shown to change prior to changes in blood pressure. Humoral factors which have been discussed include PTH and 1,25-dihydroxyvitamin D_3, both of which should decrease in the presence of an increased dietary consumption. It should not be overlooked, however, that low serum phosphate is also a strong signal to the renal tubule cell for the production of 1,25-dihydroxyvitamin D_3. A low level of phosphate concurrent with elevated Ca^{2+} may favorably alter calcium homeostasis at the cell level through an effect mediated by 1,25-dihydroxyvitamin D_3. Substantial work remains until these various parameters are sorted out in both normal and hypertensive states.

F. Conclusions

It has become clear that defects in calcium metabolism of vascular smooth muscle are present in the SHR (a genetic model of essential hypertension) and in specific cell types in humans with essential hypertension. Furthermore, it has been shown that increased consumption of calcium by people with essential hypertension results in a blood pressure lowering response in approximately 30%–40% of individuals. Further clinical work is necessary to better identify the physiologic parameters that define these calcium-sensitive people. Laboratory investigation has identified several areas which may explain the effect of dietary calcium in the SHR. Chief among these is an effect on vascular smooth muscle function produced by modulators of whole animal calcium homeostasis on an action on the peripheral nervous system. The observed age-dependent sensitivity of the blood pressure lowering effect of dietary calcium has led to several questions pertaining to the effect of exposure of an animal to calcium at critical periods in its development. In addition, these results have led to the question of whether the vascular smooth muscle cell can respond to its extracellular calcium load with a change in its ability to extrude or sequester calcium.

Acknowledgment. The authors gratefully acknowledge Gloria Ellis for her expert secretarial assistance in preparation of this manuscript.

References

Ackley S, Barrett-Connor E, Suarez L (1983) Dairy products, calcium and blood pressure. Am J Clin Nutr 38:457–461

Aksoy MO, Mras S, Kamm KE, Murphy A (1983) Ca^{2+}, cAMP, and changes in myosin phosphorylation during contraction of smooth muscle. Am J Physiol 245:C255–C270

Aoki K, Yamashita Y, Tomita N, Tazumi K, Hotta K (1974) ATPase activity and Ca binding ability of subcellular membranes of arterial smooth muscle in spontaneously hypertensive rat. Jpn Heart J 15:180–181

Aoki K, Sato K, Kawaguchi Y (1985) Increased cardiovascular responses to norepinephrine and calcium antagonists in essential hypertension compared with normotension in humans. J Cardiovasc Pharmacol [Suppl 6] 7:S182–S186

Ayachi S (1979) Increased dietary calcium lowers blood pressure in the spontaneously hypertensive rat. Metabolism 28:1234–1238
Baran DT, Milne ML (1986) 1,25 dihydroxyvitamin D increases hepatocyte cytosolic calcium levels. A potential regulator of vitamin D-25-hydroxylase. J Clin Invest 27:1622–1626
Belizan JM, Villar J (1980) The relationship between calcium intake and edema-, proteinuria-, and hypertension gestosis: an hypothesis. Am J Clin Nutr 33:2202–2210
Belizan JM, Villar J, Pineda O, Gonzalez AE, Sainz E, Garrera G, Sidrian R (1983) Reduction of blood pressure with calcium supplementation in young adults. JAMA 249:1161–1165
Bhalla RC, Webb RC, Ashley T, Brock T (1978) Calcium fluxes, calcium binding and adenosine 3′,5′-monophosphate dependent protein kinase activity in aorta of spontaneously hypertensive and Wistar normotensive rats. Mol Pharmacol 14:468–477
Bindels RJM, Geertsen JAM, Os CH van, Slegers JFG (1984) Renal calcium and phosphate handling in the spontaneously hypertensive rat and normotensive Wistar Kyoto control rat. In: Brenner F, Peterlik M (eds) Epithelial calcium and phosphate transport: molecular and cellular aspects. Liss, New York, p 369
Bohr DF, Webb RC (1985) Vascular smooth muscle and its changes in hypertension. Am J Med 5:3–16
Boland AR, Boland R (1985) Suppression of 1,25 dihydroxyvitamin D_3-dependent calcium transport by protein synthesis inhibitors and changes in phospholipids in skeletal muscle. Biochim Biophys Acta 845:237–241
Boland AR, Gallego S, Boland R (1983) Effects of vitamin D-3 in phosphate and calcium transport across and composition of skeletal muscle plasma membranes. Biochim Biophys Acta 733:264–273
Bolton TB (1979) Mechanisms of action of transmitters and other substances on smooth muscle. Physiol Rev 59:606–718
Braunwald E (1982) Mechanism of action of calcium-channel blocking agents. N Engl J Med 307:1618–1626
Brody MJ, Barron KW, Berechek KH, Faber JE, Lappe RW (1983) Neurogenic mechanisms of experimental hypertension. In: Genast J, Kuchad O, Hamet P, Cantin M (eds) Hypertension: physiopathology and treatment, 2nd edn. McGraw-Hill, New York, p 117
Brushi G, Brushi ME, Caroppo M, Orlandini G, Spaggiari M, Cavatorta A (1985) Cytoplasmic free [Ca^{2+}] is increased in the platelets of spontaneously hypertensive rats and essential hypertensive patients. Clin Sci 68:179–184
Buhler FR, Hulthen UL, Kiowski W, Muller FB, Bolli P (1982) The place of the calcium antagonist verapamil in antihypertensive therapy. J Cardiovasc Pharmacol 4:S350–S357
Bukoski RD, McCarron DA (1986) Altered aortic reactivity and lowered blood pressure associated with high Ca^{2+} intake in the SHR. Am J Physiol 251:H976–H983
Bukoski RD, Plant SB, McCarron DA (1986) Altered aortic calcium metabolism and blood pressure are altered by Ca^{2+} intake in the SHR. Am J Physiol (to be published)
Campbell MD, Deth RC, Payne RA, Honeyman TW (1985) Phosphoinositide hydrolysis is correlated with agonist-induced calcium flux and contraction in the rabbit aorta. Eur J Pharmacol 116:129–136
Cauvin C, van Breemen C (1985) Altered ^{45}Ca fluxes in isolated mesenteric resistance vessels from SHR (Abstr). Fed Proc 44(4):1008
Cauvin C, Loutzenhiser R, van Breemen C (1983) Mechanism of calcium antagonists-induced vasodilation. Annu Rev Pharmacol Toxicol 23:373–386
Cauvin C, Lukeman S, Cambron J, Henry O, Meisheri K, Yamamoto H, van Breemen C (1984a) Theoretical basis for vascular sensitivity of Ca^{2+} antagonists. J Cardiovasc Pharmacol 6:S630–S638
Cauvin C, Saida K, van Breemen C (1984b) Extracellular Ca^{2+} dependence and diltiazem inhibition of contraction in rabbit conduit arteries and mesenteric resistance vessels. Blood Vessels 21:23–31

Chacko S, Rosefeld A, Thomm G (1985) Calcium regulation of smooth muscle actomyosin. In: Grover AK, Daniel EG (eds) Calcium and contractility. Humana, Clifton, p 175
Chatterjee M, Murphy RA (1983) Calcium-dependent stress maintenance without myosin phosphorylation in skinned smooth muscle. Science 221:464–466
Daniel EE, Kwan CY (1981) Control of contraction of vascular muscle-relation to hypertension. Trends Pharmacol Sci 2:207–223
Daniel EE, Grover AK, Kwan CY (1982) Isolation and properties of plasma membrane from smooth muscle. Fed Proc 41:2895–2904
DeQuattro V, Chan S (1972) Raised plasma-catecholamines in some patients with primary hypertension. Lancet 1:806–809
Dillon PF, Aksoy MO, Driska SP, Murphy RA (1981) Myosin phosphorylation and the cross-bridge cycle in smooth muscle. Science 24:495–497
Edelman IS (1981) Receptors and effectors in hormone action on the kidney. Am J Physiol 241:F333–F339
Engelman K, Portnoy B, Sjoerdsma A (1970) Plasma catecholamine concentrations in patients with hypertension. Circ Res [Suppl 1] 26:141–146
Erne P, Bolli P, Bertel O, Hulthien L, Kiowski W, Muller FB, Buhler F (1983) Factors influencing the hypotensive effects of calcium antagonists. Hypertension [Suppl 2] 5:97–102
Erne P, Bolli P, Burgisser E, Buttler FR (1984) Correction of platelet calcium with blood pressure: effect of antihypertensive therapy. N Engl J Med 310:1084–1088
Feinleib M, Lenfant C, Miller SA (1984) Hypertension and calcium. Science 226:384–386
Fogh-Anderson N, Hedegaard L, Thode J, Siggaard-Andersen O (1984) Sex-dependent relation between ionized calcium in serum and blood pressure. Clin Chem 30:116–118
Folkow B (1982) Physiological aspects of primary hypertension. Physiol Rev 62:347–504
Folsom AR, Smith CL, Prineas RJ, Grimm RH (1984) Serum calcium fractions in essential hypertensive and matched normotensive subjects. Hypertension 8:11–15
Forder J, Scriabine A, Rasmussen H (1985) Plasma membrane calcium flux, protein kinase C activation and smooth muscle contraction. J Pharmacol Exp Ther 235:267–273
Frishman WH, Charlap S (1984) Nifedipine in the treatment of systemic hypertension. Arch Intern Med 144:2335–2336
Frishman W, Charlap S, Kimmel B, Saltzburg S, Stroh J, Weinberg P, Monasko E, Wiezner J, Dorsa F, Pollack S, Strom J (1986) Twice daily administration of oral verapamil in the treatment of essential hypertension. Arch Intern Med 146:561–565
Fuller TJ, Nichols WW, Brenner BJ, Peterson JC (1978) Reversible depression in myocardial performance in changes with experimental phosphorus deficiency. J Clin Invest 62:1194–1200
Garcia-Palmieri MR, Costas R, Cruz-Vidal M, Sorlie PD, Tillotson J, Havlik RJ (1984) Milk consumption, calcium intake, and decreased hypertension in Puerto Rico. Hypertension 6:322–328
Grover AK (1985) Ca-handling studies using isolated smooth muscle membranes. In: Grover AK, Daniel EE (eds) Calcium and contractility. Humana, Clifton, p 245
Grover AK, Daniel EE (eds) (1985) Calcium and contractility. Humana, Clifton
Harlan WR, Hull AL, Schmouder RL, Landis JR, Thompson FE, Larkin FA (1984) Blood pressure and nutrition in adults. Am J Epidemiol 120:17–28
Harlan WR, Landis JR, Schmouder RL, Goldstein NG, Harlan LC (1985) Blood lead and blood pressure. JAMA 253:530–534
Hatton DC, Huie PE, Muntzel MS, Metz JA, McCarron DA (1986) Stress-induced blood pressure responses in the SHR: effect of dietary calcium. Am J Physiol 252:R48–R54
Hunt SC, McCarron DA, Smith JB, Ash KO, Bristow MR, Williams RR (1984) The relationship of plasma ionized calcium to cardiovascular disease endpoint and family history of hypertension. Clin Exp Hypertens [A] A6(8):1397–1414
Johnson NF, Smith EL, Freudenheim JL (1985) Effects on blood pressure of calcium supplementation of women. Am J Clin Nutr 42:12–17
Jones AW (1974) Altered ion transport in large and small arteries from spontaneously hypertensive rats and the influence of calcium. Circ Res 435:I-117–I-122

Katz AI (1982) Renal Na-K-ATPase: its role in tubular sodium and potassium transplant. Am J Physiol 242:F207–F219
Kwan CY, Daniel EE (1982) Arterial muscle abnormalities of hydralazine treated spontaneously hypertensive rats. Eur J Pharmacol 82:187–190
Kwan CY, Garfield R, Daniel EE (1979) An improved procedure for the isolation of plasma membranes from rat mesenteric arteries. J Mol Cell Cardiol 11:639–659
Kwan CY, Belbeck L, Daniel EE (1980) Abnormal biochemistry of vascular smooth muscle plasma membrane isolated from hypertensive rats. Mol Pharmacol 77:137–140
Lau K, Chen S, Eby B (1984a) Evidence for the role of PO_4 deficiency in antihypertensive actions of a high-Ca diet. Am J Physiol 246:H324–H329
Lau K, Chen S, Spirnak J, Eby B (1984b) Evidence for an intestinal mechanism in hypercalciuria of spontaneously hypertensive rat. Am J Physiol 247:E625–E633
Lew PD, Favre L, Waldvogel FA, Vallotton MB (1985) Cytosolic free calcium and intracellular calcium stores in neutrophils from hypertensive subjects. Clin Sci 69:227–230
Louis WJ, Boyle AE, Anavekar S (1973) Plasma norepinephrine levels in essential hypertension. N Engl J Med 288:599–601
Luft FC, Aronoff GR, Sloan RS, Fineberg NS, Weinberger MH (1986) Short-term augmented calcium intake has no effect on sodium homeostasis. Clin Pharmacol Ther 39:414–419
Macgregor GA (1985) Sodium is more important than calcium in essential hypertension. Hypertension 7:628–637
Marver D, Kokko JP (1983) Renal target sites and the mechanism of action of aldosterone. Miner Electrolyte Metab 9:1–18
Matlib MA, Schwartz A, Yamori Y (1985) A Na^+-Ca^{2+} exchange process in isolated sarcolemmal membranes of mesenteric arteries from WKY and SHR rats. Am J Physiol 249:C166–C172
McCarron DA (1982a) Low serum concentrations of ionized calcium in patients with hypertension. N Engl J Med 307:226–228
McCarron DA (1982b) Blood pressure and calcium balance in the Wistar Kyoto rat. Life Sci 30:683–689
McCarron DA (1985) Is calcium more important than sodium in the pathogenesis of essential hypertension? Hypertension 7:607–627
McCarron DA, Morris CD (1985) Blood pressure response to oral calcium in persons with mild to moderate hypertension. Ann Intern Med 103:825–831
McCarron DA, Pingree PA, Rubin RJ, Gaucher SM, Molitch M, Krutzik S (1980) Enhanced parathyroid function in essential hypertension: response to a urinary calcium leak. Hypertension 2:162–168
McCarron DA, Yung NN, Ugoretz BA, Krutzik S (1981) Disturbances of calcium metab olism in the spontaneously hypertensive rat. Hypertension 3:I162–I167
McCarron DA, Morris CD, Cole C (1982) Dietary calcium in human hypertension. Science 217:267–269
McCarron DA, Morris CD, Henry HJ, Stanton JL (1984) Blood pressure and nutrient intake in the United States. Science 224:1392–1398
McCarron DA, Lucas PA, Shneidman RJ, LaCour B, Drüeke T (1985) Blood pressure development of the spontaneously hypertensive rat after concurrent manipulations of dietary Ca^{2+} and Na^+: relation to intestinal Ca^{2+} fluxes. J Clin Invest 76:1147–1154
Mikkelsen E, Andersson KE, Bengtsson B (1978a) Effects of verapamil and nitroglycerin on contractile responses to potassium and noradrenaline in isolated human peripheral veins. Acta Pharmacol Toxicol (Copenh) 92:14–22
Mikkelson E, Andersson KE, Pedersen OL (1978b) The effect of nifedipine on isolated human peripheral vessels. Acta Pharmacol Toxicol (Copenh) 43:291–298
Nagaoka A, Lovenberg W (1976) Plasma norepinephrine and dopamine hydroxylase in genetic hypertensive rats. Life Sci 19:28–34
Nichols GA, Metz MA, Cline WH (1986) Vasodilation of the rat mesenteric vasculature by parathyroid hormone. J Pharmacol Exp Ther 236:419–423
Nojima H, Kishi K, Sokabe H (1986) Organization of calmodulin genes in the spontaneously hypertensive rat. J Hypertens (in press)

Noon JR, Rice JP, Baldessarini R (1978) Calcium leakage as a cause of high resting tension in vascular smooth muscle from the spontaneously hypertensive rat. Proc Natl Acad Sci USA 75:1605–1607
Nyborg NCB, Byg-Hansen J, Mulvany MJ (1985) Effect of felodepine on resistance vessels from spontaneously hypertensive and normotensive rats. J Cardiovasc Pharmacol [Suppl 6] 7:S43–S46
Ozaki H, Urakawa N (1979) Na-Ca exchange and tension development in guinea pig aorta. Naunyn Schmiedebergs Arch Pharmacol 309:171–178
Page LB (1983) Epidemiology of hypertension. In: Genest J, Kuchel O, Hamet P, Cantin M (eds) Hypertension, 2nd edn. McGraw-Hill, New York, p 683
Pedersen OL, Mikkelsen E, Andersson KE (1978) Effects of extracellular calcium on potassium and noradrenaline induced contractions in the aorta of spontaneously hypertensive rats – increased sensitivity to nifidipine. Acta Pharmacol Toxicol (Copenh) 43:137–144
Pokudin NI, Orlov SN, Ryazhskv GG, Menshikov NY, Tkachuk VA, Postnov YV (1985) Isolation and characteristics of calmodulin from the brain of rats with spontaneous genetic hypertension. Kardiologiia 25(1):72–77
Postnov YV, Orlov SN (1985) Ion transport across plasma membrane in primary hypertension. Physiol Rev 65:904–945
Postnov YV, Orlov SN, Shevchenko A, Alder AM (1977) Altered sodium permeability, calcium binding and Na K ATPase activity in red blood cell membrane in essential hypertension. Pflugers Arch 371:263–269
Postnov YV, Orlov SN, Reznikova MB, Rjazlisky GG, Pokdin NI (1984) Calmodulin distribution and Ca^{2+} transport in the erythrocytes of patients with essential hypertension. Clin Sci 66:459–463
Resink TJ, Tkachuk VA, Erne P, Buhler FR (1986) Platelet membrane calmodulin-stimulated calcium-adenosine triphosphatase. Altered activity in essential hypertension. Hypertension 8:159–166
Resnick LM, Laragh JH, Sealey JE, Alderman MH (1983) Divalent cations in hypertension. Relations between serum ionized calcium, magnesium, and plasma renin activity. N Engl J Med 309:888–891
Resnick LM, Nicholson JP, Laragh JH (1984) Outpatient therapy of essential hypertension with dietary calcium supplementation (Abstr). J Am Coll Cardiol 3:616
Resnick LM, Nicholson JP, Laragh JH (1985a) Calcium metabolism and the renin-aldosterone system in essential hypertension. J Cardiovasc Pharmacol 6:S187–S193
Resnick LM, Sealey JE, Laragh JM (1985b) Short and long-term oral calcium alters blood pressure in essential hypertension (Abstr). Fed Proc 44:300
Reuter H, Blaustein MP, Haesler G (1973) Na-Ca exchange and tension development in arterial smooth muscle. Philos Trans R Soc Lond [Biol] 265:87–94
Robinson BF, Dobbs RJ, Bayley S (1982) Response of forearm resistance vessels to verapamil and sodium nitroprusside in normotensive and hypertensive men, evidence for a functional abnormality of vascular smooth muscle in primary hypertension. Clin Sci 63:33–42
Sagelikes Y, Massry SG, Iseki K, Brautbar N, Barndt B, Brunton LL, Buxton ILO, Vlachakis N, Campese VM (1985) Effect of phosphate depletion on blood pressure and vascular reactivity to norepinephrine and angiotensin II in the rat. Am J Physiol 248:F93–F99
Saida K, Breemen C van (1983) A possible Ca^{2+}-induced Ca^{2+} release mechanism mediated by norepinephrine in vascular smooth muscle. Pflugers Arch 397:166–177
Sang KHL, Montenay-Garestier J, Devynck MA (1986) Platelet cytosolic free calcium concentration in essential hypertension. Nouv Rev Fr Hematol 27:279–283
Schedl HP, Miller DL, Pape JM, Horst RL, Wilson HD (1984) Calcium and sodium transport and vitamin D metabolism in the spontaneously hypertensive rat. J Clin Invest 73:980–986
Schedl HP, Miller DL, Horst RL, Wilson HD, Natarajan K, Conway T (1986) Intestinal calcium transport in the spontaneously hypertensive rat: response to calcium depletion. Am J Physiol 250:G412–G419

Sempos CT, Cooper RS, Johnson C, Drizd TA, Yetley EA (1985) Association of dietary and serum calcium to blood pressure in adults 40–74 years (Abstr). Fed Proc 44(3):817

Simpson RU, Thomas GA, Arnold AJ (1985) Identification of 1,25-dihydroxy vitamin D_3 receptors and activities in muscle. J Biol Chem 260:8882–8891

Skuherska R, Themblay S, Ray C, Amer V, Hamet P (1986) Intracellular levels of free calcium in platelets (Abstr). Fed Proc 45(3):184

Somlyo AV, Bond M, Somlyo AP, Scarpa A (1985) Inositol triphosphate-induced calcium release and contraction in vascular smooth muscle. Proc Natl Acad Sci USA 82:5231–5235

Sowers MR, Wallace RB, Lemke JH (1985) The association of intakes of vitamin D and calcium with blood pressure among women. Am J Clin Nutr 42:135–142

Spivak C, Ocken S, Frishman WH (1983) Calcium antagonists: clinical use in the treatment of systemic hypertension. Drugs 25:154–177

Stern N, Lee DBN, Silis V, Beck FWJ, Deftos L, Manolagos SC, Sowers JR (1984) Effect of high calcium intake on blood pressure and calcium metabolism in young SHR. Hypertension 6:639–646

Stern N, Golub M, Nyby M, Berger M, Tuck ML, Trujillo A, Carnazzola AE, Lee DBN (1986) The hypotensive effect of high calcium intake in the SHR is not mediated by decreased vascular reactivity (Abstr). Clin Res 34(1):44A

Strazzullo P, Nunziata V, Cirillo M, Gianattasio R, Mancini M (1983) Abnormalities of calcium metabolism in essential hypertension. Clin Sci 65:359–363

Suematsu E, Harata M, Hashimoto T, Kuoiyama H (1984) Inositol-1,4,5-triphosphate releases Ca^{2+} from intracellular store sites in skinned single cells of porcine coronary artery. Biochim Biophys Acta 120:481–485

Tenner TE, Buddingh F, Yang ML, Pang PKT, Patel D, Savjani G (1985) The influence of dietary calcium on the development of spontaneous hypertension in the rat (Abstr). Fed Proc 44:1748

Triggle DJ, Janis RA (1985) Calcium antagonists and ionophores. In: Grover AK, Daniel EE (eds) Calcium and contractility. Humana, New Jersey, p 37

Vallieres J, Scarpa A, Somlyo AP (1975) Subcellular fractions of smooth muscle. Isolation, substrate utilization and calcium transport by main pulmonary aorta and mesenteric vein mitochondria. Arch Biochem Biophys 170:659–669

van Breemen C, Aaronson P, Loutzenhiser R (1979) Sodium-calcium interactions in mammalian smooth muscle. Pharmacol Rev 30:167–208

van Breemen C, Hwang OK, Meisheri KD (1981) The mechanism of the inhibitory action of diltiazem on vascular smooth muscle contractility. J Pharmacol Exp Ther 218:459–463

Vincenzi FS, Morris CD, Kinsel L, Kenny M, McCarron DA (1986) Decreased in vitro Ca^{2+} pump ATPase in red blood cells of hypertensive subjects. Hypertension 8:1058–1066

Webb RC, Bhalla RC (1976) Altered calcium sequestration by subcellular fractions of vascular smooth muscle from spontaneously hypertensive rats. J Mol Cell Cardiol 8:651–661

Webb RC, Bohr DF (1978) Mechanism of membrane stabilization by calcium in vascular smooth muscle. Am J Physiol 235:C227–C232

Williamson JR (1986) Inositol lipid metabolism and intracellular signaling mechanisms. News Physiol Sci 1:72–76

Wright GL, Rankin GO (1982) Concentrations of ionic and total calcium in plasma of four models of hypertension. Am J Physiol 243:H365–H390

Wuytack F, Casteels R (1980) Demonstration of a (Ca^{2+}-Mg^{2+})-ATPase activity probably related to Ca-transport in the microsomal fractions of porcine coronary artery smooth muscle. Biochim Biophys Acta 595:257–263

Yamamoto H, van Breemen C (1985) Inositol-1,4,5-triphosphate releases calcium from skinned cultured smooth muscle cells. Biochem Biophys Res Commun 130:270–274

Drugs and Toxicological Agents that Either Mimic Calcium or Elements of Intracellular Calcium Metabolism

CHAPTER 23

Calcium Chelators and Calcium Ionophores

D. M. Bers and K. T. MacLeod

A. Introduction

The preceding chapters in this volume have demonstrated that Ca ions play critical roles in the regulation of a very large number of physiologic, pharmacologic, and biochemical processes. In the study of these processes, it is invariably necessary to control and measure or monitor the free [Ca^{2+}] with a high degree of precision. Many of these processes are sensitive to submicromolar levels of Ca^{2+}, where endogenous or contaminant Ca^{2+} can become a significant problem. Therefore, in these instances it becomes especially important to control the concentration of free calcium with buffers. Thus, calcium chelators have played a crucial role in allowing progress in the understanding of Ca^{2+}-regulated processes. Chelators are extensively utilized to buffer free [Ca^{2+}] just as pH buffers are utilized to buffer low concentrations of protons. Calcium chelators which exhibit changes in their optical properties can be used as buffers and indicators of free [Ca^{2+}] and those which are hydrophobic in nature can be used as Ca ionophores to alter (or control) free [Ca^{2+}] inside cells, organelles, or vesicles. Additionally, these ionophores can be incorporated into synthetic membranes and can be used in Ca^{2+}-selective electrodes (and microelectrodes) to measure free [Ca^{2+}]. The present chapter will discuss some salient properties of these Ca^{2+} chelators which are important in their experimental use.

B. Calcium Chelators

I. EGTA

Ethylene glycol bis(β-aminoethylether)-*N,N,N′,N′*-tetraacetic acid (EGTA) has been, and continues to be, the workhorse of biologic Ca^{2+} chelators and as such deserves some special attention. Considerations of some aspects of EGTA use are directly relevant to some of the other Ca^{2+} chelators which are less broadly utilized. EGTA is an especially useful Ca chelator for biologic systems for two principal reasons:

1. At neutral pH, it has an apparent dissociation constant (K_d) around 0.4 μM Ca^{2+} which allows Ca^{2+} buffering over a range of free [Ca^{2+}] found intracellularly and over which many cellular systems are Ca^{2+} sensitive.

2. EGTA has a much higher affinity for Ca^{2+} than Mg^{2+} ($\sim 3.8 \times 10^5$ times higher).

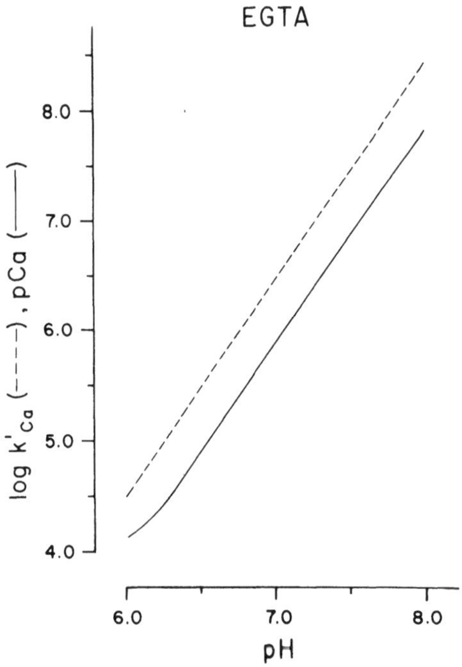

Fig. 1. The effect of pH variation on the apparent association constant of EGTA for Ca^{2+} (K'_{Ca}) and $-\log$ (free[Ca^{2+}]) (i.e., pCa) in a solution containing 1 m*M* total EGTA and 0.8 m*M* $CaCl_2$ calculated using the constants in Table 1

Since intracellular [Mg^{2+}] is often 1000- to 10000-fold higher than [Ca^{2+}], this degree of specificity is extremely valuable. These major advantages are the reason why EGTA is commonly chosen as a Ca^{2+} buffer. However, there are several characteristics and limitations of EGTA of which the user should be aware.

The apparent Ca association constant for EGTA (K'_{Ca}) is very sensitive to pH around neutral pH. Figure 1 demonstrates this pH dependence of log K'_{Ca} as well as the pH dependence of pCa in a solution containing 1 m*M* EGTA and 0.8 m*M* Ca^{2+} (using values for 0.1 *M* ionic strength at 25 °C). Thus, if the pH of a Ca–EGTA solution is not very well controlled or there is any systematic error in pH, the free [Ca^{2+}] in the solution can be substantially different from the expected value. As pointed out by Illingworth (1981), systematic errors in pH are often as large as 0.2 pH units and this error would translate into a two- to threefold error in free [Ca^{2+}] (Fig. 1). Even a pH error of 0.05 changes the pCa by 0.1. In addition to experimental errors, some erroneous calculations have been made in the past owing to the use of stoichiometric (concentration) acid association constants and pH values, which are $-\log(H^+$ activity). This fundamental oversight was pointed out by Tsien and Rink (1980) and hopefully this has increased awareness of this potential pitfall. To obtain the K'_{Ca} at a given pH (knowing the absolute association constants K_{Ca} and K_{Ca_2} and the four acid association constants K_1, K_2, K_3, and K_4) one can take advantage of the readily derived equation

$$K'_{Ca} = K_{Ca}/(1+[H^+]K_1+[H^+]^2K_1K_2+[H^+]^3K_1K_2K_3+[H^+]^4K_1K_2K_3K_4)+K_{Ca_2}/((1/[H^+]K_1)+1+[H^+]K_2+[H^+]^2K_2K_3+[H^+]^3K_2K_3K_4) \quad (1)$$

Where the second term representing Ca binding to the monoprotonated form of EGTA can, as a practical matter, be neglected. The error so produced in K'_{Ca} will be about 1 part in 10^5. This same equation can be used for correcting Mg^{2+} affinity, but in that case the second term can not be neglected. The association constants as usually reported (e.g., in MARTELL and SMITH 1974) are concentration constants. One can either use proton concentration (rather than pH) or convert the concentration constants to "mixed constants" by dividing by the proton activity coefficient for a particular ionic strength (HARNED and OWEN 1958) and use pH. The protonation constants presented in this paper are mixed constants. The K'_{Ca} obtained from Eq. 1 can then be used

$$K'_{Ca} = \frac{[\text{Ca–EGTA}]_t}{[Ca^{2+}][\text{EGTA}]_{tf}} \quad (2)$$

where $[\text{Ca–EGTA}]_t$ is the total of all forms of Ca^{2+} bound to EGTA and $[\text{EGTA}]_{tf}$ is the total of all non-Ca-bound forms of EGTA.

The purity of EGTA is a variable that can also cause substantial errors in the free $[Ca^{2+}]$ of Ca–EGTA solutions. The purity of commercially available EGTA has been carefully assessed by BERS (1982) and MILLER and SMITH (1984) using Ca electrode titration, pH-metric, and oxalate precipitation methods. The magnitudes of error incurred by assumption of nominal purity of EGTA in solutions typically used by numerous investigators were also discussed (BERS 1982; MILLER and SMITH 1984). These errors could be greater than 0.2 pCa units, even at $[Ca^{2+}]$ at which it may be expected that EGTA is a reasonable Ca buffer. This problem is greatly magnified when the total $[Ca^{2+}]$ begins to approach the nominal [EGTA]. These problems can be largely overcome by confining the use of EGTA to a narrow buffer range around $1/K'_{Ca}$ and by estimating the purity of EGTA stock solutions as described by BERS (1982) and MILLER and SMITH (1984).

The K'_{Ca} value and the proton activity coefficient are sensitive to ionic strength. HARAFUJI and OGAWA (1980) and MILLER and SMITH (1984) have developed expressions based upon the Debye–Hückel limiting law which allow prediction of the effect of ionic strength on K'_{Ca}. The effect of temperature changes on K'_{Ca} can also be estimated using the individual K values, the enthalpy of complexation (ΔH) (BOYD et al. 1965), and the following relation; all given by MARTELL and SMITH (1974).

$$\log K_2 = \log K_1 + \Delta H(T_2 - T_1)/(2.303\ RT^2) \quad (3)$$

Rather than make a whole series of adjustments to the relevant constants, it may sometimes be more valuable to evaluate the total [EGTA], K'_{Ca}, and free $[Ca^{2+}]$ using a simple method like that described by BERS (1982). The actual ionic strength, temperature, ionic composition, and pH desired can be set up and the K'_{Ca}, total [EGTA], and free $[Ca^{2+}$) can be evaluated under the precise conditions one is using experimentally. This method circumvents many of the potential errors due to the assumption of binding parameters. HARRISON and BERS (1987) have recently evaluated K'_{Ca} for EGTA, 1,2-bis (*O*-aminophenoxy) ethane-*N,N,N',N'*-tetraacetic acid (BAPTA), and dibromo-BAPTA as functions of temperature (1°–36 °C) and ionic strength (0.104–0.304 *M*).

EGTA shows much higher selectivity for Ca^{2+} over Mg^{2+} than all of the related chelators in Table 1 (and the Ca^{2+} ionophores in Table 3) with the exception of five new synthetic chelators developed by Tsien and co-workers (TSIEN 1980; GRYNKIEWICZ et al. 1985). It should be noted that all of these chelators also exhibit high affinities for other cations which may be endogenous in biologic tissues or utilized experimentally. For example, the Ca^{2+} affinity of EGTA is 20- to 320-fold lower than that for Mn^{2+}, Co^{2+}, Zn^{2+}, and Ni^{2+}, 10^3- to 10^7-fold lower than that for Al^{3+}, Pb^{2+}, La^{3+}, and Co^{3+} and $\sim 10^{12}$-fold lower than that for Mn^{3+} and Hg^{2+}. The affinity of EGTA for Ca is about 300 times that for Ba^{2+} and Sr^{2+} (MARTELL and SMITH 1974).

Another characteristic of EGTA is that it has a slow rate of binding around neutral pH. Association rate constants for Ca^{2+}–EGTA have been estimated to be about 2×10^6 M^{-1} s^{-1} at pH 7.0 and 25 °C (HELLAM and PODOLSKY 1969; SMITH et al. 1977; HARAFUJI and OGAWA 1980). Corresponding dissociation rate constants are about 0.4 s^{-1}. Part of this slow reaction may be because at pH 7, 99% of EGTA is in the diprotonated form and both protons must be removed before Ca can bind. BAPTA is one of the newly synthesized tetracarboxylate Ca chelators developed by TSIEN (1980) which exists primarily in the unprotonated state at neutral pH and is more than 100 times faster than EGTA. The dibromo derivative of BAPTA described by TSIEN (1980) has been reported to have an association rate constant at least 100 times higher and a dissociation rate constant at least 1000 times higher than EGTA (FABIATO 1985).

Ideally, a Ca^{2+} chelator should be inert, but for its Ca^{2+}-chelating properties. Is is usually tacitly assumed that when one uses a Ca^{2+} chelator, it does not affect the binding or kinetic behavior of the system being studied. EGTA (or Ca–

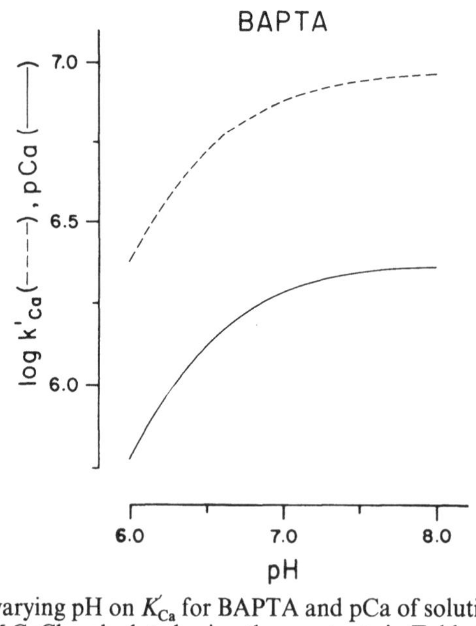

Fig. 2. The effect of varying pH on K'_{Ca} for BAPTA and pCa of solutions containing 1 mM BAPTA and 0.8 mM $CaCl_2$ calculated using the constants in Table 1

EGTA) has been reported to stimulate Ca^{2+} pumping (SAKARDI et al. 1979) and to increase Ca^{2+} affinity (AL-JABORE and ROUFOGALIS 1981; WAISMAN et al. 1981) of erythrocyte, pancreatic islet plasma membrane (KOTAGAL et al. 1983), and sarcoplasmic reticulum Ca^{2+}-dependent ATPase pumps (BERMAN 1982a, b). TROSPER and PHILIPSON (1984) demonstrated that EGTA, EDTA, *trans*-1,2-cyclohexylenedinitrilo-*N,N,N',N'*-tetraacetic acid (CDTA), and arsenazo III all increase the apparent Ca affinity of cardiac sarcolemmal Na,Ca exchange. SHIMOMURA and SHIMOMURA (1984) have also reported that free EGTA and EDTA strongly inhibit the Ca^{2+}-triggered luminescence of aequorin. The physical basis for these effects is not clear. Nevertheless, it is becoming clear that Ca^{2+} chelator inertness should not be taken for granted.

II. Carboxylate Ca^{2+} Chelators and Fluorescent Ca^{2+} Probes

Table 1 shows proton, Ca^{2+}, and Mg^{2+} association constants for several di-, tri-, and tetraacetic acids frequently used as Ca^{2+} chelators. The K'_{Ca} and K'_{Mg} values at pH 7.00 and 7.40 are also tabulated (and the indicated constants are for 100 mM KCl at 25 °C where available). The apparent dissociation constants for these chelators at pH 7.0 vary from 21 nM to 0.34 mM. The pH sensitivity of the chelators at physiologic pH values can be appreciated in the column where the ratio of the K'_{Ca} at pH 7.40 to that at pH 7.00 is listed. All of the first six chelators are rather pH sensitive (especially EGTA). The last column in Table 1 (K'_{Ca}/K'_{Mg}) is indicative of the selectivity for Ca over Mg. In this regard, EGTA stands out among the first seven chelators. The new Ca chelators developed by TSIEN (1980) and GRYNKIEWICZ et al. (1985), namely BAPTA, dibromo-BAPTA, quin2, and fura-2 are both relatively pH insensitive (Fig. 2) and exhibit a great selectivity for Ca over Mg. As noted in Sect. B.I, BAPTA and dibromo-BAPTA have much faster rates of association and dissociation than EGTA. These qualities make these new compounds excellent Ca chelators. In addition, these compounds exhibit Ca^{2+}-sensitive fluorescence, which allows them to be used as Ca^{2+} indicators as well (particularly quin2, fura-2, and the related chelator indo-1 also developed by GRYNKIEWICZ et al. 1985).

A major hurdle in using optical indicators for intracellular free [Ca^{2+}] is to get the indicator into the cell. BAPTA, quin2, fura-2, and indo-1 are all available commercially (Molecular Probes, Junction City, Oregon and Calbiochem, San Diego, California) in esterified form as acetoxymethylesters which can pass readily through biologic membranes. Once inside most cells, the esters are cleaved by endogenous esterases, leaving the membrane-impermeant indicator (or buffer) trapped in the cell. By virtue of the cleavage, the cells can concentrate these chelators substantially (up to several mM) above the extracellular concentration of the esterified compound (20–50 μM, e.g., TSIEN et al. 1982a, b). Quin2 has already been used to examine intracellular Ca^{2+} changes in a large variety of cells (see TSIEN 1983). Fura-2, the focus of a second generation of Ca^{2+} indicators developed by Tsien and co-workers (GRYNKIEWICZ et al. 1985), has several substantial advantages over quin2 which make it even more attractive as an intracellular Ca^{2+} indicator. These improvements include enhanced quantum efficiency, photochemical stability, 30-fold brighter fluorescence, slightly decreased Ca^{2+} af-

Table 1. Stability constants[a] (as $\log_{10}$) for Ca^{2+} chelators

	K_1	K_2	K_3	K_4	K_{Ca}	K_{Ca_2}	K_{Mg}	K_{Mg_2}	pH	K'_{Ca}	$[K'_{Ca}(pH\ 7.4)/K'_{Ca}(pH\ 7.0)]$	K'_{Mg}	K'_{Ca}/K'_{Mg}
EGTA	9.51	8.89	2.77[c]	2.11[c]	10.86	5.3[d]	5.28	3.4[c]	7	6.45	6.19	1.60	72202
									7.4	7.25		2.10	140904
CDTA	12.4	6.23	3.64[c]	2.53[c]	13.15		11.07		7	7.68	2.75	5.60	120
									7.4	8.12		6.04	120
EDTA	10.28	6.22	2.79	2.1	10.61	3.2	8.83	3.2	7	7.26	2.74	5.49	60
									7.4	7.70		5.92	60
HEDTA	9.92	5.48	2.7		8.2		7.0		7	5.27	2.55	4.07	16
									7.4	5.67		4.47	16
NTA[b]	9.76	2.59	1.9		6.39		5.47		7	3.63	2.50	2.71	8
									7.4	4.03		3.11	8
ADA	6.78	2.42	1.7		4.01		2.51		7	3.80	1.29	2.31	32
									7.4	3.92		2.42	32
Citrate	5.8	4.46	2.98		3.5	2.1	3.37	1.92	7	3.47	1.04	3.34	1.3
									7.4	3.49		3.36	1.3
BAPTA	6.36	5.47			6.97		1.77		7	6.88	1.13	1.68	157875
									7.4	6.93		1.73	158244
Dibromo–BAPTA	5.6	4.57			5.8		1		7	5.78	1.02	0.98	62846
									7.4	5.79		0.99	62996
Quin2	6.4	5.47			7.1		2.7		7	7.00	1.14	2.60	25106
									7.4	7.05		2.66	25114
Fura-2	6.4	5.47			6.87		2.01		7	6.77	1.14	1.91	72266
									7.4	6.83		1.97	72373

[a] Mixed stability constant (M^{-1}) are from MARTELL and SMITH (1974, 1977; 0.1 M ionic strength, 25 °C) except for ADA (NAKON 1979; 100 mM KNO_3, 25 °C), BAPTA, dibromo–BAPTA, and quin2 (TSIEN 1980; 100 mM KCl, 22 °C) fura-2 (GRYNKIEWICZ et al. 1985; 100 mM, 22 °C). The K_1 and K_2 values for quin2 and fura-2 were assumed to be the same as BAPTA.
[b] NTA also forms a Ca–$(NTA)_2$ complex with $K_{Ca} = 10^{8.76}$ M^{-2} and this complex is not considered in K'_{Ca}.
[c] Stability constants at 20 °C are from MARTELL and SMITH (1974).
[d] Stability constant is from SCHWARZENBACH et al. (1957). Note that the stability constant for K_{Ca_2} quoted in MARTELL and SMITH (1974) is in error.

finity, greater selectivity for Ca^{2+} over Mg^{2+} and a Ca^{2+}-dependent wavelength shift in the excitation spectrum. In the 10 months since this Ca^{2+} chelator/indicator was first described there have already been a large number of studies utilizing fura-2.

Chlortetracycline (CTC) has been shown to form stable complexes with a number of divalent metal cations (ALBERT 1953). CTC is a lipophilic fluorescent Ca^{2+} chelator which also displays fluorescence changes upon binding Ca^{2+} (or Mg^{2+}). This makes the compound a potentially useful Ca^{2+} indicator. However, great uncertainty exists regarding the information which the fluorescence signal provides about free [Ca^{2+}] or membrane surface [Ca^{2+}] (BLINKS et al. 1982). WHITE and PEARCE (1982) have demonstrated that CTC is a highly specific Ca^{2+} ionophore, with maximum Ca^{2+} binding at about pH 8.0. The relevance of this to the behavior of CTC as an ionophore can be appreciated in the discussion of ionophores which follows. With a pK_a of 7.7 (MATHEW and BALARAM 1980), CTC would be expected to display substantial pH dependence at physiologic pH values. CTC forms a 1:1 complex with Ca^{2+} and shows a selectivity for Ca^{2+} over Mg^{2+} of about 10- to 20-fold (WHITE and PEARCE 1982) which is high compared with other ionophores (see Table 3), but not very high when compared with some of the Ca^{2+} chelators in Table 1. Most lanthanide cations also effectively displace Ca^{2+} from CTC (WHITE and PEARCE 1982).

III. Metallochromic Dyes

A group of four Ca^{2+}-chelating dyes, arsenazo III, antipyralazo III, murexide, and tetramethylmurexide, have been used extensively as indicators of free [Ca^{2+}]. All of these dyes change their absorption spectra when they bind Ca^{2+}. The optical properties of these dyes will not be discussed per se (see, e.g., BLINKS et al. 1982; SCARPA 1979). Table 2 shows characteristics of these Ca^{2+} chelators, such as K'_{Ca} at pH 7.0, 0.1 M ionic strength, and 25 °C assuming 1:1 stoichiometry of binding. This assumption is controversial and Ca:dye stoichiometries of 1:1, 2:1, 1:2, and 2:2 have been suggested for both arsenazo III and antipyralazo III

Table 2. Binding constants of metallochromic dyes

	K'_{Ca} (M^{-1})	K'_{Ca}/K'_{Mg}	Ca/Mg sensitivity	[K'_{Ca}(pH 7.5)/K'_{Ca}(pH 7.0)]
Arsenazo III	50000[a]	48[a]	39[a]	1.41[e, a]
Antipyralazo III	5000[b]	33[c]	High[d]	1.43[b, f]
Murexide	280[a]	35[a]	1840[a]	1.3[a, g]
Tetramethylmurexide	360[a]	23[a]	513[a]	1.0[a]

[a] OHNISHI (1978, 1979b).
[b] SCARPA et al. (1978); OGAWA et al. (1980).
[c] RIOS and SCHNEIDER (1981).
[d] BLINKS et al. (1982).
[e] CHIU and HAYNES (1980).
[f] K'_{Ca}(pH 7.5)/K'_{Ca}(pH 6.8).
[g] Extrapolated from Fig. 7 in OHNISHI (1978).

(PALADE and VERGARA 1983; DOROGI et al. 1981), although murexide and tetramethylmurexide appear to interact with a 1:1 stoichiometry (OHNISHI 1978). It has also been reported that arsenazo III and antipyralazo III bind significantly to cell constituents (BEELER et al. 1980; DOROGI et al. 1981) and murexide and tetramethylmurexide are sequestered by sarcoplasmic reticulum (OHNISHI 1979 a). All of these dyes also bind Mg^{2+} and the ratio of apparent affinities is also shown in Table 2. By choosing an appropriate wavelength pair for measurement, the apparent sensitivity of the absorption signal to Mg^{2+} can be substantially reduced (Ca/Mg sensitivity in Table 2). For example, antipyralazo III can be used at a wavelength pair where the absorbance is Ca^{2+} sensitive, but Mg^{2+} insensitive (720–790 nm; SCARPA et al. 1978). However, Mg^{2+} can still bind to antipyralazo III, decreasing the effective amount of dye which is available for Ca^{2+} binding (i.e., Ca^{2+} sensitive at these wavelengths). Looking at it another way, 5 m*M* Mg^{2+} decreases the apparent Ca^{2+} affinity of antipyralazo III by 2.6-fold (SCARPA et al. 1978). Except for tetramethylmurexide, all these dyes are sensitive to pH around physiologic levels (see Table 2).

IV. Photoproteins

Several Ca^{2+}-sensitive bioluminescent proteins have been isolated and the properties of the best known of these, aequorin, have been carefully documented over the years by Blinks and his co-workers (see, e.g., BLINKS et al. 1982; ALLEN and BLINKS 1979). The light emitted by aequorin is very steeply dependent upon free [Ca^{2+}] over the intracellular physiologic range (10^{-7}–10^{-5} *M*). This steep dependence (slope = 2.5 on log–log plots) suggests that a minimum of three Ca^{2+} ions must bind to aequorin to activate luminescence. It also implies that local regions of elevated free [Ca^{2+}] will dominate light emission such that spatially averaged free [Ca^{2+}] will be very difficult to obtain if there are any regional variations in [Ca^{2+}] (BLINKS et al. 1982). The Ca^{2+} sensitivity of aequorin is not especially pH sensitive around pH 7, but is rather sensitive to ionic strength and [Mg^{2+}]. Mg^{2+} shifts the Ca^{2+} dependence of luminescence strongly to higher [Ca^{2+}], thus depressing luminescence. In contrast, Ba^{2+}, Sr^{2+}, and lanthanide ions are all capable of inducing aequorin luminescence (BLINKS et al. 1982).

C. Calcium Ionophores

Calcium ionophores are presently used for (a) allowing rapid equilibration of Ca^{2+} across biologic membranes and, in so doing, changing and/or controlling [Ca^{2+}] in cells, organelles, or vesicles and (b) making Ca^{2+}-selective electrodes to measure changes in free extracellular [Ca^{2+}] or free intracellular [Ca^{2+}]. Ionophores are molecules which can form liposoluble complexes with ions and so promote the transfer of hydrophilic ions into and across a hydrophobic region. Therefore, Ca^{2+} ionophores increase the permeability of biologic membranes to Ca^{2+}, causing changes in cellular [Ca^{2+}]. As outlined by SIMON et al. (1977) ionophores must have several properties:

1. Ionophore molecules should provide a site which is surrounded by polar groups and suitable for the binding of an ion. These groups are usually oxygen, or sometimes nitrogen atoms which replace the hydration sphere around the ion. The polar groups should be surrounded by nonpolar groups, which form a hydrophobic shell and must confer a large lipid solubility on the ligand and complex. This explains why complexing compounds such as EGTA or EDTA can not act as carriers in membrane systems.

2. The polar groups should contain (for Ca^{2+}) between five and eight coordination sites. The effect of coordination site number on complex stability is considered in detail by MORF and SIMON (1971). Suffice it to say that coordination spheres of eight, six, and four oxygen atoms will form minimal site radii for the binding of a cation of 0.1, 0.06, and 0.03 nm, respectively. This is consistent with observed coordination numbers of eight for Ca^{2+} (ionic radius ~0.099 nm) and Mg^{2+} (ionic radius ~0.065 nm). If one could fix a coordination number of eight, then this would discriminate heavily against Mg^{2+} and Li^{+} and at least destabilize the binding of other ions, for example Na^{+} (ionic radius ~0.095 nm). This together with achieving precise coordination site arrangement around the cavity of ligation will govern the selectivity of the ionophore. These same considerations regarding selectivity apply to the calcium chelators discussed in the previous section. The thickness of the ligand shell around the cation is important for discrimination between cations of the same radius but different charge.

3. Given that the necessary ligand architecture can be achieved, there must be a compromise between the binding energy of the ion to its cavity and its hydration energy in solution. If the hydration energy is much greater than the binding energy, then the ion would not lose its hydration shell. Conversely, if the hydration energy is much less than the binding energy, then the ion would tend not to go back into solution. The selectivity of the ionophore will therefore also be influenced by the final balance between the respective binding and hydration energies of it and the ions for which it exhibits affinities.

4. The ligand should be able to undergo a degree of rapid conformational change to allow fast ion exchange. Thus, there must be a balance between the rigidity of structure, which assures stability and selectivity, and flexibility of structure, which promotes high exchange rates.

Calcium ionophores can be divided into two groups: (a) those which are naturally occurring and (b) those which are synthetically produced. In group (a) are X-537A, A23187, and ionomycin. In group (b) are DPP^{-}, $DOPP^{-}$, and ETH 1001. They will be discussed in that order.

I. Naturally Occurring Calcium Ionophores

The compounds in this class, produced by the *Streptomyces* species, are carboxylic acids and form cationic complexes ($M^{+}L^{-}$) in their deprotonated form. Complexing, transport, and subsequent solution of the cation takes the form shown in Fig. 3. At physiologic pH, the carboxyl group of the compounds is usually ionized, giving the molecules a negative charge. The carrier and ion cross the membrane as a neutral zwitterionic complex. The equilibrium of ion concentrations set up across the membrane will therefore be pH sensitive and the iono-

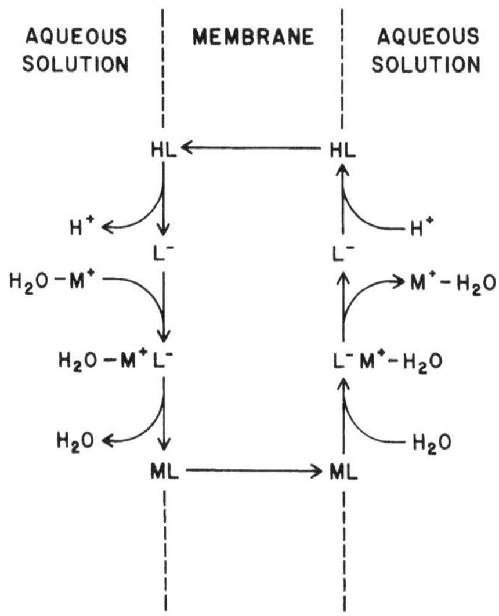

Fig. 3. For details see text

phores will also act as protonophores. The ionophore may also transport a different counter-ion, depending upon the relative affinities of the ligand for other ions (discussed in Sects. C.I.2 and C.II.1). The structures of these compounds have been published elsewhere (e.g., DOBLER 1981; TRUTER 1976; TOEPLITZ et al. 1979). Oxygen atoms generally form the ligands in the compounds, but A23187 differs from the other two in that it also contains two ring nitrogen atoms which are used for coordination.

1. X-537A

This compound was first isolated by BERGER et al. (1951) from an unknown *Streptomyces* species. Identification of the organism producing X-537A as *Streptomyces lasaliensis* led to the alternative name for the antibiotic, lasalacid A. X-537A has a molecular weight of 590. PRESSMAN (1976) points out that models of its structure indicate that it is more "shell-shaped" than the classical ionophore "doughnut-shape"; thus, complexed ions sit on, rather than within the structure and this may explain the wide range of ion complexes of the compound. In addition to binding a large number of different cations, it can also transport catecholamines across lipid membranes (DOBLER 1981). PRESSMAN (1973) assessed the ionophore properties of X-537A and some of its derivatives. Acylation of the phenolic hydroxyl group increases the pK_a above that of the parent ionophore. Substitution of the aromatic ring R group with Cl^-, Br^-, I^-, or NO_2 gives a successive decrease of pK_a (see Table 3). Although Ca^{2+} affinity changes little as pK_a changes, and Mg^{2+} affinity decreases as pK_a decreases, in general, cation affinities increase with decreasing pK_a (PRESSMAN 1973; see also Table 3). The general affinity series for X-537A is $Ba^{2+} > Sr^{2+} > Ca^{2+} > Mg^{2+}$ for divalent cations and $Cs > Rb \sim K > Na > Li$ for monovalent cations. The compound's relative affinities for mono- and divalent cations (M^Z) with respect to Ca^{2+} are shown in

Table 3. X-537A will facilitate the passive movement of Ca^{2+} across a membrane by forming neutral Ca–(X-537A)$_2$ complexes. However, its selectivity for Ca^{2+} is questionable and care should be used in attributing effects of the ionophore to changes in [Ca^{2+}] (see, e.g., DEVORE and NASTUK 1975). Available evidence suggests that it is not as selective for Ca^{2+} over monovalent cations as is ionophore A23187 (discussed in Sect. C.I.2). It is difficult to make exact comparisons between the relative affinities of the ionophores as their relative association constants have been calculated in different ways, and sometimes assuming different binding stoichiometries. CORNELIUS et al. (1974) show that Na^+, K^+, and Rb^+ are preferentially bound to X-537A about five times more effectively than Ca^{2+} or Mg^{2+} while the results of PRESSMAN (1973) and DEGANI et al. (1973) suggest that monovalent cations are not as preferred over Ca^{2+} (see Table 3). Physiologically useful concentrations of the compound appear to be in the range of 1–20 μ*M*.

2. A23187

This compound was first used by REED and LARDY (1972) who found that it uncoupled oxidative phosphorylation in rat liver mitochondria, probably by perturbing the mitochondrial membranes and allowing Ca^{2+} and Mg^{2+} to reach equilibrium. A23187 (free acid) has a molecular weight of 523 and, at pH 7.4, binds divalent cations specifically. It does not appear to act as an ionophore for K^+ or Na^+ at physiologic pH and [Ca^{2+}] (PFEIFFER et al. 1979). The compound appears to transport Li^+ although the concentrations of Li^+ required to saturate the carrier are 100 times greater than Mg^{2+}. However, this may be a relevant consideration if Li^+ is being used as a replacement ion for Na^+ in physiologic saline. FLATMAN and LEW (1977) have demonstrated that the ionophore can transport Na^+ across red cell membranes when the level of divalent cations in the superfusing medium is low (<50 μ*M*) owing to the presence of EDTA. PFEIFFER et al. (1979) also found evidence for the antibiotic forming complexes with La^{3+} and ruthenium red, but the nature of these complexes remains unclear. The ionophore has $pK_a = 6.9$ (in dimethylsulfoxide) (TRUTER 1976). The relative selectivity series for A23187 is: $Mn^{2+} \gg Ca^{2+} \sim Mg^{2+} \gg Sr^{2+} > Ba^{2+}$ for divalent cations and, for monovalent cations: $Li^+ > Na^+ > K^+$. Li^+ has less affinity for the ionophore than Sr^{2+}, but greater affinity than Ba^{2+}. The relative affinities of the ionophore for other cations with respect to Ca^{2+} are given in Table 3. Disagreement exists as to the complexing stoichiometry of the ionophore. It appears that generally two molecules of ionophore complex with each calcium atom (LIU and HERMANN 1978; PFEIFFER et al. 1979; REED and LARDY 1972). CASE et al. (1974), however, detected only 1:1 binding in biologic membranes.

Efforts have been made to produce derivatives of A23187 which possess enhanced transport selectivity for Ca^{2+} over Mg^{2+} (DEBONO et al. 1981). 4-Bromo-A23187 seems to increase the transport selectivity for Ca^{2+} over Mg^{2+} by between four and ten times (DEBONO et al. 1981). Affinities for *both* Ca^{2+} and Mg^{2+} are approximately four times greater than those of the parent ionophore. Physiologic experiments have generally used concentrations of A23187 between 0.5 and 15 μ*M*.

Table 3. Selectivities of ionophores X-537A, A23187 and ionomycin

		pK_a	$K'_{M^z}/K'_{Ca^{2+}}$							
			Na^+	K^+	Rb^+	Cs^+	Mg^{2+}	Sr^{2+}	Ba^{2+}	Mn^{2+}
Acetyl	X-537A[a]	6.30	8.1×10^{-4}			5.4×10^{-4}	0.12	1.05		
	X-537A[a]	5.80	1.3×10^{-3}			0.01	0.091	2.27		
NO_2–	X-537A[a]	4.20	1.7×10^{-3}			0.015	0.091	3.34		
	X-537A[b]		0.01	0.102	0.1	0.073	0.181	8.108	78.37	
	X-537A[c]		5.55	5.19	4.44	3.52	1.259	0.519	1.70	
	A23187[e]	6.9[d]					1.2	0.134		
	A23187[f]						0.38	0.0046		80.8
	A23187[g]						0.43	0.078		108.7
	A23187[h]						0.36	0.043	4.6×10^{-6}	357.1
	A23187[j]		0.012	0.012	0.012	0.009	0.953	0.011	0.009	
	Ionomycin[j]	~8.4	0.023	0.028	0.021	0.023	0.507	0.023	0.031	

[a] PRESSMAN (1973).
[b] DEGANI et al. (1973).
[c] CORNELIUS et al. (1974).
[d] TRUTER (1976).
[e] CASWELL and PRESSMAN (1972).
[f] PFEIFFER et al. (1979).
[g] PUSKIN and GUNTER (1975).
[h] PFEIFFER and LARDY (1976).
[j] LIU and HERMANN (1978).

3. Ionomycin

The ionophore ionomycin was isolated by LIU et al. (1978) from *Streptomyces conglobatus*. It has a molecular weight of about 708 (free acid) (LIU et al. 1978) and is a dibasic acid having a carboxylic acid group and an enolized β-diketone group (TOEPLITZ et al. 1979). Thus, in the unprotonated state (i.e., near pH 9) the compound has a double charge. This explains why only one ionomycin molecule is required to complex a Ca^{2+} ion, in contrast to the other ionophores of this group. It is highly selective for divalent cations and exhibits very little affinity for monovalent metal ions. The general divalent selectivity sequence is $Ca > Mg \gg Sr = Ba$ (LIU and HERMANN 1978) which is similar to A23187, but the antibiotic appears to bind Mg^{2+} about half as effectively as A23187 (Table 3).

Complexing between ionomycin and the cation is dependent upon the pH of the aqueous phase. Essentially, complexing ceases below pH 7.0 and reaches maximum at pH 9.5. In comparison, A23187 complexing ceases below pH 5.0 and reaches maximum at far more physiologic levels of pH 7.5. It remains to be seen whether or not this new ionophore can be a useful tool in allowing changes of intracellular $[Ca^{2+}]$ to be made without altering intracellular $[Mg^{2+}]$ as markedly as A23187.

II. Synthetically Produced Calcium Ionophores

1. DDP^- and $DOPP^-$

Didecylphosphoric acid (DDP^-) and di(*n*-octylphenyl)phosphoric acid ($DOPP^-$) are electrically charged ion exchangers belonging to the dialkylphosphate group of compounds. They have been used in a variety of ion-sensitive electrode systems as sensors for Ca^{2+} and seem to have good discrimination for Ca^{2+} over Na^+ and K^+. DDP^- has a poorer selectivity for Mg^+ than systems based on $DOPP^-$. However, both are affected by other divalent cations (Sr^{2+} and Ba^{2+}) and H^+ (SIMON et al. 1978). It is difficult to give a selectivity sequence for these ionophores as their selectivities have been calculated in different ways and under different circumstances. The selectivities are generally described by K^{Pot}_{CaM} values. K^{Pot}_{CaM} is the potentiometric selectivity coefficient for Ca^{2+} with respect to an interferring ion M with a valency z. This is used in the Nicolsky-Eisenman equation

$$E = E_0 + \frac{RT}{2F} \ln [a_{Ca} + K^{Pot}_{CaM} \quad (a_M)^{2/Z}] \tag{4}$$

(where E, R, T, and F have their usual thermodynamic meanings) to describe the electrical potential difference arising across a Ca^{2+}-selective ligand interspersed between two aqueous solutions. For systems based on DDP^-, SIMON et al. (1979) give log K^{Pot}_{CaM} values for M^Z shown in Table 4. The use of $DOPP^-$ generally gives better selectivity for Ca^{2+} over Na^+, K^+, and Mg^{2+}, though the interfering properties of Mg^{2+} and H^+ are still large. Careful calibration is required for electrodes based on these ligands.

Table 4. Selectivities for Ca^{2+}-selective sensors

	$\log K^{Pot^{+}}_{CaM}$					
	H^{+}	Na^{+}	K^{+}	Mg^{2+}	Sr^{2+}	Ba^{2+}
DDP^{-} [a]	5.0	−4.2 to −0.7	−4.5 to −0.3	−3.2 to −1.3	−1.8 to −1.5	−2.7 to −1.4
$DOPP^{-}$ [a, b]	4.2	−4.7	−5.7	−3.6	−1.77	−3.6
ETH 1001 [a]	−4.39	−6.1	−5.22	−5.1	−2.09	−3.23
ETH 1001 [c]	−4.4	−6.1	−6.6	−4.9		
ETH 129 [c]	−1.6	−7.4	−8.0	−4.4		

[a] SIMON et al. (1978).
[b] RUZICKA et al. (1973).
[c] SCHEFER et al. (1986).

2. ETH 1001

Advances in selectivities have been achieved through the synthesis of an electrically neutral Ca^{2+} carrier (ETH 1001) by Simon and co-workers. This neutral carrier *N,N'*-di(1,1-ethoxycarbonyl)undecyl-*N,N'*-4,5-tetramethyl-3,6-dioxaoctane diacid diamide was originally described as an ion-sensitive component for microelectrodes which showed very high Ca^{2+} selectivity (OEHME et al. 1976). It has much better discrimination against Mg^{2+} than the two previous compounds. Values of log K^{Pot}_{CaM} for M^{Z} are shown in Table 4. It is highly sensitive to changes in pH, but only below about pH 6.0. Around physiologic pH values the carrier is not significantly affected by protons.

In addition to its extensive use in Ca^{2+}-selective electrodes, this ligand has been used as a very specific Ca^{2+} ionophore to promote Ca^{2+} flux across human erythrocytes (HINDS and VINCENZI 1985) without changing intracellular [Mg^{2+}] markedly. A drawback of ETH 1001 is the slow transport time (turnover value for ETH 1001 $= 2.9 \times 10^{-3}$ ions/s per ionophore molecule) compared with X-537A and A23187 (8.3×10^{-2} and 5.0 ions/s per ionophore molecule, respectively; CASWELL and PRESSMAN 1972). The action of ETH 1001 also seems to be dependent upon the solvent used. SCHEFER et al. (1986) have recently described Ca-selective electrodes based on a newly synthesized neutral Ca carrier (ETH 129). The superior selectivity of ETH 129 (see Table 4) allows construction of Ca-selective electrodes with detection limits in the subnanomolar range.

D. Conclusion

We have discussed a number of Ca^{2+} chelators which are used as tools in the study of a wide variety of Ca^{2+}-sensitive processes in biology. Undoubtedly, the Ca^{2+} dependence of all of these biologic processes is attributable to integral or endogenous Ca^{2+} chelators which may be on proteins, phospholipids (or other lipids), or carbohydrates. The same principles discussed in this chapter must also be considered to apply to the regulation of these biologic systems.

References

Albert A (1953) Avidity of terramycin and aureomycin for metallic cations. Nature 172:201

Al-Jabore A, Roufogalis BD (1981) Influence of EGTA on the apparent Ca^{2+} affinity of Mg^{2+}-dependent, Ca^{2+}-stimulated ATPase in the human erythrocyte membrane. Biochim Biophys Acta 645:1–9

Allen DG, Blinks JR (1979) The interpretation of light signals from aequorin-injected skeletal and cardiac muscle a new method of calibration. In: Ashley CC, Campbell AK (eds) Detection and measurements of free Ca^{2+} in cells. Elsevier/North Holland, Amsterdam, pp 159–174

Beeler TJ, Schibeci A, Martonosi A (1980) The binding of arsenazo III to cell components. Biochim Biophys Acta 629:317–327

Berger J, Rachlin AI, Scott WE, Sternbach LH, Goldberg MW (1951) The isolation of three new crystalline antibiotics from streptomyces. J Am Chem Soc 73:5295–5298

Berman MC (1982a) Stimulation of calcium transport of sarcoplasmic reticulum vesicles by the calcium complex of ethylene glycol bis(β-aminoethylether)N,N,N′,N′-tetraacetic acid. J Biol Chem 257:1953–1957

Berman MC (1982b) Energy coupling and uncoupling of active calcium transport by sarcoplasmic reticulum membrane. Biochim Biophys Acta 694:95–121

Bers DM (1982) A simple method for the accurate determination of free [Ca] in Ca-EGTA solutions. Am J Physiol 242:C404–C408

Blinks JR, Weir WG, Hess P, Prendergast FG (1982) Measurement of Ca^{2+} concentrations in living cells. Prog Biophys Mol Biol 40:1–114

Boyd S, Bryson A, Nancollas GH, Torrance K (1965) Thermodynamics of ion association. XII. EGTA complexes with divalent metal ions. J Chem Soc London 5:7353–7358

Case GD, Vanderkooi JM, Scarpa A (1974) Physical properties of biological membranes determined by the fluorescence of the calcium ionophore A23187. Arch Biochem Biophys 162:174–185

Caswell AH, Pressman BC (1972) Kinetics of transport of divalent cations across sarcoplasmic reticulum vesicles induced by ionophores. Biochem Biophys Res Commun 49:292–298

Chiu VCK, Haynes D (1980) The pH dependence and binding equilibria of the calcium indicator-arsenazo III. Membr Biochem 3:169–183

Cornelius A, Gartner W, Haynes DH (1974) Cation complexation by valinomycin- and nigericin-type ionophores registered by the fluorescence signal of Tl^{+}. Biochemistry 13:3052–3057

Debono M, Molloy RM, Dorman DE, Paschal JW, Babcock DF, Deber CM, Pfeiffer DR (1981) Synthesis and characterization of halogenated derivatives of the ionophore A23187: enhanced calcium ion transport specificity by the 4-bromo derivative. Biochemistry 20:6865–6872

Degani H, Friedman HL, Navon G, Kosower EM (1973) Fluorimetric complexing constants and circular dichroism measurements for antibiotic X-537A with univalent and bivalent cations. J Chem Soc Chem Comm 1973:431–432

Devore DI, Nastuk WL (1975) Effects of "calcium ionophore" X537A on frog skeletal muscle. Nature 253:644–646

Dobler M (1981) Ionophores and their structure. Wiley Interscience, New York

Dorogi PL, Moss K, Neumann E (1981) Spectrophotometric determination of reaction stoichiometry and equilibrium constants of metallochromic indicators. Biophys Chem 14:91–100

Fabiato A (1985) Time and calcium dependence of activation and inactivation of calcium-induced release of calcium from the sarcoplasmic reticulum of a skinned canine cardiac Purkinje cell. J Gen Physiol 85:247–289

Flatman PW, Lew VL (1977) Does ionophore A23187 mediate Na transport in the absence of divalent cations? Nature 270:444–445

Grynkiewicz G, Poenie M, Tsien RY (1985) A new generation of Ca^{2+} indicators with greatly improved fluorescent properties. J Biol Chem 260:3440–3450

Harafuji H, Ogawa Y (1980) Re-examination of the apparent binding constant of ethylene glycol bis(β-aminoethyl ethylene)-N,N,N',N'-tetraacetic acid with calcium around neutral pH. J Biochem 87:1305–1312
Harned HS, Owen BB (1958) The physical chemistry of electrolytic solutions, 2nd edn. Reinhold, New York, p 575
Harrison SM, Bers DM (1987) The effect of temperature and ionic strength on the apparent Ca-affinity of EGTA and the analogous Ca-Chelators BAPTA and dibromo-BAPTA. Biochim Biohpys Acta 925:133–143
Hellam DC, Podolsky RJ (1969) Force measurements in skinned muscle fibres. J Physiol (Lond) 200:807–819
Hinds TR, Vincenzi FF (1985) The effect of ETH 1001 on ion fluxes across red blood cell membranes. Cell Calcium 6:265–279
Illingworth JA (1981) A common source of error in pH measurements. Biochem J 195:259–262
Kotagel N, Colca JR, McDaniel ML (1983) Activation of an islet cell plasma membrane ($Ca^{2+}+Mg^{2+}$)-ATPase by calmodulin and Ca-EGTA. J Biol Chem 258:4808–4813
Liu C, Hermann TE (1978) Characterization of ionomycin as a calcium ionophore. J Biol Chem 253:5892–5894
Liu WC, Shesarchyk DS, Ashe G, Trejo WH, Brown WE, Meyers E (1978) Ionomycin, a new polyether antibiotic. J Antibiot (Tokyo) 31:815–819
Martell AE, Smith RM (1974) Critical stability constants, vol 1. Plenum, New York
Martell AE, Smith RM (1977) Critical stability constants, vol 3. Plenum, New York
Mathew MK, Balaram P (1980) A reinvestigation of chlortetracycline fluorescence: effect of pH, metal ions and environment. J Inorgan Biochem 13:339–346
Meier PC, Lanter F, Ammann D, Steiner RA, Simon W (1982) Applicability of available ion-selective liquid-membrane microelectrodes to intracellular ion-activity measurements. Pflügers Arch 393:23–30
Miller DJ, Smith GL (1984) EGTA purity and the buffering of calcium ions in physiological solutions. Am J Physiol 246:C160–C166
Morf WE, Simon W (1971) Berechnung von freien Hydratationsenthalpien und Koordinationszahlen für Kationen aus leicht zugänglichen Parametern. Helv Chem Acta 54:794–810
Nakon R (1979) Free metal ion depletion by Good's buffers. Anal Biochem 95:527–532
Oehme M, Kessler M, Simon W (1976) Neutral carrier Ca^{2+}-microelectrode. Chimia 30:204–206
Ogawa Y, Harafjuji H, Kurebayashi N (1980) Comparison of the characteristics of four metallochromic dyes as potential calcium indicators for biological experiments. J Biochem 87:1293–1303
Ohnishi ST (1978) Characterization of the murexide method: dual wavelength spectrophotometry of cations under physiological conditions. Anal Biochem 85:165–179
Ohnishi ST (1979a) Interaction of metallochromic indicators with calcium sequestering organelles. Biochim Biophys Acta 585:315–319
Ohnishi ST (1979b) A method of estimating the amount of calcium bound to the metallochromic indicator arsenazo III. Biochim Biophys Acta 586:217–230
Palade P, Vergara J (1983) Stoichiometries of arsenazo III-Ca complexes. Biophys J 43:355–369
Pfeiffer DR, Lardy HA (1976) Ionophore A23187: the effect of H^+ concentration on complex formation with divalent and monovalent cations and the demonstration of K^+ transport in mitochondria mediated by A23187. Biochemistry 15:935–943
Pfeiffer DR, Reed PW, Lardy HA (1979) Ultraviolet and fluorescent spectral properties of the divalent cation ionophore A23187 and its metal ion complexes. Biochemistry 13:4007–4014
Pressman BC (1973) Properties of ionophores with broad range cation selectivity. Fed Proc 32:1698–1703
Pressman BC (1976) Biological applications of ionophores. Ann Rev Biochem 45:501–530

Puskin JS, Gunter TE (1975) Electron paramagnetic resonance of copper ion and manganese ion complexes with the ionophore A23187. Biochemistry 14:187–191
Reed PW, Lardy HA (1972) A23187: a divalent cation ionophore. J Biol Chem 247:6970–6977
Rios E, Schneider MF (1981) Stoichiometry of the reactions of calcium with the metallochromic indicator dyes antipyralazo III and arsenazo III. Biophys J 36:607–621
Ruzicka J, Hanse EH, Tjell JC (1973) Selectrode – the universal ion-selective electrode. Part IV: The calcium (II) selectrode employing a new ion exchanger in a non-porous membrane and a solid state reference system. Anal Chim Acta 67:155–178
Sarkadi B, Shubert A, Gardos G (1979) Effects of Ca-EGTA buffers on active calcium transport in inside-out red cell membrane vesicles. Experientia 35:1045–1047
Scarpa A (1979) Measurement of calcium ion concentrations with metallochromic indicators. In: Ashley CC, Campbell AK (eds) Detection and measurement of free Ca^{2+} in cells. Elsevier/North Holland, Amsterdam, pp 85–115
Scarpa A, Brinley FJ, Dubyak G (1978) Antipyralazo III, a middle range Ca^{2+} metallochromic indicator. Biochemistry 17:1378–1386
Schefer U, Ammann D, Pretsch E, Oesch U, Simon W (1986) Neutral carrier based Ca^{2+}-selective electrode with detection limit in the sub-nanomolar range. Anal Chem 58:2282–2285
Schwarzenbach VG, Seen H, Anderegg G (1957) Komplexone XXIX. Ein grosser Chelateffekt besonderer Art. Helv Chim Acta 40:1886–1900
Shimomura O, Shimomura A (1984) Effect of calcium chelators on the Ca^{2+}-dependent luminescence of aequorin. Biochem J 221:907–910
Simon W, Morf WE, Ammann D (1977) Calcium ionophores. In: Wasserman RH, Corradino RA, Carafoli E, Kretsinger RH, MacLennan DH, Siegel FL (eds) Calcium binding proteins and calcium function. Elsevier North-Holland, Amsterdam, pp 50–62
Simon W, Ammann D, Oehme M, Morf WE (1978) Calcium-selective electrodes. Ann NY Acad Sci 307:52–69
Smith PD, Berger RL, Podolsky RJ (1977) Stopped-flow study of the rate of calcium binding by EGTA. Biophys J 17:159a
Toeplitz BA, Cohen AI, Funke PT, Parker WL, Gougaitas JZ (1979) Structure of ionomycin – a novel diacidic polyether antibiotic having high affinity for calcium ions. J Am Chem Soc 101:3344–3353
Trosper TL, Philipson KD (1984) Stimulatory effect of calcium chelators on Na^{+}-Ca^{2+} exchange in cardiac sarcolemmal vesicles. Cell Calcium 5:211–222
Truter MR (1976) Chemistry of calcium ionophores. Symp Soc Exp Biol 30:19–40
Tsien RY (1980) New calcium indicators and buffers with high selectivity against magnesium and protons: design, synthesis and properties of prototype structures. Biochemistry 19:2396–2404
Tsien RY (1983) Intracellular measurements of ion activities. Annu Rev Biophys Bioeng 12:91–116
Tsien RY, Rink TJ (1980) Neutral carrier ion-selective microelectrodes for measurement of intracellular free calcium. Biochim Biophys Acta 599:623–638
Tsien RY, Pozzan T, Rink TJ (1982a) T-cell mitogens cause early changes in cytoplasmic free Ca^{2+} and membrane potential in lymphocytes. Nature 295:68–71
Tsien RY, Pozzan T, Rink TJ (1982b) Calcium homeostasis in intact lymphocytes: cytoplasmic free calcium monitored with a new intracellularly trapped fluorescent indicator. J Cell Biol 94:325–334
Waisman DM, Gimble JM, Goodman DBP, Rasmussen H (1981) Studies of the Ca^{2+} transport mechanism of human erythrocyte inside-out plasma membrane vesicles. J Biol Chem 256:409–414
White JR, Pierce FL (1982) Characterization of chlortetracycline (aureomycin) as a calcium ionophore. Biochemistry 21:6309–6312

CHAPTER 24

Lead–Calcium Interactions and Lead Toxicity

T. J. B. SIMONS

A. Overview

The aim of this chapter is to bring together information on the interactions between Ca and Pb at the cellular level, and to discuss them in relation to possible mechanisms for the toxicity of Pb. Interactions between Pb and Ca in the whole organism have been known for many years. For example, Ca inhibits Pb uptake in the gastrointestinal tract, and long-term storage of Pb occurs in the bones (SOBEL et al. 1940; SMITH and HURSH 1977; MAHAFFEY 1980). These processes are now being studied at the cellular and molecular level. Interactions with binding proteins, enzymes, membrane transport, and secretory mechanisms will be discussed in later sections of this chapter. Consideration of the organism as a whole will be excluded, as will the interaction of Pb with other metals. (For interactions with Fe, Cu and Zn, see, for example MAHAFFEY and RADER 1980; PETERING 1980.)

The traditional view of the toxicology of Pb and other heavy metals is that they inhibit enzymes by interaction with SH groups (VALLEE and ULMER 1972). An alternative hypothesis has recently been suggested, based upon the observation that Pb and other heavy metals interact with calmodulin. Calmodulin is found in all cells and acts as a Ca^{2+} receptor, mediating many of the intracellular effects of Ca^{2+} ions. Lead can replace Ca^{2+} and causes activation of calmodulin at low concentrations, and inhibition at high concentrations (CHAO et al. 1984). It has been suggested that the toxic effects of Pb and other heavy metals might be brought about by: (a) occupying Ca-binding sites on calmodulin (GOLDSTEIN and AR 1983); (b) inhibiting calmodulin (COX and HARRISON 1983); or (c) activating calmodulin (CHEUNG 1984). It will be shown that this view is oversimplified, because there are other Pb–Ca interactions at the cellular and molecular level, and because Pb has other actions that are unlike those of Ca. These include enzyme inhibition, binding to specific proteins, and a novel mechanism of membrane transport. The conclusion will be that Pb toxicity is due to a combination of effects, some like and some unlike those of Ca.

B. Relevant Chemistry of Lead and Calcium

I. Introduction

Lead, atomic number 82, is in group IVB of the periodic table. It exists in two oxidation states, Pb(II) and Pb(IV), but only Pb(II) is normally encountered in nature. Inorganic Pb(IV) compounds are unstable under physiological conditions. One organic Pb(IV) compound, tetraethyl lead, is widely used as a petroleum additive, but it is converted to Pb(II) during or after combustion. Pb occurs naturally as the minerals PbS and $PbCO_3$. Lead oxide (PbO) forms salts with many acids. These salts contain the Pb^{2+} ion, which is the form in which Pb usually interacts with biological systems.

Calcium, atomic number 20, is an alkaline earth (group IIA). It only exists in one oxidation state, Ca(II), and is widely distributed in nature as $CaCO_3$ and $CaSO_4$. It is an essential constituent of living matter, where it is always found as the Ca^{2+} ion, either in solution, or in complexed or mineral form.

II. Chemistry of the Ions in Solution

1. Complexes with Simple Anions

The Pb^{2+} ions has two 6*s* electrons in its outer valence shell, and also low-lying unfilled *p* and *d* orbitals. It can use these to form covalent or partial ionic/covalent bonds with other ions. Complexes are formed with many inorganic anions in solution, e.g. with OH^-, Cl^- and NO_3^-. Successive addition of an anion (X^-) leads to formation of PbX^+, PbX_2, PbX_3^- and PbX_4^{2-} species (SILLEN and MARTELL 1964). Other, more complex ions can also be formed. The neutral species often has a low water solubility, and this can limit the range of Pb^{2+} concentrations that can be achieved experimentally. Addition of Pb in increasing quantities to physiological solutions that mimic mammalian extracellular fluid leads to precipitation in the sequence $Pb_3(PO_4)_2$, $PbCO_3$ then $Pb(OH)_2$ (MAXWELL and BISCHOFF 1929; SIMONS 1986a). The solubility product of $Pb(OH)_2$ (1.7×10^{-19} M^3; BIRRAUX et al. 1977) imposes an upper limit of about $(5–10) \times 10^{-6}$ M as the maximum Pb^{2+} concentration that can be achieved in vitro at neutral pH. Many publications report the use of much higher concentrations of lead nitrate or acetate. These experiments should be interpreted with considerable caution. All the valence shells on Ca^{2+} are full, and it does not form significant complexes with inorganic anions under normal conditions.

2. Complexes with Organic Ligands

Ca^{2+} and Pb^{2+} both form complexes with chelating ligands, such as ethylenediamine tetraacetic acid and nitrilotriacetic acid. The binding of Pb^{2+} is always stronger than that of Ca^{2+}, because of its orbital structure. The difference in affinity depends upon the donor atoms present at the binding site. Pb^{2+} has a much greater tendency to form bonds to N and S ligands than Ca^{2+}, and the presence of these ligands leads to a greatly increased binding of Pb^{2+} compared with Ca^{2+}.

Ca^{2+} is always coordinated by oxygen atoms in proteins containing specific Ca-binding sites. The coordination number is 6, 7 or 8, and the oxygens are in carbonyl or carboxyl groups, or water molecules (KRETSINGER 1976; MARTIN 1984). It seems reasonable to assume that Pb^{2+} binds at these sites when it produces Ca-like effects. In view of its tendency to form bonds to N and S, it probably also binds to other sites in proteins, where these may be present. The use of Pb derivatives in the determination of crystal structures provides some information on this point. In two of three proteins, the Pb atom is bound at sites with N ligands (imidazole, amide, amine), in addition to carboxylate groups (see Table 8.V in BLUNDELL and JOHNSON 1976).

3. Ionic Radius

The effective radius of an ion varies with its coordination number. The values for Ca^{2+} at 6- and 8-coordinated sites are 1.00 and 1.18 Å, respectively. Corresponding values for Pb^{2+} are 1.18 and 1.29 Å (SHANNON and PREWITT 1969). It is hard to draw any conclusions from these numbers. If Pb^{2+} interacts with a single negatively charged ion, the electrostatic attraction would be less than for Ca^{2+}. At a protein-binding site, the larger Pb^{2+} ion might be disfavoured because of steric factors, but the ability to form partially covalent bonds with the ligands would be energetically favourable.

III. Measurement of Pb^{2+} Concentration

Pb^{2+} ions can be measured with a commercially available Pb electrode (KIVALO et al. 1976). In the author's hands, this gives a linear voltage–log concentration response down to 10^{-8} M Pb^{2+} in unbuffered solutions, and to 10^{-11} M Pb^{2+} in well-buffered solutions. The units of concentration have to be defined arbitrarily. The authors has used the convention that the concentration measured by the electrode is the same as the total concentration of Pb for dilute solutions of $Pb(NO_3)_2$ in 0.1 M KNO_3 (SIMONS 1985). These measurements are not Pb^{2+} activities: the activity of Pb^{2+} is considerably lower, because an appreciable fraction of the Pb in the standard solutions is $PbNO_3^+$, and other species are present.

Most studies of Pb action report merely the total concentration of Pb present. Pb is added as the soluble nitrate or acetate salt. If measurements are made in a saline solution at neutral pH, most of the Pb will be in the form of $PbCl^+$ or $PbOH^+$. For example, in a solution containing 80 mM Cl^- at pH 7.4, the Pb^{2+} concentration measured with the electrode is about 25% of the total Pb concentration (SIMONS 1986a). In theory, the binding of Pb by anions should be taken into account when literature values for affinity constants are compared. In practice, this is very difficult, because variations in ionic strength affect the corrections that need to be applied, and the fact that differences in ionic strength have a considerable effect on affinity constants, even after correction for the binding of Pb by anions. All values for Pb^{2+} affinities should be considered with these difficulties in mind.

IV. Pb^{2+} Buffers

The use of H^+ and Ca^{2+} buffers to maintain fixed ion concentrations is well established. Pb^{2+} buffers work on the same principles. Their use is desirable for in vitro work with low concentrations of Pb^{2+}, because Pb^{2+} is strongly absorbed or bound by most biological material. This binding can easily amount to over 90% of the lead added to a preparation. Pb buffers provide a supply of Pb which maintains the Pb^{2+} concentration in the face of extensive binding. They are made by mixing a Pb salt with a chelating anion, so that a very small fraction of the Pb is present as free Pb^{2+} ions (SIMONS 1985). If Pb buffers are used, one should check that the buffer does not produce artifacts, e.g. by acting as a carrier for the movement of Pb across a membrane. On the other hand, the failure to use Pb buffers may lead to the overestimation of affinity constants for Pb-dependent actions, because of appreciable Pb binding to tissues. The general conclusion is that there are a number of reasons why literature estimates of Pb affinities are likely to be overestimates, if Pb electrodes or Pb buffers have not been used.

C. Nonenzymic Actions of Lead

Pb^{2+} catalyses the hydrolysis of ATP (ROSENTHAL et al. 1966) and the depolymerisation of RNA (FARKAS 1968). The mechanism of RNA breakdown has been elucidated and it has even been suggested as a possible explanation for the toxicity of Pb (BROWN et al. 1983). It is hard to assess the plausibility of this hypothesis as the reaction has been studied only at very high Pb concentrations, in the range 10^{-3}–10^{-4} *M*.

D. Interactions Between Lead and Binding Proteins

I. Calmodulin

Calmodulin has four Ca-binding sites, from which Pb can readily displace bound Ca (HABERMANN et al. 1983; CHAO et al. 1984). Experiments show that Pb^{2+} has a higher affinity than Ca^{2+} for binding to these sites. When calmodulin is equilibrated with ^{45}Ca at a Ca^{2+} concentration of 10^{-5} *M*, 50% of the ^{45}Ca is displaced by 10^{-6} *M* Pb^{2+} (FULLMER et al. 1985). Dialysis of calmodulin with 1.1×10^{-5} *M* Pb^{2+} and Ca^{2+} (together) leads to Pb binding of 4.7 mol/mol, with no Ca bound (FULLMER et al. 1985). The binding of more than 4 mol/mol may be explained by the recent discovery of regulatory metal-binding sites on calmodulin (MILLS and JOHNSON 1985). Many drugs can bind to calmodulin, among them felodipine, a fluorescent dihydropyridine. Some metals (La, Pb, Zn, Cd) can increase its fluorescence by allosterically increasing the affinity of calmodulin for felodipine. This action requires the occupation of secondary sites, in addition to the four Ca-binding sites. The secondary sites have a different metal selectivity from the primary Ca-binding sites. They probably have a relatively low affinity for Pb^{2+}, as 50% activation of fluorescence occurs at 3×10^{-5} *M* Pb^{2+} (MILLS and JOHNSON 1985). Their number and function is unknown. The interaction of Pb with calmodulin has also been studied in connection with its ability to replace Ca and activate Ca-dependent processes (Sect. E.V).

II. Troponin C

The competition of Pb^{2+} for the Ca^{2+} sites on troponin C appears to be very similar to the competition for Ca^{2+} sites on calmodulin (FULLMER et al. 1985). When troponin C is equilibrated with ^{45}Ca at 10^{-5} *M* Ca^{2+}, 50% of the ^{45}Ca

is displaced by 6×10^{-6} M Pb^{2+}. Dialysis of tropinin C with 1.1×10^{-5} M Ca^{2+} and Pb^{2+} (together) leads to Pb binding of 4.9 mol/mol and Ca binding of 0.2 mol/mol (FULLMER et al. 1985). These experiments suggest that the affinity for Pb^{2+} is higher than for Ca^{2+}. Contrary results have recently been obtained by flow dialysis (RICHARDT et al. 1986). The IC_{50} for displacement of ^{45}Ca was 18 μM for Ca and 52 μM for Pb.

III. Intestinal Calcium-Binding Proteins

The binding of Pb to bovine and chick intestinal Ca-binding proteins has been studied by equilibrium dialysis (FULLMER et al. 1985). The chick protein binds Ca 4 mol/mol, with $K_a = 2 \times 10^6$ M^{-1}. These sites also bind Pb, with a higher affinity, $K_a = 1.6 \times 10^7$ M^{-1}. The bovine protein behaves similarly, but binds only two Ca or Pb ions per molecule. Sulfhydryl groups are thought not to be involved in the binding of Pb to these high affinity sites. The bovine Ca-binding protein contains no sulfhydryl groups, while the binding of Pb to the chick protein is unaffected by the blockage of all the sulfhydryl groups by the formation of an *S*-β-pyridylethylcysteine derivative (FULLMER et al. 1985). This work is the clearest study which shows that Pb^{2+} binds to the Ca^{2+}-binding sites in proteins with higher affinity than Ca^{2+}, and that sulfhydryl groups are not involved in the binding of Pb^{2+} to these sites. However, flow dialysis of porcine intestinal Ca-binding protein showed a higher affinity for Ca than Pb (RICHARDT et al. 1986). The discrepancy with the findings of FULLMER et al. (1985) may be due to insufficient time being allowed for equilibration during flow dialysis. Pb was also shown to bind to parvalbumin and a 28 kdalton Ca-binding protein from rat kidney, in both cases with a lower affinity than Ca (RICHARDT et al. 1986).

IV. Lead-Binding Proteins

Two lead-binding proteins have been isolated from rat kidney cytosol (GOERING and FOWLER 1984; MISTRY et al. 1985). They are 11.5 and 63 kdalton proteins, and are also found in brain, but not liver or lung. Binding studies show the presence of both high and low affinity sites on the isolated proteins. Nonlinear curve fitting gives K_d for the high affinity sites as 1.3×10^{-8} M Pb^{2+} (63 kdalton protein) and 4×10^{-8} M Pb^{2+} (11.5 kdalton protein). These affinities must be open to doubt, because of the nonlinear Scatchard plots, and the fact that the calculated number of Pb-binding sites with these affinities is much less than one per molecule. ^{203}Pb is displaced from these proteins when cytosol is incubated with a 100-fold excess of Pb^{2+}, Zn^{2+} or Cd^{2+}, but not Ca^{2+}. This observations suggests a fundamental distinction from the Ca-binding proteins discussed already. The function of these renal Pb-binding proteins is not certain: they are involved in the accumulation of Pb by the kidney. The fact that they reverse the inhibition of δ-aminolevulinic acid dehydratase by Pb, partly through the donation of Zn, suggests that their main function may be as Zn-binding proteins (GOERING and FOWLER 1985; GOERING et al. 1986). They are distinct from metallothionein, and it is not known if they are induced by Pb or other metals. A 10 kdalton Pb-binding protein has been characterised in human red cells (RAGHAVEN and GANICK 1977).

Unlike the kidney protein, this one is only found in subjects who have previously been exposed to high levels of Pb. Detailed binding studies are not available.

V. Summary

Lead binds to Ca-binding sites in all Ca-binding proteins that have been examined, and to Pb-binding sites in "Pb-binding proteins", that do not bind Ca. All of these are high affinity sites. In addition, Pb seems to bind to all proteins with low affinity, e.g. to haemoglobin (BARLTROP and SMITH 1971; SIMONS 1986a), albumin (SIMONS 1986a) as well as the intestinal Ca-binding proteins (FULLMER et al. 1985) and renal Pb-binding proteins (MISTRY et al. 1985) discussed previously. The nature of these low affinity sites is not known.

E. Lead-Enzyme Interactions

I. Ca^{2+}-ATPase

The Ca^{2+}-ATPase of human red cell membranes is associated with the active transport mechanism for the extrusion of Ca from cells. It is stimulated by a variety of divalent cations, including Pb^{2+}, which gives a maximum velocity 70% of that with Ca^{2+}, and has $K_m = 2 \times 10^{-12}$ M (determined with Pb buffers; PFLEGER and WOLF 1975). This observation has not been followed by any more detailed study.

II. δ-Aminolevulinic Acid Dehydratase

This enzyme has attracted much attention, because it is on the pathway for haem biosynthesis, and is extremely sensitive to inhibition by Pb. Its activity in red blood cells correlates inversely with the blood Pb concentration (HERNBERG 1980). It is an example of an enzyme that is inhibited by Pb, but not by Ca. In vitro K_i for Pb^{2+} is about 10^{-6} M (BONSIGNORE et al. 1965; WEISSBERG and VOYTEK 1974; GOERING and FOWLER 1984; GOERING et al. 1986). Contamination by other proteins would be expected to increase the concentration of Pb needed to cause inhibition. In fact, the only study using purified enzyme (WEISSBERG and VOYTEK 1974) also showed the lowest affinity for inhibition by Pb ($K_i = 3–4 \times 10^{-6}$ M). The presence of cysteine reduces the inhibitory effect of Pb (BONSIGNORE et al. 1965), while reduced glutathione has little effect (WEISSBERG and VOYTEK 1974). The effect of cysteine may be due to its ability to chelate Pb^{2+} ions, but the lack of effect of reduced glutathione is surprising, because it should also bind Pb^{2+}. It is not possible to draw any conclusions about the nature of the Pb-binding site (or sites) on this enzyme.

III. Adenylate Cyclase

Lead and other heavy metals inhibit this enzyme with high affinity (NATHANSON and BLOOM 1976). K_i for Pb^{2+} is the range $1–2 \times 10^{-6}$ M, using enzyme from rat heart, salivary gland or cerebellum. 2-Mercaptoethanol is more successful than EGTA in reversing inhibition after the enzyme has been exposed to Pb. This, together with the lack of effect of Ca on the enzyme, and the high affinity for inhibition by Zn, Cd and Hg (NATHANSON and BLOOM 1976), suggests the presence of N ligands and/or an SH group at the inhibitory site.

IV. Na^+, K^+-ATPase

Both Ca and Pb inhibit the Na^+, K^+-ATPase and K^+-phosphatase activities associated with the Na pump, but by different mechanisms. Calcium inhibits mainly by slowing the rate of enzyme phosphorylation from ATP (TOBIN et al. 1973). K_i for Ca^{2+} inhibition of the ATPase can vary from 10^{-5} to 6×10^{-4} M, depending upon the Mg and ATP concentrations (BEAUGE and COOPER 1983). K_i for the inhibition of Na,K exchange by intracellular Ca^{2+} in intact red blood cells is about 2.5×10^{-5} M (BROWN and LEW 1983). Pb^{2+} inhibits the ATPase with $K_i = 4 \times 10^{-6}$ M (SIEGEL and FOGT 1977), but stimulates the phosphorylation of the enzyme from ATP, even in the absence of Na (SIEGEL et al. 1978). It may be presumed that Pb inhibits the dephosphorylation of the enzyme, but this has not been shown directly. The difference in behaviour of the enzyme with Ca and Pb suggests that the ions have different effects when bound to the enzyme. There is evidence that Ca^{2+} may bind at a transport site normally occupied by K^+ – the "occluded" site (POST and STEWART 1984; FORBUSH 1986), but not evidence at which site Pb^{2+} binds.

V. Calmodulin-Dependent Actions

The binding of Pb to calmodulin is discussed in Sect. D.I. Ca-calmodulin stimulates a wide variety of intracellular processes. Pb can substitute for Ca, for example in the stimulation of bovine brain 3′,5′-cyclic AMP phosphodiesterase, the stimulation of rat brain membrane-bound protein kinase, and in promoting the binding of calmodulin to brain membranes (HABERMANN et al. 1983). Pb^{2+} gives the same maximum rate of enzyme activity as Ca^{2+} does, and produces its effects at slightly lower concentrations than Ca^{2+} (HABERMANN et al. 1983; CHAO et al. 1984). This probably corresponds to the higher affinity that Pb^{2+} has for binding to the Ca^{2+} sites (Sect. D.I). Much higher concentrations of Pb^{2+} inhibit 3′,5′-cyclic AMP phosphodiesterase (10^{-3} M rather than 10^{-6} M) (COX and HARRISON 1983).

VI. Summary

In some cases (Ca-ATPase and calmodulin-stimulated phosphodiesterase), Pb and Ca produce the same activating effect. In others (δ-aminolevulinic acid dehydratase and adenylate cyclase), the enzyme is sensitive to heavy metals, but not Ca. A third pattern is shown by Na^+, K^+-ATPase, which is inhibited by Pb and Ca in mechanistically distinct ways.

F. Transport of Lead, and Its Effects Upon Ion Transport

I. Transport Across the Plasma Membrane

1. Human Red Blood Cells

Human red cells maintain a low intracellular free Ca^{2+} concentration, probably about 3×10^{-8} *M* (LEW et al. 1982). The membrane has a low Ca permeability, and the cells maintain a steady state by a balance between passive influx and the active extrusion of Ca, brought about by the Ca^{2+}-ATPase. The membrane also possesses a Ca-dependent K permeability (GARDOS 1959), and if intracellular free Ca^{2+} rises above a threshold of about 10^{-7} *M*, net loss of KCl and cell shrinkage occurs (LEW and FERREIRA 1978). Pb interacts with these transport mechanisms, sometimes by mimicking the actions of Ca^{2+} ions, but also in other ways.

The passive permeability of the red cell membrane to Pb^{2+} is about 10^5 times the permeability to Ca^{2+} ions. This is because Pb^{2+} is able to take advantage of the anion exchange pathway, which is unavailable to Ca^{2+}. Pb transport is dependent upon the presence of HCO_3^- ions, which are present in sufficient quantities, even in nominally HCO_3^--free solutions. Pb crosses the membrane by virtue of its ability to form anion complexes, either as $PbCO_3$ (neutral) in exchange for Cl^-, or as the ternary complex $PbCO_3Cl^-$ (SIMONS 1986a, b).

The red cell can readily take up millimolar quantities of Pb, because of the high membrane permeability, and the binding of Pb to cell constituents, such as haemoglobin. If cells are loaded with Pb in the cold, they can pump out much of it on rewarming to 37 °C, provided ATP is present. The transport of Pb out of the cells is active, as it occurs against a concentration gradient for Pb^{2+} (SIMONS 1984). These observations suggest that Pb^{2+} may be able to replace Ca^{2+} and be transported by the Ca pump, as well as stimulating the Ca^{2+}-ATPase (Sect. E.I).

Pb also affects the permeability of the membrane to monovalent cations. Low concentrations of extracellular Pb increase the permeability to K (ØRSKOV 1935) and Na (ØRSKOV 1947). It was later realised that the increase in permeability to K has very similar properties to the Ca-dependent K permeability (PASSOW 1981). The effects of Pb have now been characterised in detail (SIMONS 1985). Intracellular Pb^{2+} can replace Ca^{2+} in the activation of the K permeability mechanism, but has a considerably higher affinity. It is found that 50% activation of ^{86}Rb efflux occurs at 2×10^{-9} *M* Pb^{2+}, compared with 5×10^{-7} *M* Ca^{2+} (measured with buffers). The maximum rate is the same for both Pb^{2+} and Ca^{2+} (SIMONS 1985). There is also evidence that higher concentrations of Pb (up to 10^{-4} *M*) can block K movement through the channels (SHIELDS et al. 1985; ALVAREZ et al. 1986). In addition to effects of Pb^{2+} on the Ca-dependent K permeability, extracellular Pb^{2+} at concentrations above 6×10^{-7} *M* increases the membrane permeability of red blood cells to both Na^+ and K^+ (SIMONS 1985). This effect is not produced by Ca^{2+}.

2. Ca Channels

Ca entry through Ca channels is of considerable physiological significance, because of its importance in neurotransmission and secretion (Sect. H). Ca channels may be opened by a change in the voltage across a cell membrane, or the binding of a hormone to a receptor. Pb has been known to interact with Ca channels since the discovery that Pb inhibits transmission, competitively with Ca, in the cat superior cervical ganglion (KOSTIAL and VOUK 1957). Neurotransmission requires both the entry of Ca through Ca channels, and the elevation of intracellular free Ca^{2+}, which leads to the release of transmitter. This can complicate the interpretation of experiments in which only transmitter release is measured. The inhibition of ^{45}Ca uptake into bullfrog sympathetic ganglia, in response to presynaptic nerve stimulation, provided direct evidence that Pb blocks Ca entry (KOBER and COOPER 1976). Subsequent work with rat brain synaptosomes has shown that ^{45}Ca uptake through Ca channels is inhibited by Pb^{2+} with $K_i = 4 \times 10^{-7}$ M (NACHSHEN 1984) or 1.1×10^{-6} M (SUSZKIW et al. 1984). Seemingly contrary results have been obtained with dopaminergic synaptosomes prepared from rat caudate nucleus (SILBERGELD 1977). In these, Pb increases Ca uptake, with a maximal effect at 5×10^{-5} M Pb. It is doubtful whether this represents entry via Ca channels, as the experiments measured basal uptake, rather than uptake stimulated by depolarisation in high K solutions.

The interaction of Pb with Ca channels has now been studied in bovine adrenal medullary cells, using Pb buffers to control free Pb^{2+} levels (POCOCK and SIMONS 1987; SIMONS and POCOCK 1987). Pb^{2+} inhibits Ca entry with a $K_i = 1.2 \times 10^{-6}$ M at 1×10^{-3} M Ca^{2+}. Pb^{2+} was found to enter the cells via Ca channels with an apparent permeability about 30 times the permeability to Ca^{2+}. (An exact comparison is not possible from flux measurements, because of very different inactivation kinetics for Pb and Ca.) Pb entry through Ca channels occurs with K_m for external Pb^{2+} equal to 2.6×10^{-6} M, and is inhibited by Ca^{2+} with $K_i = 1.4 \times 10^{-3}$ M. It shows similar pharmacological properties to Ca entry (SIMONS and POCOCK 1987). Pb entry through Ca channels at the frog neuromuscular junction has been inferred from the observation that Cd antagonises the stimulation of miniature end-plate potential frequency by Pb (COOPER and MANALIS 1984; see Sect. H).

Pb has a second action on membrane channels in bovine adrenal medullary cells. It inhibits Na entry through channels linked to nicotinic cholinergic receptors, with $K_i \sim 5 \times 10^{-6}$ M (POCOCK and SIMONS 1987). A similar effect may occur at the frog neuromuscular junction, where 10^{-4} M Pb causes a small reduction in the postsynaptic depolarisation induced by iontophoresis of acetylcholine (MANALIS and COOPER 1973).

II. Transport Across Epithelia

1. Intestinal Absorption

Pb is incompletely absorbed by the small intestine in vivo. The presence of Ca in the lumen inhibits the uptake of Pb, with $K_i \sim 10^{-3}$ M for a 10^{-6} M dose of Pb in rats (BARTON et al. 1978). Lead competes with Ca for intestinal Ca-binding proteins (Sect. D.III), but their function is unknown. It has recently been shown that everted gut sacs from rat duodenum can bring about the active transport of ^{210}Pb from the mucosal to the serosal surface. This requires an energy supply, and is blocked by metabolic inhibitors, such as 2,4-dinitrophenol, or cyanide (BARTON 1984). These observations would fit with the hypothesis that Pb enters the enterocytes passively at the mucosal border, and is actively transported out of the cells by the Ca pump at the basolateral surface. More work is needed to establish the mechanism of Pb absorption, and to investigate the nature of the competition between Pb, Fe and Zn, as well as Ca (MAHAFFEY and RADER 1980; PETERING 1980).

2. Renal Absorption and Excretion

The handling of Pb by the kidney is quite complex. Only 10%–20% of the Pb in blood is ultrafiltrable, and both tubular absorption and secretion occur (VANDER et al. 1977; VICTERY et al. 1979). Pb is accumulated in tubular cells, apparently by uptake mechanisms at both luminal and basolateral surfaces. There are several pieces of evidence which suggest that Pb and Ca are handled differently by the kidney. Mannitol and furosemide inhibit distal tubular Na and Ca reabsorption, but have no effect on Pb reabsorption (VANDER ct al. 1977). The uptake of Pb into cortical slices is markedly reduced by metabolic inhibitors, and may be due to active transport (VANDER et al. 1979). It is thought mainly to reflect transport at the basolateral surface of the tubular cells. Elevated Ca levels do not inhibit this uptake (VANDER et al. 1979). A 100-fold excess of Ca also has no effect on the accumulation of Pb by brush border membrane vesicles (VICTERY et al. 1984). Inhibition of Pb accumulation is brought about by other metals, such as Sn(IV) and Fe(III). It is possible that Pb may be transported in the kidney by pathways specific for heavy metals.

III. Mitochondria

Most of the Ca inside cells is not in the cytoplasm, but is found in organelles, such as mitochondria, endoplasmic reticulum, secretory granules and microvesicles. Of these, only mitochondria have been investigated in relation to the handling of Pb. An extensive study has been performed, using beef heart mitochondria (SCOTT et al. 1971). Lead is taken up by both passive and energy-dependent routes. The passive uptake depends upon the anions present, and Pb affects both the anion and cation permeability of the mitochondria. It causes both activation and inhibition of respiration and respiration-dependent ion movements, dependent upon the amount of Pb taken up (SCOTT et al. 1971). Later work has concentrated on the interaction of Pb with the Ca uptake mechanism. Pb competitively inhibits Ca uptake, and it does this as concentrations lower than those which inhibit respiration (PARR and HARRIS 1976). Ca uptake is half-inhibited by 4×10^{-7} *M* Pb^{2+} in mitochondria from rat heart (PARR and HARRIS 1976 and 5×10^{-6} *M* Pb^{2+} in those from rat brain (GOLDSTEIN 1977). Ruthenium red is a relatively specific inhibitor of mitochondrial Ca uptake, and causes partial inhibition of Pb uptake into rat kidney mitochondria (KAPOOR et al. 1985). Most of the

Pb uptake occurs via a ruthenium red-sensitive pathway at Pb^{2+} concentrations up to 10^{-5} *M*, and this is blocked by other agents which inhibit Ca uptake (N_3^-, La^{3+}). The ruthenium red-sensitive fraction shows a sigmoid variation with Pb concentration, and is half-maximal at 5×10^{-6} *M* Pb^{2+}. It is inhibited at concentrations above 10^{-5} *M*, possibly because of the inhibition of respiration, and Pb then enters by another route (KAPOOR et al. 1985). Pb also displaces Ca from isolated mitochondria (KAPOOR and VAN ROSSUM 1984). It is not clear whether this is because it inhibits Ca influx, or accelerates Ca efflux. Nothing is known about Pb exit from mitochondria.

G. Cellular Homeostasis

Pb–Ca interactions have been studied with three types of cells. The most extensive work has utilised rat hepatocytes in primary culture, and this has been reviewed (POUNDS 1984). Tracer washout experiments allowed the identification of three cellular compartments for Ca. Incubation with Pb for 20 h increases the size of all three compartments, but most notably the slowest exchanging one, which includes the mitochondria (POUNDS et al. 1982a). Lead itself distributes into three compartments, the largest of which is thought to include the mitochondria, as it is sharply reduced in size by cyanide and 2,4-dinitrophenol, and increased by phosphate (POUNDS et al. 1982a, b). Attempts to identify the compartments by cellular fractionation were inconclusive (MITTELSTAEDT and POUNDS 1984). Some physiological stimuli cause Ca mobilisation in hepatocytes, and this can be detected by an increased rate of ^{45}Ca efflux. When the cells are loaded with ^{210}Pb, α-adrenergic agents, angiotensin and vasopressin all stimulate ^{45}Ca efflux and ^{210}Pb efflux in parallel (POUNDS and MITTELSTAEDT 1983). This is a further indication that Ca and Pb are handled by similar mechanisms.

Tracer Pb movements have also been studied with cultured mouse osteoblasts and osteoclasts (ROSEN 1983). Lead uptake is stimulated by parathyroid hormone. A more detailed examination of ^{210}Pb equilibration in osteoclasts shows that it is distributed between three cellular compartments. The largest and slowest exchanging component was identified with mitochondria, as in hepatocytes (POUNDS and ROSEN 1986).

When kidney cortex slices are cooled to 1 °C, they gain Ca, and this extra Ca is lost on warming to 25 °C. Parallel changes occur in the mitochondria. Addition of 2×10^{-4} *M* Pb during the cooling phase reduces the amount of Ca uptake by the slices, and the amount of Ca taken up by the mitochondria within the slices. Addition of Pb during rewarming reduces the amount of Ca finally left in the mitochondria, thus seeming to displace Ca from them (KAPOOR and VAN ROSSUM 1984). Incubation of slices with Pb leads to Pb uptake into both slices and the mitochondria within them. Mitochondrial Pb uptake (using slices) is increased in the absence of Ca from the medium, suggesting Ca–Pb competition at either the plasma or the mitochondrial membrane (KAPOOR et al. 1985).

H. Neurotransmission and Neurosecretion

Lead–calcium interactions at the nerve terminal have attracted much interest, because lead intoxication causes both peripheral neuropathy and encephalopathy. There are two possible presynaptic sites for Pb–Ca interaction: the Ca channel where Ca enters the nerve terminal, and the intracellular site where Ca causes transmitter release. Lead competitively inhibits Ca entry, and there is some evidence that Pb actually enters the terminal through Ca channels (Sect. F.II). The effects on transmitter release are rather variable.

Lead has two presynaptic effects at the frog neuromuscular junction (MANALIS et al. 1984). It inhibits acetylcholine release evoked by nerve stimulation, seemingly by blocking Ca entry into the nerve terminal. The effect is competitive with respect to Ca, and characterised by $K_d = 9.9 \times 10^{-7}$ M between Pb and the Ca receptor site (MANALIS et al. 1984). Lead also stimulates miniature end-plate potential (m.e.p.p.) frequency. This has a slower onset than the inhibition of evoked release, and is probably produced by Pb acting within the nerve terminal. The reason for thinking this is that Cd, like Pb, inhibits transmitter release, but fails to stimulate the m.e.p.p. frequency. Cadmium does antagonise the stimulatory effect of Pb, showing similar competition to its effects on nerve-evoked transmitter release (COOPER and MANALIS 1984). The stimulation of secretion by Pb is thought to result from the inhibition of mitochondrial and other Ca uptake mechanisms, leading to a rise in intracellular Ca^{2+} (MANALIS et al. 1984). This hypothesis would fit with what is known about mitochondrial function (Sect. F.III), but it would also require that the secretory mechanism itself is less susceptible to inhibition by Pb than mitochondrial Ca uptake. Alternatively it might arise from a direct stimulatory effect of Pb on the secretory mechanism.

The isolated bovine adrenal medullary cell responds slightly differently to Pb (POCOCK and SIMONS 1987). Carbachol-evoked secretion (equivalent to nerve-evoked secretion) is blocked, because of the inhibition of Ca entry into the cell. Basal secretion is analogous to the stimulation of m.e.p.p. frequency, and is unaffected by Pb, even though Pb enters the cells (SIMONS and POCOCK 1987). The lack of stimulation of basal secretion could either be because intracellular Ca^{2+} is not increased, or because intracellular Pb antagonises the secretory effects of Ca.

Lead–calcium interactions have also been studied with rat brain synaptosomes (SILBERGELD 1977). Forebrain synaptosomes take up ^{3}H-labelled choline, and Pb inhibits both choline uptake and release. Synaptosomes derived from caudate nucleus are dopaminergic, and Pb inhibits dopamine uptake, but stimulates its release (SILBERGELD and ADLER 1978). This has been confirmed in more recent work with striatal synaptosomes (MINNEMA et al. 1986). Here 1–30 μM Pb stimulates basal dopamine release in a graded fashion, whilst inhibiting transmitter release induced by K depolarisation. The latter effect shows Pb–Ca competition, and presumably arises from the blockade of Ca entry through Ca channels. The stimulation of basal dopamine release by Pb shows similarities to its effect at the neuromuscular junction. A difference is that dopamine release is not inhibited by Ca channel blockers, suggesting that Pb enters the synaptosomes by a different route, or that the stimulation of secretion is brought about by extracellular, rather than intracellular Pb. Thus, the three exocytotic systems which have been examined in some detail all show significant differences in the effects of Pb on basal secretion. This makes it difficult to draw any general conclusions about the interactions of Pb with secretory systems.

J. Summary and Conclusions

Two important conclusions follow from the experiments discussed in this chapter:

1. Lead interacts in some way with most sites of action of Ca. It interacts competitively at all well-defined sites, and in most cases it can replace Ca not only in binding, but also in a further step, such as activation of an enzyme or transport across a membrane. Where the affinity can be measured, it is higher for Pb^{2+} than for Ca^{2+}. Pb does not interact with Ca in some cases, for example in renal Ca transport, or at the intracellular site in neurosecretion. These should repay further investigation.

2. Lead has other effects, unlike those of Ca. Some of these have affinities comparable with the effects of Pb at Ca-specific sites (for example the inhibition of enzymes discussed in Sect. D).

It is hard to draw conclusions about the mechanism of Pb toxicity. There is very little information about free Pb^{2+} concentrations, as opposed to total Pb, in vivo. The total Pb in human whole blood is typically $1–2 \times 10^{-6}$ *M*, but increases with environmental exposure to Pb. Serious haematological disorders (anaemia, shortening of red cell life, excretion of haem precursors) appear at around 5×10^{-6} *M*, and gross disturbances of other body systems (brain, nerve, kidney, gut) at around 10^{-5} *M* (HERNBERG 1980). The free Pb^{2+} concentration in plasma in many orders of magnitude lower than these values, perhaps in the range $10^{-12}–10^{-11}$ *M* (SIMONS 1986a), although there are many uncertainties. Intracellular Pb^{2+} levels may be similar, although this is pure speculation. The majority of "high affinity" Pb effects occur at $10^{-7}–10^{-6}$ *M* free Pb^{2+}. This would lead one to expect little or no disturbance of function, or displacement of Ca, at toxic doses of Pb in vivo. This may not be so, because of cumulative effects of Pb on a combination of reactions, or because of longer-term effects associated with the induction or repression of protein synthesis. The red cell enzyme δ-aminolevulinic acid dehydratase is a specific example which should repay further study. When its activity is correlated with blood Pb levels, K_i is 2×10^{-6} *M* total Pb (HERNBERG 1980), of which 2×10^{-11} *M* at most is free Pb^{2+}. K_i for Pb added to enzyme in vitro is around 10^{-6} *M* (Sect. E.II), although the free Pb^{2+} level in these experiments is rather uncertain. This is preliminary evidence for a massive discrepancy between the inhibition constants in vivo and in vitro. With this in mind, extrapolation from other studies seems fraught with difficulties. There is no reason at present for favouring one mechanism of Pb toxicity over any other.

References

Alvarez J, García-Sancho J, Herreros B (1986) Inhibition of Ca^{2+}-dependent K^+ channels by lead in one-step inside-out vesicles from human red cell membranes. Biochem Biophys Acta 857:291–294
Barltrop D, Smith A (1971) Interaction of lead with erythrocytes. Experientia 27:92–93
Barton JC (1984) Active transport of lead-210 by everted segments of rat duodenum. Am J Physiol 247:G193–G198
Barton JC, Conrad ME, Harrison L, Nuby S (1978) Effects of calcium on the absorption and retention of lead. J Lab Clin Med 91:366–376
Beaugé L, Campos MA (1983) Calcium inhibition of the ATPase and phosphatase activities of the $(Na^+ + K^+)$-ATPase. Biochem Biophys Acta 729:137–149
Birraux C, Landry JCl, Haerdi W (1977) Etude de l'equilibre plomb-hydroxide par électrodes selectives. Anal Chim Acta 93:281–285
Blundell TL, Johnson LN (1976) Protein crystallography. Academic, New York
Bonsignore D, Calissano P, Cartasegna C (1965) Sul meccanismo di inhibizione da piombo dell'ala-deidratasi eritrocitaria. Med Lavoro 56:727–731
Brown AM, Lew VL (1983) The effect of intracellular calcium on the sodium pump of human red cells. J Physiol (Lond) 343:455–493
Brown RS, Hingerty BE, Dewan JC, Klug A (1983) Pb(II)-catalysed cleavage of the sugar-phosphate backbone of yeast $tRNA^{Phe}$ – implications for lead toxicity and self-splicing RNA. Nature 303:543–546
Chao SH, Suzuki Y, Zysk JR, Cheung WY (1984) Activation of calmodulin by various metal cations as a function of ionic strength. Mol Pharmacol 26:75–82
Cheung WY (1984) Calmodulin: its potential role in cell proliferation and heavy metal toxicity. Fed Proc 43:2995–2999
Cooper GP, Manalis RS (1984) Interactions of lead and cadmium on acetylcholine release at the frog neuromuscular junction. Toxicol Appl Pharmacol 74:411–416
Cox JL, Harrison SD (1983) Correlation of metal toxicity with in vitro calmodulin inhibition. Biochem Biophys Res Commun 115:106–111
Farkas WR (1968) Depolymerization of ribonucleic acid by plumbous ion. Biochem Biophys Acta 155:401–409
Forbush B (1986) Calcium and the occluded state of the Na/K pump. J Gen Physiol 88:22a–23a
Fullmer CS, Edelstein S, Wasserman RH (1985) Lead-binding properties of intestinal calcium-binding proteins. J Biol Chem 260:6816–6819
Gárdos G (1959) The role of calcium in the potassium permeability of human erythrocytes. Acta Physiol Hung 15:121–125
Goering PL, Fowler BA (1984) Regulation of lead inhibition of δ-aminolevulinic acid dehydratase by a low molecular weight, high affinity renal lead-binding protein. J Pharmacol Exp Ther 231:66–71
Goering PL, Fowler BA (1985) Mechanism of renal lead-binding protein reversal of δ-aminolevulinic acid dehydratase inhibition by lead. J Pharmacol Exp Ther 234:365–371
Goering PL, Mistry P, Fowler BA (1986) A low molecular weight lead-binding protein in brain attenuates lead inhibition of δ-aminolevulinic acid dehydratase: comparison with a renal lead-binding protein. J Pharmacol Exp Ther 237:220–225
Goldstein GW (1977) Lead encephalopathy: the significance of lead inhibition of calcium uptake by brain mitochondria. Brain Res 136:185–188
Goldstein GW, Ar D (1983) Lead activates calmodulin-sensitive processes. Life Sci 33:1001–1006
Habermann E, Crowell K, Janicki P (1983) Lead and other metals can substitute for Ca^{2+} in calmodulin. Arch Toxicol 54:61–70
Hernberg S (1980) Biochemical and clinical effects and responses as indicated by blood concentration. In: Singhal RL, Thomas A (eds) Lead toxicity. Urban and Schwarzenberg, Baltimore, pp 367–399
Kapoor SC, Van Rossum GDV (1984) Effects of Pb^{2+} added in vitro on Ca^{2+} movements in isolated mitochondria and slices of rat kidney cortex. Biochem Pharmacol 33:1771–1778

Kapoor SC, Van Rossum GDV, O'Neill KJ, Mercorella I (1985) Uptake of inorganic lead in vitro by isolated mitochondria and tissue slices of rat renal cortex. Biochem Pharmacol 34:1439–1448
Kivalo P, Virtanen R, Wickstrom K, Wilson M, Pungor E, Horvai G, Toth K (1976) An evaluation of some commercial lead(II)-selective electrodes. Anal Chim Acta 87:401–409
Kober TE, Cooper GP (1976) Lead competitively inhibits calcium-dependent synaptic transmission in the bullfrog sympathetic ganglion. Nature 262:704–705
Kostial K, Vouk VB (1957) Lead ions and synaptic transmission in the superior cervical ganglion of the cat. Br J Pharmacol 12:219–222
Kretsinger RH (1976) Calcium-binding proteins. Annu Rev Biochem 45:239–266
Lew VL, Ferreira HG (1978) Calcium transport and the properties of a calcium-activated potassium channel in red cell membranes. Curr Top Membr Transp 10:217–271
Lew VL, Tsien RY, Miner C, Bookchin RM (1982) Physiological $[Ca^{2+}]_i$ level and pump-leak turnover in intact red cells measured using an incorporated Ca chelator. Nature 298:478–481
Mahaffey KR (1980) Nutrient-lead interactions. In: Singhal RL, Thomas JA (eds) Lead toxicity. Urban and Schwarzenberg, Baltimore, pp 425–460
Mahaffey KR, Rader JI (1980) Metabolic interactions: lead, calcium and iron. Ann NY Acad Sci 355:285–297
Manalis RS, Cooper GP (1973) Presynaptic and postsynaptic effects of lead at the frog neuromuscular junction. Nature 243:354–356
Manalis RS, Cooper GP, Pomeroy SL (1984) Effects of lead on neuromuscular transmission in the frog. Brain Res 294:95–109
Martin BR (1984) Bioinorganic chemistry of calcium. In: Sigel H (ed) Metal ions in biological systems, vol 17. Calcium and its role in biology. Dekker, New York, pp 1–49
Maxwell LC, Bischoff F (1929) The stability of lead salts in physiological salt solution. J Pharmacol Exp Ther 36:279–293
Mills JS, Johnson JD (1985) Metal ions as allosteric regulators of calmodulin. J Biol Chem 260:15100–15105
Minnema DJ, Greenland RD, Michaelson IA (1986) Effect of in vitro inorganic lead on dopamine release from superfused rat striatal synaptosomes. Toxicol Appl Pharmacol 84:400–411
Mistry P, Lucier GW, Fowler BA (1985) High-affinity lead-binding proteins in rat kidney cytosol mediate cell-free nuclear translocation of lead. J Pharmacol Exp Ther 232:462–469
Mittelstaedt RA, Pounds JG (1984) Subcellular distribution of lead in cultured rat hepatocytes. Environ Res 35:188–196
Nachshen DA (1984) Selectivity of the Ca binding site in synaptosome Ca channels. Inhibition of Ca influx by multivalent metal cations. J Gen Physiol 83:941–967
Nathanson JA, Bloom FE (1976) Heavy metals and adenosine cyclic 3′5′-monophosphate metabolism: possible relevance to heavy metal toxicity. Mol Pharmacol 12:390–398
Ørskov SL (1935) Untersuchungen über den Einfluß von Kohlensäure und Blei auf die Permeabilität der Blutkörperchen für Kalium und Rubidium. Biochem Z 279:250–261
Ørskov SL (1947) The volume of the erythrocytes at different osmotic pressure. Further experiments on the influence of lead on the permeability of cations. Acta Physiol Scand 12:202–212
Parr DR, Harris EJ (1976) The effect of lead on the calcium-handling capacity of rat heart mitochondria. Biochem J 158:289–294
Passow H (1981) Selective enhancement of potassium efflux from red blood cells by lead. In: Wallach DFH (ed) The function of red blood cells: erythrocyte pathobiology. Liss, New York, pp 79–104
Petering HG (1980) The influence of dietary zinc and copper on the biologic effects of orally ingested lead in the rat. Ann NY Acad Sci 355:298–308
Pfleger H, Wolf HU (1975) Activation of membrane-bound high-affinity calcium ion-sensitive adenosine triphosphatase of human erythrocytes by bivalent metal ions. Biochem J 147:359–361

Pocock G, Simons TJB (1987) The effects of Pb^{2+} ions on events associated with exocytosis in isolated bovine adrenal medullary cells. J Neurochem 48:376–382
Post RL, Stewart HB (1984) Effect of Ca^{2+} on phosphorylation of (Na,K)-ATPase from P_i. Biophys J 45:78a
Pounds JG (1984) Effect of lead intoxication on calcium homeostasis and calcium-mediated cell function: a review. Neurotoxicology 5:295–332
Pounds JG, Mittelstaedt RA (1983) Mobilization of cellular calcium-45 and lead-210: effect of physiological stimuli. Science 220:308–310
Pounds JG, Rosen JF (1986) Cellular metabolism of lead: a kinetic analysis in cultured osteoclastic bone cells. Toxicol Appl Pharmacol 83:531–545
Pounds JG, Wright R, Morrison D, Casciano DA (1982a) Effect of lead on calcium homeostasis in the isolated rat hepatocyte. Toxicol Appl Pharmacol 63:389–401
Pounds JG, Wright R, Kodell RL (1982b) Cellular metabolism of lead: a kinetic analysis in the isolated rat hepatocyte. Toxicol Appl Pharmacol 66:88–101
Raghaven SRV, Ganick HC (1977) Isolation of low-molecular-weight lead-binding proteins from human erythrocytes. Proc Soc Exp Biol Med 155:164–167
Richardt G, Federolf G, Habermann E (1986) Affinity of heavy metal ions to intracellular Ca^{2+}-binding proteins. Biochem Pharmacol 35:1331–1335
Rosen JF (1983) The metabolism of lead in isolated bone-cell populations: interactions between Pb and Ca. Toxicol Appl Pharmacol 71:101–112
Rosenthal AS, Moses HL, Beaver DL, Schuffman SS (1966) Lead ion and phosphatase histochemistry I. Nonenzymatic hydrolysis of nucleoside phosphates by lead ion. J Histochem Cytochem 14:698–701
Scott KM, Hwang KM, Jurkowitz M, Brierley GP (1971) Ion transport by heart mitochondria. XXIII The effects of lead on mitochondrial reactions. Arch Biochem Biophys 147:557–567
Shannon RD, Prewitt CT (1969) Effective ionic radii in oxides and fluorides. Acta Cryst B25:925–946
Shields M, Grygorczyk R, Fuhrmann GF, Schwarz W, Passow (1985) Lead-induced activation and inhibition of potassium-selective channels in the human red blood cell. Biochem Biophys Acta 815:223–232
Siegel GJ, Fogt SM (1977) Inhibition by lead ion of electrophorus electroplax (Na^+ + K^+)-adenosine triphosphatase and K^+-p-nitrophenylphosphatase. J Biol Chem 252:5201–5206
Siegel GJ, Fogt SM, Iyengar S (1978) Characteristics of lead ion-stimulated phosphorylation of electrophorus electricus electroplax ($Na^+ + K^+$)-adenosine triphosphatase and inhibition of ATP-ADP exchange. J Biol Chem 253:7207–7211
Silbergeld EK (1977) Interactions of lead and calcium on the synaptosomal uptake of dopamine and calcium. Life Sci 20:309–318
Silbergeld EK, Adler HS (1978) Subcellular mechanisms of lead neurotoxicity. Brain Res 294:95–109
Sillen LG, Martell AE (1964) Stability constants of metal-ion complexes. Chem Soc Spec Publ 17. The Chemical Society, London
Simons TJB (1984) Active transport of lead by human red blood cells. FEBS Lett 172:250–253
Simons TJB (1985) Influence of lead ions on cation permeability in human red cell ghosts. J Membr Biol 84:61–71
Simons TJB (1986a) Passive transport and binding of lead by human red blood cells. J Physiol (Lond) 378:267–286
Simons TJB (1986b) The role of anion transport in the passive movement of lead across the human red cell membrane. J Physiol (Lond) 378:287–312
Simons TJB, Pocock G (1987) Lead enters bovine adrenal medullary cells through calcium channels. J Neurochem 48:383–389
Smith FA, Hursh JB (1977) Bone storage and release. In: Lee DHK (ed) Handbook of physiology, sect 9. Reactions to environmental agents. American Physiological Society, Bethesda, pp 469–482

Sobel AE, Yuska H, Peters DD, Kramer B (1940) The biochemical behavior of lead. I. Influences of calcium, phosphorus, and vitamin D on lead in blood and bone. J Biol Chem 132:239–265
Suszkiw J, Toth G, Murawsky M, Cooper GP (1984) Effects of Pb^{2+} and Cd^{2+} on acetylcholine release and Ca^{2+} movements in synaptosomes and subcellular fractions from rat brain and torpedo electric organ. Brain Res 323:31–46
Tobin T, Akera T, Baskin SI, Brody TM (1973) Calcium ion and sodium- and potassium-dependent adenosine triphosphatase: its mechanism of inhibition and identification of the E_1-P intermediate. Mol Pharmacol 9:336–349
Vallee BT, Ulmer DD (1972) Biochemical effects of mercury, cadmium and lead. Annu Rev Biochem 41:91–128
Vander AJ, Taylor DL, Kalitis K, Mouw DR, Victery W (1977) Renal handling of lead in dogs: clearance studies. Am J Physiol 233:F532–F538
Vander AJ, Mouw DR, Cox J, Johnson R (1979) Lead transport by renal slices and its inhibition by tin. Am J Physiol 236:F373–378
Victery W, Vander AJ, Mouw DR (1979) Effect of acid-base status on renal excretion and accumulation of lead in dogs and rats. Am J Physiol 237:F398–F407
Victery W, Miller CR, Fowler BA (1984) Lead accumulation by rat renal brush border membrane vesicles. J Pharmacol Exp Ther 231:589–596
Weissberg JB, Voytek PE (1974) Liver and red-cell porphobilinogen synthase in the adult and fetal guinea pig. Biochem Biophys Acta 364:304–319

CHAPTER 25

Alkaline Earths, Transition Metals, and Lanthanides

C. H. EVANS

A. Introduction

As a ubiquitous, versatile, and crucial component of biologic systems, calcium is without peer. Its repertoire of functions is remarkable, including its presence in the structural supporting systems of various invertebrate and vertebrate animals, its role as an essential cofactor for many enzymes, and its involvement in modulating cellular physiologic responses. In these functions, Ca^{2+} can be replaced, to a greater or lesser degree, by other metal ions. The deliberate substitution of Ca^{2+} by another metal ion has proved a fruitful experimental approach to the investigation of various aspects of the inorganic biochemistry of calcium. This chapter will review some of the results obtained by this strategy. In view of the likely readership of this volume, the emphasis will be on the cellular pharmacologic aspects of the alkaline earths, transition metals, and lanthanides (Ln^{3+}). However, it should not be forgotten that Ln^{3+}, in particular, have served as highly successful probes of the interactions of Ca^{2+} with a number of enzymes and other proteins (reviewed in MARTIN and RICHARDSON 1979; HORROCKS 1982; EVANS 1983).

Organic Ca^{2+} blockers have an actual or potential use in the treatment of several diseases, including angina, arrhythmia, atherosclerosis, and hypertension. There exist a small number of animal studies suggesting that inorganic blockers, Ln^{3+} in particular, could play a future therapeutic role in certain instances. For reasons which will become apparent later, I would like to suggest that Ln^{3+} are worth investigating as possible anti-inflammatory agents with particular relevance to arthritis.

B. Interactions of Alkaline Earths, Transition Metals, and Lanthanides with Calcium Channels

There are at least four ways in which Ca^{2+} can enter a cell. Two of these are through so-called channels, one voltage-operated (VOC) and the other receptor-operated (ROC). In addition, there is a Na^+/Ca^{2+} exchange which can operate in either direction, and passive leaking of Ca^{2+} into the cell down the concentration gradient. Of these, only the slow VOCs have been studied in detail.

I. Slow, Voltage-Operated Calcium Channels

Slow, voltage-operated Ca^{2+} channels show rather poor metal ion (M^{n+}) selectivity. They transport Ba^{2+} and Sr^{2+} at rates equal to or better than Ca^{2+}, while transport of Cd^{2+}, Mg^{2+}, Mn^{2+}, and Zn^{2+} has been reported in certain cases. When the extracellular concentration of Ca^{2+} is very low, the VOP may permit entry of Li^+, Na^+, K^+, Rb^+, and Cs^+ (ALMERS et al. 1984). In addition, a number of M^{n+}, although not permeant through the slow VOC, bind to the external surface of the channel, thereby blocking the passage of Ca^{2+}. Such blockers include Ln^{3+}, Co^{2+}, and Ni^{2+} (Table 1). Physiologic selectivity for Ca^{2+} is presumably maintained by the high extracellular concentration of Ca^{2+} relative to its serious competitors. Thus, the only M^{n+} present at millimolar concentrations in human plasma are Ca^{2+} (3.2 m*M*), Mg^{2+} (1.1 m*M*), Na^+ (140 m*M*), and K^+ (5.3 m*M*). In seawater, the respective concentrations are Ca^{2+} (10 m*M*), Mg^{2+} (54 m*M*), Na^+ (490 m*M*), and K^+ (9.8 m*M*), (CARAFOLI and CROMPTON 1978). Obviously, inside the cell where the cytosolic concentration of free Ca^{2+} is around 10^{-8}–10^{-7} *M*, nature has had to devise other ways of ensuring Ca^{2+} specificity.

Before examining specificity more closely, it is worth pointing out one semantic issue. Ions, such as La^{3+}, which bind to Ca^{2+} channels and inhibit the influx of Ca^{2+} without being themselves transported, can be safely referred to in all instances as channel blockers. For ions such as Mn^{2+} and Mg^{2+} which are, in several cases, weakly permeant, the nomenclature is a little more complicated. In the absence of high levels of extracellular Ca^{2+}, Mn^{2+} or Mg^{2+} may carry current into the cell, in which case they can not be truly referred to as blockers. However, under these circumstances they will competitively reduce the influx of Ca^{2+} through VOCs, and hence function as blockers of the inward Ca^{2+} current. Thus, the designation as a current "carrier" or "blocker" is context dependent in instances like this.

For barnacle muscle, HAGIWARA and TAKAHASHI (1967) have shown that the rise in Ca^{2+} current as a function of the extracellular concentration of Ca^{2+} can be described by Michaelis–Menten kinetics.

$$I_{Ca} = I_{Ca}^{max}\left(\frac{C}{C+K'_{Ca}}\right)$$

where C = external [Ca^{2+}]; I_{Ca}^{max} = maximum Ca^{2+} current; and K'_{Ca} is a dissociation constant, equal to the concentration of Ca^{2+} at half I_{Ca}^{max}. I_{Ca}^{max} and K'_{Ca} are equivalent to the V_{max} and K_m of standard enzyme kinetics. As HAGIWARA and BYERLY (1981a) have pointed out, to measure specifically the Ca^{2+} current through the calcium channel, it is necessary to eliminate complications such as the stabilizing effect of Ca^{2+} on membranes. One way to do so is to use a high external concentration of Mg^{2+}, although the apparent affinity of Ca^{2+} for the channel is consequently reduced by competition from the large excess of Mg^{2+} ions. Under these conditions, apparent dissociation constants (K'_{Ca}) of 25–40 m*M* (HAGIWARA and TAKAHASHI 1967) and 21 m*M* (BEIRAO and LAKSHMINARAYANAIAH 1979) have been measured for the Ca^{2+} current of barnacle muscle. The dissociation constant for Mg^{2+} can be independently determined. When the necessary correction is made on this basis, the true dissociation constant for Ca^{2+}

Table 1. Some representative values for the affinity of various M^{n+} for the exterior surface of Ca^{2+} channels

M^{n+}	Ionic radius	K_d (mM)		
	(Å)[a]	1[b]	2[c]	3[d]
Alkaline earths				
Ba^{2+}	1.34		15.0	
Sr^{2+}	1.12	25.0	10.0	2.0
Ca^{2+}	0.99	< 15.0	5.4	0.3
Mg^{2+}	0.66	176.0		5.7
Transition elements				
Mn^{2+}	0.80	9.4	0.36	0.05
Co^{2+}	0.72	3.8	0.74	0.06
Ni^{2+}	0.69	14.0	0.74	0.05
Zn^{2+}	0.74	3.8		0.04
Lanthanum				
La^{3+}	1.06	< 3.8		0.0002

[a] Ionic radii (WEAST 1984) vary with coordination number. The given values are indicative only.
[b] HAGIWARA and TAKAHASHI (1967). Values are for barnacle muscle fiber. Value for Ca^{2+} is calculated from the apparent K_d after allowance for the effects of 100 mM Mg^{2+}. Values for the transition metals were measured in the presence of 42 mM Ca^{2+} and 12 mM Mg^{2+}.
[c] KOSTYUK (1981). Values for the alkaline earths are for *Helix* nerve cells. Values for the transition elements were measured in the presence of 4 mM Sr^{2+}.
[d] NACHSHEN (1984). Values are for the fast Ca^{2+} channels of rat brain synaptosomes. These are K_i values derived from Lineweaver–Burk analysis. K_{Ca} is derived from competition experiments.

(K_{Ca}) can be calculated to be <15 mM (HAGIWARA and TAKAHASHI 1967). The dissociation constant for Ca^{2+} has been experimentally determined in a number of different tissues under various conditions, yielding K'_{Ca} values in the millimolar range (Table 1).

In all systems yet tested, slow Ca^{2+} VOCs are also permeant to Ba^{2+} and Sr^{2+}, usually in the sequence $Ba^{2+} > Sr^{2+} \geqq Ca^{2+}$. For a permeant M^{n+}, HAGIWARA and BYERLY (1981 b) have derived the expression

$$\frac{I_M}{I_{Ca}} = \frac{I_M^{max}}{I_{Ca}^{max}} \left(\frac{C + K'_{Ca}}{C + K'_M} \right)$$

The ratio I_M^{max}/I_{Ca}^{max} compares the theoretical current produced by the permeant cation with the Ca^{2+} current at infinite concentration. For all lower concentrations (C) of M^{n+}, it is necessary to take into account the dissociation constants (K'_{Ca}, K'_M) as shown. For example, Ba^{2+} has a higher I^{max} than Ca^{2+}, but also a higher dissociation constant. Thus, it carries a lower current than Ca^{2+} at low concentration, but a higher current at high M^{n+} concentration.

The effects of blocking ions on the inward Ca current can be mathematically described by the equation (HAGIWARA and BYERLY 1981b)

$$I_{Ca} = I_{Ca}^{max} \frac{C}{C + \left(1 + \frac{[M]}{K_M}\right) C'_{Ca}}$$

where [M] is the concentration of the blocker, and K_M is its dissociation constant. It should be pointed out that this only holds true if the blocker is a bona fide competitive inhibitor of the binding of Ca^{2+} to the channel and its transport through it. (In this sense, the K_M term is equivalent to the K_i in enzyme kinetics.) Competitive kinetics have indeed been demonstrated by a Lineweaver–Burk plot of the inhibition of inward Ca^{2+} currents by Mn^{2+} in, for example, rat ventricular trabecular cells (PAYET et al. 1980). In this system, Mn^{2+} cannot transport current through the slow VOC, but, as shown in Fig. 1, it displays competitive inhibition kinetics in suppressing Ca^{2+} influx. Of interest is the observation that inhibition of Ca^{2+} transport by verapamil and D600 in these cells is noncompetitive. Further evidence of the accuracy of this equation is given by the good correlation between observed and theoretical inhibitions of Ca currents in barnacle muscle for a number of M^{n+} (Fig. 2).

One complication of using whole cells to investigate Ca channels is the fact that M^{n+} bind to more than one class of sites on cell surfaces, reduce the surface charge, and act as membrane stabilizers. To study the properties of channels in partial isolation from these complications, rat brain VOC have been incorporated into planar lipid bilayers (NELSON et al. 1984). In agreement with studies using intact cells, the order of ionic conductance was $Ba^{2+} > Sr^{2+} = Ca^{2+}$, with Ba^{2+} transport being 1.7 times that of Ca^{2+} or Sr^{2+}. Channel blockers also functioned in this system, in the order $La^{3+} > Cd^{2+} \gg Mn^{2+}$. Interestingly, the channel open time was shorter for Ba^{2+} (0.4 μs) than Ca^{2+} or Sr^{2+} (0.7 μs).

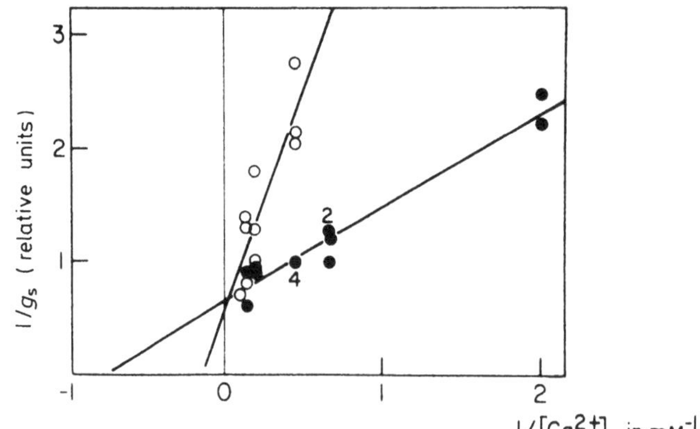

Fig. 1. Lineweaver–Burk analysis of inhibition by 1 mM Mn^{2+}. Control *full circles;* 1 mM Mn^{2+} *open circles*. Rat ventricular muscle. Ordinate: reciprocal of slow inward conductance. (PAYET et al. 1980)

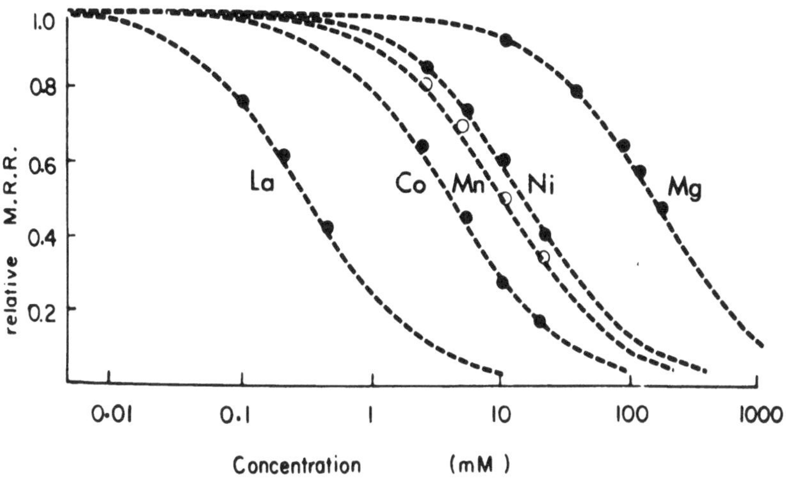

Fig. 2. Blocking of Ca currents by various polyvalent cations. Barnacle muscle. *Broken curves* show the theoretical inhibitions predicted by the equation described in the text; *circles* show experimental data points. *Ordinate:* relative maximum rate of rise of the action potential. (HAGIWARA and BYERLY 1981 a)

Binding of Ca^{2+} or competing M^{n+} to Ca^{2+} VOCs is most likely to occur through ligands with 0 donor atoms, as Ca^{2+} and Ln^{3+} have poor affinity for N or S donors. It has proved possible to titrate the putative ligands on the external surface of certain channels, and thereby calculate pK values of around 6.0–6.5 (SPITZER 1979; NACHSHEN and BLAUSTEIN 1979). These are consistent with the possibility that the bond site contains one or more carboxyl groups. All the M^{n+} under discussion bind readily to such groups (Table 1), and their affinities for these ligands are in roughly the same order as their affinities for the Ca^{2+} channel. From the data in Table 1, it is also apparent that the permeability of M^{n+} through the slow Ca^{2+} VOC is strongly determined by its K_d, the most permeant M^{n+} tending to have the highest K_d values. The exception is Mg^{2+}, which is poorly transported and has low affinity. Transport is thus generally a compromise between having an affinity for the Ca^{2+} channel sufficient to permit recognition and binding, yet a sufficiently high dissociation constant to permit subsequent release. Thus, the most permeant M^{n+} are paradoxically the poorest competitors for the channel. Hence, HAGIWARA and BYERLY (1981 a, b) have shown that, whereas the normal relative rates of transport of M^{n+} by barnacle muscle fibers is $Ba^{2+} > Sr^{2+} > Ca^{2+}$, this order is reversed in the presence of the competing channel blocker, Co^{2+}.

Although Ba^{2+}, Sr^{2+}, and Ca^{2+} are invariably transported through slow VOC and La^{3+}, Co^{2+}, and Ni^{+} are usually considered to be impermeant blockers, there are several M^{n+} whose status depends on the type of cell, the species of animal, and the experimental conditions. Thus, Zn^{2+} is transported through the VOCs in the larval muscle fibers of the beetle *Xylotrupes dichotomus* (FUKUDA and KAWA 1977) and in snail neuron (KAWA 1979), but not in barnacle muscle fibers (HAGIWARA and BYERLY 1981 a) nor *Aplysia juliana* (ADAMS and

GAGE 1979). Similar examples can be found for Mg^{2+} and Mn^{2+}. Such discrepancies are consistent with suggestions of considerable heterogeneity among slow Ca^{2+} VOCs. Further evidence of this comes from differences in the order of relative effectiveness of different M^{n+} as channel blockers. Thus, in *Helix* neurons, the Ca^{2+} current is blocked by M^{n+} in the order $Ni^{2+} \gg La^{3+} \gg Cd^{2+} > Co^{2+} \gg Mg^{2+}$ (AKAIKE et al. 1978), whereas for smooth muscle it is $La^{3+} > Ni^{2+} > Co^{2+} > Mg^{2+}$ (VAN BREEMAN et al. 1973). In these respects, the slow Ca channels are much more variable than the Na channel.

II. Other Types of Calcium Transport Mechanisms

Calcium entry into nerve terminals occurs through separate "fast phase" and "slow phase" processes (GRIPENBERG et al. 1980; NACHSHEN and BLAUSTEIN 1980). The ion selectivity of voltage-dependent influx of ^{45}Ca through *fast* Ca^{2+} channels in presynaptic nerve terminals isolated from rat brain has been subjected to close examination by NACHSHEN (1984). These channels are associated with voltage- and Ca^{2+}-dependent dopamine release from striated synaptosomes. Using high K^+ to open VOC, NACHSHEN (1984) measured inhibition of ^{45}Ca influx by a range of M^{n+}, whose half-inhibition constants (K_I) fell into three groups. The weakest inhibitors, Mg^{2+}, Sr^{2+}, and Ba^{2+}, had K_I valucs in excess of 1 m*M* in the presence of 20 m*M* Ca^{2+}. Most of the transition metals, including Mn^{2+}, Co^{2+}, Ni^{2+}, Cu^{2+}, and Zn^{2+} had intermediate K_I values of 30–100 µ*M*, while Ln^{3+} were very strong blockers with K_I values <1 µ*M*. These channels are permeable to Sr^{2+}, Ba^{2+}, and Mn^{2+} (NACHSHEN and BLAUSTEIN 1982; DRAPEAU and NACHSHEN 1984), but not Na^+ (KRUEGER and NACHSHEN 1980). Lineweaver–Burk plots were performed to confirm that inhibition was competitive with regard to Ca^{2+} and to derive K_i values. In most cases the I_{50} values and K_i values corresponded well, confirming that competitive inhibition by these M^{n+} indeed accounts for the inhibition in Ca^{2+} flux.

The potency of the Ln^{3+} as channel blockers increases with ionic radius. Based on this and other evidence, NACHSHEN (1984) suggested a model for the Ca-binding site in synaptosome channels with two anionic sites, possibly carboxyl groups, separated by a distance which permits simultaneous binding of the larger anions Ca^{2+}, Ba^{2+}, and Ln^{3+} to both ligands. Small M^{n+}, such as Mg^{2+}, can not span the gap and thus bind only to one anionic group at a time. An attraction of this model is that small changes in the spacing of the two anionic ligands will alter the theoretical relative affinities of the binding site for different M^{n+}. Such modifications could underly the differences in relative Ca^{2+}-blocking activities, noted already, for different species and different types of cell.

Far less is known of the cation selectivity of ROC for Ca^{2+}. Several of the VOC blockers, such as Ln^{3+}, Co^{2+}, and Ni^{2+} appear to block at least some of the ROC. Most of the evidence is indirect, coming from inhibition studies on Ca^{2+}-requiring, stimulus-coupled cellular responses. Examples include the inhibition of receptor-activated Rb^+ efflux in rat parotid gland by La^{3+}, Co^{2+}, and Ni^+ (MARIER et al. 1978) and the extensive inhibitory properties of La^{3+} (see Table 2). Recent data on the inhibition of prolactin secretion by Co^{2+} (THAW et al. 1984; see Sect. C) caution careful interpretation of such data. It has been sug-

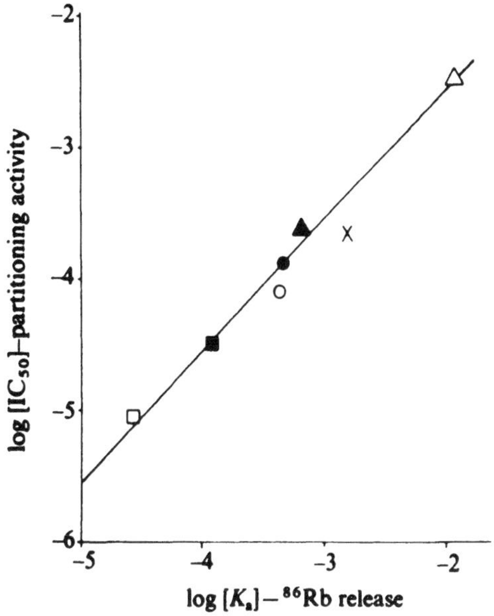

Fig. 3. Relationship between ability of various M^{n+} and neomycin to inhibit Ca^{2+}-dependent ^{86}Rb release and to inhibit ^{45}Ca partitioning by phosphatidic acid. La^{3+} *open squares;* Tm^{3+} *full squares;* neomycin *open circles;* Co^{2+} *full circles;* Ni^{2+} *full triangles;* Mg^{2+} *open triangles;* Ca^{2+} *crosses.* (PUTNEY et al. 1980)

gested that phosphatidic acid actis as an ionophore in the transport of Ca^{2+} into the cell following the occupation of a receptor by an agonist (PUTNEY et al. 1980). This links Ca^{2+} influx to increased phospholipid metabolism in activated cells and suggests that it may not occur through an actual "channel" in the strict sense of the word. In support of the phosphatidic acid hypothesis, there is a good correlation between the ability of various M^{n+} to inhibit Ca^{2+} influx through ROC and to inhibit in vitro Ca^{2+} transport into $CHCl_3$, by phosphatidic acid (Fig. 3).

Ca^{2+} uptake into isolated, cardiac sarcolemmal vesicles through Na^+/Ca^{2+} exchange is inhibited by M^{n+} in the order $Cd^{2+} > La^{3+} > Y^{3+} > Mn^{2+} > Co^{2+}$ (BERS et al. 1980). Mg^{2+} does not inhibit this process. Thus, all of the best-studied routes through which Ca^{2+} enters cells are inhibited by M^{n+}. The physiologic effects of this blockade will be considered next.

C. Cellular Physiologic Effects of Inorganic Blockers of Calcium Channels

As expected, impermeant, inorganic channel blockers suppress those cellular responses upon which Ca^{2+} influx depends. The earliest recorded demonstration of this is that of MINES (1910) who showed that La^{3+} and Ce^{3+} reversibly inhibited contraction of the frog's heart. Ln^{3+} also inhibit contraction of skeletal

(ANDERSSON and EDMAN 1974) and smooth (WEISS and GOODMAN 1969) muscle, with cardiac and smooth muscle being a little more sensitive to their effects than skeletal muscle. A differential response such as this may reflect differences in the properties of the channels, or in the source of the additional cytosolic Ca^{2+}.

It is now known that Ca^{2+} can be released from intracellular stores as an alternative or supplement to Ca^{2+} entering the cell from the extracellular milieu (BERRIDGE 1983). As agents which block Ca^{2+} channels should not affect intracellular release of Ca^{2+} in the short term, they have found use as discriminators between these two sources of cytosolic Ca^{2+}. BERS (1985), for instance, has used La^{3+} and Co^{2+} in this way to investigate the relative contributions of Ca^{2+} from extracellular and intracellular sources during the generation of twitch tension in isolated rabbit ventricles. The first post-rest beat is much less susceptible to inhibition by Co^{2+} or La^{3+} than is the second beat, suggesting that the contribution of intracellular reservoirs to increases in cytosolic Ca^{2+} is greater in the first beat. An advantage of inorganic blockers in experiments of this kind is that, unlike organic blockers, their effectiveness is not beat dependent. Experiments of this type have revealed interesting differences in the Ca^{2+} source upon which the cell draws for different purposes. For example, in polymorphonuclear leukocytes, La^{3+} inhibits chemotaxis (O'FLAHERTY et al. 1978; BOUCEK and SNYDERMAN 1976), but has a smaller effect on degranulation (SHOWELL et al. 1977).

If added early enough, inorganic Ca^{2+} blockers also prevent many of the destructive cellular changes that characterize the "calcium paradox." These occur when Ca^{2+} is readministered to adult mammalian hearts following a period of perfusion in the absence of Ca^{2+}, and are associated with a rapid, uncontrolled influx of Ca^{2+}. Both Co^{2+} and Mn^{2+} protect against these changes in a dose-dependent manner, if present during the entire period of Ca^{2+} depletion and repletion (NAYLOR et al. 1983). However, if added at the time of Ca^{2+} repletion, they are ineffective, suggesting that Ca^{2+} does not enter through normal Ca^{2+} channels under these conditions.

In several studies, the inhibition of physiologic function has been linked to blocking of specific Ca^{2+} channels. For instance, there exists a correlation between the ability of various M^{n+} to inhibit voltage-dependent transmitter release at the neuromuscular junction (GINSBORG and JENKINSON 1976), and their ability to block Ca^{2+} VOCs (NACHSHEN 1984). A note of caution in the interpretation of results obtained from certain types of inhibitor studies has been introduced by recent results of THAW et al. (1984). In rat pituitary cells, Co^{2+} inhibits Ca^{2+}-dependent action potentials whose frequency of firing is stimulated by thyrotropin-releasing hormone (TRH) and inhibits prolactin secretion in response to TRH (LORENSON et al. 1983). However, THAW et al. (1984) have shown that Co^{2+} ions can enter the cytoplasm of such cells and act as intracellular antagonists of Ca^{2+}. In a similar vein, the glucose-induced release of insulin by pancreatic islet cells, although inhibited by Co^{2+}, Mn^{2+}, and Ni^{2+} (HENQUIN and LAMBERT 1975), is inhibited at low concentrations (0.25, 0.5 m*M*) of Co^{2+} which do not affect the rate of glucose-stimulated $^{45}Ca^{2+}$ uptake (WOLLHEIM and JANJIC 1984). In this regard, it is interesting to note that low concentrations of Co^{2+} appear to stimulate the production of collagenase by synoviocytes (OKAZAKI et al. 1981; FERGUSON and EVANS 1987).

Table 2. Examples of the inhibition of stimulus-coupled cellular responses by lanthanum[a]

Type of cells	Stimulus	Inhibited response	Reference
Mast cells	Antigen	Histamine release	FOREMAN and MONGAR (1972)
Polymorphonuclear leukocytes	Chemotactic factors	Chemotaxis	BOUCEK and SNYDERMAN (1976)
Adrenal glomerulosa	Angiotensin, K^+, ACTH	Aldosterone production	SCHIFFRIN et al. (1981)
Pancreatic acinar cells	Various secretagogues	Sustained release of trypsin	SCHULZ et al. (1981)
Pancreatic acinar cells	Cholecystokinin octapeptide	Incorporation of [^{3}H]phenylalanine into protein	KORC (1983)
Neuron	K^+	Neurotransmitter release	MILEDI (1971)
Neuron	Acetylcholine	Activation	COLTON (1976)
Cockroach fat body	Trehalagon	Activation of glycogen phosphorylase	McCLURE and STEELE (1981)
Mouse skin	12-O-Tetradecanoylphorbol-13-acetate	Induction of ornithine decarboxylase	VERMA and BOUTWELL (1981)
Granulosa cells	Various agonists	Induction of ornithine decarboxylase	VELDHUIS 1982)
Salivary glands	Serotonin	Fluid secretion	BERRIDGE et al. (1975)

[a] This table includes examples of both VOC and ROC Ca^{2+} channel function.

La^{3+} is probably a more reliable, impermeant, inorganic Ca^{2+} channel blocker than Co^{2+}, as there is a considerable literature attesting to its inability to penetrate the cellular membranes of living, intact cells. Examples of the inhibition of stimulus-coupled cellular events by La^{3+} are given in Table 2.

Several psychoactive drugs, including opiates, alcohol, barbiturates, and cannabinols, appear to act as Ca^{2+} antagonists in the central nervous system (HARRIS 1981). La^{3+}, injected into the brains of laboratory rodents, shares a number of properties with the opiates. These include the production of analgesia and the suppression of morphine withdrawal signs. In addition, animals tolerant to morphine are also tolerant to La^{3+}.

The effects of the freely permeant M^{n+}, Sr^{2+} and Ba^{2+}, often mimic those of Ca^{2+} in short-term experiments. For instance, Sr^{2+} can substitute for Ca^{2+} in the release of histamine from mast cells (FOREMAN and MONGAR 1972) and in stimulus–secretion coupling in the adrenal medulla; indeed, Ba^{2+} causes secretion of catecholamines by itself (CORCORAN and KIRSHNER 1983; DOUGLAS and RUBIN 1964a, b). As long ago as 1911, MINES showed that Sr^{2+} or Ba^{2+}, but not Mg^{2+}, could substitute for Ca^{2+} in the contraction of the frog gastrocnemius muscle.

Although Sr^{2+} can substitute for Ca^{2+} in transmitter release at the neuromuscular junction (MILEDI 1966), it is less efficient than Ca^{2+} (MEIRI and RAHAMIMOFF 1971). Ba^{2+} is actually better than Ca^{2+} in supporting the evoked discharge of miniature end-plate potentials (SILINSKY 1978). However, Ba^{2+}-supported discharge is physiologically useless as the effects of the Ba^{2+} signal are too long-lived, thus destroying the "impulsive" nature of the release. This may occur because Ba^{2+}, unlike Sr^{2+} and Ca^{2+}, is not taken up by mitochondria (LEHNINGER 1970). Ba^{2+} also has cardiac dysrhythmogenic properties (SPEDDING and CAVERO 1984). TRIGGLE et al. (1975) have transported various M^{n+} into guinea pig ileal muscle cells with A23187. Ca^{2+}, Mn^{2+}, Sr^{2}, and Ba^{2+}, but not Co^{2+}, Ni^{2+}, or Tm^{3+}, produced a contractive response.

D. Metal Ions as Drugs

Several metals form the basis for drugs in present use (SADLER 1982). The oldest is gold, which has been used therapeutically since ancient times, and now forms the basis for a variety of antiarthritic drugs. Platinum, in the form of cisplatin, is used in the treatment of certain cancers, while manic depression has been treated with lithium salts for a number of years. Of relevance to the present discussion is the possibility that Li^{+} acts by inhibiting inositol-1-phosphatase (HALLCHER and SHERMAN 1980), and thus influences the intracellular release of Ca^{2+}.

Zn^{2+}, as an inhibitor of histamine release from mast cells, has been discussed as a topical treatment for allergic conditions (KASIMIERCZAK and DIAMANT 1978). In a double-blind trial, SIMKIN (1976) demonstrated the beneficial effects of oral $ZnSO_4$ in rheumatoid arthritis, although its effectiveness has been challenged (MATTINGLY and MOWATT 1982). Zinc may be of clinical use in wound healing (CLAYTON 1970).

Apart from the treatment of deficiency diseases, none of the metals under discussion find clinical therapeutic application. However, copper complexes have remarkable anticonvulsant, antiarthritic, anti-inflammatory, and antidiabetic activities in experimental animals (SORENSON 1984). I would like to draw attention to potential anti-inflammatory and antiarthritic properties of the lanthanides. Before doing so, it is first necessary to consider briefly their wider metabolism.

E. Metabolism of Lanthanides by Whole Animals

Interest in the toxicology of the lanthanides was born along with the atomic age, with the rare earth elements being generated as a by-product of the fission of uranium. Most of this work is now over 20 years old. Interest in this subject has recently returned through the possible use of lanthanides in bone scanning and as contrast reagents in NMR imaging. However, the modern literature remains scant. HALEY (1965) has comprehensively reviewed the older literature.

As shown in Table 3, the acute toxicity of aqueous solutions of simple, inorganic lanthanide salts depends critically upon their mode of administration. Orally administered lanthanides are very well tolerated, with LD_{50} values of several grams per kilogram body weight. This is due to poor absorption of Ln^{3+} from the alimentary canal. HAMILTON (1949), for instance, reported that only 0.3% of orally administered $^{140}La\ Cl_3$ was retained by rats 4 days later. Oral in-

Table 3. Acute lethal toxicity of selected lanthanide salts (HALEY 1965)

Lanthanide	Route of administration	LD_{50} (mg/kg)	Animal
LaO_2CCH_3	p.o.	10000	Rat
$LaCl_3$	s.c.	3500[a]	Mouse
	i.p.	372	Mouse
	i.v.	200–250[a]	Rabbit
$SmCl_3$	p.o	> 2000	Mouse
	s.c.	750–1000	Guinea pig
	i.p	585	Mouse
$Sm(NO_3)_3$	i.v.	59.1	Rat
$ErCl_3$	p.o.	6200	Mouse
	i.p.	535	Mouse
$Er(NO_3)_3$	i.v.	52.4	Rat

[a] Minimum lethal dose.
p.o. oral; s.c. subcutaneous; i.p. intraperitoneal; i.v. intravenous.

gestion of $LuCl_3$, $SmCl_3$, or $GdCl_3$ has no effect on the growth of rats nor upon various haematologic parameters (HALEY et al. 1961, 1964).

Lanthanide salt solutions are a little more toxic when given to experimental animals by subcutaneous injection. This probably reflects their greater rates of absorption. However, these rates are still low. For example, 6 days after subcutaneous injection into mice of an aqueous solution of $^{140}La\ Cl_3$, 54% of the radioactivity was recoverable from the injection site, with only 0.5% present in the liver (LASZLO et al. 1952). Excretion occurs very slowly by the intestinal route. Direct contact of the rare earths with the skin is benign. However, where the skin is abraded, lanthanides produce ulceration, epilation, and eventually scar formation. Granulomas result from intradermal injection of rare earths (HALEY and UPHAM 1963).

Clearance of $^{169}Yb^{3+}$ following intra-articular injection into rabbits' knees was shown to be biphasic (MCCARTY et al. 1979). About 50% of the radioactivity was eliminated from the knee in 10 days. After 28 days, by which time around one-third of the original radioactivity remained in the knee, most of the Yb^{3+} which had not been excreted remained intra-articular, with little accumulation in liver, kidney, or bone marrow.

Intraperitoneally injected $^{140}La\ Cl_1$ coats the organs of the abdominal cavity. Autoradiography of tissue sections of mice killed 24 h after i.p. administration of this salt confirms that the radioactive material deposits upon, rather than within, these organs. Excretion is again slow, with 70% of the radioactivity remaining in the abdominal cavity 1 week after i.p. injection (LASZLO et al. 1952).

Thus, in each of the examples discussed so far, the lanthanides have tended to stay close to the site of injection. Absorption and excretion are slow, and toxicity is consequently low to moderate. It is only after intravenous injection that solutions of simple inorganic lanthanide salts travel quickly to other organs and show moderately high toxicity. A number of studies using a variety of different Ln^{3+} agree that the liver, spleen and, to a lesser degree, the kidney, sequester the bulk of the injected metal. Values for uptake by bone tend to vary, but it is worth noting that once bone has sequestered Ln^{3+}, it releases it again only slowly. This stability probably results from incorporation of Ln^{3+} into the bone mineral as a Ca^{2+} substitute at sites of new bone synthesis. Indeed, Ln^{3+} are normally included in the category of "bone-seeking elements."

About half the radioactivity from an i.v. injection of $^{140}La^{3+}$ in mice is removed from the circulation in 1 h (LASZLO et al. 1952). None is detectable after 20 h. At least part of this transport is mediated by albumin and unidentified, low molecular weight carriers (EKMAN et al. 1961). Nearly 75% of the radioactivity accumulates in the liver after 20 h, with slow elimination through the bile. This has been estimated to have a half-life of 15 days in rats. Total excretion is also slow, with only 8% being cleared 96 h after the injection.

The organ distribution does not vary over a 25-fold range of dose injected, except for the lung, whose sequestration of $^{140}La^{3+}$ increased from 0.75% to 8.6% as the dose increased from 0.45 to 0.85 mg. It has been suggested that, above a threshold concentration, $^{140}La^{3+}$ may be associated with serum components in such a way as to form aggregates whose size exceeds a certain critical value. For all organs other than the lung, the uptake was proportional to the size of the injected dose (LASZLO et al. 1952).

The most pronounced pathologic effect of i.v. injected rare earth salt solutions is the production of a fatty liver. This only occurs with the series members La^{3+} to Sm^{3+}. It is seen in rats, mice, and hamsters, occasionally in rabbits, and never in guinea pigs, chickens, or dogs (SNYDER et al. 1960). Although its mechanism is unknown, it may be related to the ability of Ce^{3+} to increase plasma concentrations of free fatty acids (SNYDER and STEPHENS 1961). The fatty liver is a self-limiting condition, which spontaneously recovers, provided there is no further challenge with lanthanides. In view of the hepatotoxicity of Ln^{3+}, it is interesting to note that Ce^{3+} protects rats against liver damage by CCl_4 (SANNA et al. 1976). This is thought to occur by inhibition of the conversion of CCl_4 to toxic derivatives by liver microsomes.

All the rare earths lower the blood pressure of animals after i.v. injection (HALEY 1965), possibly through a Ca^{2+}-blocking mechanism. There is a concomitant change in hemoglobin content and leukocyte, erythrocyte, and differential counts. Hypoglycemia has also been reported following i.v. injection of Ce^{3+}, La^{3+}, Nd^{3+}, or Pr^{3+}. Symptoms of acute lanthanide toxicity have been described as writhing, ataxia, labored respiration, sedation, and walking on the toes with arched back (HALEY 1965). Death usually occurs some 48–96 h after injection of a lethal dose, by cardiovascular collapse and respiratory paralysis. Atropine or epinephrine are without effect on this. A number of neurophysiologic parameters change in this time. Animals which survive to experience chronic toxicity show a generalized peritonitis and adhesions with hemorrhagic ascitic fluid. The latter may result from the known anticoagulant properties of Ln^{3+} ions. Also present are a granulomatous peritonitis and hepatic necrosis.

For obvious reasons, there are few studies of lanthanide metabolism in humans. In one instance, $^{140}LaCl_3$ was injected into the right femoral artery of a patient prior to amputation of this leg for an osteogenic sarcoma (LASZLO et al. 1952). In agreement with other animal studies, most of the radioactivity accumulated in the liver, and there was little excretion over the 31-h period of the experiment. Autopsy analysis showed ^{140}La in bone, tumor, and soft tissue. In bone, autoradiography showed preferential uptake in the epiphyseal line, as shown in animals.

Administration of complexes which contain Ln^{3+} gives a more even tissue distribution, lower toxicity, and greater rate of excretion. In mice, the proportion of ^{140}La accumulating in the liver is much lower following i.v. administration of ^{140}La–EDTA than with $^{140}LaCl_3$ (LASZLO et al. 1952). There is a corresponding increase in radioactivity associated with muscle, and a far higher rate of excretion, 33% being excreted within 24 h. Similar results are seen with i.p. injected ^{140}La–EDTA, with over 20% excreted within 48 h. Gd–DTPA (diethylenetriaminepentaacetic acid) has been tested as a possible contrast agent in NMR imaging (WEINMANN et al. 1984). When given i.v., its molar toxicity is 50 times less than that of $GdCl_3$ in rats. Furthermore, it is rapidly and completely excreted. The half-life in the blood is 20 min. By 3 h, over 80% has been excreted in the urine. This contrasts with only 2% excretion of $GdCl_3$ 7 days after injection.

F. Possible Therapeutic Uses of the Lanthanides

From the foregoing discussion, it seems likely that any specific therapeutic role for nonradioactive Ln^{3+} would involve their ability to modulate Ca^{2+}-dependent metabolic functions. Prominent among these are the suppression of certain types of stimulus-coupled cell activation processes and alteration of the kinetics of Ca^{2+}-dependent enzymes. However, certain difficulties arise. Simple lanthanide salts are too poorly absorbed from the gastrointestinal tract to be administered orally, yet are too toxic to be given by i.v. injection. Furthermore, when given parenterally outside the circulation, Ln^{3+} remain close to the site of injection. Specificity is also of concern, as Ln^{3+} appear to interact indiscriminately with Ca^{2+}-dependent biochemical processes. Certain complexes which contain rare earths have reduced toxicity and, as we shall see, have some potential as anti-inflammatory agents. Another approach is to introduce the Ln^{3+} directly into the proposed site of action. Some parts of the body obviously lend themselves to this approach better than others, but if it can be done there are certain advantages. Several studies have confirmed that Ln^{3+} administered in this way tend to remain close to the site of injection, thus concentrating the agent where it is needed. And as Ln^{3+} do not accumulate in the liver under these conditions, toxicity is very low. Possible complications include the tendency of Ln^{3+} to form insoluble hydroxides at physiologic pH values and to form insoluble phosphates. In addition, the target ligands (e.g., surfaces of specific cells; certain enzymes) will have to compete for the Ln^{3+} with other ligands in the microenvironment around the site of injection.

As long ago as 1920, rare earth salts were injected into humans as possible therapeutic agents for tuberculosis (GRENET and DROUIN 1920). A decade or so later, their anticoagulant properties were being investigated in humans (DYCKERHOFF and GOOSSENS 1939). Although their anticoagulant mechanism is unclear, it may well be related to the ability of Ln^{3+} to inhibit the activation of factor X (FURIE and FURIE 1975) and to lower the rate of conversion of prothrombin to thrombin (FURIE et al. 1976). While they were found to be potent anticoagulants in dogs, rabbits, and humans, severe side effects proscribed their clinical use (BEASER et al. 1942).

On the basis that agents which reduce excessive calcium deposition in the arteries retard the progress of atherosclerotic changes in experimental animals, KRAMSCH et al. (1980) attempted to suppress experimental atherosclerosis in rabbits with $LaCl_3$. Oral supplementation of aqueous $LaCl_3$ (20, 30, or 40 mg/kg), while having no effect on serum levels of cholesterol or Ca^{2+}, prevented the rise in arterial Ca^{2+} and other related atherosclerotic changes, including reduced cholesterol associated with the aortic intima media. The rabbits remained healthy during the 8-week experiment. GINSBURG et al. (1983) confirmed the suppression of atherogenesis in rabbit aorta, but found that La^{3+} had no effect on this process in the intramural coronary artery. An effect on the aortic wall is most surprising in view of the poor oral absorption of lanthanides, unless the La^{3+} formed carrier complexes with certain substances in the animal's food. In support of the suggestion that the Ca^{2+}-blocking properties of La^{3+} are responsible for its antiatherosclerotic properties is the observation that the organic blockers verapamil (ROULEAU et al. 1982) and nifedipine (HENRY and BENTLEY 1981) have the same effect.

Our own interest is in the biochemistry of arthritis, particularly the erosion of articular cartilage that occurs in arthritic joints. At least part of the degradation of cartilage is probably mediated by a small number of neutral proteinases, including collagenase, which require Ca^{2+}. These are inhibited by Ln^{3+}, with the inhibition of collagenase by La^{3+} being uncompetitive (Fig. 4), although the molecular details of the mode of inhibition are as yet unknown (EVANS and RIDELLA 1985). A suppressor of these enzymic activities in vivo should theoretically have antierosive properties (EVANS 1979; BROWN 1981). In addition, these neutral metalloproteinases are released from cells in and around the arthritic joint as a result of stimulus-coupled cellular activation. Synoviocytes are important resident cells in this regard, while in inflammation, these are joined by polymorphonuclear leukocytes and lymphocytes. In vitro studies have shown that La^{3+} inhibits synovial cell activation (M. E. BARATZ and C. H. EVANS 1986, unpublished work), lymphocyte activation (Fig. 5; M. YAMAGE and C. H. EVANS 1986, unpublished work), and aggregation of neutrophils (O'FLAHERTY et al. 1978). We have thus been attracted to the idea that Ln^{3+} may have anti-inflammatory properties of relevance to arthritis.

The joint lends itself to putative therapeutic intervention of this type. It is a straightforward matter to introduce a lanthanide into the joint by intra-articular injection. Secondly, as MCCARTY'S (1979) studies have shown (Sect. E), intra-articularly injected Ln^{3+} are retained within the synovium, cartilage, and other tissues of the joint, with little transport to the liver. Furthermore, there are sugges-

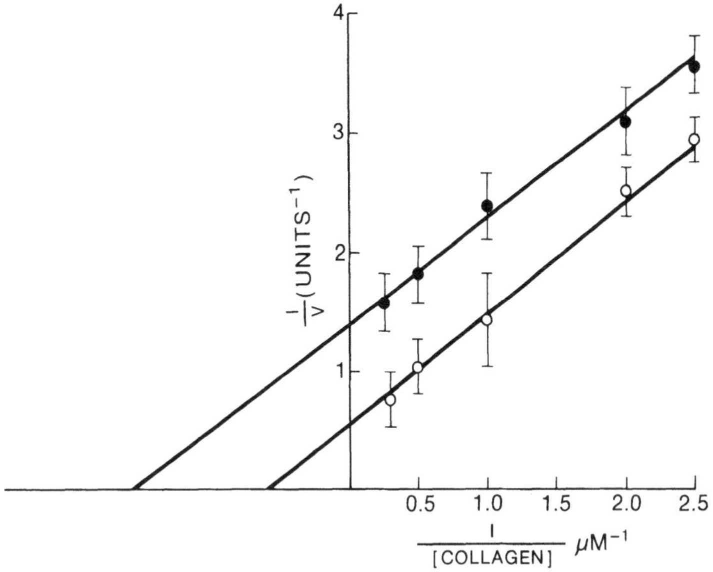

Fig. 4. Lineweaver–Burk analysis of the inhibition of collagenase by La^{3+}. Control *open circles;* 150 μ*M* La^{3+} *full circles.* (EVANS and RIDELLA 1985)

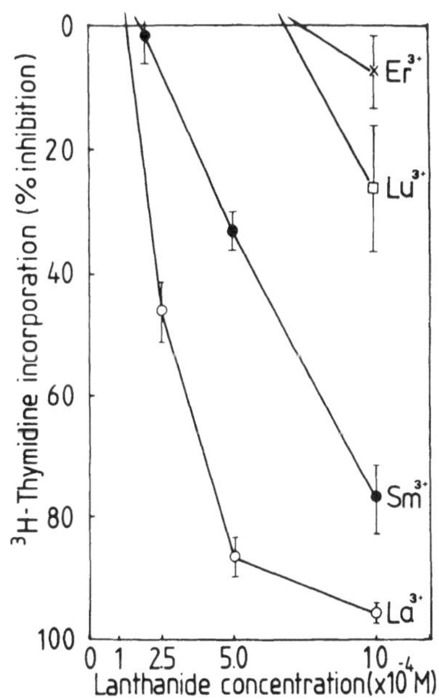

Fig. 5. Inhibition of lymphocyte activation by Ln^{3+}. Lymphocytes were isolated from venous blood of human volunteers and stimulated with antigen in the presence or absence of lanthanides as indicated. (M. YAMAGE and C. H. EVANS 1986, unpublished work)

tions of a correlation between the uptake of $^{169}Y^{3+}$ into articular cartilage and the severity of osteoarthritic, cartilaginous lesions (LIPIELLO et al. 1984). From what is known of the toxicology of the rare earths (Sect. E), intra-articular injections of Ln^{3+} ions would be expected to be of low toxicity.

A literature search revealed experimental evidence of anti-inflammatory properties of the rare earths. JANCSO (1962), examining the hypothesis that inflammation was closely linked to blood clotting, used inorganic salts of La^{3+}, Ce^{3+}, Nd^{3+}, Pr^{3+}, and Sm^{3+} as experimental anticoagulants. Anti-inflammatory effects were indeed observed which, I would like to suggest, were more likely to have resulted from direct suppression of certain types of inflammatory cell in the manner alluded to already. Heparin's lack of effect in JANCSO's experimental system supports this conclusion.

ELLIS (1977) drew attention to results in the Hungarian literature demonstrating that various complexes which contain Ln^{3+}, such as Nd^{3+} pyrocatechol disulfonate, have anti-inflammatory properties suitable for topical use. Examples include mouthwashes for gingivitis, and ointments for treating eczema and skin lesions resulting from radiation. ELLIS (1977) refers to the anti-inflammatory properties of Nd^{3+}, and its complexes, in carrageenan-induced paw inflammation in rats. In a similar manner, i.p. administration of $PrCl_3$, $GdCl_3$, and $YbCl_3$ (15–75 mg/kg) reduced carrageenan-induced inflammation in rats, as well as the nystatin-induced edema and vascular permeability response to histamine and serotonin. Repeated doses of these salts also inhibited arthritis development and granuloma formation (BASILE et al. 1984).

$GdCl_3$ (10 mg/kg) injected i.v. into mice 24 or 48 h prior to challenge greatly reduced the symptoms of anaphylaxis, including anaphylactic death (LAZAR et al. 1982). The authors explained this as depression of macrophage function by $GdCl_3$, although suppression of mast cells or other activatable cell types could have been responsible.

G. Summary

Transport of Ca^{2+} from the extracellular milieu through the plasmalemma is competitively inhibited by other alkaline earths, a number of transition elements, and the lanthanides. In all cells yet studied, the lanthanides appear to be impermeant, while the larger alkaline earths, Sr^{2+} and Ba^{2+} are permeant. Mg^{2+} and most transition elements are weakly permeant in some types of cell, but impermeant in others.

Inhibition of Ca^{2+} transport into the cell by these inorganic blockers suppresses stimulus-coupled cellular responses which depend upon an influx of extracellular Ca^{2+}. Those responses which can occur through Ca^{2+} release from intracellular stores are not strongly inhibited in the short term. Sr^{2+} and Ba^{2+} can substitute for Ca^{2+} in a variety of cellular stimulus-coupled responses in vitro, but Ba^{2+} in particular is unable to maintain a sustained, physiologically useful response. Selective blocking of specific Ca^{2+} channels is therapeutically useful. From what is known of the biochemical, toxicologic, and physiologic properties of the lanthanides, they seem to merit consideration as forming the basis for future anti-inflammatory and antiarthritic agents.

Acknowledgments. I would like to thank Mrs. Diana Montgomery for considerable assistance in preparing this manuscript.

References

Adams DJ, Gage PW (1979) Ionic currents in response to membrane depolarization in Aplysia neurone. J Physiol (Lond) 289:115–141

Akaike N, Lee KS, Brown AM (1978) The calcium content of helix neuron. J Gen Physiol 71:509–531

Almers W, McCleskey EW, Palade PT (1984) A non-selective cation conductance in frog muscle membrane blocked by micromolar external calcium ions. J Physiol (Lond) 353:565–583

Andersson KE, Edman KAP (1974) Effects of La on potassium contractures of isolated twitch muscle fibres of the frog. Acta Physiol Scand 90:124–131

Basile AC, Hanada S, Sertie JAA, Oga S (1984) Anti-inflammatory effects of praseodymium, gadolinium and ytterbium chlorides. J Pharmacobiodyn 7:94–100

Beaser SB, Segel A, Vandam L (1942) The anticoagulant effects in rabbits and man of the intravenous injection of salts of the rare earths. J Clin Invest 21:447–454

Beirao PS, Lakshminarayanaiah N (1979) Calcium carrying system in the giant muscle fibre of the barnacle species, *Balanus nubilis*. J Physiol (Lond) 293:319–327

Berridge MJ (1983) Calcium-mobilizing receptors. Membrane phosphoinositides and signal transduction. In: Rubin RP, Weiss GB, Putney JW (eds) Calcium in biological systems. Plenum, New York, pp 37–44

Berridge MJ, Oschman JL, Wall BJ (1975) Intracellular calcium reservoirs in *Calliphora* salivary glands. In: Carafoli E (ed) Calcium transport in contraction and secretion. North Holland, Amsterdam, pp 131–138

Bers DM (1985) Ca influx and sarcoplasmic reticulum Ca release in cardiac muscle activation during postrest recovery. Am J Physiol 248:H366–381

Bers DM, Philipson KD, Nishimoto AY (1980) Sodium-calcium exchange and sidedness of isolated cardiac sarcolemmal vesicles. Biochim Biophys Acta 601:358–371

Boucek MM, Synderman R (1976) Calcium influx requirements for human neutrophil chemotaxis: inhibition by lanthanum chloride. Science 193:905–907

Brown SI (1981) Prevention of collagenase-induced disease by treatment with collagenase inhibitors. US Patent No 4,276,284

Carafoli E, Crompton M (1978) The regulation of intracellular calcium. In: Bronner F, Kleinzeller A (eds) Current topics in membranes and transport, vol 10. Academic, New York, pp 151–216

Clayton RJ (1970) Zinc and healing. Lancet 2:1254

Colton CA (1976) Post-synaptic effect of La at the frog neuromuscular junction. J Neurobiol 7:87–91

Corcoran JJ, Kirshner N (1983) Effects of manganese and other divalent cations on calcium uptake and catecholamine secretion by primary cultures of bovine adrenal medulla. Cell Calcium 4:127–137

Douglas WW, Rubin RP (1964a) The effects of alkaline earths and other divalent cations on adrenal medullary secretion. J Physiol (Lond) 175:231–241

Douglas WW, Rubin RP (1964b) Stimulant action of barium on the adrenal medulla. Nature 203:305–307

Drapeau P, Nachshen DA (1984) Manganese fluxes and manganese-dependent neurotransmitter release in presynaptic nerve endings isolated from rat brain. J Physiol (Lond) 348:493–510

Dyckerhoff H, Goossens N (1939) Über die thromboseverhütende Wirkung des Neodyms. Z Ges Exp Med 106:181–190

Ekman L, Valmet A, Aberg B (1961) Behaviour of Yttrium-91 and some lanthanides towards serum proteins in paper electrophoresis, density gradient electrophoresis and gel filtration. Int J Appl Radiat Isot 12:32–41

Ellis KJ (1977) The lanthanide elements in biochemistry, biology and medicine. Inorg Persp Biol Med 1:101–135

Evans CH (1979) Enzyme inhibitors as possible therapeutic agents for arthritis. Orthop Survey 3:63–69
Evans CH (1983) Interesting and useful biochemical properties of the lanthanides. Trends Biochem Sci 8:445–449
Evans CH, Ridella JD (1985) Inhibition, by lanthanides, of neutral proteinases secreted by human, rheumatoid synovium. Eur J Biochem 151:29–32
Ferguson GM, Evans CH (1987) The possible role of implant materials in promoting the aseptic loosening of prosthetic joints. In: Crawford N, Taylor DEM (eds) Interactions of cells with natural and foreign surfaces. Plenum, New York, pp 279–286
Foreman JC, Mongar JL (1972) The role of alkaline earths in anaphylactic histamine secretion. J Physiol (Lond) 244:753–769
Fukada J, Kawa K (1977) Permeation of manganese, cadmium, zinc and beryllium through calcium channels of an insect muscle membrane. Science 196:309–311
Furie BC, Furie B (1975) Interaction of lanthanide ions with bovine factor X and their use in the affinity chromatography of the venom coagulant protein of vipera russelli. J Biol Chem 250:601–608
Furie BC, Mann KG, Furie B (1976) Substitution of lanthanide ions for calcium ions in the activation of bovine prothrombin by activated factor X. High affinity metal-binding sites of prothrombin and the derivatives of prothrombin activation. J Biol Chem 254:3235–3241
Ginsborg BL, Jenkinson DH (1976) Transmission of impulse from nerve to muscle. In: Zamis E (ed) Neuromuscular junction. Springer, Berlin Heidelberg New York, pp 229–364 (Handbook of experimental pharmacology, vol 42)
Ginsburg R, Davis K, Bristow MR, McKennett K, Kodsi SR, Billingham ME, Schroeder JS (1983) Calcium antagonists suppress atherogenesis in aorta but not in the intramural coronary arteries of cholesterol-fed rabbits. Lab Invest 49:154–158
Grenet H, Drouin H (1920) Les sels de terres de la serie du cerium dans le traitement de la tuberculose pulmonaire chronique. Gaz Hop 93:789–792
Gripenberg J, Heinonen E, Jansson SE (1980) Uptake of radiocalcium by nerve endings isolated from rat brain: kinetic studies. Br J Pharmacol 71:265–271
Hagiwara S, Byerly L (1981a) Membrane biophysics of calcium currents. Fed Proc 40:2220–2225
Hagiwara S, Byerly L (1981b) Calcium channel. Annu Rev Neurosci 4:69–125
Hagiwara S, Takahashi K (1967) Surface density of calcium ions and calcium spikes in the barnacle muscle fiber membrane. J Gen Physiol 50:583–601
Haley TJ (1965) Pharmacology and toxicology of the rare earth elements. J Pharm Sci 54:663–670
Haley TJ, Upham HC (1963) Skin reaction to intradermal injection of rare earths. Nature 200:271
Haley TJ, Raymond K, Komesu N, Upham HC (1961) Toxicological and pharmacological effects of gadolinium and samarium chlorides. Br J Pharmacol 17:526–532
Haley TJ, Komesu N, Efros M, Koste L, Upham HC (1964) Pharmacology and toxicology of lutetium chloride. J Pharm Sci 53:1186–1188
Hallcher LM, Sherman WR (1980) The effects of lithium ion and other agents on the activity of *myo*-inositol-1-phosphatase from bovine brain. J Biol Chem 255:10896–10901
Hamilton JG (1949) The metabolism of radioactive elements created by nuclear fission. N Engl J Med 240:863–870
Harris RA (1981) Psychoactive drugs as antagonists of actions of calcium. In: Weiss GB (ed) New perspectives on calcium antagonists. American Physiological Society, Bethesda, pp 223–231
Henquin JL, Lambert AE (1975) Cobalt inhibition of insulin secretion and calcium uptake by isolated rat islets. Am J Physiol 228:1669–1677
Henry PD, Bentley KI (1981) Suppression of atherogenesis in cholesterol-fed rabbit treated with nifedipine. J Clin Invest 68:1366–1396
Horrocks WDeW (1982) Lanthanide ion probes of biomolecular structure. In: Eichhorn GL, Marzilli LG (eds) Advances in inorganic biochemistry, vol 4. Elsevier, New York, pp 201–260

Jancso H (1962) Inflammation and the inflammatory mechanisms. J Pharm Pharmacol 13:577–594
Kawa K (1979) Zinc-dependent action potentials in giant neurones of the snail, *Euhadra quaestia*. J Membr Biol 49:325–344
Kazimierczak W, Diamant B (1978) Mechanisms of histamine release in anaphylactic and anaphylactoid reactions. Prog Allergy 24:295–365
Korc M (1983) Effect of lanthanum on pancreatic protein synthesis in streptozotocin-diabetic rats. Am J Physiol 244:G321–326
Kostyuk PG (1981) Calcium channels in the neuronal muscle. Biochim Biophys Acta 650:128–150
Kramsch DM, Aspen AJ, Rozler LJ (1980) Suppression of experimental atherosclerosis by the Ca-antagonist lanthanum: possible role of calcium in atherogenesis. J Clin Invest 65:967–981
Krueger BK, Nachshen DA (1980) Selectivity of Na^+ and Ca^+ channels in synaptosomes. Fed Proc 39:2038
Laszlo D, Ekstein DM, Lewin R, Stern KG (1952) Biological studies on stable and radioactive rare earth compounds. I. On the distribution of lanthanum in the mammalian organism. J Natl Cancer Inst 13:559–571
Lazar G, Husztik E, Somfay A (1982) Effect of Kupffer cell phagocytosis blockade induced by gadolinium chloride on the humoral immune response and sensitivity to anaphylaxis. 9th Int Cong Reticuloendothelial Society, Davos, Abstract No 147
Lehninger AL (1970) Mitochondria and Ca^{2+} ion transport. Biochem J 119:129–138
Lippiello L, Prellwitz J, Schmelter R, Connolly J (1984) Investigation of ytterbium-169 as a predictor of joint degeneration in osteoarthritis. Trans Orthop Res Soc 9:68
Lorenson MY, Robson DL, Jacobs LS (1983) Divalent cation inhibition of hormone release from isolated adenohypophysial secretory granules. J Biol Chem 258:8618–8620
Marier SH, Putney JW, Van de Walle CM (1978) Control of calcium channels by membrane receptors in the rat parotid gland. J Physiol (Lond) 279:141–151
Martin RB, Richardson FS (1979) Lanthanides as probes for calcium in biological systems. Q Rev Biophys 12:181–209
Mattingly PC, Mowat AG (1982) Zinc sulphate in rheumatoid arthritis. Ann Rheum Dis 41:456–457
McCarty DJ, Palmer DW, Halverson PG (1979) Clearance of calcium pyrophosphate dihydrate crystals in vivo. I. Studies using ^{169}Yb labeled triclinic crystals. Arthritis Rheum 22:718–727
McClure JB, Steele JE (1981) The role of extracellular calcium in hormonal activation of glycogen phosphorylase in cockroach fat body. Insect Biochem 11:605–613
Meiri U, Rahamimoff R (1971) Activation of transmitter release by strontium and calcium ions at the neuromuscular junction. J Physiol (Lond) 215:709–726
Miledi R (1966) Strontium as a substitute for calcium in the process of transmitter release at the neuromuscular junction. Nature 212:1233–1234
Miledi R (1971) La ions abolish the "calcium response" of nerve terminals. Nature 229:410–411
Mines GR (1910) The action of beryllium, lanthanum, yttrium and cerium on the frog's heart. J Physiol (Lond) 40:327–346
Mines GR (1911) On the replacement of calcium in certain neuromuscular mechanisms by allied substances. J Physiol (Lond) 42:251–266
Nachshen DA (1984) Selectivity of the Ca binding site in synaptosome Ca channels. Inhibition of Ca influx by multivalent metal cations. J Gen Physiol 83:941–967
Nachshen DA, Blaustein MP (1979) Regulation of nerve terminal calcium channel selectivity by a weak acid site. Biophys J 26:329–334
Nachshen DA, Blaustein MP (1980) Some properties of potassium-stimulated calcium influx in presynaptic nerve endings. J Gen Physiol 76:709–728
Nachshen DA, Blaustein MP (1982) The influx of calcium, strontium, and barium in presynaptic nerve endings. J Gen Physiol 79:1065–1087
Nayler WG, Elz JS, Perry SE, Daly MJ (1983) The biochemistry of uncontrolled calcium entry. Eur Heart J 4 (Suppl H):29–41

Nelson MT, French RJ, Krueger BK (1984) Voltage-dependent calcium channels from brain incorporated into planar lipid bilayers. Nature 308:77–80
O'Flaherty JT, Showell HJ, Becker EL, Ward PA (1978) Substances which aggregate neutrophils. Mechanism of action. Am J Pathol 92:155–166
Okazaki I, Brinckerhoff CE, Sinclair JF, Sinclair PR, Bonkowsky HL, Harris ED (1981) Iron increases collagenase production by rabbit synovial fibroblasts. J Lab Clin Med 97:396–402
Payet MD, Schanne OF, Ruiz-Ceretti E (1980) Competition for slow channel of Ca^{2+}, Mn^{2+}, verapamil and D-600 in rat ventricular muscle. J Mol Cell Cardiol 12:635–638
Putney JW, Weiss SJ, Van de Walle CM, Haddas RA (1980) Is phosphatidic acid a calcium ionophore under neurohumoral control? Nature 284:345–347
Rouleau JL, Parmley WW, Stevens J, Wilkman-Coffelt J, Sievers R, Mahley R, Havel RJ (1982) Verapamil suppresses atherosclerosis in cholesterol-fed rabbits. Am J Cardiol 49:889
Sadler PJ (1982) Inorganic pharmacology. Chem Br 18:182–188
Sanna A, Mascia A, Pani P, Congui L (1976) Protection by cerium chloride on CCl_4-induced hepatotoxicity. Experientia 32:91–92
Schiffrin EL, Lis M, Grutkowska J, Genest J (1981) Role of Ca^{2+} in response of adrenal glomerulosa cells to angiotensin II, ACTH, K^+ and ouabain. Am J Physiol 241:E42–46
Schulz I, Wakasugi H, Stolze H, Kribben A, Haase W (1981) Analysis of Ca^{2+} fluxes and their relation to enzyme secretion in dispersed pancreatic acinar cells. Fed Proc 40:2503–2510
Showell HJ, Naccache PH, Sha'afi RI, Becker EL (1977) The effects of extracellular K^+, Na^+, and Ca^{++} on lysosomal enzyme secretion from polymorphonuclear leukocytes. J Immunol 119:804–811
Silinsky EM (1978) On the role of barium in supporting the asynchronous release of acetylcholine quanta by motor nerve impulses. J Physiol (Lond) 274:157–171
Simkin PA (1976) Oral zinc sulphate in rheumatoid arthritis. Lancet 2:539–542
Snyder F, Stephens N (1961) Plasma free fatty acids and the rare-earth fatty liver. Proc Soc Exp Biol Med 106:202–204
Snyder F, Cress EA, Kyker GC (1960) Rare-earth fatty liver. Nature 185:480–481
Sorenson JRJ (1984) Copper complexes in biochemistry and pharmacology. Chem Br 20:1110–1113
Spedding M, Cavero I (1984) Calcium antagonists: a class of drugs with a bright future, part II. Determination of basic pharmacological properties. Life Sci 35:575–587
Spitzer NC (1979) Low pH selectively blocks calcium action potentials in amphibian neurons developing in culture. Brain Res 161:555–559
Thaw CN, Raaka EG, Gershengorn MC (1984) Evidence that cobalt ion inhibition of prolactin secretion occurs at an intracellular locus. Am J Physiol 247:C150–155
Triggle CR, Grant WF, Triggle DJ (1975) Intestinal smooth muscle contraction and the effects of cadmium and A23187. J Pharmacol Exp Ther 194:182–190
Van Breeman C, Farinas BR, Casteels R, Gerba P, Wuytack F, Deth R (1973) Factors controlling cytoplasmic Ca^{2+} concentration. Philos Trans R Soc Lond 265B:57–71
Veldhuis JD (1982) Role of calcium ions in enzyme induction in isolated swine granulosa cells in vitro. Biochim Biophys Acta 720:211–216
Verma AK, Boutwell RK (1981) Intracellular calcium and skin tumor promotion: calcium regulation of the induction of epidermal ornithine decarboxylase activity by the tumor promotor 12-O-tetradecanoylphorbol-13-acetate. Biochem Biophys Res Commun 101:375–383
Weast RC (ed) (1984) Handbook of chemistry and physics, 64th edn. CRC Press, Boca Raton
Weinmann HJ, Brasch RC, Press WR, Wesbey GE (1984) Characteristics of gadolinium-DTPA complex: a potential NMR contrast agent. Am J Rad 142:619–624
Weiss GB, Goodman FR (1969) Effects of lanthanum on contraction, calcium distribution and Ca^{2+} movements in intestinal smooth muscle. J Pharmacol Exp Ther 169:46–55
Wollheim CB, Janjic D (1984) Cobalt inhibition of insulin release: evidence for an action not related to Ca^{2+} uptake. Am J Physiol 246:C57–62

Subject Index

MIX
Papier aus verantwortungsvollen Quellen
Paper from responsible sources
FSC® C105338

If you have any concerns about our products,
you can contact us on
ProductSafety@springernature.com

In case Publisher is established outside the EU,
the EU authorized representative is:
Springer Nature Customer Service Center GmbH
Europaplatz 3, 69115 Heidelberg, Germany

Printed by Libri Plureos GmbH
in Hamburg, Germany